SCHAUM'S OUTLINE OF

THEORY AND PROBLEMS

OF

ELECTRIC CIRCUITS IN SI UNITS

•

FIRST EDITION

BY

JOSEPH A EDMINISTER, MSE

Assistant Professor of Electrical Engineering
The University of Akron

SI EDITION adapted by

J E SWANN, BSc, CEng, MIEE

Senior Lecturer in Electrical Engineering
Slough College of Technology

•

McGraw-Hill International Book Company, New York
co-published with
McGraw-Hill Book Company (UK) Limited, London
McGraw-Hill Book Company GmbH, Düsseldorf
McGraw-Hill Book Company Australia Pty, Limited
McGraw-Hill Book Company (SA) (Pty) Limited, Johannesburg

07 084397 X

Contents

CONTENTS

Preface

This book is designed for use as a supplement to all current standard texts or as a textbook for a first course in circuit analysis. Emphasis is placed on the basic laws, theorems and techniques which are common to the various approaches found in other texts.

The subject matter is divided into chapters covering duly-recognized areas of theory and study. Each chapter begins with statements of pertinent definitions, principles and theorems together with illustrative and other descriptive material. This is followed by graded sets of solved and supplementary problems. The solved problems serve to illustrate and amplify the theory, present methods of analysis, provide practical examples, and bring into sharp focus those fine points which enable the student to apply the basic principles correctly and confidently. The large number of supplementary problems serve as a complete review of the material of each chapter.

Topics covered include fundamental circuit responses, analysis of waveforms, the complex number system, phasor notation, series and parallel circuits, power and power factor correction, and resonance phenomena. Considerable use of matrices and determinants is made in the treatment of mesh current and node voltage methods of analysis. Matrix methods are also employed in the development of wye-delta transformations and network theorems such as superposition and reciprocity. Mutually coupled circuits are very carefully explained. Polyphase circuits of all types are covered, with emphasis on the one-line equivalent circuit which has important applications. The trigonometric and exponential Fourier series are treated simultaneously, and the coefficients of one are frequently converted to coefficients of the other to show their relationship. Direct and alternating current transients are treated using classical differential equations so that this topic can precede the phasor notation of Chapter 5, and this is recommended for those whose proficiency in mathematics will permit this arrangement. The Laplace transform method is introduced and applied to many of the same problems treated in Chapter 16 by differential equations. This permits a convenient comparison of the two methods and emphasizes the strong points of the Laplace method.

I wish to avail myself of this opportunity to express my gratitude to the staff of the Schaum Publishing Company, especially to Mr. Nicola Miracapillo, for their valuable suggestions and helpful co-operation. Thanks and more are due my wife, Nina, for her unfailing assistance and encouragement in this endeavour.

JOSEPH A. EDMINISTER

The University of Akron
August 21, 1965

CHAPTER 1

Definitions and Circuit Parameters

BASIC UNITS

The SI system is now used in electrical engineering. In this system the fundamental mechanical units are the metre (m) for length, the kilogram (kg) for mass, and the second (s) for time. The basic electrical unit is the ampere (A) for electric current, which is defined in terms of the force (F) per unit length between two current-carrying conductors using the relationship

$$F = k\frac{II'}{d}$$

where d is the distance apart of the two straight parallel conductors of infinite length, I and I' are the two currents and k is a dimensional constant given the value $\mu_0/2\pi$ in the SI system where $\mu_0 = 4\pi \times 10^{-7}$ N/A^2 = $0{\cdot}4\pi$ μH/m. μ_0 is known as the permeability of free space and for practical purposes is the same as the permeability of air. In general, $F = \mu II'/2\pi d$ newtons per metre where $\mu = \mu_r\mu_0$ where μ_r is a dimensionless constant called the relative permeability of the material between the conductors, μ is the permeability of the material and μ_0 the permeability of free space.

The unit of current, the ampere, may be defined as that magnitude of current which, when flowing in two infinitely long straight parallel conductors placed in a vacuum, causes a force between the conductors of 2×10^{-7} N/m. Convenient submultiples of the ampere are

$$1 \text{ mA} = 1 \text{ milliampere} = 10^{-3} \text{ amperes}$$
$$1\ \mu\text{A} = 1 \text{ microampere} = 10^{-6} \text{ amperes}$$

The positive current direction is, by convention, opposite to the direction in which electrons move. See Fig. 1-1.

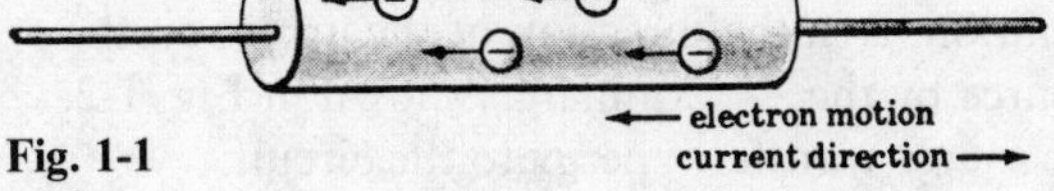

electron motion
current direction

Fig. 1-1

DERIVED UNITS

CHARGE q

An electric current flows in a conductor when charge q is transferred from one point to another in that conductor. Current is the rate of flow of charge with time so that

$$i \text{ (amperes)} = \frac{dq \text{ (coulombs)}}{dt \text{ (seconds)}}$$

Some authorities consider the coulomb (C) as the fundamental electrical unit, in which case the force (F) between two point charges q and q' distance d apart is expressed in the Coulomb's Law formula

$$F = k\frac{qq'}{d^2}$$

where k is a dimensional constant given the value $1/4\pi\varepsilon_0$ in the SI system where $\varepsilon_0 = 8{\cdot}854\ 10^{-12}$ C^2/N m^2 $= 8{\cdot}854$ pF/m.

When the surrounding medium is not a vacuum, forces caused by the charges induced in the medium reduce the resultant force between the free charges immersed in the medium. The net force is now given by $F = qq'/4\pi\varepsilon d^2$ newtons. For air ε is only just greater than ε_0 and for most purposes is taken equal to ε_0. For other materials ε is given by $\varepsilon = \varepsilon_r\varepsilon_0$ where ε_r is a dimensionless constant called the relative permittivity of the material between the charges, ε is called the permittivity of the material and ε_0 the permittivity of free space. For a vacuum $\varepsilon_r = 1$ and $\varepsilon = \varepsilon_0$.

The unit of charge, the coulomb, may be defined as that magnitude of charge which, when placed one metre from an equal and similar charge in a vacuum, repels it with a force of $1/4\pi\varepsilon_0$ newtons. Convenient submultiples of the coulomb are

$$1\ \mu C = 1 \text{ microcoulomb} = 10^{-6} \text{ coulombs}$$
$$1\ pC = 1 \text{ picocoulomb} = 10^{-12} \text{ coulombs}$$

The charge carried by an electron $(-e)$ or by a proton $(+e)$ is $e = 1{\cdot}602 \times 10^{-19}$ coulombs.

POTENTIAL DIFFERENCE v

The potential difference v between two points is measured by the work required to transfer unit charge from one point to the other. The *volt* is the potential difference (p.d.) between two points when 1 joule of work is required to transfer 1 coulomb of charge from one point to the other: 1 volt = 1 joule/coulomb.

If two points of an external circuit have a potential difference v, then a charge q in passing between the two circuit points does an amount of work qv as it moves from the higher to the lower potential point.

An agent such as a battery or generator has an electromotive force (emf) if it does work on the charge moving through it, the charge receiving electrical energy as it moves from the lower to the higher potential side. Emf is measured by the p.d. between the terminals when the generator is not delivering current.

POWER p

Electrical power p is the product of impressed voltage v and resulting current i.

$$p\,(\text{watts}) = v\,(\text{volts}) \times i\,(\text{amperes})$$

Positive current, by definition, is in the direction of the arrow on the voltage source; it leaves the source by the + terminal as shown in Fig. 1-2. When p has a positive value the source transfers energy to the circuit.

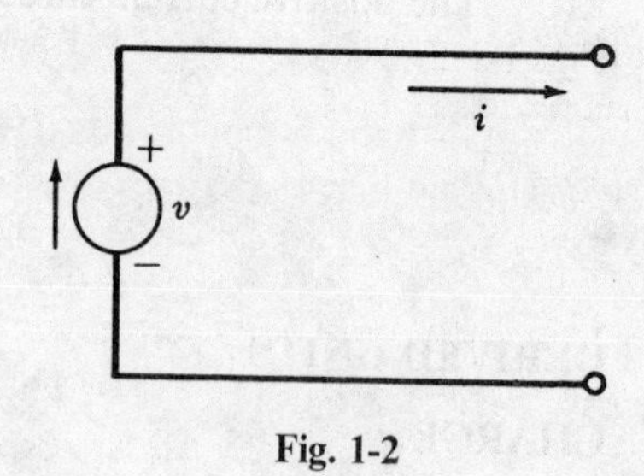

Fig. 1-2

If power p is a periodic function of time t with period T, then the

$$\text{Average power} \quad P = \frac{1}{T}\int_0^T p\,dt$$

ENERGY w

Since power p is the time rate of energy transfer,

$$p = \frac{dw}{dt} \quad \text{and} \quad W = \int_{t_1}^{t_2} p\,dt$$

where W is the energy transferred during the time interval.

RESISTOR, INDUCTOR, CAPACITOR

When electrical energy is supplied to a circuit element, it will respond in one or more of the following three ways. If the energy is consumed, then the circuit element is a pure *resistor*. If the energy is stored in a magnetic field, the element is a pure *inductor*. And if the energy is stored in an electric field, the element is a pure *capacitor*. A practical circuit device exhibits more than one of the above and perhaps all three at the same time, but one may be predominant. A coil may be designed to have a high inductance, but the wire with which it is wound has some resistance; hence the coil has both properties.

RESISTANCE R

The potential difference $v(t)$ across the terminals of a pure resistor is directly proportional to the current $i(t)$ in it. The constant of proportionality R is called the resistance of the resistor and is expressed in volts/ampere or ohms.

$$v(t) = R\,i(t) \qquad \text{and} \qquad i(t) = \frac{v(t)}{R}$$

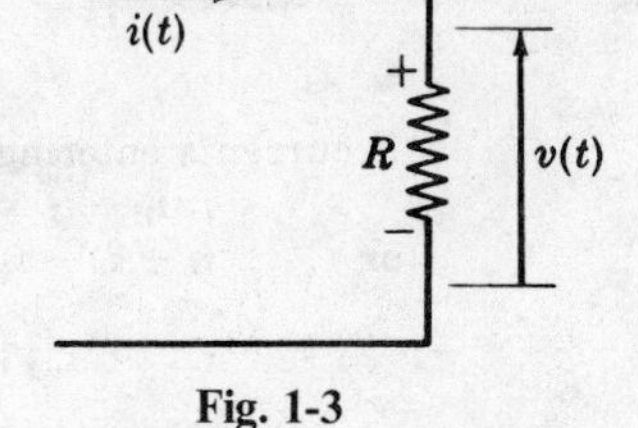

Fig. 1-3

No restriction is placed on $v(t)$ and $i(t)$; they may be constant with respect to time, as in D.C. circuits, or they may be sine or cosine functions, etc.

Lower case letters (v, i, p) indicate general functions of time. Capital letters (V, I, P) denote constant quantities, and peak or maximum values carry a subscript (V_m, I_m, P_m).

INDUCTANCE L

When the current in a circuit is changing, the magnetic flux linking the same circuit changes. This change in flux causes an emf v to be induced in the circuit. The induced emf v is proportional to the time rate of change of current if the permeability is constant. The constant of proportionally is called the *self-inductance* or *inductance* of the circuit.

$$v(t) = L\frac{di}{dt} \qquad \text{and} \qquad i(t) = \frac{1}{L}\int v\,dt$$

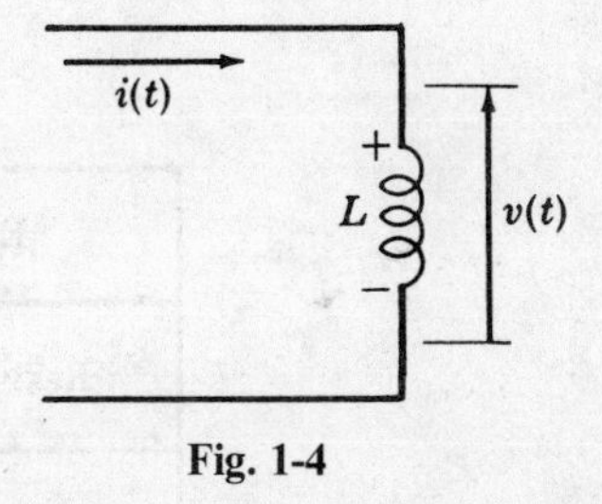

Fig. 1-4

When v is in volts and di/dt in amperes/sec, L is in volt seconds/ampere or *henrys*. The self-inductance of a circuit is 1 henry (1 H) if an emf of 1 volt is induced in it when the current changes at the rate of 1 ampere/second.

CAPACITANCE C

The potential difference v between the terminals of a capacitor is proportional to the charge q on it. The constant of proportionality C is called the *capacitance* of the capacitor.

$$q(t) = C\,v(t), \quad i = \frac{dq}{dt} = C\frac{dv}{dt}, \quad v(t) = \frac{1}{C}\int i\,dt$$

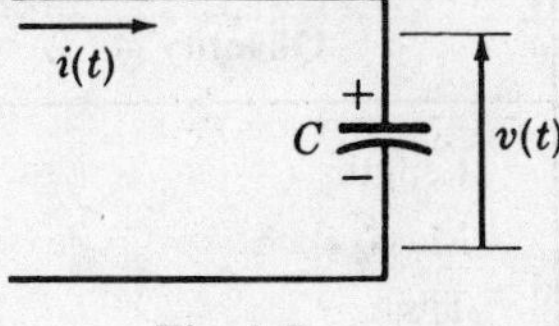

Fig. 1-5

When q is in coulombs and v in volts, C is in coulombs/volt or *farads*. A capacitor has capacitance 1 farad (1 F) if it requires 1 coulomb of charge per volt of potential difference between its conductors. Convenient submultiples of the farad are

$$1\ \mu\text{F} = 1 \text{ microfarad} = 10^{-6}\text{ F and } 1\text{ pF} = 1\text{ picofarad} = 10^{-12}\text{ F}$$

KIRCHHOFF'S LAWS

1. The sum of the currents entering a junction is equal to the sum of the currents leaving the junction. If the currents toward a junction are considered positive and those away from the same junction negative, then this law states that the algebraic sum of all the currents meeting at a common junction is zero.

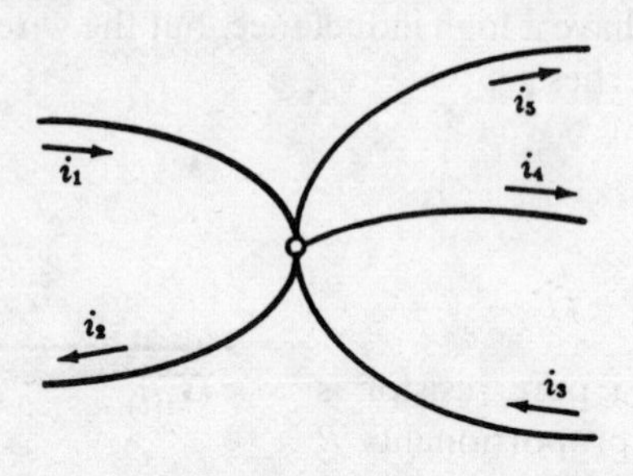

Σ currents entering = Σ currents leaving

$$i_1 + i_3 = i_2 + i_4 + i_5$$

or $$i_1 + i_3 - i_2 - i_4 - i_5 = 0$$

Fig. 1-6

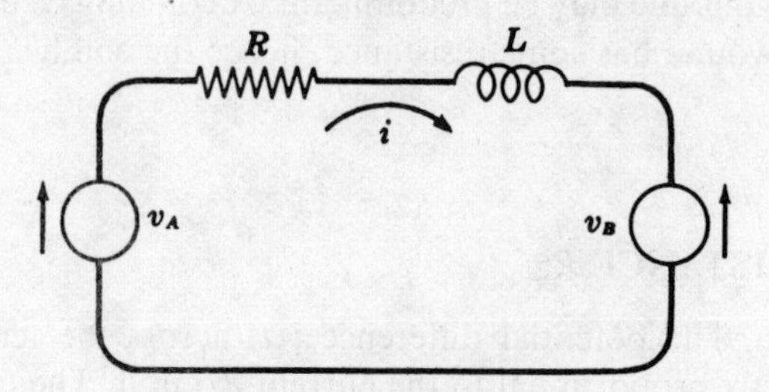

Σ potential rises = Σ potential drops

$$v_A - v_B = Ri + L(di/dt)$$

or $$v_A - v_B - Ri - L(di/dt) = 0$$

Fig. 1-7

2. The sum of the rises of potential around any closed circuit equals the sum of the drops of potential in that circuit. In other words, the algebraic sum of the potential differences around a closed circuit is zero. With more than one source when the directions do not agree, the voltage of the source is taken as positive if it is in the direction of the assumed current.

Circuit Response of Single Elements

Element	Voltage across element	Current in element
Resistance R	$v(t) = R\,i(t)$	$i(t) = \dfrac{v(t)}{R}$
Inductance L	$v(t) = L\dfrac{di}{dt}$	$i(t) = \dfrac{1}{L}\int v\,dt$
Capacitance C	$v(t) = \dfrac{1}{C}\int i\,dt$	$i(t) = C\dfrac{dv}{dt}$

Units in the SI System

Quantity		Unit		Quantity		Unit	
Length	l	metre	m	Charge	Q, q	coulomb	C
Mass	m	kilogramme	kg	Potential	V, v	volt	V
Time	t	second	s	Current	I, i	ampere	A
Force	F, f	newton	N	Resistance	R	ohm	Ω
Energy	W, w	joule	J	Inductance	L	henry	H
Power	P, p	watt	W	Capacitance	C	farad	F

Solved Problems

1.1. In the circuit shown in Fig. 1-8 the applied constant voltage is $V = 45$ volts. Find the current, the voltage drop across each resistor, and the power in each resistor.

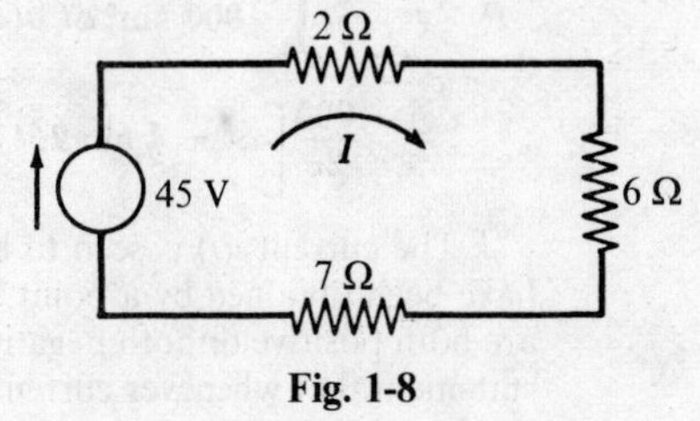

Fig. 1-8

The sum of the voltage rises equals the sum of the voltage drops around any closed loop; thus

$$V = I(2) + I(6) + I(7), \quad 45 = 15I, \quad I = 3\text{ A}$$

The voltage drop across the 2 ohm resistor is $V_2 = IR_2 = 3(2) = 6$ V. Similarly, $V_6 = 3(6) = 18$ V, and $V_7 = 21$ V.

The power in the 2 ohm resistor is $P_2 = V_2 I = 6(3) = 18$ W or $P_2 = I^2R_2 = 3^2(2) = 18$ W. Similarly, $P_6 = V_6 I = 54$ W, and $P_7 = V_7 I = 63$ W.

1.2. A current I_T divides between two parallel branches having resistances R_1 and R_2 respectively as shown in Fig. 1-9. Develop formulae for the currents I_1 and I_2 in the parallel branches.

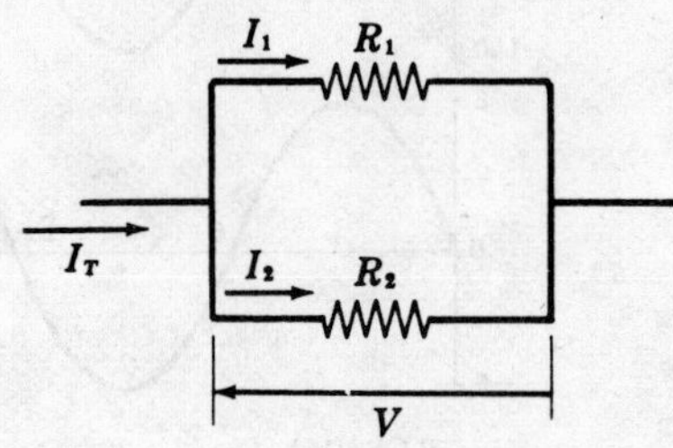

Fig. 1-9

The voltage drop in each branch is the same, i.e. $V = I_1R_1 = I_2R_2$. Then

$$I_T = I_1 + I_2 = \frac{V}{R_1} + \frac{V}{R_2} = V\left(\frac{1}{R_1} + \frac{1}{R_2}\right)$$

$$= I_1R_1\left(\frac{R_2 + R_1}{R_1R_2}\right) = I_1\left(\frac{R_2 + R_1}{R_2}\right)$$

from which $I_1 = I_T\left(\dfrac{R_2}{R_1 + R_2}\right)$. Similarly, $I_2 = I_T\left(\dfrac{R_1}{R_1 + R_2}\right)$.

1.3. Three resistors R_1, R_2, R_3 are in parallel as shown in Fig. 1-10. Derive a formula for the equivalent resistance R_e of the network.

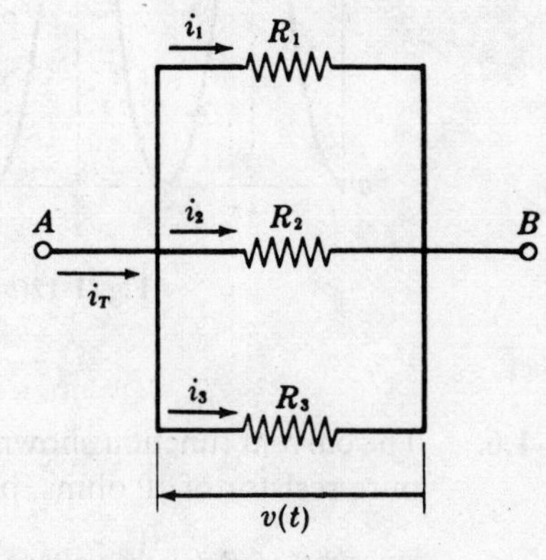

Fig. 1-10

Assume a voltage A to B of $v(t)$, and let the currents in R_1, R_2, R_3 be $i_1(t)$, $i_2(t)$, $i_3(t)$ respectively. The current in R_e must be the total current $I_T(t)$. Then $v(t) = R_1i_1(t) = R_2i_2(t) = R_3i_3(t) = R_ei_T(t)$ and

$$i_T(t) = i_1(t) + i_2(t) + i_3(t) \quad \text{or} \quad \frac{v(t)}{R_e} = \frac{v(t)}{R_1} + \frac{v(t)}{R_2} + \frac{v(t)}{R_3}$$

or $$\frac{1}{R_e} = \frac{1}{R_1} + \frac{1}{R_2} + \frac{1}{R_3}$$

For a two-branch parallel circuit, $\dfrac{1}{R_e} = \dfrac{1}{R_1} + \dfrac{1}{R_2}$ or

$$R_e = \frac{R_1R_2}{R_1 + R_2}.$$

1.4. The two constant voltage sources V_A and V_B act in the same circuit as shown in Fig. 1-11. What power does each deliver?

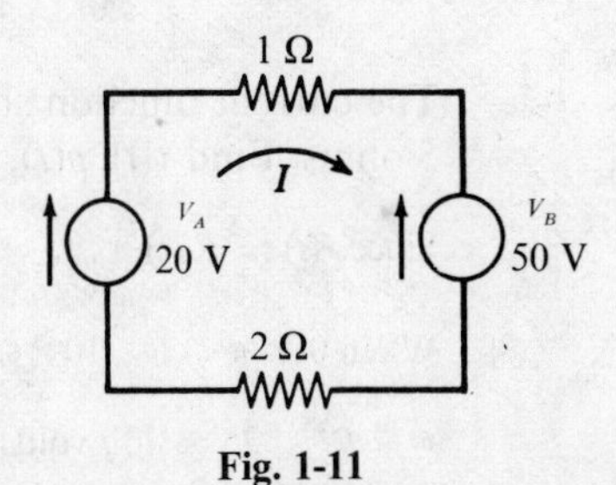

Fig. 1-11

The sum of the potential rises is equal to the sum of the potential drops around a closed circuit; hence

$$20 - 50 = I(1) + I(2), \quad I = -10\text{ A}$$

Power delivered by $V_A = V_AI = 20(-10) = -200$ W.
Power delivered by $V_B = V_BI = 50(10) = 500$ W.

1.5. In the circuit shown in Fig. 1-12(a), the voltage function is $v(t) = 150 \sin \omega t$. Find the current $i(t)$, the instantaneous power $p(t)$, and the average power P.

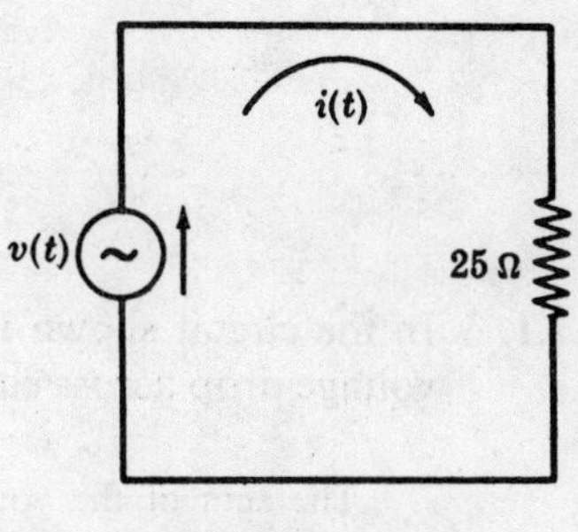

Fig. 1-12(a)

$$i(t) = \frac{1}{R}v(t) = \frac{150}{25}\sin \omega t = 6 \sin \omega t \text{ amperes}$$

$$p(t) = v(t)\,i(t) = (150 \sin \omega t)(6 \sin \omega t) = 900 \sin^2 \omega t \text{ watts}$$

$$P = \frac{1}{\pi}\int_0^\pi 900 \sin^2 \omega t \, d(\omega t) = \frac{900}{\pi}\int_0^\pi \tfrac{1}{2}(1 - \cos 2\omega t)\, d(\omega t)$$

$$= \frac{900}{2\pi}\left[\omega t - \tfrac{1}{2}\sin 2\omega t\right]_0^\pi = 450 \text{ W}$$

The current $i(t)$ is seen to be related to the voltage $v(t)$ by the constant R. The instantaneous power plot could have been obtained by a point by point product of the v and i plots shown in Fig. 1-12(b) below. Note that v and i are both positive or both negative at any instant; the product must therefore always be positive. This agrees with the statement that whenever current flows through a resistor, electrical energy is delivered by the source.

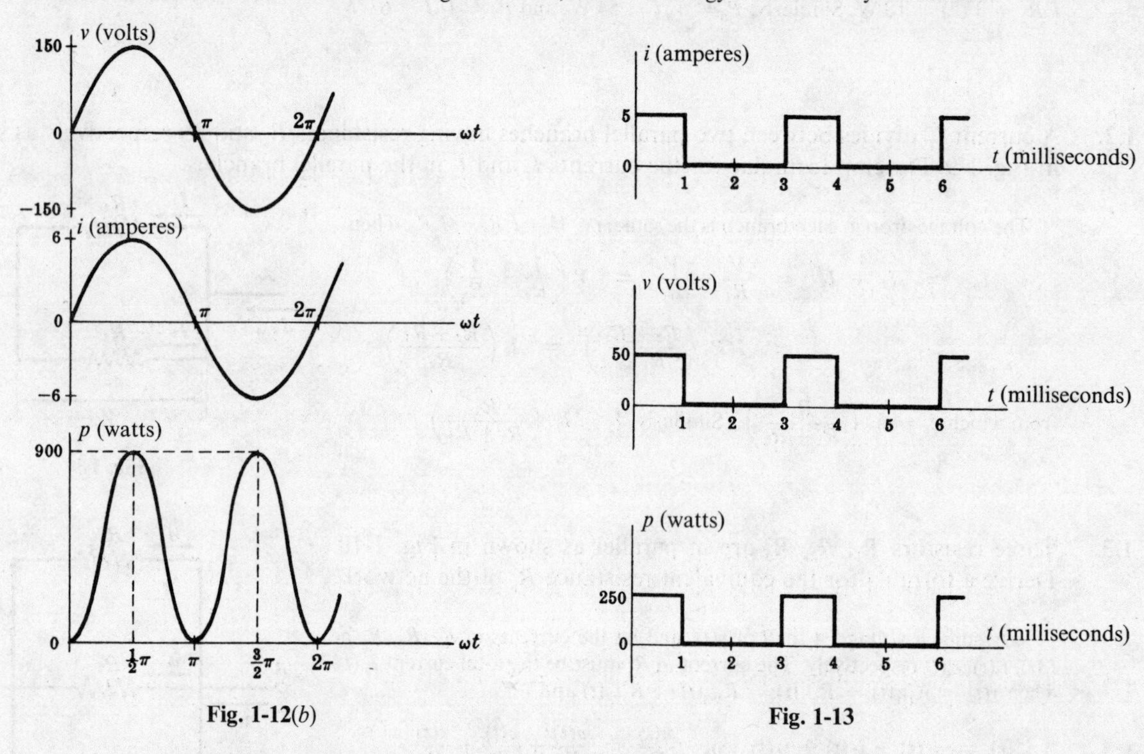

Fig. 1-12(b) Fig. 1-13

1.6. The current function shown in Fig. 1-13 above is a repeating square wave. With this current existing in a pure resistor of 10 ohms, plot voltage $v(t)$ and power $p(t)$.

Since $v(t) = R\,i(t)$, the voltage varies directly as the current. The maximum value is $Ri_{max} = 5(10) = 50$ V.
Since $p = vi$, the power plot is a point by point product. The maximum value is $i_{max}\,v_{max} = 50(5) = 250$ W.

1.7. The current function shown in Fig. 1-14 below is a repeating sawtooth and exists in a pure resistor of 5 ohms. Find $v(t)$, $p(t)$, and average power P.

Since $v(t) = Ri(t)$, $v_{max} = Ri_{max} = 5(10) = 50$ V.

When $0 < t < 2 \times 10^{-3}$ s, $i = \dfrac{10}{2 \times 10^{-3}}t = 5 \times 10^3 t$ amperes. Then

$$v = Ri = 25 \times 10^3 t \text{ volts}, \quad p = vi = 125 \times 10^6 t^2 \text{ watts}, \quad P = \frac{1}{2 \times 10^{-3}}\int_0^{2\times 10^{-3}} 125 \times 10^6 t^2\, dt = 167 \text{ W}$$

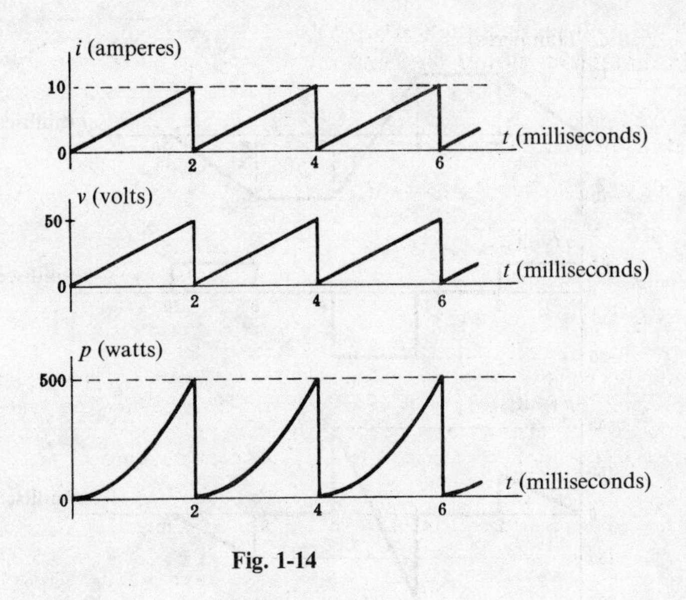

Fig. 1-14

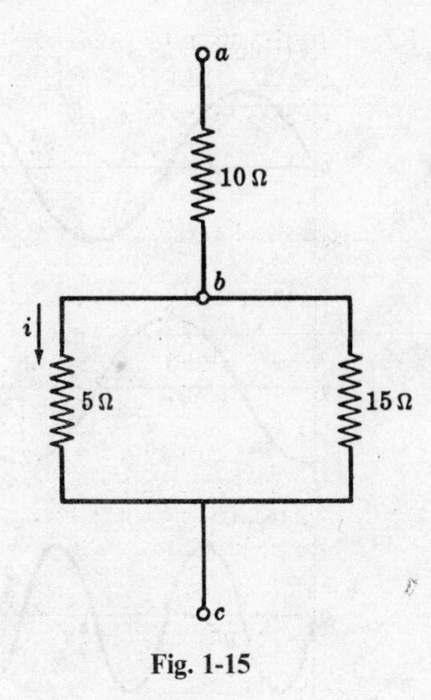

Fig. 1-15

1.8. In the circuit shown in Fig. 1-15 above, the current in the 5 ohm resistor is $i(t) = 6 \sin \omega t$ amperes. (*a*) Determine the current in the 15 and 10 ohm resistors and the voltages *a* to *b* and *b* to *c*. (*b*) Find the instantaneous and average power consumed in each resistor.

(*a*) The same voltage v_{bc} is across the 5 and 15 ohm resistors; then

$$v_{bc} = i_5 R_5 = (6 \sin \omega t)(5) = 30 \sin \omega t \text{ volts} \quad \text{and} \quad i_{15} = v_{bc}/R_{15} = 2 \sin \omega t \text{ amperes}$$

Now $\quad i_{10} = i_{15} + i_5 = 8 \sin \omega t$ amperes and $v_{ab} = i_{10}R_{10} = 80 \sin \omega t$ volts

(*b*) Instantaneous power $p = vi$. Thus $p_5 = (30 \sin \omega t)(6 \sin \omega t) = 180 \sin^2 \omega t$ watts. Similarly, $p_{15} = 60 \sin^2 \omega t$ watts and $p_{10} = 640 \sin^2 \omega t$ watts.

Average power in 5 ohm resistor is

$$P_5 = \frac{1}{\pi}\int_0^\pi 180 \sin^2 \omega t \, d(\omega t) = \frac{1}{\pi}\int_0^\pi 180[\tfrac{1}{2}(1 - \cos 2\omega t)] \, d(\omega t) = 90 \text{ W}$$

Similarly, $P_{15} = 30$ W and $P_{10} = 320$ W.

1.9. A pure resistor of 2 ohms has an applied voltage $v(t)$ given by

$$v(t) = 50\left[1 - \frac{(\omega t)^2}{2!} + \frac{(\omega t)^4}{4!} - \frac{(\omega t)^6}{6!} + \cdots\right] \text{ volts}$$

Determine the current and power for this single circuit element.

Expanding $\cos x$ as a power series in x, $\cos x = 1 - \frac{x^2}{2!} + \frac{x^4}{4!} - \frac{x^6}{6!} + \cdots$.

Hence $v(t) = 50 \cos \omega t$ volts, $i(t) = 25 \cos \omega t$ amperes, $p(t) = 1250 \cos^2 \omega t$ watts, and $P = 625$ W.

1.10. A pure inductance $L = 0.02$ henrys has an applied voltage $v(t) = 150 \sin 1000t$ volts. Determine the current $i(t)$, the instantaneous power $p(t)$, and the average power P.

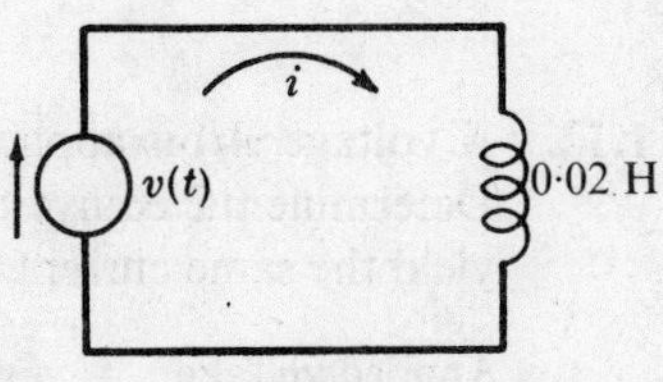

Fig. 1-16(*a*)

$$i(t) = \frac{1}{L}\int v(t)\,dt = \frac{1}{0{\cdot}02}\int 150 \sin 1000t \, dt$$

$$= \frac{150}{0{\cdot}02}\left(\frac{-\cos 1000t}{1000}\right) = -7{\cdot}5 \cos 1000t \text{ amperes}$$

$p = vi = -150(7{\cdot}5)(\tfrac{1}{2} \sin 2000t) = -562{\cdot}5 \sin 2000t$ watts. [$\sin x \cos x = \tfrac{1}{2} \sin 2x$.] The average power P is obviously zero, as shown in Fig. 1-16(*b*) below.

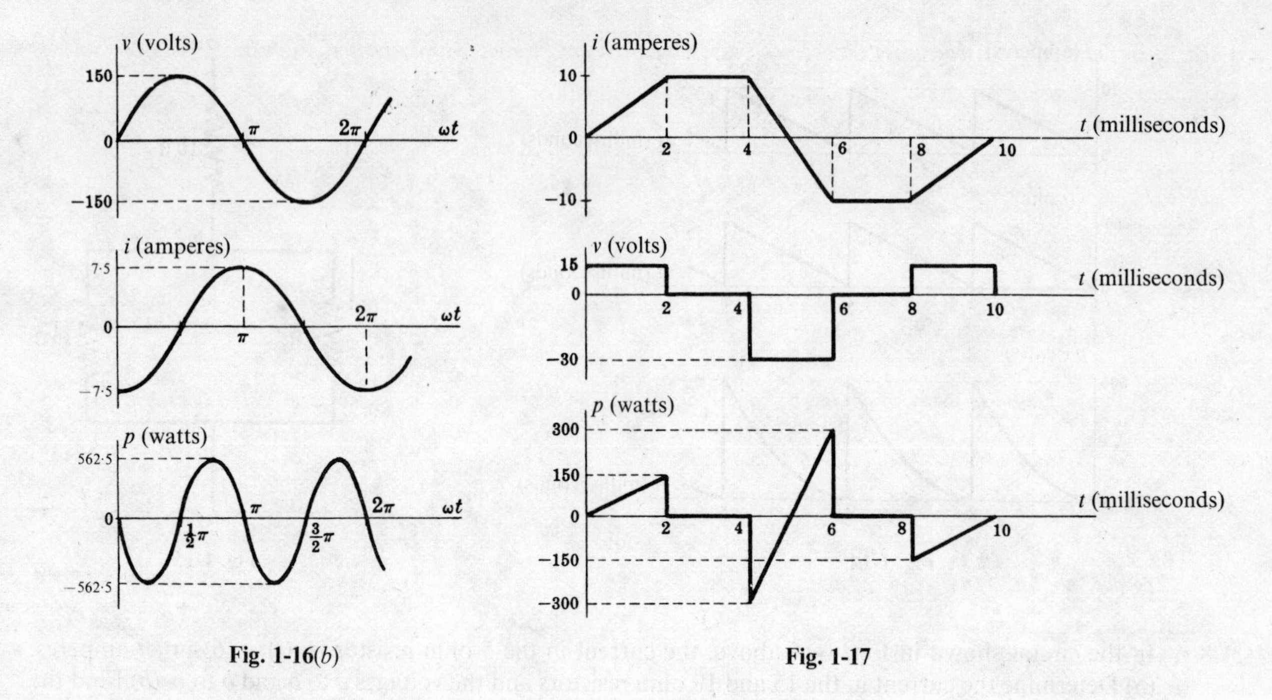

Fig. 1-16(*b*) Fig. 1-17

1.11. A single pure inductance of 3 millihenrys passes a current of the waveform shown in Fig. 1-17 above. Determine and sketch the voltage $v(t)$ and the instantaneous power $p(t)$. What is the average power P?

The instantaneous current $i(t)$ is given by (see Fig. 1-17 above):

(1) $0 < t < 2$ ms $\quad i = 5 \times 10^3 t$ amperes
(2) $2 < t < 4$ ms $\quad i = 10$ A
(3) $4 < t < 6$ ms $\quad i = 10 - 10 \times 10^3(t - 4 \times 10^{-3}) = 50 - 10 \times 10^3 t$ amperes
(4) $6 < t < 8$ ms $\quad i = -10$ A
(5) $8 < t < 10$ ms $\quad i = -10 + 5 \times 10^3(t - 8 \times 10^{-3}) = -50 + 5 \times 10^3 t$ amperes

The corresponding voltages are:

(1) $v_L = L\dfrac{di}{dt} = 3 \times 10^{-3}\dfrac{d}{dt}(5 \times 10^3 t) = 15$ V

(2) $v_L = L\dfrac{di}{dt} = 3 \times 10^{-3}\dfrac{d}{dt}(10) = 0$

(3) $v_L = L\dfrac{di}{dt} = 3 \times 10^{-3}\dfrac{d}{dt}(50 - 10 \times 10^3 t) = -30$ V, etc.

The corresponding instantaneous power values are:

(1) $p = vi = 15(5 \times 10^3 t) = 75 \times 10^3 t$ watts
(2) $p = vi = 0(10) = 0$
(3) $p = vi = -30(50 - 10 \times 10^3 t) = -1500 + 300 \times 10^3 t$ watts, etc.

The average power P is evidently zero.

1.12. A voltage $v(t)$ is applied across two inductances L_1 and L_2 in series. Determine the equivalent inductance L_e which can replace them and yield the same current.

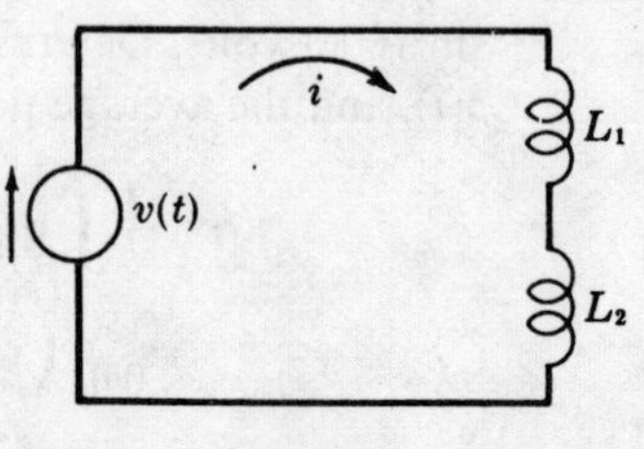

Fig. 1-18

Applied voltage = voltage drop across L_1 + drop across L_2

$$v(t) = L_e\frac{di}{dt} = L_1\frac{di}{dt} + L_2\frac{di}{dt}$$

from which $L_e = L_1 + L_2$.

1.13. Find the equivalent inductance L_e of two inductances L_1 and L_2 in parallel as shown in Fig. 1-19 below.

Assume a voltage $v(t)$ exists across the parallel combination and let the currents in L_1 and L_2 be i_1 and i_2 respectively. Since the total current i_T is the sum of the branch currents,

$$i_T = i_1 + i_2 \quad \text{or} \quad \frac{1}{L_e}\int v\,dt = \frac{1}{L_1}\int v\,dt + \frac{1}{L_2}\int v\,dt$$

Then

$$\frac{1}{L_e} = \frac{1}{L_1} + \frac{1}{L_2} \quad \text{or} \quad L_e = \frac{L_1 L_2}{L_1 + L_2}$$

The reciprocal of the equivalent inductance of any number of inductors connected in series is the sum of the reciprocals of the individual inductances.

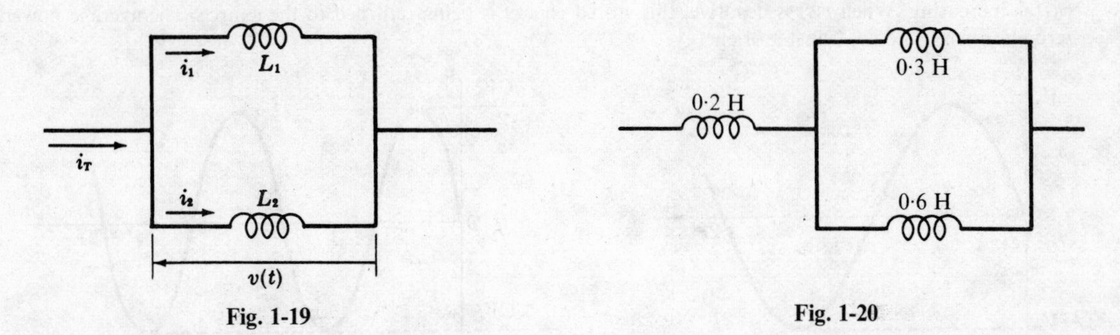

Fig. 1-19

Fig. 1-20

1.14. Three pure inductances are connected as shown in Fig. 1-20 above. What equivalent inductance L_e may replace this circuit?

Equivalent inductance of parallel combination is $L_p = \dfrac{L_1 L_2}{L_1 + L_2} = \dfrac{0{\cdot}3(0{\cdot}6)}{0{\cdot}3 + 0{\cdot}6} = 0{\cdot}2$ H

The required equivalent inductance $L_e = 0{\cdot}2 + L_p = 0{\cdot}4$ H.

1.15. A pure inductor carries a current $i(t) = I_m \sin \omega t$ amperes. Assuming the stored energy in the magnetic field is zero at $t = 0$, derive and sketch the energy function $w(t)$.

$$v(t) = L\frac{d}{dt}(I_m \sin \omega t) = \omega L I_m \cos \omega t \text{ volts}$$

$$p(t) = vi = \omega L I_m^2 \sin \omega t \cos \omega t = \tfrac{1}{2}\omega L I_m^2 \sin 2\omega t \text{ watts}$$

$$w(t) = \int_0^t \tfrac{1}{2}\omega L I_m^2 \sin 2\omega t\,dt = \tfrac{1}{4}L I_m^2[-\cos 2\omega t + 1] = \tfrac{1}{2}L I_m^2 \sin^2 \omega t \text{ joules}$$

At $\omega t = \pi/2, 3\pi/2, 5\pi/2$, etc, the stored energy is maximum and equals $\frac{1}{2}LI_m^2$. At $\omega t = 0, \pi, 2\pi, 3\pi$, etc, the stored energy is zero. See Fig. 1-21 below.

When $p(t)$ is positive the flow of energy is toward the load and the stored energy increases. When $p(t)$ is negative the energy is returning from the magnetic field of the inductor to the source. In a pure inductor no energy is consumed. The average power is zero and there is no net transfer of energy.

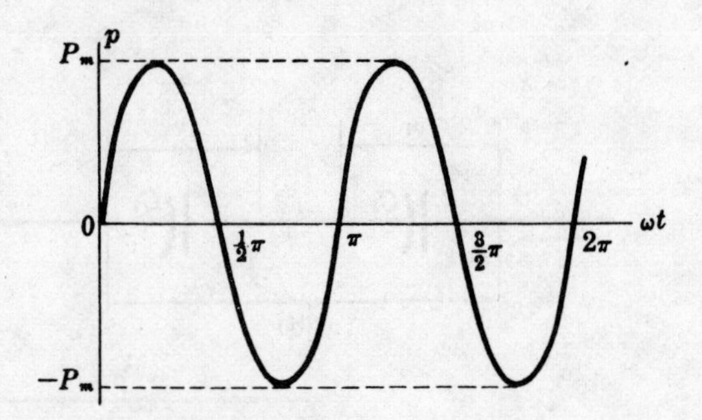

Fig. 1-21

1.16. Consider a pure capacitor with an applied voltage $v(t) = V_m \sin \omega t$ volts. Find the current $i(t)$, the power $p(t)$, the charge $q(t)$, and the stored energy $w(t)$ in the electric field assuming $w(t) = 0$ at $t = 0$.

$$i(t) = C\,dv/dt = \omega C V_m \cos \omega t \text{ amperes}$$

$$p(t) = vi = \tfrac{1}{2}\omega C V_m^2 \sin 2\omega t \text{ watts}$$

$$q(t) = Cv = CV_m \sin \omega t \text{ coulombs}$$

$$w(t) = \int_0^t p\,dt = \tfrac{1}{4}CV_m^2(1 - \cos 2\omega t) = \tfrac{1}{2}CV_m^2 \sin^2 \omega t \text{ joules}$$

At $\omega t = \pi/2, 3\pi/2, 5\pi/2$, etc., the stored energy is maximum and equals $\frac{1}{2}CV_m^2$. At $\omega t = 0, \pi, 2\pi, 3\pi$, etc, the stored energy is zero. See Fig. 1-22 below.

When $p(t)$ is positive the flow of energy is from the source to the electric field of the capacitor and the stored energy $w(t)$ is increasing. When $p(t)$ is negative, this stored energy is being returned to the source. The average power P is zero and there is no net transfer of energy.

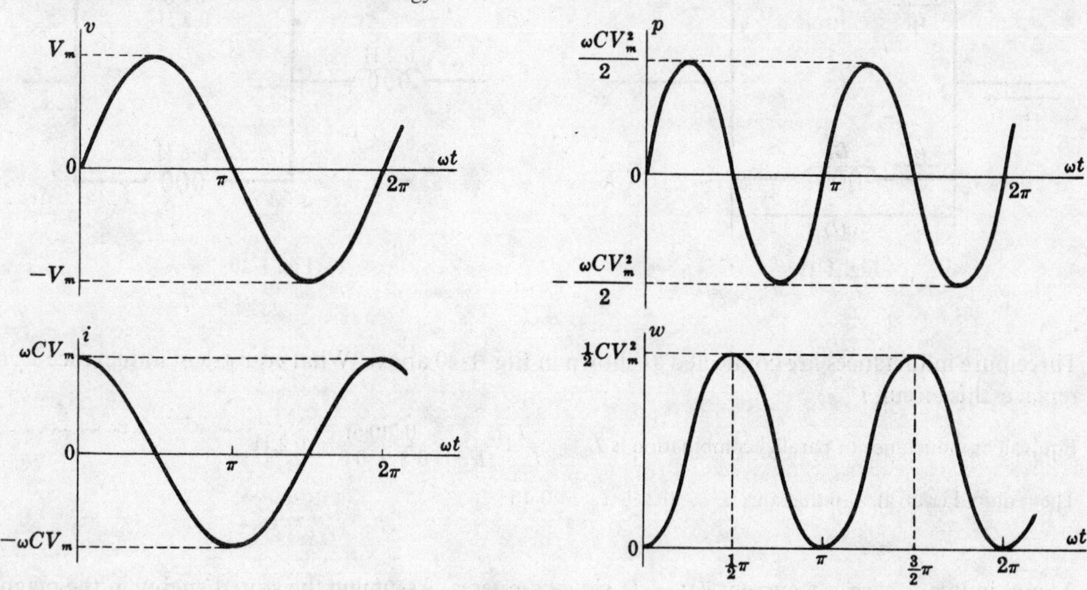

Fig. 1-22

1.17. Determine the equivalent capacitance C_e of the parallel combination of two capacitors C_1 and C_2 shown in Fig. 1-23 below.

Assume a voltage $v(t)$ exists across the parallel combination and let the currents in C_1 and C_2 be i_1 and i_2 respectively. Then, if the total current is i_T,

$$i_T = i_1 + i_2 \quad \text{or} \quad C_e\frac{d}{dt}v(t) = C_1\frac{d}{dt}v(t) + C_2\frac{d}{dt}v(t) \quad \text{or} \quad C_e = C_1 + C_2$$

The resultant (equivalent) capacitance of any number of capacitors connected in parallel is the sum of their individual capacitances.

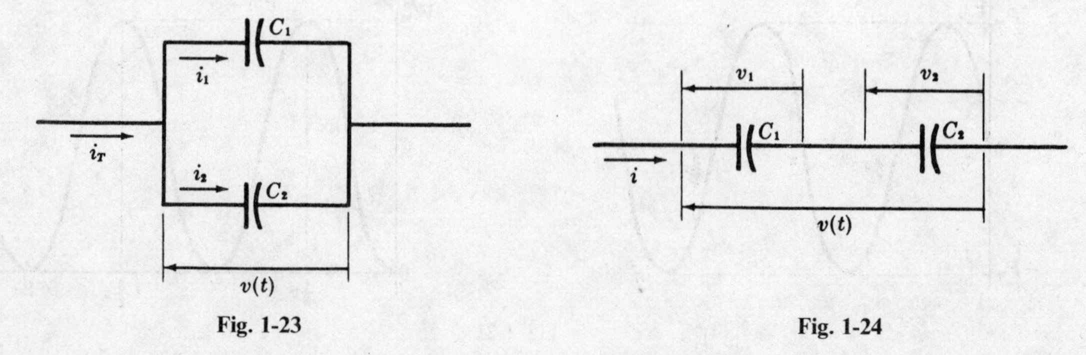

Fig. 1-23

Fig. 1-24

1.18. Determine the equivalent capacitance C_e of the series combination of two capacitors C_1 and C_2 shown in Fig. 1-24 above.

Assume a voltage exists across the series circuit. Then

Applied voltage = voltage drop across C_1 + voltage drop across C_2

$$\frac{1}{C_e}\int i(t)\,dt = \frac{1}{C_1}\int i(t)\,dt + \frac{1}{C_2}\int i(t)\,dt$$

Then $$\frac{1}{C_e} = \frac{1}{C_1} + \frac{1}{C_2} \quad \text{or} \quad C_e = \frac{C_1C_2}{C_1+C_2}$$

The reciprocal of the resultant (equivalent) capacitance of any number of capacitors connected in series is the sum of the reciprocals of the individual capacitances.

1.19. Find the equivalent capacitance C_e of the combination of capacitors shown in Fig. 1-25.

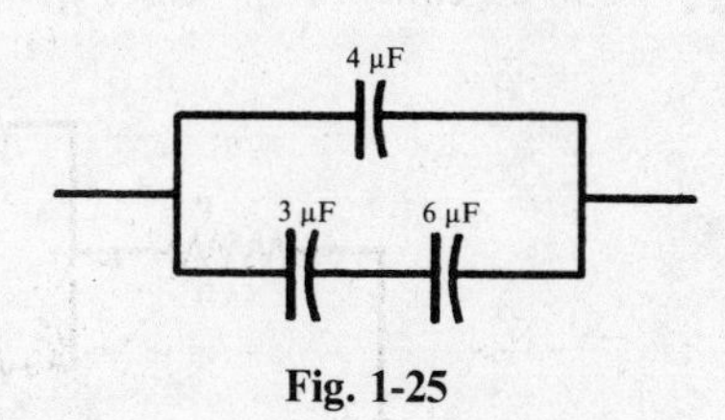

Fig. 1-25

Equivalent capacitance of series branch is

$$Cs = \frac{C_1C_2}{C_1+C_2} = \frac{3(6)}{3+6} = 2\ \mu\text{F}$$

The required equivalent capacitance is

$$C_e = 4 + C_s = 6\ \mu\text{F}$$

1.20. The given series circuit passes a current $i(t)$ of waveform shown in Fig. 1-26. Find the voltage across each element and sketch each voltage to same time scale. Also sketch $q(t)$, the charge on the capacitor.

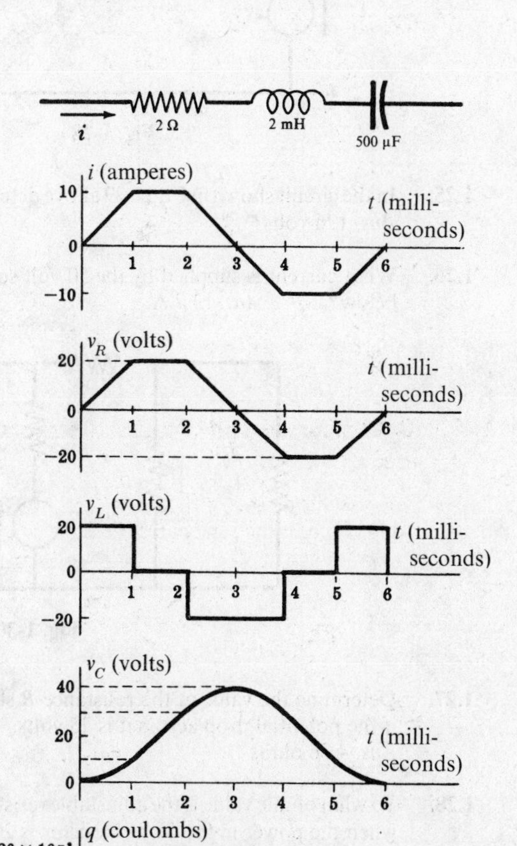

Fig. 1-26

Across Resistor: $v_R = Ri$

The plot of v_R is a duplicate of the current function plot, with a peak value of 2(10) = 20 V.

Across Inductor: $v_L = L\,di/dt$

(1) $0 < t < 1$ ms $\quad i = 10 \times 10^3 t$ amperes

$$v_L = (2 \times 10^{-3})(10 \times 10^3) = 20\ \text{V}$$

(2) $1 < t < 2$ ms $\quad i = 10$ A

$$v_L = (2 \times 10^{-3})(0) = 0$$

etc.

Across Capacitor: $$v_C = \frac{1}{C}\int i\,dt$$

(1) $0<t<1$ ms $$v_C = \frac{1}{500\times10^{-6}}\int_0^t (10\times10^3 t)\,dt = 10\times10^6 t^2 \text{ volts}$$

(2) $1<t<2$ ms $$v_C = 10 + \frac{1}{500\times10^{-6}}\int_{10^{-3}}^t (10)\,dt = 10 + 20\times10^3(t-10^{-3}) \text{ volts}$$

etc.

The plot of q is easily made using the relationship $q = Cv_C$. Note that when i is positive, both q and v_C increase, i.e. both the charge on the capacitor and the voltage across the capacitor increase; when i is negative, both decrease.

Supplementary Problems

1.21. Three resistors R_1, R_2, R_3 are in series with a constant voltage V. The voltage across R_1 is 20 volts, the power in R_2 is 25 watts, and R_3 has a resistance of 2 ohms. Find the voltage V if the current is 5 amperes. *Ans.* 35 volts

1.22. Two resistors, R_1 and R_2 in parallel have an equivalent resistance R_e of 10/3 ohms. When a current enters the parallel circuit it divides between the two resistors in the ratio of 2 to 1. Determine R_1 and R_2. *Ans.* $R_1 = 5$ ohms, $R_2 = 10$ ohms.

1.23. (*a*) Determine R_e for the four resistors in the circuit shown in Fig. 1-27.
(*b*) If a constant voltage $V = 100$ volts is applied, which resistor contains the most power?
Ans. (*a*) $R_e = 5{\cdot}42$ ohms (*b*) The 5 ohm resistor with $P = 957$ watts

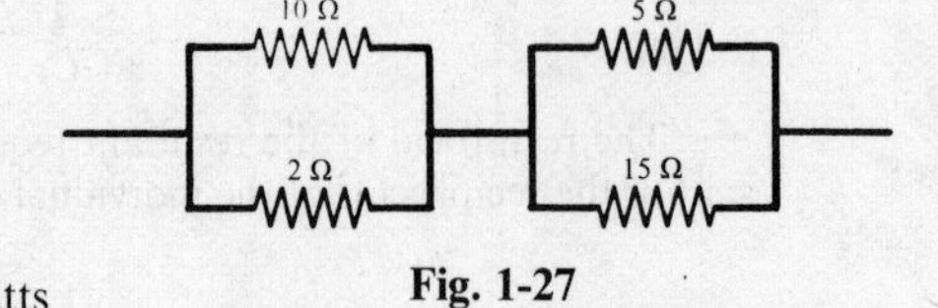

Fig. 1-27

1.24. Two constant voltage sources act in the circuit shown in Fig. 1-28 below. Find the power P supplied by each source to the circuit. *Ans.* $P_{25} = 75$ watts, $P_5 = 15$ watts

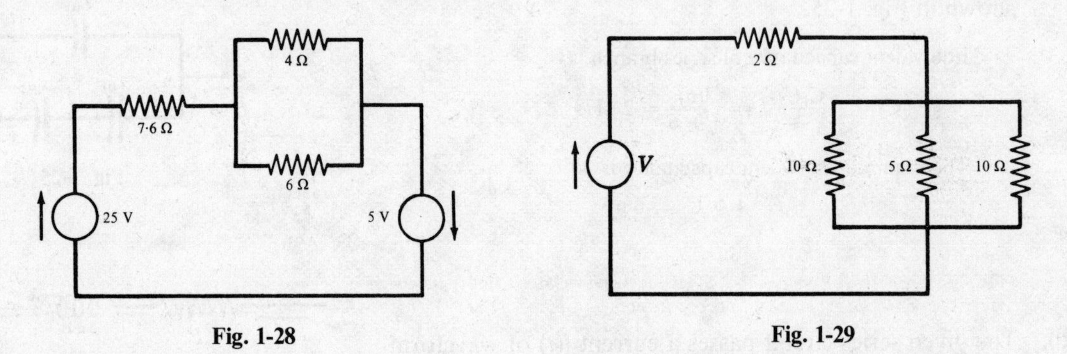

Fig. 1-28 **Fig. 1-29**

1.25. In the circuit shown in Fig. 1-29 above determine the constant voltage V if the current in the 5 ohm resistor is 14 amperes. *Ans.* 126 volts

1.26. What current is supplied by the 50 volt source to the connected network of resistors in the circuit shown in Fig. 1-30 below? *Ans.* 13·7 A

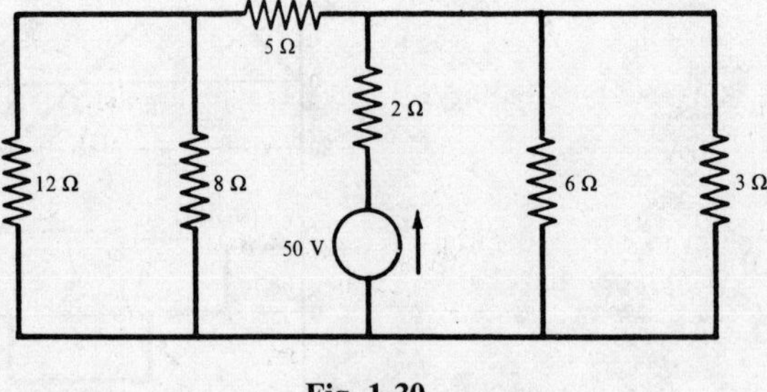

Fig. 1-30

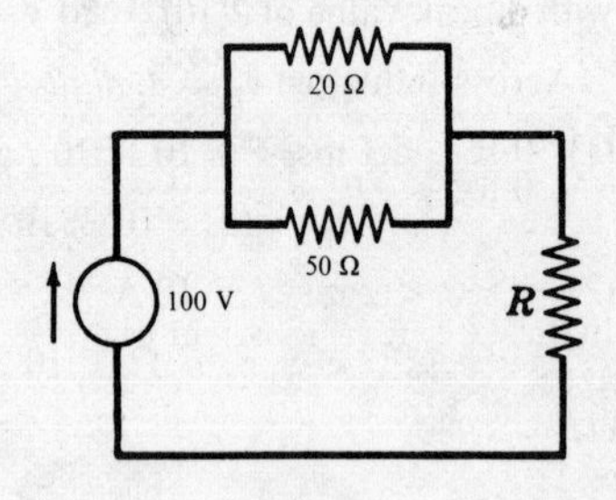

Fig. 1-31

1.27. Determine the value of the resistance R shown in Fig. 1-31 above if the potential drop across it is 25 volts.
Ans. 4·76 ohms

1.28. To what ohmic value is the adjustable resistor shown in Fig. 1-32 set when the power in the 5 ohm resistor is 20 watts?
Ans. 16 ohms

1.29. A 10 ohm resistor is in series with a parallel combination of two resistors of 15 and 5 ohms. If the constant current in the 5 ohm resistor is 6 amperes, what total power is dissipated in the three resistors? *Ans.* 880 watts

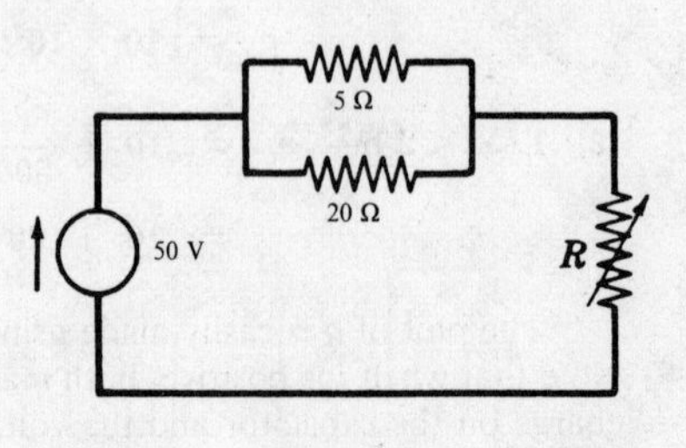

Fig. 1-32

1.30. Inductances L_1 and L_2 shown in Fig. 1-33 below are in the ratio of 2 to 1. If the equivalent inductance L_e of all three is 0·7 H, what are the inductances L_1 and L_2? *Ans.* $L_1 = 0{\cdot}6$ H, $L_2 = 0{\cdot}3$ H

1.31. The three inductances in parallel shown in Fig. 1-34 below have an equivalent inductance L_e of 0·0755 H. (*a*) Find the unknown inductance L. (*b*) Is there a value of L which can make L_e equal to 0·5 H? (*c*) What is the maximum value of L_e if inductance L is adjustable without limit? *Ans.* (*a*) $L = 0{\cdot}1$ H, (*b*) No, (*c*) 0·308 H

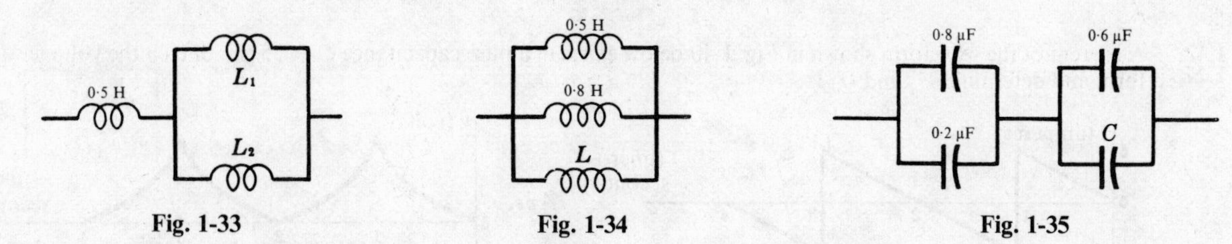

Fig. 1-33 Fig. 1-34 Fig. 1-35

1.32. What capacitance C gives an equivalent capacitance $C_e = 0{\cdot}5\mu$F for the combination of the four capacitors shown in Fig. 1-35 above? *Ans.* 0·4μF

1.33. A constant voltage $V = 100$ V is applied across the combination of four capacitors shown in Fig. 1-36. Find the charge q coulombs on each capacitor.
Ans. $q_{0\cdot8} = 40$ μC, $q_{0\cdot2} = 10$ μC,
$q_{0\cdot3} = 15$ μC, $q_{0\cdot7} = 35$ μC

1.34. The two capacitors shown in Fig. 1-37 are charged by momentary connection to a constant voltage of 50 volts at terminals AB. Then the terminals A and B are connected without the 50 volt source. Determine the final charge on each capacitor.
Ans. $q_{20} = 444\frac{1}{3}$ μC, $q_{40} = 888\frac{2}{3}$ μC.

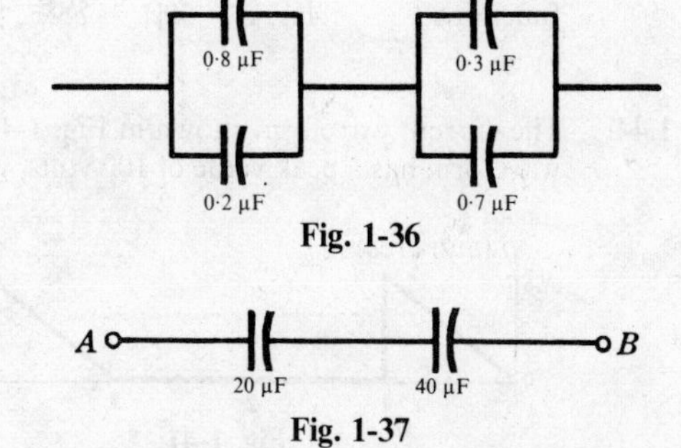

Fig. 1-36

Fig. 1-37

1.35. Show that when a pure resistance R has an applied voltage $v = V_m \sin \omega t$, the energy is given by $w = \dfrac{V_m^2}{2R}\left(t - \dfrac{\sin 2\omega t}{2\omega}\right)$.

1.36. An inductance L contains a current $i = I_m[1 - e^{-\frac{R}{L}t}]$. Show that the maximum stored energy W_m in the magnetic field is given by $W_m = \frac{1}{2}LI_m^2$. ($i = 0$ for $t < 0$)

1.37. If the current in a capacitor is $i = \dfrac{V_m}{R}e^{-\frac{t}{RC}}$, show that the maximum stored energy in the electric field is $W_m = \frac{1}{2}CV_m^2$. ($i = 0$ for $t < 0$).

1.38. In the RC circuit shown in Fig. 1-38 a total energy of $3{\cdot}6 \times 10^{-3}$ joules is dissipated in the 10 ohm resistor when the switch is closed. What initial charge q_0 was on the capacitor? *Ans.* $q_0 = 120$ μC

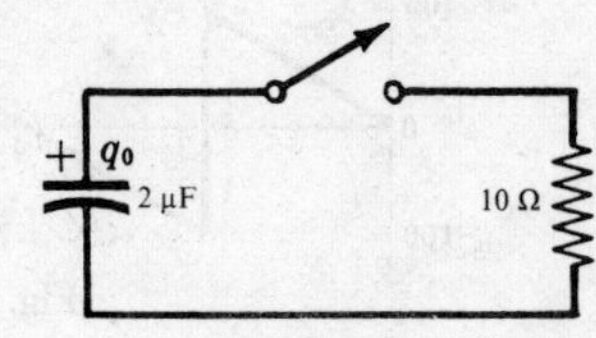

Fig. 1-38

1.39. Show that both $\frac{1}{2}CV^2$ and $\frac{1}{2}LI^2$ have the same units as energy w.

1.40. The voltage waveform shown in Fig. 1-39 is applied to a pure capacitor of 60 μF. Sketch $i(t)$, $p(t)$, and determine I_m and P_m. *Ans.* $I_m = 1{\cdot}5$ A, $P_m = 75$ W

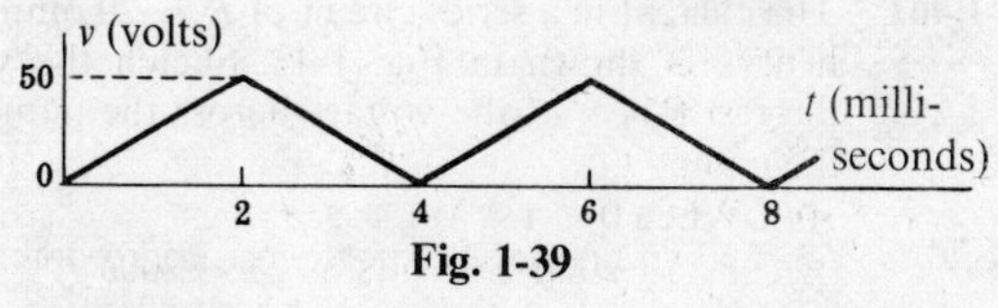

Fig. 1-39

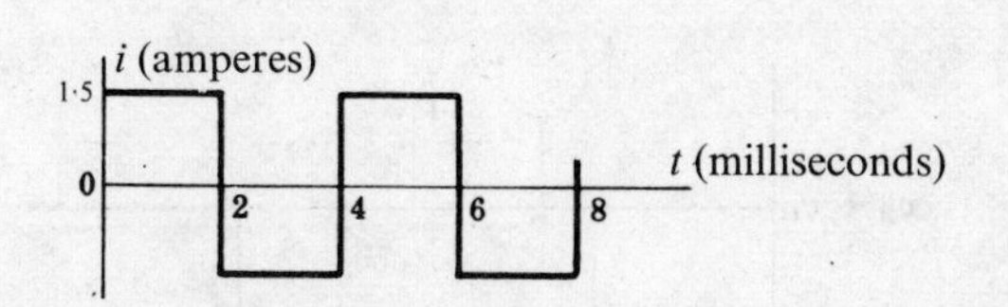

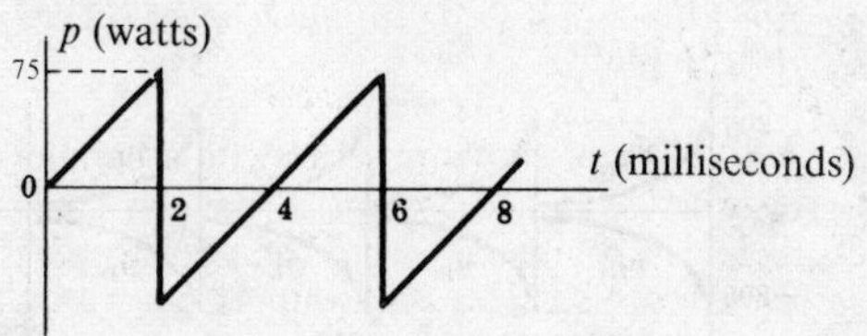

1.41. Determine an expression for the current if the voltage across a pure capacitor is given as

$$v = V_m\left[\omega t - \frac{(\omega t)^3}{3!} + \frac{(\omega t)^5}{5!} - \frac{(\omega t)^7}{7!} + \cdots\right].$$

Ans. $i = \omega C V_m\left[1 - \frac{(\omega t)^2}{2!} + \frac{(\omega t)^4}{4!} - \frac{(\omega t)^6}{6!} + \cdots\right]$ or $i = \omega C V_m \cos \omega t$

1.42. A current of the waveform shown in Fig. 1-40 below exists in a pure capacitance $C = 25\ \mu F$. Sketch the voltage waveform and determine V_m and Q_m.

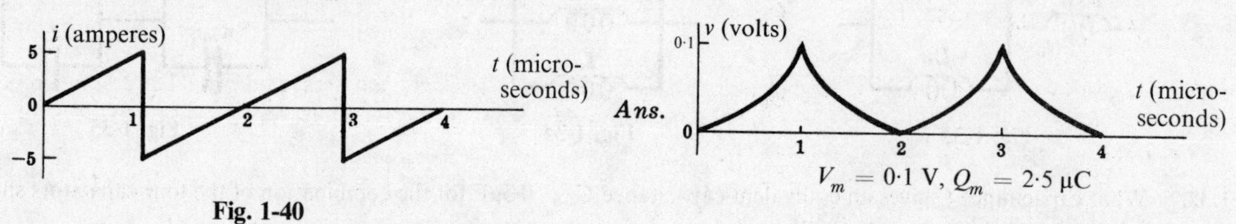

$V_m = 0{\cdot}1$ V, $Q_m = 2{\cdot}5\ \mu C$

Fig. 1-40

1.43. A 2 μF capacitor has a charge function $q = 100[1 \times e^{-5 \times 10^{4t}}]$ μC. Determine the corresponding voltage and current functions. *Ans.* $v = 50[1 + e^{-5 \times 10^{4t}}]$ volts, $i = -5e^{-5 \times 10^{4t}}$ amperes

1.44. The current waveform shown in Fig. 1-41 below is the current in a pure inductance L. If the corresponding voltage waveform has a peak value of 100 volts, what is the magnitude of L? Sketch the voltage waveform.

Ans. $L = 0{\cdot}5$ H

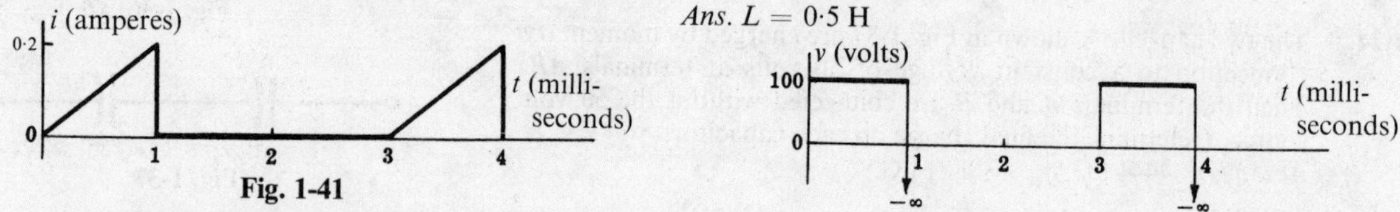

Fig. 1-41

Note. In the practical case it is not possible for the current in an inductance to be a discontinuous function as shown in the current waveform at $t = 1$ ms and $t = 4$ ms. Since the voltage is equal to the first derivative of the current function times the inductance L, and this derivative has an infinite negative value at the points of discontinuity, there will be negative infinite spikes on the voltage waveform at these points.

1.45. A pure inductance of 0·05 H has an applied voltage with the waveform shown in Fig. 1-42 below. Sketch the corresponding current waveform and determine the expression for i in the first interval, i.e. $0 < t < 2$ ms.

Ans. $i = 5 \times 10^5 t^2$ amperes

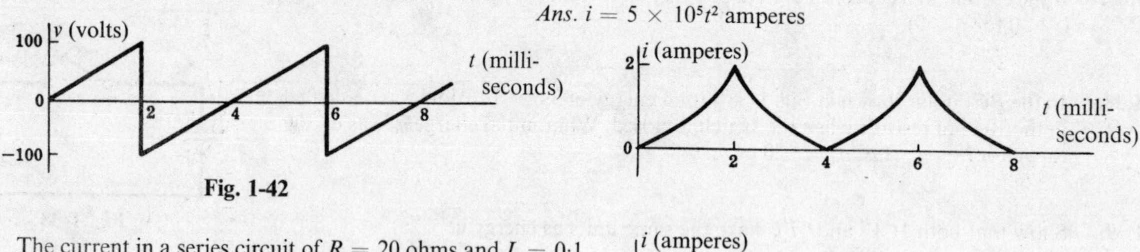

Fig. 1-42

1.46. The current in a series circuit of $R = 20$ ohms and $L = 0{\cdot}1$ henrys is shown in Fig. 1-43. Sketch the voltage across the resistance v_R, the voltage across the inductance v_L and their sum.

Ans. When $0 < t < 0{\cdot}1$ s:

$V_R = 200e^{-200t}$ volts, $v_L = -200e^{-200t}$ volts, $v_T = 0$

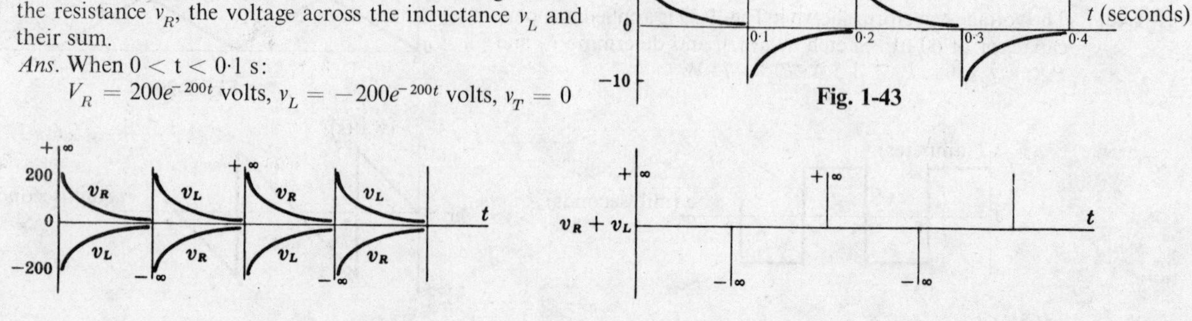

Fig. 1-43

1.47. A series RL circuit with $R = 5$ ohms and $L = 0{\cdot}004$ henrys contains a current with a waveform as shown in Fig. 1-44 below. Sketch v_R and v_L.

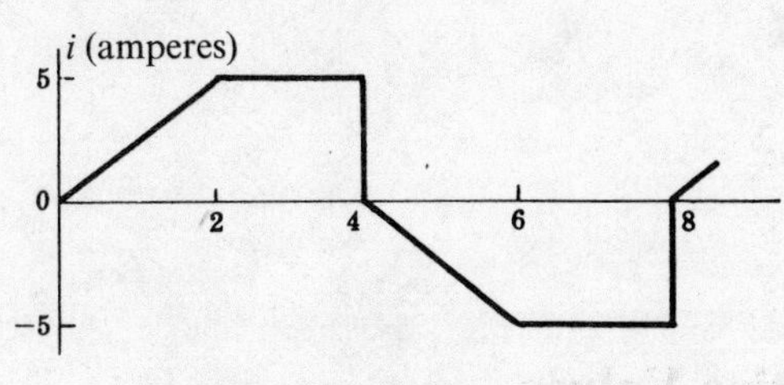

Fig. 1-44

Ans.

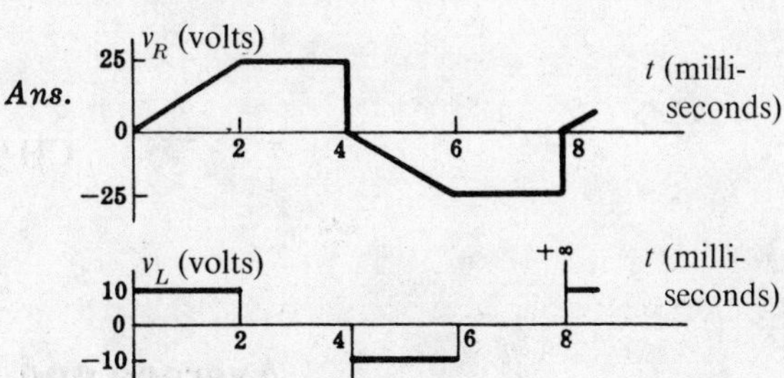

1.48. A series RL circuit with $R = 10$ ohms and $L = 0{\cdot}5$ henrys has a sinusoidal applied voltage. The resulting current is $i = 0{\cdot}822e^{-20t} + 0{\cdot}822 \sin(377t - 86{\cdot}96°)$ amperes. Determine the corresponding voltages v_R, v_L, and v_T.

Ans. $v_R = 8{\cdot}22e^{-20t} + 8{\cdot}22 \sin(377t - 86{\cdot}96°)$ volts
$v_L = -8{\cdot}22e^{-20t} + 155 \cos(377t - 86{\cdot}96°)$ volts
$v_T = 155 \sin 377t$ volts

1.49. A series RL circuit with $R = 100$ ohms and $L = 0{\cdot}05$ henrys has a current function as described below. Determine v_M and v_L in each interval.
(1) $0 < t < 10 \times 10^{-3}$ s, $i = 5[1 - e^{-2000t}]$ amperes
(2) $10 \times 10^{-3} < t$, $i = 5\,e^{-2000(t - 10 \times 10^{-3})}$ amperes

Ans. (1) $v_R = 500[1 - e^{-2000t}]$ volts, $v_L = 500\,e^{-2000t}$ volts
(2) $v_R = 500\,e^{-2000(t - 10 \times 10^{-3})}$ volts, $v_L = -500\,e^{-2000(t - 10 \times 10^{-3})}$ volts

1.50. The current in a series RC circuit is $i = 10\,e^{-500t}$ amperes. There was no initial charge on the capacitor. After the current transient the capacitor has a charge of 0·02 coulombs. If the applied voltage is $V = 100$ volts, and $v_C = 100[1 - e^{-500t}]$ volts, find C and v_R.

Ans. $C = 200\ \mu$F, $v_R = 100\,e^{-500t}$ volts

1.51. A series LC circuit with $L = 0{\cdot}02$ H and $C = 30\ \mu$F contains a current $i = 1{\cdot}5 \cos 1000t$ amperes. Find the total voltage v_T. *Ans.* $v_T = 20 \sin 1000t$ volts

1.52. The parallel RL circuit has a square wave of voltage as shown in Fig. 1-45. Determine the total current.

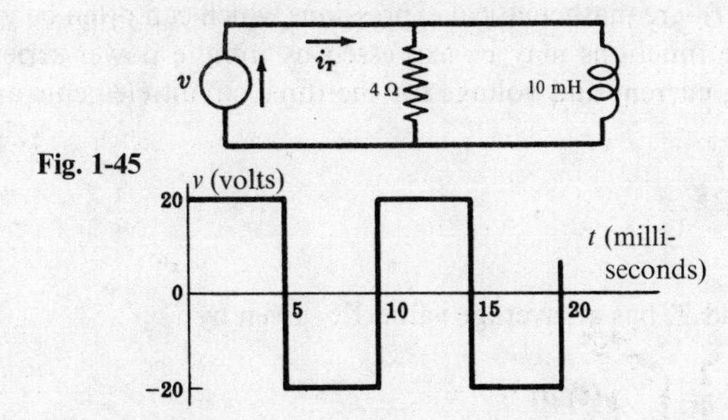

Fig. 1-45

Ans.

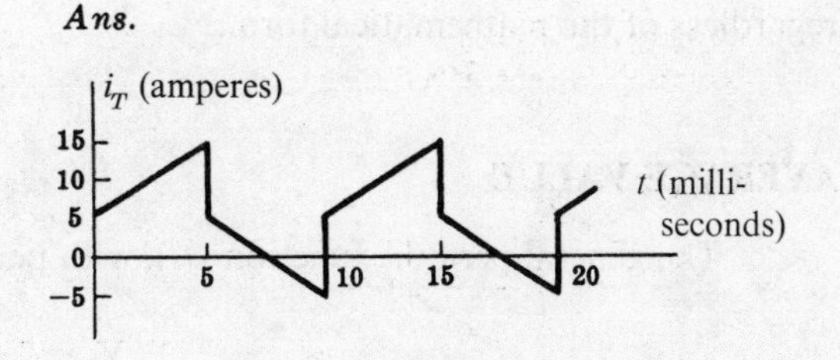

1.53. The parallel RC circuit has an applied voltage of waveform shown in Fig. 1-46 below. Determine the total current i_T.

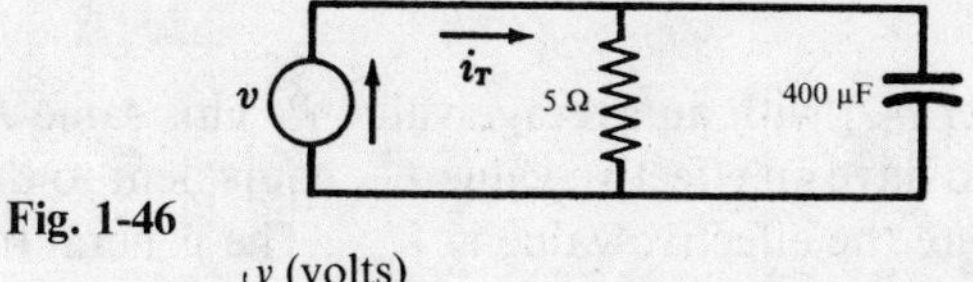

Fig. 1-46

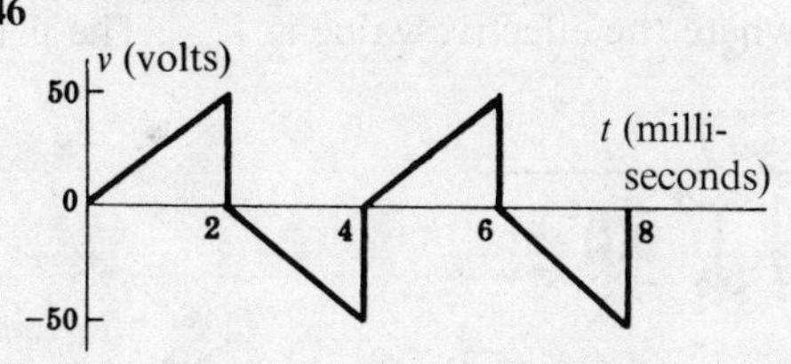

Ans.

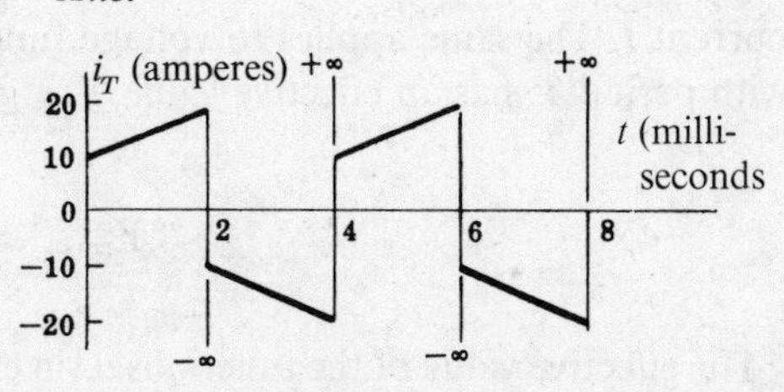

CHAPTER 2

Average and Effective Values

WAVEFORMS

The pictorial representations or graphs of $v(t)$, $i(t)$, $p(t)$, etc., are the waveforms of voltage, current and power respectively. Introductory circuit analysis deals only with periodic functions, i.e. those for which $f(t) = f(t + nT)$, where n is an integer and T is the period, as shown in Fig. 2-1. If the function is periodic, at least one period is required in the plot to be called a waveform.

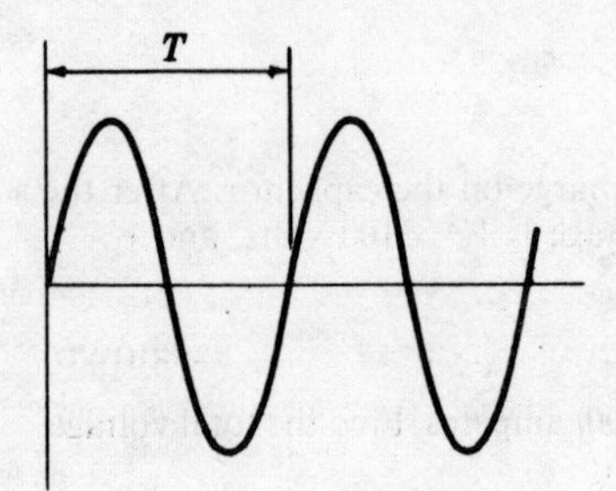

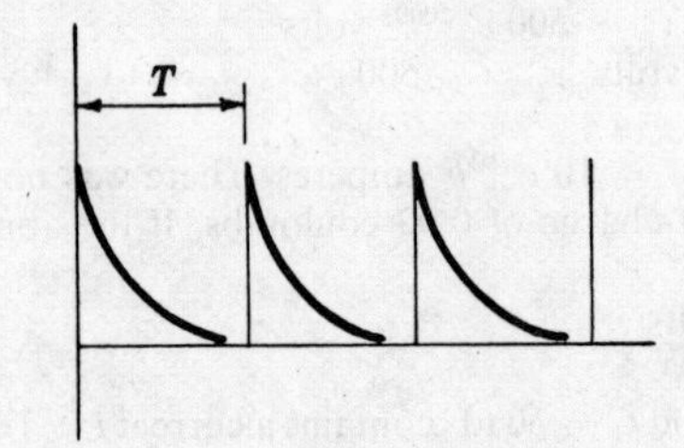

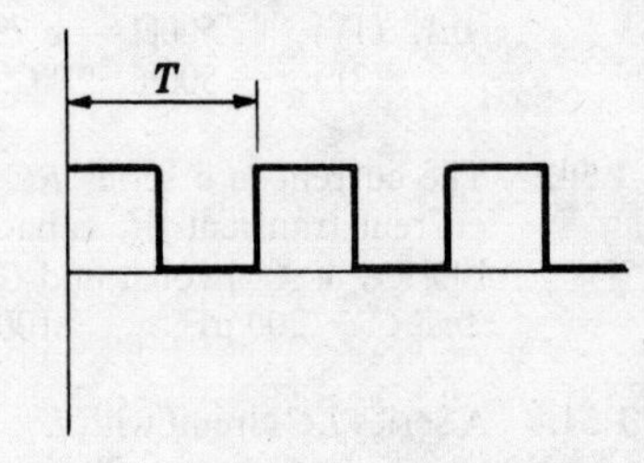

Fig. 2-1. Periodic Waveforms

The voltage and current functions, $v(t)$ and $i(t)$, are mathematical expressions which can often be given in several forms. For example, the sine and cosine functions may be expressed by infinite power series. It should be stressed that the basic equations relating current and voltage for the three circuit elements apply regardless of the mathematical form.

AVERAGE VALUE

The general periodic function $y(t)$, with period T, has an average value Y_{av} given by

$$Y_{av} = \frac{1}{T}\int_0^T y(t)\,dt$$

ROOT MEAN SQUARE OR EFFECTIVE VALUE

A current $i(t)$ in a pure resistor R results in a power $p(t)$ with an average value P. This same P could be produced in R by a constant current I. Then $i(t)$ is said to have an effective value I_{rms} equivalent to this constant current I. The same applies to voltage functions where the effective value is V_{rms}. The general function $y(t)$, with period T, has an effective value Y_{rms} given by

$$Y_{rms} = \sqrt{\frac{1}{T}\int_0^T \overline{y(t)}^2\,dt}$$

The effective value of the functions $a \sin \omega t$ and $a \cos \omega t$ is $a/\sqrt{2}$. See Problem 2.2.

RMS VALUE FOR SEVERAL SINE AND COSINE TERMS

The function $y(t) = a_0 + (a_1 \cos \omega t + a_2 \cos 2\omega t + \cdots) + (b_1 \sin \omega t + b_2 \sin 2\omega t + \cdots)$ has an effective value given by

$$Y_{\text{rms}} = \sqrt{a_0^2 + \tfrac{1}{2}(a_1^2 + a_2^2 + \cdots) + \tfrac{1}{2}(b_1^2 + b_2^2 + \cdots)}$$

Also, if A_1 is the effective value of $a_1 \cos \omega t$, then $A_1 = \dfrac{a_1}{\sqrt{2}}$, $A_1^2 = \dfrac{a_1^2}{2}$, and

$$Y_{\text{rms}} = \sqrt{a_0^2 + (A_1^2 + A_2^2 + \cdots) + (B_1^2 + B_2^2 + \cdots)}$$

FORM FACTOR

The ratio of effective value to average value is the form factor F of the waveform. It has use in voltage generation and instrument correction factors.

$$\text{Form Factor} = \frac{Y_{\text{rms}}}{Y_{\text{av}}} = \frac{\sqrt{\dfrac{1}{T}\displaystyle\int_0^T \overline{y(t)}^2\, dt}}{\dfrac{1}{T}\displaystyle\int_0^T y(t)\, dt}$$

Waveforms with half wave symmetry, i.e. $f(t) = -f(t + \frac{1}{2}T)$, have an average value of zero, as shown in Fig. 2-2 below. For these waveforms, of which the sine wave is an example, Y_{av} is computed over the positive half of the period. This is sometimes called the half cycle average.

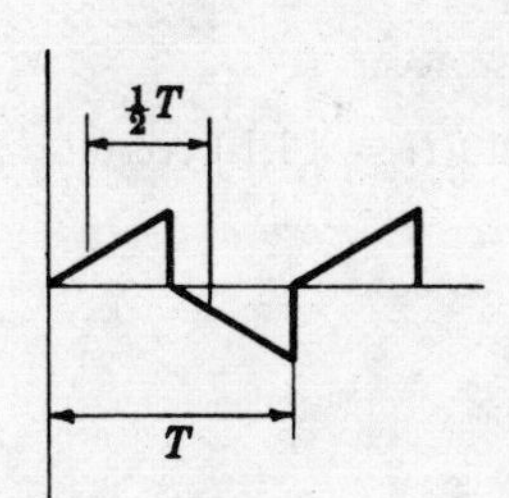

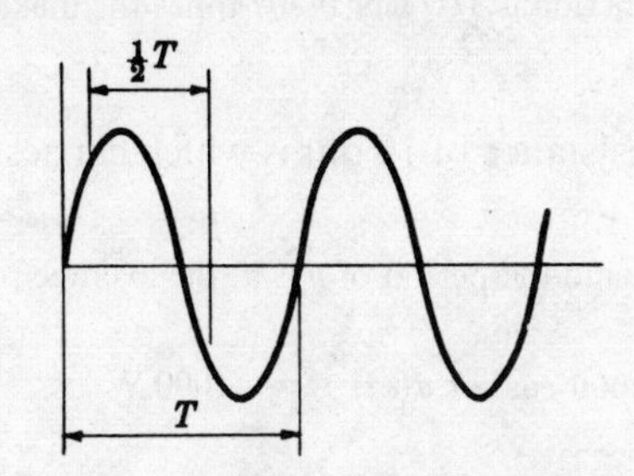

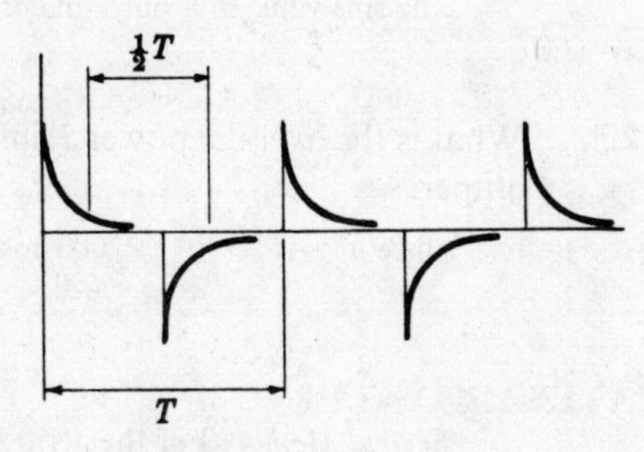

Fig. 2-2. Half Wave Symmetry

Other waveforms will have an average value of zero but do not have half wave symmetry, as shown in Fig. 2-3 below. In calculation of Y_{av} for use in the form factor, one-half the period is chosen just as with waveforms showing half wave symmetry.

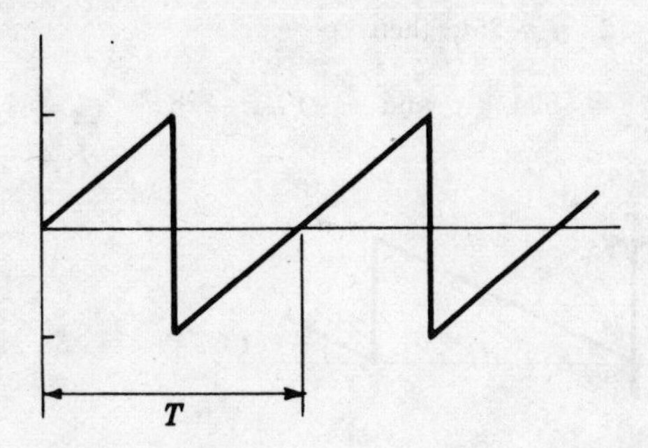

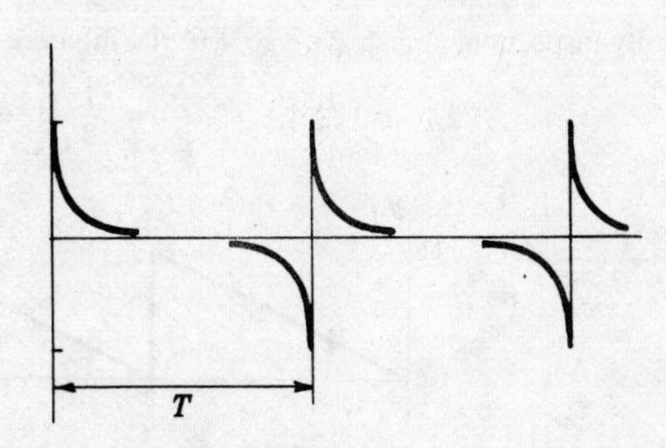

Fig. 2-3

Solved Problems

2.1. A resistor carries (*a*) a constant current I, (*b*) a periodic current $i(t)$ with period T. See Fig. 2-4 below. Show that the average power P is the same in each case if $I_{\text{rms}} = I$.

For constant current I: $\quad P = VI = I^2R$

For periodic current $i(t)$: $\quad p = vi = i^2R \quad$ and $\quad P = \left(\frac{1}{T}\int_0^T i^2\,dt\right)R = I_{\text{rms}}^2 R$

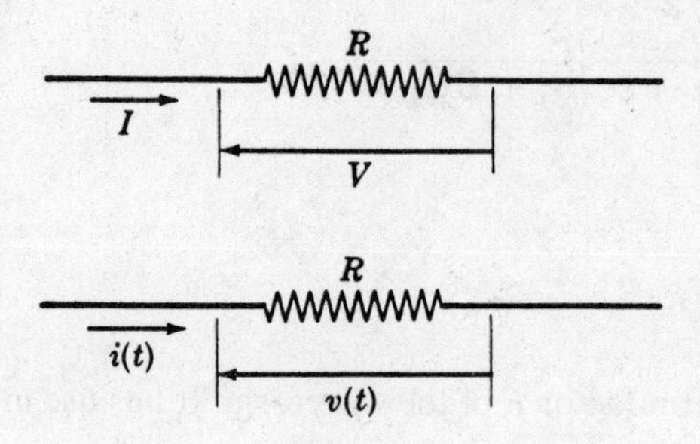

Fig. 2-4

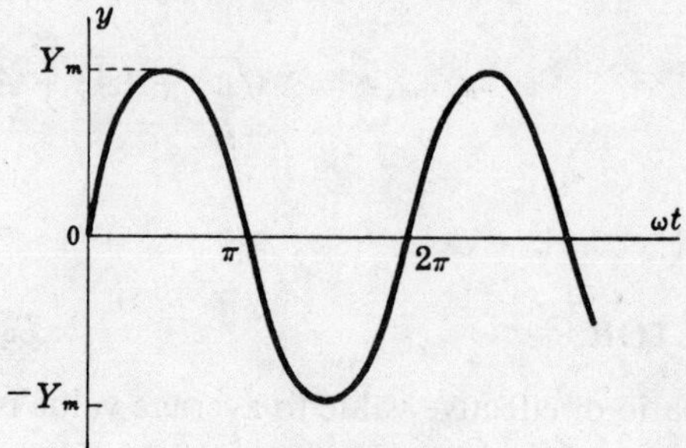

Fig. 2-5

2.2. Find the average and effective (root mean square) values for the function $y(t) = Y_m \sin \omega t$.

The period is 2π. The plot is given with ωt as the independent variable as shown in Fig. 2-5 above.

$$Y_{\text{av}} = \frac{1}{T}\int_0^T y(t)\,dt = \frac{1}{2\pi}\int_0^{2\pi} Y_m \sin \omega t\, d(\omega t) = \frac{1}{2\pi}Y_m\Big[-\cos \omega t\Big]_0^{2\pi} = 0$$

$$Y_{\text{rms}} = \sqrt{\frac{1}{T}\int_0^T y^2\,dt} = \sqrt{\frac{1}{2\pi}\int_0^{2\pi}(Y_m \sin \omega t)^2\,d(\omega t)} = \frac{Y_m}{\sqrt{2}} = 0{\cdot}707\,Y_m$$

The rms value of a pure sine or cosine function is $1/\sqrt{2}$ or 0·707 times the maximum value.

2.3. What is the average power P in a pure resistance of 10 ohms which carries a current $i(t) = 14.14 \cos \omega t$ amperes?

Since $p = vi = Ri^2 = 2000 \cos^2 \omega t$ watts and the period of p is π, the average power is

$$P = \frac{1}{\pi}\int_0^{\pi} 2000 \cos^2 \omega t\, d(\omega t) = 1000\text{ W}$$

Second Method. For the periodic current $i(t)$ in a pure resistor R, the average power is

$$P = I_{\text{rms}}^2 R = \left\{\frac{1}{2\pi}\int_0^{2\pi}(14.14 \cos \omega t)^2\,d(\omega t)\right\}10 = (14.14/\sqrt{2})^2(10) = 1000\text{ W}$$

2.4. Find the average and effective values of the sawtooth waveform shown in Fig. 2-6 below.

By inspection, $Y_{\text{av}} = 25$. For the interval $0 < t < 2$, $y = 25t$; then

$$Y_{\text{rms}}^2 = \frac{1}{T}\int_0^T y^2\,dt = \frac{1}{2}\int_0^2 625t^2\,dt = 834, \quad\text{and}\quad Y_{\text{rms}} = 28{\cdot}9$$

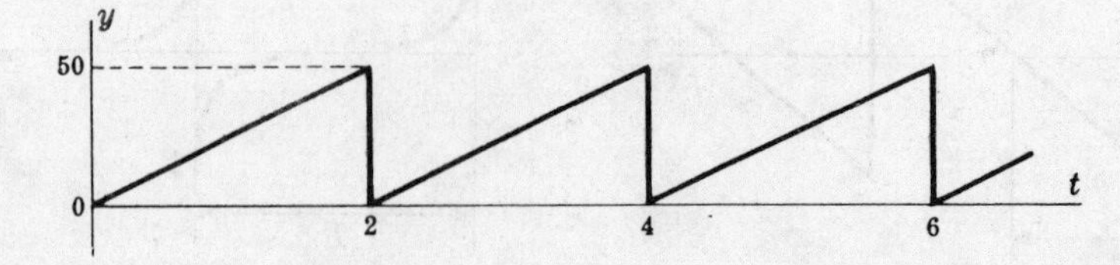

Fig. 2-6

2.5. Determine the average and rms values of the waveform shown in Fig. 2-7 below, where in the first interval $y = 10e^{-200t}$.

$$Y_{av} = \frac{1}{T}\int_0^T y\,dt = \frac{1}{0.05}\int_0 10\,e^{-200t}\,dt = \frac{10}{0.05(-200)}\Big[e^{-200t}\Big]_0^{0.05}$$

$$= -1[e^{-10} - e^0] = 1.00$$

$$Y^2_{rms} = \frac{1}{T}\int_0^T y^2\,dt = \frac{1}{0.05}\int_0^{0.05} 100\,e^{-400t}\,dt = 5.00, \quad \text{and} \quad Y_{rms} = 2.24$$

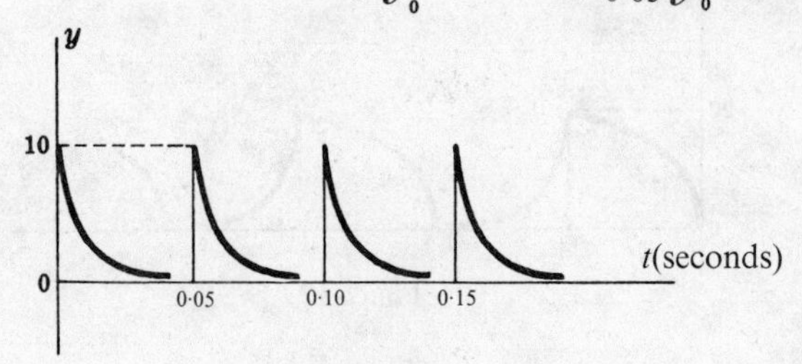

Fig. 2-7

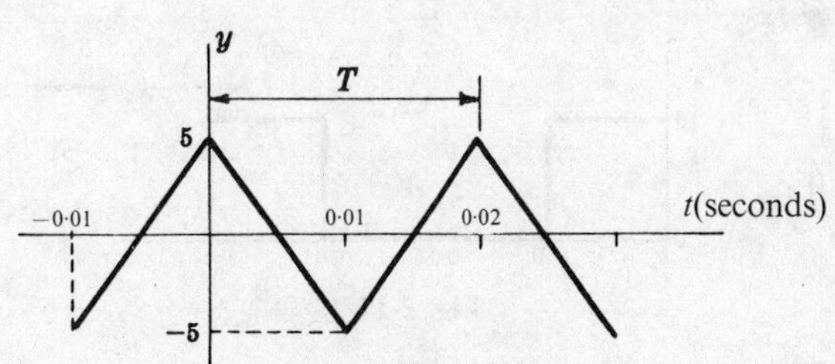

Fig. 2-8

2.6. Find the form factor of the triangular wave shown in Fig. 2-8 above.

$$-0.01 < t < 0: \quad y(t) = 1000t + 5; \quad \overline{y(t)}^2 = 10^6t^2 + 10^4t + 25$$

$$0 < t < 0.01: \quad y(t) = -1000t + 5; \quad \overline{y(t)}^2 = 10^6t^2 - 10^4t + 25$$

$$Y^2_{rms} = \frac{1}{0.02}\left\{\int_{-0.01}^0 (10^6t^2 + 10^4t + 25)\,dt + \int_0^{0.01} (10^6t^2 - 10^4t + 25)\,dt\right\} = 8.33, \qquad Y_{rms} = 2.89$$

Since the waveform has half-wave symmetry the average is taken over the positive portion; thus,

$$Y_{av} = \frac{1}{0.01}\left\{\int_{-0.005}^0 (1000t + 5)\,dt + \int_0^{-0.005} (-1000t + 5)\,dt\right\} = 2.5$$

$$\text{Form factor} = \frac{Y_{rms}}{Y_{av}} = \frac{2.89}{2.5} = 1.16$$

2.7. Find the average and effective values of the half-wave rectified sine wave shown in Fig. 2-9 below.

For $0 < \omega t < \pi$, $y = Y_m \sin \omega t$; for $\pi < \omega t < 2\pi$, $y = 0$. The period is 2π.

$$Y_{av} = \frac{1}{2\pi}\left\{\int_0^\pi Y_m \sin \omega t\,d(\omega t) + \int_\pi^{2\pi} 0\,d(\omega t)\right\} = 0.318Y_m$$

$$Y^2_{rms} = \frac{1}{2\pi}\int_0^\pi (Y_m \sin \omega t)^2\,d(\omega t) = \tfrac{1}{4}Y^2_m, \qquad Y_{rms} = \tfrac{1}{2}Y_m$$

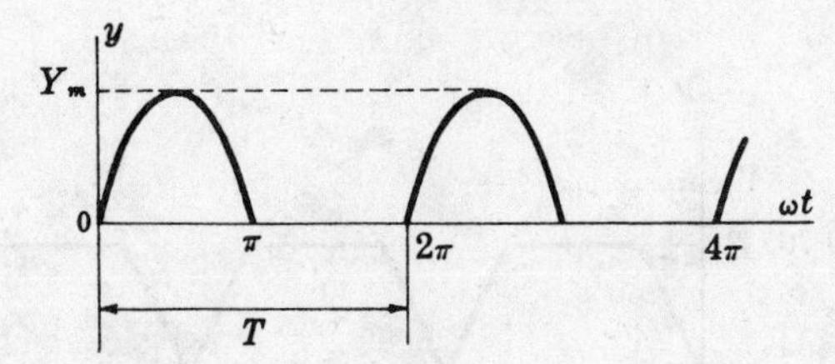

Fig. 2-9

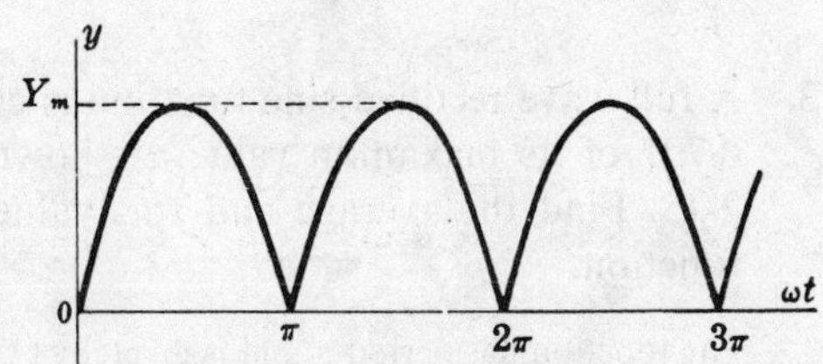

Fig. 2-10

2.8. Find the average and rms values of the full-wave rectified sine wave shown in Fig. 2-10 above. The period is π.

$$Y_{av} = \frac{1}{\pi}\int_0^\pi Y_m \sin \omega t\,d(\omega t) = 0.637Y_m$$

$$Y^2_{rms} = \frac{1}{\pi}\int_0^\pi (Y_m \sin \omega t)^2\,d(\omega t) = \frac{Y^2_m}{2}, \qquad Y_{rms} = 0.707Y_m$$

2.9. Find the average and effective values of the square wave shown in Fig. 2-11 below.

For $0 < t < 0{\cdot}01$, $y = 10$; for $0{\cdot}01 < t < 0{\cdot}03$, $y = 0$. The period is 0·03 s.

$$Y_{av} = \frac{1}{0{\cdot}03}\int_0^{0{\cdot}01} 10\,dt = \frac{10(0{\cdot}01)}{0{\cdot}03} = 3{\cdot}33$$

$$Y_{rms}^2 = \frac{1}{0{\cdot}03}\int_0^{0{\cdot}01} 10^2\,dt = 33.3, \qquad Y_{rms} = 5{\cdot}77$$

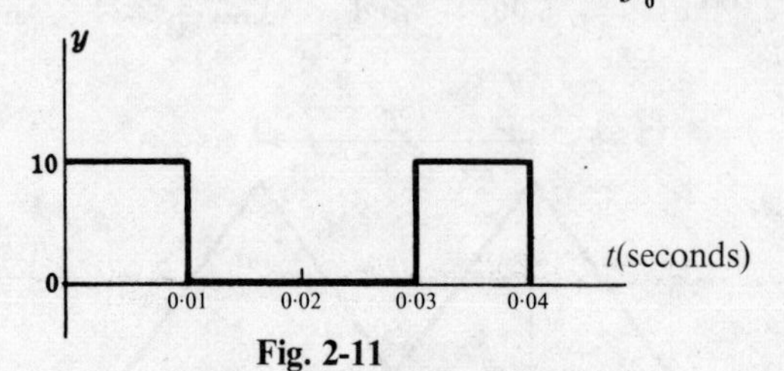

Fig. 2-11

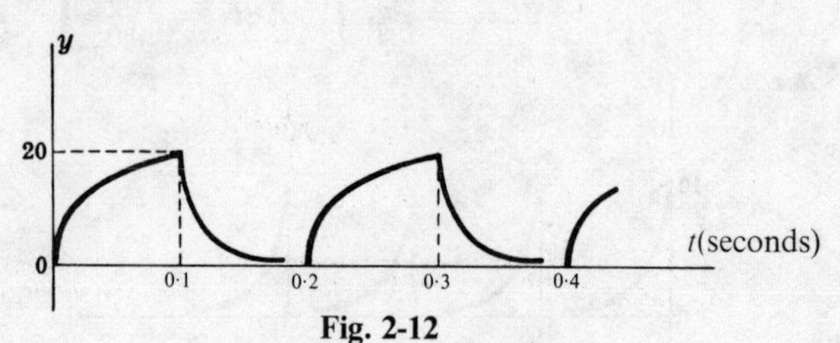

Fig. 2-12

2.10. Find the rms value of the function shown in Fig. 2-12 above and described as follows:

$0 < t < 0{\cdot}1 \quad y = 20(1 - e^{-100t})$; $\qquad 0{\cdot}1 < t < 0{\cdot}2 \quad y = 20\,e^{-50(t-0{\cdot}1)}$

$$Y_{rms}^2 = \frac{1}{0{\cdot}2}\left\{\int_0^{0{\cdot}1} 400(1 - 2\,e^{-100t} + e^{-200t})\,dt + \int_{0{\cdot}1}^{0{\cdot}2} 400\,e^{-100(t-0{\cdot}1)}\,dt\right\}$$

$$= 2000\left\{\left[t + 0{\cdot}02\,e^{-100t} - 0{\cdot}005\,e^{-200t}\right]_0^{0{\cdot}1} + \left[-0{\cdot}01\,e^{-100(t-0{\cdot}1)}\right]_{0{\cdot}1}^{0{\cdot}2}\right\}$$

$= 190$, and $Y_{rms} = 13{\cdot}78$. (The terms in e^{-10} and e^{-20} are not significant.)

2.11. Find the effective value of the function $y = 50 + 30\sin\omega t$.

$$Y_{rms}^2 = \frac{1}{2\pi}\int_0^{2\pi}(2500 + 3000\sin\omega t + 900\sin^2\omega t)\,d(\omega t)$$

$$= \frac{1}{2\pi}[2500(2\pi) + 0 + 900\pi] = 2950, \qquad Y_{rms} = 54{\cdot}3$$

Another method: $\quad Y_{rms} = \sqrt{(50)^2 + \frac{1}{2}(30^2)} = \sqrt{2950} = 54{\cdot}3$

2.12. Find the effective value of the voltage function $v = 50 + 141.4\sin\omega t + 35.5\sin 3\omega t$ volts.

$$V_{rms} = \sqrt{(50)^2 + \tfrac{1}{2}(141{\cdot}4)^2 + \tfrac{1}{2}(35{\cdot}5)^2} = 114{\cdot}6\text{ V}$$

2.13. A full wave rectified sine function is clipped at 0·707 of its maximum value as shown in Fig. 2-13. Find the average and rms values of the function.

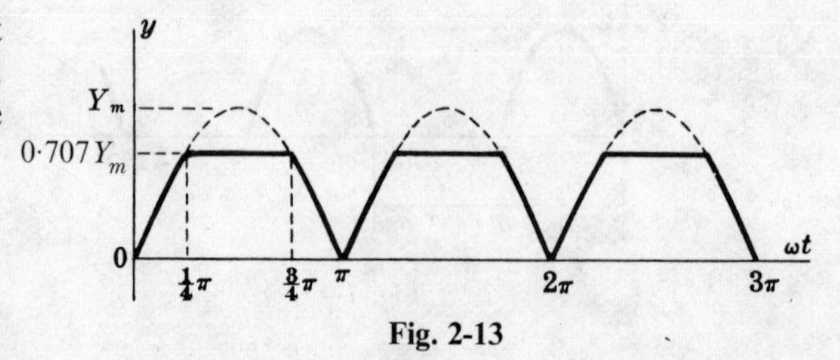

Fig. 2-13

The function has period π and is given by

$0 < \omega t < \pi/4 \quad y = Y_m \sin\omega t$

$\pi/4 < \omega t < 3\pi/4 \quad y = 0{\cdot}707Y_m$

$3\pi/4 < \omega t < \pi \quad y = Y_m \sin\omega t$

$$Y_{av} = \frac{1}{\pi}\left\{\int_0^{\pi/4} Y_m \sin\omega t\,d(\omega t) + \int_{\pi/4}^{3\pi/4} 0{\cdot}707Y_m\,d(\omega t) + \int_{3\pi/4}^{\pi} Y_m\sin\omega t\,d(\omega t)\right\} = 0{\cdot}54Y_m$$

$$Y_{rms}^2 = \frac{1}{\pi}\left\{\int_0^{\pi/4}(Y_m\sin\omega t)^2\,d(\omega t) + \int_{\pi/4}^{3\pi/4}(0{\cdot}707Y_m)^2\,d(\omega t) + \int_{3\pi/4}^{\pi}(Y_m\sin\omega t)^2\,d(\omega t)\right.$$

$$= 0{\cdot}341Y_m^2, \qquad Y_{rms} = 0{\cdot}584Y_m$$

2.14. A delayed full wave rectified sine wave has an average value of half the maximum value as shown in Fig. 2-14. Find the angle θ.

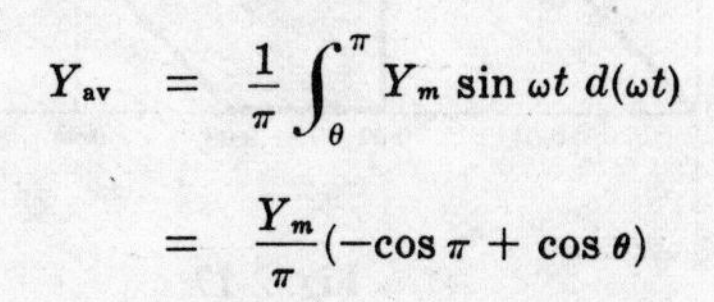

$$Y_{av} = \frac{1}{\pi}\int_{\theta}^{\pi} Y_m \sin \omega t \, d(\omega t)$$

$$= \frac{Y_m}{\pi}(-\cos \pi + \cos \theta)$$

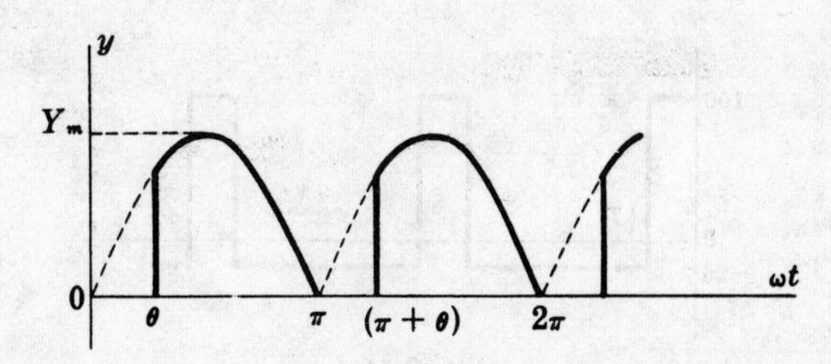

Fig. 2-14

Then $0{\cdot}5Y_m = (Y_m/\pi)(1 + \cos\theta)$, $\cos\theta = 0{\cdot}57$, $\theta = 55{\cdot}25°$.

2.15. The current in a 2 ohm resistor has the waveform given in Problem 2.14, with a maximum value of 5 amperes. The average power in the resistor is 20 watts. Find the angle θ.

$$P = I_{rms}^2 R, \quad 20 = I_{rms}^2 (2), \quad I_{rms}^2 = 10. \quad \text{Then}$$

$$I_{rms}^2 = 10 = \frac{1}{\pi}\int_{\theta}^{\pi} (5 \sin \omega t)^2 \, d(\omega t) = \frac{25}{\pi}\left[\frac{\omega t}{2} - \frac{\sin 2\omega t}{4}\right]_{\theta}^{\pi} = \frac{25}{\pi}\left(\frac{\pi}{2} - \frac{\sin 2\pi}{4} - \frac{\theta}{2} + \frac{\sin 2\theta}{4}\right)$$

from which $\sin 2\theta = 2\theta - 10\pi/25$ and $\theta = 60.5°$ (graphical solution).

Supplementary Problems

2.16. A 25 ohm resistor has an average power of 400 watts. Determine the maximum value of the current function if it is (*a*) sinusoidal, (*b*) triangular. *Ans.* (*a*) 5·66 A, (*b*) 6·93 A

2.17. Determine the effective value V_{rms} of the voltage function given by $v(t) = 100 + 25 \sin 3\omega t + 10 \sin 5\omega t$ volts.
Ans. 101·8 V

2.18. What average power results in a 25 ohm resistor when it passes a current $i(t) = 2 + 3 \sin \omega t + 2 \sin 2\omega t + 1 \sin 3\omega t$ amperes? *Ans.* 275 W

2.19. Calculate Y_{rms} if $y(t) = 50 + 40 \sin \omega t$. *Ans.* 57·4

2.20. Calculate Y_{rms} if $y(t) = 150 + 50 \sin \omega t + 25 \sin 2\omega t$. *Ans.* 155·3

2.21. The effective value of $y(t) = 100 + A \sin \omega t$ is known to be 103·1. Determine the amplitude A of the sine term.
Ans. 35·5

2.22. A certain function contains a constant term, a fundamental and a third harmonic. The maximum value of the fundamental is 80% of the constant term and the maximum value of the third harmonic is 50% of the constant term. If the effective value of this function is 180·3, find the magnitude of the constant term and the two harmonics.
Ans. 150, 120, 75

2.23. If a half-wave rectified sine wave has an effective value of 20, what is its average value? *Ans.* 12·7

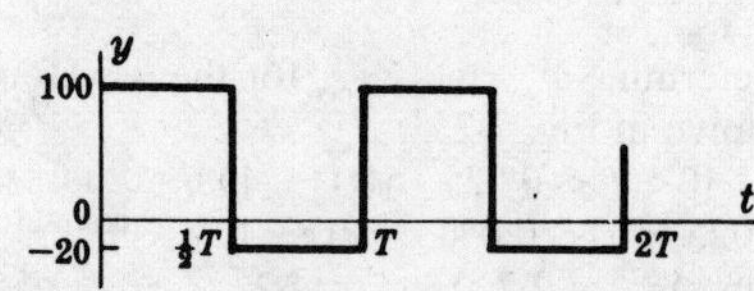

Fig. 2-15

2.24. Calculate Y_{av} and Y_{rms} for the waveform shown in Fig. 2-15.
Ans. $Y_{av} = 40$, $Y_{rms} = 72{\cdot}1$

2.25. Calculate Y_{av} and Y_{rms} for the waveform shown in Fig. 2-16 below. *Ans.* $Y_{av} = 10$, $Y_{rms} = 52{\cdot}9$

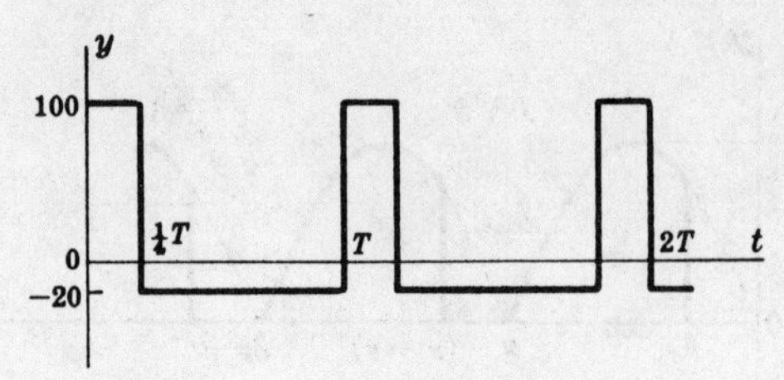

Fig. 2-16

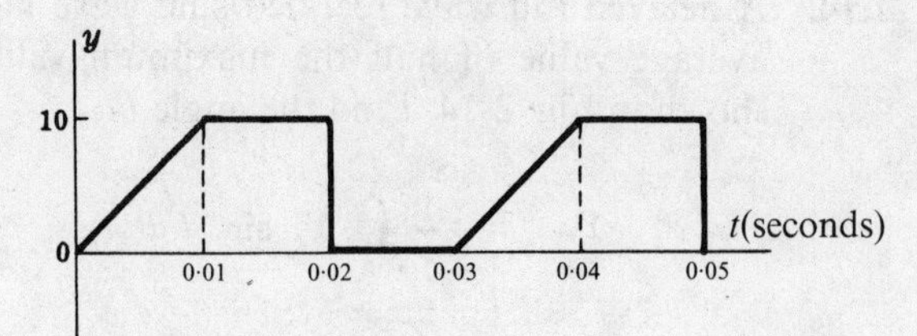

Fig. 2-17

2.26. Determine Y_{rms} of the waveform shown in Fig. 2-17 above. *Ans.* $Y_{rms} = 6{\cdot}67$

2.27. Determine Y_{rms} of the waveform shown in Fig. 2-18 below. *Ans.* $Y_{rms} = Y_m/\sqrt{3} = 0{\cdot}577\,Y_m$

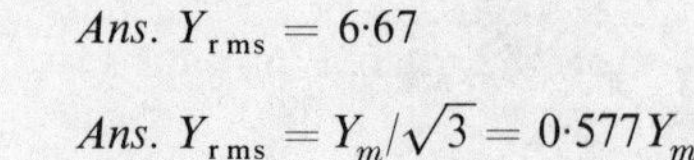

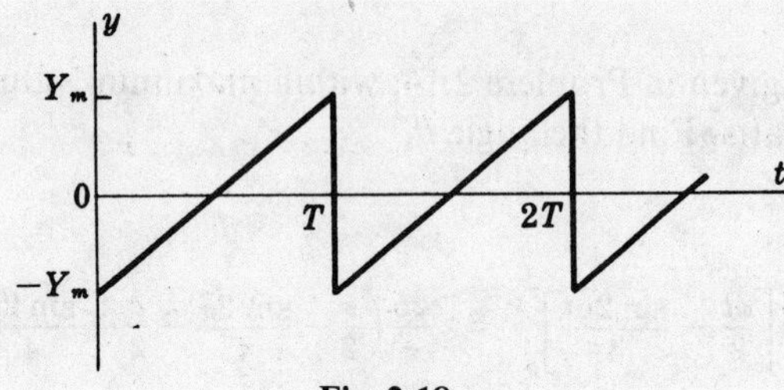

Fig. 2-18

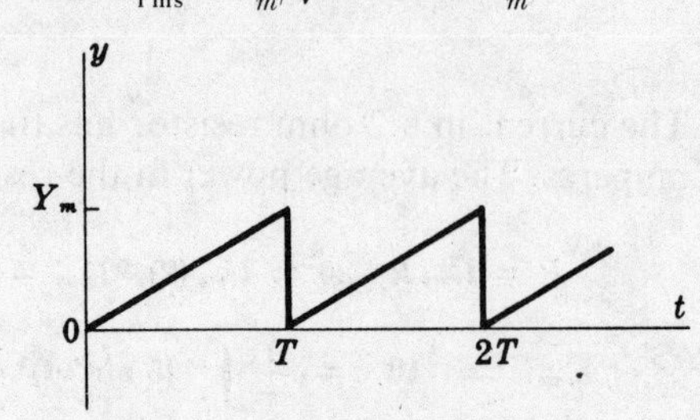

Fig. 2-19

2.28. Calculate the effective value of the waveform shown in Fig. 2-19 above and compare with Prob. 2.27.

2.29. Determine the effective value of the triangular wave shown in Fig. 2-20 below and compare with Problem 2.27.

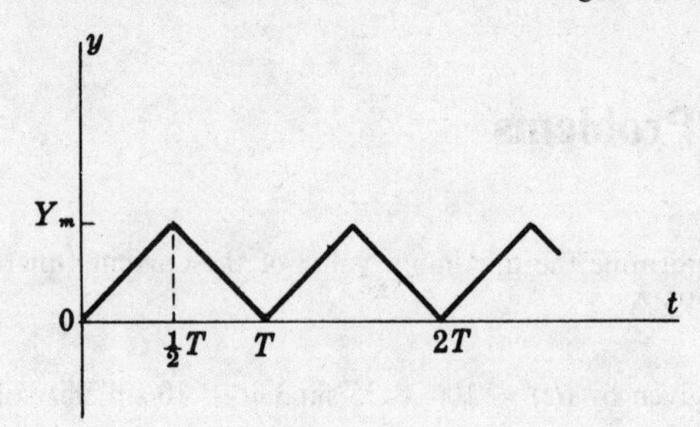

Fig. 2-20

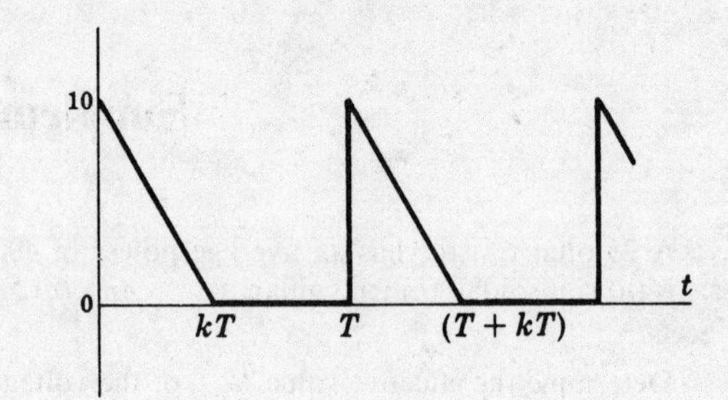

Fig. 2-21

2.30. Determine k in the waveform shown in Fig. 2-21 above where k is some fraction of the period T, such that the effective value is (*a*) 2, (*b*) 5. What is the highest possible effective value for the waveform as k varies?
Ans. (*a*) 0·12, (*b*) 0·75. 5·77 for $k = 1$

2.31. Find V_{av} and V_{rms} of the waveform shown in Fig. 2-22.
Ans. $V_{av} = 21{\cdot}6$ V, $V_{rms} = 24{\cdot}75$ V

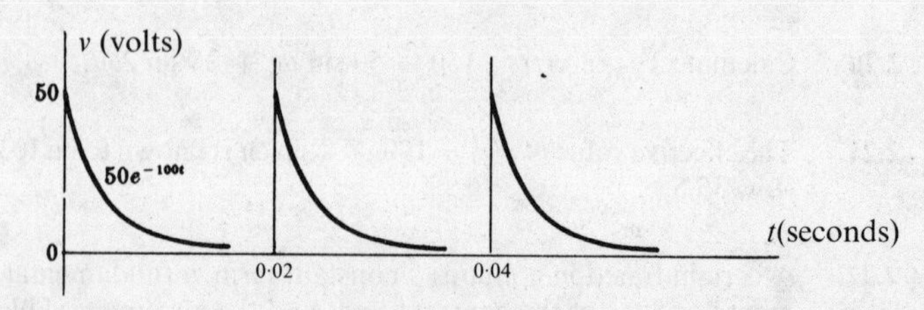

Fig. 2-22

2.32. Referring to Problem 2.31, determine V_{av} and V_{rms} if the function is described in the first interval by (*a*) $50\,e^{-200t}$ volts, (*b*) $50\,e^{-500t}$ volts.
Ans. (*a*) $V_{av} = 12{\cdot}25$ V, $V_{rms} = 17{\cdot}67$ V
(*b*) $V_{av} = 5{\cdot}0$ V, $V_{rms} = 11{\cdot}18$ V

2.33. Determine Y_{av} and Y_{rms} for the waveform shown in Fig. 2-23.

$0 < t < 0{\cdot}025 \quad y(t) = 400t$

$0{\cdot}025 < t < 0{\cdot}050 \quad y(t) = 10\,e^{-1000(t-0{\cdot}025)}$

Ans. $Y_{av} = 2{\cdot}7$, $Y_{rms} = 4{\cdot}2$

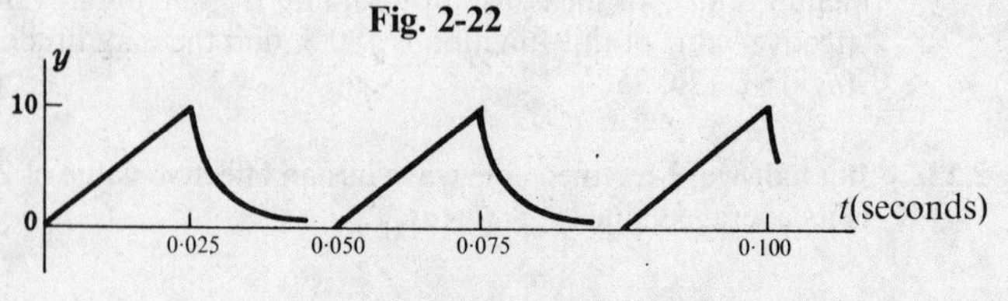

Fig. 2-23

2.34. The waveform shown in Fig. 2-24 is similar to that of Problem 2.33, but with a shorter rise time. Find Y_{av} and Y_{rms}.

$$0 < t < 0{\cdot}01 \quad y(t) = 1000t$$
$$0{\cdot}01 < t < 0{\cdot}05 \quad y(t) = 10\, e^{-1000\, t0{\cdot}01}$$

Ans. $Y_{av} = 1{\cdot}2$, $Y_{rms} = 2{\cdot}77$

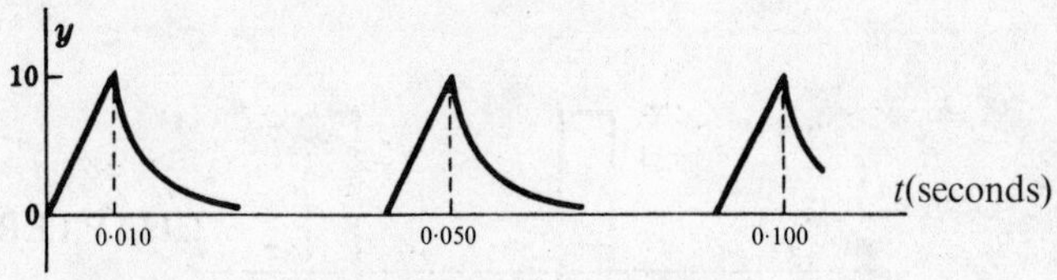

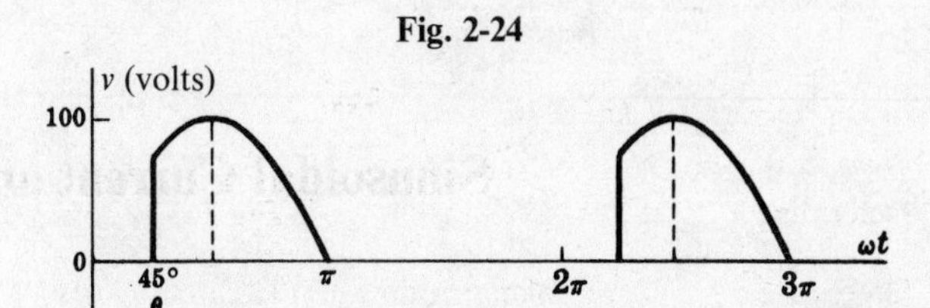

Fig. 2-24

2.35. Find V_{av} and V_{rms} of the delayed half-wave rectified sine wave of voltage shown in Fig. 2-25 when the delay angle is 45°.
Ans. $V_{av} = 27{\cdot}2$ V, $V_{rms} = 47{\cdot}7$ V

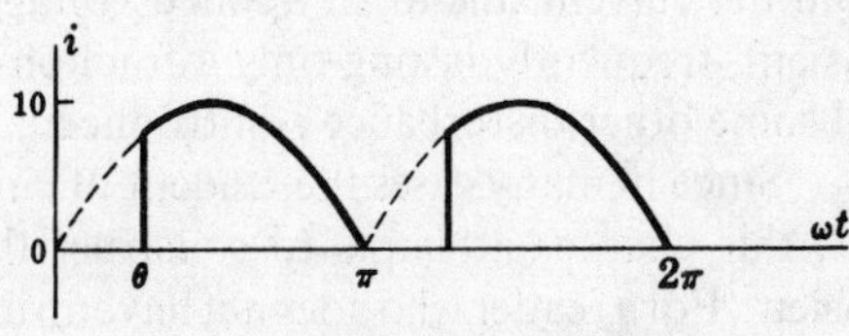

Fig. 2-25

2.36. Referring to the waveform of Problem 2.35, determine V_{av} and V_{rms} if the delay angle is (*a*) $\theta = 90°$, (*b*) $\theta = 135°$
Ans. (*a*) $V_{av} = 15{\cdot}95$ V, $V_{rms} = 35{\cdot}4$ V
(*b*) $V_{av} = 4{\cdot}66$ V, $V_{rms} = 15{\cdot}06$ V

2.37. The full-wave rectified sine wave shown in Fig. 2-26 below has a delay angle of 60°. Calculate V_{av} and V_{rms} in terms of V_m. *Ans.* $V_{av} = 0{\cdot}478\ V_m$, $V_{rms} = 0{\cdot}633\ V_m$

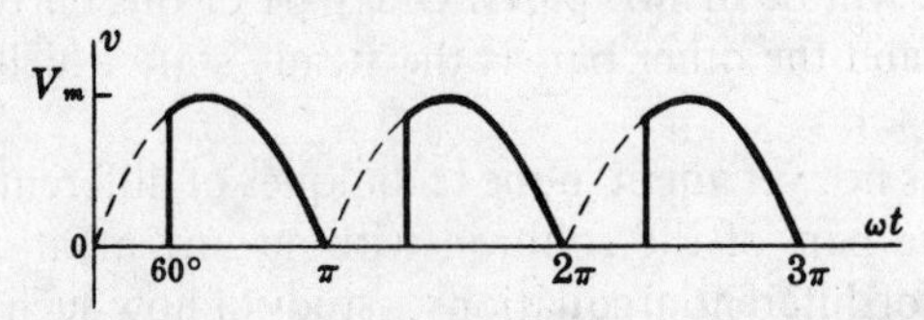

Fig. 2-26

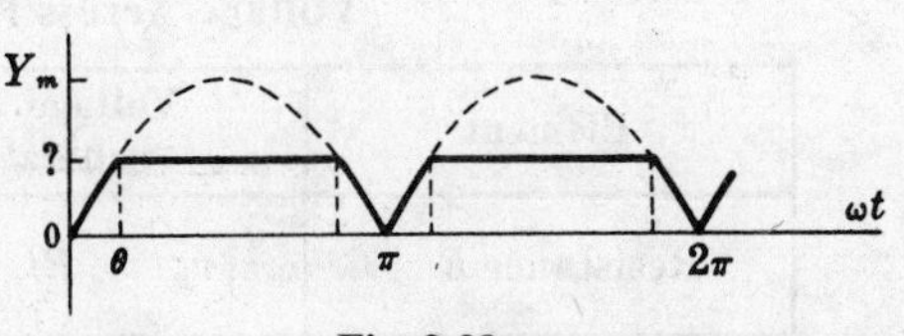

Fig. 2-27

2.38. A control circuit makes it possible to vary the delay angle in the current waveform shown in Fig. 2-27 above, such that the effective value has lower and upper limits of 2·13 and 7·01 amperes. Find the angles.
Ans. $\theta_1 = 135°$, $\theta_2 = 25°$

2.39. Determine the effective value of a full-wave rectified sine wave which is clipped at one-half of its peak as shown in Fig. 2-28 below. *Ans.* $Y_{rms} = 0{\cdot}422\ Y_m$

2.40. Referring to the waveform of Problem 2.39, find the effective value if the wave is clipped at 60° or $\pi/3$ radians.
Ans. $Y_{rms} = 0{\cdot}668\ Y_m$

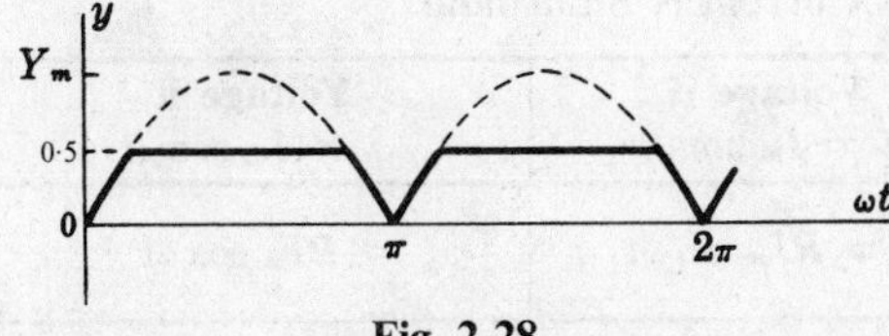

Fig. 2-28

Fig. 2-29

2.41. A full-wave rectified sine wave is clipped such that the effective value is $0{\cdot}5\ Y_m$ as shown in Fig. 2-29 above. Determine the amplitude at which the waveform is clipped. *Ans.* $0{\cdot}581\ Y_m$ or $\theta = 35{\cdot}5°$

2.42. Find the average effective values of the waveform which results from a three phase half-wave rectifier circuit as shown in Fig. 2·30 below. *Ans.* $V_{av} = 0{\cdot}827\ V_m$, $V_{rms} = 0{\cdot}0840\ V_m$

2.43. The waveform resulting from a six phase half-wave rectifier circuit is shown in Fig. 2-31 below. Calculate V_{av} and V_{rms}. *Ans.* $V_{av} = 0{\cdot}955\ V_m$, $V_{rms} = 0{\cdot}956\ V_m$

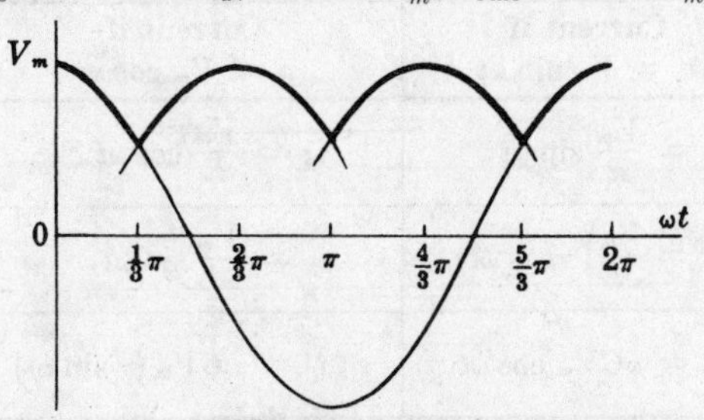

Fig. 2-30

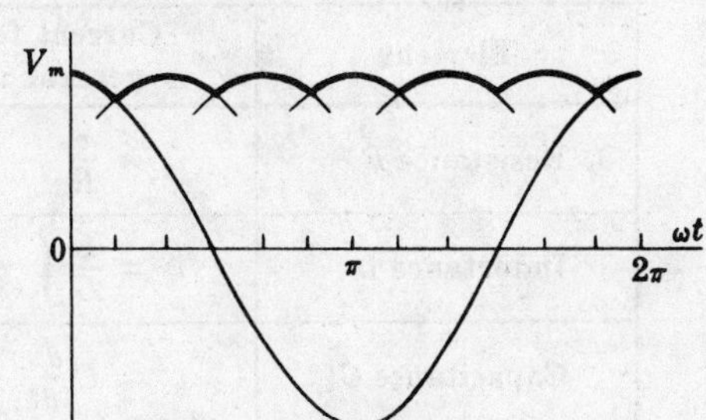

Fig. 2-31

CHAPTER 3

Sinusoidal Current and Voltage

INTRODUCTION

When Kirchhoff's Laws are applied to a circuit the result is usually an integrodifferential equation. The methods of classical differential equations will yield the required solutions. When these methods are used to obtain the current due to an applied voltage, the current will be in two parts. One part of this current is the transient, frequently lasting only a fraction of a second, and the other part is the steady state which remains until some other disturbance is introduced.

Since in many cases the student of circuit analysis is not yet adept in the techniques of differential equations, this chapter attempts to construct the steady state part of the solution without any mention of the transient. For a reader who does not have an understanding of differential equations, a study of how such methods apply to circuit analysis would be valuable at this time. Chapter 16 discusses classical differential equations and gives a number of examples illustrating the transient and steady state parts of the solutions.

SINUSOIDAL CURRENTS

When the current in the pure elements R, L, and C is sinusoidal, the voltages across the elements are as given in Table 3-1.

Table 3-1

Voltage Across Pure Element if Current is Sinusoidal

Element	Voltage for general i	Voltage if $i = I_m \sin \omega t$	Voltage if $i = I_m \cos \omega t$
Resistance R	$v_R = Ri$	$v_R = RI_m \sin \omega t$	$v_R = RI_m \cos \omega t$
Inductance L	$v_L = L\dfrac{di}{dt}$	$v_L = \omega L I_m \cos \omega t$	$v_L = \omega L I_m (-\sin \omega t)$
Capacitance C	$v_C = \dfrac{1}{C}\int i\,dt$	$v_C = \dfrac{I_m}{\omega C}(-\cos \omega t)$	$v_C = \dfrac{I_m}{\omega C} \sin \omega t$

Table 3-2

Current in Pure Element if Voltage is Sinusoidal

Element	Current for general v	Current if $v = V_m \sin \omega t$	Current if $v = V_m \cos \omega t$
Resistance R	$i_R = \dfrac{v}{R}$	$i_R = \dfrac{V_m}{R} \sin \omega t$	$i_R = \dfrac{V_m}{R} \cos \omega t$
Inductance L	$i_L = \dfrac{1}{L}\int v\,dt$	$i_L = \dfrac{V_m}{\omega L}(-\cos \omega t)$	$i_L = \dfrac{V_m}{\omega L} \sin \omega t$
Capacitance C	$i_C = C\dfrac{dv}{dt}$	$i_C = \omega C V_m \cos \omega t$	$i_C = \omega C V_m (-\sin \omega t)$

SINUSOIDAL VOLTAGES

When the voltages across the three elements are sinusoidal, the currents in the elements are as given in Table 3-2 above.

IMPEDANCE

The impedance of an element, of a branch, or of a complete circuit is the ratio of the voltage to the current

$$\text{Impedance} = \frac{\text{Voltage function}}{\text{Current function}}$$

With sinusoidal voltages and currents this ratio will have a magnitude and an angle. In Chapter 5, impedance is treated in much more detail and there the angle will be considered. For the present, only the magnitude of the impedance will be considered. The angle between v and i is discussed below as the phase angle.

PHASE ANGLE

If voltage and current are both sinusoidal functions of time, a plot of both to the same time scale will show a displacement between them except for the case of pure resistance. This displacement is the phase angle and never exceeds 90° or $\pi/2$ radians. By agreement this phase angle is always described as "what the current i does with respect to the voltage v"; e.g. i leads v by 90° in a pure capacitor, i lags v by 45° in an RL series circuit with R and ωL equal, i is in phase with v in a pure resistance, etc. The following sketches should clarify both impedance and phase angle.

Pure R. The current and voltage are in phase in a pure reistor. See Fig. 3-1 below. The impedance magnitude is R.

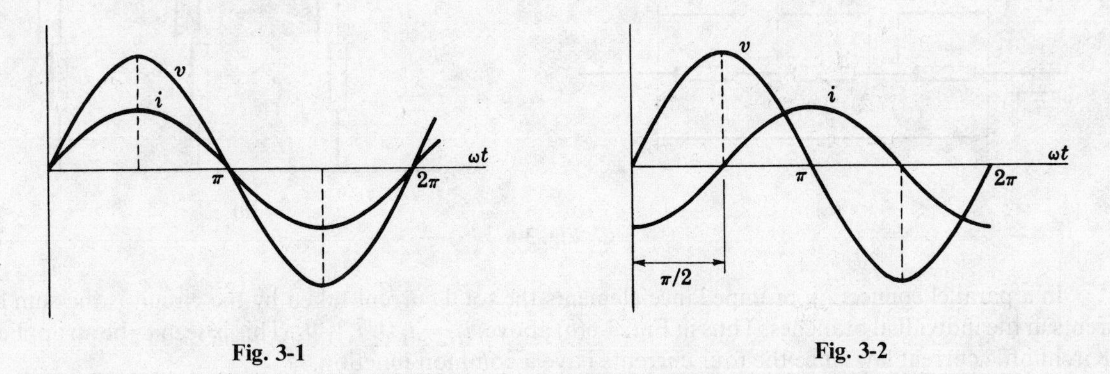

Fig. 3-1

Fig. 3-2

Pure L. The current lags the voltage by 90° or $\pi/2$ in a pure inductor. See Fig. 3-2 above. The magnitude of the impedance is ωL.

Pure C. The current leads the voltage by 90° or $\pi/2$ in a pure capacitor. See Fig. 3-3 below. The impedance magnitude is $\frac{1}{\omega C}$.

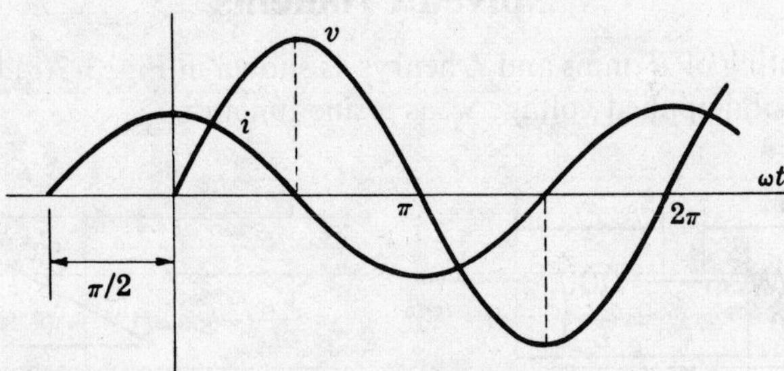

Fig. 3-3

Series RL. The current lags the voltage by $\tan^{-1}(\omega L/R)$ in a series RL circuit. See Fig. 3-4 below. The impedance magnitude is $\sqrt{R^2 + (\omega L)^2}$.

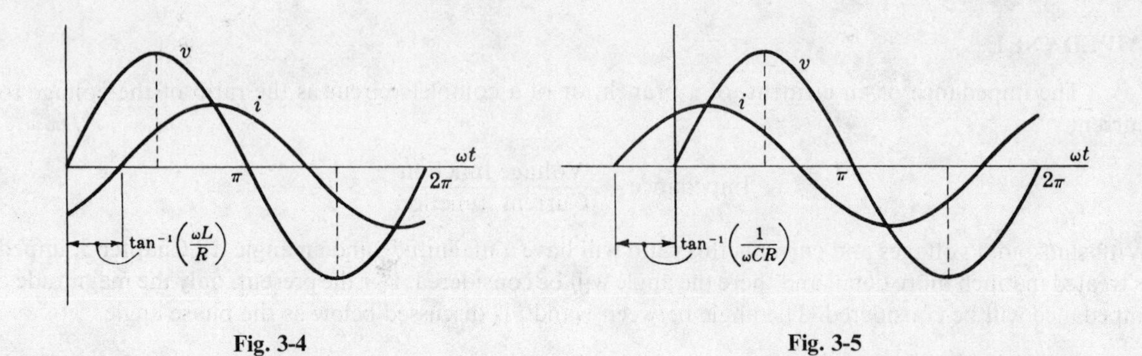

Fig. 3-4 **Fig. 3-5**

Series RC. The current leads the voltage by $\tan^{-1}(1/\omega CR)$ in a series RC circuit. See Fig. 3-5 above. The impedance magnitude is $\sqrt{R^2 + (1/\omega C)^2}$.

SERIES AND PARALLEL CIRCUITS

For circuit elements in a series connection the total voltage is the sum of the voltages across the individual elements. Thus in Fig. 3-6(*a*) below, $v_T = v_1 + v_2 + v_3$.

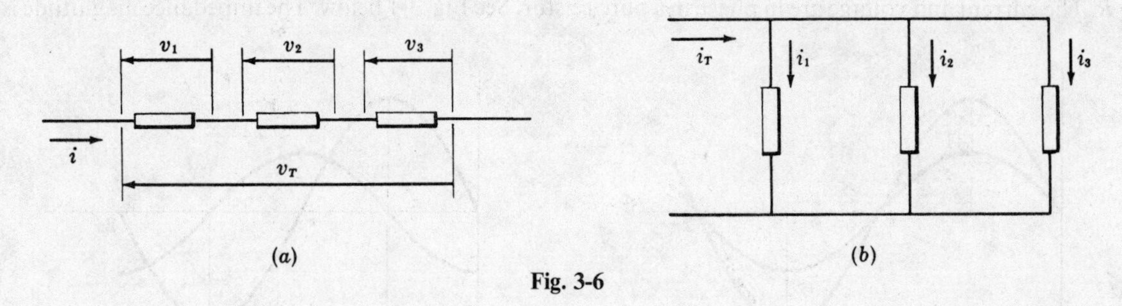

Fig. 3-6

In a parallel connection of impedance elements the total current taken by the circuit is the sum of the currents in the individual branches. Thus in Fig. 3-6(*b*) above, $i_T = i_1 + i_2 + i_3$. This is seen to be an application of Kirchhoff's current law since the four currents have a common junction.

Solved Problems

3.1. A series circuit consisting of R ohms and L henrys as shown in Fig. 3-7(*a*) below passes a current $i = I_m$ sine ωt. Express the total applied voltage v_T as a sine function.

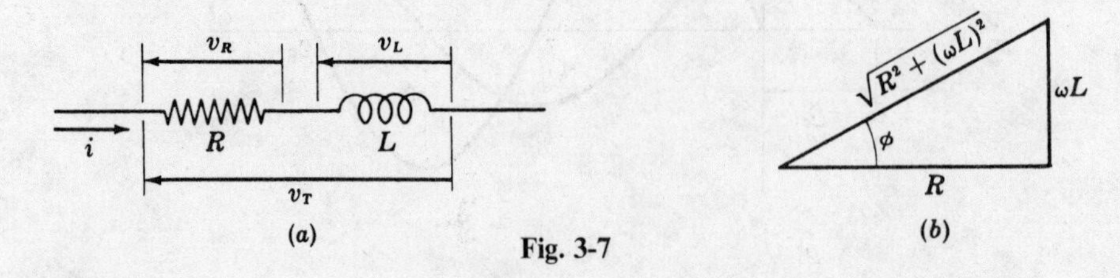

Fig. 3-7

$$v_T = v_R + v_L = RI_m \sin \omega t + \omega L I_m \cos \omega t \qquad (1)$$

Any number of sine and cosine terms of the same frequency can be combined into a single sine or cosine term with amplitude A and phase angle φ; thus we may write

$$v_T = A \sin(\omega t + \varphi) = A \sin \omega t \cos \varphi + A \cos \omega t \sin \varphi \qquad (2)$$

In (1) and (2) equate coefficients of $\sin \omega t$ and then of $\cos \omega t$ to obtain

$$RI_m = A \cos \varphi, \qquad \omega L I_m = A \sin \varphi$$

Now (see Fig. 3-7(b)) $\tan \varphi = \dfrac{\sin \varphi}{\cos \varphi} = \dfrac{\omega L}{R}$, $\cos \varphi = \dfrac{R}{\sqrt{R^2 + (\omega L)^2}}$, $A = \dfrac{RI_m}{\cos \varphi} = \sqrt{R^2 + (\omega L)^2}\, I_m$,

and $v_T = A \sin(\omega t + \varphi) = \sqrt{R^2 + (\omega L)^2}\, I_m \sin(\omega t + \tan^{-1} \omega L/R)$

which indicates that the current lags the voltage by the phase angle $\varphi = \tan^{-1} \omega L/R$.

The magnitude of the impedance is $\sqrt{R^2 + (\omega L)^2}$.

If $R \gg \omega L$, then $\omega L/R \to 0$ and $\varphi \to 0$, i.e. the result obtained with pure resistance, If $\omega L \gg R$, then $\omega L/R \to \infty$ and $\varphi \to \pi/2$, i.e. the result obtained with pure inductance.

In a series combination of R and L, the current will lag the voltage by some angle from 0° to 90° depending on the relative magnitudes of R and ωL.

3.2. The given circuit shown in Fig. 3-8 below passes a current $i = 2 \sin 500t$ amperes. Calculate the total applied voltage v_T.

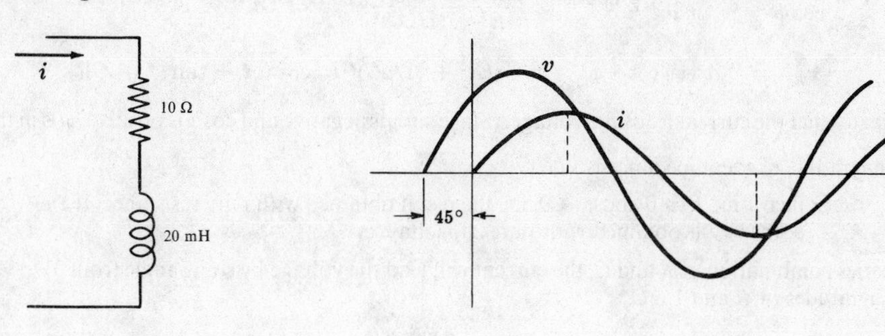

Fig. 3-8

$$v_T = \sqrt{R^2 + (\omega L)^2}\, I_m \sin(\omega t + \tan^{-1} \omega L/R) = 28{\cdot}28 \sin(500t + 45°) \text{ volts}$$

where $R = 10\ \Omega$, $\omega L = 500(0{\cdot}02) = 10\ \Omega$, $\tan^{-1} \omega L/R = 45°$, $I_m = 2$ A.

Since here $R = \omega L$, the current lags the voltage by 45°.

3.3. In a series circuit of $R = 20$ ohms and $L = 0{\cdot}06$ henrys the current lags the voltage by 80°. Determine ω.

$$\tan \varphi = \omega L/R, \ \tan 80° = 5{\cdot}68 = \omega(0{\cdot}06)/20, \ \omega = 1893 \text{ rad/s}$$

3.4. A series RL circuit has $L = 0{\cdot}02$ henrys and an impedance of 17·85 ohms. With a sinusoidal voltage applied, the current lags the voltage by 63·4°. Calculate ω and R.

$$\tan \varphi = \omega L/R, \ \tan 63{\cdot}4° = 2 = 0{\cdot}02\omega/R, \ R = 0{\cdot}01\omega$$

$$17{\cdot}85 = \sqrt{R^2 + (\omega L)^2} = \sqrt{(0{\cdot}01\omega)^2 + (0{\cdot}02\omega)^2}$$

$$\omega = 800 \text{ rad/s and } R = 0{\cdot}01\omega = 8 \text{ ohms}$$

3.5. A series circuit consisting of R ohms and C farads as shown in Fig. 3-9 below passes a current $i = I_m \cos \omega t$. Express the total applied voltage v_T as a single cosine function.

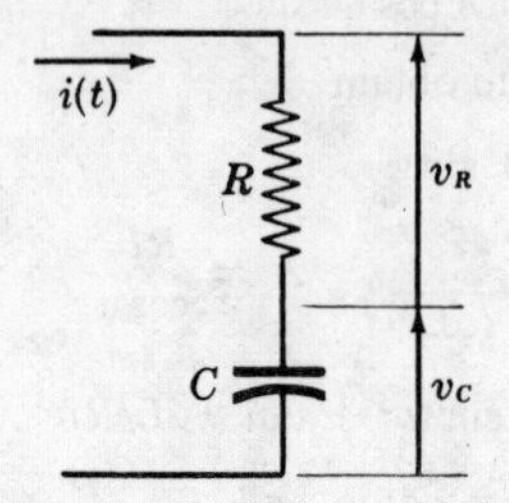

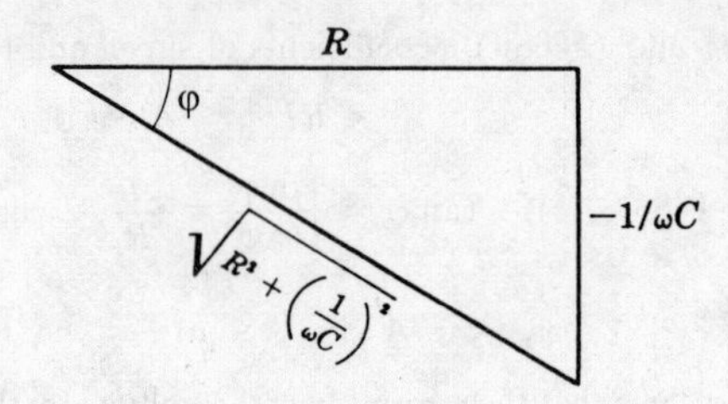

Fig. 3-9

$$v_T = v_R + v_C = RI_m \cos \omega t + (1/\omega C)I_m \sin \omega t \tag{1}$$

Expressing v_T by a single cosine term with amplitude A and phase angle φ,

$$v_T = A \cos (\omega t + \varphi) = A \cos \omega t \cos \varphi - A \sin \omega t \sin \varphi \tag{2}$$

In (*1*) and (*2*) equate coefficients of $\cos \omega t$ and then of $\sin \omega t$ to obtain

$$RI_m = A \cos \varphi, \quad (1/\omega C)I_m = -A \sin \varphi$$

Now $\tan \varphi = \dfrac{\sin \varphi}{\cos \varphi} = -\dfrac{1}{\omega CR}$, $\cos \varphi = \dfrac{R}{\sqrt{R^2 + (1/\omega C)^2}}$, $A = \sqrt{R^2 + (1/\omega C)^2}\, I_m$, and

$$v_T = A \cos (\omega t + \varphi) = \sqrt{R^2 + (1/\omega C)^2}\, I_m \cos (\omega t - \tan^{-1} 1/\omega CR)$$

which indicates that the current leads the voltage. (Since $\sin \varphi$ is negative and $\cos \varphi$ is positive, φ is in the fourth quadrant.)

The magnitude of the impedance is $\sqrt{R^2 + (1/\omega C)^2}$.

If $R \gg 1/\omega C$, then $1/\omega CR \to 0$ and $\varphi \to 0$, i.e. the result obtained with pure resistance. If $1/\omega C \gg R$, then $1/\omega CR \to \infty$ and $\varphi \to \pi/2$, i.e. the result obtained with pure capacitance.

In a series combination of R and C, the current will lead the voltage by some angle from 0° to 90° depending on the relative magnitudes of R and $1/\omega C$.

3.6. The series circuit shown in Fig. 3-10 below passes a current $i = 2 \cos 5000t$ amperes. Find the total applied voltage v_T.

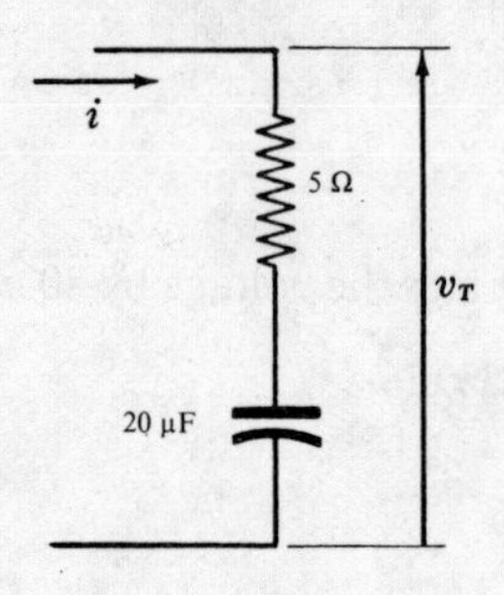

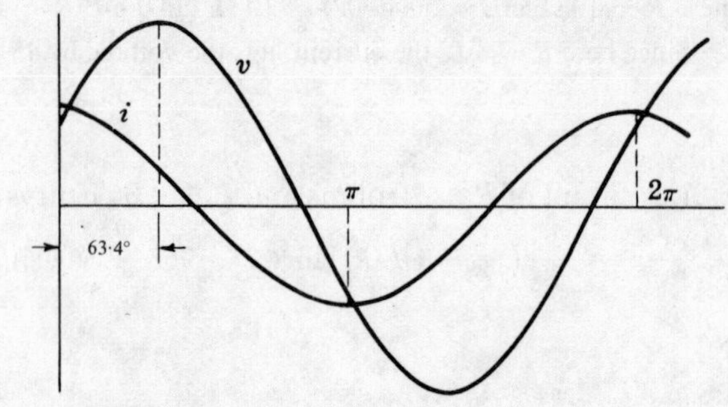

Fig. 3-10

$$v_T = \sqrt{R^2 + (1/\omega C)^2}\, I_m \cos (\omega t - \tan^{-1} 1/\omega CR) = 22{\cdot}4 \cos (5000t - 63{\cdot}4°) \text{ volts}$$

where $R = 5\,\Omega$, $1/\omega C = 1/(5000 \times 20 \times 10^{-6}) = 10\,\Omega$, $\tan^{-1} 1/\omega CR = \tan^{-1} 10/5 = 63{\cdot}4°$, $I_m = 2$ A.

The current leads the voltage by the phase angle 63·4°. The absolute value of the impedance is 11·18 ohms.

3.7. A series circuit consisting of R, L and C passes a current $i = I_m \sin \omega t$. Determine the voltage across each element. Refer to Fig. 3-12.

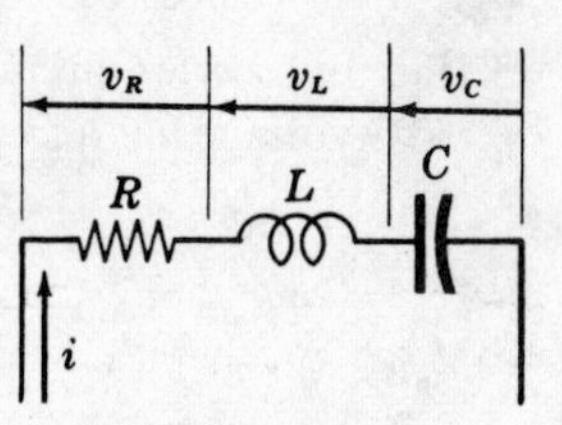

Fig. 3-11

$$v_R = Ri = RI_m \sin \omega t$$

$$v_L = L\frac{d}{dt}(I_m \sin \omega t) = \omega L I_m \cos \omega t$$

$$v_C = \frac{1}{C}\int I_m \sin \omega t \, dt = \frac{1}{\omega C} I_m(-\cos \omega t)$$

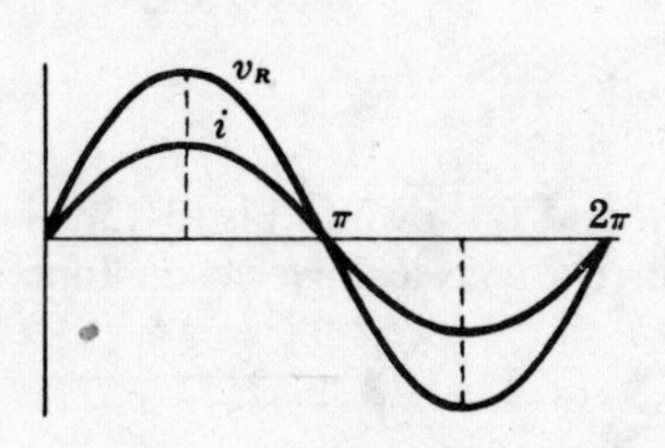

v_R and i
(i in phase with v_R)

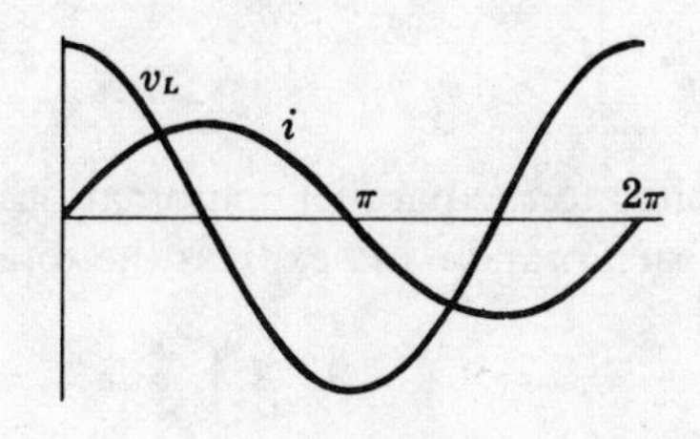

v_L and i
(i lags v_L by 90°)

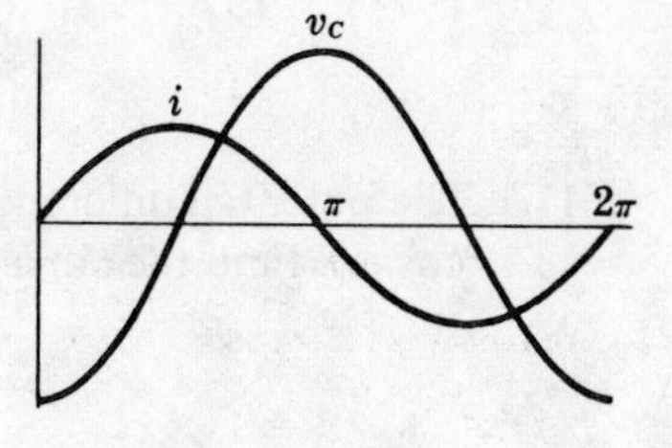

v_C and i
(i leads v_C by 90°)

Fig. 3-12

3.8. Referring to Problem 3.7, express the total voltage v_T across the three circuit elements as a single sine function.

$$v_T = v_R + v_L + v_C = RI_m \sin \omega t + (\omega L - 1/\omega C)I_m \cos \omega t \tag{1}$$

Expressing v_T as a single sine function with amplitude A and phase angle φ,

$$\begin{aligned} v_T &= A \sin(\omega t + \varphi) \\ &= A \sin \omega t \cos \varphi + A \cos \omega t \sin \varphi \end{aligned} \tag{2}$$

In (1) and (2) equate coefficients of $\sin \omega t$ and then of $\cos \omega t$ to obtain

$$RI_m = A \cos \varphi, \qquad I_m(\omega L - 1/\omega C) = A \sin \varphi$$

Now $\tan \varphi = \dfrac{\omega L - 1/\omega C}{R}$, $\cos \varphi = \dfrac{R}{\sqrt{R^2 + (\omega L - 1/\omega C)^2}}$, $A = \dfrac{RI_m}{\cos \varphi} = \sqrt{R^2 + (\omega L - 1/\omega C)^2}\, I_m$,

and $\quad v_T = A \sin(\omega t + \varphi) = \sqrt{R^2 + (\omega L - 1/\omega C)^2}\, I_m \sin[\omega t + \tan^{-1}(\omega L - 1/\omega C)/R]$

where $\sqrt{R^2 + (\omega L - 1/\omega C)^2}$ is the absolute value of the impedance, and $\tan^{-1}(\omega L - 1/\omega C)/R$ is the phase angle.

If $\omega L > 1/\omega C$, the phase angle φ is positive; the current *lags* the voltage and the circuit has an overall effect which is inductive.

If $1/\omega C > \omega L$, the phase angle φ is negative; the current *leads* the voltage and the circuit has an overall effect which is capacitive.

If $\omega L = 1/\omega C$, the phase angle φ is zero; the current and voltage are said to be in phase and the impedance has the value R. This condition is called series resonance.

3.9. Show that ωL and $1/\omega C$ are given in ohms when ω is in rad/s, L in henrys, and C in farads.

$$\omega L = \frac{\text{radians}}{\text{seconds}} \cdot \text{henrys} = \frac{1}{\text{seconds}} \cdot \frac{\text{volt seconds}}{\text{amperes}} = \frac{\text{volts}}{\text{amperes}} = \text{ohms}$$

$$\frac{1}{\omega C} = \frac{\text{seconds}}{\text{radians}} \cdot \frac{1}{\text{farads}} = \text{seconds} \cdot \frac{\text{volts}}{\text{amperes seconds}} = \frac{\text{volts}}{\text{amperes}} = \text{ohms}$$

Note that the radian measure of an angle is a pure (dimensionless) number.

3.10. In a series circuit consisting of $R = 15$ ohms, $L = 0{\cdot}08$ henrys and $C = 30$ microfarads, the applied voltage has a frequency of 500 rad/s. Does the current lead or lag the applied voltage and by what angle?

$$\omega L = 500(0{\cdot}08) = 40 \text{ ohms}, \quad \frac{1}{\omega C} = \frac{1}{500(30 \times 10^{-6})} = 66{\cdot}7 \text{ ohms}$$

$$\tan^{-1}\frac{\omega L - 1/\omega C}{R} = \tan^{-1}\frac{-26{\cdot}7}{15} = -60{\cdot}65^\circ$$

The capacitive reactance $1/\omega C$ is greater than the inductive reactance ωL. The current leads the applied voltage by $60{\cdot}65^\circ$ and the circuit has an overall effect which is capacitive. The magnitude of the impedance is $\sqrt{R^2 + (\omega L - 1/\omega C)^2} = 30{\cdot}6$ ohms.

3.11. The potential difference applied to the parallel combination of R and L shown in Fig. 3-13 is $v = V_m \cos \omega t$. Find the current in each branch and express the total current i_T as a single cosine function.

$$i_T = i_R + i_L = \frac{1}{R}v + \frac{1}{L}\int v\,dt = \frac{V_m}{R}\cos\omega t + \frac{V_m}{\omega L}\sin\omega t$$

Then $i_T = \sqrt{(1/R)^2 + (1/\omega L)^2}\; V_m \cos(\omega t - \tan^{-1} R/\omega L)$

The current lags the applied voltage by an angle $\varphi = \tan^{-1} R/\omega L$.

If $R \gg \omega L$, then $\varphi \rightarrow 90^\circ$ and $i_T \approx (V_m/\omega L)\cos(\omega t - 90^\circ)$. With this relatively high resistance the current drawn by the resistive branch is quite low. Hence i_T is essentially given by i_L, i.e. the inductive current controls in fixing total current.

If $\omega L \gg R$, then $\varphi \rightarrow 0^\circ$ and $i_T \approx (V_m/R)\cos\omega t$. In this case the inductive branch has a high reactance and hence draws a small current compared to that taken by the resistive branch. Here the resistive current controls in setting total current.

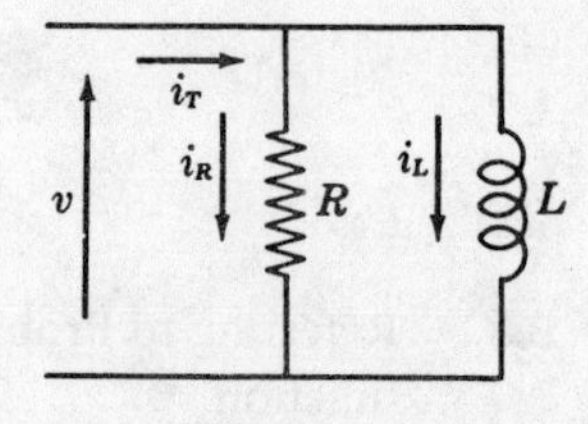

Fig. 3-13

3.12. The voltage applied to the parallel combination of R and C shown in Fig. 3-14 is $v = V_m \sin \omega t$. Find the current in each branch and express the total current i_T as a single sine function.

$$i_T = i_R + i_C = \frac{v}{R} + C\frac{dv}{dt} = \frac{V_m}{R}\sin\omega t + \omega C V_m \cos\omega t$$

Then $i_T = \sqrt{(1/R)^2 + (\omega C)^2}\; V_m \sin(\omega t + \tan^{-1}\omega CR)$

The current leads the voltage by the angle $\varphi = \tan^{-1}\omega CR$.

If $R \gg 1/\omega C$, then $\varphi \rightarrow 90^\circ$ and $i_T \approx i_C = \omega C V_m \sin(\omega t + 90^\circ)$, i.e. the capacitive branch controls in setting the total current.

If $1/\omega C \gg R$, then $\varphi \rightarrow 0^\circ$ and $i_T \approx i_R = (V_m/R)\sin\omega t$, (i.e. the resistive branch controls in setting the total current.

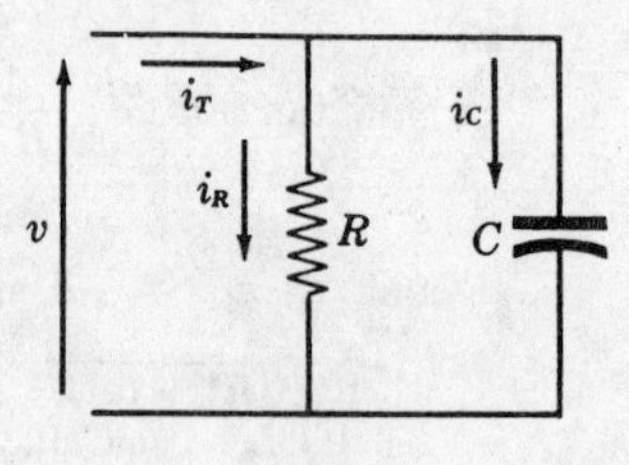

Fig. 3-14

3.13. The potential difference applied to the parallel combination of R, L and C shown in Fig. 3-15 is $v = V_m \sin \omega t$. Find the current in each branch and express the total current i_T as a single sine function.

$$i_T = i_R + i_L + i_C = \frac{v}{R} + \frac{1}{L}\int v\,dt + C\frac{dv}{dt}$$

$$= \frac{V_m}{R}\sin\omega t - \frac{V_m}{\omega L}\cos\omega t + \omega C V_m \cos\omega t \qquad (1)$$

Expressing v_T as a single sine function with amplitude A and phase angle φ,

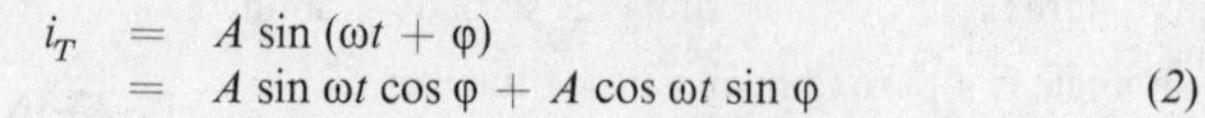

$$i_T = A\sin(\omega t + \varphi)$$

$$= A\sin\omega t\cos\varphi + A\cos\omega t\sin\varphi \qquad (2)$$

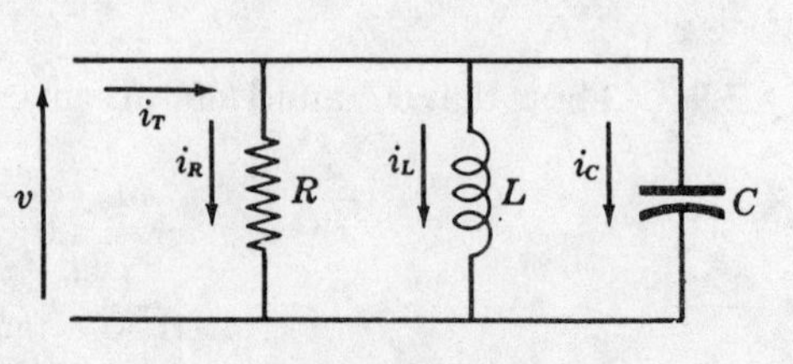

Fig. 3-15

In (*1*) and (*2*) equate coefficients of sin ωt and then of cos ωt to get

$$V_m/R = A\cos\varphi, \qquad (\omega C - 1/\omega L)V_m = A\sin\varphi$$

Then $\tan\varphi = \dfrac{\omega C - 1/\omega L}{1/R}$, $\cos\varphi = \dfrac{1/R}{\sqrt{(1/R)^2 + (\omega C - 1/\omega L)^2}}$, $A = \sqrt{(1/R)^2 + (\omega C - 1/\omega L)^2}\,V_m$, and

$$i_T = \sqrt{(1/R)^2 + (\omega C - 1/\omega L)^2}\,V_m \sin[\omega t + \tan^{-1}(\omega C - 1/\omega L)R]$$

Evidently the sign of the phase angle depends on the relative values of ωC and $1/\omega L$.

The inductive branch draws a current which lags the applied voltage by 90°. The capacitive branch draws a current which leads the applied voltage by 90°. In combination these two currents could with equal magnitudes cancel out. If the inductive branch current is the greater, then the total current lags the applied voltage; if the capacitive branch current is the greater, then the total current leads the applied voltage.

3.14. Two pure circuit elements in a series connection have the following current and applied voltage:

$$v = 150\sin(500t + 10°)\text{ volts}, \quad i = 13{\cdot}42\sin(500t - 53{\cdot}4°)\text{ amperes}$$

Find the elements comprising the circuit.

By inspection, the current lags the voltage by $53{\cdot}4° + 10° = 63{\cdot}4°$; hence the circuit must contain R and L.

$$\tan 63{\cdot}4° = 2 = \omega L/R, \ \omega L = 2R$$

$$V_m/I_m = \sqrt{R^2 + (\omega L)^2}, \quad 150/13{\cdot}42 = \sqrt{R^2 + (2R)^2}, \quad R = 5\text{ ohms}$$

and $L = 2R/\omega = 0{\cdot}02$ henrys. The circuit contains a resistor of 5 Ω and an inductor of 0·02 H.

3.15. A series circuit containing two pure elements has the following current and applied voltage:

$$v = 200\sin(2000t + 50°) \quad i = 4\cos(2000t + 13{\cdot}2°)\text{ amperes}$$

Find the elements comprising the circuit.

Since $\cos x = \sin(x + 90°)$, we may write $i = 4\sin(2000t + 103{\cdot}2°)$ amperes. Hence the current leads the voltage by $103{\cdot}2° - 50° = 53{\cdot}2°$ and the circuit must contain R and C.

$$\tan 53{\cdot}2° = 1{\cdot}33 = 1/\omega CR, \ 1/\omega C = 1{\cdot}33R$$

$$V_m/I_m = \sqrt{R^2 + (1/\omega C)^2}, \quad 200/4 = \sqrt{R^2 + (1.33R)^2}, \quad R = 30\text{ ohms}$$

and $C = 1/(1{\cdot}33\omega R) = 1{\cdot}25 \times 10^{-5}$ farads $= 12{\cdot}5\ \mu$F.

3.16. In the series circuit shown in Fig. 3-16 the voltage and current are given by

$$v = 353{\cdot}5\cos(3000t - 10°)\text{ volts},$$
$$i = 12{\cdot}5\cos(3000t - 55°)\text{ amperes}$$

and the inductance is 0·01 henrys. Find R and C.

The current lags the voltage by $55° - 10° = 45°$. Hence the inductive reactance ωL is greater than the capacitive reactance $1/\omega C$.

$$\tan 45° = 1 = (\omega L - 1/\omega C)/R, \quad (\omega L - 1/\omega C) = R$$

$$V_m/I_m = \sqrt{R^2 + (\omega L - 1/\omega C)^2}, \quad 353{\cdot}5/12{\cdot}5 = \sqrt{2R^2}$$

$$R = 20\text{ ohms}$$

and from $(\omega L - 1/\omega C) = R$ we find

$$C = 3{\cdot}33 \times 10^{-5}\text{ farads} = 33{\cdot}3\ \mu\text{F}$$

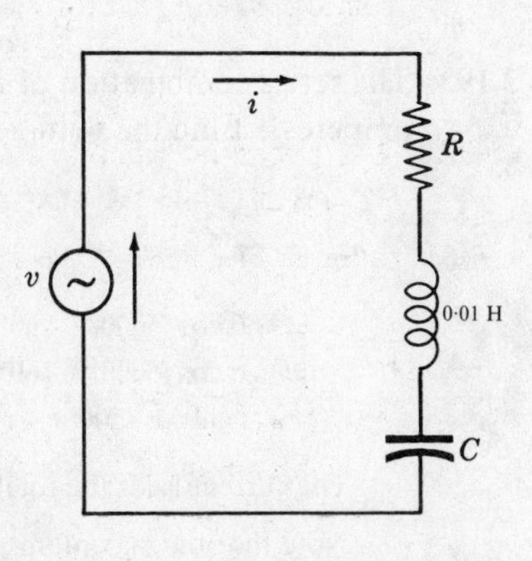

Fig. 3-16

3.17. In the parallel circuit, as shown in Fig. 3-17 below, the voltage function is $v = 100 \sin(1000t + 50°)$ volts. Express the total current as a single sine function.

$$
\begin{aligned}
i_T &= i_R + i_L = \frac{v}{R} + \frac{1}{L}\int v\,dt \\
&= 20 \sin(1000t + 50°) - 5 \cos(1000t + 50°) \\
&= A \sin(1000t + 50°) \cos\varphi + A \cos(1000t + 50°) \sin\varphi
\end{aligned}
$$

from which $20 = A\cos\varphi$ and $-5 = A\sin\varphi$. Then $\tan\varphi = -5{:}20$, $\varphi = -14{\cdot}05°$; and $A = 20/(\cos\varphi) = 20{\cdot}6$. Thus

$$i_T = 20{\cdot}6 \sin(1000t + 50° - 14{\cdot}05°) = 20{\cdot}6 \sin(1000t + 35{\cdot}95°) \text{ amperes}$$

The current lags the applied voltage by 14·05°.

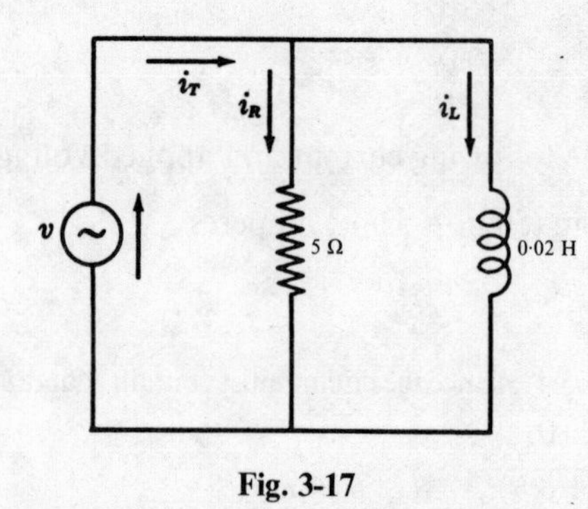

Fig. 3-17

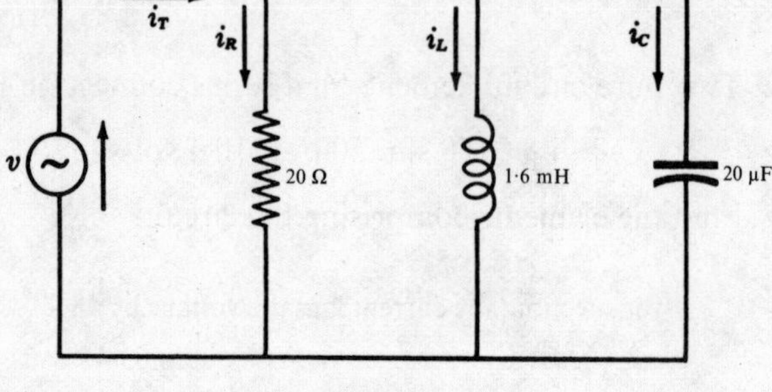

Fig. 3-18

3.18. The applied voltage in the given circuit shown in Fig. 3-18 is $v = 50 \sin(5000t + 45°)$ volts. Find all branch currents and the total current.

$$
\begin{aligned}
i_T &= i_R + i_L + i_C = \frac{v}{R} + \frac{1}{L}\int v\,dt + C\frac{dv}{dt} \\
&= 2{\cdot}5 \sin(5000t + 45°) - 6{\cdot}25 \cos(5000t + 45°) + 5 \cos(5000t + 45°) \\
&= 2{\cdot}5 \sin(5000t + 45°) - 1{\cdot}25 \cos(5000t + 45°) \\
&= 2{\cdot}8 \sin(5000t + 18{\cdot}4°) \text{ amperes, using the methods of this chapter.}
\end{aligned}
$$

The current lags the applied voltage by $45° - 18{\cdot}4° = 26{\cdot}6°$.

Note that the total current has a maximum value of 2·8 amperes. This is less than either the maximum inductive or capacitive branch currents which are 6·25 and 5 amperes respectively. The explanation is obvious from the plots of these three branch currents to the same scale.

3.19. The series combination of R, L and C shown in Fig. 3-19 passes a current $i = 3 \cos(5000t - 60°)$ amperes. Find the voltage across each element and the total voltage.

$$
\begin{aligned}
v_T &= v_R + v_L + v_C = Ri + L\frac{di}{dt} + \frac{1}{C}\int i\,dt \\
&= 6 \cos(5000t - 60°) - 24 \sin(5000t - 60°) + 30 \sin(5000t - 60°) \\
&= 6 \cos(5000t - 60°) + 6 \sin(5000t - 60°) \\
&= 8{\cdot}49 \cos(5000t - 105°) \text{ volts, using the methods of this chapter.}
\end{aligned}
$$

The current leads the total voltage by $105° - 60° = 45°$.

Note that the maximum applied voltage is 8·49 volts. The voltage across the individual circuit elements is greater than this for the inductive and capacitive elements. A carefully scaled plot will demonstrate what is taking place.

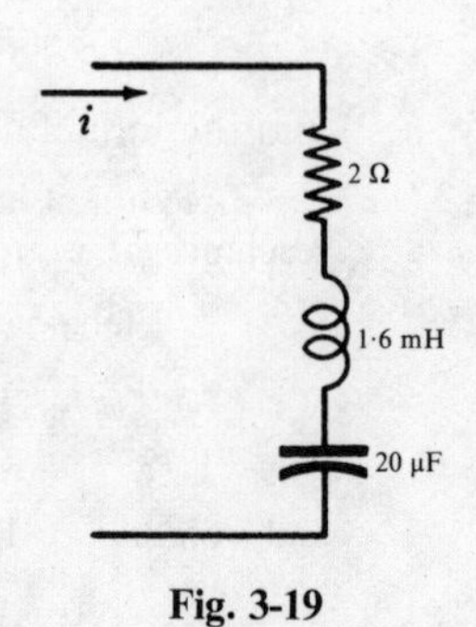

Fig. 3-19

Supplementary Problems

3.20. A pure inductance of $L = 0{\cdot}01$ H passes a current $i = 5 \cos 2000t$ amperes. What is the voltage across the element?
Ans. $100 \cos (2000t + 90°)$ volts

3.21. A pure capacitance of $C = 30\ \mu F$ passes a current $i = 12 \sin 2000t$ amperes. Find the voltage across the element.
Ans. $200 \sin (2000t - 90°)$ volts

3.22. In a series of $R = 5$ ohms and $L = 0{\cdot}06$ henrys the voltage across the inductance is $v_L = 15 \sin 200t$ volts. Find the total voltage, the current, the angle by which i lags v_T and the magnitude of the impedance.
Ans. $i = 1{\cdot}25 \sin (200t - 90°)$ amperes, $v_T = 16{\cdot}25 \sin (200t - 22{\cdot}65°)$ volts, $67{\cdot}35°$, $V_m/I_m = 13$ ohms

3.23. In the same series circuit as Problem 3.22 the voltage across the resistance is $v_R = 15 \sin 200t$ volts. Find the total voltage, the current, the angle by which i lags v_T and the magnitude of the impedance.
Ans. $i = 3 \sin 200t$ amperes, $v_T = 39 \sin (200t + 67{\cdot}35°)$ volts, $67{\cdot}35°$, $V_m/I_m = 13$ ohms

3.24. A series circuit of two pure elements has the following applied voltage and resulting current:

$$v_T = 255 \sin (300t + 45°) \text{ volts}, \quad i = 8{\cdot}5 \sin (300t + 15°) \text{ amperes}$$

Find the elements comprising the circuit. *Ans.* $R = 26$ ohms, $L = 0{\cdot}05$ henrys.

3.25. A series circuit of two pure elements has the following applied voltage and resulting current:

$$v_T = 150 \cos (200t - 30°) \text{ volts}, \quad i = 4{\cdot}48 \cos (200t - 56{\cdot}6°) \text{ amperes}$$

Find the elements comprising the circuit. *Ans.* $R = 30$ ohms, $L = 0{\cdot}075$ henrys

3.26. Two pure elements, $R = 12$ ohms and $C = 31{\cdot}3\ \mu F$ are connected in series with an applied voltage $v = 100 \cos (2000t - 20°)$ volts. The same two elements are then connected in parallel with the same applied voltage. Find the total current for each connection.
Ans. Series: $i = 5 \cos (2000t + 33{\cdot}2°)$ amperes, parallel: $i = 10{\cdot}4 \cos (2000t + 16{\cdot}8°)$ amperes

3.27. A resistor of $R = 27{\cdot}5$ ohms and a capacitor of $C = 66{\cdot}7\ \mu F$ are in series. The capacitor voltage is $v_C = 50 \cos 1500t$ volts. Find the total voltage v_T, the angle by which the current leads the voltage, and the impedance magnitude.
Ans. $v_T = 146{\cdot}3 \cos (1500t + 70°)$ volts, $20°$, $V_m/I_m = 29{\cdot}3$ ohms

3.28. A resistor of $R = 5$ ohms and an unknown capacitor are in series. The voltage across the resistor is $v_R = 25 \sin (2000t + 30°)$ volts. If the current leads the applied voltage by $60°$ what is the unknown capacitance C? *Ans.* $57{\cdot}7\ \mu F$

3.29. A series circuit of $L = 0{\cdot}05$ H and an unknown capacitance has the following applied voltage and resulting current:

$$v_T = 100 \sin 5000t \text{ volts}, \quad i = 2 \sin (5000t + 90°) \text{ amperes}$$

Find the capacitance C. *Ans.* $C = 0{\cdot}667\ \mu F$

3.30. An *RLC* series circuit has a current which lags the applied voltage by $30°$. The voltage across the inductance has a maximum value which is twice the maximum value of the voltage across the capacitor and $v_L = 10 \sin 1000t$ volts. If $R = 20$ ohms, determine the values of L and C. *Ans.* $L = 23{\cdot}1$ mH, $C = 86{\cdot}5\ \mu F$

3.31. A series circuit consisting of $R = 5$ ohms, $L = 0{\cdot}02$ henrys and $C = 80\ \mu F$ has a variable frequency sinusoidal voltage applied. Find the values of ω for which the current (*a*) will lead the voltage by $45°$, (*b*) be in phase, (*c*) lag by $45°$.
Ans. (*a*) 675, (*b*) 790, (*c*) 925 rad/s

3.32. A two branch parallel circuit with one branch of $R = 50$ ohms and a single unknown element in the other branch has the following applied voltage and total current:

$$v = 100 \cos (1500t + 45°) \text{ volts}, \quad i_T = 12 \sin (1500t + 135°) \text{ amperes}$$

Find the unknown element. *Ans.* $R = 10$ ohms

3.33. Find the total current to the parallel circuit of $L = 0{\cdot}05$ H and $C = 0{\cdot}667\ \mu F$ with an applied voltage $v = 100 \sin 5000t$ volts. *Ans.* $i_T = 0{\cdot}067 \sin (5000t - 90°)$ amperes

3.34. A resistor of $R = 10$ ohms and an inductor of $L = 0{\cdot}005$ henrys are in parallel. The current in the inductive branch is $i_L = 5 \sin(2000t - 45°)$ amperes. Find the total current and the angle between i_T and the applied voltage.
Ans. $i_T = 7{\cdot}07 \sin(2000t + 0°)$ amperes, 45° (i_T lags v)

3.35. A parallel circuit with one branch of $R = 5$ ohms and a single unknown element in the other branch has the following applied voltage and total current:

$$v = 10 \cos(50t + 60°) \text{ volts}, \quad i = 5{\cdot}38 \cos(50t - 8{\cdot}23°) \text{ amperes}$$

Find the unknown element. *Ans.* $L = 0{\cdot}04$ henrys

3.36. Two pure elements, $R = 10$ ohms and $C = 100$ μF, in a parallel connection have an applied voltage $v = 150 \cos(5000t - 30°)$ volts. Find the total current. *Ans.* $i_T = 76{\cdot}5 \cos(5000t + 48{\cdot}7°)$ amperes

3.37. A pure capacitor of $C = 35$ μF is in parallel with another single circuit element. If the applied voltage and resulting total current are $v = 150 \sin 3000t$ volts and $i_T = 16{\cdot}5 \sin(3000t + 72{\cdot}4°)$ amperes respectively, find the other element.
Ans. $R = 30$ ohms

3.38. An LC parallel circuit has an applied voltage $v = 50 \cos(3000t + 45°)$ volts and a total current $i_T = 2 \cos(3000t - 45°)$ amperes. It is also known that the current in the L branch is five times greater than the current in the C branch. Find L and C. *Ans.* $L = 6{\cdot}67$ mH, $C = 3{\cdot}33$ μF

3.39. Three parallel branches, each containing one pure element, have an applied voltage $v = 200 \sin 1000t$ volts. The branches contain $R = 300$ ohms, $L = 0{\cdot}5$ H and $C = 10$ μF respectively. Find the total current, the angle between i_T and the applied voltage, and the magnitude of the impedance.
Ans. $i_T = 1{\cdot}74 \sin(1000t + 67{\cdot}4°)$ amperes, 67·4° (i_T leads v), $V_m/I_m = 115$ ohms

3.40. Find L in the parallel circuit shown in Fig. 3-20 below if the applied voltage and total current are $v = 100 \sin 500t$ volts and $i_T = 2{\cdot}5 \sin 500t$ amperes respectively. *Ans.* $L = 0{\cdot}08$ H

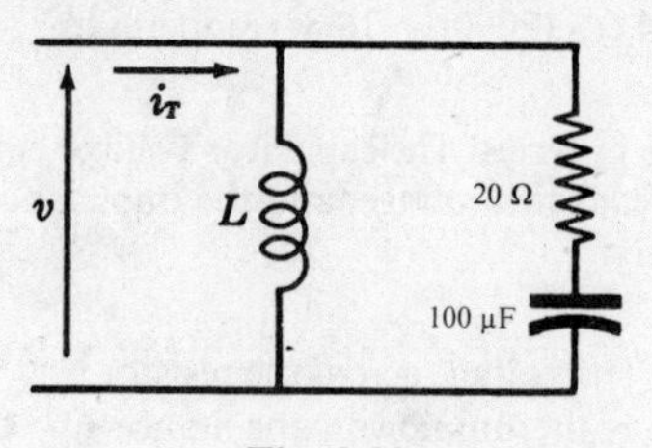

Fig. 3-20

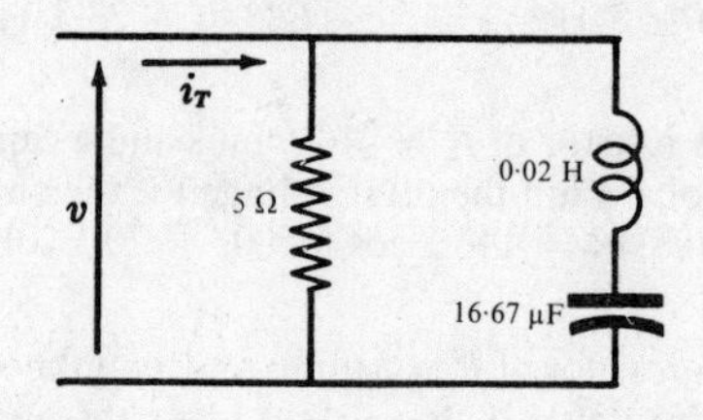

Fig. 3-21

3.41. In the parallel circuit shown in Fig. 3-21 above the applied voltage is $v = 50 \sin(2000t - 90°)$ volts. Find the total current. *Ans.* $i_T = 11{\cdot}2 \sin(2000t - 116{\cdot}6°)$ amperes

3.42. In the parallel circuit shown in Fig. 3-22 below the applied voltage is $v = 100 \sin 5000t$ volts. Find the currents i_1, i_2, i_T.
Ans. $i_1 = 7{\cdot}07 \sin(5000t - 45°)$ amperes, $i_2 = 7{\cdot}07 \sin(5000t + 45°)$ amperes, $i_T = 10 \sin 5000t$ amperes

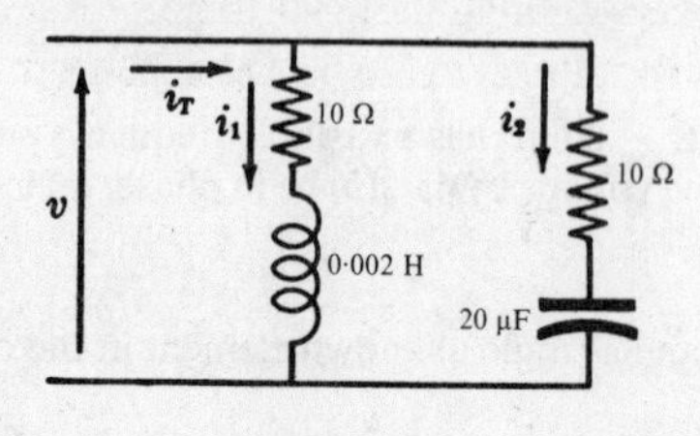

Fig. 3-22

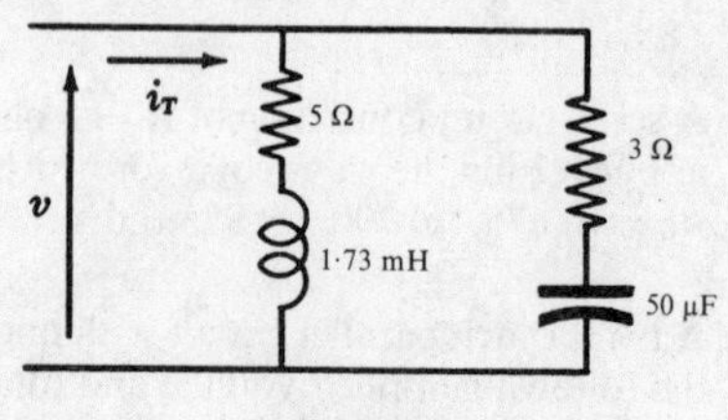

Fig. 3-23

3.43. In the parallel circuit shown in Fig. 3-23 the applied voltage is $v = 100 \cos(5000t + 45°)$ volts. (*a*) Find the total current. (*b*) What two elements in a series connection would result in the same current and would therefore be equivalent to the parallel circuit for the same frequency?
Ans. (*a*) $i_T = 18{\cdot}5 \cos(5000t + 68{\cdot}4°)$ amperes, (*b*) series circuit of $R = 4{\cdot}96$ ohms and $C = 93$ μF

CHAPTER 4

Complex Numbers

REAL NUMBERS

The *real number system* consists of the rational and irrational numbers. The set of all real numbers may be placed in one-to-one correspondence with the set of all points on a straight line, called the *real number line*, such that each point represents a unique real number and each real number is represented by a unique point on the line, as shown in Fig. 4-1. The operations of addition, subtraction, multiplication and division can be performed with any of the numbers in this system. Square roots of positive real numbers may be represented on the real number line, but the square root of a negative number does not exist in the real number system.

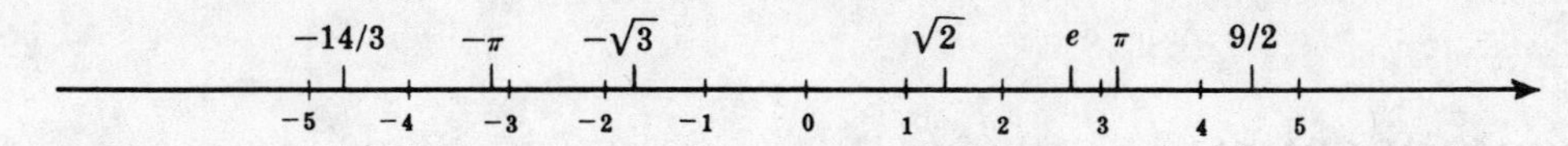

Fig. 4-1. Real Number Line

IMAGINARY NUMBERS

The square root of a negative real number is called a *pure imaginary number*, e.g. $\sqrt{-1}$, $\sqrt{-2}$, $\sqrt{-5}$, $\sqrt{-16}$.

If we designate $j = \sqrt{-1}$, then $\sqrt{-2} = j\sqrt{2}$, $\sqrt{-4} = j2$, $\sqrt{-5} = j\sqrt{5}$, etc. Also, it follows that

$$j^2 = -1, \quad j^3 = j^2 \cdot j = (-1)j = -j, \quad j^4 = (j^2)^2 = 1, \quad j^5 = j, \quad \ldots$$

All pure imaginary numbers can be represented by points on a straight line called the *imaginary number line*, as shown in Fig. 4-2.

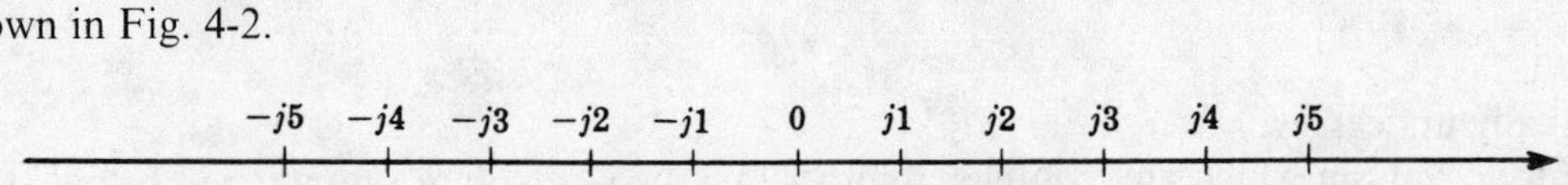

$-j5 \quad -j4 \quad -j3 \quad -j2 \quad -j1 \quad 0 \quad j1 \quad j2 \quad j3 \quad j4 \quad j5$

Fig. 4-2. Imaginary Number Line.

The choice of the word imaginary is an unfortunate one since imaginary numbers exist as surely as real numbers. The term merely signifies that such numbers cannot be represented on the real number line but are located on a second number line, the imaginary number line.

COMPLEX NUMBERS

A complex number $\mathbf{z}$ is a number of the form $x + jy$ where x and y are real numbers and $j = \sqrt{-1}$. In a complex number $x + jy$, the first term x is called the real part and the second term jy the imaginary part. When $x = 0$, the complex number is a pure imaginary and corresponds to a point on the j axis. Similarly, if $y = 0$, the complex number is a real number and corresponds to a point on the real axis. Thus complex numbers include all real numbers and all pure imaginary numbers.

Two complex numbers, $a + jb$ and $c + jd$, are equal if and only if $a = c$ and $b = d$.

If, as in Fig. 4-3, the axis of reals is perpendicular to the axis of imaginaries (or j axis) at their common point 0, then each point in the resulting complex plane represents a unique complex number, and conversely. Six complex numbers ($\mathbf{z}_1, \ldots, \mathbf{z}_6$) are plotted in Fig. 4-3.

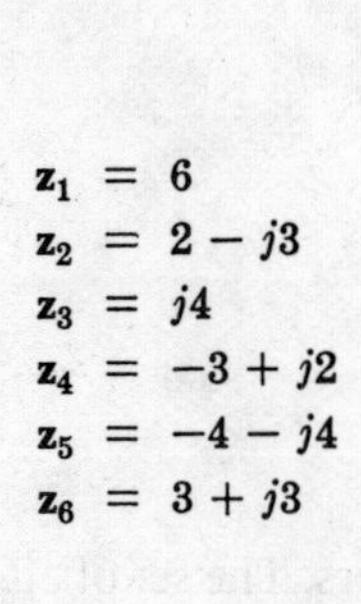

$$\mathbf{z}_1 = 6$$
$$\mathbf{z}_2 = 2 - j3$$
$$\mathbf{z}_3 = j4$$
$$\mathbf{z}_4 = -3 + j2$$
$$\mathbf{z}_5 = -4 - j4$$
$$\mathbf{z}_6 = 3 + j3$$

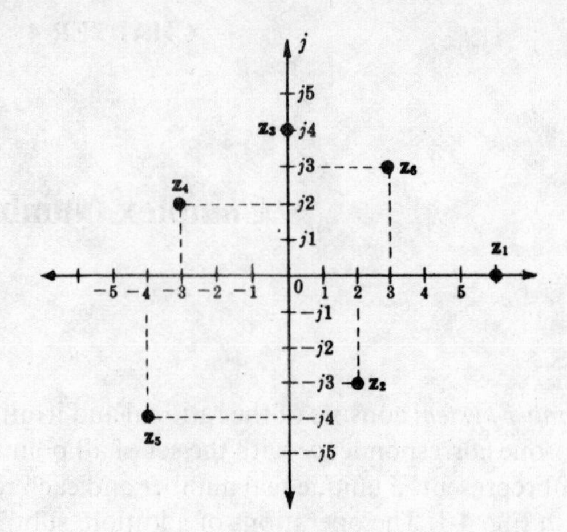

Fig. 4-3.

OTHER FORMS OF COMPLEX NUMBERS

In Fig. 4-4, $x = r\cos\theta$, $y = r\sin\theta$, and the complex number $\mathbf{z}$ is

$$\mathbf{z} = x + jy = r(\cos\theta + j\sin\theta)$$

where $r = \sqrt{x^2 + y^2}$ is called the *modulus* or *absolute value* of $\mathbf{z}$, and the angle $\theta = \tan^{-1} y/x$ is called the *amplitude* or *argument* of $\mathbf{z}$.

Euler's formula, $e^{j\theta} = (\cos\theta + j\sin\theta)$, permits another form of a complex number, called the exponential form (see Problem 4.1):

$$\mathbf{z} = r\cos\theta + jr\sin\theta = r\,e^{j\theta}$$

Polar representation of a complex number $\mathbf{z}$

Fig. 4-4.

The polar or Steinmetz form of a complex number $\mathbf{z}$ is widely used in circuit analysis and is written

$$r\underline{/\theta}$$

where θ is usually in degrees.

These four ways in which any complex number may be written are summarized below. The one employed will depend upon the operation which is to be performed.

Rectangular form	$\mathbf{z} = x + jy$
Polar or Steinmetz form	$\mathbf{z} = r\underline{/\theta}$
Exponential form	$\mathbf{z} = r\,e^{j\theta}$
Trigonometric form	$\mathbf{z} = r(\cos\theta + j\sin\theta)$

CONJUGATE OF A COMPLEX NUMBER

The conjugate $\mathbf{z}^*$ of a complex number $\mathbf{z} = x + jy$ is the complex number $\mathbf{z}^* = x - jy$. For example, two pairs of conjugate complex numbers are: (*1*) $3 - j2$ and $3 + j2$, (*2*) $-5 + j4$ and $-5 - j4$.

In the polar form, the conjugate of $\mathbf{z} = r\,\angle\theta$ is $\mathbf{z}^* = r\,\angle{-\theta}$. Since $\cos(-\theta) = \cos\theta$ and $\sin(-\theta) = -\sin\theta$, the conjugate of $\mathbf{z} = r(\cos\theta + j\sin\theta)$ is $\mathbf{z}^* = r(\cos\theta - j\sin\theta)$. For example, the conjugate of $\mathbf{z} = 7\,\angle 30°$ is $\mathbf{z}^* = 7\,\angle{-30°}$.

The conjugate $\mathbf{z}^*$ of a complex number $\mathbf{z}$ is always the image of $\mathbf{z}$ with respect to the axis of reals, as indicated in Fig. 4-5.

Thus, four ways of writing a complex number $\mathbf{z}$ and its conjugate are:

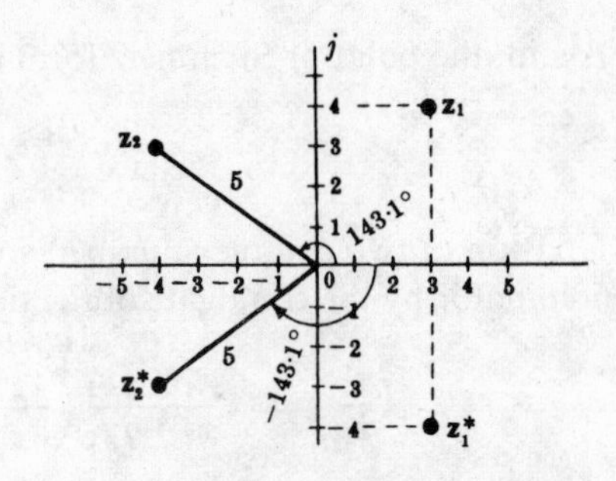

$z_1 = 3 + j4, \quad z_1^* = 3 - j4$

$z_2 = 5\underline{/143{\cdot}1°}, \quad z_2^* = 5\underline{/-143{\cdot}1°}$

Fig. 4-5. Complex Numbers and Their Conjugates

$$\mathbf{z} = x + jy \qquad \mathbf{z} = r\underline{/\theta} \qquad \mathbf{z} = r\,e^{j\theta} \qquad \mathbf{z} = r(\cos\theta + j\sin\theta)$$
$$\mathbf{z}^* = x - jy \qquad \mathbf{z}^* = r\underline{/-\theta} \qquad \mathbf{z}^* = r\,e^{-j\theta} \qquad \mathbf{z}^* = r(\cos\theta - j\sin\theta)$$

SUM AND DIFFERENCE OF COMPLEX NUMBERS

To add two complex numbers, add the real parts and the imaginary parts separately. To subtract two complex numbers, subtract the real parts and the imaginary parts separately. From the practical standpoint, addition and subtraction of complex numbers can be performed conveniently only when both numbers are in the rectangular form.

Example 1. Given $\mathbf{z}_1 = 5 - j2$ and $\mathbf{z}_2 = -3 - j8$. Then

$$\mathbf{z}_1 + \mathbf{z}_2 = (5-3) + j(-2-8) = 2 - j10$$
$$\mathbf{z}_2 - \mathbf{z}_1 = (-3-5) + j(-8+2) = -8 - j6$$

MULTIPLICATION OF COMPLEX NUMBERS

The product of two complex numbers when both are in the exponential form follows directly from the laws of exponents.

$$\mathbf{z}_1\mathbf{z}_2 = (r_1 e^{j\theta_1})(r_2 e^{j\theta_2}) = r_1 r_2 e^{j(\theta_1+\theta_2)}$$

The polar or Steinmetz form product is evident from reference to the exponential form.

$$\mathbf{z}_1\mathbf{z}_2 = (r_1\underline{/\theta_1})(r_2\underline{/\theta_2}) = r_1 r_2\underline{/\theta_1 + \theta_2}$$

The rectangular form product can be found by treating the two complex numbers as binomials.

$$\mathbf{z}_1\mathbf{z}_2 = (x_1 + jy_1)(x_2 + jy_2) = x_1x_2 + jx_1y_2 + jy_1x_2 + j^2y_1y_2$$
$$= (x_1x_2 - y_1y_2) + j(x_1y_2 + y_1x_2)$$

Example 2. If $\mathbf{z}_1 = 5e^{j\pi/3}$ and $\mathbf{z}_2 = 2e^{-j\pi/6}$, then $\mathbf{z}_1\mathbf{z}_2 = (5e^{j\pi/3})(2e^{-j\pi/6}) = 10e^{j\pi/6}$.

Example 3. If $\mathbf{z}_1 = 2\underline{/30°}$ and $\mathbf{z}_2 = 5\underline{/-45°}$, then $\mathbf{z}_1\mathbf{z}_2 = (2\underline{/30°})(5\underline{/-45°}) = 10\underline{/-15°}$.

Example 4. If $\mathbf{z}_1 = 2 + j3$ and $\mathbf{z}_2 = -1 - j3$, then $\mathbf{z}_1\mathbf{z}_2 = (2+j3)(-1-j3) = 7 - j9$.

DIVISION OF COMPLEX NUMBERS

For two complex numbers in the exponential form, the quotient follows directly from the laws of exponents.

$$\frac{\mathbf{z}_1}{\mathbf{z}_2} = \frac{r_1 e^{j\theta_1}}{r_2 e^{j\theta_2}} = \frac{r_1}{r_2}\,e^{j(\theta_1-\theta_2)}$$

Again, the polar or Steinmetz form of division is evident from reference to the exponential form.

$$\frac{\mathbf{z}_1}{\mathbf{z}_2} = \frac{r_1\underline{/\theta_1}}{r_2\underline{/\theta_2}} = \frac{r_1}{r_2}\underline{/\theta_1 - \theta_2}$$

Division of two complex numbers in the rectangular form is performed by multiplying the numerator and denominator by the conjugate of the denominator.

$$\frac{\mathbf{z}_1}{\mathbf{z}_2} = \frac{x_1 + jy_1}{x_2 + jy_2}\left(\frac{x_2 - jy_2}{x_2 - jy_2}\right) = \frac{(x_1x_2 + y_1y_2) + j(y_1x_2 - y_2x_1)}{x_2^2 + y_2^2}$$

Example 5. Given $\mathbf{z}_1 = 4e^{j\pi/3}$ and $\mathbf{z}_2 = 2e^{j\pi/6}$; then $\dfrac{\mathbf{z}_1}{\mathbf{z}_2} = \dfrac{4e^{j\pi/3}}{2e^{j\pi/6}} = 2e^{j\pi/6}$.

Example 6. Given $\mathbf{z}_1 = 8\underline{/-30^\circ}$ and $\mathbf{z}_2 = 2\underline{/-60^\circ}$; then $\dfrac{\mathbf{z}_1}{\mathbf{z}_2} = \dfrac{8\underline{/-30^\circ}}{2\underline{/-60^\circ}} = 4\underline{/30^\circ}$.

Example 7. Given $\mathbf{z}_1 = 4 - j5$ and $\mathbf{z}_2 = 1 + j2$; then $\dfrac{\mathbf{z}_1}{\mathbf{z}_2} = \dfrac{4 - j5}{1 + j2}\left(\dfrac{1 - j2}{1 - j2}\right) = \dfrac{-6 - j13}{5}$.

ROOTS OF COMPLEX NUMBERS

Any complex number $\mathbf{z} = r\,e^{j\theta}$ may be written $\mathbf{z} = r\,e^{j(\theta + 2\pi n)}$, where $n = 0, \pm 1, \pm 2, \ldots$. Similarly, $\mathbf{z} = r\underline{/\theta}$ may be written $\mathbf{z} = r\underline{/(\theta + n360^\circ)}$. Thus

$$\mathbf{z} = r\,e^{j\theta} = r\,e^{j(\theta + 2\pi n)} \quad \text{and} \quad \sqrt[k]{\mathbf{z}} = \sqrt[k]{r}\,e^{j(\theta + 2\pi n)/k}$$

$$\mathbf{z} = r\underline{/\theta} = r\underline{/(\theta + n360^\circ)} \quad \text{and} \quad \sqrt[k]{\mathbf{z}} = \sqrt[k]{r}\underline{/(\theta + n360^\circ)/k}$$

Now the k distinct kth roots of the complex number can be obtained by assigning to n the values 0, 1, 2, 3, . . ., $k - 1$.

Example 8.

If $\mathbf{z} = 8\underline{/60^\circ}$, then $\sqrt[3]{\mathbf{z}} = \sqrt[3]{8}\underline{/(60^\circ + n360^\circ)/3} = 2\underline{/(20^\circ + n120^\circ)}$. As n assumes the values 0, 1 and 2, the three cube roots obtained are $2\underline{/20^\circ}$, $2\underline{/140^\circ}$ and $2\underline{/260^\circ}$.

Example 9. Find the five fifth roots of 1.

Since $1 = 1e^{j2\pi n}$, then $\sqrt[5]{1} = \sqrt[5]{1}\,e^{j2\pi n/5} = 1\,e^{j2\pi n/5}$. As n assumes the values 0, 1, 2, 3 and 4, the five fifth roots obtained are $1\underline{/0^\circ}$ or 1, $1\underline{/72^\circ}$, $1\underline{/144^\circ}$, $1\underline{/216^\circ}$ and $1\underline{/288^\circ}$.

LOGARITHM OF A COMPLEX NUMBER

The natural logarithm of a complex number may be readily found when the number is expressed in exponential form.

$$\ln \mathbf{z} = \ln r\,e^{j(\theta + 2\pi n)} = \ln r + \ln e^{j(\theta + 2\pi n)} = \ln r + j(\theta + 2\pi n)$$

The result is not unique. The principal value, when $n = 0$, is most frequently used.

Example 10. If $\mathbf{z} = 3e^{j\pi/6}$, then $\ln \mathbf{z} = \ln z\,e^{j\pi/6} = \ln 3 + j\pi/6 = 1{\cdot}099 + j0{\cdot}523$.

USE OF SLIDE RULE WITH COMPLEX NUMBERS

Introduction

In the phasor notation of Chapter 5, voltage, current and impedance are all complex numbers. The rectangular and polar forms of these quantities are used most frequently. There is need for a rapid and dependable conversion from one form to the other, since the polar form leads to the most direct multiplication or division and the rectangular form is required for addition or subtraction.

Any slide rule of the deci-trig type may be used for these conversions. Most rules have the tangent scale folded so that the tangents of angles greater than 45° appear on the inverted C scale. The instructions which follow apply to such a rule. For any slide rule not of this type, the reader should consult the instruction book accompanying the rule.

Since the sole purpose of this discussion is to arrive at a rapid and dependable conversion in either direction, the trigonometric explanations are reduced to a minimum.

POLAR TO RECTANGULAR FORM

Example 11. Express $50\angle 53{\cdot}1°$ in rectangular form, $x + jy$.

1. Make a sketch, exaggerating the fact that the angle is greater than 45°.
2. Set the C scale index to 50 on the D scale.
3. Move the cursor to place the hairline at 53·1° on the sine and cosine readings of the S scale. Write down both numbers read from the D scale: 40 and 30.
4. Refer to the sketch, noting that the j part is larger than the real part and that both are positive.
5. $50\angle 53{\cdot}1° = 30 + j40$.

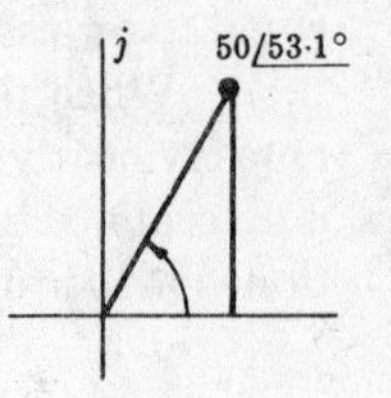

Example 12. Express $100\angle -120°$ in rectangular form, $x + jy$.

1. Make the sketch. The reference angle is 60°.
2. Set the C scale index to 100 on the D scale.
3. Move the cursor to place the hairline at 60° on the sine cosine scale. The two numbers read on the D scale are 86·6 and 50·0.
4. Refer to the sketch, noting that the j part is larger than the real part and that both are negative.
5. $100\angle -120° = -50{\cdot}0 - j86{\cdot}6$.

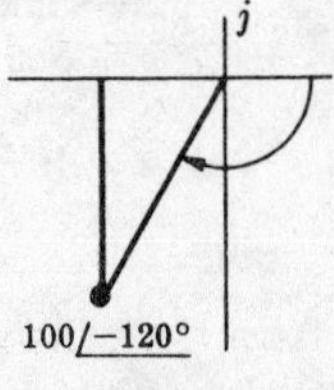

RECTANGULAR TO POLAR FORM

Example 13. Express $4 + j3$ in polar form, $r\angle\theta$.

1. Make a sketch, exaggerating the fact that the real part is larger than the j part.
2. Set the hairline to the *smaller* of the two numbers (3) on the D scale.
3. Position the C scale index at the *larger* number (4) on the D scale.
4. On the tangent scale there are two angles at the hairline: 53·1° and 36·9°. Refer to the sketch and note that the smaller of the two applies in this problem. Place this angle in an angle bracket: $4 + j3 = \ldots\ldots \angle 36{\cdot}9°$.
5. With the hairline fixed, move the sliding part of the rule to bring 36·9° on the sine cosine scale under the hairline. The 36·9° appears twice on this scale. To obtain the correct one, note that on the tangent scale the 36·9° was on the *right side* of the markings. Therefore 36·9° on the *right side* of the sine cosine scale markings is used in this step. At the C scale index the value of r is read, 5. (Always use the leftmost number. Note that this causes the index to move to the right.)
6. $4 + j3 = 5\angle 36{\cdot}9°$.

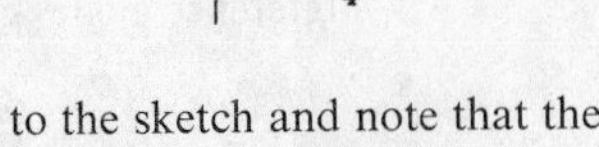

Example 14. Express $-10 + j20$ in polar form, $r\angle\theta$.

1. Make a sketch.
2. Set the hairline to the *smaller* of the two numbers (10) on the D scale.
3. Position the C scale at the *larger* number (20) on the D scale.
4. The sketch show that the reference angle is greater than 45°. Read 63·4° on the tangent scale. This is not θ but a reference angle. Write it below the angle bracket for use later in the problem, and write $\theta = 180° - 63{\cdot}4° = 116{\cdot}6°$ inside the bracket: $-10 + j20 = \ldots\ldots \underset{(63{\cdot}4°)}{\angle 116{\cdot}6°}$.

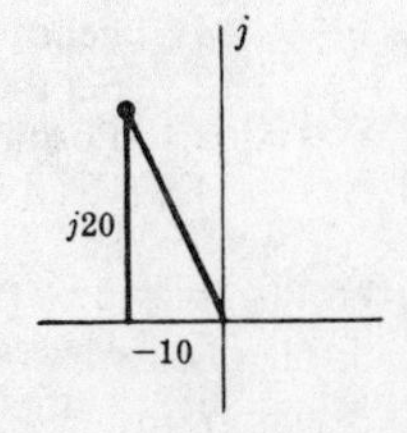

5. With the hairline fixed, move the slider until 63·4° (on the left side of the sine cosine scale markings) appears under the hairline. At the C scale index read 22·4 on the D scale.
6. $-10 + j20 = 22{\cdot}4\angle\underline{116{\cdot}6^\circ}$.

OPERATIONS WITH THE EXTENDED SINE TANGENT SCALE

As the numerical value of the angle θ in the polar form of a complex number becomes very small, r and the numerical value of the real part x of the rectangular form become very nearly equal. When $|\theta| \leqq 5{\cdot}73^\circ$, r and x are assumed equal. The imaginary part jy, equal to $jr \sin \theta$, is determined by using the extended sine tangent scale. The same assumption is made for values of θ close to 180° where the reference angle is numerically equal to or less than 5·73°.

When the numerical value of θ is close to 90°, r and the numerical value of y of the rectangular form are very nearly equal. When $84{\cdot}27^\circ \leqq \theta \leqq 95{\cdot}73^\circ$, r and y are assumed equal. The real part x, equal to $r \cos \theta$, is determined by using the extended sine tangent scale, considering that $\cos \theta = \sin (90^\circ - \theta)$. The same assumption is made for values of θ near 270° where the reference angle equals or exceeds 84·27°.

Example 15. Express $10\angle\underline{3{\cdot}5^\circ}$ in rectangular form, $x + jy$.

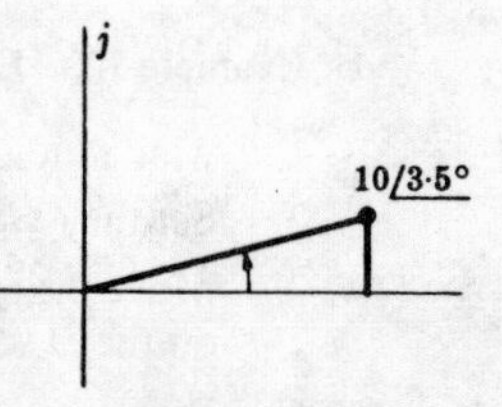

1. Make a sketch.
2. Since the angle is less than 5·73°, the real part x is equal to 10.
3. Set the C scale index to 10 on the D scale. Move the hairline to 3·5° on the extended sine scale. Read 0·61 on the D scale.
4. $10\angle\underline{3{\cdot}5^\circ} = 10 + j0{\cdot}61$. The decimal point is determined by noting that for small angles the real and j parts are in a ratio greater than 10 to 1.

Example 16. Express $450\angle\underline{94^\circ}$ in rectangular form, $x + jy$.

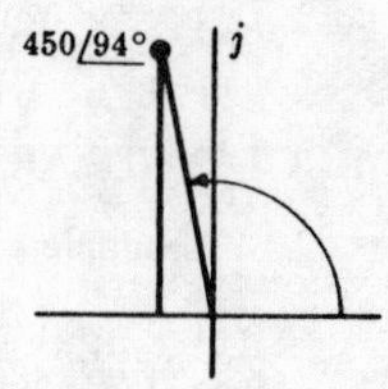

1. Make a sketch. The reference angle is 86°.
2. Since the reference angle is greater than 84·27°, the j part is equal to r, 450.
3. Set the C scale index to 450 on the D scale. Move the hairline to 86° on the extended sine scale. Read 31·4 on the D scale. For those rules without the complementary angles marked on the extended scale, move the hairline to (90° − 86°) or 4° and read 31·4 on the D scale.
4. $450\angle\underline{94^\circ} = -31{\cdot}4 + j450$. Just as with the ordinary angles, the sketch aids in determining the signs of the components. The ratio of the j part to the real part exceeds 10 to 1 for reference angles greater than 84·27°, and the decimal point is thereby set at 31·4.

Example 17. Express $20 + j500$ in polar form, $r\angle\underline{\theta}$.

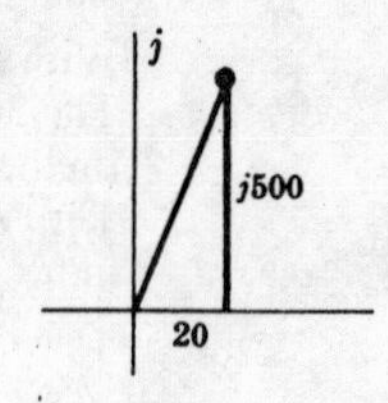

1. Make a sketch. The ratio greater than 10 to 1 indictaes an angle greater than 84·27°; therefore $r = 500$.
2. Set the hairline to the *smaller* part (20) on the D scale.
3. Position the C scale at the *larger* number (500) on the D scale.
4. Read from the extended tangent scale 87·7°.
5. $20 + j500 = 500\angle\underline{87{\cdot}7^\circ}$.

Example 18. Express $-4 - j85$ in polar form, $r\angle\underline{\theta}$.

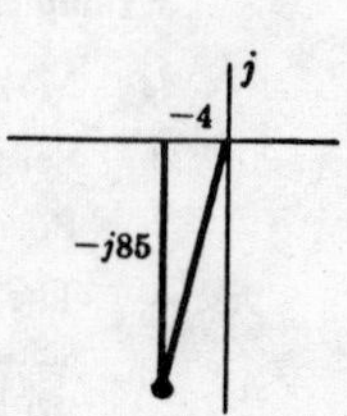

1. Make a sketch. The ratio greater than 10 to 1 indicates a reference angle greater than 84·27°; hence $r = 85$.
2. Set the hairline to the *smaller* part (4) on the D scale.
3. Position the C scale index at the *larger* number (85) on the D scale.
4. Read from the extended scale 87·3°. This is not θ but the reference angle. $-4 - j85 =$ $85\angle\underline{\quad}$ (87·3°)

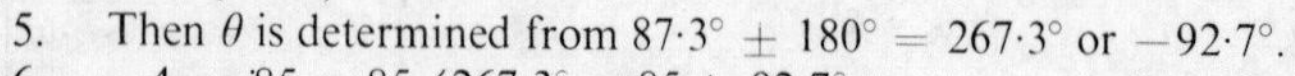

5. Then θ is determined from $87{\cdot}3^\circ \pm 180^\circ = 267{\cdot}3^\circ$ or $-92{\cdot}7^\circ$.
6. $-4 - j85 = 85\angle\underline{267{\cdot}3^\circ}$ or $85\angle\underline{-92{\cdot}7^\circ}$.

Problems

4.1. Prove Euler's formula.

Assuming a function $f(x)$ can be represented by a power series in x, that series must be of the form of the Maclaurin series

$$f(x) = f(0) + x\,f'(0) + \frac{x^2}{2!}f''(0) + \frac{x^3}{3!}f'''(0) + \cdots + \frac{x^{(n-1)}}{(n-1)!}f^{(n-1)}(0) + \cdots$$

where the function and all its derivatives exist at $x = 0$.

The Maclaurin series expansions of $\cos\theta$, $\sin\theta$ and $e^{j\theta}$ in powers of θ are

$$\cos\theta = 1 - \frac{\theta^2}{2!} + \frac{\theta^4}{4!} - \frac{\theta^6}{6!} + \cdots \qquad \sin\theta = \theta - \frac{\theta^3}{3!} + \frac{\theta^5}{5!} - \frac{\theta^7}{7!} + \cdots$$

$$e^{j\theta} = 1 + j\theta - \frac{\theta^2}{2!} - j\frac{\theta^3}{3!} + \frac{\theta^4}{4!} + j\frac{\theta^5}{5!} - \frac{\theta^6}{6!} - j\frac{\theta^7}{7!} + \cdots$$

Rearranging the terms in the Maclaurin series of $e^{j\theta}$, we have

$$e^{j\theta} = \left(1 - \frac{\theta^2}{2!} + \frac{\theta^4}{4!} - \frac{\theta^6}{6!} + \cdots\right) + j\left(\theta - \frac{\theta^3}{3!} + \frac{\theta^5}{5!} - \frac{\theta^7}{7!} + \cdots\right) = \cos\theta + j\sin\theta$$

4.2. Sketch the complex plane and locate the following complex numbers. Convert each number into polar form and repeat the sketch. A comparison of the two sketches will show if the conversion was performed correctly.

(*a*) $2 - j2$ (*b*) $3 + j8$ (*c*) $-5 + j3$ (*d*) $-4 - j4$ (*e*) $5 + j0$ (*f*) $j6$ (*g*) -4 (*h*) $-j5$

4.3. Express each complex number in the polar form.

(*a*) $15e^{j\pi/4}$ (*b*) $5e^{-j2\pi/3}$ (*c*) $-4e^{j5\pi/6}$ (*d*) $-2e^{-j\pi/2}$ (*e*) $10e^{-j7\pi/6}$ (*f*) $-18e^{-j3\pi/2}$

Ans. (*a*) $15\angle 45^\circ$, (*b*) $5\angle -120^\circ$, (*c*) $4\angle -30^\circ$, (*d*) $2\angle 90^\circ$, (*e*) $10\angle -210^\circ$ or $10\angle 150^\circ$, (*f*) $18\angle -90^\circ$

4.4. Perform the indicated operation.

(*a*) $z = 3 - j4$. Find zz^*.
(*b*) $z = 10\angle -40^\circ$. Find zz^*.
(*c*) $z = 20\angle 53{\cdot}1^\circ$. Find $z + z^*$.
(*d*) $z = 2{\cdot}5e^{-j\pi/3}$. Find zz^*.
(*e*) $z = 2 + j8$. Find $z - z^*$.
(*f*) $z = 10 - j4$. Find $z + z^*$.
(*g*) $z = 95\angle 25^\circ$. Find $z - z^*$.
(*h*) $z = r\angle\theta$. Find z/z^*.

Ans. (*a*) 25, (*b*) 100, (*c*) 24, (*d*) 6·25, (*e*) $j16$, (*f*) 20, (*g*) $j80{\cdot}2$, (*h*) $1\angle 2\theta$.

4.5. Determine the indicated roots of each complex number.

(*a*) $\sqrt{5 + j8}$ (*b*) $\sqrt{150\angle -60^\circ}$ (*c*) $\sqrt[3]{6{\cdot}93 - j4}$ (*d*) $\sqrt[3]{27e^{j3\pi/2}}$ (*e*) $\sqrt[4]{1}$ (*f*) $\sqrt{4}$

Ans. (*a*) $3{\cdot}07\angle 29^\circ$, $3{\cdot}07\angle 209^\circ$, (*b*) $12{\cdot}25\angle -30^\circ$, $12{\cdot}25\angle 150^\circ$, (*c*) $2\angle -10^\circ$, $2\angle 110^\circ$, $2\angle 230^\circ$, (*d*) $3e^{j\pi/2}$, $3e^{j7\pi/6}$, $3e^{j11\pi/6}$, (*e*) $1\angle 0$, $1\angle 90^\circ$, $1\angle 180^\circ$, $1\angle 270^\circ$, (*f*) $2\angle 0$, $2\angle 180^\circ$, i.e. ± 2.

4.6. In (*a*)-(*d*), find the natural logarithm of the complex number. In part (*e*) use logarithms to determine the product.

(*a*) $20\angle 45^\circ$ (*b*) $6\angle -60^\circ$ (*c*) $0{\cdot}5\angle 120^\circ$ (*d*) $0{\cdot}3\angle 180^\circ$ (*e*) $(0{\cdot}3\angle 180^\circ)(20\angle 45^\circ)$

Ans. (*a*) $3 + j\pi/4$, (*b*) $1{\cdot}79 - j\pi/3$, (*c*) $-0{\cdot}693 + j2\pi/3$, (*d*) $-1{\cdot}2 + j\pi$, (*e*) $6\angle 225^\circ$

4.7. Use the slide rule to convert each complex number from polar to rectangular form.

(*a*) $12{\cdot}3\angle 30^\circ$ *Ans.* $10{\cdot}63 + j6{\cdot}15$
(*b*) $53\angle 160^\circ$ $-49{\cdot}8 + j18{\cdot}1$
(*c*) $25\angle -45^\circ$ $17{\cdot}7 - j17{\cdot}7$
(*d*) $86\angle -115^\circ$ $-36{\cdot}3 - j78$
(*e*) $0{\cdot}05\angle -20^\circ$ *Ans.* $0{\cdot}047 - j0{\cdot}0171$
(*f*) $0{\cdot}003\angle 80^\circ$ $0{\cdot}00052 + j0{\cdot}00295$
(*g*) $0{\cdot}013\angle 260^\circ$ $-0{\cdot}00226 - j0{\cdot}0128$
(*h*) $0{\cdot}156\angle -190^\circ$ $-0{\cdot}1535 + j0{\cdot}0271$

4.8. Use the slide rule to convert each complex number from rectangular to polar form.

(*a*) $-12 + j16$ *Ans.* $20\angle 126{\cdot}8^\circ$
(*b*) $2 - j4$ $4{\cdot}47\angle -63{\cdot}4^\circ$
(*c*) $-59 - j25$ $64\angle 203^\circ$
(*d*) $700 + j200$ $727\angle 16^\circ$
(*e*) $0{\cdot}048 - j0{\cdot}153$ *Ans.* $0{\cdot}160\angle -72{\cdot}55^\circ$
(*f*) $0{\cdot}0171 + j0{\cdot}047$ $0{\cdot}05\angle 70^\circ$
(*g*) $-69{\cdot}4 - j40$ $80\angle 210^\circ$
(*h*) $-2 + j2$ $28{\cdot}3\angle 135^\circ$

4.9. Use the slide rule to convert the complex numbers from polar to rectangular form.

(*a*) $10\angle 3^\circ$ *Ans.* $10 + j0{\cdot}523$
(*b*) $25\angle 88^\circ$ $0{\cdot}871 + j25$
(*c*) $50\angle -93^\circ$ $-2{\cdot}62 - j50$
(*d*) $45\angle 179^\circ$ $-45 + j0{\cdot}785$
(*e*) $0{\cdot}02\angle 94^\circ$ *Ans.* $-0{\cdot}00139 + j0{\cdot}02$
(*f*) $0{\cdot}70\angle 266^\circ$ $-0{\cdot}0488 - j0{\cdot}70$
(*g*) $0{\cdot}80\angle -5^\circ$ $0{\cdot}8 - j0{\cdot}0696$
(*h*) $200\angle 181^\circ$ $-200 - j3{\cdot}49$

4.10. Use the slide rule to convert the complex numbers from rectangular to polar form.

(*a*) $540 + j40$ *Ans.* $540\angle 4{\cdot}25^\circ$
(*b*) $-10 - j250$ $250\angle -92{\cdot}29^\circ$
(*c*) $8 - j0{\cdot}5$ $8\angle -3{\cdot}58^\circ$
(*d*) $25 + j717$ $717\angle 88^\circ$
(*e*) $0{\cdot}8 - j0{\cdot}0696$ *Ans.* $0{\cdot}8\angle -5^\circ$
(*f*) $10 + j0{\cdot}253$ $10\angle 3^\circ$
(*g*) $-200 - j3{\cdot}49$ $200\angle 181^\circ$
(*h*) $0{\cdot}02 - j0{\cdot}001$ $0{\cdot}02\angle 2{\cdot}87^\circ$

4.11. The following is an exercise in the use of the slide rule. Convert the following numbers written in the polar form to the rectangular form, and those given in rectangular form to polar form. Then convert the answers back to the original form.

(*a*) $40\angle 10^\circ$
(*b*) $18 - j9$
(*c*) $0{\cdot}03 + j0{\cdot}80$
(*d*) $0{\cdot}06\angle -100^\circ$
(*e*) $5{\cdot}0 + j0{\cdot}3$
(*f*) $0{\cdot}50\angle 174^\circ$
(*g*) $180 + j55$
(*h*) $25\angle 88^\circ$
(*i*) $-0{\cdot}05 - j0{\cdot}80$
(*j*) $150\angle -5^\circ$
(*k*) $0{\cdot}002\angle -178^\circ$
(*l*) $-1080 + j250$
(*m*) $80\angle -98^\circ$
(*n*) $-15 - j30$
(*o*) $5\angle 233{\cdot}1^\circ$
(*p*) $-26 + j15$
(*q*) $0{\cdot}85\angle 1^\circ$
(*r*) $3 + j4$
(*s*) $20\angle -143{\cdot}1^\circ$
(*t*) $-5 - j8{\cdot}66$

4.12. Determine the indicated sum or difference.

(*a*) $(10\angle 53{\cdot}1^\circ) + (4 + j2)$ *Ans.* $10 + j10$
(*b*) $(10\angle 90^\circ) + (8 - j2)$ $8 + j8$
(*c*) $(-4 - j6) + (2 + j4)$ $-2 - j2$
(*d*) $(2{\cdot}83\angle 45^\circ) - (2 - j8)$ $j10$
(*e*) $(-5 + j5) - (7{\cdot}07\angle 135^\circ)$ *Ans.* 0
(*f*) $(2 - j10) - (1 - j10)$ 1
(*g*) $(10 + j1) + 6 - (13{\cdot}45\angle -42^\circ)$ $6 + j10$
(*h*) $-(5\angle 53{\cdot}1^\circ) - (1 - j6)$ $-4 + j2$

4.13. Calculate the product of the following complex numbers. As an additional exercise they may be converted to polar form and the product again determined and checked.

(*a*) $(3 - j2)(1 - j4)$ *Ans.* $-5 - j14$
(*b*) $(2 + j0)(3 - j3)$ $6 - j6$
(*c*) $(-1 - j1)(1 + j1)$ $-j2$
(*d*) $(j2)(4 - j3)$ $6 + j8$
(*e*) $(j2)(j5)$ *Ans.* -10
(*f*) $(-j1)(j6)$ 6
(*g*) $(2 + j2)(2 - j2)$ 8
(*h*) $(x + jy)(x - jy)$ $x^2 + y^2$

4.14. In the following problems find the quotient by multiplying the numerator and denominator by the conjugate of the denominator. Convert the numbers to the polar form and determine the quotient from this form.

(*a*) $(5 + j5)/(1 - j1)$ *Ans.* $j5$
(*b*) $(4 - j8)/(2 + j2)$ $-1 - j3$
(*c*) $(5 - j10)/(3 + j4)$ $-1 - j2$
(*d*) $(8 + j12)/(j2)$ $6 - j4$
(*e*) $(3 + j3)/(2 + j2)$ *Ans.* $1{\cdot}5$
(*f*) $(-5 - j10)/(2 + j4)$ $-2{\cdot}5$
(*g*) $10/(6 + j8)$ $0{\cdot}6 - j0{\cdot}8$
(*h*) $j5/(2 - j2)$ $-1{\cdot}25 + j1{\cdot}25$

4.15. Find each indicated product.

(*a*) $(2{\cdot}5 + j10)(-0{\cdot}85 + j4{\cdot}3)$ *Ans.* $45\angle 177{\cdot}1^\circ$
(*b*) $(3{\cdot}8 - j\,1{\cdot}5)(6 - j2{\cdot}3)$ $26{\cdot}2\angle -42{\cdot}6^\circ$
(*c*) $(72 - j72)(1{\cdot}3 + j4{\cdot}8)$ $506\angle 29{\cdot}8^\circ$
(*d*) $(3\angle 20^\circ)(2\angle -45^\circ)$ $6\angle -25^\circ$
(*e*) $(2 + j6)(18\angle 21^\circ)$ *Ans.* $113{\cdot}5\angle 92{\cdot}5^\circ$
(*f*) $1\angle 80^\circ\,(25\angle -45^\circ)(0{\cdot}2\angle -15^\circ)$ $5\angle 20^\circ$
(*g*) $(1{\cdot}2 - j16)(0{\cdot}23 + j0{\cdot}75)$ $15{\cdot}66\angle 19{\cdot}7^\circ$
(*h*) $(j1{\cdot}63)(2{\cdot}6 + j1)$ $4{\cdot}53\angle 111{\cdot}1^\circ$

4.16. Express each ratio as a single complex number.

(*a*) $(23{\cdot}5 + j8{\cdot}55)/(4{\cdot}53 - j2{\cdot}11)$ *Ans.* $5\angle 45^\circ$
(*b*) $(21{\cdot}2 - j21{\cdot}2)/(3{\cdot}54 - j3{\cdot}54)$ $6\angle 0^\circ$
(*c*) $(-7{\cdot}07 + j7{\cdot}07)/(4{\cdot}92 + j0{\cdot}868)$ $2\angle 125^\circ$
(*d*) $(-j45)/(6{\cdot}36 - j6{\cdot}36)$ $5\angle -45^\circ$
(*e*) $(6{\cdot}88\angle 12^\circ)/(2 + j1)$ *Ans.* $3{\cdot}08\angle -14{\cdot}6^\circ$
(*f*) $(5 + j5)/5\angle 80^\circ$ $1{\cdot}414\angle -35^\circ$
(*g*) $1/(6 + j8)$ $0{\cdot}1\angle -53{\cdot}1^\circ$
(*h*) $(-10 + j20)/(2 - j1)$ $10\angle 143{\cdot}2^\circ$

4.17. In each case, evaluate $z_1z_2/(z_1 + z_2)$.

(*a*) $z_1 = 10 + j5$, $z_2 = 20\angle 30^\circ$ *Ans.* $7{\cdot}18\angle 27{\cdot}8^\circ$
(*b*) $z_1 = 5\angle 45^\circ$, $z_2 = 10\angle -70^\circ$ $5{\cdot}5\angle 15{\cdot}2^\circ$
(*c*) $z_1 = 6 - j2$, $z_2 = 1 + j8$ *Ans.* $5{\cdot}52\angle 23{\cdot}81^\circ$
(*d*) $z_1 = 20$, $z_2 = j40$ $17{\cdot}9\angle 26{\cdot}6^\circ$

CHAPTER 5

Complex Impedance and Phasor Notation

INTRODUCTION

Analysis of circuits in the sinusoidal steady state is of importance not only because the voltages supplied by alternating current generators are very nearly pure sine functions, but also because any periodic waveform may be replaced by a constant term and a series of sine and cosine terms. This is called the *Fourier method of waveform analysis* and is the subject of a later chapter.

Many simple circuits were analyzed in Chapter 3 in which the voltage and current were sinusoidal functions. However, the work became cumbersome even though the circuits were relatively simple. With *phasors* used to represent voltages and currents, and with a *complex impedance* to account for the circuit elements, analysis in the sinusoidal steady state will be greatly simplified.

COMPLEX IMPEDANCE

Consider a series RL circuit with an applied voltage $v(t) = V_m e^{j\omega t}$, as shown in Fig. 5-1. By Euler's formula this function includes a cosine term and a sine term, $V_m \cos \omega t + jV_m \sin \omega t$. Kirchhoff's voltage law for the closed loop is

$$Ri(t) + L\frac{di(t)}{dt} = V_m e^{j\omega t}$$

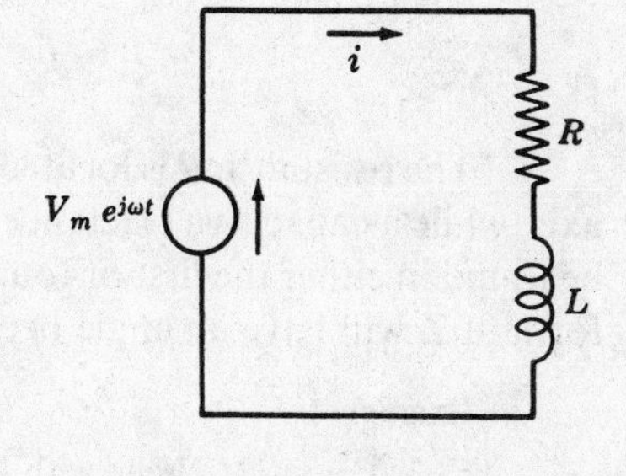

Fig. 5-1

This first order linear differential equation has a particular solution of the form $i(t) = Ke^{j\omega t}$. Substituting this current function, we obtain

$$RKe^{j\omega t} + j\omega LKe^{j\omega t} = V_m e^{j\omega t}$$

from which $K = \dfrac{V_m}{R + j\omega L}$ and $i(t) = \dfrac{V_m}{R + j\omega L} e^{j\omega t}$. The ratio of the voltage function to the current function shows that the impedance is a complex number with a real part R and a j part ωL:

$$\mathbf{Z} = \frac{v(t)}{i(t)} = \frac{V_m e^{j\omega t}}{\dfrac{V_m}{R + j\omega L} e^{j\omega t}} = R + j\omega L$$

Now consider an RC series circuit with the same applied voltage as shown in Fig. 5-2. Then

$$Ri(t) + \frac{1}{C}\int i(t)\,dt = V_m e^{j\omega t} \qquad (1)$$

Let $i(t) = Ke^{j\omega t}$ and substitute in (*1*) to obtain

$$RKe^{j\omega t} + \frac{1}{j\omega C}Ke^{j\omega t} = V_m e^{j\omega t}$$

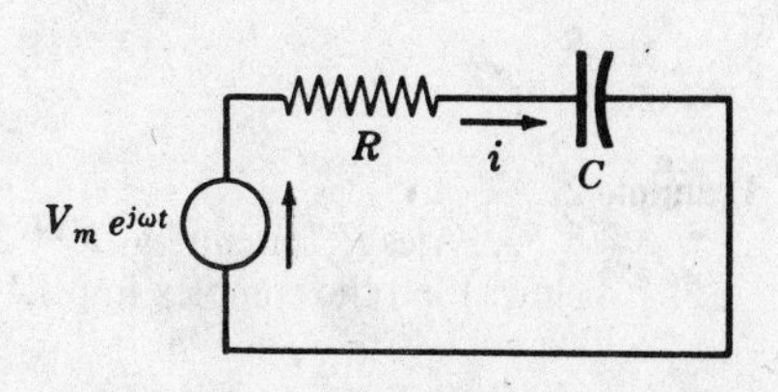

Fig. 5-2

from which $K = \dfrac{V_m}{R + 1/j\omega C} = \dfrac{V_m}{R - j(1/\omega C)}$ and $i(t) = \dfrac{V_m}{R - j(1/\omega C)} e^{j\omega t}$. Then

$$\mathbf{Z} = \frac{V_m e^{j\omega t}}{\dfrac{V_m}{R - j(1/\omega C)} e^{j\omega t}} = R - j(1/\omega C)$$

The impedance is again a complex number with a real part R and a j part $-1/\omega C$.

The above suggests that the circuit elements can be expressed in terms of their complex impedance **Z**, which can be placed directly on the circuit diagram as shown in Fig. 5-3.

Since the impedance is a complex number it may be displayed on the complex plane. However, since resistance is never negative, only the first and fourth quadrants are required. The resulting display is called an *impedance diagram*. See Fig. 5-4 below.

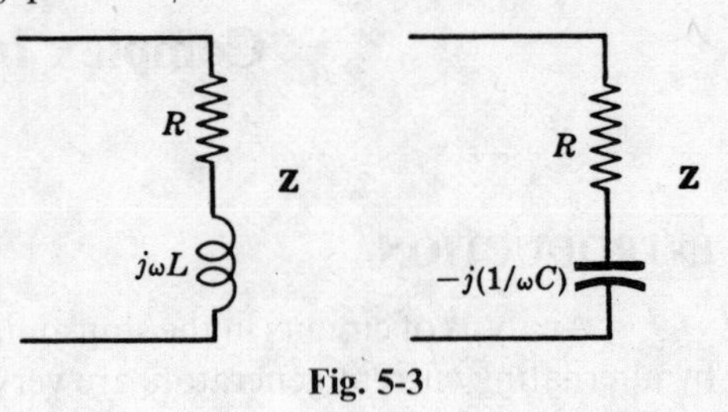

Fig. 5-3

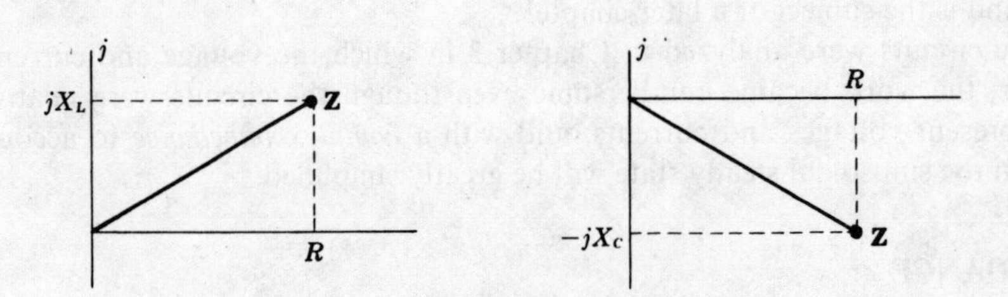

Fig. 5-4. Impedance Diagrams

The resistance R is located on the positive real axis. An inductive reactance X_L is located on the positive j axis, while a capacitive reactance X_C is plotted on the negative j axis. In general the complex impedance **Z** may be found in either the first or fourth quadrants, depending upon the elements comprising the circuit. The polar form of **Z** will have an angle between $\pm 90°$.

Example 1.

A series *RL* circuit with $R = 5$ ohms and $L = 2$ mH has an applied voltage $v = 150 \sin 5000t$ volts. Refer to Fig. 5-5 below. Find the complex impedance **Z**.

Inductive reactance $X_L = \omega L = 5000(2 \times 10^{-3}) = 10$ ohms; then $\mathbf{Z} = 5 + j10\,\Omega$. In polar form, $\mathbf{Z} = 11{\cdot}16\angle 63{\cdot}4°\,\Omega$

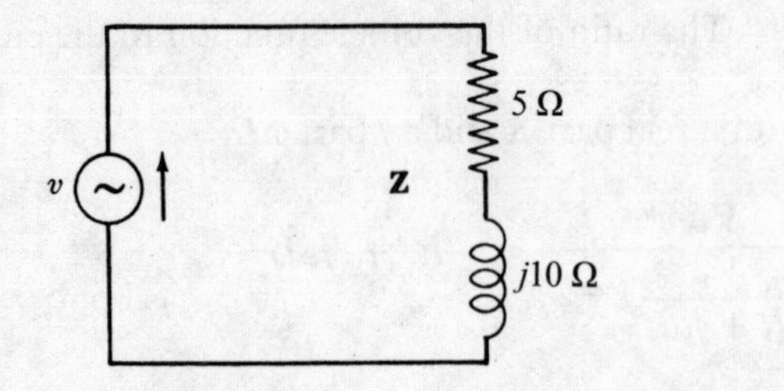

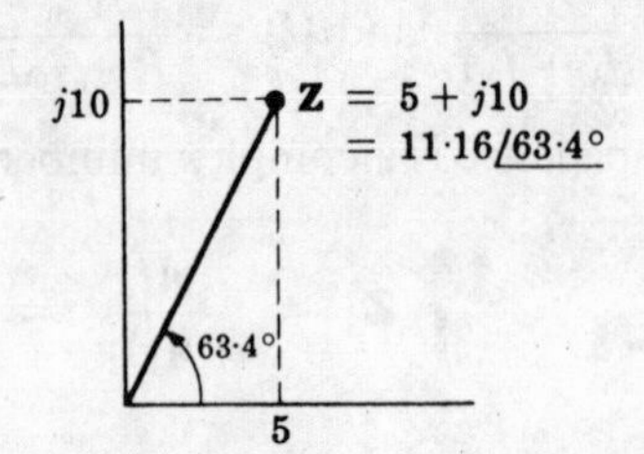

Fig. 5-5

Example 2.

A series *RC* circuit has $R = 20$ ohms, $C = 5\,\mu$F, and an applied voltage $v = 150 \cos 10{,}000t$ volts. Refer to Fig. 5-6 below. Find the complex impedance **Z**.

Capacitive reactance $X_C = \dfrac{1}{\omega C} = \dfrac{1}{10{,}000(5 \times 10^{-6})} = 20$ ohms; then $\mathbf{Z} = 20 - j20\,\Omega$. In polar form $\mathbf{Z} = 28{\cdot}3\angle -45°\,\Omega$.

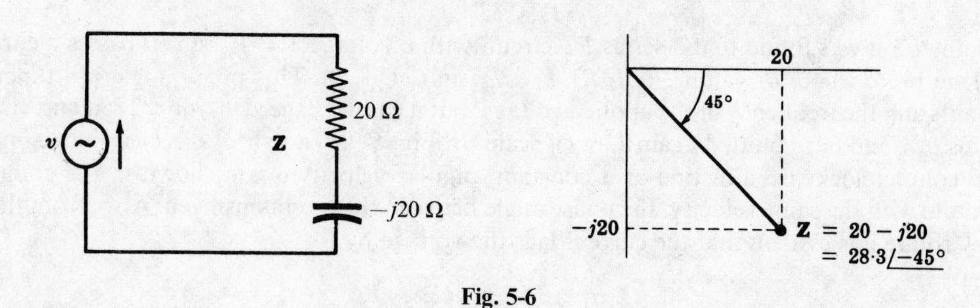

Fig. 5-6

Impedance is a function of ω in all circuits except pure resistance, since both X_L and X_C are functions of ω. Therefore any complex impedance is valid only for that frequency at which it was computed.

PHASOR NOTATION

Consider the function $f(t) = r\,e^{j\omega t}$. This is a complex number which includes the variable t. However, the absolute value is fixed at r. If a few sketches are made, for instance at $t = 0$, $\pi/4\omega$ and $\pi/2\omega$, as shown in Fig. 5-7, the nature of the function becomes apparent.

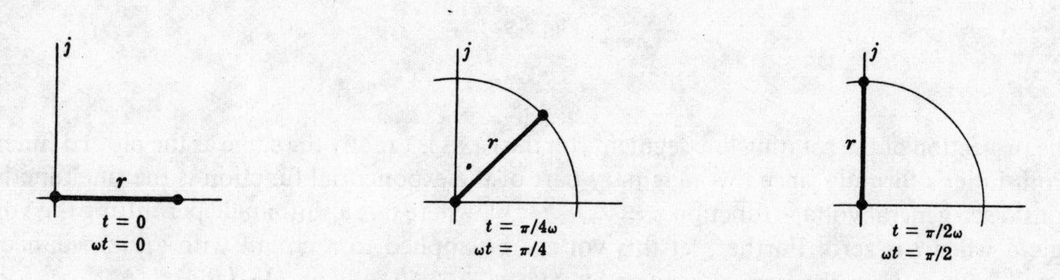

Fig. 5-7. The Function $r\,e^{j\omega t}$

With ω constant, the line segment will rotate in the counterclockwise direction at constant angular velocity. If the projections of this rotating line segment on both the real axis and j axis are examined, the cosine and sine parts of $e^{j\omega t}$ as given in Euler's formula result.

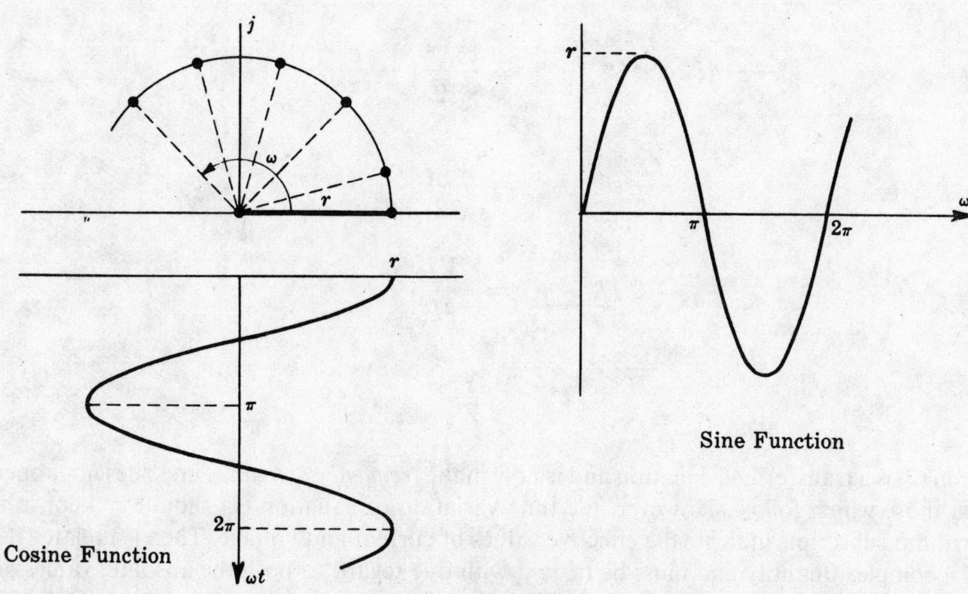

Fig. 5-8

In Chapter 3 it was found that a series RL circuit with a voltage $v = V_m \sin \omega t$ passes a current which lags the voltage by θ, where $\theta = \tan^{-1}(\omega L/R)$, $i = I_m \sin(\omega t - \theta)$. This phase angle is a function of the circuit constants and the frequency of the applied voltage, but it cannot exceed 90° or $\pi/2$ radians. In Fig. 5-9(b) the waveforms of v and i are plotted against an ωt scale. In Fig. 5-9(a) a pair of directed line segments which rotate in the counterclockwise direction at a constant angular velocity ω are shown in the complex plane. Since both rotate with the same velocity, the phase angle between them remains fixed. Also, from the direction in which they rotate it is evident that the current lags the voltage by θ.

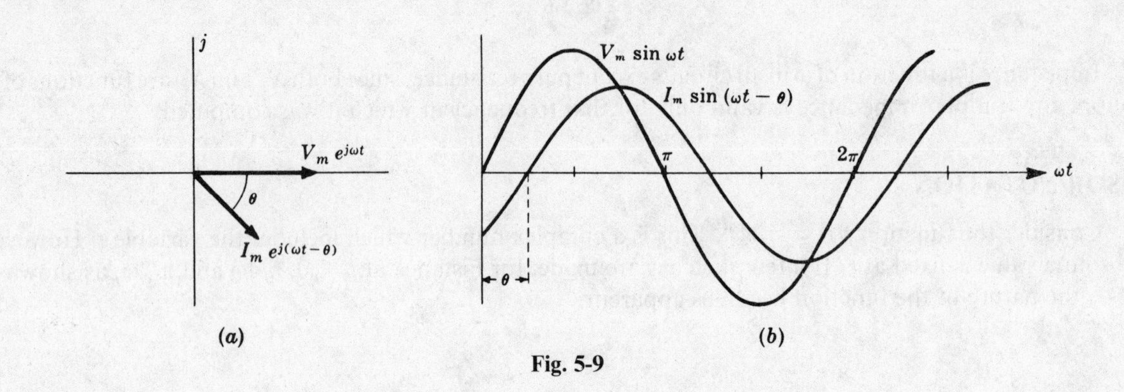

Fig. 5-9

The projection of the rotating line segments on the j axis is exactly the same as the plotted functions. This follows from Euler's formula since the imaginary part of the exponential function is the sine function.

Consider a general voltage function $v = V_m\, e^{j(\omega t + \alpha)}$, where α is a shift angle permitting the voltage to be at an angle α when t is zero. Further, let this voltage be applied to a circuit with an impedance $\mathbf{Z} = ze^{j\theta}$, $(-\pi/2 \leq \theta \leq \pi/2)$. Then the current is given by $(V_m\, e^{j(\omega t + \alpha)})/(ze^{j\theta}) = (V_m/z)\, e^{j(\omega t + \alpha - \theta)} = I_m\, e^{j(\omega t + \alpha - \theta)}$,

$$\text{i.e.,}\quad I_m\, e^{j(\omega t + \alpha - \theta)} = \frac{V_m\, e^{j(\omega t + \alpha)}}{ze^{j\theta}} \tag{1}$$

The above equality is in the *time domain* since time appears explicitly in both the current and voltage expressions. Two changes are now made to yield *phasors*. First, the equality is multiplied by $e^{-j\omega t}$ to remove the time function. Secondly, the equality is multiplied by $1/\sqrt{2}$ to yield effective values of current and voltage.

$$\frac{e^{-j\omega t}}{\sqrt{2}}\left(I_m\, e^{j(\omega t + \alpha - \theta)}\right) = \frac{e^{-j\omega t}}{\sqrt{2}}\left(\frac{V_m\, e^{j(\omega t + \alpha)}}{ze^{j\theta}}\right)$$

$$\frac{I_m}{\sqrt{2}}\, e^{j(\alpha - \theta)} = \frac{V_m}{\sqrt{2}} \cdot \frac{e^{j\alpha}}{ze^{j\theta}} \tag{2}$$

$$I\underline{/\alpha - \theta} = \frac{V\underline{/\alpha}}{z\underline{/\theta}} \tag{3}$$

$$\mathbf{I} = \frac{\mathbf{V}}{\mathbf{Z}} \tag{4}$$

Equation (2) is a transformed equation and is now in the *frequency domain*. Time does not appear in this equation or in those which follow. However, the time variation of equation (1) should be kept in mind. In (3), I and V without subscripts indicate the effective values of current and voltage. Then (4) relates **I**, **V** and **Z** where each is a complex quantity and must be treated with due regard to both the absolute values and arguments. This is a *phasor equivalent* of Ohm's law sometimes referred to as the *complex form* of Ohm's law.

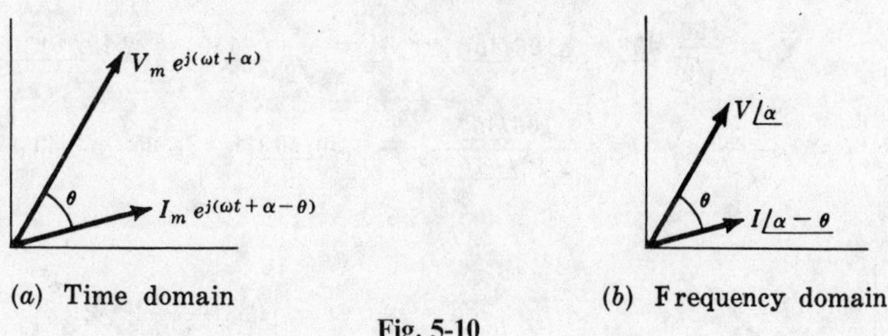

Fig. 5-10

In Fig. 5-10(*a*) the voltage and current functions are shown in the complex plane, expressed in the exponential form. This is a *time domain* plot since t is shown explicity. In Fig. 5-10(*b*) the *phasor voltage* and *phasor current* are shown. Here the line segments are $1/\sqrt{2}$ times those in Fig. 5-10(*a*), and there is no evidence of time. But the angle θ and the absolute value of the current are functions of frequency, and thus Fig. 5-10(*b*) is said to be in the *frequency domain*.

Solved Problems

5.1. Show how X_L and X_C vary with frequency by plotting each against ω over the range from 400 to 4000 radians per second. $L = 400$ mH and $C = 25$ μF.

Substituting in $X_L = \omega L$ and $X_C = 1/\omega C$ convenient values of ω over the given range, the corresponding values of X_L and X_C are tabulated as in Fig. 5-11(*a*). Fig. 5-11(*b*) shows graphs of X_L and X_C.

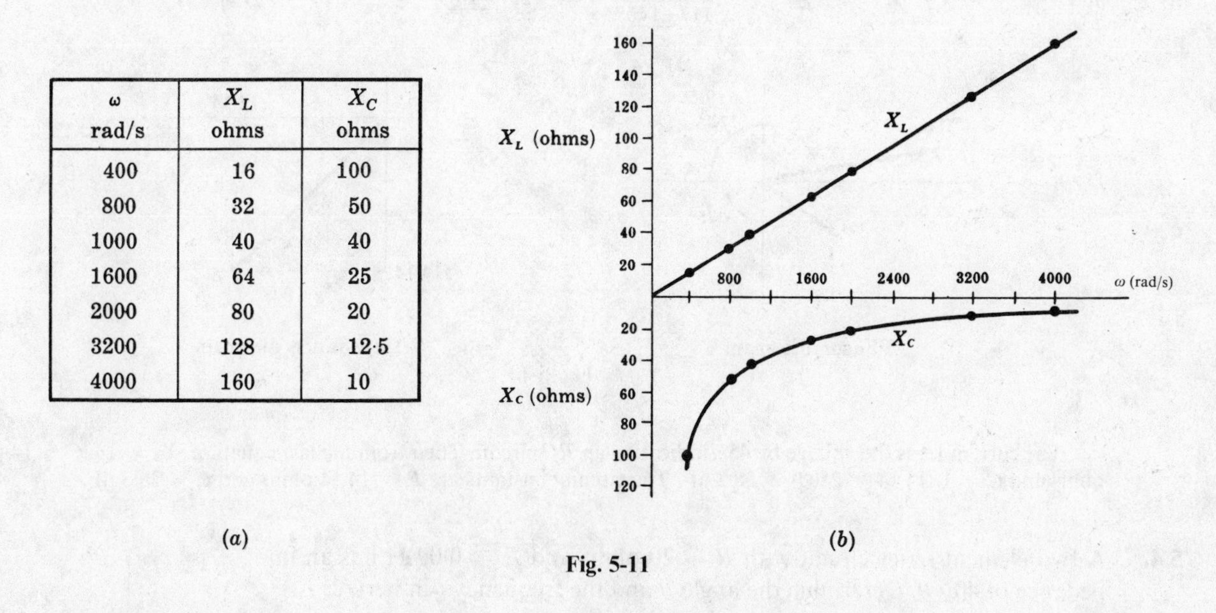

ω rad/s	X_L ohms	X_C ohms
400	16	100
800	32	50
1000	40	40
1600	64	25
2000	80	20
3200	128	12·5
4000	160	10

(*a*) (*b*)

Fig. 5-11

Any circuit containing L or C will have an impedance which is a function of frequency. Therefore any impedance diagram constructed for a given frequency is valid only for that specific frequency.

5.2. Construct the phasor and impedance diagrams and determine the circuit constants for the following voltage and current: $v = 150 \sin(5000t + 45°)$ volts, $i = 3 \sin(5000t - 15°)$ amperes.

The phasor quantities have absolute values of $1/\sqrt{2}$ times the maximum values. Thus

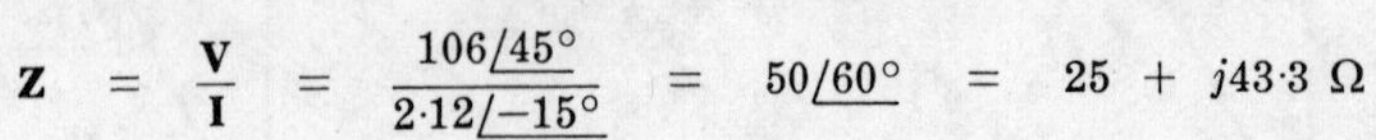

$$\mathbf{V} = \frac{150}{\sqrt{2}}\underline{/45^\circ} = 106\underline{/45^\circ}, \qquad \mathbf{I} = \frac{3}{\sqrt{2}}\underline{/-15^\circ} = 2{\cdot}12\underline{/-15^\circ}\ \text{A}$$

and
$$\mathbf{Z} = \frac{\mathbf{V}}{\mathbf{I}} = \frac{106\underline{/45^\circ}}{2{\cdot}12\underline{/-15^\circ}} = 50\underline{/60^\circ} = 25 + j43{\cdot}3\ \Omega$$

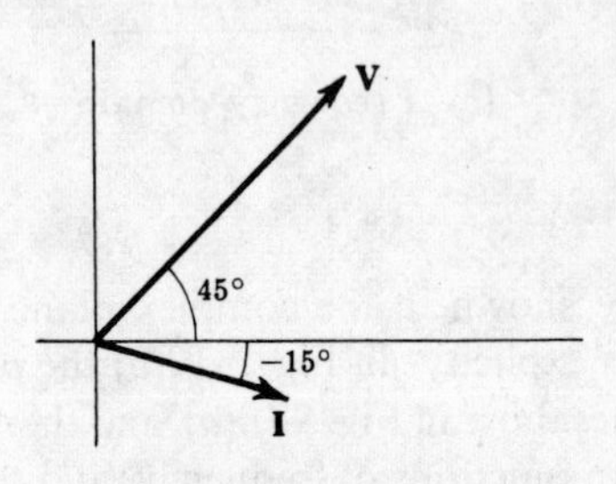

Phasor diagram

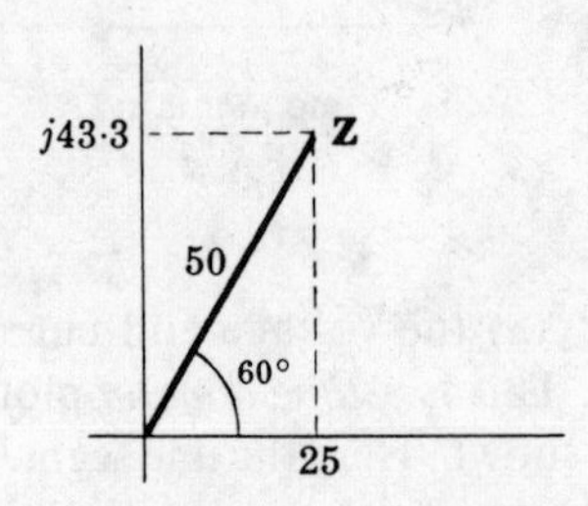

Impedance diagram

Fig. 5-12

The current lags the voltage by 60°, indicating an RL series circuit. Then from the last equation, $\omega L = 43{\cdot}3$ ohms and $L = 43{\cdot}3/5000 = 8{\cdot}66$ mH. The circuit constants are $R = 25$ ohms and $L = 8{\cdot}66$ mH.

5.3. Construct the phasor and impedance diagrams and determine the circuit constants for the following voltage and current: $v = 311 \sin(2500t + 170^\circ)$ volts and $i = 15{\cdot}5 \sin(2500t - 145^\circ)$ amperes.

$$\mathbf{V} = \frac{311}{\sqrt{2}}\underline{/170^\circ} = 220\underline{/170^\circ}\ \text{V}, \qquad \mathbf{I} = \frac{15.5}{\sqrt{2}}\underline{/-145^\circ} = 11\underline{/-145^\circ}\ \text{A}$$

and
$$\mathbf{Z} = \frac{\mathbf{V}}{\mathbf{I}} = \frac{220\underline{/170^\circ}}{11\underline{/-145^\circ}} = 20\underline{/-45^\circ} = 14\,14 - j14{\cdot}14\ \Omega$$

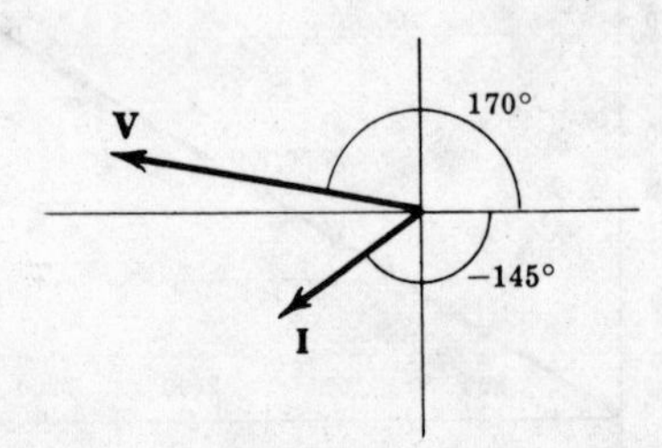

Phasor diagram

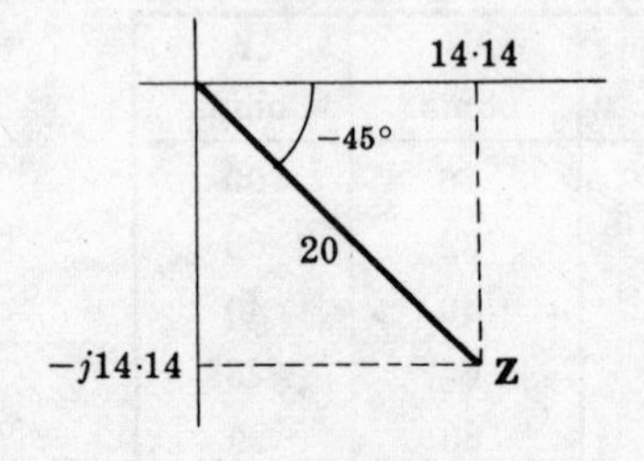

Impedance diagram

Fig. 5-13

The current leads the voltage by 45°, indicating an RC circuit. Then from the last equation, $X_C = 1/\omega C = 14{\cdot}14$ ohms and $C = 1/(14{\cdot}14 \times 2500) = 28{\cdot}3\ \mu$F. The circuit constants are $R = 14{\cdot}14$ ohms and $C = 28{\cdot}3\ \mu$F.

5.4. A two element series circuit with $R = 20$ ohms and $L = 0{\cdot}02$ H has an impedance of $40\underline{/\theta}$. Determine the angle θ and the frequency f in hertz.

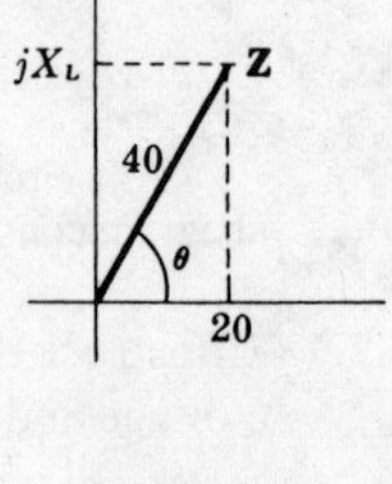

Impedance of circuit $= 20 + jX_L = 40\underline{/\theta}$.

From Fig. 5-14, $\theta = \cos^{-1} 20/40 = 60^\circ$; then

$$X_L = 40 \sin 60^\circ = 34{\cdot}6 \text{ ohms}$$

Now $X_L = \omega L = 2\pi f L$ and $f = \dfrac{X_L}{2\pi L} = \dfrac{34{\cdot}6}{2\pi(0{\cdot}02)} = 275$ Hz.

Fig. 5-14

5.5. A series circuit with $R = 10$ ohms and $C = 50\ \mu F$ has an applied voltage with a frequency such that the current leads by 30°. What change in frequency would be necessary to cause the current to lead by 70°?

From Fig. 5-15, $\tan -30° = -X_{C_1}/10 = -0{\cdot}576$ or $X_{C_1} = 5{\cdot}76$ ohms. Then $X_{C_1} = 1/2\pi f_1 C$ and

$$f_1 = \frac{1}{2\pi C X_{C_1}} = \frac{1}{2\pi(50 \times 10^{-6}\ f)(5{\cdot}76\ \text{ohms})} = 553\ \text{Hz}$$

At the new frequency f_2 the current leads by 70°. Now $\tan -70° = -X_{C_2}/10 = -2{\cdot}74$ or $X_{C_2} = 27{\cdot}4$ ohms. Then $f_2/f_1 = X_{C_1}/X_{C_2}$, $f_2/553 = 5{\cdot}76/27{\cdot}4$ and $f_2 = 116$ Hz.

Since X_C varies inversely with ω, the series RC circuit has a larger phase angle at a lower frequency.

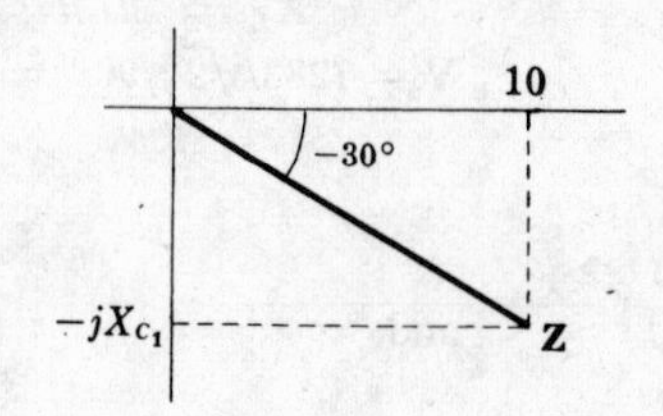

Fig. 5-15

5.6. With $f = 500$ Hz, find the pure element in series with $R = 25$ ohms which causes the current to lag the voltage by 20°. Repeat for an angle of 20° leading.

An angle of 20° lagging requires an inductive reactance X_L in series with R. But the capacitive reactance X_C which yields the same angle of lead has the same ohmic value as X_L.

For the current lagging, $\tan 20° = X_L/25$ or $X_L = 9{\cdot}1$ ohms. Then $L = X_L/2\pi f = 9{\cdot}1/2\pi(500) = 2{\cdot}9$ mH.

For the current leading, $C = 1/2\pi f X_C = 1/2\pi(500)(9{\cdot}1) = 35\ \mu F$.

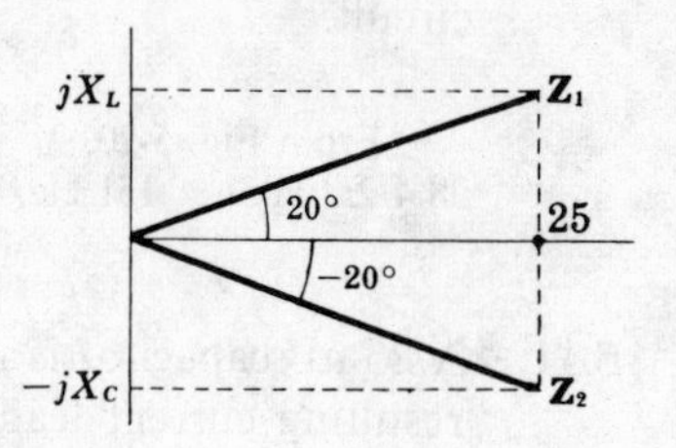

Fig. 5-16

5.7. A series circuit of $R = 25$ ohms and $L = 0{\cdot}01$ H is to be used at frequencies of 100, 500 and 1000 Hz. Find the impedance **Z** at each of these frequencies.

At $f = 100$ Hz, $X_L = 2\pi f L = 2\pi(100)(0{\cdot}01) = 6{\cdot}28$ ohms. Similarly, $f = 500$ Hz gives $X_L = 31{\cdot}4$ ohms and $f = 1000$ Hz gives $X_L = 62{\cdot}8$ ohms. The corresponding **Z** values are given in Fig. 5-17 below.

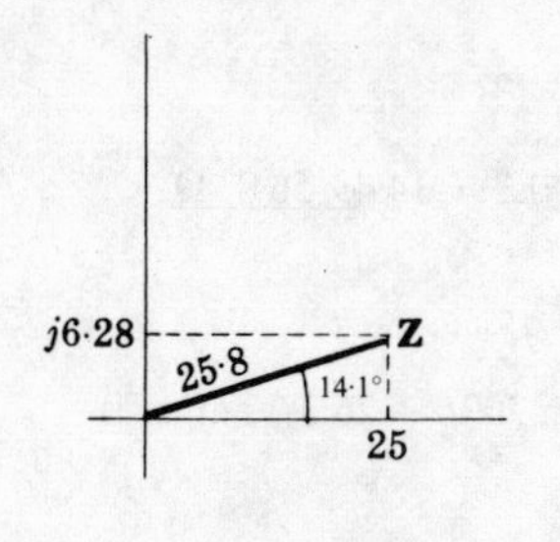

$f = 100$ Hz

$\mathbf{Z} = 25 + j6{\cdot}28 = 25{\cdot}8\underline{/14{\cdot}1°}\ \Omega$

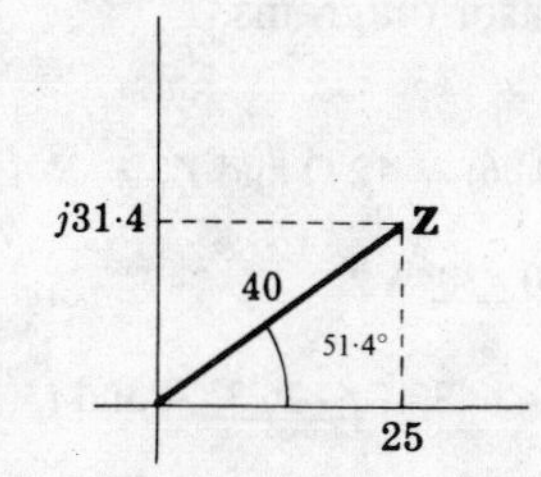

$f = 500$ Hz

$\mathbf{Z} = 25 + j31{\cdot}4 = 40\underline{/51{\cdot}4°}\ \Omega$

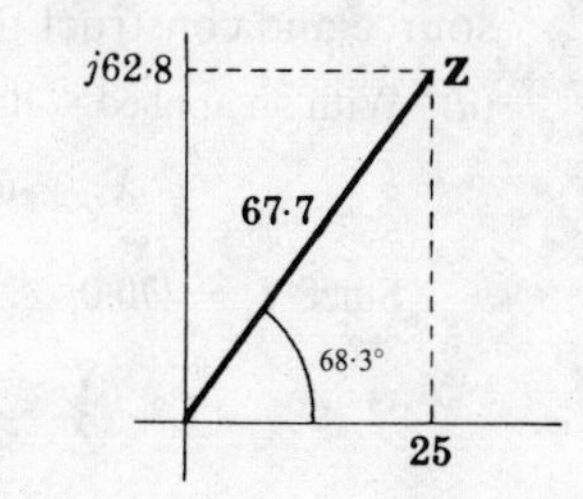

$f = 1000$ Hz

$\mathbf{Z} = 25 + j62{\cdot}8 = 67{\cdot}7\underline{/68{\cdot}3°}\Omega$

Fig. 5-17

5.8. The series circuit of $R = 10$ ohms and $C = 40\ \mu F$ has an applied voltage $v = 500 \cos (2500t - 20°)$ volts. Find the current i.

$X_C = 1/\omega C = 1/2500(40 \times 10^{-6}) = 10$ ohms and the complex impedance $\mathbf{Z} = 10 - j10 = 10\sqrt{2}\underline{/-45°}\ \Omega$. Converting the voltage to phasor notation, $\mathbf{V} = (500/\sqrt{2})\underline{/-20°}$ V. Then

$$\mathbf{I} = \frac{\mathbf{V}}{\mathbf{Z}} = \frac{(500/\sqrt{2})\underline{/-20°}}{(10\sqrt{2})\underline{/-45°}} = 25\underline{/25°}\ \text{A}$$

and $$i = 25\sqrt{2}\cos(2500t + 25°)\ \text{amperes}$$

The phasor diagram in Fig. 5-18 shows the current **I** leading **V** by the angle on the impedance, 45°.

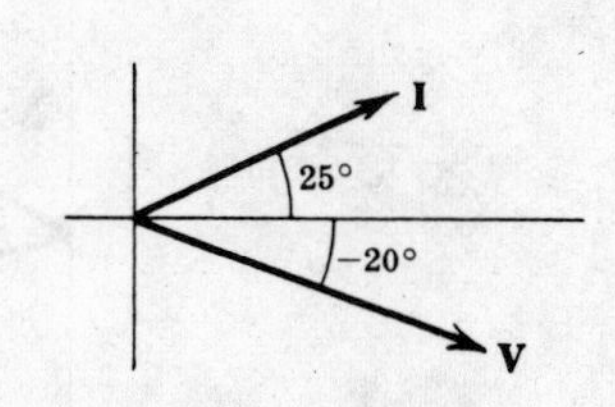

Fig. 5-18

5.9. The series circuit of $R = 8$ ohms and $L = 0{\cdot}02$ H has an applied voltage $v = 283 \sin(300t + 90°)$ volts. Find the current i.

$X_L = \omega L = 300(0{\cdot}02) = 6$ ohms, $\mathbf{Z} = 8 + j6 = 10\angle 36{\cdot}9°\ \Omega$, and $\mathbf{V} = (283/\sqrt{2})\angle 90° = 200\angle 90°$ V. Then

$$\mathbf{I} = \frac{200\angle 90°}{10\angle 36{\cdot}9°} = 20\angle 53{\cdot}1°\ \text{A}$$

and $$i = 20\sqrt{2}\sin(300t + 53{\cdot}1°) \text{ amperes}$$

Fig. 5-19

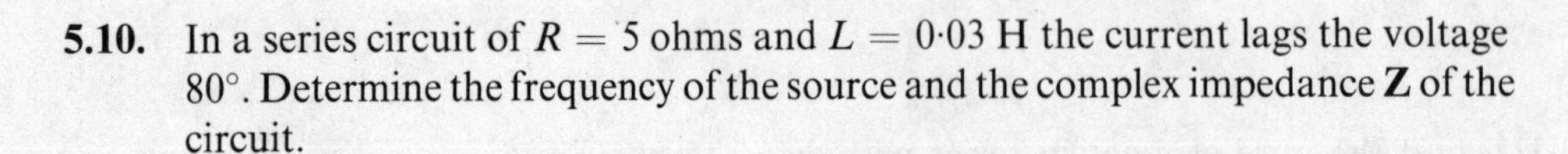

5.10. In a series circuit of $R = 5$ ohms and $L = 0{\cdot}03$ H the current lags the voltage 80°. Determine the frequency of the source and the complex impedance $\mathbf{Z}$ of the circuit.

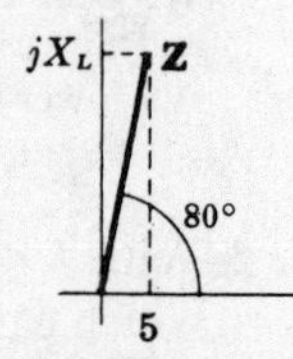

Fig. 5-20

From Fig. 5-20, $X_L = 5 \tan 80° = 28{\cdot}4$ ohms. Since $X_L = 2\pi fL$, $f = X_L/2\pi L = 28{\cdot}4/2\pi(0{\cdot}03) = 151$ Hz. The complex impedance $\mathbf{Z} = 5 + j28{\cdot}4 = 28{\cdot}8\angle 80°\ \Omega$.

5.11. A 25 μF capacitor is in series with a resistor R at a frequency of 60 Hz. The resulting current leads the voltage by 45°. Determine the magnitude of R.

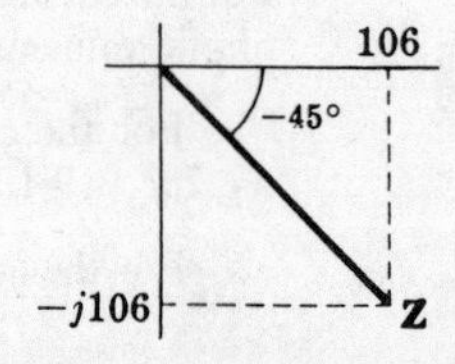

Fig. 5-21

$X_C = \dfrac{1}{2\pi fC} = \dfrac{1}{2\pi(60)(25 \times 10^{-6})} = 106$ ohms. Since the phase angle is 45°, $R = X_C = 106$ ohms.

5.12. A series circuit of $R = 8$ ohms and $L = 0{\cdot}06$ H has an applied voltage $v_1 = 70{\cdot}7 \sin(200t + 30°)$ volts. Later a second voltage $v_2 = 70{\cdot}7 \sin(300t + 30°)$ volts is applied in place of the first. Find i for each source and construct the two phasor diagrams.

(*a*) With an applied voltage v_1,

$$X_L = \omega L = 200(0{\cdot}06) = 12\ \Omega \text{ and } \mathbf{Z}_1 = R + jX_L = 8 + j12 = 14{\cdot}4\angle 56{\cdot}3°\ \Omega$$

Since $\mathbf{V}_1 = (70{\cdot}0/\sqrt{2})\angle 30° = 50\angle 30°$ V,

$$\mathbf{I}_1 = \frac{\mathbf{V}_1}{\mathbf{Z}_1} = \frac{50\angle 30°}{14{\cdot}4\angle 56{\cdot}3°} = 3{\cdot}47\angle -26{\cdot}3° \text{ A and } i_1 = 3{\cdot}47\sqrt{2}(\sin 200t - 26{\cdot}3°) \text{ amperes}$$

(*b*) With an applied voltage v_2,

$$X_L = \omega L = 300(0{\cdot}06) = 18\ \Omega \text{ and } \mathbf{Z}_2 = 8 + j18 = 19{\cdot}7\angle 66°\ \Omega$$

Since $\mathbf{V}_2 = 50\angle 30°$ V,

$$\mathbf{I}_2 = \frac{\mathbf{V}_2}{\mathbf{Z}_2} = \frac{50\angle 30°}{19{\cdot}7\angle 66°} = 2{\cdot}54\angle -36° \text{ A and } i_2 = 2{\cdot}54\sqrt{2}\sin(300t - 36°) \text{ amperes}$$

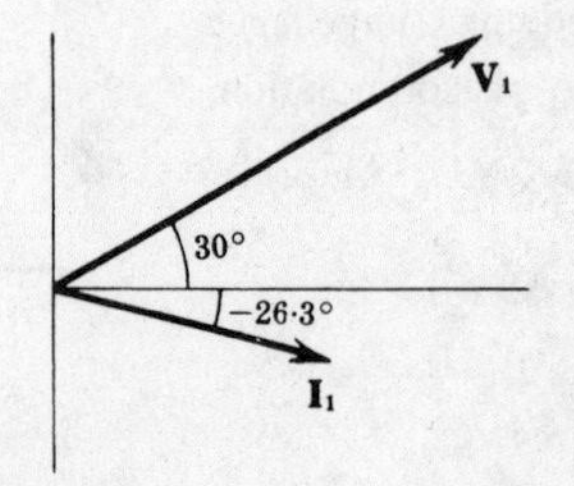

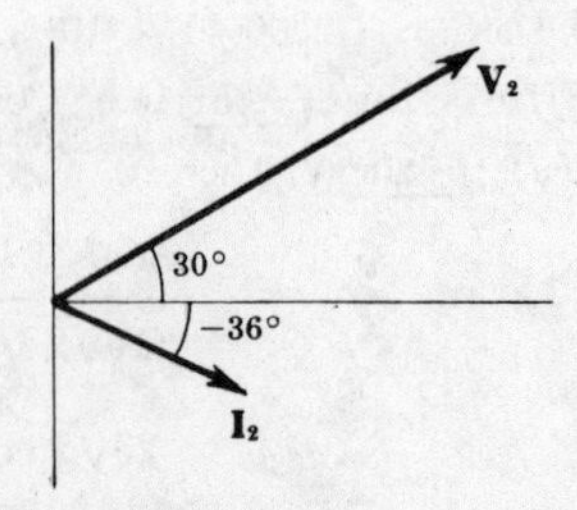

Phasor diagram ω = 200 rad/s   Phasor diagram, ω = 300 rad/s

Fig. 5-22

5.13. Use phasors to find the sum of the two currents $i_1 = 14{\cdot}14 \sin(\omega t + 13{\cdot}2°)$ amperes and $i_2 = 8{\cdot}95 \sin(\omega t + 121{\cdot}6°)$ amperes. Refer to Fig. 5-23 below.

$$\mathbf{I}_1 = (14{\cdot}14/\sqrt{2})\underline{/13{\cdot}2°} = 10\underline{/13{\cdot}2°} = 9{\cdot}73 + j2{\cdot}28 \text{ A}$$
$$\mathbf{I}_2 = (8{\cdot}95)/\sqrt{2})\underline{/121{\cdot}6°} = 6{\cdot}33\underline{/121{\cdot}6°} = -3{\cdot}32 + j5{\cdot}39 \text{ A}$$
$$\mathbf{I}_1 + \mathbf{I}_2 = 6{\cdot}41 + j7{\cdot}67 = 10\underline{/50°} \text{ A}$$

Then $i_1 + i_2 = 10\sqrt{2} \sin(\omega t + 50°)$ amperes.

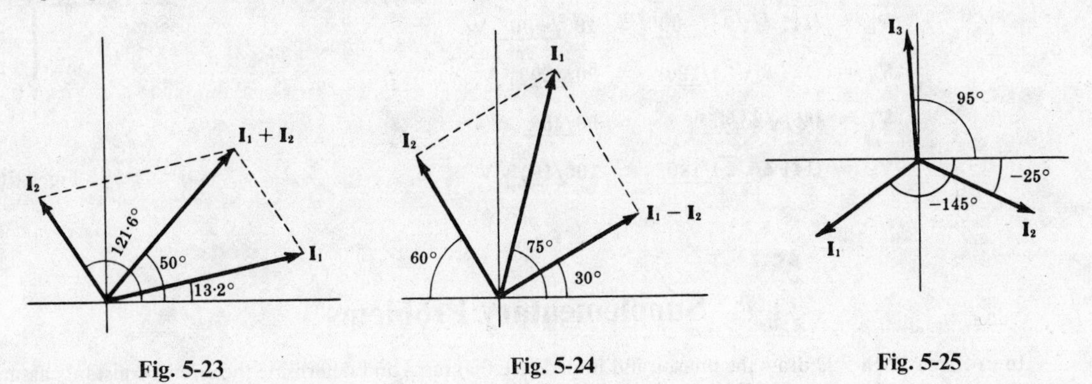

Fig. 5-23 Fig. 5-24 Fig. 5-25

5.14. Find the difference $i_1 - i_2$ where $i_1 = 50 \cos(\omega t + 75°)$ amperes and $i_2 = 35{\cdot}4 \cos(\omega t + 120°)$ amperes. Refer to Fig. 5-24 above.

$$\mathbf{I}_1 = (50/\sqrt{2})\underline{/75°} = 35{\cdot}4\underline{/75°} = 9{\cdot}16 + j34{\cdot}2 \text{ A}$$
$$\mathbf{I}_2 = (35{\cdot}4/\sqrt{2})\underline{/120°} = 25\underline{/120°} = -12{\cdot}5 + j21{\cdot}7 \text{ A}$$
$$\mathbf{I}_1 - \mathbf{I}_2 = 21{\cdot}7 + j12{\cdot}5 = 25\underline{/30°} \text{ A}$$

Then $i_1 - i_2 = 25\sqrt{2} \cos(\omega t + 30°)$ amperes.

5.15. Find the sum of the three currents $i_1 = 32{\cdot}6 \sin(\omega t - 145°)$ amperes, $i_2 = 32{\cdot}6 \sin(\omega t - 25°)$ amperes and $i_3 = 32{\cdot}6 \sin(\omega t + 95°)$ amperes.

$$\mathbf{I}_1 = (32{\cdot}6/\sqrt{2})\underline{/-145°} = 23\underline{/-145°} = -18\,8 - j13{\cdot}2 \text{ A}$$
$$\mathbf{I}_2 = (32{\cdot}6/\sqrt{2})\underline{/-25°} = 23\underline{/-25°} = 20\,8 - j9{\cdot}71 \text{ A}$$
$$\mathbf{I}_3 = (32{\cdot}6/\sqrt{2})\underline{/95°} = 23\underline{/95°} = -2 + j23 \text{ A}$$
$$\mathbf{I}_1 + \mathbf{I}_2 + \mathbf{I}_3 = j0{\cdot}09 \text{ A}$$

Within the limits of slide rule accuracy the sum is zero. The phasor diagram, Fig. 5-25 above, shows that the three currents are spaced 120° apart. This, together with the equal magnitudes, obviously results in a sum of zero.

5.16. Find the sum of the two voltages $v_1 = 126{\cdot}5 \sin(\omega t + 63{\cdot}4°)$ volts and $v_2 = 44{\cdot}7 \cos(\omega t - 161{\cdot}5°)$ volts. Express the sum as a sine function and then also as a cosine function.

Converting v_2 into a sine function, $v_2 = 44{\cdot}7 \sin(\omega t - 161{\cdot}5° + 90°) = 44{\cdot}7 \sin(\omega t - 71{\cdot}5°)$ volts.

Now

$$\mathbf{V}_1 = (126{\cdot}5/\sqrt{2})\underline{/63{\cdot}4°} = 89\,5\underline{/63{\cdot}4°} = 40 + j80 \text{ V}$$
$$\mathbf{V}_2 = (44{\cdot}7/\sqrt{2})\underline{/-71{\cdot}5°} = 31\,6\underline{/-71{\cdot}5°} = 10 - j30 \text{ V}$$
$$\mathbf{V}_1 + \mathbf{V}_2 = 50 + j50 = 50\sqrt{2}\underline{/45°} \text{ V}$$

and $v_1 + v_2 = 100 \sin(\omega t + 45°)$ volts.

Also, since $\sin x = \cos(x - 90°)$, $v_1 + v_2 = 100 \cos(\omega t - 45°)$ volts.

5.17. Express each of the following voltages in phasor notation and locate them on a phasor diagram; $v_1 = 212 \sin(\omega t + 45°)$ volts, $v_2 = 141{\cdot}4 \sin(\omega t - 90°)$ volts, $v_3 = 127{\cdot}3 \cos(\omega t + 30°)$ volts, $v_4 = 85 \cos(\omega t - 45°)$ volts, $v_5 = 141{\cdot}4 \sin(\omega t + 180°)$ volts.

The voltages must all be of the same function, either sine or cosine, before they can be expressed as phasors on the same phasor diagram. Change v_3 and v_4 to sine functions: $v_3 = 127{\cdot}3 \sin(\omega t + 120°)$ volts, $v_4 = 85 \sin(\omega t + 45°)$ volts.

$$\mathbf{V}_1 = (212/\sqrt{2})\angle 45° = 150\angle 45° \text{ V}$$

$$\mathbf{V}_2 = (141{\cdot}4/\sqrt{2})\angle -90° = 100\angle -90° \text{ V}$$

$$\mathbf{V}_3 = (127{\cdot}3/\sqrt{2})\angle 120° = 90\angle 120° \text{ V}$$

$$\mathbf{V}_4 = (85/\sqrt{2})\angle 45° = 60\angle 45° \text{ V}$$

$$\mathbf{V}_5 = (141{\cdot}4/\sqrt{2})\angle 180° = 100\angle 180° \text{ V}$$

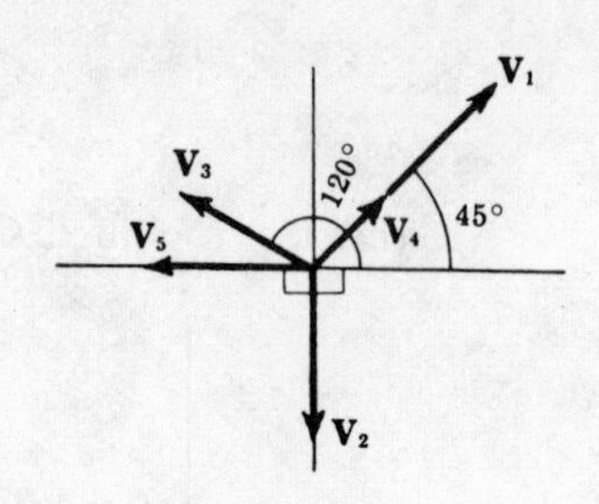

Fig. 5-26

Supplementary Problems

In Problems 5.18-5.22 draw the phasor and impedance diagrams and determine the circuit constants assuming a two element series circuit.

5.18. $v = 283 \cos(800t + 150°)$ volts, $i = 11{\cdot}3 \cos(800t + 140°)$ amperes. *Ans.* R $= 24{\cdot}6\ \Omega$, $L = 5{\cdot}43$ mH

5.19. $v = 50 \sin(2000t - 25°)$ volts, $i = 8 \sin(2000t + 5°)$ amperes. *Ans.* $R = 5{\cdot}41\ \Omega$, $C = 160\ \mu$F

5.20. $v = 10 \cos(5000t - 160°)$ volts, $i = 1{\cdot}333 \cos(5000t - 73{\cdot}82°)$ amperes. *Ans.* $R = 0{\cdot}5\ \Omega$, $C = 26{\cdot}7\ \mu$F

5.21. $v = 80 \sin(1000t + 45°)$ volts, $i = 8 \cos(1000t - 90°)$ amperes. *Ans.* $R = 7{\cdot}07\ \Omega$, $L = 7{\cdot}07$ mH

5.22. $v = 424 \cos(2000t + 30°)$ volts, $i = 28{\cdot}3 \cos(2000t + 83{\cdot}2°)$ amperes. *Ans.* $R = 9\ \Omega$, $C = 41{\cdot}6\ \mu$F

5.23. A series circuit has $R = 8$ ohms and $C = 30\ \mu$F. At what frequency will the current lead the voltage by 30°?
Ans. $f = 1155$ Hz

5.24. An *RL* series circuit has $L = 21{\cdot}2$ mH. At a frequency of 60 Hz the current lags the voltage by 53·1°. Find R.
Ans. $R = 6$ ohms

5.25. A two element series circuit has a voltage $\mathbf{V} = 240\angle 0°$ V and a current $\mathbf{I} = 50\angle -60°$ A. Determine the phasor current which would result from the same applied voltage if the circuit resistance is reduced to (*a*) 60%, (*b*) 30% of its former value. *Ans.* (*a*) $54{\cdot}7\angle -70{\cdot}85°$ A, (*b*) $57{\cdot}1\angle -80{\cdot}15°$ A

5.26. The voltage and current in a two element series circuit are $\mathbf{V} = 150\angle -120°$ V and $\mathbf{I} = 7{\cdot}5\angle -90°$ A. What per cent change in the resistance will result in a phasor current of 12 amperes, and what is the angle associated with this current?
Ans. 56·8% reduction, $\angle -66{\cdot}8°$

5.27. An *RC* series circuit with $R = 10$ ohms has an impedance with an angle of $-45°$ at a frequency $f_1 = 500$ Hz. Find the frequency for which the absolute value of the impedance is (*a*) twice that at f_1, (*b*) one-half that at f_1.
Ans. (*a*) 189 Hz, (*b*) Not possible since the lower limit of $\mathbf{Z}$ is $10 + j0\ \Omega$

5.28. An *RL* series circuit with $R = 10$ ohms has an angle of 30° on the impedance at a frequency $f_1 = 100$ Hz. At what frequency will the absolute value of the impedance be twice the value at f_1? *Ans.* 360 Hz

5.29. In a two element series circuit with $R = 5$ ohms the current lags the applied voltage by 75° at a frequency of 60 Hz. (*a*) Determine the second element in the circuit. (*b*) Find the phase angle which results from a third harmonic $f = 180$ Hz.
Ans. (*a*) 0·0496 H, (*b*) $\theta = 84{\cdot}88°$

5.30. A series circuit consists of $R = 5$ ohms and $C = 50\ \mu$F. Two voltage sources are applied one at a time, $v_1 = 170 \cos(1000t + 20°)$ volts and $v_2 = 170 \cos(2000t + 20°)$ volts. Find the current which results from each source.
Ans. $i_1 = 8{\cdot}25 \cos(1000t + 95{\cdot}85°)$ amperes, $i_2 = 15{\cdot}2 \cos(2000t + 83{\cdot}4°)$ amperes

5.31. A two element series circuit has the following voltage and current for $\omega = 2000$ rad/s: $\mathbf{V} = 150\angle{-45°}$ V and $\mathbf{I} = 4{\cdot}74\angle{-116{\cdot}6°}$ A. A second voltage source results in an angle of 30° between the voltage and current. Determine ω of this second source. *Ans.* 385 rad/s

5.32. Referring to Problem 5·31, what change in source frequency would result in a phasor current of 6 A? With unlimited variation in frequency, what is the maximum possible phasor current? *Ans.* 23·6% reduction in f, 15·0 A

5.33. Find the sum of the two voltages $v_1 = 50 \sin(\omega t + 90°)$ volts and $v_2 = 50 \sin(\omega t + 30°)$ volts shown in Fig. 5-27 below. What voltage would be indicated on a voltmeter across the two outer terminals? *Ans.* $86{\cdot}6 \sin(\omega t + 60°)$ volts, 61·2 V

5.34. Find the sum of the two voltages $v_1 = 35 \sin(\omega t + 45°)$ volts and $v_2 = 100 \sin(\omega t - 30°)$ volts shown in Fig. 5-28 below. Let the positive sense of the sum be in agreement with v_1. *Ans.* $97 \sin(\omega t + 129{\cdot}6°)$ volts

5.35. Referring to Problem 5.34, repeat for the reversed direction of v_2. *Ans.* $114 \sin(\omega t - 12{\cdot}75°)$ volts

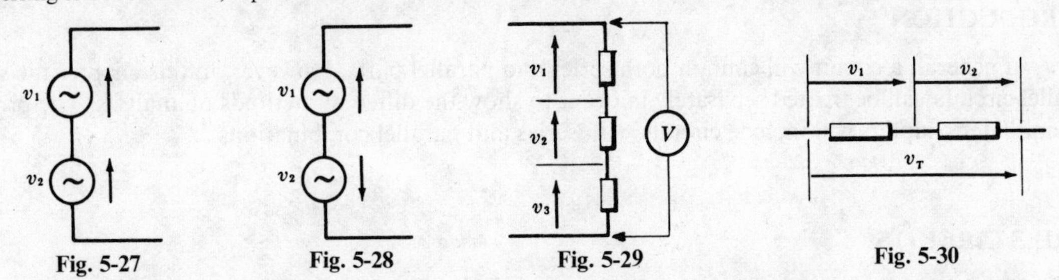

Fig. 5-27 Fig. 5-28 Fig. 5-29 Fig. 5-30

5.36. Find the voltmeter reading across the three impedances shown in Fig. 5-29 above when the individual voltages are $v_1 = 70{\cdot}7 \sin(\omega t + 30°)$ volts, $v_2 = 28{\cdot}3 \sin(\omega t + 120°)$ volts, $v_3 = 14{\cdot}14 \cos(\omega t + 30°)$ volts. *Ans.* 58·3 V

5.37. In Fig. 5-30 above, determine v_1 if the other voltages are $v_2 = 31{\cdot}6 \cos(\omega t + 73{\cdot}4°)$ volts and $v_T = 20 \cos(\omega t - 35°)$ volts. *Ans.* $v_1 = 42{\cdot}4 \cos(\omega t - 80°)$ volts

5.38. Referring to Problem 5.37, find the reading on a voltmeter applied across each impedance and then across both impedances. How can this result be explained? *Ans.* $V_1 = 30$ V, $V_2 = 22{\cdot}4$ V, $V_T = 14{\cdot}14$ V

5.39. Determine the indication on the ammeter of Fig. 5-31 below when the two currents are $i_1 = 14{\cdot}14 \sin(\omega t - 20°)$ amperes and $i_2 = 7{\cdot}07 \sin(\omega t + 60°)$ amperes. *Ans.* 11·9 A

5.40. In Fig. 5-32 below, determine i_T when the three currents are $i_1 = 14{\cdot}14 \sin(\omega t + 45°)$ amperes, $i_2 = 14{\cdot}14 \sin(\omega t - 75°)$ amperes, $i_3 = 14{\cdot}14 \sin(\omega t - 195°)$ amperes. *Ans.* $i_T = 0$

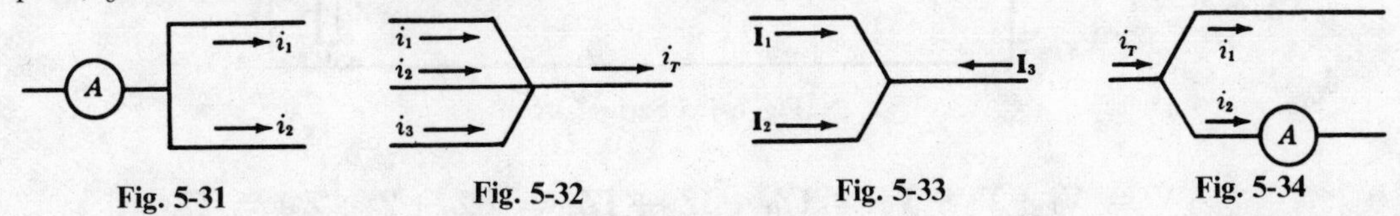

Fig. 5-31 Fig. 5-32 Fig. 5-33 Fig. 5-34

5.41. Find the phasor current $\mathbf{I}_3$ with the direction shown on the diagram, Fig. 5-33 above, if $\mathbf{I}_1 = 25\angle{70°}$ A and $\mathbf{I}_2 = 25\angle{-170°}$ A. *Ans.* $\mathbf{I}_3 = 25\angle{-50°}$ A

5.42. Find the current i_2 and the indication on the ammeter in Fig. 5-34 above when the other currents are $i_T = 13{\cdot}2 \sin(\omega t - 31°)$ amperes and $i_1 = 3{\cdot}54 \sin(\omega t + 20°)$ amperes. *Ans.* $i_2 = 11{\cdot}3 \sin(\omega t - 45°)$ amperes, 8 A

5.43. With fixed frequency and fixed circuit elements, the impedance is a point on the impedance diagram. However, if an element or the frequency is variable there results an impedance locus instead of the single point. For each of the following figures, discuss what may be variable to produce the impedance locus.

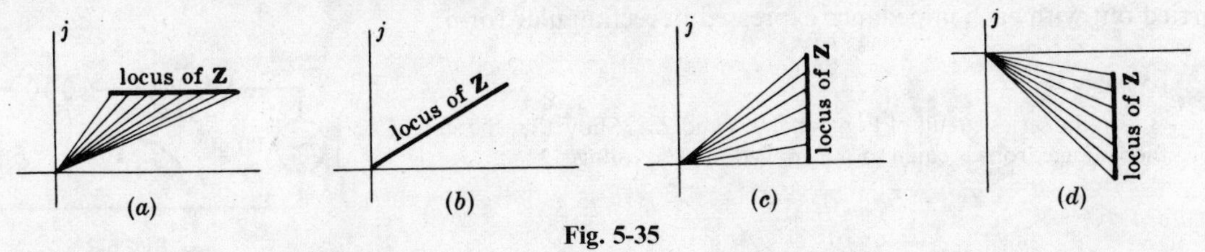

Fig. 5-35

CHAPTER 6

Series and Parallel Circuits

INTRODUCTION

In general, a circuit will contain both series and parallel parts. However, in this chapter the series and parallel circuits will be treated separately in order to show the different methods of analysis. The problems in this and later chapters will include circuits with series and parallel combinations.

SERIES CIRCUITS

Fig. 6-1 below shows a series circuit consisting of one voltage source and three impedances. The source voltage is assumed constant and is a *potential rise*. The phasor current **I** develops a voltage across each impedance through which it passes; these voltages are the *potential drops*. Kirchhoff's voltage law states that the *sum of the potential rises is equal to the sum of the potential drops around any closed path*. This law permits a solution to the series circuit problem.

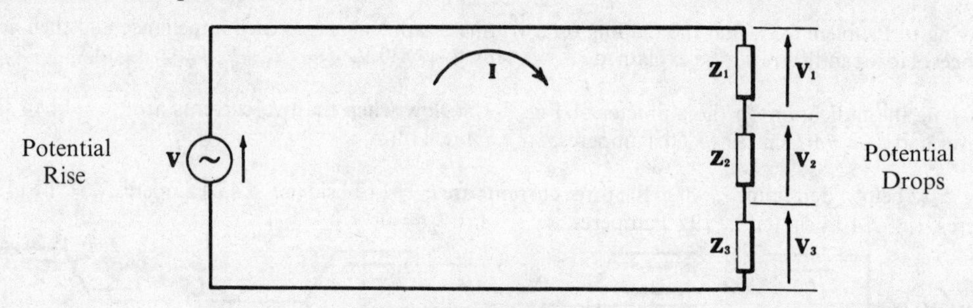

Fig. 6-1. Series Circuit

$$\mathbf{V} = \mathbf{V}_1 + \mathbf{V}_2 + \mathbf{V}_3 = \mathbf{IZ}_1 + \mathbf{IZ}_2 + \mathbf{IZ}_3 = \mathbf{I}(\mathbf{Z}_1 + \mathbf{Z}_2 + \mathbf{Z}_3) = \mathbf{IZ}_{eq}$$

from which $\mathbf{I} = \mathbf{V}/\mathbf{Z}_{eq}$ and $\mathbf{Z}_{eq} = \mathbf{Z}_1 + \mathbf{Z}_2 + \mathbf{Z}_3$

The voltage drop in an impedance is given by the product of phasor current **I** and the complex impedance **Z**. Thus in the circuit of Fig. 6-1, $\mathbf{V}_1 = \mathbf{IZ}_1$, $\mathbf{V}_2 = \mathbf{IZ}_2$ and $\mathbf{V}_3 = \mathbf{IZ}_3$. The arrow establishes a reference direction for this voltage and it points to the terminal by which the phasor current **I** enters.

The equivalent impedance $\mathbf{Z}_{eq}$ of any number of impedances in series is the sum of the individual impedances, $\mathbf{Z}_{eq} = (\mathbf{Z}_1 + \mathbf{Z}_2 + \mathbf{Z}_3 + \cdots)$. These impedances are complex numbers, and the indicated sum is carried out with each impedance expressed in rectangular form.

Example 1.

In the series circuit of Fig. 6-2, find **I** and $\mathbf{Z}_{eq}$. Show that the sum of the voltage drops is equal to the applied phasor voltage.

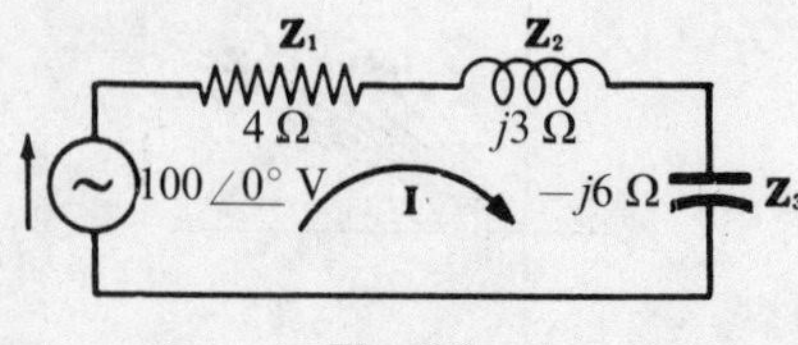

Fig. 6-2

$$\mathbf{Z}_{eq} = \mathbf{Z}_1 + \mathbf{Z}_2 + \mathbf{Z}_3 = 4 + j3 - j6$$
$$= 4 - j3 = 5\angle -36.9^\circ\ \Omega$$

and $$\mathbf{I} = \frac{\mathbf{V}}{\mathbf{Z}_{eq}} = \frac{100\angle 0^\circ}{5\angle -36{\cdot}9^\circ} = 20\angle 36{\cdot}9^\circ \text{ A}$$

Then $\mathbf{V}_1 = \mathbf{IZ}_1 = 20\angle 36{\cdot}9^\circ\,(4) = 80\angle 36{\cdot}9^\circ$ V, $\mathbf{V}_2 = 60\angle 126{\cdot}9^\circ$ V, $\mathbf{V}_3 = 120\angle -53{\cdot}1^\circ$ V

and $$\mathbf{V}_1 + \mathbf{V}_2 + \mathbf{V}_3 = (64 + j48) + (-36 + j48) + (72 - j96) = 100 + j0 = \mathbf{V}$$

as shown graphically in the voltage phasor diagram, Fig. 6-3(c).

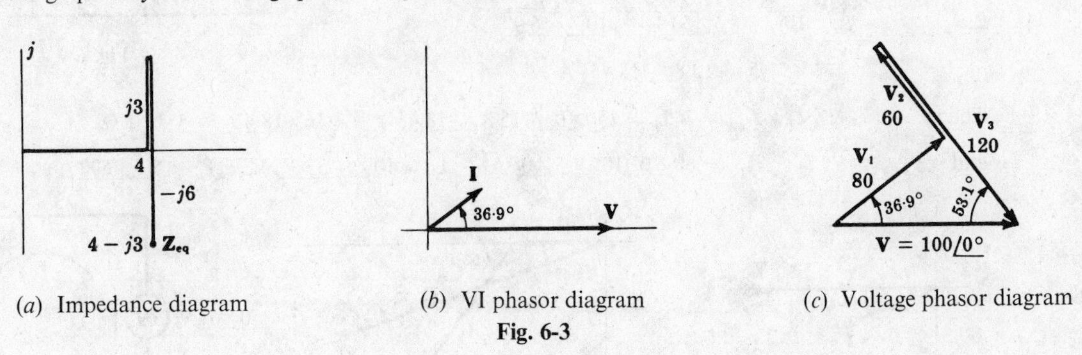

(a) Impedance diagram (b) VI phasor diagram (c) Voltage phasor diagram

Fig. 6-3

The equivalent impedance is capacitive, resulting in a current **I** which *leads* **V** by the angle on the impedance, 36·9°, as shown in Fig. 6-3(b) above. Note that $\mathbf{V}_1$, the voltage drop across the pure resistor, is in phase with the current. The current **I** lags $\mathbf{V}_2$ by 90° and leads $\mathbf{V}_3$ by 90°.

If a voltmeter were placed across each of the impedances $\mathbf{Z}_1$, $\mathbf{Z}_2$ and $\mathbf{Z}_3$ it would indicate 80, 60 and 120 volts respectively. This would seem to indicate a total voltage of 260 volts. But the meter would indicate 100 volts when placed across all three. It must be remembered that in the sinusoidal steady state analysis *all voltages and currents are phasors* and as such must be added *vectorially*.

PARALLEL CIRCUITS

In Fig. 6-4(a) below, a single voltage source is applied to a parallel connection of three impedances. The circuit is redrawn in Fig. 6-4(b) to emphasize the fact that the source and the three impedances have two common junctions. It is at one of these junctions (either may be used) that we apply Kirchhoff's current law, i.e. *the sum of the currents entering a junction is equal to the sum of the currents leaving the junction.*

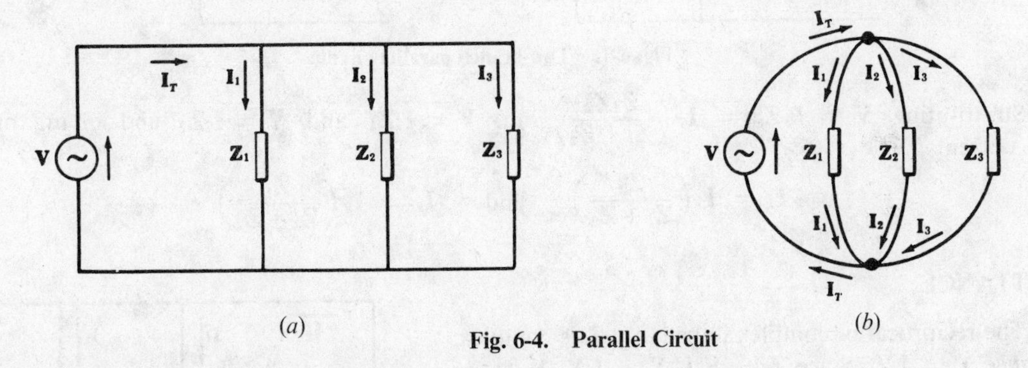

(a) **Fig. 6-4. Parallel Circuit** (b)

The constant potential of the source appears directly across each of the branch impedances. Hence the branch currents may be computed independently.

$$\mathbf{I}_T = \mathbf{I}_1 + \mathbf{I}_2 + \mathbf{I}_3 = \mathbf{V}/\mathbf{Z}_1 + \mathbf{V}/\mathbf{Z}_2 + \mathbf{V}/\mathbf{Z}_3 = \mathbf{V}(1/\mathbf{Z}_1 + 1/\mathbf{Z}_2 + 1/\mathbf{Z}_3) = \mathbf{V}/\mathbf{Z}_{eq}$$

Then $$\mathbf{I}_T = \mathbf{V}/\mathbf{Z}_{eq} \quad \text{and} \quad 1/\mathbf{Z}_{eq} = (1/\mathbf{Z}_1 + 1/\mathbf{Z}_2 + 1/\mathbf{Z}_3)$$

Thus the equivalent impedance of any number of impedances in parallel is given by

$$1/\mathbf{Z}_{eq} = 1/\mathbf{Z}_1 + 1/\mathbf{Z}_2 + 1/\mathbf{Z}_3 + \cdots$$

Example 2.

Find the total current and the equivalent impedance in the parallel circuit shown in Fig. 6-5, and sketch the **VI** phasor diagram.

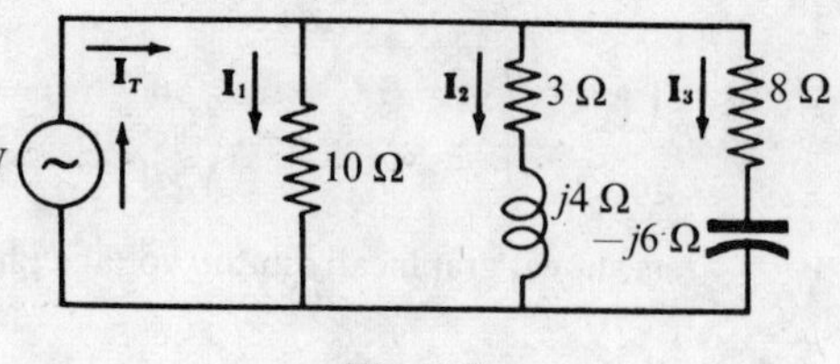

Fig. 6-5

$$\mathbf{I}_T = \mathbf{I}_1 + \mathbf{I}_2 + \mathbf{I}_3$$

$$= \frac{50\angle 0^\circ}{10} + \frac{50\angle 0^\circ}{5\angle 53{\cdot}1^\circ} + \frac{50\angle 0^\circ}{10\angle -36{\cdot}9^\circ}$$

$$= 15 - j5 = 15{\cdot}8\angle -18{\cdot}45^\circ \text{ A}$$

Then $$\mathbf{Z}_{eq} = \mathbf{V}/\mathbf{I}_T = (50\angle 0^\circ)/(15{\cdot}8\angle -18{\cdot}45^\circ) = 3{\cdot}16\angle 18{\cdot}45^\circ = 3 + j1\ \Omega$$

and $$\mathbf{I}_1 = 50\angle 0^\circ/10 = 5\angle 0^\circ \text{ A}, \quad \mathbf{I}_2 = 10\angle -53{\cdot}1^\circ \text{ A}, \quad \mathbf{I}_3 = 5\angle 36{\cdot}9^\circ \text{ A}$$

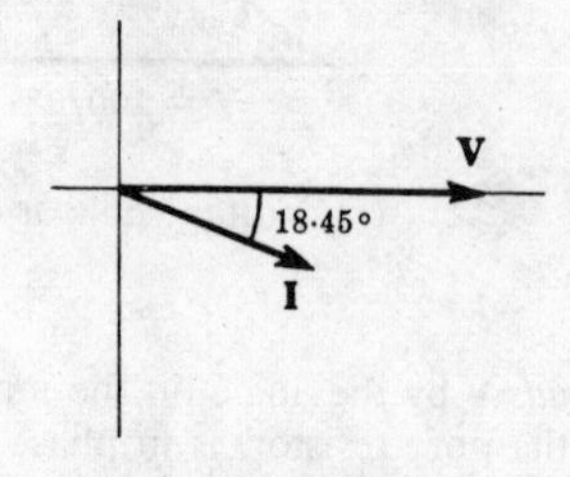

(a) VI phasor diagram

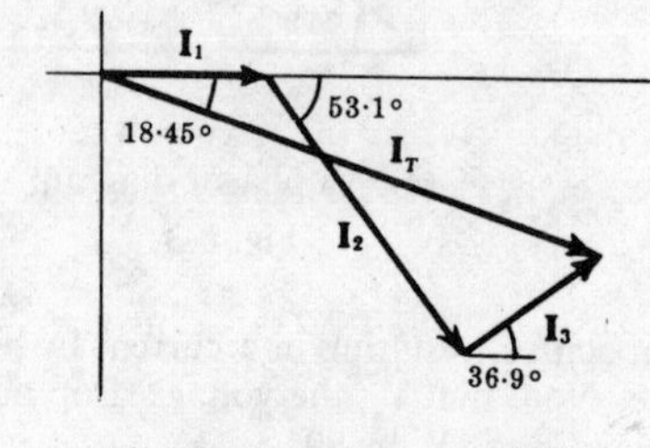

(b) Sum of phasor currents

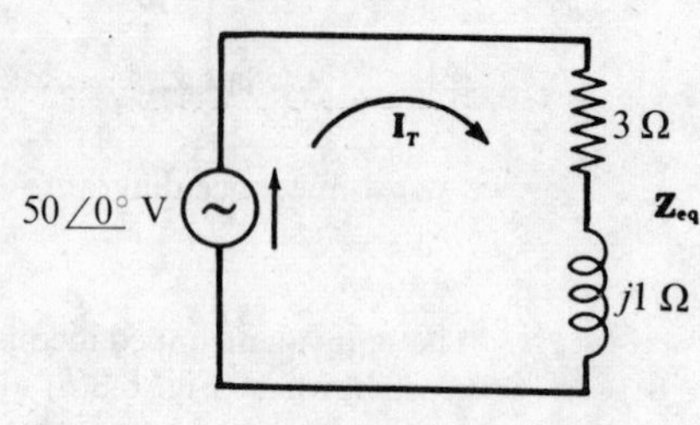

(c) Equivalent circuit

Fig. 6-6

TWO-BRANCH PARALLEL CIRCUIT

The case of two impedances in a parallel connection occurs frequently in circuit work and therefore deserves further consideration. In Fig. 6-7(a) below, impedances $\mathbf{Z}_1$ and $\mathbf{Z}_2$ in parallel have an applied voltage **V**. The equivalent impedance is given by $1/\mathbf{Z}_{eq} = 1/\mathbf{Z}_1 + 1/\mathbf{Z}_2$ or $\mathbf{Z}_{eq} = \mathbf{Z}_1\mathbf{Z}_2/(\mathbf{Z}_1 + \mathbf{Z}_2)$.

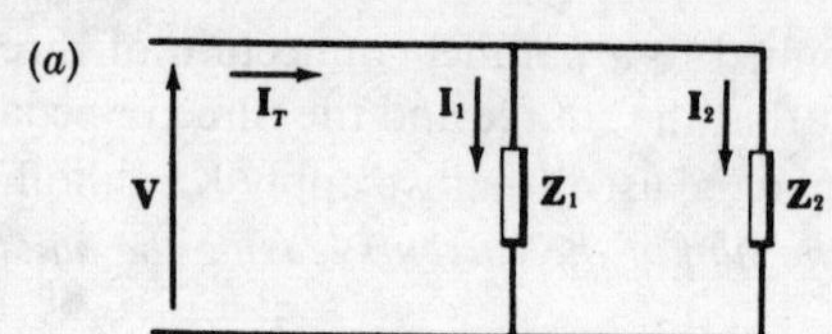

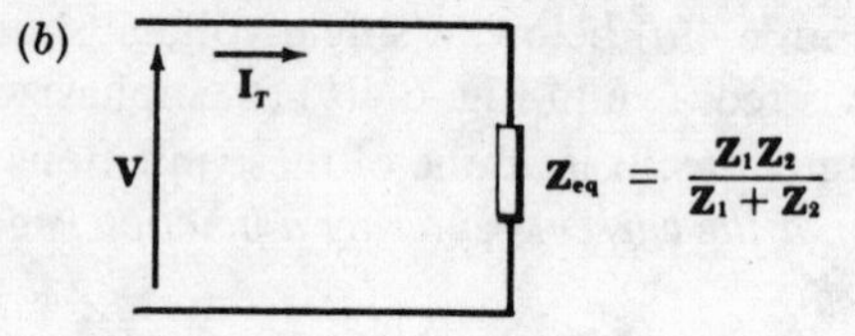

Fig. 6-7. Two-branch parallel circuit

Substituting $\mathbf{V} = \mathbf{I}_T\mathbf{Z}_{eq} = \mathbf{I}_T\left(\frac{Z_1Z_2}{Z_1+Z_2}\right)$ into $\mathbf{V} = \mathbf{I}_1\mathbf{Z}_1$ and $\mathbf{V} = \mathbf{I}_2\mathbf{Z}_2$ and solving for the two branch currents,

$$\mathbf{I}_1 = \mathbf{I}_T\left(\frac{\mathbf{Z}_2}{\mathbf{Z}_1+\mathbf{Z}_2}\right) \quad \text{and} \quad \mathbf{I}_2 = \mathbf{I}_T\left(\frac{\mathbf{Z}_1}{\mathbf{Z}_1+\mathbf{Z}_2}\right)$$

ADMITTANCE

The reciprocal of complex impedance **Z** is complex *admittance* $\mathbf{Y} = 1/\mathbf{Z}$. Since $\mathbf{Z} = \mathbf{V}/\mathbf{I}$, $\mathbf{Y} = \mathbf{I}/\mathbf{V}$. **Y** is expressed in siemens (S). The admittance concept is well suited to parallel circuits such as shown in Fig. 6-8.

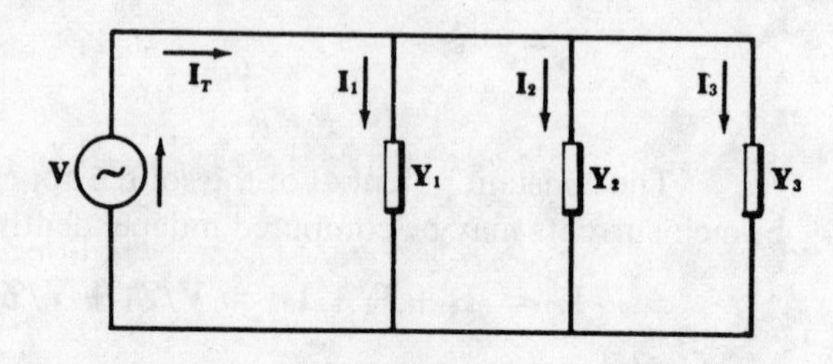

Fig. 6-8

$$\mathbf{I}_T = \mathbf{I}_1 + \mathbf{I}_2 + \mathbf{I}_3 = \mathbf{V}\mathbf{Y}_1 + \mathbf{V}\mathbf{Y}_2 + \mathbf{V}\mathbf{Y}_3$$

$$= \mathbf{V}(\mathbf{Y}_1 + \mathbf{Y}_2 + \mathbf{Y}_3) = \mathbf{V}\mathbf{Y}_{eq}$$

and $$\mathbf{Y}_{eq} = \mathbf{Y}_1 + \mathbf{Y}_2 + \mathbf{Y}_3$$

Thus the equivalent admittance of any number of admittances in parallel is the sum of the individual admittances.

In rectangular form, $\mathbf{Z} = R \pm jX$. The positive sign indicates inductive reactance $X_L = \omega L$, and the negative sign signifies capacitive reactance $X_C = 1/\omega C$.

Similarly, $\mathbf{Y} = G \pm jB$ where G is called *conductance* and B is called *susceptance*. The positive sign indicates capacitive susceptance B_C, and the negative sign implies inductive susceptance B_L.

Consider a general phasor voltage **V** and the resulting current **I**. Now the current **I** may lead, lag or be in phase with **V**, but in no case can the angle between them exceed 90°. Accordingly, three cases occur.

Case 1. The phasor current and voltage are in phase as in Fig. 6-9.

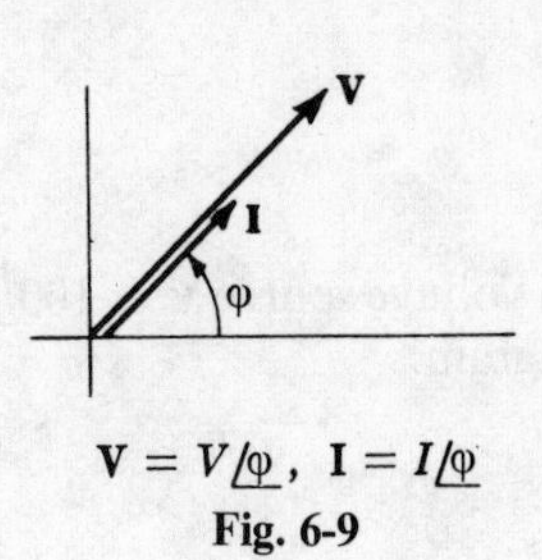

$\mathbf{V} = V\angle\varphi$, $\mathbf{I} = I\angle\varphi$

Fig. 6-9

Impedance

$$\mathbf{Z} = V\angle\varphi / I\angle\varphi = Z\angle 0° = R$$

On the basis of impedance the circuit consists of a pure resistance R ohms.

Admittance

$$\mathbf{Y} = I\angle\varphi / V\angle\varphi = Y\angle 0° = G$$

On the basis of admittance the circuit consists of a pure conductance G siemens.

Case 2. The phasor current lags the voltage by an angle θ as shown in Fig. 6-10.

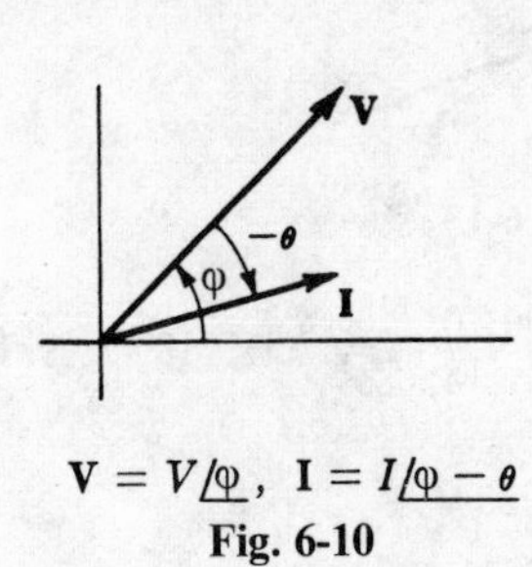

$\mathbf{V} = V\angle\varphi$, $\mathbf{I} = I\angle\varphi - \theta$

Fig. 6-10

Impedance

$$\mathbf{Z} = V\angle\varphi / I\angle\varphi - \theta = Z\angle\theta = R + jX_L$$

On the basis of impedance the circuit consists of resistance and inductive reactance in series.

Admittance

$$\mathbf{Y} = I\angle\varphi - \theta / V\angle\varphi = Y\angle -\theta = G - jB_L$$

On the basis of admittance the circuit consists of conductance and inductive susceptance in parallel.

Case 3. The phasor current leads the voltage by an angle θ as shown in Fig. 6-11.

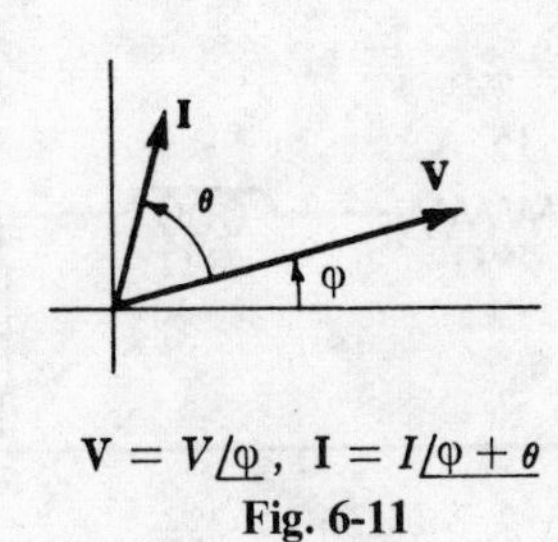

$\mathbf{V} = V\angle\varphi$, $\mathbf{I} = I\angle\varphi + \theta$

Fig. 6-11

Impedance

$$\mathbf{Z} = V\angle\varphi / I\angle\varphi + \theta = Z\angle -\theta = R - jX_C$$

On the basis of impedance the circuit consists of resistance and capacitive reactance in series.

Admittance

$$\mathbf{Y} = I\angle\varphi + \theta / V\angle\varphi = Y\angle\theta = G + jB_C$$

On the basis of admittance the circuit consists of conductance and capacitive susceptance in parallel.

ZY CONVERSION

Using the polar form, it is simple to convert **Z** into **Y**, and vice versa, since $\mathbf{Y} = 1/\mathbf{Z}$. However, at times it will be necessary to work with the relationships between the rectangular components as developed below.

$$\mathbf{Y} = 1/\mathbf{Z}$$

$$G + jB = \frac{1}{R + jX} = \frac{R - jX}{R^2 + X^2}$$

$$\therefore \quad G = \frac{R}{R^2 + X^2} \quad \text{and} \quad B = \frac{-X}{R^2 + X^2}$$

$$\mathbf{Z} = 1/\mathbf{Y}$$

$$R + jX = \frac{1}{G + jB} = \frac{G - jB}{G^2 + B^2}$$

$$\therefore \quad R = \frac{G}{G^2 + B^2} \quad \text{and} \quad X = \frac{-B}{G^2 + B^2}$$

Example 3.

Given $\mathbf{Z} = 3 + j4\ \Omega$, find the equivalent admittance **Y**.

$$\mathbf{Y} = 1/\mathbf{Z} = 1/5\angle 53{\cdot}1^\circ = 0{\cdot}2\angle -53{\cdot}1^\circ = 0{\cdot}12 - j0{\cdot}16\ \text{S}$$

from which $G = 0{\cdot}12$ siemens conductance and $B = 0{\cdot}16$ siemens inductive susceptance.

Alternate method.

$G = R/(R^2 + X^2) = 3/(9 + 16) = 0{\cdot}12$ and $B = -X/(R^2 + X^2) = -4/25 = -0{\cdot}16$. Thus $\mathbf{Y} = 0{\cdot}12 - j0{\cdot}16$ S.

Solved Problems

6.1. The two impedances $\mathbf{Z}_1$ and $\mathbf{Z}_2$ shown in Fig. 6-12 below are in series with a voltage source $\mathbf{V} = 100\angle 0^\circ$ V. Find the voltage across each impedance and draw the voltage phasor diagram.

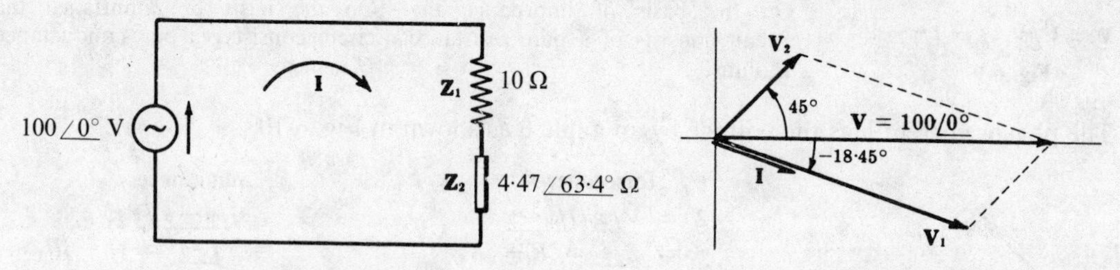

Fig. 6-12 Fig. 6-13

$\mathbf{Z}_{eq} = \mathbf{Z}_1 + \mathbf{Z}_2 = 10 + (2 + j4) = 12 + j4 = 12{\cdot}65\angle 18{\cdot}45^\circ\ \Omega$ and $\mathbf{I} = \dfrac{\mathbf{V}}{\mathbf{Z}_{eq}} = \dfrac{100\angle 0^\circ}{12{\cdot}65\angle 18{\cdot}45^\circ} = 7{\cdot}9\angle -18{\cdot}45^\circ$ A. Then

Then

$$\mathbf{V}_1 = \mathbf{I}\mathbf{Z}_1 = 7{\cdot}9\angle -18{\cdot}45^\circ\ (10) = 79\angle -18{\cdot}45^\circ = 74{\cdot}9 - j25\ \text{V}$$
$$\mathbf{V}_2 = \mathbf{I}\mathbf{Z}_2 = (7{\cdot}9\angle -18{\cdot}45^\circ)(4{\cdot}47\angle 63{\cdot}4^\circ) = 35{\cdot}3\angle 45^\circ = 25 + j25\ \text{V}$$

Now $\mathbf{V}_1 + \mathbf{V}_2 = (74{\cdot}9 - j25) + (25 + j25) = 99{\cdot}9 + j0 \approx 100\angle 0^\circ = \mathbf{V}$ as shown graphically in the voltage phasor diagram of Fig. 6-13 above.

6.2. Calculate the impedance $\mathbf{Z}_2$ in the series circuit of Fig. 6-14.

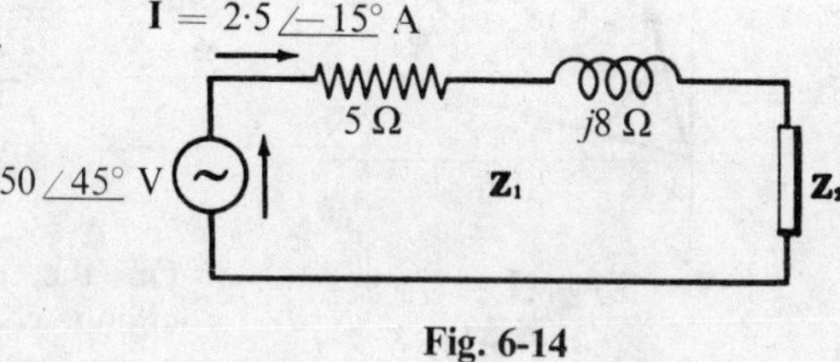

Fig. 6-14

For the given circuit, $\mathbf{Z}_{eq} = \dfrac{\mathbf{V}}{\mathbf{I}} = \dfrac{50\angle 45^\circ}{2{\cdot}5\angle -15^\circ} = 20\angle 60^\circ = 10 + j17{\cdot}3\ \Omega$. Since $\mathbf{Z}_{eq} = \mathbf{Z}_1 + \mathbf{Z}_2$, $10 + j17{\cdot}3 = (5 + j8) + \mathbf{Z}_2$ and $\mathbf{Z}_2 = 5 + j9{\cdot}3\ \Omega$.

6.3. In the circuit shown in Fig. 6-15 below, at a frequency $\omega = 400$ rad/s the current leads the voltage by $63{\cdot}4^\circ$. Find R and the voltage across each circuit element. Draw the voltage phasor diagram.

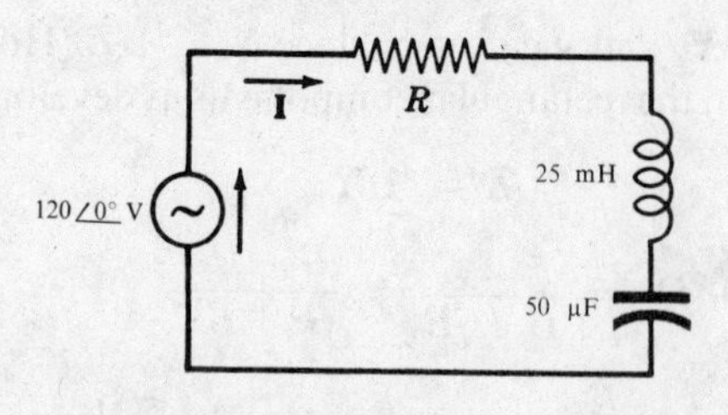

Fig. 6-15

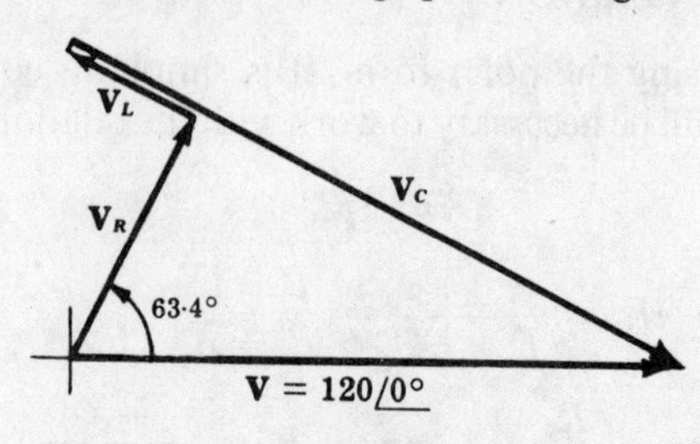

Fig. 6-16

$X_L = \omega L = 400(25 \times 10^{-3}) = 10$ ohms, $X_C = 1/\omega C = 1/400(50 \times 10^{-6}) = 50$ ohms, and $\mathbf{Z} = R + j(X_L - X_C) = R - j40$. Also, $\mathbf{Z} = Z\angle{-63{\cdot}4^\circ}\ \Omega$. Since $\tan -63{\cdot}4^\circ = (X_L - X_C)/R$, $R = -40/(\tan -63{\cdot}4^\circ) = 20$ ohms.

The impedance $\mathbf{Z} = 20 - j40 = 44{\cdot}7\angle{-63{\cdot}4^\circ}\ \Omega$ and the current $\mathbf{I} = \dfrac{\mathbf{V}}{\mathbf{Z}} = \dfrac{120\angle{0^\circ}}{44{\cdot}7\angle{-63{\cdot}4^\circ}} = 2{\cdot}68\angle{63{\cdot}4^\circ}$ A. Then

$$\mathbf{V}_R = 53{\cdot}6\angle{63{\cdot}4^\circ}\text{ V},\quad \mathbf{V}_L = 26{\cdot}8\angle{153{\cdot}4^\circ}\text{ V, and } V_C = 134\angle{-26{\cdot}6^\circ}\text{ V}$$

The voltage phasor diagram in Fig. 6-16 above shows $\mathbf{V}_R + \mathbf{V}_L + \mathbf{V}_C = \mathbf{V}$.

6.4. The circuit constants R and L of a coil are to be found by placing the coil in series with a 10 ohm standard resistor and reading the voltages across R_s, across the coil and across the complete series circuit. Determine R and L if the following 60 Hz voltages were read: $V_{Rs} = 20$ V, $V_{\text{coil}} = 22{\cdot}4$ V, $V_T = 36$ V.

The voltage $\mathbf{V}_{Rs}$ across the standard resistor and the current $\mathbf{I}$ are in phase. Let $\mathbf{V}_{Rs} = 20\angle{0^\circ}$ V; then $\mathbf{I} = \mathbf{V}_{Rs}/R_s = 2\angle{0^\circ}$ A.

In Fig. 6-17, from the tail end of phasor $\mathbf{V}_{Rs}$ we swing an arc of radius 36, and from the arrow end of $\mathbf{V}_{Rs}$ we swing an arc of radius 22·4. The intersection of the two arcs gives the arrow end of phasors $\mathbf{V}_T$ and $\mathbf{V}_{\text{coil}}$, thus satisfying the relation $\mathbf{V}_T = \mathbf{V}_{Rs} + \mathbf{V}_{\text{coil}}$.

We calculate the angle of phasor $\mathbf{V}_T$ by applying the law of cosines.

$$\cos\alpha = \frac{(36)^2 + (20)^2 - (22{\cdot}4)^2}{2(36)(20)} = 0{\cdot}831,\ \alpha = 33{\cdot}7^\circ$$

Then $\mathbf{V}_T = 36\angle{33{\cdot}7^\circ} = 30 + j20$ V and $\mathbf{V}_{\text{coil}} = \mathbf{V}_T - \mathbf{V}_{Rs} = 10 + j20 = 22{\cdot}4\angle{63{\cdot}4^\circ}$ V. The impedance of the coil is $\mathbf{Z}_{\text{coil}} = \mathbf{V}_{\text{coil}}/\mathbf{I} = (10 + j20)/2 = 5 + j10\ \Omega$, from which $R = 5$ ohms.

At $f = 60$ Hz, $X_L = 2\pi fL = 2\pi(60)L = 10\ \Omega$ and $L = 26{\cdot}5$ mH.

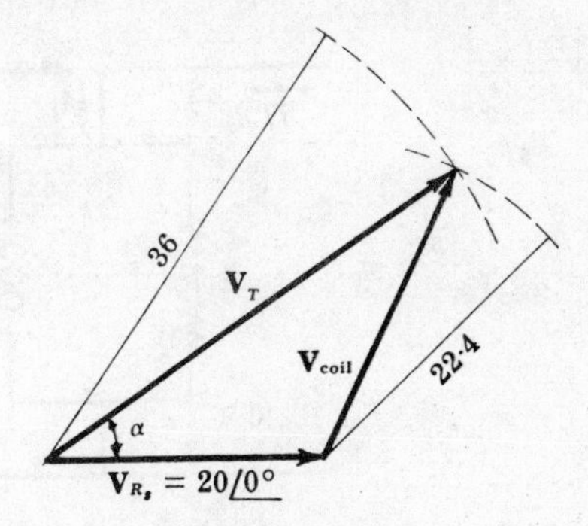

Fig. 6-17

6.5. For the parallel circuit of Fig. 6-18 below, find the branch currents and the total current. Construct the phasor diagram. Find $\mathbf{Z}_{\text{eq}}$ from $\mathbf{V}/\mathbf{I}$ and compare with $\mathbf{Z}_1\mathbf{Z}_2/(\mathbf{Z}_1 + \mathbf{Z}_2)$.

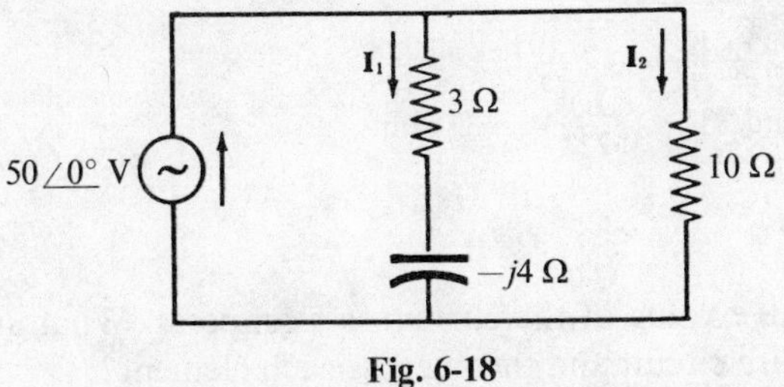

Fig. 6-18

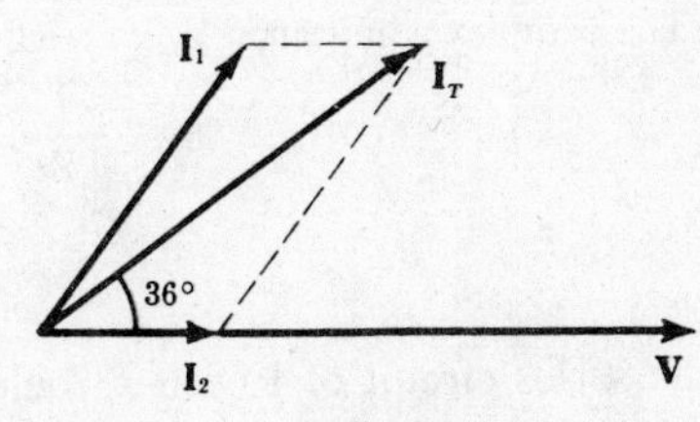

Fig. 6-19

$\mathbf{Z}_1 = 3 - j4 = 5\angle{-53{\cdot}1^\circ}\ \Omega$ and $\mathbf{Z}_2 = 10\ \Omega$. Then

$$\mathbf{I}_1 = \frac{\mathbf{V}}{\mathbf{Z}_1} = \frac{50\angle{0^\circ}}{5\angle{-53{\cdot}1^\circ}} = 10\angle{53{\cdot}1^\circ} = 6 + j8\text{ A}$$

$$\mathbf{I}_2 = \frac{\mathbf{V}}{\mathbf{Z}_2} = \frac{50\angle{0^\circ}}{10} = 5\angle{0^\circ} = 5\text{ A}$$

$$\mathbf{I}_T = \mathbf{I}_1 + \mathbf{I}_2 = 11 + j8 = 13{\cdot}6\angle{36^\circ}\text{ A}$$

$$\mathbf{Z}_{\text{eq}} = \frac{\mathbf{V}}{\mathbf{I}_T} = \frac{50\angle{0^\circ}}{13{\cdot}6\angle{36^\circ}} = 3{\cdot}67\angle{-36^\circ}\ \Omega,\quad \mathbf{Z}_{\text{eq}} = \frac{\mathbf{Z}_1\mathbf{Z}_2}{\mathbf{Z}_1 + \mathbf{Z}_2} = \frac{5\angle{-53{\cdot}1^\circ}(10)}{(3 - j4) + 10} = \frac{50\angle{-53{\cdot}1^\circ}}{13{\cdot}6\angle{-17{\cdot}1^\circ}} = 3{\cdot}67\angle{-36^\circ}\ \Omega$$

The phasor diagram is shown in Fig. 6-19 above.

6.6. For the series-parallel circuit of Fig. 6-20, find the current in each element.

$$\mathbf{Z}_{eq} = 10 + \frac{5(j10)}{5 + j10} = 14 + j2 = 14{\cdot}14\angle 8{\cdot}14^\circ\ \Omega \text{ and}$$

$$\mathbf{I}_T = \frac{\mathbf{V}}{\mathbf{Z}_{eq}} = \frac{100\angle 0^\circ}{14{\cdot}14\angle 8{\cdot}14^\circ} = 7{\cdot}07\angle -8{\cdot}14^\circ \text{ A. Then}$$

$$\mathbf{I}_{10} = \mathbf{I}_T = 7{\cdot}07\angle -8{\cdot}14^\circ \text{ A}$$

$$\mathbf{I}_{j10} = \mathbf{I}_T\left(\frac{5}{5 + j10}\right) = 7{\cdot}07\angle -8{\cdot}14^\circ\left(\frac{5}{5 + j10}\right) = 3{\cdot}16\angle -71{\cdot}54^\circ \text{ A}$$

$$\mathbf{I}_5 = \mathbf{I}_T\left(\frac{j10}{5 + j10}\right) = 7{\cdot}07\angle -8{\cdot}14^\circ\left(\frac{j10}{5 + j10}\right) = 6{\cdot}32\angle 18{\cdot}46^\circ \text{ A}$$

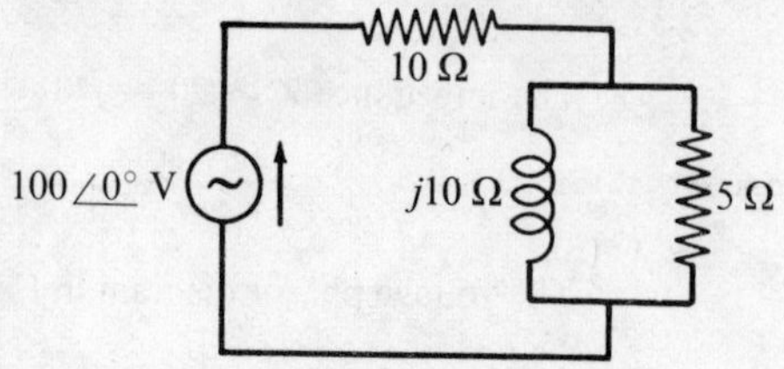

Fig. 6-20

6.7. In the parallel circuit of Fig. 6-21 below, the effective values of currents $\mathbf{I}_1$, $\mathbf{I}_2$ and $\mathbf{I}_T$ are 18, 15 and 30 amperes respectively. Determine the unknown impedances R and X_L.

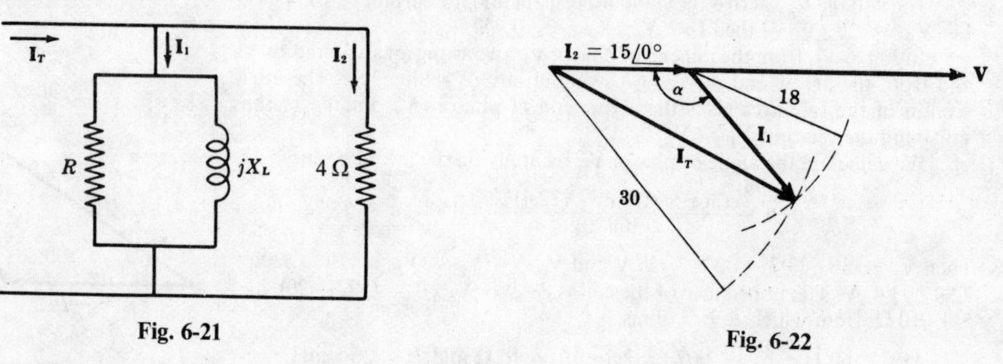

Fig. 6-21

Fig. 6-22

Applying Kirchhoff's current law, $\mathbf{I}_1 + \mathbf{I}_2 = \mathbf{I}_T$. $\mathbf{I}_2$ is in phase with the applied voltage $\mathbf{V}$. Let $\mathbf{I}_2 = 15\angle 0^\circ$ A; then $\mathbf{V} = 15\angle 0^\circ\,(4) = 60\angle 0^\circ$ V. With an inductive reactance present, $\mathbf{I}_1$ lags the applied voltage. Using the same construction as in Problem 6.4, Fig. 6-22 above is drawn. Then

$$\cos\alpha = \frac{(15)^2 + (18)^2 - (30)^2}{2(15)(18)} = -0{\cdot}65 \text{ and } \alpha = 130{\cdot}5^\circ$$

From the diagram, $\mathbf{I}_1 = 18\angle -49{\cdot}5^\circ$ A. Then $\mathbf{Z}_1 = \dfrac{\mathbf{V}}{\mathbf{I}_1} = \dfrac{60\angle 0^\circ}{18\angle -49{\cdot}5^\circ} = 3{\cdot}33\angle 49{\cdot}5^\circ\ \Omega$.

The complex admittance $\mathbf{Y}_1 = 1/R + 1/jX_L = 1/3{\cdot}33\angle 49{\cdot}5^\circ = 0{\cdot}195 - j0{\cdot}228$ S. Hence

$$R = \frac{1}{0{\cdot}195} = 5{\cdot}13 \text{ ohms and } X_L = \frac{1}{0{\cdot}228} = 4{\cdot}39 \text{ ohms}$$

6.8. In the series circuit of Fig. 6-23 below, the effective value of the current is 5 amperes. What are the readings on a voltmeter placed first across the entire circuit and then across each element?

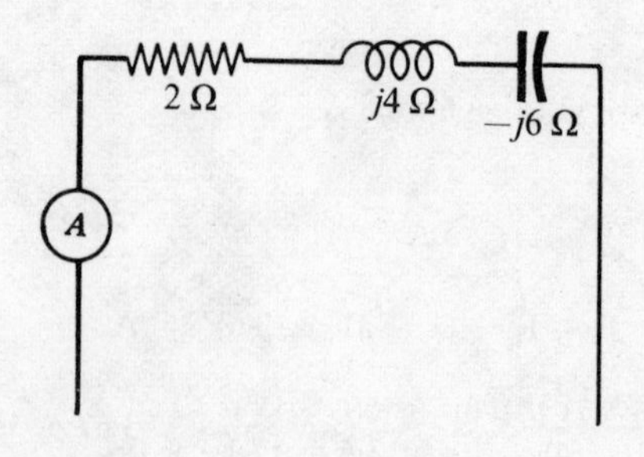

Fig. 6-23

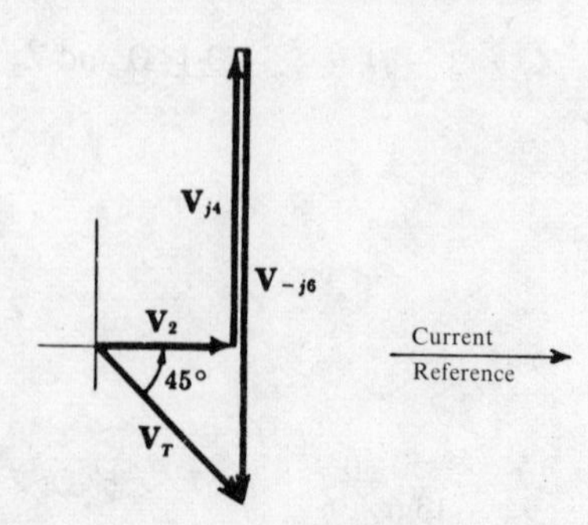

Fig. 6-24

$\mathbf{Z}_{eq} = 2 + j4 - j6 = 2{\cdot}83\angle{-45^\circ}\ \Omega$. Then

$$V_T = 5(2{\cdot}83) = 14{\cdot}14 \text{ volts} \qquad V_{j4} = 5(4) = 20 \text{ volts}$$
$$V_2 = 5(2) = 10 \text{ volts} \qquad V_{-j6} = 5(6) = 30 \text{ volts}$$

The voltage phasor diagram, Fig. 6-24, shows the addition of the voltage phasors across each circuit element.

6.9. In the parallel circuit of Fig. 6-25 the voltmeter reads 45 volts across the 3 ohm resistor. What is the indication on the ammeter?

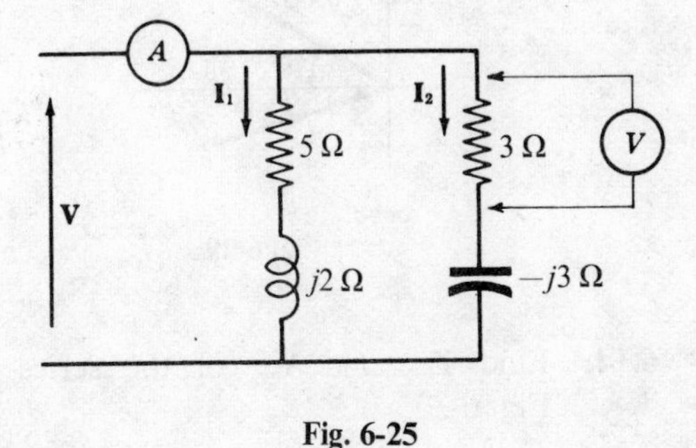

Fig. 6-25

$I_2 = 45/3 = 15$ A. Assuming an angle of 0°, $\mathbf{I}_2 = 15\angle 0^\circ$ A. Then $\mathbf{V} = 15\angle 0^\circ\,(3 - j3) = 63{\cdot}6\angle{-45^\circ}$ V and $\mathbf{I}_1 = 63{\cdot}6\angle{-45^\circ}/(5 + j2) = 11{\cdot}8\angle{-66{\cdot}8^\circ} = 4{\cdot}64 - j10{\cdot}85$ A. Now

$$\mathbf{I}_T = \mathbf{I}_1 + \mathbf{I}_2 = (4{\cdot}64 - j10{\cdot}85) + 15$$
$$= 19{\cdot}64 - j10{\cdot}85 = 22{\cdot}4\angle{-29^\circ} \text{ A}$$

The ammeter reads 22·4 A.

6.10. In the series-parallel circuit shown in Fig. 6-26 the effective value of the voltage across the parallel part of the circuit is 50 volts. Find the corresponding magnitude of **V**.

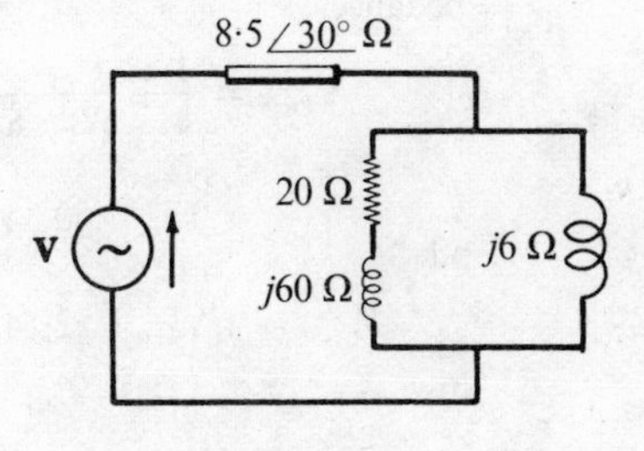

Fig. 6-26

$$\mathbf{Z}_p = \frac{(20 + j60)j6}{20 + j60 + j6} = 5{\cdot}52\angle 88{\cdot}45^\circ = 0{\cdot}149 + j5{\cdot}52\ \Omega$$

$$\mathbf{Z}_{eq} = 8{\cdot}5\angle 30^\circ + (0{\cdot}149 + j5{\cdot}52) = 12{\cdot}3\angle 52{\cdot}4^\circ\ \Omega$$

Since $\mathbf{V} = \mathbf{I}\mathbf{Z}_{eq}$ and $\mathbf{V}_p = \mathbf{I}\mathbf{Z}_p$, $V_p/Z_p = V/Z_{eq}$. Then $V = V_p(Z_{eq}/Z_p) = 50(12{\cdot}3/5{\cdot}52) = 111{\cdot}5$ volts

6.11. For the four branch parallel circuit of Fig. 6-27, find the total current and the equivalent impedance.

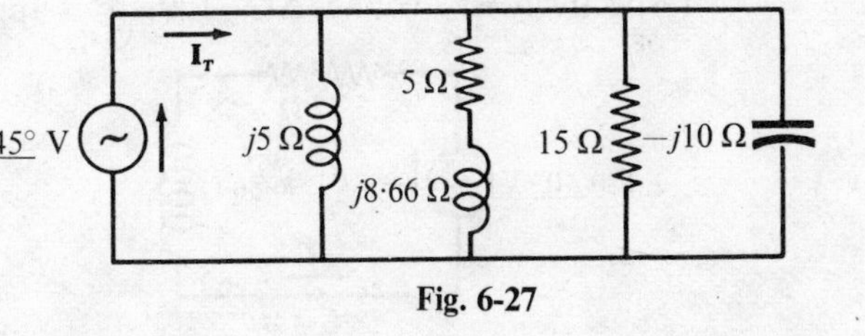

Fig. 6-27

$$\mathbf{Y}_1 = 1/j5 = -j0{\cdot}2 \text{ S}$$
$$\mathbf{Y}_2 = 1/10\angle 60^\circ = 0{\cdot}05 - j0{\cdot}0866 \text{ S}$$
$$\mathbf{Y}_3 = 1/15 = 0{\cdot}067 \text{ S}$$
$$\mathbf{Y}_4 = 1/-j10 = j0{\cdot}1 \text{ S}$$
$$\mathbf{Y}_{eq} = 0{\cdot}117 - j0{\cdot}1866 = 0{\cdot}22\angle{-58^\circ} \text{ S}$$

Then $\mathbf{I}_T = \mathbf{V}\mathbf{Y}_{eq} = (150\angle 45^\circ)(0{\cdot}22\angle{-58^\circ}) = 33\angle{-13^\circ}$ A and $\mathbf{Z}_{eq} = 1/\mathbf{Y}_{eq} = 1/(0{\cdot}22\angle{-58^\circ}) = 4{\cdot}55\angle 58^\circ\ \Omega$

6.12. In the three branch parallel circuit of Fig. 6-28, determine the impedance $\mathbf{Z}_1$.

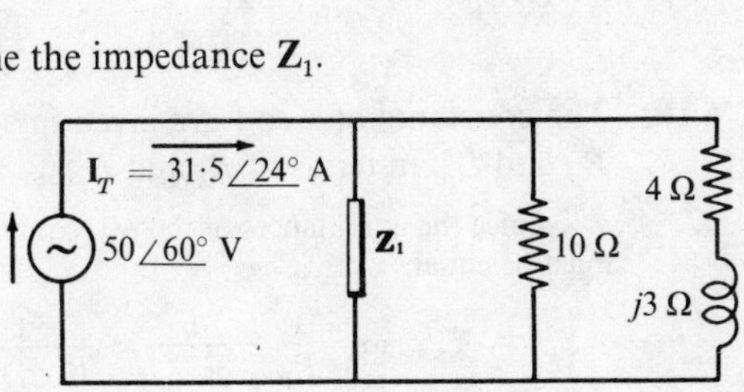

Complex admittance of circuit, $\mathbf{Y}_{eq} = \dfrac{\mathbf{I}_T}{\mathbf{V}} = \dfrac{31{\cdot}5\angle 24^\circ}{50\angle 60^\circ} = 0{\cdot}63\angle{-36^\circ} = 0{\cdot}51 - j0{\cdot}37$ S. Since $\mathbf{Y}_{eq} = \mathbf{Y}_1 + \mathbf{Y}_2 + \mathbf{Y}_3 = \mathbf{Y}_1 + (0{\cdot}1) + (0{\cdot}16 - j0{\cdot}12) = 0{\cdot}51 - j0{\cdot}37$ S, $\mathbf{Y}_1 = 0{\cdot}25 - j0{\cdot}25 = 0{\cdot}25\sqrt{2}\angle{-45^\circ}$ S. Then

$$\mathbf{Z}_1 = 1/\mathbf{Y}_1 = 2\sqrt{2}\angle 45^\circ = 2 + j2\ \Omega$$

Fig. 6-28

Alternate method.

$$\mathbf{I}_T = \mathbf{I}_1 + \mathbf{I}_2 + \mathbf{I}_3 = \mathbf{I}_1 + \frac{50\angle 60^\circ}{10} + \frac{50\angle 60^\circ}{5\angle 36{\cdot}9^\circ} = 31{\cdot}5\angle 24^\circ \text{ A from which } \mathbf{I}_1 = 17{\cdot}7\angle 15^\circ \text{ A.}$$

Hence $\mathbf{Z}_1 = \dfrac{\mathbf{V}}{\mathbf{I}_1} = \dfrac{50\angle 60^\circ}{17{\cdot}7\angle 15^\circ} = 2\sqrt{2}\angle 45^\circ = 2 + j2\ \Omega$.

6.13. Given the voltage phasor diagram of Fig. 6-29, determine the corresponding equivalent impedance and admittance.

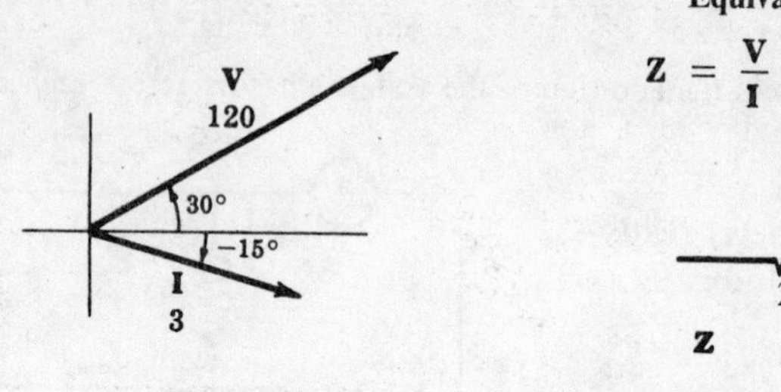

Fig. 6-29

Equivalent impedance

$$\mathbf{Z} = \frac{\mathbf{V}}{\mathbf{I}} = \frac{120\angle 30^\circ}{3\angle -15^\circ} = 40\angle 45^\circ = 28{\cdot}3 + j28{\cdot}3\ \Omega$$

Equivalent admittance

$$\mathbf{Y} = \frac{\mathbf{I}}{\mathbf{V}} = \frac{3\angle -15^\circ}{120\angle 30^\circ} = 0{\cdot}025\angle -45^\circ = 0{\cdot}0177 - j0{\cdot}0177\ \text{S}$$

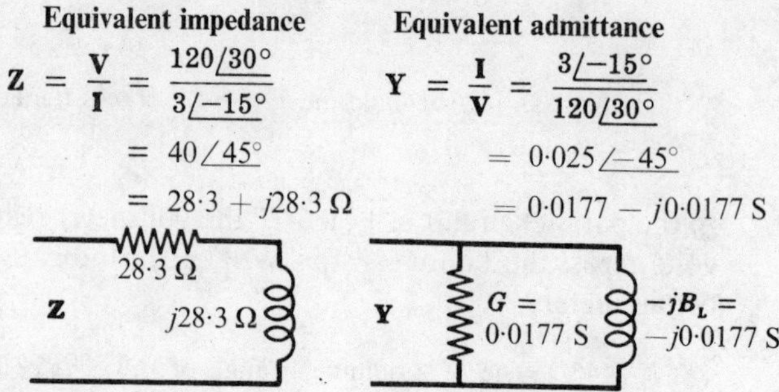

6.14. Find $\mathbf{Z}_{eq}$ and $\mathbf{Y}_{eq}$ of the series-parallel circuit in Fig. 6-30.

First calculate the equivalent admittance of the three branch parallel part of the circuit and then convert it to impedance.

$$\mathbf{Y}_{p_{eq}} = \frac{1}{5} + \frac{1}{j2} + \frac{1}{5\angle -53{\cdot}1^\circ}$$

$$= 0{\cdot}32 - j0{\cdot}34 = 0{\cdot}467\angle -46{\cdot}7^\circ\ \text{S}$$

Fig. 6-30

and

$$\mathbf{Z}_{peq} = 1/\mathbf{Y}_{peq} = 2{\cdot}14\angle 46{\cdot}7^\circ = 1{\cdot}47 + j1{\cdot}56\ \Omega$$

Now

$$\mathbf{Z}_{eq} = (2 + j5) + (1{\cdot}47 + j1{\cdot}56) = 3{\cdot}47 + j6{\cdot}56 = 7{\cdot}42\angle 62{\cdot}1^\circ\ \Omega$$

$$\mathbf{Y}_{eq} = 1/(7{\cdot}42\angle 62{\cdot}1^\circ) = 0{\cdot}135\angle -62{\cdot}1^\circ = 0{\cdot}063 - j0{\cdot}119\ \text{S}$$

6.15. Convert the series-parallel circuit of Problem 6.14 to two equivalent circuits containing $\mathbf{Z}_{eq}$ and $\mathbf{Y}_{eq}$ respectively. If a voltage $\mathbf{V} = 120\angle 0^\circ$ is applied to each circuit, find the currents.

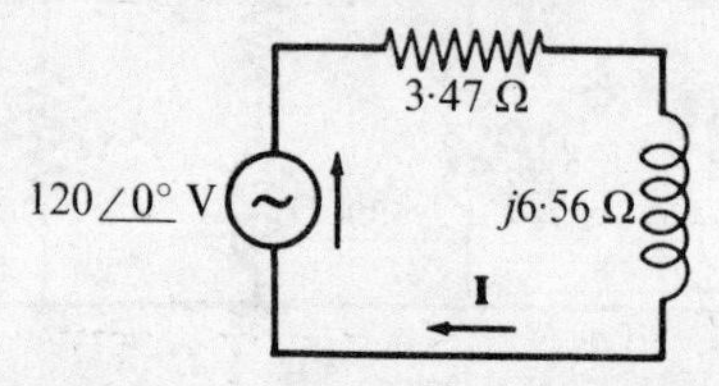

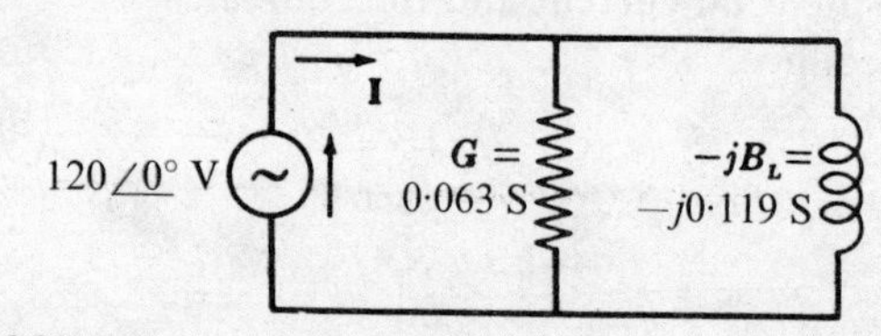

Fig. 6-31

(*a*) $\mathbf{Z} = 7{\cdot}42\angle 62{\cdot}1^\circ\ \Omega$

$$\mathbf{I} = \frac{\mathbf{V}}{\mathbf{Z}} = \frac{120\angle 0^\circ}{7{\cdot}42\angle 62{\cdot}1^\circ} = 16{\cdot}2\angle -62{\cdot}1^\circ\ \text{A}$$

(*b*) $\mathbf{Y} = 0{\cdot}135\angle -62{\cdot}1^\circ$ S

$$\mathbf{I} = \mathbf{VY} = (120\angle 0^\circ)(0{\cdot}135\angle -62{\cdot}1^\circ) = 16{\cdot}2\angle -62{\cdot}1^\circ\ \text{A}$$

6.16. The constants of a coil are given in series as R_s and L_s. Determine the parallel equivalent constants, R_p and L_p, in terms of R_s and L_s.

Since the admittances of the two equivalent circuits shown in Fig. 6-32 must be equal,

$$\mathbf{Y}_p = \mathbf{Y}_s \quad \text{or} \quad \frac{1}{R_p} + \frac{1}{j\omega L_p} = \frac{1}{R_s + j\omega L_s} = \frac{R_s - j\omega L_s}{(R_s)^2 + (\omega L_s)^2}$$

Equating real parts and j parts of the two admittances,

$$\frac{1}{R_p} = \frac{R_s}{(R_s)^2 + (\omega L_s)^2} \quad \text{and} \quad \frac{1}{j\omega L_p} = \frac{-j\omega L_s}{(R_s)^2 + (\omega L_s)^2}$$

from which $R_p = R_s + (\omega L_s)^2/R_s$ and $L_p = L_s + R_s^2/\omega^2 L_s$.

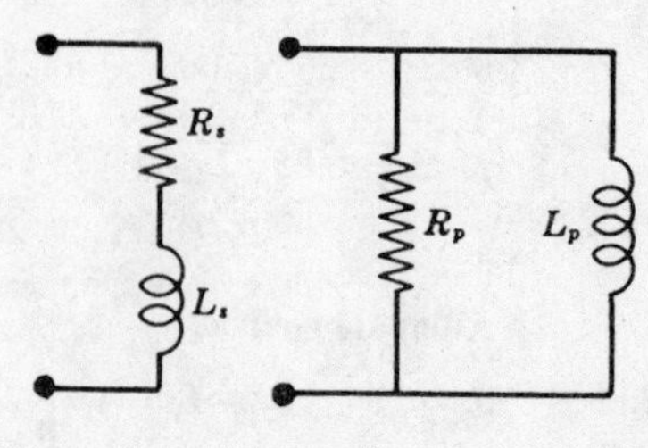

Fig. 6-32

6.17. Find the equivalent impedance of the series-parallel circuit shown in Fig. 6-33.

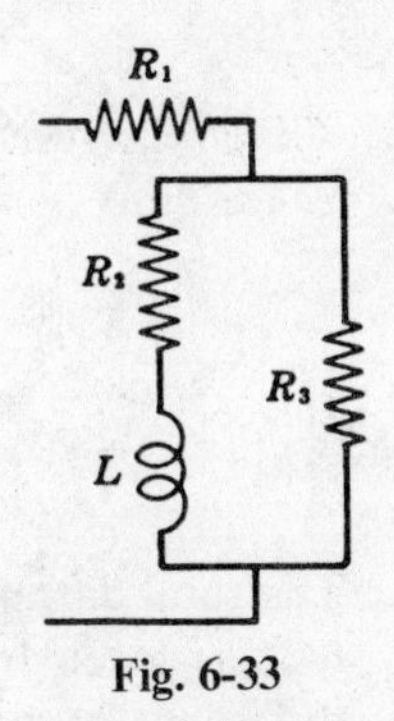

Fig. 6-33

$$\begin{aligned}
\mathbf{Z}_{eq} &= R_1 + \frac{(R_2 + j\omega L)R_3}{R_2 + R_3 + j\omega L} = R_1 + \frac{(R_2R_3 + j\omega LR_3)[(R_2+R_3) - j\omega L]}{(R_2+R_3)^2 + (\omega L)^2} \\
&= R_1 + \frac{R_2R_3(R_2+R_3) + \omega^2L^2R_3 + j\omega LR_3(R_2+R_3) - j\omega L(R_2R_3)}{(R_2+R_3)^2 + (\omega L)^2} \\
&= \left[R_1 + \frac{R_3(R_2^2 + R_2R_3 + \omega^2L^2)}{(R_2+R_3)^2 + (\omega L)^2}\right] + j\left[\frac{\omega LR_3^2}{(R_2+R_3)^2 + (\omega L)^2}\right] \\
&= R_{eq} + j\omega L_{eq}
\end{aligned}$$

6.18. In the parallel circuit of Fig. 6-34 the first branch contains two equal resistors R in series and the second branch contains a resistor R_1 in series with a variable inductance L. Show how the voltage between A and B varies as L is changed.

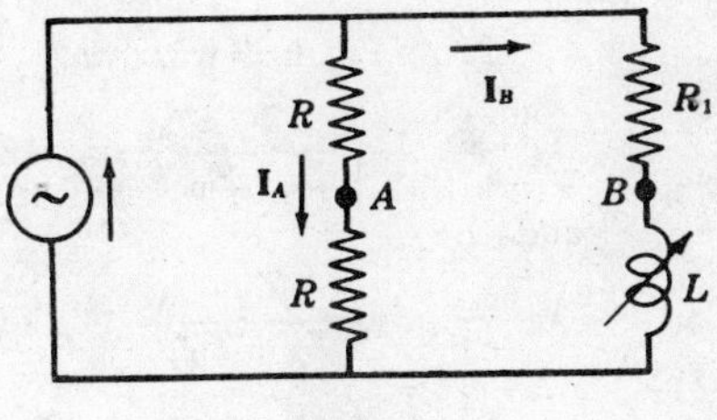

Fig. 6-34

In the first branch, the current $\mathbf{I}_A = \mathbf{V}/2R$ and the potential across the lower resistor is $\mathbf{I}_A R = \frac{1}{2}\mathbf{V}$.

In the second branch, the current

$$\mathbf{I}_B = \mathbf{V}/(R_1 + j\omega L)$$

and the potential across the inductance is

$$\mathbf{I}_B j\omega L = \frac{\mathbf{V}}{(R_1 + j\omega L)}(j\omega L)$$

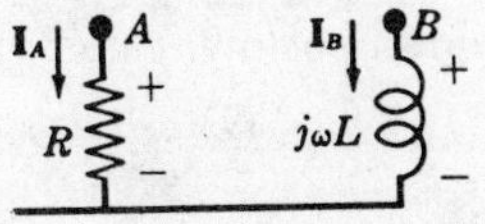

Fig. 6-35

Since the polarities are as shown in Fig. 6-35,

$$\mathbf{V}_{AB} = \mathbf{I}_A R - \mathbf{I}_B(j\omega L) = \tfrac{1}{2}\mathbf{V} - \frac{\mathbf{V}}{(R_1 + j\omega L)}(j\omega L)$$

Rationalizing the right hand term of the above expression and separating the real and j parts, we obtain

$$\mathbf{V}_{AB} = \mathbf{V}\left[\left(\frac{1}{2} - \frac{\omega^2L^2}{R_1^2 + (\omega L)^2}\right) - j\left(\frac{\omega LR_1}{R_1^2 + (\omega L)^2}\right)\right]$$

The expression in brackets is a complex number which, converted to polar form, has an absolute value r and an amplitude φ as follows.

$$r = \sqrt{\left(\frac{1}{2} - \frac{\omega^2L^2}{R_1^2 + (\omega L)^2}\right)^2 + \left(\frac{\omega LR_1}{R_1^2 + (\omega L)^2}\right)^2} = \frac{1}{2}$$

$$\varphi = \tan^{-1}\frac{-\omega LR_1/[R_1^2 + (\omega L)^2]}{\frac{1}{2} - \omega^2L^2/[R_1^2 + (\omega L)^2]} = \tan^{-1}\frac{-2\omega LR_1}{R_1^2 - (\omega L)^2} = \tan^{-1}\frac{-2(\omega L/R_1)}{1 - (\omega L/R_1)^2}$$

Thus the absolute value of $\mathbf{V}_{AB}$ is constant, i.e. $V_{AB} = \frac{1}{2}V$; and since $\tan 2x = (2\tan x)/(1 - \tan^2 x)$ and $\omega L/R = \tan\theta$, $\varphi = -2\theta$ where θ is the angle on the complex impedance of the second branch.

6.19. In the network of Fig. 6-36, two active loops are connected by a 10 ohm resistor. Find the potential difference between A and B.

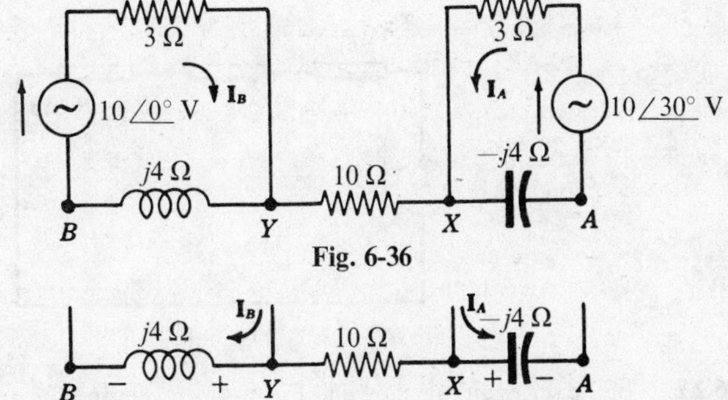

Fig. 6-36

Fig. 6-37

From Fig. 6-36 we calculate

$$\mathbf{I}_A = \frac{10\angle 30°}{3 - j4} = \frac{10\angle 30°}{5\angle -53{\cdot}1°} = 2\angle 83{\cdot}1°\ \text{A}$$

and

$$\mathbf{I}_B = \frac{10\angle 0°}{3 + j4} = \frac{10\angle 0°}{5\angle 53{\cdot}1°} = 2\angle -53{\cdot}1°\ \text{A}$$

To calculate $\mathbf{V}_{AB}$, the voltages across the elements shown in Fig. 6-37 are needed. Thus, considering the proper polarities, we have

$$\mathbf{V}_{AX} = -\mathbf{I}_A(-j4) = -2\underline{/83{\cdot}1^\circ}\,(-j4) = -8\underline{/-6{\cdot}9^\circ} = -7{\cdot}94 + j0{\cdot}96 \text{ V}$$

$$\mathbf{V}_{XY} = 0 \text{ (no current is passing through the 10 ohm resistor)}$$

$$\mathbf{V}_{YB} = \mathbf{I}_B(j4) = 2\underline{/-53{\cdot}1^\circ}\,(j4) = 8\underline{/36{\cdot}9^\circ} = 6{\cdot}4 + j4{\cdot}8 \text{ V}$$

Hence $\mathbf{V}_{AB} = \mathbf{V}_{AX} + \mathbf{V}_{XY} + \mathbf{V}_{YB} = -1{\cdot}54 + j5{\cdot}76 = 5{\cdot}95\underline{/105^\circ}$ V

6.20. The total current entering the parallel circuit shown in Fig. 6-38 is given by $\mathbf{I}_T = 18\underline{/45^\circ}$ A. Determine the potential difference between points A and B.

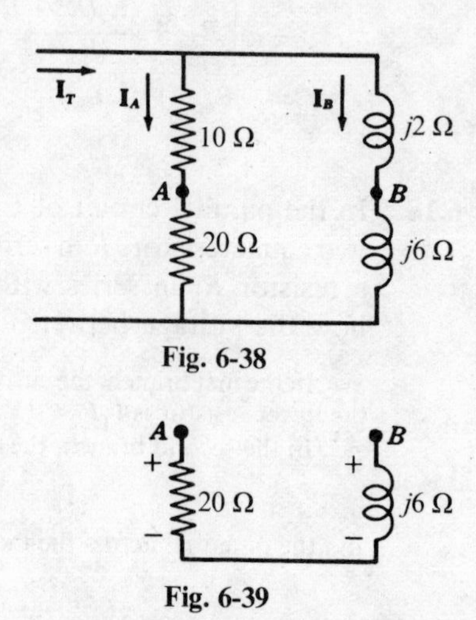

Fig. 6-38

From Fig. 6-38 we have

$$\mathbf{I}_A = \mathbf{I}_T\left(\frac{\mathbf{Z}_B}{\mathbf{Z}_A + \mathbf{Z}_B}\right) = 18\underline{/45^\circ}\left(\frac{j8}{30 + j8}\right) = 4{\cdot}66\underline{/120^\circ} \text{ A}$$

and

$$\mathbf{I}_B = \mathbf{I}_T\left(\frac{\mathbf{Z}_A}{\mathbf{Z}_A + \mathbf{Z}_B}\right) = 18\underline{/45^\circ}\left(\frac{30}{30 + j8}\right) = 17{\cdot}5\underline{/30^\circ} \text{ A}$$

The voltages across the 20 ohm resistance and the $j6$ ohm reactance and $\mathbf{V}_{20} = \mathbf{I}_A(20) = 93{\cdot}2\underline{/120^\circ}$ V and $\mathbf{V}_{j6} = \mathbf{I}_B(j6) = 105\underline{/120^\circ}$ V respectively.

The sketch in Fig. 6-39 permits the two voltages to be added with the correct polarity. Thus

$$\mathbf{V}_{AB} = (93{\cdot}2\underline{/120^\circ}) - (105\underline{/120^\circ}) = 11{\cdot}8\underline{/-60^\circ} \text{ V}$$

Fig. 6-39

6.21. Determine the equivalent impedance between terminals A and B in the bridge circuit of Fig. 6-40.

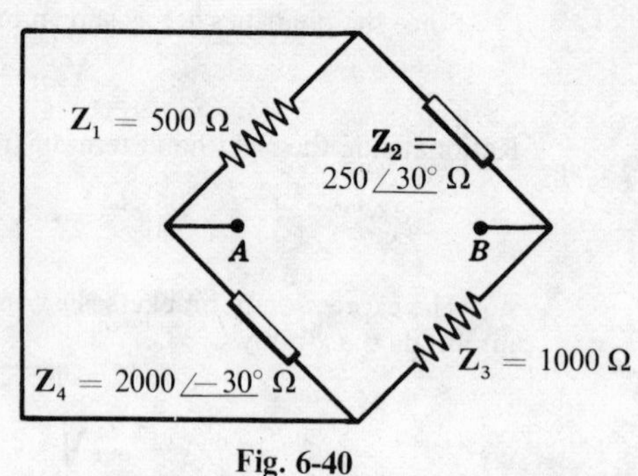

Fig. 6-40

The parallel combination of $\mathbf{Z}_1$ and $\mathbf{Z}_4$ is in series with the parallel combination of $\mathbf{Z}_2$ and $\mathbf{Z}_3$. Then

$$\mathbf{Z}_{eq} = \frac{\mathbf{Z}_1\mathbf{Z}_4}{\mathbf{Z}_1 + \mathbf{Z}_4} + \frac{\mathbf{Z}_2\mathbf{Z}_3}{\mathbf{Z}_2 + \mathbf{Z}_3}$$

$$= \frac{500(2000\underline{/-30^\circ})}{500 + 2000\underline{/-30^\circ}} + \frac{250\underline{/30^\circ}\,(1000)}{250\underline{/30^\circ} + 1000}$$

$$= 596\underline{/4{\cdot}05^\circ}\ \Omega$$

Supplementary Problems

6.22. In the series circuit of Fig. 6-41 below, find the voltage across each impedance. Show on a phasor diagram that the sum $\mathbf{V}_1 + \mathbf{V}_2 + \mathbf{V}_3$ is equal to the applied voltage $\mathbf{V} = 100\underline{/0^\circ}$ V.
Ans. $31{\cdot}4\underline{/20{\cdot}8^\circ}$ V, $25{\cdot}1\underline{/50{\cdot}8^\circ}$ V, $62{\cdot}9\underline{/-29{\cdot}2^\circ}$ V

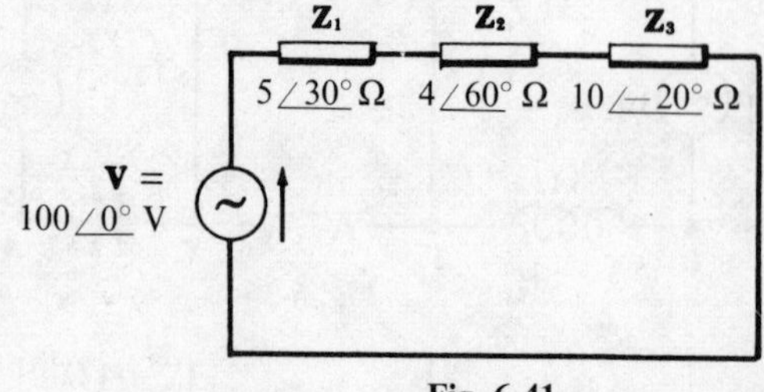

Fig. 6-41

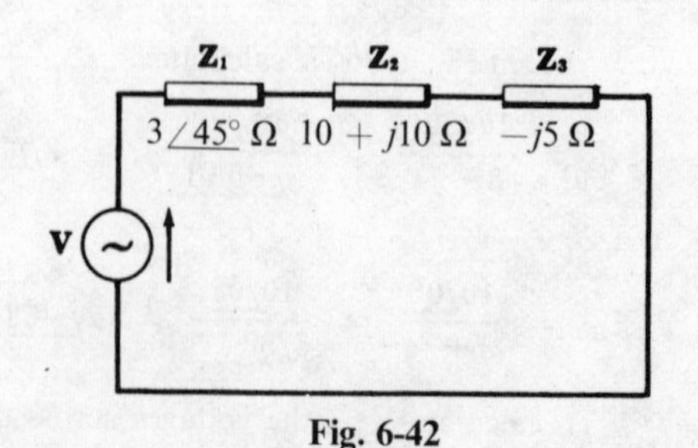

Fig. 6-42

6.23. In the series circuit shown in Fig. 6-42 above, find the applied voltage $\mathbf{V}$ if the voltage drop across $\mathbf{Z}_1$ is $27\underline{/-10^\circ}$ volts.
Ans. $126{\cdot}5\underline{/-24{\cdot}6^\circ}$ V

6.24. Three impedances $\mathbf{Z}_1 = 5 + j5\ \Omega$, $\mathbf{Z}_2 = -j8\ \Omega$ and $\mathbf{Z}_3 = 4\ \Omega$ are connected in series to an unknown voltage source $\mathbf{V}$. Find $\mathbf{I}$ and $\mathbf{V}$ if the voltage drop across $\mathbf{Z}_3$ is $63{\cdot}2\angle 18{\cdot}45^\circ$ volts. *Ans.* $\mathbf{I} = 15{\cdot}8\angle 18{\cdot}45^\circ$ A, $\mathbf{V} = 150\angle 0^\circ$ V

6.25. A voltage source $\mathbf{V} = 25\angle 180^\circ$ V is connected to a series circuit of fixed R and variable X_L. The inductive reactance is set to an arbitrary value and the resulting current is $\mathbf{I} = 11{\cdot}15\angle 153{\cdot}4^\circ$ A. Then X_L is adjusted to cause the current to lag the voltage by 60°. Under this second condition, what is the effective value of the current? *Ans.* 6·25 A

6.26. In the series circuit of Fig. 6-43 below, there is a voltage drop across the $j2\ \Omega$ reactance of $\mathbf{V}_{j2} = 13{\cdot}04\angle 15^\circ$ V. Determine $\mathbf{Z}$. *Ans.* $R = 4$ ohms, $X_C = 15$ ohms

6.27. A series circuit consists of a resistance $R = 1$ ohm, an inductive reactance $jX_L = j4$ ohms and a third impedance $\mathbf{Z}$. If the applied voltage and resulting current are $\mathbf{V} = 50\angle 45^\circ$ V and $\mathbf{I} = 11{\cdot}2\angle 108{\cdot}4^\circ$ A, what is the impedance $\mathbf{Z}$? *Ans.* $\mathbf{Z} = 1 - j8\Omega$

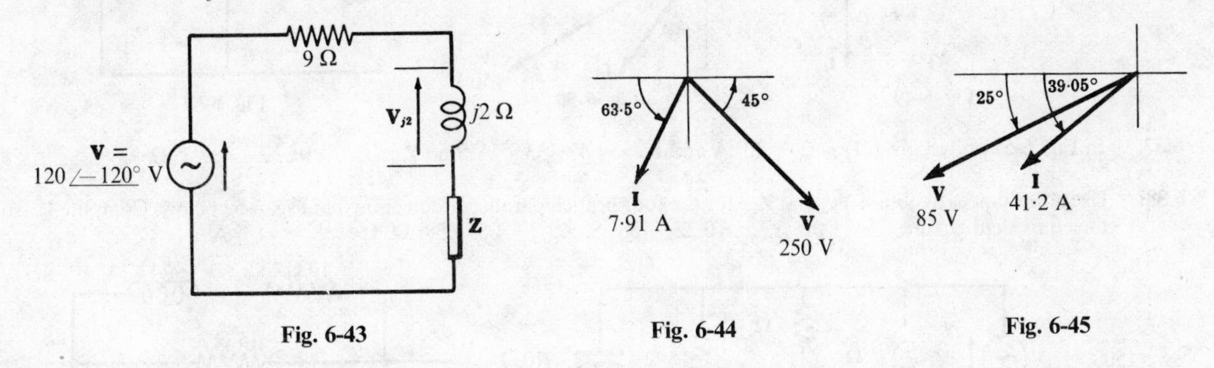

Fig. 6-43 Fig. 6-44 Fig. 6-45

6.28. A three element series circuit contains one inductance $L = 0{\cdot}02$ H. The applied voltage and resulting current are shown on the phasor diagram in Fig. 6-44 above. If $\omega = 500$ rad/s, what are the other two circuit elements?
Ans. $R = 10$ ohms, $L = 0{\cdot}04$ H

6.29. Find $\mathbf{Z}$ and $\mathbf{Y}$ which correspond to the phasor diagram in Fig. 6-45 above.
Ans. $\mathbf{Z} = 2 - j0{\cdot}5\ \Omega$, $\mathbf{Y} = 0{\cdot}47 + j0{\cdot}1175$ S

6.30. The circuit constants R and L of a coil are to be determined by connecting the coil in series with a resistor of 25 ohms and applying a 120 volt, 60 Hz source and then reading the voltages across the resistor and the coil. Find R and L if $V_R = 70{\cdot}8$ volts and $V_{\text{coil}} = 86$ volts.
Ans. $R = 5$ ohms, $L = 79{\cdot}6$ mH

6.31. A series combination of R and C is connected in series with a resistance of 15 ohms. When a source of 120 volts at 60 Hz is applied to the complete circuit, the effective voltages across the RC combination and the pure resistor are 87·3 and 63·6 volts respectively. Find R and C. *Ans.* $R = 5$ ohms, $C = 132{\cdot}5\ \mu$F

6.32. Find $\mathbf{Z}_{\text{eq}}$ and $\mathbf{Y}_{\text{eq}}$ for the two branch parallel circuit shown in Fig. 6-46 below. Compute the current from each equivalent circuit. *Ans.* $\mathbf{Z}_{\text{eq}} = 18{\cdot}6\angle 7{\cdot}15^\circ\ \Omega$, $\mathbf{Y}_{\text{eq}} = 0{\cdot}0538\angle -7{\cdot}15^\circ$ S, $\mathbf{I}_T = 10{\cdot}75\angle -7{\cdot}15^\circ$ A

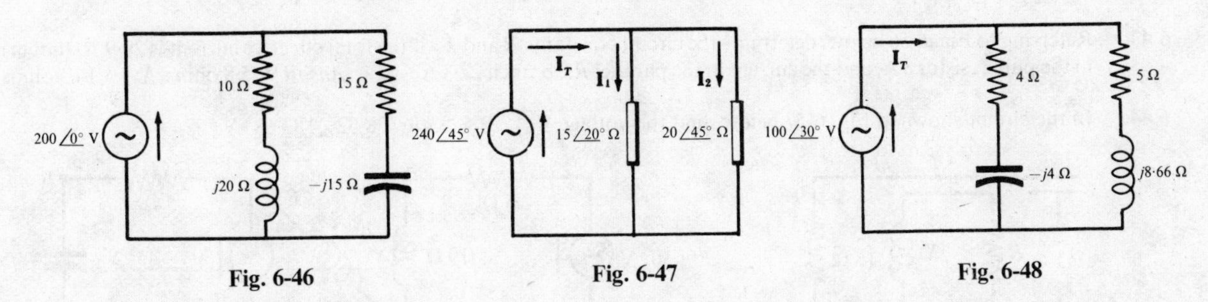

Fig. 6-46 Fig. 6-47 Fig. 6-48

6.33. In the parallel circuit shown in Fig. 6-47 above, find the two branch currents and the total current. Construct the current phasor diagram showing $\mathbf{I}_1$, $\mathbf{I}_2$ and $\mathbf{I}_T$, *Ans.* $16\angle 25^\circ$ A, $12\angle 0^\circ$ A, $27{\cdot}4\angle 14{\cdot}3^\circ$ A

6.34. Find $\mathbf{I}_T$ in the two branch parallel circuit shown in Fig. 6-48 above. Find $\mathbf{Z}_{eq}$ from the ratio $\mathbf{V}/\mathbf{I}_T$ and compare this with $\mathbf{Z}_{eq} = \mathbf{Z}_1\mathbf{Z}_2/(\mathbf{Z}_1 + \mathbf{Z}_2)$. *Ans.* $\mathbf{I}_T = 17{\cdot}9\angle 42{\cdot}4^\circ$ A, $\mathbf{Z}_{eq} = 5{\cdot}59\angle -12{\cdot}4^\circ\ \Omega$

6.35. A two branch parallel circuit has a corresponding phasor diagram as shown in Fig. 6-49 below. Find the branch impedances $\mathbf{Z}_1$ and $\mathbf{Z}_2$. *Ans.* $\mathbf{Z}_1 = 2{\cdot}5 + j20\ \Omega$, $\mathbf{Z}_2 = 15\angle -90^\circ\ \Omega$

6.36. A two branch parallel circuit has an applied voltage and resulting currents as given in the phasor diagram shown in Fig. 6-50 below. Find the branch impedances $\mathbf{Z}_1$ and $\mathbf{Z}_2$. *Ans.* $\mathbf{Z}_1 = 11{\cdot}55 - j20\ \Omega$, $\mathbf{Z}_2 = 27{\cdot}6 + j11{\cdot}75\ \Omega$

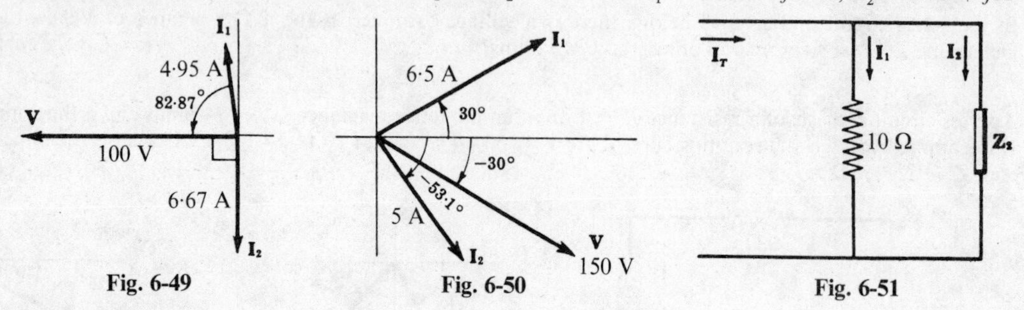

Fig. 6-49 Fig. 6-50 Fig. 6-51

6.37. In Fig. 6-51 above, given $\mathbf{I}_1 = 2\angle -30^\circ$ A and $\mathbf{I}_T = 4{\cdot}47\angle 33{\cdot}4^\circ$ A, find $\mathbf{Z}_2$. *Ans.* $\mathbf{Z}_2 = -j5\ \Omega$

6.38. Use admittances to obtain $\mathbf{Y}_{eq}$ and $\mathbf{Z}_{eq}$ for the four branch parallel circuit shown in Fig. 6-52 below. Compute $\mathbf{I}_T$ from the equivalent circuit. *Ans.* $\mathbf{Y}_{eq} = 0{\cdot}22\angle -58^\circ$ S, $\mathbf{Z}_{eq} = 4{\cdot}55\angle 58^\circ\ \Omega$, $\mathbf{I}_T = 33\angle -13^\circ$ A

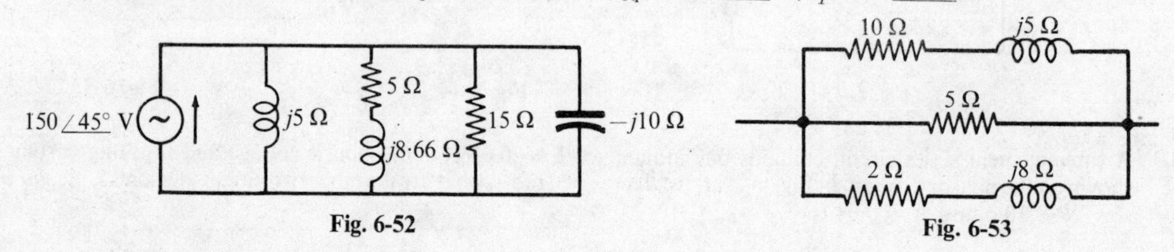

Fig. 6-52 Fig. 6-53

6.39. Find $\mathbf{Z}_{eq}$ and $\mathbf{Y}_{eq}$ of the three branch parallel circuit of Fig. 6-53 above.
Ans. $\mathbf{Z}_{eq} = 2{\cdot}87\angle 27^\circ\ \Omega$, $\mathbf{Y}_{eq} = 0{\cdot}348\angle -27^\circ$ S

6.40. In Fig. 6-54 below, given $\mathbf{V} = 50\angle 30^\circ$ V and $\mathbf{I}_T = 27{\cdot}9\angle 57{\cdot}8^\circ$ A, determine $\mathbf{Z}$. *Ans.* $\mathbf{Z} = 5\angle -30^\circ\ \Omega$

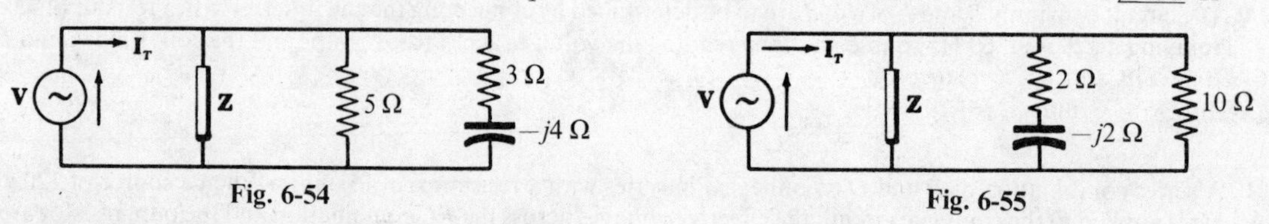

Fig. 6-54 Fig. 6-55

6.41. In Fig. 6-55 above, given $\mathbf{V} = 100\angle 90^\circ$ V and $\mathbf{I}_T = 50{\cdot}2\angle 102{\cdot}5^\circ$ A, determine $\mathbf{Z}$. *Ans.* $\mathbf{Z} = 5\angle 45^\circ\ \Omega$

6.42. A series combination of R and C is in parallel with a 20 ohm resistor. A 60 Hz source results in a total current of 7·02 A, a current through the 20 ohm resistor of 6 Ω, and a current in the RC branch of 2·3 A. Determine R and C.
Ans. $R = 15$ ohms, $C = 53{\cdot}1\ \mu$F

6.43. Referring to Fig. 6-56 below, determine the circuit constants R and X_L if the total effective current is 29·9 A, the current in the pure resistor 8 A, and the current in the parallel RL branch 22·3 A. *Ans.* $R = 5{\cdot}8$ ohms, $X_L = 14{\cdot}5$ ohms

6.44. In the circuit shown in Fig. 6-57 below, find the voltage $\mathbf{V}_{AB}$. *Ans.* $28{\cdot}52\angle 183{\cdot}68^\circ$ V

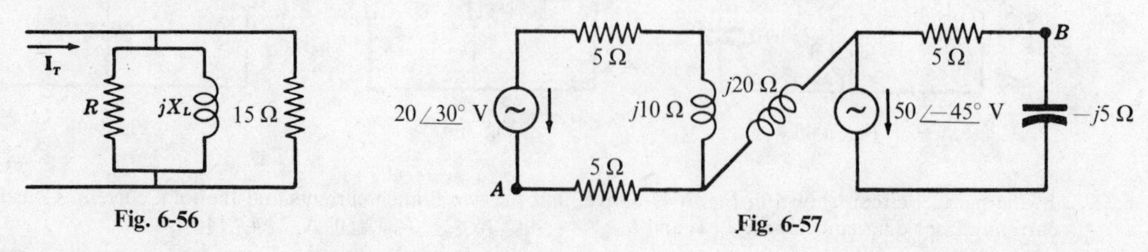

Fig. 6-56 Fig. 6-57

6.45. A voltmeter placed across the 3 ohm resistor shown in Fig. 6-58 reads 45 volts. What is the ammeter reading?
Ans. 19·4 A

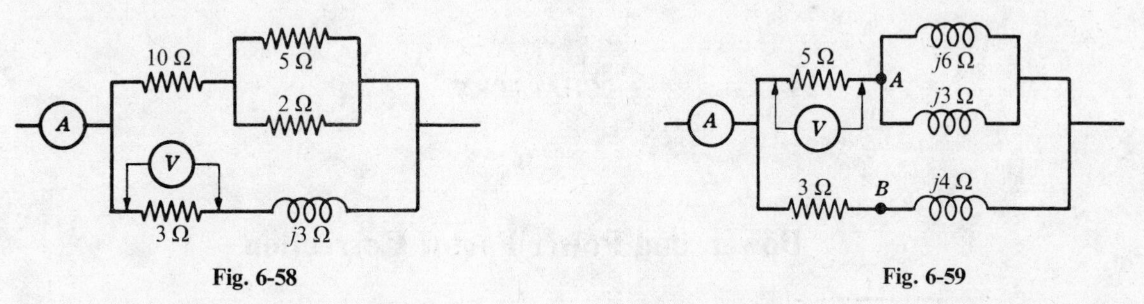

Fig. 6-58 **Fig. 6-59**

6.46. The voltmeter reads 45 volts across the 5 ohm resistor in the circuit shown in Fig. 6-59 above. Find the ammeter reading.
Ans. 18 A

6.47. Referring to the circuit of Problem 6.46, find the effective voltage between the points *A* and *B*. Ans. 25·2 V

6.48. In the circuit of Fig. 6-60 below, the effective voltage between points *A* and *B* is 25 volts. Find the corresponding effective values of **V** and $\mathbf{I}_T$. *Hint.* Assume any convenient **V**′ and determine the corresponding $\mathbf{V}'_{AB}$. Then $V/25 = V'/V'_{AB}$.
Ans. 54·3 V, 14·2 A

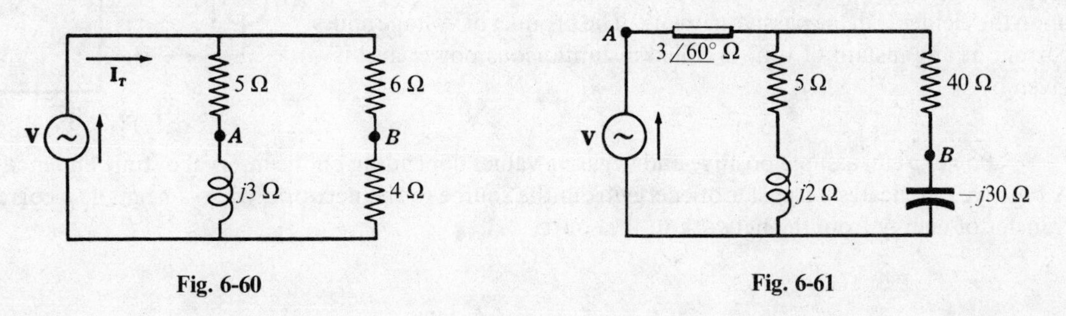

Fig. 6-60 **Fig. 6-61**

6.49. In the parallel circuit shown in Fig. 6-61 above, find the effective value of the voltage source if the potential between *A* and *B* is 50 volts. *Ans.* 54·6 V

6.50. Referring to Fig. 6-62 below, select any convenient values for *R* and X_L. Verify that, for any values given to *R* and X_L, the effective value of $\mathbf{V}_{AB}$ is 50 volts.

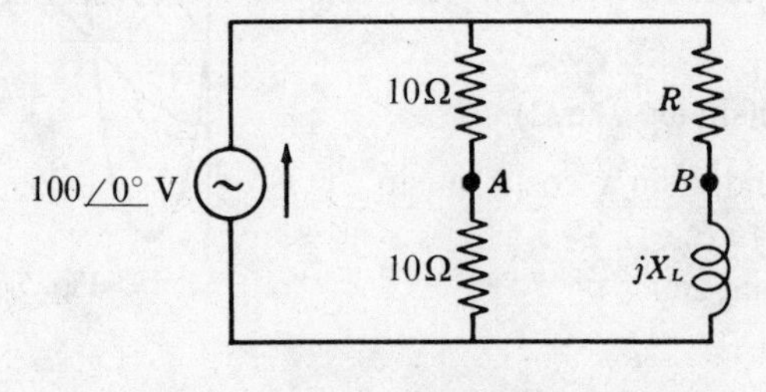

Fig. 6-62

CHAPTER 7

Power, and Power Factor Correction

INTRODUCTION

In many electric devices we are mainly interested in the power. For example, we are interested in the power generated by an alternator, the power input to an electric motor drive, or the power delivered by a radio or television transmitter.

In Fig. 7-1, let the voltage be a time function. The resulting current will also be a time function, and its magnitude will depend upon the elements in the passive network. The product of voltage and current at any instant of time is called instantaneous power and is given by

$$p = vi$$

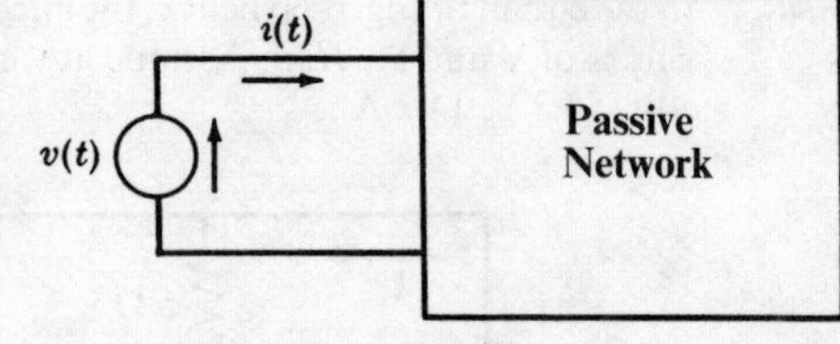

Fig. 7-1

Power p can assume positive and negative values depending on the interval of time under consideration. A positive p indicates a transfer of energy from the source to the network, while a negative p corresponds to a transfer of energy from the network to the source.

POWER IN THE SINUSOIDAL STEADY STATE. AVERAGE POWER (P)

Consider the ideal case where the passive network consists only of an inductive element, and apply to the network a sinusoidal voltage of the form $v = V_m \sin \omega t$. The resulting current will have the form $i = I_m \sin (\omega t - \pi/2)$. Then the power at any instant of time is

$$p = vi = V_m I_m (\sin \omega t)(\sin \omega t - \pi/2)$$

Since $\sin (\omega t - \pi/2) = -\cos \omega t$ and $2 \sin x \cos x = \sin 2x$, we have

$$p = -\tfrac{1}{2} V_m I_m \sin 2\omega t$$

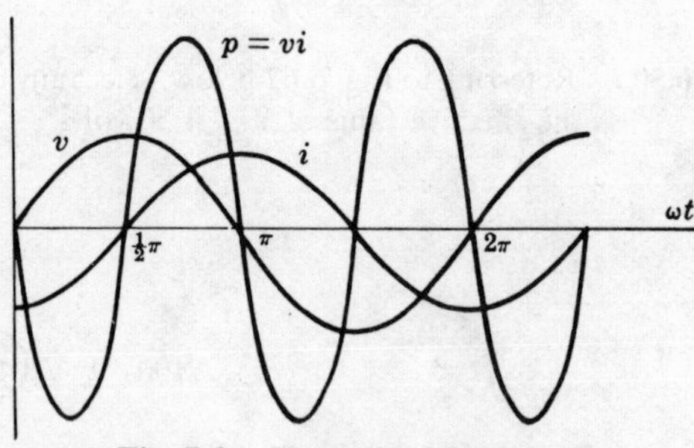

Fig. 7-2. Network of Pure L

The above result is illustrated in Fig. 7-2. When v and i are both positive, the power p is positive and energy is delivered from the source to the inductance. When v and i have opposite sign, the power is negative and energy is returning from the inductance to the source. Power has a frequency twice that of the voltage or current. The average value of the power, indicated with the symbol P, is zero when calculated over a complete cycle.

In the ideal case of a pure capacitive network, analogous results are obtained. Fig. 7-3 illustrates such case.

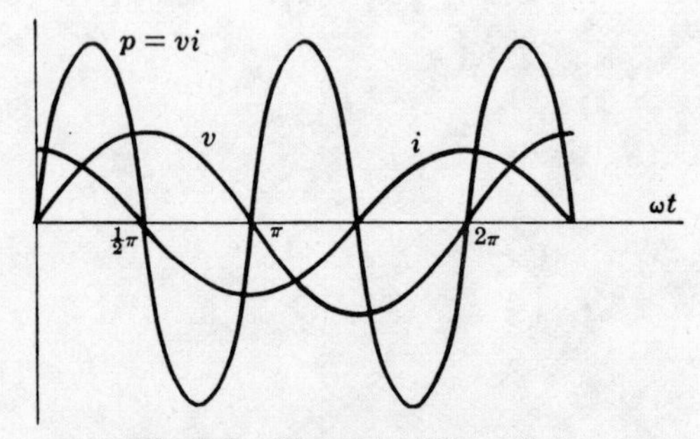

Fig. 7-3. Network of Pure C

Apply now a voltage $v = V_m \sin \omega t$ to a network containing only resistance. The resulting current is $i = I_m \sin \omega t$ and the corresponding power is

$$p = vi = V_m I_m \sin^2 \omega t$$

Since $\sin^2 x = \frac{1}{2}(1 - \cos 2x)$, we have

$$p = \tfrac{1}{2} V_m I_m (1 - \cos 2\omega t)$$

This result is illustrated in Fig. 7-4. Here also we notice that the power has a frequency twice that of the voltage or current. Moreover, the power is always positive and varies from zero to a maximum value of $V_m I_m$. The average value of the power is $\frac{1}{2} V_m I_m$.

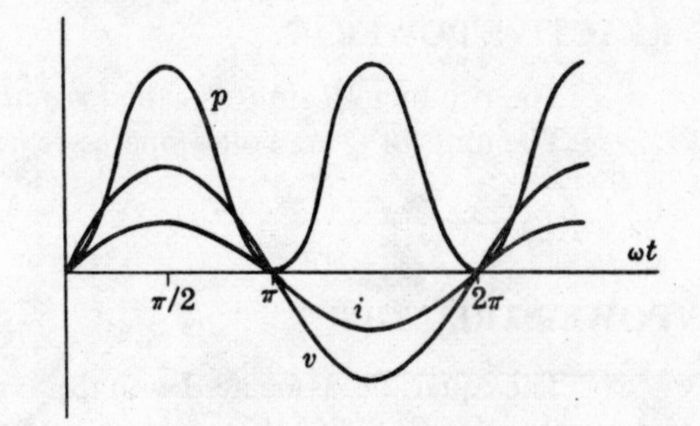

Fig. 7-4. Network of Pure R

Finally consider the case of a general passive network. For an applied sinusoidal voltage $v = V_m \sin \omega t$, we have a resulting current $i = I_m \sin(\omega t + \theta)$. The phase angle θ will be positive or negative depending on the capacitive or inductive character of the network. Then

$$p = vi = V_m I^m \sin \omega t \sin(\omega t + \theta)$$

Since $\sin \alpha \sin \beta = \frac{1}{2}[\cos(\alpha - \beta) - \cos(\alpha + \beta)]$ and $\cos - \alpha = \cos \alpha$,

$$p = \tfrac{1}{2} V_m I_m [\cos \theta - \cos(2\omega t + \theta)]$$

The instantaneous power p consists of a sinusoidal term $-\frac{1}{2} V_m I_m \cos(2\omega t + \theta)$ which has an average value of zero and a constant term $\frac{1}{2} V_m I_m \cos \theta$. Then the average value of p is

$$P = \tfrac{1}{2} V_m I_m \cos \theta = VI \cos \theta$$

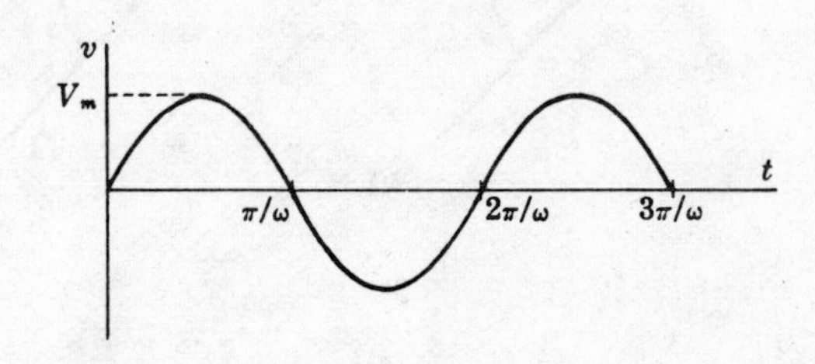

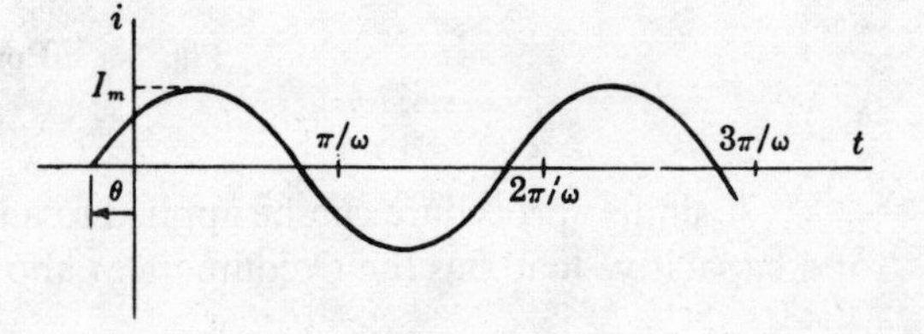

where $V = V_m/\sqrt{2}$ and $I = I_m/\sqrt{2}$ are the effective values of the phasors **V** and **I** respectively. The term $\cos \theta$ is called *power factor*, abbreviated pf. The angle θ is the angle between **V** and **I**, and its value is always between $\pm 90°$. Hence, $\cos \theta$ and consequently P are always positive. However, to indicate the sign of θ, an *inductive circuit*, where the current lags the voltage, has a *lagging power factor*. In a *capacitive circuit* the current leads the voltage, and the circuit has a *leading power factor*.

The average power P can also be obtained from the relation $P = \dfrac{1}{T} \displaystyle\int_0^T p\, dt$.

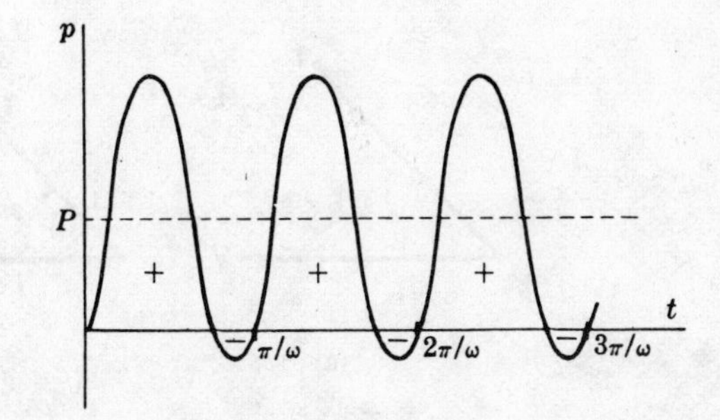

Fig. 7-5

Units of average power are the watt (W) and kilowatt (kW) = 1000 W.

APPARENT POWER (S)

The product VI is called *apparent power* and is indicated by the symbol S. The units of S are volt ampere (VA) and kilovolt ampere (kVA) = 1000 VA.

REACTIVE POWER (Q)

The product $VI \sin \theta$ is called *reactive power* and is indicated by the symbol Q.
The units of Q are volt amperes reactive (var) and kilovar (kvar) $=$ 1000 var.

POWER TRIANGLE

The equations associated with the average, apparent and reactive power can be developed geometrically on a right triangle called the *power triangle*.

Given an inductive circuit, sketch the lagging current and the terminal voltage as shown below in Fig. 7-6(a) with **V** as reference. In Fig. 7-6(b), redraw the current with its in-phase and quadrature components. The in-phase component is in phase with **V** and the quadrature or reactive component is normal to **V** or 90° out of phase. Repeat the diagram in Fig. 7-6(c), multiplying **I**, $I \sin \theta$ and $I \cos \theta$ by the effective voltage V. Then

Average power P $=$ voltage $\times$ in-phase component of current $= VI \cos \theta$

Apparent power S $=$ voltage $\times$ current $= VI$

Reactive power Q $=$ voltage $\times$ quadrature component of current $= VI \sin \theta$

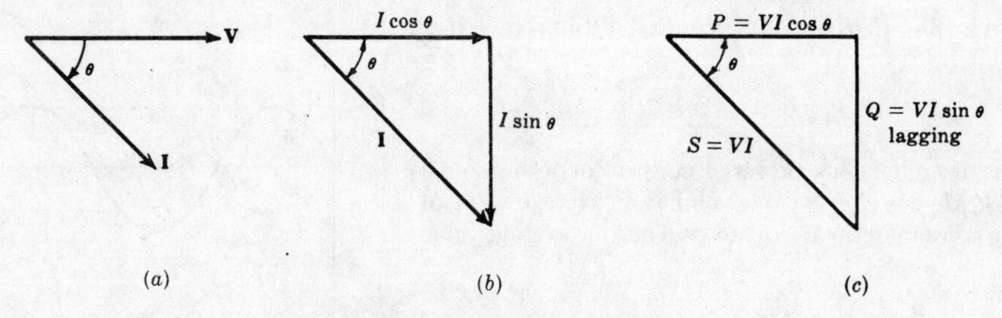

Fig. 7-6. Power triangle, Inductive load

A similar procedure can be applied to a leading current as shown in Fig. 7-7 below. The power triangle for a capacitive load has the Q component above the horizontal.

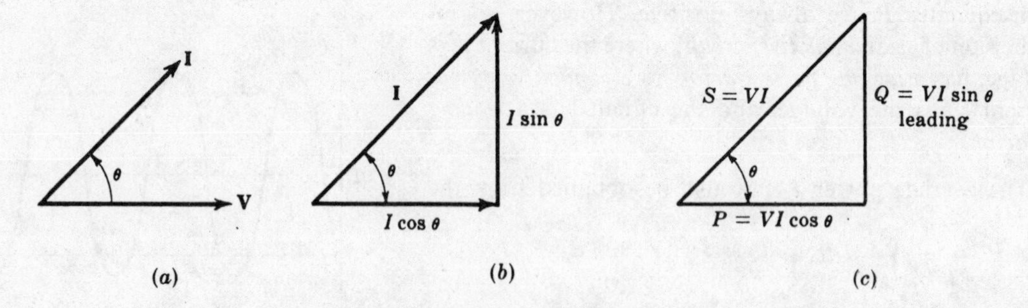

Fig. 7-7. Power triangle, Capacitive load

COMPLEX POWER

The three sides S, P and Q of the power triangle can be obtained from the product $\mathbf{VI}^*$. The result of this product is a complex number called the *complex power* **S**. Its real part equals the average power P and its imaginary part is equal to the reactive power Q.

Consider $\mathbf{V} = Ve^{j\alpha}$ and $\mathbf{I} = Ie^{j(\alpha+\theta)}$. Then

$$\mathbf{S} = \mathbf{VI}^* = Ve^{j\alpha}Ie^{-j(\alpha+\theta)} = VIe^{-j\theta} = VI\cos\theta - jVI\sin\theta = P - jQ$$

The absolute value of **S** is the apparent power $S = VI$. A leading phase angle (**I** leading **V**) determines a leading Q while a lagging phase angle indicates a lagging Q. This consideration should be kept in mind in the construction of the power triangle.

A summary of the equations that can be used to determine the components of the power triangle is given below.

$$\text{Average power } P = VI\cos\theta = I^2R = V_R^2/R = \text{Re } \mathbf{VI}^*$$

$$\text{Reactive power } Q = VI\sin\theta = I^2X = V_X^2/X = \text{Im } \mathbf{VI}^*$$

$$\text{Apparent power } S = VI = I^2Z = V^2/Z = \text{absolute value of } \mathbf{VI}^*$$

$$\text{Power factor pf} = \cos\theta = R/Z = P/S$$

Example 1.

Given a circuit with an impedance $\mathbf{Z} = 3 + j4\ \Omega$ and an applied phasor voltage $\mathbf{V} = 100\angle 30°$ V, determine the power triangle.

The resulting current is $\mathbf{I} = \mathbf{V}/\mathbf{Z} = (100\angle 30°)/(5\angle 53{\cdot}1°) = 20\angle -23{\cdot}1°$ A.

Method 1.

$P = I^2R = (20)^2 3 = 1200$ W
$Q = I^2X = 1600$ var lagging
$S = I^2Z = 2000$ VA
p.f. $= \cos 53{\cdot}1° = 0{\cdot}6$ lagging

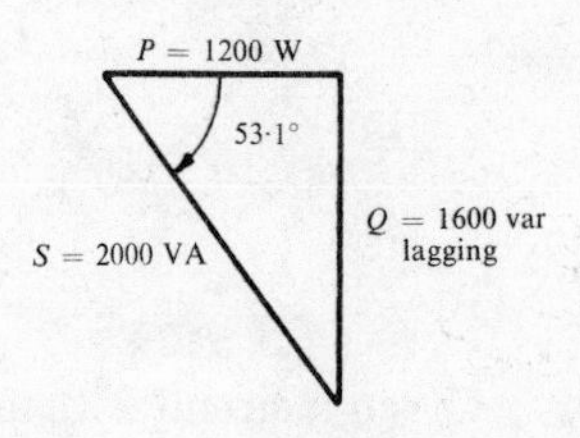

Fig. 7-8

Method 2.

$S = VI = 100(20) = 2000$ VA
$P = VI\cos\theta = 2000\cos 53{\cdot}1° = 1200$ W
$Q = VI\sin\theta = 2000\sin 53{\cdot}1° = 1600$ var lagging
p.f. $= \cos\theta = \cos 53{\cdot}1° = 0{\cdot}6$ lagging

Method 3.

$\mathbf{S} = \mathbf{VI}^* = (100\angle 30°)(20\angle 23{\cdot}1°) = 2000\angle 53{\cdot}1° = 1200 + j1600$ VA from which
$P = 1200$ W, $Q = 1600$ var lagging, $S = 2000$ VA, and p.f. $= \cos 53{\cdot}1° = 0{\cdot}6$ lagging

Method 4.

$$V_R = \mathbf{I}R = 20\angle -23{\cdot}1°(3) = 60\angle -23{\cdot}1°\text{ V},\quad V_X = (20\angle -23{\cdot}1°)(4\angle 90°) = 80\angle 66{\cdot}9°\text{ V}$$

Hence

$$P = V_R^2/R = 60^2/3 = 1200\text{ W}$$
$$Q = V_X^2/X = 80^2/4 = 1600\text{ var lagging}$$
$$S = V^2/Z = 100^2/5 = 2000\text{ VA}$$
$$\text{p.f.} = P/S = 0{\cdot}6\text{ lagging}$$

Care should be taken in substituting values in the equation $P = V_R^2/R$. The most frequent error is to replace V_R, the voltage across the resistor alone, by V, the total voltage across the impedance **Z**.

POWER FACTOR CORRECTION

In ordinary residential and industrial applications, the loads appear inductive, and the current lags the applied voltage. The average power, P, delivered to a load is a measure of the useful work per unit time that the load can perform. This power is usually transmitted through distribution lines and transformers.

Since a transformer, rated in kVA, is often used at a fixed voltage, the kVA rating is merely an indication of the maximum current permitted. Theoretically, if a pure inductive or capacitive load were connected, the transformer could be fully loaded while the average power delivered would be zero.

Referring to the power triangle, the hypotenuse S is a measure of the loading on the distribution system, and the side P is a measure of the useful power delivered. It is therefore desirable to have S as close to P as possible, that is, to make the angle θ approach zero. Since p.f. $= \cos\theta$, the power factor should approach unity. For the usual case of an inductive load, it is often possible to improve the power factor by placing capacitors in parallel with the load. Note that since the voltage across the load remains the same, the useful power P also does not change. Since the power factor is increased, the current and apparent power decrease, and a more efficient utilization of the power distribution system is obtained.

Example 2.

In the circuit of Example 1, correct the power factor to 0·9 lagging by addition of parallel capacitors. Find S' after the correction is introduced, and the var of capacitors required to obtain such correction.

Redraw the power triangle determined in Example 1. Now, $0{\cdot}9 = \cos\theta'$ and $\theta' = 26°$; then

$$S' = P/\cos\theta' = 1200/\cos 26° = 1333 \text{ VA}$$

Since $Q' = S' \sin\theta' = 1333 \sin 26° = 585$ var lagging,

$$\text{Capacitor var} = Q - Q' = 1600 - 585 = 1015 \text{ leading}$$

Since P remains unchanged, the work also remains unchanged after correction of the power factor. However, the value of S has been reduced from 2000 to 1333 VA.

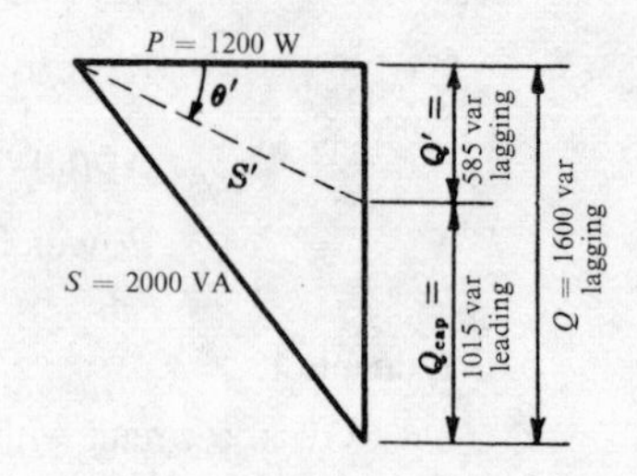

Fig. 7-9

Solved Problems

7.1. Given a circuit with an applied voltage $v = 150 \sin(\omega t + 10°)$ volts and a resulting current $i = 5 \sin(\omega t - 50°)$ amperes, determine the power triangle.

$$\mathbf{V} = (150/\sqrt{2})\angle 10° = 106\angle 10° \text{ V and } \mathbf{I} = (5/\sqrt{2})\angle -50° = 3{\cdot}54\angle -50° \text{ A.}$$

Then

$$\mathbf{S} = \mathbf{VI}^* = (106\angle 10°)(3{\cdot}54\angle 50°) = 375\angle 60° = 187{\cdot}5 + j325 \text{ VA from which}$$

$$P = \text{Re } \mathbf{VI}^* = 187{\cdot}5 \text{ W}$$

$$Q = \text{Im } \mathbf{VI}^* = 325 \text{ var lagging}$$

$$S = |\mathbf{VI}^*| = 375 \text{ VA}$$

$$\text{p.f.} = \cos 60° = 0{\cdot}5 \text{ lagging}$$

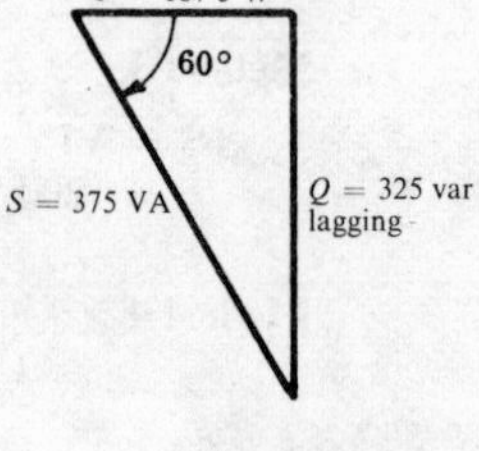

P = 187·5 W; 60°; S = 375 VA; Q = 325 var lagging

Fig. 7-10

7.2. A two element series circuit has a power of 940 watts and a power factor of 0·707 leading. If the applied voltage $v = 99 \sin(6000t + 30°)$ volts, determine the circuit constants.

The phasor form of the applied voltage is $\mathbf{V} = (99/\sqrt{2})\angle 30° = 70\angle 30°$ V. Now the power $P = VI\cos\theta$, $940 = 70I(0{\cdot}707)$ and $I = 19$ amperes. Since the power factor is 0·707 leading, the phasor current must lead the voltage by $\cos^{-1} 0{\cdot}707 = 45°$. Then $\mathbf{I} = 19\angle 75°$ A. The impedance of the circuit $\mathbf{Z} = \mathbf{V}/\mathbf{I} = (70\angle 30°)/(19\angle 75°) = 3{\cdot}68\angle -45° = 2{\cdot}6 - j2{\cdot}6\,\Omega$. Since $\mathbf{Z} = R - jX_C$ and $X_C = 1/\omega C$, we have

$$R = 2{\cdot}6 \text{ ohms and } C = \frac{1}{6000(2{\cdot}6)} = 64{\cdot}1\ \mu\text{F}$$

Alternate method.

Put $I = 19$ amperes in $P = I^2R$ and obtain $940 = (19)^2R$, $R = 2{\cdot}6$ ohms.
Then $\mathbf{Z} = Z\angle -45° = 2{\cdot}6 - jX_C$ and $X_C = 2{\cdot}6\,\Omega$. Hence $C = 1/\omega X_C = 64{\cdot}1\ \mu$F.

7.3. Given the series circuit of Fig. 7-11, determine the power triangle.

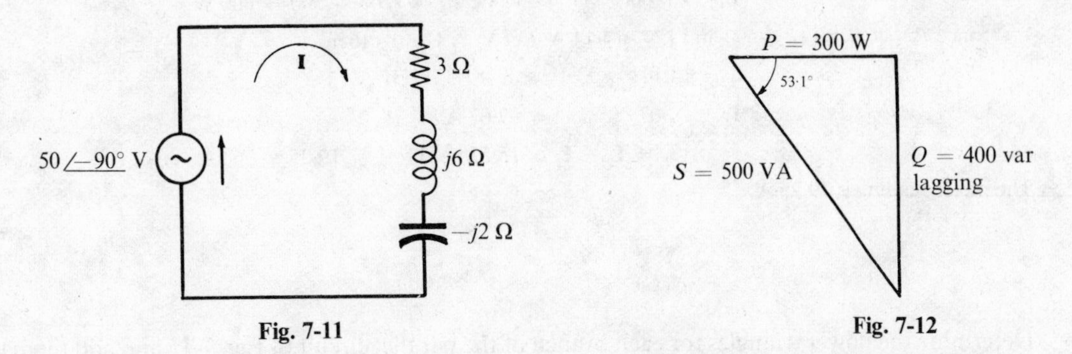

Fig. 7-11 Fig. 7-12

From Fig. 7-11, $\mathbf{Z} = 3 + j6 - j2 = 5\angle 53{\cdot}1^\circ\ \Omega$ and $\mathbf{I} = \mathbf{V}/\mathbf{Z} = (50\angle -90^\circ)/(5\angle 53{\cdot}1^\circ) = 10\angle -143{\cdot}1^\circ$ A. Then

$$\mathbf{S} = \mathbf{VI}^* = (50\angle -90^\circ)(10\angle 143{\cdot}1^\circ) = 500\angle 53{\cdot}1^\circ = 300 + j400 \text{ VA}$$

The components of the power triangle shown in Fig. 7-12 are

$$P = 300 \text{ W},\ Q = 400 \text{ var lagging},\ S = 500 \text{ VA},\ \text{p.f.} = \cos 53{\cdot}1^\circ = 0{\cdot}6 \text{ lagging}$$

Alternate method.

Substituting $I = 10$ amperes in the power equation of each element, $P = I^2R = 10^2(3) = 300$ W, $Q_{j6} = 10^2(6) = 600$ var lagging, $Q_{-j2} = 10^2(2) = 200$ var leading and $Q = Q_{j6} + Q_{-j2} = 600 - 200 = 400$ var lagging.

7.4. In the circuit shown in Fig. 7-13 the total effective current is 30 amperes. Determine the power relations.

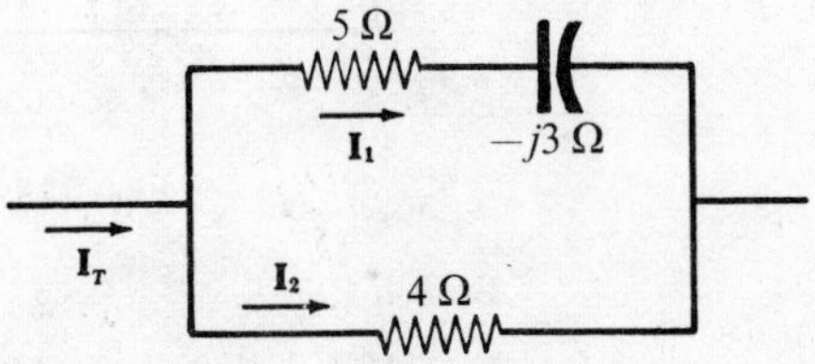

Letting $\mathbf{I}_T = 30\angle 0^\circ$ A, $\mathbf{I}_2 = 30\angle 0^\circ \left(\dfrac{5 - j3}{9 - j3}\right) = 18{\cdot}45\angle -12{\cdot}55^\circ$ A and $\mathbf{I}_1 = 30\angle 0^\circ \left(\dfrac{4}{9 - j3}\right) = 12{\cdot}7\angle 18{\cdot}45^\circ$ A. Then

Fig. 7-13

$$\begin{aligned} P &= I_2^2R_4 + I_1^2R_5 = (18{\cdot}45)^2\,(4) + (12{\cdot}7)^2\,(5) = 2165 \text{ W} \\ Q &= I_1^2X = (12{\cdot}7)^2(3) = 483 \text{ var leading} \\ \mathbf{S} &= P - jQ = 2165 - j483 = 2210\angle -12{\cdot}6^\circ,\ S = 2210 \text{ VA} \\ \text{p.f.} &= P/S = 2165/2210 = 0{\cdot}98 \text{ leading} \end{aligned}$$

The above results can also be found by calculating the equivalent impedance $\mathbf{Z}_{eq} = \dfrac{(5 - j3)4}{9 - j3} = 2{\cdot}4 - j0{\cdot}533\ \Omega$.

Then

$$P = I_T^2R = 30^2(2{\cdot}4) = 2160 \text{ W and } Q = 30^2(0{\cdot}533) = 479{\cdot}7 \text{ var leading}$$

7.5. In the parallel circuit of Fig. 7-14, the total power is 1100 watts. Find the power in each resistor and the reading on the ammeter.

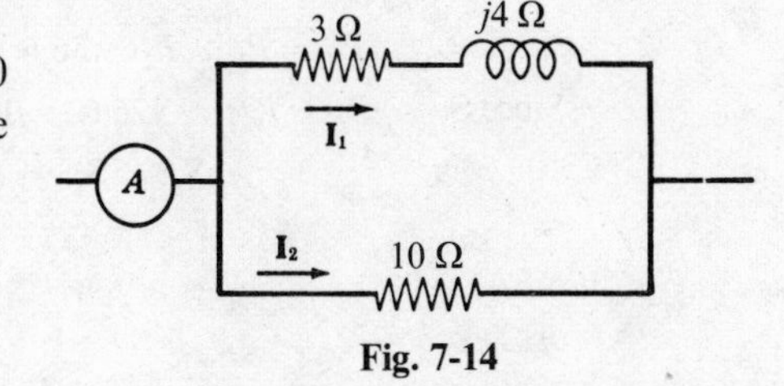

Fig. 7-14

From Fig. 7-14 obtain

$$\mathbf{I}_1 = \frac{\mathbf{V}}{\mathbf{Z}_1} = \frac{\mathbf{V}}{3 + j4} = \frac{\mathbf{V}}{5\angle 53{\cdot}1^\circ},\quad \mathbf{I}_2 = \frac{\mathbf{V}}{\mathbf{Z}_2} = \frac{\mathbf{V}}{10}$$

The ratio of the magnitudes of the currents is $\dfrac{I_1}{I_2} = \dfrac{V/5}{V/10} = \dfrac{2}{1}$. Using the relation $P = I^2R$, the ratio of the powers in the 3 and 10 ohm resistors is

$$\frac{P_3}{P_{10}} = \frac{I_1^2R_1}{I_2^2R_2} = \left(\frac{2}{1}\right)^2 \frac{3}{10} = \frac{6}{5}$$

Now $P_T = P_3 + P_{10}$, from which, dividing both terms by P_{10}, $P_T/P_{10} = P_3/P_{10} + 1$ and

$$P_{10} = 1100(5/11) = 500 \text{ W}, \; P_3 = 1100 - 500 = 600 \text{ W}$$

Since $P = I^2R$, $I_1^2(3) = 600$ and $I_1 = 14{\cdot}14$ A. Let $\mathbf{V} = V\angle 0^\circ$; then

$$\mathbf{I}_1 = 14{\cdot}14\angle -53{\cdot}1^\circ = 8{\cdot}48 - j11{\cdot}31 \text{ A}$$

$$\mathbf{I}_2 = 7{\cdot}07\angle 0^\circ = 7{\cdot}07 \text{ A}$$

and

$$I_T = \mathbf{I}_1 + \mathbf{I}_2 = 15{\cdot}55 - j11{\cdot}31 = 19{\cdot}25\angle -36^\circ \text{ A}$$

The meter reading is 19·25 A.

7.6. Determine the power triangles for each branch of the parallel circuit of Fig. 7-15 and add them to obtain the power triangle for the entire circuit.

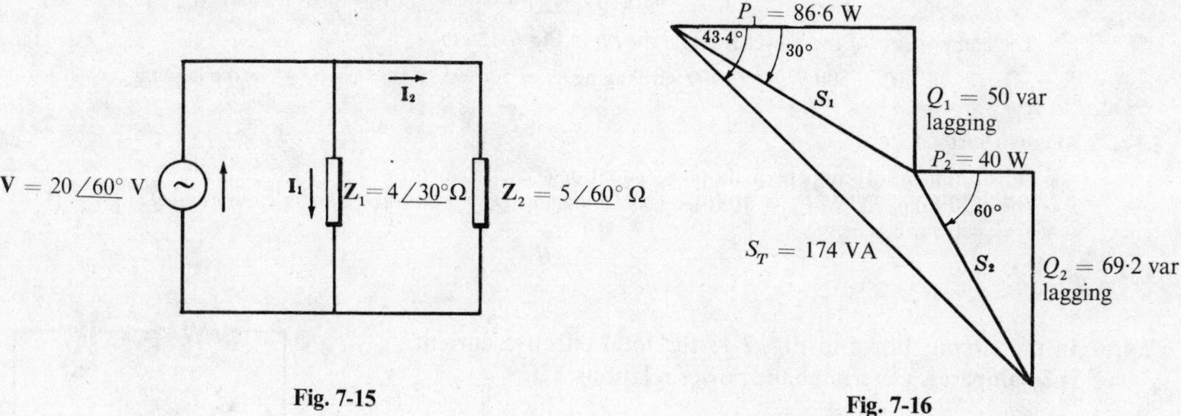

Fig. 7-15

Fig. 7-16

Branch 1.

$$\mathbf{I}_1 = \mathbf{V}/\mathbf{Z}_1 = (20\angle 60^\circ)/(4\angle 30^\circ) = 5\angle 30^\circ \text{ A}$$

$$\mathbf{S}_1 = \mathbf{VI}_1^* = (20\angle 60^\circ)(5\angle -30^\circ) = 100\angle 30^\circ \text{ VA} = 86{\cdot}6 + j50 \text{ VA}$$

Then

$P_1 = \text{Re } \mathbf{VI}_1^* = 86{\cdot}6$ W
$Q_1 = \text{Im } \mathbf{VI}_1^* = 50$ var lagging
$S_1 = |\mathbf{VI}_1^*| = 100$ VA
p.f.$_1 = P_1/S_1 = 0{\cdot}866$ lagging

Branch 2.

$$\mathbf{I}_2 = \mathbf{V}/\mathbf{Z}_2 = (20\angle 60^\circ)/(5\angle 60^\circ) = 4\angle 0^\circ \text{ A}$$

$$\mathbf{S}_2 = \mathbf{VI}_2^* = (20\angle 60^\circ)(4\angle 0^\circ) = 80\angle 60^\circ \text{ VA} = 40 + j69{\cdot}2 \text{ VA}$$

Then

$P_2 = 40$ W
$Q_2 = 69{\cdot}2$ var lagging
$S_2 = 80$ VA
p.f.$_2 = 0{\cdot}5$ lagging

From the above results and referring to Fig. 7-16, obtain the total power triangle as follows:

$$P_T = P_1 + P_2 = 86{\cdot}6 + 40 = 126{\cdot}6 \text{ W}, \; Q_T = Q_1 + Q_2 = 50 + 69{\cdot}2 = 119{\cdot}2 \text{ var lagging}$$

Since $\mathbf{S}_T = P_T + jQ_T = 126{\cdot}6 + j119{\cdot}2 = 174\angle 43{\cdot}4^\circ$ VA,

$$S_T = |\mathbf{S}_T| = 174 \text{ VA and p.f.}_T = P_T/S_T = 126{\cdot}6/174 = 0{\cdot}727 \text{ lagging}$$

7.7. An induction motor with a 2 hp output has an efficiency of 85%. At this load the power factor is 0·8 lagging. Determine the complete input power information.

Since 1 hp = 746 watts, $P_{in} = 2(746)/0{\cdot}85 = 1755$ W. Then

$$S = 1755/0{\cdot}8 = 2190 \text{ VA}, \; \theta = \cos^{-1}(0{\cdot}8) = 36{\cdot}9^\circ, \; Q = 2190 \sin 36{\cdot}9^\circ = 1315 \text{ var lagging}$$

7.8. Determine the total power triangle for the parallel circuit of Fig. 7-17 if the power in the 2 ohm resistor is 20 watts.

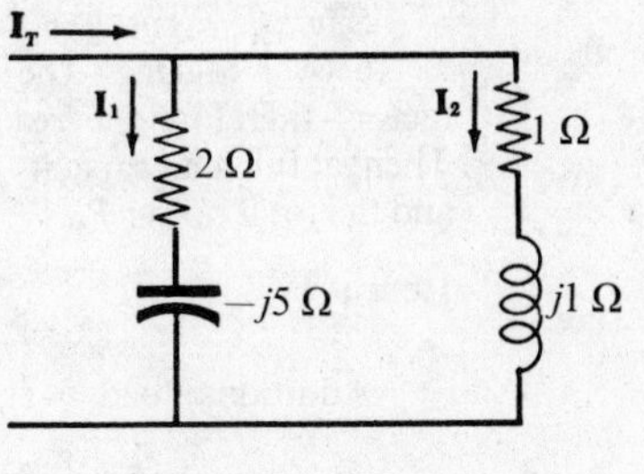

Fig. 7-17

From $P = I^2R$ we have $I_1^2(2) = 20$ A and $I_1 = 3{\cdot}16$ A. Since $\mathbf{Z}_1 = 2 - j5 = 5{\cdot}38\angle{-68{\cdot}2^\circ}\ \Omega$, $V = I_1Z = 3{\cdot}16(5{\cdot}38) = 17$ volts. Let $\mathbf{V} = 17\angle 0^\circ$ V; then

$$\mathbf{I}_1 = 3{\cdot}16\angle 68{\cdot}2^\circ \text{ A},\ \mathbf{I}_2 = \mathbf{V}/\mathbf{Z}_2 = (17\angle 0^\circ)/(\sqrt{2}\angle 45^\circ)\text{ A}$$

and
$$\mathbf{I}_T = \mathbf{I}_1 + \mathbf{I}_2 = 11{\cdot}1\angle{-29{\cdot}8^\circ}\text{ A}$$

To calculate the power triangle components, S_T is needed. Thus

$$S_T = \mathbf{VI}_T^* = 17\angle 0^\circ(11{\cdot}1\angle 29{\cdot}8^\circ = 189\angle 29{\cdot}8^\circ = 164 + j94\text{ VA}$$

from which

$$P_T = 164\text{ W},\ Q_T = 94\text{ var lagging},\ S_T = 189\text{ VA, p.f.} = 164/189 = 0{\cdot}868\text{ lagging}$$

7.9. Determine the power components of a combination of three individual loads specified as follows: Load 1, 250 VA, p.f. 0·5 lagging; Load 2, 180 W, p.f. 0·8 leading; Load 3, 300 VA, 100 var lagging.

Calculate the unknown average power and reactive power of each load. Thus.

Load 1. Given $S = 250$ VA, p.f. $= 0{\cdot}5$ lagging.
$P = S\,\text{p.f.} = 250(0{\cdot}5) = 125$ W, $\theta = \cos^{-1}0{\cdot}5 = 60^\circ$, $Q = S\sin\theta = 250\sin 60^\circ = 216$ var lagging

Load 2. Given $P = 180$ W, p.f. $= 0{\cdot}8$ leading.
$S = P/\text{p.f.} = 180/0{\cdot}8 = 225$ VA, $\theta = \cos^{-1}0{\cdot}8 = 36{\cdot}9^\circ$, $Q = 225\sin 36{\cdot}9^\circ = 135$ var leading

Load 3. Given $S = 300$ VA, $Q = 100$ var lagging
$\theta = \sin^{-1}(Q/S) = \sin^{-1}(100/300) = 19{\cdot}5^\circ$, $P = S\cos\theta = 300\cos 19{\cdot}5^\circ = 283$ W

Then $$P_T = 125 + 180 + 283 = 588\text{ W},\ Q_T = 216 - 135 + 100 = 181\text{ var lagging}$$

Since $\mathbf{S}_T = P_T + jQ_T = 588 + j181 = 616\angle 17{\cdot}1^\circ$ VA,

$$S_T = 616\text{ VA}\quad\text{and}\quad\text{p.f.} = P/S = 588/616 = 0{\cdot}955\text{ lagging}$$

Fig. 7-18 shows the power triangles of the individual and combined loads.

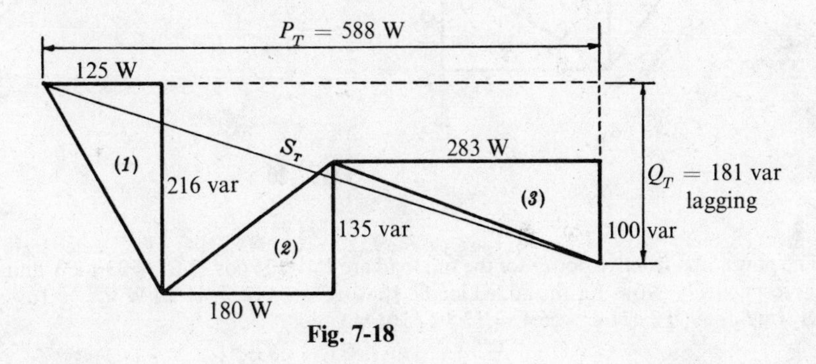

Fig. 7-18

7.10. A 25 kVA transformer supplies a load of 12 kW at a power factor of 0·6 lagging. Find the percentage of full load the transformer is carrying. If additional unity power factor loads are to be served with the same transformer, how many kW may be added before the transformer is at full load?

For the 12 kW load, $S = P/\text{p.f.} = 12/0{\cdot}6 = 20$ kVA. Then

$$\%\text{ full load} = (20/25)100 = 80\%$$

Since $\theta = \cos^{-1}0{\cdot}6 = 53{\cdot}1°$, $Q = S \sin\theta = 20 \sin 53{\cdot}1° = 16$ kvar lagging. The additional loads have unity power factor; therefore the reactive power Q remains unchanged. Then, at full load capacity, the angle $\theta' = \sin^{-1}(16/25) = 39{\cdot}8°$ and the total power $P_T = S' \cos\theta' = 25 \cos 39{\cdot}8° = 19{\cdot}2$ kW.

Hence

$$\text{Additional load} = P_T - P = 19{\cdot}2 - 12 = 7{\cdot}2 \text{ kW}$$

The above result can also be obtained graphically as shown in Fig. 7-19.

Notice that with the addition of loads at unity power factor the overall power factor is improved, i.e. p.f. = cos 39·8° 0·768 lagging.

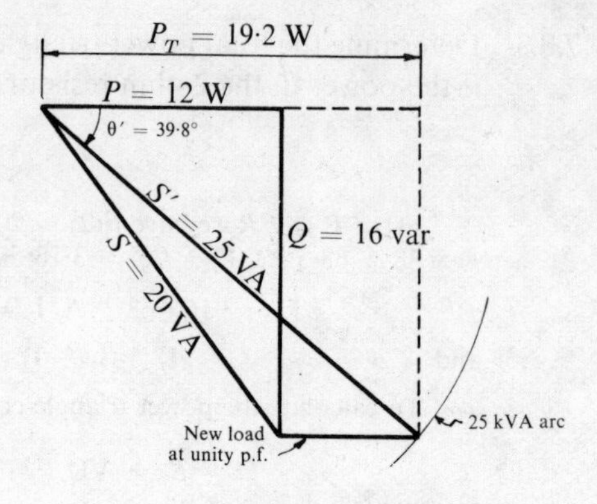

Fig. 7-19

7.11. Referring to Problem 7.10, if the additional loads have a power factor of 0·866 leading, how many kVA of these loads can be added to bring the transformer to its full load capacity?

From Problem 7.11, $S = 20$ kVA, $\theta = 53{\cdot}1°$, $Q = 16$ kvar lagging. Draw the power triangle as shown in Fig. 7-20(a). With the S_2 of the new loads added to an angle $\theta_2 = \cos^{-1}0{\cdot}866 = 30°$ the angle θ is needed. Referring to Fig. 7-20(b), we have

$$25/\sin 96{\cdot}9° = 20/\sin\beta,\ \sin\beta = 0{\cdot}795,\ \beta = 52{\cdot}6°$$

Then $\gamma = 180° - (96{\cdot}9° + 52{\cdot}6°) = 30{\cdot}5°$ and $\theta' = 53{\cdot}1° - 30{\cdot}5° = 22{\cdot}6°$.

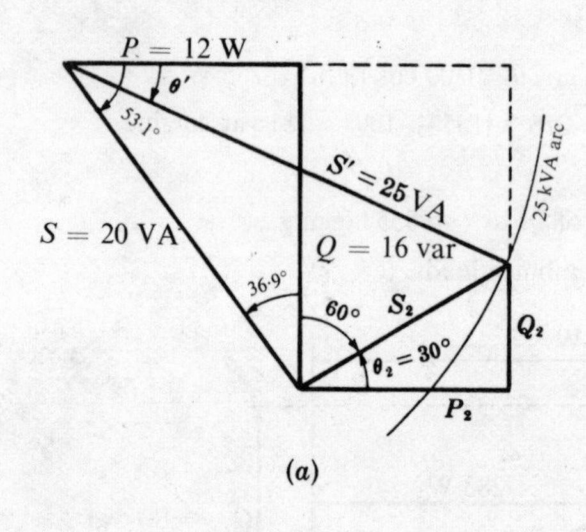

(a)

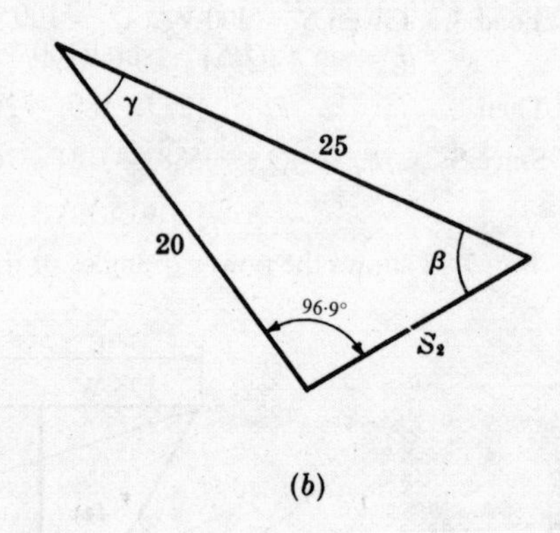

(b)

Fig. 7-20

The power and reactive power for the full load are $P_T = 25 \cos 22{\cdot}6° = 23{\cdot}1$ kW and $Q_T = 25 \sin 22{\cdot}6° = 9{\cdot}6$ kvar lagging respectively. Now, for the added loads, $P_2 = 23{\cdot}1 - 12 = 11{\cdot}1$ kW, $Q_2 = 16 - 9{\cdot}6 = 6{\cdot}4$ kvar leading and, since $\mathbf{S}_2 = P_2 + jQ_2 = 11{\cdot}1 - j6{\cdot}4 = 12{\cdot}8\angle{-30°}$ VA,

$$S_2 = 12{\cdot}8 \text{ kVA}$$

Thus 12·8 kVA of new loads with a power factor of 0·866 leading may be added to the 12 kW at 0·6 p.f. lagging to bring the transformer up to its rated 25 kVA.

Another method. From Fig. 7-20(a), for an angle $\theta_2 = 30°$,

$$P_2 = S_2 \cos 30° = (\sqrt{3/2})S_2,\ Q_2 = S_2 \sin 30° = \tfrac{1}{2}S_2$$

Now

$$(S')^2 = (P + P_2)^2 + (Q - Q_2)^2$$

Substituting, $(25)^2 = (12 + \sqrt{3/2}\, S_2)^2 + (16 - \tfrac{1}{2}S_2)^2$ and $S_2 = 12{\cdot}8$ kVA

7.12. A 500 kVA transformer is at full load with an overall power factor of 0·6 lagging. The power factor is improved by adding capacitors until the overall power factor becomes 0·9 lagging. Determine the kvar of capacitors required. After correction of the power factor, what percentage of full load is the transformer carrying?

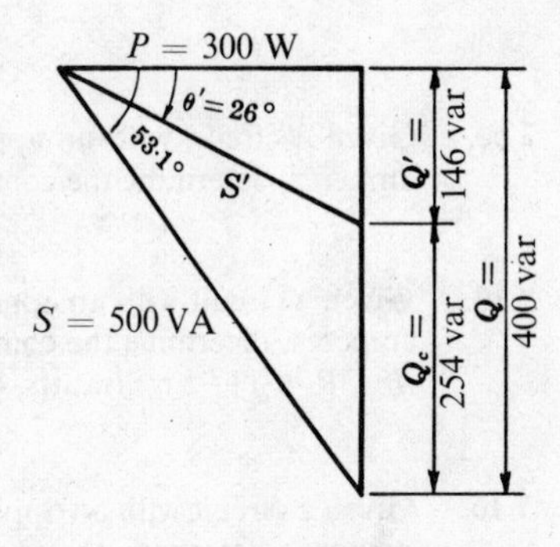

Fig. 7-21

For the transformer at full load (see Fig. 7-21),

$P = VI\cos\theta = 500(0{\cdot}6) = 300$ kW

$\theta = \cos^{-1}0{\cdot}6 = 53{\cdot}1°$

$Q = VI\sin\theta = 500\sin 53{\cdot}1° = 400$ kvar lagging

When p.f. = 0·9 lagging,

$$\theta' = \cos^{-1}0{\cdot}9 = 26°,\ S' = 300/0{\cdot}9 = 333\text{ kVA},\ Q' = 333\sin 26° = 146\text{ kvar lagging}$$

Hence $\qquad$ Capacitor kvar $= Q - Q' = 400 - 146 = 254$ leading

and $\qquad$ % full load $= (333/500)100 = 66{\cdot}7\%$

7.13. A group of induction motors with a total of 500 kW and a power factor of 0·8 lagging is to be partially replaced with synchronous motors of the same efficiency but leading power factor of 0·707. As the replacement programme continues, the overall power factor is constantly improving. What percentage of the load will have been replaced when the system power factor reaches 0·9 lagging?

Since the synchronous motors have the same efficiency as the induction motors, the total average power remains constant at 500 kW. Before replacement of the motors,

$$S = 500/0{\cdot}8 = 625\text{ kVA},\ \theta = \cos^{-1}0{\cdot}8 = 36{\cdot}9°,\ Q = 625\sin 36{\cdot}9° = 375\text{ kvar lagging}$$

When the system p.f. = 0·9 lagging,

$$\theta' = \cos^{-1}0{\cdot}9 = 26°,\ S' = 500/0{\cdot}9 = 556\text{ kVA},\ Q' = 556\sin 26° = 243\text{ kvar lagging}$$

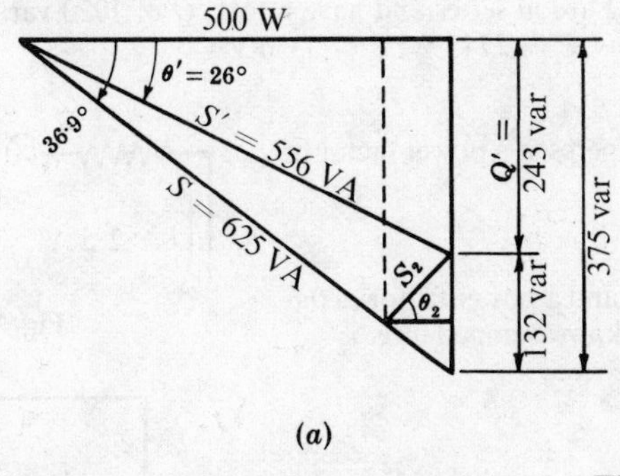

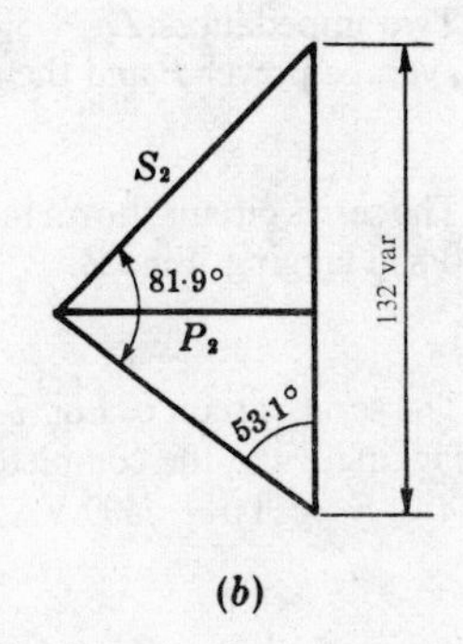

Fig. 7-22

With the power factor of the replacing motors equal to 0·707 leading, $\theta_2 = \cos^{-1}0{\cdot}707 = 45°$. Referring to Fig. 7-22(*b*), apply the law of sines to obtain

$$S_2/\sin 53{\cdot}1° = 132/\sin 81{\cdot}9°,\ S_2 = 106{\cdot}5\text{ kVA}$$

Then $P_2 = 106{\cdot}5\cos 45° = 75{\cdot}3$ kW and

$$\%\text{ load replaced} = (75{\cdot}3/500)100 = 15\%$$

Supplementary Problems

7.14. Given a circuit with an applied voltage $v = 200 \sin(\omega t + 110°)$ volts and a resulting current $i = 5 \sin(\omega t + 20°)$ amperes, determine the complete power triangle. *Ans.* $P = 0$, $Q = 500$ var lagging

7.15. Given a circuit with an applied voltage $v = 14{\cdot}14 \cos \omega t$ volts and resulting current $i = 17{\cdot}1 \cos(\omega t - 14{\cdot}05°)$ milliamperes, determine the complete power triangle.
Ans. $P = 117{\cdot}5$ milliwatts, $Q = 29{\cdot}6$ mvar lagging, p.f. $= 0{\cdot}97$ lagging

7.16. Given a circuit with an applied voltage $v = 340 \sin(\omega t - 60°)$ volts and a resulting current $i = 13{\cdot}3 \sin(\omega t - 48{\cdot}7°)$ amperes, determine the complete power triangle. *Ans.* $P = 2215$ W, $Q = 442$ var leading, p.f. $= 0{\cdot}98$ leading

7.17. A two element series circuit of $R = 10$ ohms and $X_C = 5$ ohms has an effective applied voltage of 120 V. Determine the power triangle. *Ans.* $\mathbf{S} = 1154 - j577$ VA, p.f. $= 0{\cdot}894$ leading

7.18. A two element series circuit of $R = 5$ ohms and $X_L = 15$ ohms has an effective voltage across the resistor of 31·6 V. Determine the power triangle. *Ans.* $\mathbf{S} = 200 + j600$ VA, p.f. $= 0{\cdot}316$ lagging

7.19. A series circuit of $R = 8$ ohms and $X_C = 6$ ohms has an applied phasor voltage $\mathbf{V} = 50\angle{-90°}$ V. Find the complete power information. *Ans.* $\mathbf{S} = 200 - j150$ VA, p.f. $= 0{\cdot}8$ leading

7.20. Determine the circuit impedance which takes 5040 volt amperes at a power factor of 0·894 leading from an applied phasor voltage $\mathbf{V} = 150\angle 45°$ V. *Ans.* $4 - j2\ \Omega$

7.21. An impedance carries an effective current of 18 amperes which results in 3500 volt amperes at a power factor of 0·76 lagging. Find the impedance. *Ans.* $8{\cdot}21 + j7{\cdot}0\ \Omega$

7.22. A two element series circuit with an instantaneous current $i = 4{\cdot}24 \sin(5000t + 45°)$ amperes has a power of 180 watts and a power factor of 0·8 lagging. Find the circuit constants. *Ans.* $R = 20$ ohms, $L = 3$ mH

7.23. Two impedances $\mathbf{Z}_1 = 5{\cdot}83\angle{-59°}\ \Omega$ and $\mathbf{Z}_2 = 8{\cdot}95\angle 63{\cdot}4°\ \Omega$ are in series and pass an effective current of 5 amperes. Determine the complete power information. *Ans.* $\mathbf{S}_T = 175 + j75$ VA, p.f. $= 0{\cdot}918$ lagging

7.24. Two impedances $\mathbf{Z}_1 = 5\angle 45°\ \Omega$ and $\mathbf{Z}_2 = 10\angle 30°\ \Omega$ are in series and have a total Q of 1920 var lagging. Find the average power P and the apparent power S. *Ans.* $P = 2745$ W, $S = 3350$ VA

7.25. The series circuit shown in Fig. 7-23 takes 36·4 volt amperes at a power factor of 0·856 lagging. Find **Z**. *Ans.* $\mathbf{Z} = 1\angle 90°\ \Omega$

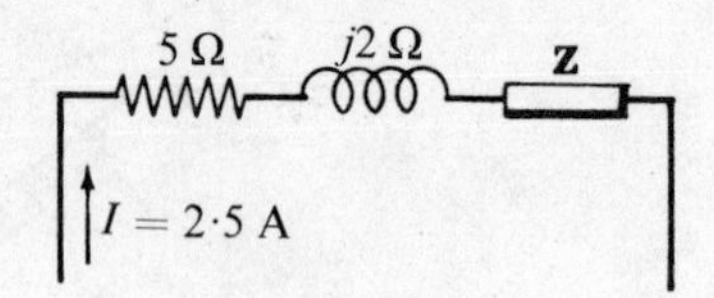

Fig. 7-23

7.26. The series circuit of Fig. 7-24 has a power of 300 watts and a power factor of 0·6 lagging. Find the complete power triangle and the unknown impedance.
Ans. $\mathbf{S} = 300 + j400$ VA, $\mathbf{Z} = 4\angle 90°\ \Omega$

7.27. Two impedances $\mathbf{Z}_1 = 4\angle{-30°}\ \Omega$ and $\mathbf{Z}_2 = 5\angle 60°\ \Omega$ are in parallel and have an applied phasor voltage $\mathbf{V} = 20\angle 0°$ V. Find the power triangle of each branch and combine them to determine the total power triangle.
Ans. $P = 126{\cdot}6$ W, $Q = 19{\cdot}3$ var lagging, p.f. $= 0{\cdot}99$ lagging

7.28. A circuit consisting of $R = 10$ ohms in parallel with $\mathbf{Z} = 8\angle{-30°}\ \Omega$ has a total effective current of 5 amperes. Find the complete power triangle.
Ans. $P = 110$ W, $Q = 33$ var leading, p.f. $= 0{\cdot}957$ leading

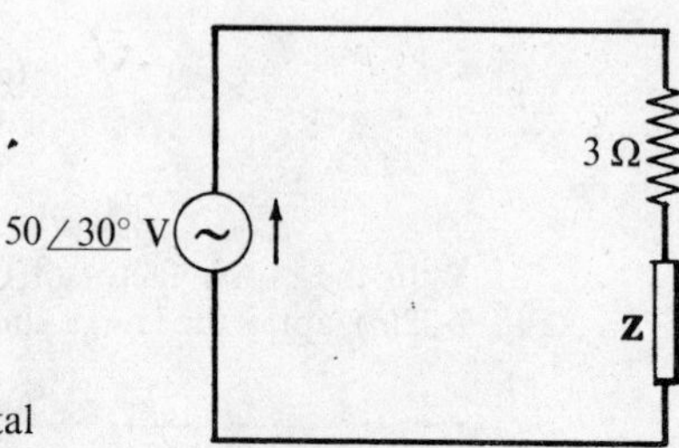

Fig. 7-24

7.29. If branch *1* of the parallel circuit of Fig. 7-25 below contains 8 kvar, find the power and the power factor of the complete circuit. *Ans.* 8 kW, p.f. = 0·555 lagging

7.30. If branch *2* of the parallel circuit of Fig. 7-26 below contains 1490 volt amperes, what will be the indication on the ammeter? Find the complete power information. *Ans.* 42·4 A, $\mathbf{S} = 2210 + j3630$ VA, p.f. = 0·521 lagging

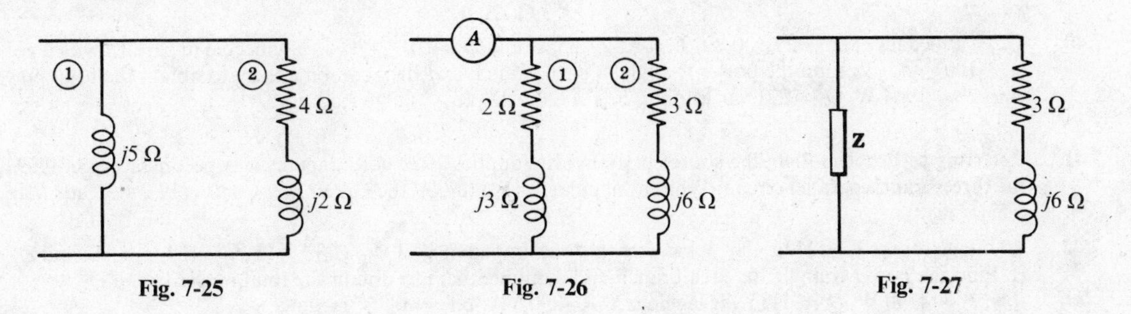

Fig. 7-25 Fig. 7-26 Fig. 7-27

7.31. In the parallel circuit of Fig. 7-27 above, the power in the 3 ohm resistor is 666 watts and the total circuit takes 3370 volt amperes at a power factor of 0·937 leading. Find **Z**. *Ans.* $\mathbf{Z} = 2 - j2\ \Omega$

7.32. The parallel circuit shown in Fig. 7-28 below has a total power of 1500 watts. Obtain the complete power triangle. *Ans.* $\mathbf{S} = 1500 + j2480$ VA, p.f. = 0·518 lagging

7.33. If the total power in the circuit of Fig. 7-29 below is 2000 watts, what is the power in each of the resistors? *Ans.* $P_{15} = 724$ W, $P_8 = 1276$ W

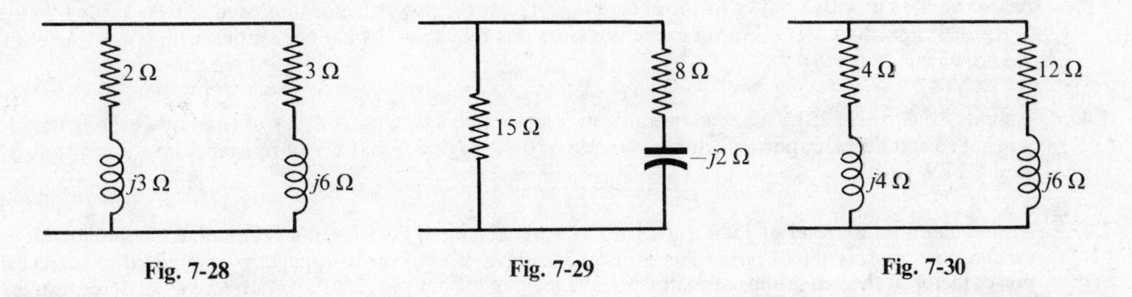

Fig. 7-28 Fig. 7-29 Fig. 7-30

7.34. The total Q in the parallel circuit shown in Fig. 7-30 above is 2500 var lagging. Find the complete power triangle. *Ans.* $S = 3920$ VA, $P = 3020$ W, p.f. = 0·771 lagging

7.35. Find the power factor of the given parallel circuit in Fig. 7-31 below. If the 6 ohm resistor is changed such that the overall power factor is 0·9 lagging, what will be its new ohmic value? *Ans.* p.f. = 0·8 lagging, $R = 3{\cdot}22$ ohms

7.36. In the circuit of Fig. 7-32 below, $\mathbf{Z} = 5 + j8{\cdot}66\ \Omega$ is the original load. With the power factor of the circuit improved by the addition of a $-j20\ \Omega$ capacitor in parallel, find the per cent reduction in total current. *Ans.* 38%

7.37. In the parallel circuit shown in Fig. 7-33 below, find the capacitance C necessary to correct the power factor to 0·95 lagging. *Ans.* $C = 28{\cdot}9\ \mu$F

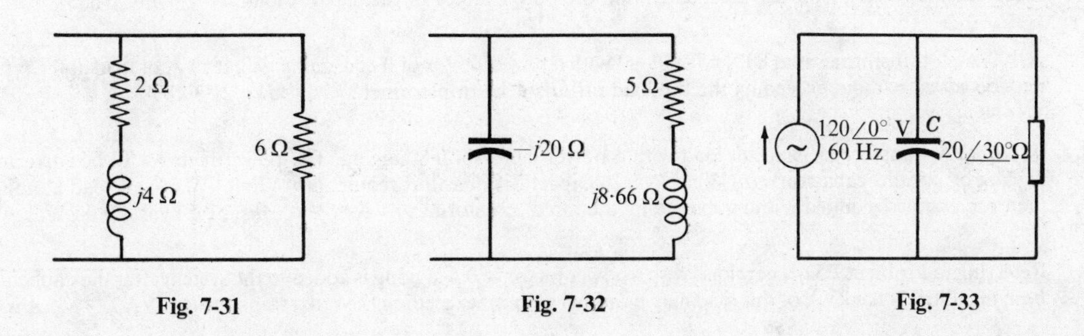

Fig. 7-31 Fig. 7-32 Fig. 7-33

7.38. A 60 Hz source with an effective voltage of 240 volts supplies 4500 volt amperes to a load with a power factor of 0·75 lagging. Determine the parallel capacitance required to improve the power factor to (*a*) 0·9 lagging and (*b*) 0·9 leading.
Ans. (*a*) 61·3 μF, (*b*) 212 μF

7.39. Referring to Problem 7.38, what per cent reduction in line current resulted in part (*a*)? Is there any further reduction in current in part (*b*)? *Ans.* 16·7%. No, the currents are the same.

7.40. Three impedances $\mathbf{Z}_1 = 20\angle 30^\circ\ \Omega$, $\mathbf{Z}_2 = 15\angle -45^\circ\ \Omega$ and $\mathbf{Z}_3 = 10\angle 0^\circ\ \Omega$ are connected in parallel with a voltage source $\mathbf{V} = 100\angle -45^\circ$ V. Find the power triangle of each branch and then combine them to obtain the total power triangle.
Ans. $P = 1904$ W, $Q = 221$ var leading, $S = 1920$ VA, p.f. $= 0{\cdot}993$ leading

7.41. Referring to Problem 7.40, the source of 100 volts supplies 1920 volt amperes at a power factor of 0·993 leading to the three-branch parallel circuit. What total current is taken by the circuit? *Ans.* 19·2 A, leads **V** by 6·62°

7.42. A voltage source $\mathbf{V} = 240\angle -30^\circ$ V has three parallel impedances $\mathbf{Z}_1 = 25\angle 15^\circ\ \Omega$, $\mathbf{Z}_2 = 15\angle -60^\circ\ \Omega$ and $\mathbf{Z}_3 = 15\angle 90^\circ\ \Omega$. Find the power triangle for each branch and combine them to obtain the total power triangle.
Ans. $P = 4140$ W, $Q = 1115$ var lagging, $S = 4290$ VA, p.f. $= 0{\cdot}967$ lagging

7.43. Obtain the total power triangle for the following three loads: load 1, 5 kW at a power factor of 0·8 lagging; load 2, 4 kVA with Q of 2 kvar leading; load 3, 6 kVA at a power factor of 0·9 lagging.
Ans. $P = 13{\cdot}86$ kW, $Q = 4{\cdot}38$ kvar lagging, $S = 14{\cdot}55$ kVA, p.f. $= 0{\cdot}965$ lagging

7.44. Obtain the total power triangle for the following three loads: load 1, 200 VA at a power factor of 0·7 lagging; load 2, 350 VA at a power factor of 0·5 lagging; load 3, 275 VA at unity power factor.
Ans. $P = 590$ W, $Q = 446$ var lagging, $S = 740$ VA, p.f. $= 0{\cdot}798$ lagging

7.45. A load of 300 kW with a power factor of 0·65 lagging has the power factor improved to 0·90 lagging by the addition of parallel capacitors. What kvar of capacitors does this require and what per cent reduction in kVA results?
Ans. 204 kvar, 28%

7.46. An industrial load of 25 kVA has an overall power factor of 0·8 lagging. A group of resistance heating units (unity p.f.) is installed and the plant power factor is found to be 0·85 lagging. What kW of resistance heat was installed?
Ans. 4·3 kW

7.47. An induction motor load of 1500 watts with a power factor of 0·75 lagging is combined with 500 volt amperes of synchronous motors with a power factor of 0·65 leading. What kvar of capacitors is required to correct the overall power factor of the two groups of motors to 0·95 lagging? What per cent reduction in volt amperes results?
Ans. 347 var, 6·3%

7.48. The power factor of a certain load is corrected to 0·9 lagging with the addition of 20 kvar of capacitors. If the final kVA is 185, determine the power triangle of the load before correction?
Ans. $P = 166{\cdot}5$ kW, $Q = 101{\cdot}0$ kvar lagging, p.f. $= 0{\cdot}856$ lagging

7.49. A 2000 volt ampere induction motor load with a power factor of 0·80 lagging is combined with 500 volt amperes of synchronous motors. If the overall power factor is 0·90 lagging, find the power factor of the synchronous motors.
Ans. 0·92 leading

7.50. A 65 kVA load with a lagging power factor is added to 25 kVA of synchronous motors with leading power factor of 0·6. If the overall power factor is 0·85 lagging, find the power factor of the 65 kVA load. *Ans.* 0·585.

7.51. A 100 kVA transformer is at 80% of full load with a power factor of 0·85 lagging. What kVA of load at 0·6 p.f. lagging may be added without exceeding the full load rating of the transformer? *Ans.* 21·3 kVA

7.52. A 250 kVA transformer is at full load with a power factor of 0·8 lagging. The power factor is to be corrected to 0·9 lagging by parallel capacitors. (*a*) What kvar of capacitors does this require? (*b*) What kW of new load at unity power factor may now be added without exceeding the rated transformer kVA? *Ans.* 52·5 kvar, 30·0 kW

7.53. Referring to Problem 7.52, a new load with a power factor of 0·5 lagging is added to the system after the capacitors have been installed. What kVA of this load may be added without exceeding the rated transformer kVA? *Ans.* 32 kVA

CHAPTER 8

Series and Parallel Resonance

INTRODUCTION

A circuit is said to be in resonance when the applied voltage **V** and the resulting current **I** are in phase. Thus at resonance, the equivalent complex impedance of the circuit consists of only resistance R.

Since **V** and **I** are in phase, the power factor of a resonant circuit is unity.

SERIES RESONANCE

The RLC series circuit of Fig. 8-1 has a complex impedance $\mathbf{Z} = R + j(\omega L - 1/\omega C) = R + jX$. The circuit is in resonance when $X = 0$, i.e. when $\omega L = 1/\omega C$ or $\omega = 1/\sqrt{LC} = \omega_0$. Then, since $\omega = 2\pi f$, the resonant frequency is given by

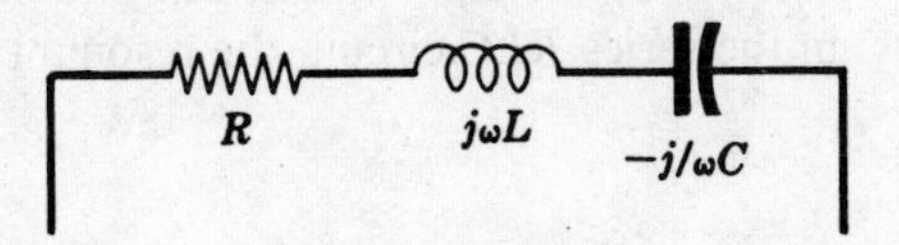

Fig. 8-1

$$f_0 = \frac{1}{2\pi\sqrt{LC}} \text{ Hz}$$

In Fig. 8-2(a) below, the absolute value of **Z** and its three components R, X_L and X_C are plotted as functions of ω. At $\omega = \omega_0$ the inductive and capacitive reactances are equal and, since $|\mathbf{Z}| = \sqrt{R^2 + X^2}$, $\mathbf{Z} = R$. Thus at resonance the impedance **Z** is a minimum. Since $\mathbf{I} = \mathbf{V}/\mathbf{Z}$, the current is a maximum.

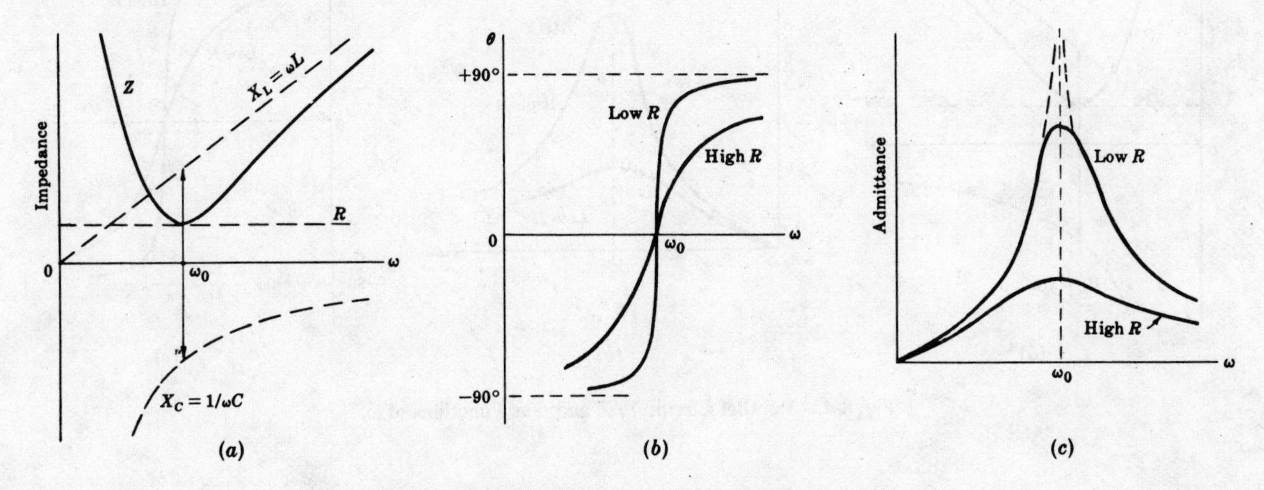

Fig. 8-2. Series Circuit Z, θ and Y as Functions of ω.

At frequencies below ω_0 the capacitive reactance is greater than the inductive reactance, and the angle of the impedance is negative. If the resistance is low, the angle changes more rapidly with frequency as shown in Fig. 8-2(b) above. As ω approaches zero the angle of **Z** approaches $-90°$.

At frequencies above ω_0 the inductive reactance exceeds the capacitive reactance and the angle of **Z** is positive and approaches $+90°$ as $\omega \gg \omega_0$.

In Fig. 8-2(*c*) above, the series circuit admittance $\mathbf{Y} = 1/\mathbf{Z}$ is plotted as a function of ω. Since $\mathbf{I} = \mathbf{VY}$, this plot is also an indication of current versus ω. Thus Fig. 8-2(*c*) shows that the maximum current occurs at ω_0 and that a low resistance results in a higher current. The dotted curve shows the limiting case where $R = 0$. The angle of admittance, not shown here, is the negative of the angle of impedance shown in Fig. 8-2(*b*).

PARALLEL RESONANCE, PURE *RLC* CIRCUIT

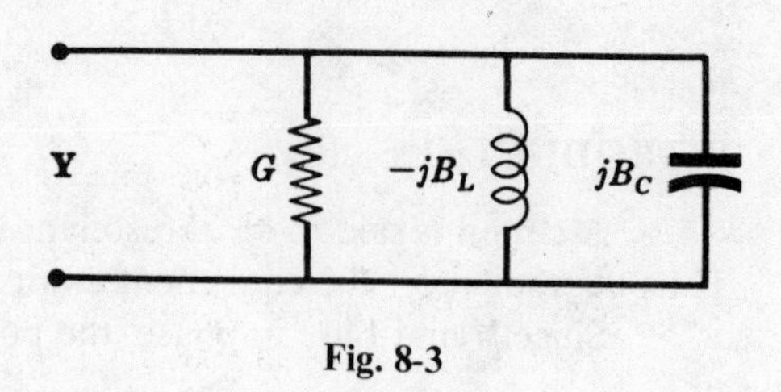

Fig. 8-3

The parallel circuit of Fig. 8-3, consisting of branches with single pure elements *R*, *L* and *C*, is an ideal circuit. However, the performance of such a circuit is of interest in the general subject of resonance. This ideal parallel circuit can be compared to the series circuit examined above and it can be seen that a duality can be established between the two circuits.

The admittance of the three elements is $\mathbf{Y} = G + j(\omega C - 1/\omega L) = G + jB$, where $B = B_C - B_L$, $B_C = \omega C$, and $B_L = 1/\omega L$. The circuit is in resonance when $B = 0$, i.e. $\omega C = 1/\omega L$ or $\omega = 1/\sqrt{LC} = \omega_0$. As in the series *RLC* circuit, the resonant frequency is

$$f_0 = \frac{1}{2\pi\sqrt{LC}} \text{ Hz}$$

In Fig. 8-4(*a*) below, the absolute value of **Y** and its three components *G*, B_C and B_L are plotted as functions of ω. At $\omega = \omega_0$ the capacitive and inductive susceptances are equal and $\mathbf{Y} = G$. Thus at resonance the admittance is a minimum and, since $\mathbf{I} = \mathbf{VY}$, the current has also a minimum value.

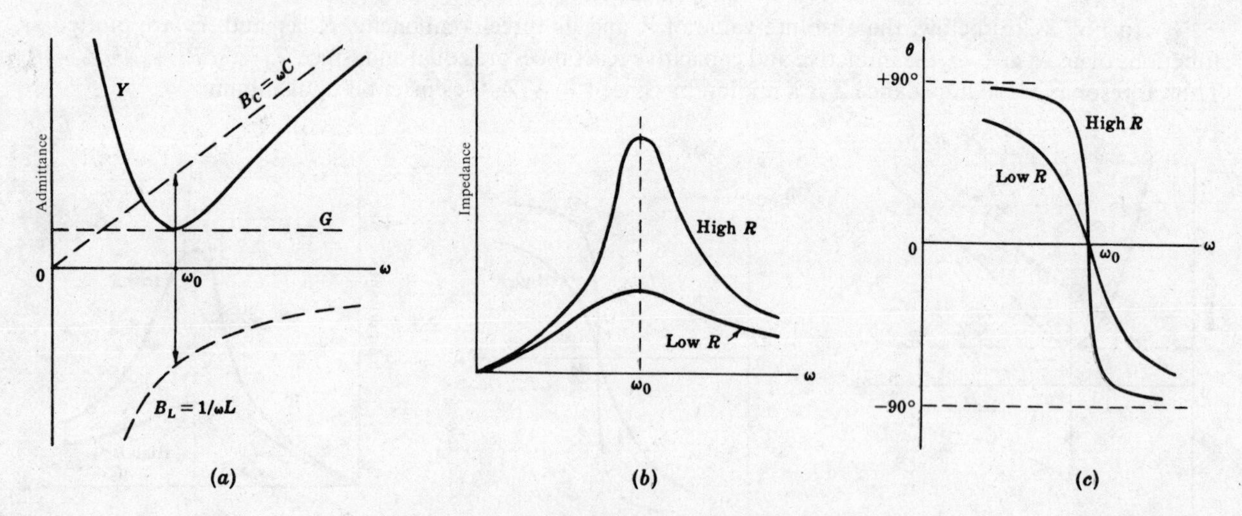

Fig. 8-4. Parallel Circuit Y, Z and θ as Functions of ω.

At frequencies below ω_0 the inductive susceptance exceeds the capacitive susceptance and the angle of **Y** is negative. Then the angle of the impedance is positive and approaches $+90°$ as ω approaches zero. See Fig. 8-4(*c*).

At frequencies above ω_0 the angle of **Z** is negative, and its variation as a function of ω is more rapid for high *R*.

PARALLEL RESONANCE, TWO-BRANCH CIRCUIT

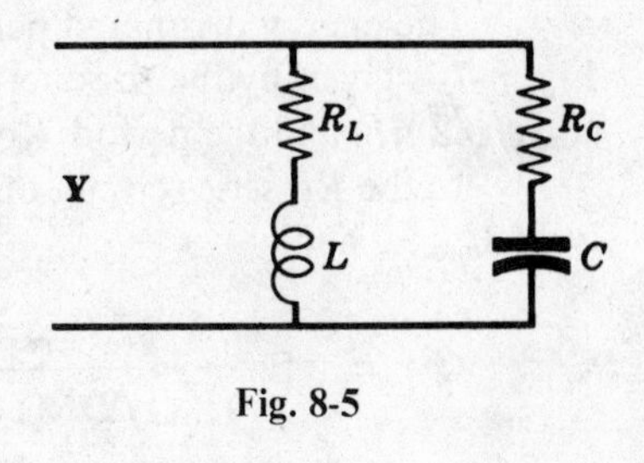

Fig. 8-5

In the two-branch parallel circuit of Fig. 8-5, the admittance **Y** is the sum of the admittances of the individual branches.

$$\mathbf{Y} = \mathbf{Y}_L + \mathbf{Y}_C = \frac{1}{R_L + jX_L} + \frac{1}{R_C - jX_C}$$

$$= \left(\frac{R_L}{R_L^2 + X_L^2} + \frac{R_C}{R_C^2 + X_C^2}\right) + j\left(\frac{X_C}{R_C^2 + X_C^2} - \frac{X_L}{R_L^2 + X_L^2}\right)$$

The circuit is at resonance when the complex admittance is a real number. Then $X_C/(R_C^2 + X_C^2) = X_L/(R_L^2 + X_L^2)$ and

$$\frac{1}{\omega_0 C}(R_L^2 + \omega_0^2 L^2) = \omega_0 L(R_C^2 + 1/\omega_0^2 C^2) \tag{1}$$

Each of the five quantities in (*1*) may be made variable in order to obtain resonance.

Solving (*1*) for ω_0, we obtain

$$\omega_0 = \frac{1}{\sqrt{LC}}\sqrt{\frac{R_L^2 - L/C}{R_C^2 - L/C}} \tag{2}$$

Thus the resonant frequency ω_0 of the two-branch parallel circuit differs from that of the pure R, L and C in parallel by the factor $\sqrt{\dfrac{R_L^2 - L/C}{R_C^2 - L/C}}$.

Frequency must be a real positive number; hence the circuit will have a resonant frequency ω_0 when $R_L^2 > L/C$ and $R_C^2 > L/C$ or $R_L^2 < L/C$ and $R_C^2 < L/C$. When $R_L^2 = R_C^2 = L/C$, the circuit is resonant at all frequencies. For this special case see Problem 8.12.

Solving (*1*) for L, we obtain

$$L = \tfrac{1}{2}C\left[(R_C^2 + X_C^2) \pm \sqrt{(R_C^2 + X_C^2)^2 - 4R_L^2 X_C^2}\right]$$

or, since $Z_C = \sqrt{R_C^2 + X_C^2}$,

$$L = \tfrac{1}{2}C\left[Z_C^2 \pm \sqrt{Z_C^4 - 4R_L^2 X_C^2}\right] \tag{3}$$

Now if in (*3*), $Z_C^4 > 4R_L^2 X_C^2$, we obtain two values of L for which the circuit is resonant. If $Z_C^4 = 4R_L^2 X_C^2$, the circuit is in resonance at $L = \frac{1}{2}CZ_C^2$. When $Z_C^4 < 4R_L^2 X_C^2$, no value of L will make the circuit resonant.

Solving (*1*) for C, we obtain

$$C = 2L\left[\frac{1}{Z_L^2 \pm \sqrt{Z_L^4 - 4R_C^2 X_L^2}}\right] \tag{4}$$

Here if $Z_L^4 > 4R_C^2 X_L^2$, we obtain two values of C for which the circuit is resonant.

Solving (*1*) for R_L, we obtain

$$R_L = \sqrt{\omega^2 LC R_C^2 - \omega^2 L^2 + L/C} \tag{5}$$

and solving for R_C,

$$R_C = \sqrt{R_L^2/(\omega^2 LC) - 1/\omega^2 C^2 + L/C} \tag{6}$$

If the radicand in (*5*) or (*6*) is positive, then we have a value for R_L or R_C for which the two-branch circuit is in resonance.

QUALITY FACTOR Q

The quality factor of coils, capacitors and circuits is defined by

$$Q = 2\pi \frac{\text{maximum stored energy}}{\text{energy dissipated per cycle}}$$

The energy dissipated per cycle in the circuits of Fig. 8-6 and Fig. 8-7 is given by the product of the average power in the resistor $(I_{max}/\sqrt{2})^2R$ and the period T or $1/f$.

In the RL series circuit of Fig. 8-6 the maximum stored energy is $\frac{1}{2}LI_{max}^2$. Then

Fig. 8-6

$$Q = 2\pi \frac{\frac{1}{2}LI_{max}^2}{(I_{max}^2/2)R(1/f)} = \frac{2\pi fL}{R} = \frac{\omega L}{R}$$

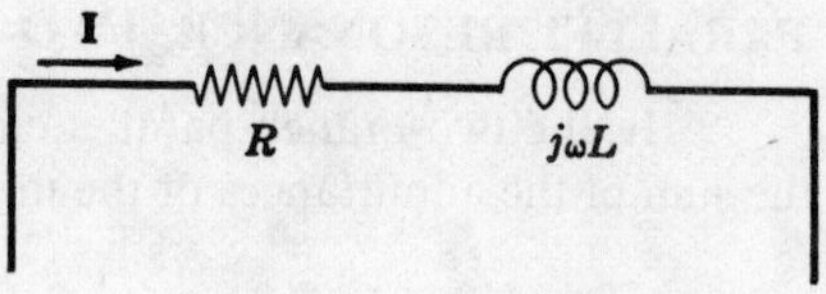

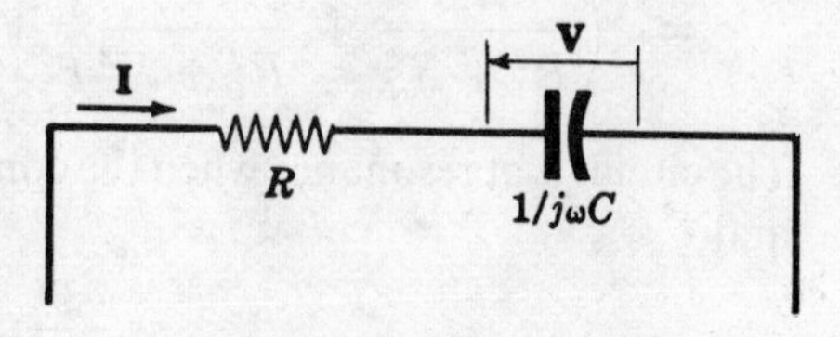

Fig. 8-7

In the RC series circuit of Fig. 8-7 the maximum stored energy is $\frac{1}{2}CV_{max}^2$ or $\frac{1}{2}I_{max}^2/\omega^2C$. Then

$$Q = 2\pi \frac{\frac{1}{2}I_{max}^2/\omega^2C}{(I_{max}^2/2)R(1/f)} = \frac{1}{\omega CR}$$

A series RLC circuit at resonance stores a constant amount of energy. Since when the capacitor voltage is maximum the inductor current is zero, and vice versa, $\frac{1}{2}CV_{max}^2 = \frac{1}{2}LI_{max}^2$. Then

$$Q_0 = \frac{\omega_0 L}{R} = \frac{1}{\omega_0 CR}$$

The RLC series circuit has a current function with respect to frequency similar to the admittance plot of Fig. 8-2(c). In Fig. 8-8 the current of the RLC circuit is plotted as a function of ω or, with an appropriate change of scale, as a function of f. At ω_0 the current $\mathbf{I}_0$ is a maximum. The points where the current is 0·707 of the maximum are indicated. The corresponding frequencies are ω_1 and ω_2.

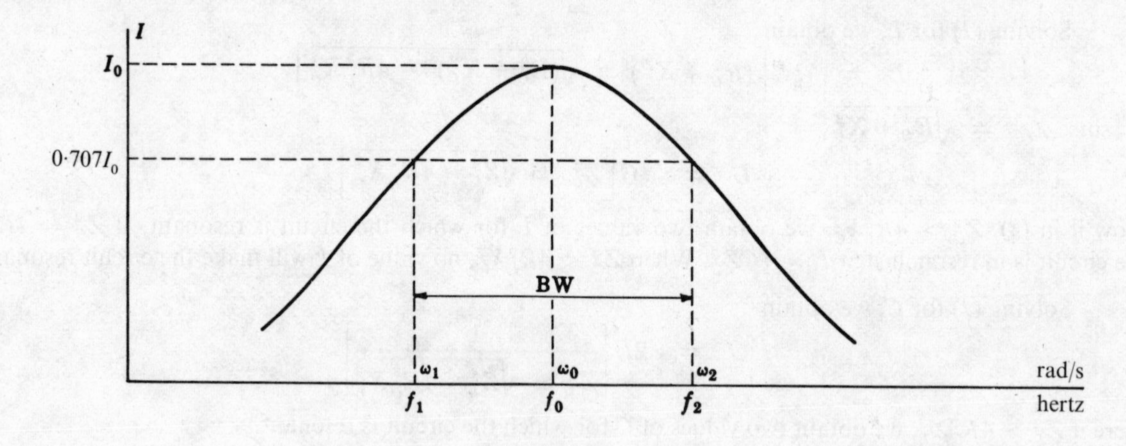

Fig. 8-8

Since the power delivered to the circuit is I^2R, at $I = 0{\cdot}707I_0$ the power is one-half of the maximum value which occurs at ω_0. The points corresponding to ω_1 and ω_2 are called the half-power points. The distance between these points, measured in hertz, is called the bandwidth BW.

Now the quality factor can be given as the ratio of the resonant frequency to bandwidth; hence (see Problem 8.13)

$$Q_0 = \frac{\omega_0}{\omega_2 - \omega_1} = \frac{f_0}{f_2 - f_1} = \frac{f_0}{\text{BW}}$$

The resonant frequency ω_0 is the geometric mean of ω_1 and ω_2 (see Problem 8.6);

$$\omega_0 = \sqrt{\omega_1\omega_2} \quad \text{and} \quad f_0 = \sqrt{f_1 f_2}$$

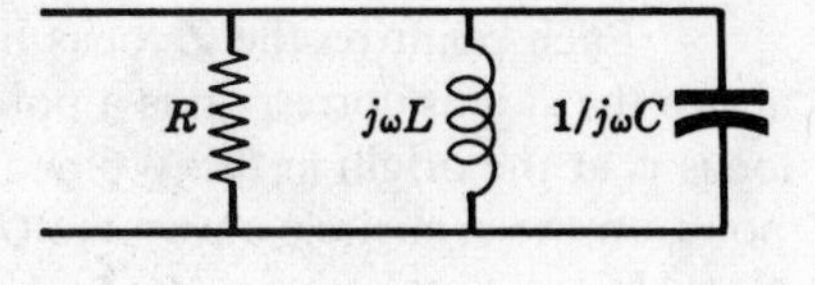

Fig. 8-9

The three branch parallel circuit of Fig. 8-9 stores a constant amount of energy at resonance. Since when the inductor current is maximum the capacitor voltage is zero, and vice versa, $\frac{1}{2}LI_{max}^2 = \frac{1}{2}CV_{max}^2$. Then

$$Q_0 = \frac{R}{\omega_0 L} = \omega_0 CR$$

LOCUS DIAGRAMS

Circuits with one variable element are conveniently analyzed by the use of admittance locus diagrams. Since $\mathbf{I} = \mathbf{VY}$, and $\mathbf{V}$ is generally constant, the $\mathbf{Y}$ locus describes the variation of $\mathbf{I}$ as the variable element is changed.

The series circuit of Fig. 8-10(a) has a fixed resistance, and a variable reactance which can assume values either positive or negative. If we consider the $\mathbf{Z}$ plane formed by a set of Cartesian axes R and X, the locus of the impedance $\mathbf{Z}$ for the given circuit is a straight line parallel with the X axis and intersecting the R axis at R_1, as shown in Fig. 8-10(b).

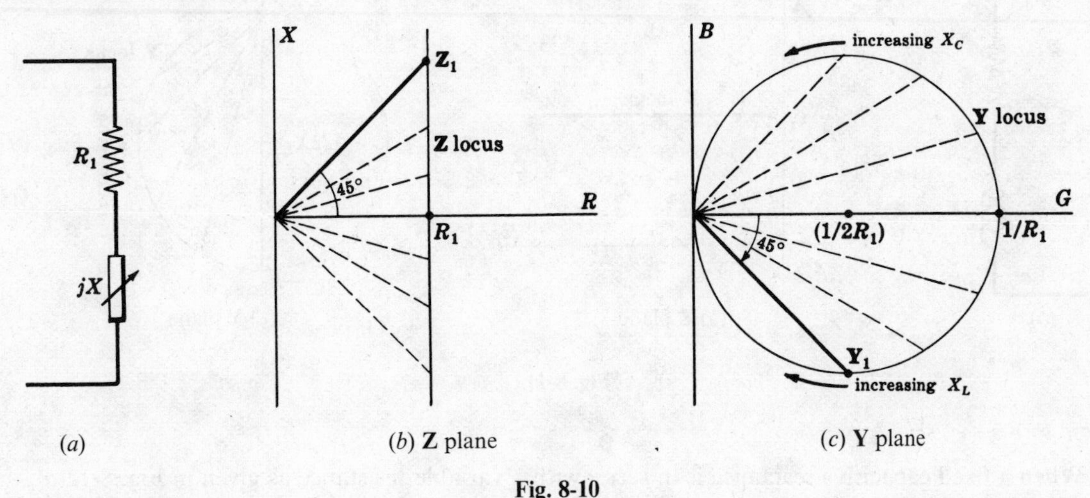

Fig. 8-10

We can determine a locus of the admittance $\mathbf{Y}$ for the given circuit in the $\mathbf{Y}$ plane formed by a set of Cartesian axes G and B.

Since $\mathbf{Z} = 1/\mathbf{Y}$,

$$R_1 + jX = \frac{1}{G + jB} \tag{1}$$

Rationalizing and equating the real parts in (1), we obtain

$$R_1 = \frac{G}{G^2 + B^2}$$

or

$$G^2 - G/R_1 + B^2 = 0 \tag{2}$$

Adding $1/4R_1^2$ to both sides in (2) and simplifying,

$$\left(G - \frac{1}{2R_1}\right)^2 + B^2 = \left(\frac{1}{2R_1}\right)^2 \tag{3}$$

If we compare the standard form of the equation of a circle as given in analytic geometry, $(x - h)^2 + (y - k)^2 = r^2$, with ($3$), we note that the $\mathbf{Y}$ plane locus is a circle with centre at $(1/2R_1, 0)$ and with radius $1/2R_1$. See Fig. 8-10(c).

Each point on the **Z** locus has a corresponding point on the **Y** locus. To each point on the **Z** locus above the R axis corresponds a point on the semicircle below the G axis in the **Y** plane. And $+\infty$ on the **Z** locus is at the origin in the **Y** plane. Similarly, to each point below the R axis on the **Z** locus corresponds a point on the semicircle above the G axis in the **Y** plane. $-\infty$ on the **Z** locus is at the origin in the **Y** plane. Note the relative positions of $\mathbf{Z}_1$ and $\mathbf{Y}_1$. The distances of $\mathbf{Z}_1$ and $\mathbf{Y}_1$ from the respective origins are different while the angles with the horizontal axes are equal but of opposite sign.

With fixed inductive reactance and variable resistance as in Fig. 8-11(a), the **Z** locus is a horizontal line in the first quadrant of the **Z** plane at $X = X_{L_1}$. Using the same method as above, the equation of the **Y** locus is

$$G^2 + (B + 1/2X_{L_1})^2 = (1/2X_{L_1})^2 \tag{4}$$

Comparing equation (4) with the standard form of the equation of a circle, we find that the **Y** locus is a circle with centre at $(0, -1/2X_{L_1})$ and radius $1/2X_{L_1}$ in the **Y** plane. See Fig. 8-11 (c). However, since the **Z** locus of Fig. 8-11(b) consists of the straight line in the first quadrant of the **Z** plane, only the semicircle in the fourth quadrant of the **Y** plane is the transformation of the **Z** locus for this circuit.

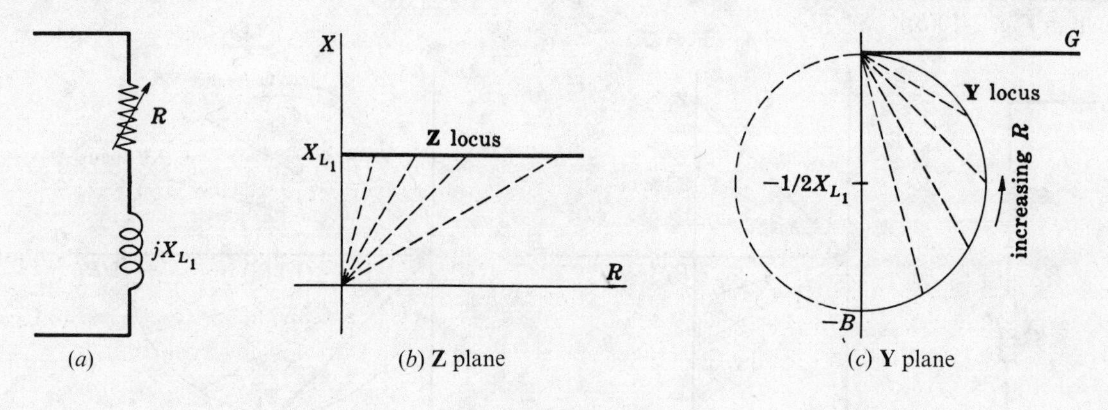

(a) (b) **Z** plane (c) **Y** plane

Fig. 8-11

When a fixed capacitive reactance is in series with a variable resistance as given in Fig. 8-12(a), the **Z** locus is a horizontal line in the fourth quadrant of the **Z** plane at $X = -X_C$. See Fig. 8-12(b). Using the same methods as above, the **Y** locus equation is

$$G^2 + (B - 1/2X_{C_1})^2 = (1/2X_{C_1})^2 \tag{5}$$

Comparing equation (5) with the standard form of the equation of a circle, we see that the **Y** locus is a semicircle with centre at $(0, 1/2X_{C_1})$ and radius $1/2X_{C_1}$ in the first quadrant of the **Y** plane. See Fig. 8-12(c).

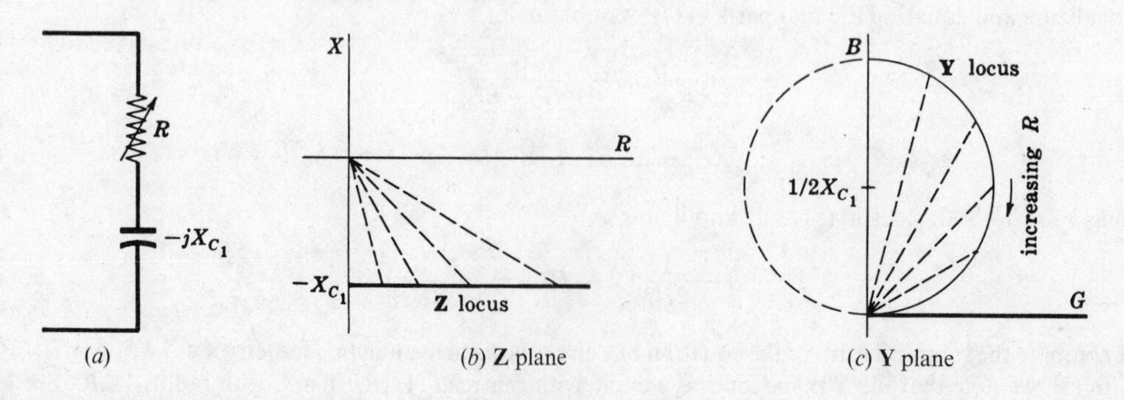

(a) (b) **Z** plane (c) **Y** plane

Fig. 8-12

CURRENT LOCUS DIAGRAMS

Consider the parallel circuit of Fig. 8-13(a) with a fixed R_1 and jX_L in series in the first branch and with a fixed R_2 and a variable $-jX_C$ in series in the second branch. The total admittance of the two branches in parallel is

$$\mathbf{Y}_T = \mathbf{Y}_1 + \mathbf{Y}_2$$

In Fig. 8-13(b), adding the second branch $\mathbf{Y}_2$ locus to the fixed point $\mathbf{Y}_1$, we obtain the $\mathbf{Y}_T$ locus.

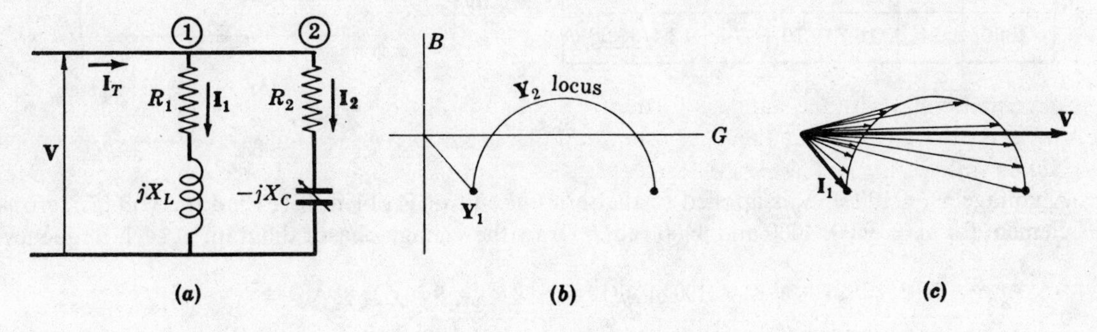

Fig. 8-13

The current is given by $\mathbf{I} = \mathbf{VY}$, and Fig. 8-13(c) shows that, as the fixed current $\mathbf{I}_1$ is added to the various values of $\mathbf{I}_2$, the result is a locus of the total current. The diagram shows further how there can be two values of C for which the total current is in phase with $\mathbf{V}$.

Further examination of Fig. 8-13(c) illustrates why under certain conditions we may not obtain a value of C which results in resonance. If the radius of the semicircular part of the locus, $1/2R_2$, is reduced in such a way that the curve does not intersect the $\mathbf{V}$ axis, then there would be no value of C which could cause resonance. Further applications of the locus diagrams are examined in the following problems.

Solved Problems

8.1. In a series RLC circuit, $R = 10$ ohms, $L = 5$ mH and $C = 12{\cdot}5$ μF. Plot the magnitude and angle of the impedance as a function of ω with ω varying from $0{\cdot}8\omega_0$ to $1{\cdot}2\omega_0$.

At resonance,

$$\omega = \omega_0 = 1/\sqrt{LC} = 1/\sqrt{(5 \times 10^{-3})(12{\cdot}5 \times 10^{-6})} = 4000 \text{ rad/s}$$

$$X_{L_0} = \omega_0 L = 4000(5 \times 10^{-3}) = 20 \text{ ohms}$$

$$X_{C_0} = 1/\omega_0 C = 1/(4000 \times 12{\cdot}5 \times 10^{-6}) = 20 \text{ ohms}$$

Then $\qquad \mathbf{Z}_0 = R + j(X_{L0} - X_{C0}) = 10 + j(20 - 20) = 10\angle 0^\circ$ ohms

Since $X_L = \omega L$ and $X_C = 1/\omega C$, then $X_L/X_{L_0} = \omega/\omega_0$ and $X_C/X_{C_0} = \omega_0/\omega$. Thus values of X_L, X_C and $\mathbf{Z}$ at other frequencies can be calculated.

Fig. 8-14(a) below gives a tabulation of reactances and impedances, and Fig. 8-14(b) below shows the required plot.

ω (rad/s)	X_L (Ω)	X_C (Ω)	**Z** (Ω)	
3200	16	25	$10 - j9$	$13{\cdot}4\angle{-42°}$
3600	18	22·2	$10 - j4{\cdot}2$	$10{\cdot}8\angle{-22{\cdot}8°}$
4000	20	20	10	$10\angle 0°$
4400	22	18·2	$10 + j3{\cdot}8$	$10{\cdot}7\angle 20{\cdot}8°$
4800	24	16·7	$10 + j7{\cdot}3$	$12{\cdot}4\angle 36{\cdot}2°$

(*a*)

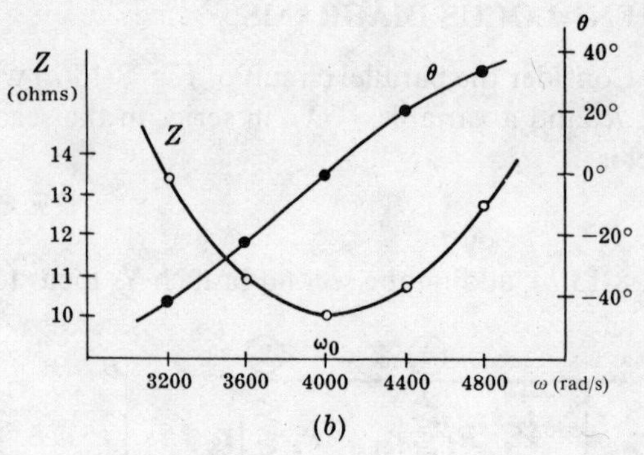

(*b*)

Fig. 8-14

8.2. A voltage $\mathbf{V} = 100\angle 0°$ V is applied to the series circuit of Problem 8.1. Find the voltage across each element for ω = 3600, 4000 and 4400 rad/s. Draw the voltage phasor diagram at each frequency.

At ω = 3600 rad/s, $\mathbf{I} = \mathbf{V}/\mathbf{Z} = (100\angle 0°)/(10{\cdot}8\angle{-22{\cdot}8°}) = 9{\cdot}26\angle 22{\cdot}8°$ A. Then

$$\mathbf{V}_R = 9{\cdot}26\angle 22{\cdot}8°(10) = 92{\cdot}6\angle 22{\cdot}8 \text{ V},\ \mathbf{V}_L = 9{\cdot}26\angle 22{\cdot}8°(18\angle 90°) = 167\angle 112{\cdot}8° \text{ V},\ \mathbf{V}_C = 206\angle{-67{\cdot}2°} \text{ V}$$

At ω = 4000 rad/s, $\mathbf{I} = (100\angle 0°)/(10\angle 0°) = 10\angle 0°$ A. Then

$$\mathbf{V}_R = 100\angle 0° \text{ V},\ \mathbf{V}_L = 10\angle 0°(20\angle 90°) = 200\angle 90° \text{ V},\ \mathbf{V}_C = 200\angle{-90°} \text{ V}$$

At ω = 4400 rad/s, $\mathbf{I} = (100\angle 0°)/(10{\cdot}7\angle 20{\cdot}8°) = 9{\cdot}34\angle{-20{\cdot}8°}$ A. Then

$$\mathbf{V}_R = 9{\cdot}34\angle{-20{\cdot}8°}(10) = 93{\cdot}4\angle{-20{\cdot}8°} \text{ V},\ \mathbf{V}_L = 9{\cdot}34\angle{-20{\cdot}8°}(22\angle 90°) = 206\angle 69{\cdot}2° \text{ V},$$
$$\mathbf{V}_C = 170\angle{-110{\cdot}8°} \text{ V}$$

The three voltage phasor diagrams are drawn in Fig. 8-15. Note that the magnitude of the voltage across each reactive element of a series circuit near resonance may exceed the magnitude of the applied voltage.

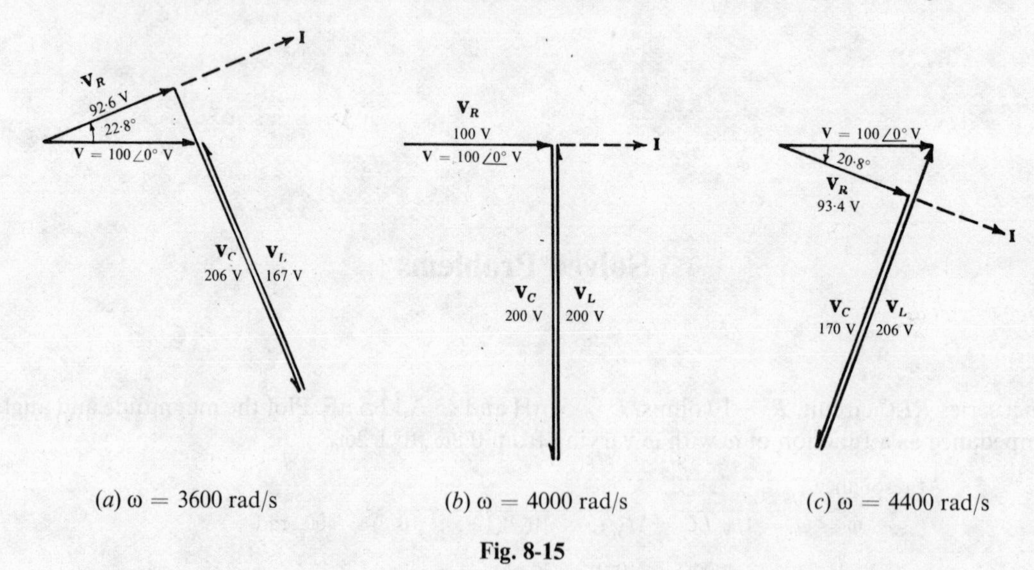

(*a*) ω = 3600 rad/s (*b*) ω = 4000 rad/s (*c*) ω = 4400 rad/s

Fig. 8-15

8.3. A series circuit with $R = 5$ ohms, $L = 20$ mH and a variable capacitance C has an applied voltage with a frequency $f = 1000$Hz. Find C for series resonance.

At resonance the reactances are equal, $2\pi fL = 1/2\pi fC$. Then

$$C = \frac{1}{L(2\pi f)^2} \quad \frac{1}{(20 \times 10^{-3})(2\pi \times 1000)^2} = 1{\cdot}27\ \mu\text{F}$$

8.4. A series circuit with $R = 5$ ohms, $C = 20\ \mu F$ and a variable inductance L has an applied voltage $\mathbf{V} = 10\angle 0^\circ$ V with a frequency of 1000 rad/s. L is adjusted until the voltage across the resistor is a maximum. Find the voltage across each element.

Since $\mathbf{V}_R = \mathbf{I}R$, the maximum voltage across the resistor occurs at resonance, when the current is a maximum. At resonance the reactances are equal; then

$$X_C = \frac{1}{\omega C} = \frac{1}{1000(20 \times 10^{-6})} = 50 \text{ ohms}, \quad X_L = 50 \text{ ohms}$$

and $\mathbf{Z} = R = 5\angle 0^\circ\ \Omega$. Now $\mathbf{I} = \mathbf{V}/\mathbf{Z} = (10\angle 0^\circ)/(5\angle 0^\circ) = 2\angle 0^\circ$ A and

$$\mathbf{V}_R = 2\angle 0^\circ(5) = 10\angle 0^\circ \text{ V}, \quad \mathbf{V}_L = (2\angle 0^\circ)(50\angle 90^\circ) = 100\angle 90^\circ \text{ V and } \mathbf{V}_C = 100\angle -90^\circ \text{ V}$$

8.5. Given a series RLC circuit with $R = 100$ ohms, $L = 0{\cdot}5$ H and $C = 40\ \mu F$, calculate the resonant, lower and upper half-power frequencies.

$$\omega_0 = 1/\sqrt{LC} = 1/\sqrt{0{\cdot}5(40 \times 10^{-6})} = 224 \text{ rad/s and } f_0 = \omega_0/2\pi = 35{\cdot}7 \text{ Hz}.$$

At the lower half-power frequency ω_1 the capacitive reactance exceeds the inductive reactance, the current is 0·707 of it maximum value and, since $\mathbf{I} = \mathbf{V}/\mathbf{Z}$, $|\mathbf{Z}|$ is 1·414 times its value at ω_0. Since $\mathbf{Z} = 100\ \Omega$ at ω_0, then $|\mathbf{Z}| = 141{\cdot}4$ ohms at ω_1. Now $\mathbf{Z} = 100 - j(X_C - X_L) = 141{\cdot}4\angle\theta$, $\cos\theta = R/Z = 100/141{\cdot}4 = 0{\cdot}707$, and $\theta = -45^\circ$. Then

$$X_C - X_L = R \text{ or } 1/\omega_1 C - \omega_1 L = \mathrm{R} \tag{1}$$

Substituting in (*1*) the given values and solving for ω_1, we obtain $\omega_1 = 145$ rad/s and $f_1 = 145/2\pi = 23{\cdot}1$ Hz.

At the upper half-power frequency ω_2 the inductive reactance exceeds the capacitive reactance, $|\mathbf{Z}|$ is also 141·4 ohms and $\theta = +45^\circ$. Then

$$X_L - X_C = R \text{ or } \omega_2 L - 1/\omega_2 C = R \tag{2}$$

Substituting in (*2*) and solving for ω_2, we have $\omega_2 = 345$ rad/s and $f_2 = 55$ Hz.

ω_0 is the geometric mean of ω_1 and ω_2; hence

$$\omega_0 = \sqrt{\omega_1\omega_2} = \sqrt{145 \times 345} = 224 \text{ rad/s}$$

8.6. Show that the resonant frequency ω_0 of an RLC series circuit is the geometric mean of ω_1 and ω_2, the lower and upper half-power frequencies respectively.

As seen in Problem 8.5, $1/\omega_1 C - \omega_1 L = R$ at ω_1, and $\omega_2 L - 1/\omega_2 C = R$ at ω_2. Then

$$1/\omega_1 C - \omega_1 L = \omega_2 L - 1/\omega_2 C \tag{1}$$

Multiplying through by C and substituting $\omega_0^2 = 1/LC$ in (*1*). obtain

$$1/\omega_1 - \omega_1/\omega_0^2 = \omega_2/\omega_0^2 - 1/\omega_2 \text{ or } 1/\omega_1 + 1/\omega_2 = (\omega_1 + \omega_2)/\omega_0^2$$

from which $\omega_0 = \sqrt{\omega_1\omega_2}$.

8.7. A series circuit with $R = 50$ ohms, $L = 0{\cdot}05$ H and $C = 20\ \mu F$ has an applied voltage $\mathbf{V} = 100\angle 0^\circ$ V with a variable frequency. Find the maximum voltage across the inductor as the frequency is varied.

The magnitude of the impedance as a function of ω is $Z = \sqrt{R^2 + (\omega L - 1/\omega C)^2}$. Then the magnitude of the current is $I = V/\sqrt{R^2 + (\omega L - 1/\omega C)^2}$.

The magnitude of the voltage across L is

$$V_L = \omega L I = \omega L V/\sqrt{R^2 + (\omega L - 1/\omega C)^2} \tag{1}$$

Setting the derivative $dV_L/d\omega$ of (*1*) equal to zero and solving for ω, we obtain the value of ω when V_L is a maximum.

$$\begin{aligned} \frac{dV_L}{d\omega} &= \frac{d}{d\omega}\omega L V(R^2 + \omega^2L^2 - 2L/C + 1/\omega^2C^2)^{-1/2} \\ &= \frac{(R^2 + \omega^2L^2 - 2L/C + 1/\omega^2C^2)^{1/2}LV - \omega LV\frac{1}{2}(R^2 + \omega^2L^2 - 2L/C + 1/\omega^2C^2)^{-1/2}(2\omega L^2 - 2/\omega^3C^2)}{R^2 + \omega^2L^2 - 2L/C + 1/\omega^2C^2} \end{aligned} \tag{2}$$

Factoring $LV(R^2 + \omega^2L^2 - 2L/C + 1/\omega^2C^2)^{-1/2}$ in (2) and setting the numerator equal to zero, we have

$$R^2 - 2L/C + 2/\omega^2C^2 = 0$$

from which
$$\omega = \sqrt{\frac{2}{2LC - R^2C^2}} = 1/\sqrt{LC}\sqrt{\frac{2}{2 - R^2C/L}} \qquad (3)$$

Since $Q_0 = \omega_0 L/R = 1/\omega_0 CR$, $Q_0^2 = L/R^2C$; substituting in (3), we obtain

$$\omega = \frac{1}{\sqrt{LC}}\sqrt{\frac{2Q_0^2}{2Q_0^2 - 1}} \qquad (4)$$

Substituting the given values in (3),

$$\omega = \sqrt{\frac{2}{2(\cdot 05)(20\times 10^{-6}) - (50\times 20\times 10^{-6})^2}} = 1414 \text{ rad/s}$$

Now $X_L = \omega L = 1414(0{\cdot}05) = 70{\cdot}7$ ohms, $X_C = 1/\omega C = 1/(1414 \times 20 \times 10^{-6}) = 35{\cdot}4$ ohms and $\mathbf{Z} = 50 + j(70{\cdot}7 - 35{\cdot}4) = 50 + j35{\cdot}4 = 61{\cdot}2\angle 35{\cdot}3°\ \Omega$. Then $I = V/Z = 100/61{\cdot}2 = 1{\cdot}635$ A and

$$V_{L(\max)} = 1{\cdot}635(70{\cdot}7) = 115{\cdot}5 \text{ V}$$

Equation (4) shows that for high Q, the maximum voltage across L accurs at $\omega_0 \approx 1/\sqrt{LC}$. If Q is high, maximum voltages are obtained also across R and C at ω_0. With low Q, V_C maximum occurs below and V_L maximum above ω_0. See Problem 8.28.

8.8. The circuit in Fig. 8-16 represents a parallel connection of a capacitor and a coil where the coil resistance is R_L. Find the resonant frequency of the circuit.

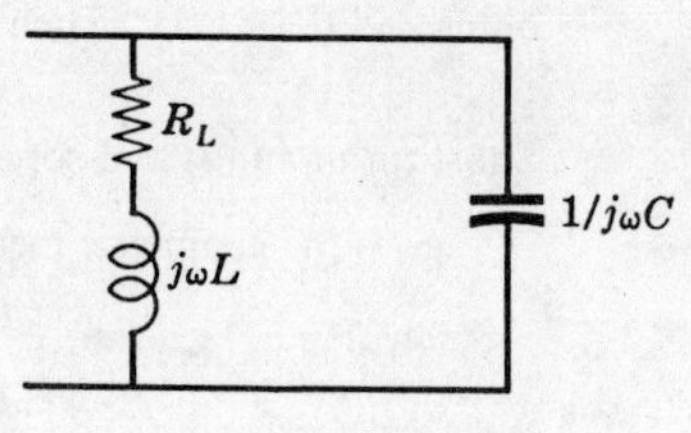

Fig. 8-16

The total admittance of the circuit is

$$\mathbf{Y}_T = \frac{1}{R_L + j\omega L} + j\omega C$$
$$= \frac{R_L}{R_L^2 + \omega^2L^2} + j\left(\omega C - \frac{\omega L}{R_L^2 + \omega^2L^2}\right)$$

At resonance the j part is zero or

$$\frac{\omega_0 L}{R_L^2 + \omega_0^2 L^2} = \omega_0 C \qquad \text{from which} \qquad \omega_0 = \frac{1}{\sqrt{LC}}\sqrt{1 - \frac{R_L^2 C}{L}}$$

If the coil resistance is small compared to $\omega_0 L$, the resonant frequency is given by $1/\sqrt{LC}$.

8.9. Find the resonant frequency ω_0 for the two-branch parallel circuit in Fig. 8-17. If the resistor in the RC branch is increased, what is its maximum value for which there is a resonant frequency?

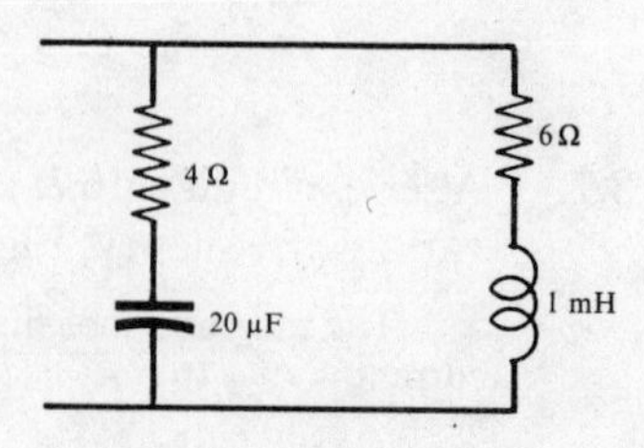

Fig. 8-17

$$\omega_0 = \frac{1}{\sqrt{LC}}\sqrt{\frac{R_L^2 - L/C}{R_C^2 - L/C}}$$
$$= \frac{1}{\sqrt{10^{-3}\times 20\times 10^{-6}}}\sqrt{\frac{6^2 - 10^{-3}/(20\times 10^{-6})}{4^2 - 10^{-3}/(20\times 10^{-6})}}$$
$$= 4540 \text{ rad/s}$$

The numerator within the radical has a value $36 - 50 = -14$. Then the radical will have a real root if the denominator is negative, i.e. if $R_C^2 < L/C$ or $R_C < 7{\cdot}07$ ohms. As the value of R_C approaches 7·07 ohms the frequency ω_0 approaches infinity.

If the value of R_L is increased, then ω_0 will approach zero as R_L approaches 7·07 ohms.

8.10. Find the values of L for which the circuit of Fig. 8-18 is resonant at a frequency of $\omega = 5000$ rad/s.

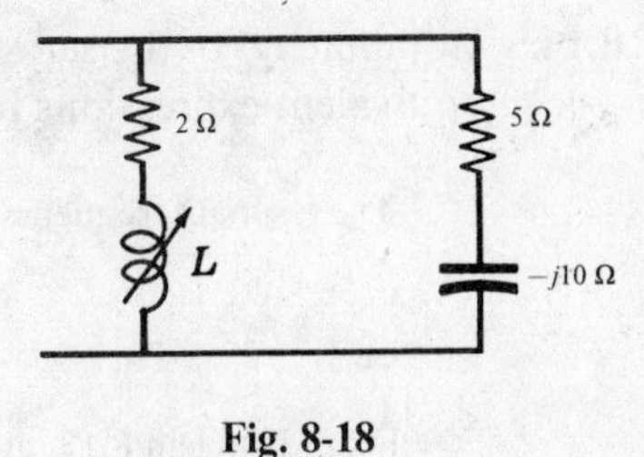

Fig. 8-18

The total admittance is

$$\mathbf{Y} = \frac{1}{2 + jX_L} + \frac{1}{5 - j10} \text{ siemens}$$

$$= \left(\frac{2}{4 + X_L^2} + \frac{5}{125}\right) + j\left(\frac{10}{125} - \frac{X_L}{4 + X_L^2}\right)$$

Setting the j part equal to zero,

$$10/125 = X_L/(4 + X_L^2) \text{ or } X_L^2 - 12{\cdot}5X_L + 4 = 0 \qquad (1)$$

The roots of (1) are $X_L = 12{\cdot}17\ \Omega$ and $X_L = 0{\cdot}33\ \Omega$. Substitute these values into the equation $X_L = \omega L$. Then the condition for a resonant circuit is that $L = 2{\cdot}43$ mH or $0{\cdot}066$ mH.

8.11. Find C which results in resonance for the circuit of Fig. 8-19 when $\omega = 5000$ rad/s.

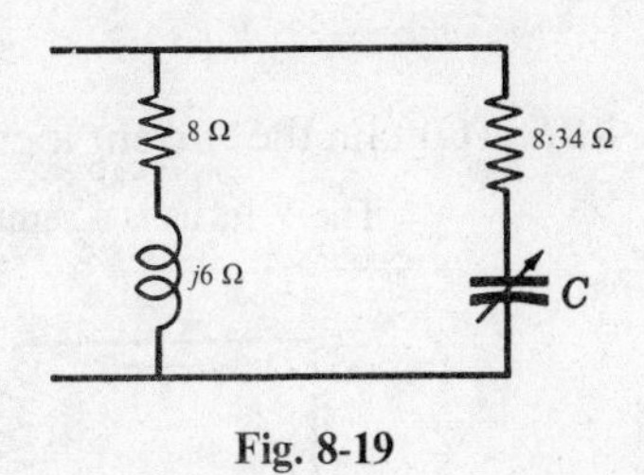

Fig. 8-19

$$\mathbf{Y} = \frac{1}{8 + j6} + \frac{1}{8{\cdot}34 - jX_C} \text{ siemens}$$

$$= \left(\frac{8}{100} + \frac{8{\cdot}34}{69{\cdot}5 + X_C^2}\right) + j\left(\frac{X_C}{69{\cdot}5 + X_C^2} - \frac{6}{100}\right)$$

At resonance the complex admittance is a real number. Then

$$X_C/(69{\cdot}5 + X_C^2) = 6/100 \text{ and } X_C^2 - 16{\cdot}7X_C + 69{\cdot}5 = 0$$

from which $X_C = 8{\cdot}35$ ohms. Substituting this value in $X_C = 1/\omega C$ and solving, we have $C = 24\ \mu$F.

8.12. Determine R_L and R_C which cause the circuit of Fig. 8-20 to be resonant at all frequencies.

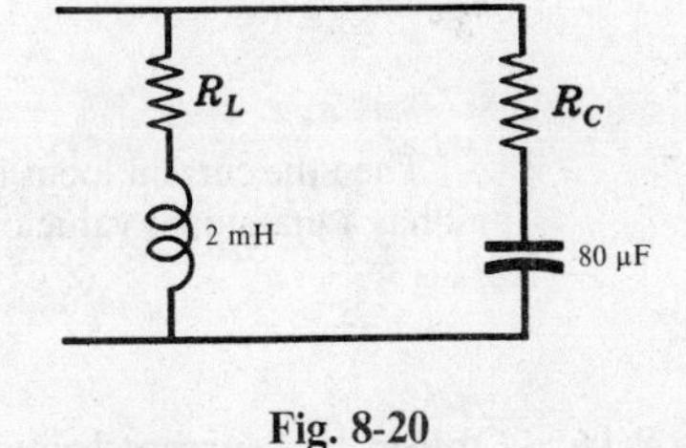

Fig. 8-20

The circuit is resonant at a frequency

$$\omega_0 = \frac{1}{\sqrt{LC}}\sqrt{\frac{R_L^2 - L/C}{R_C^2 - L/C}}$$

ω_0 can assume any value if $R_L^2 = R_C^2 = L/C$. If $L/C = (2 \times 10^{-3})/(80 \times 10^{-6}) = 25$, then

$$R_L = R_C = \sqrt{25} = 5 \text{ ohms}$$

It is left as an exercise for the student to check this result at values of $\omega = 2500$ rad/s and 5000 rad/s.

8.13. Show that $Q_0 = \omega_0 L/R = f_0/BW$ for a series RLC circuit.

At the half-power frequencies the net reactance is equal to the resistance.

At the lower half-power frequency the capacitive reactance exceeds the inductive reactance. Then

$$1/2\pi f_1 C - 2\pi f_1 L = R \qquad \text{from which} \qquad f_1 = \frac{-R + \sqrt{R^2 + 4L/C}}{4\pi L}$$

At the upper half-power frequency the inductive reactance exceeds the capacitive reactance. Then

$$2\pi f_2 L - 1/2\pi f_2 C = R \qquad \text{from which} \qquad f_2 = \frac{R + \sqrt{R^2 + 4L/C}}{4\pi L}$$

Since $\mathrm{BW} = f_2 - f_1$, $\mathrm{BW} = R/2\pi L$. Hence

$$Q_0 = f_0/\mathrm{BW} = 2\pi f_0 L/R = \omega_0 L/R$$

8.14. Compute Q_0 of the series circuit with $R = 20$ ohms, $L = 0{\cdot}05$ H and $C = 1\ \mu$F, using each of the three equivalent expressions for Q_0: $\omega_0 L/R$, $1/\omega_0 CR$ and f_0/BW.

The resonant frequency is $\omega_0 = 1/\sqrt{LC} = 1/\sqrt{0{\cdot}05 \times 10^{-6}} = 4470$ rad/s and $f_0 = \omega_0/2\pi = 712$ Hz. Then

$$Q_0 = \omega_0 L/R = 4470(0{\cdot}05)/20 = 11{\cdot}2$$

or

$$Q_0 = 1/\omega_0 CR = 1/(4470 \times 10^{-6} \times 20) = 11{\cdot}2$$

From Problem 8.13, at the lower half-power frequency, $1/2\pi f_1 C - 2\pi f_1 L = R$. Substituting,

$$1/(2\pi f_1 \times 10^{-6}) - 2\pi f_1(0{\cdot}05) = 20 \text{ and } f_1 = 681 \text{ Hz}$$

At the upper half-power frequency, $2\pi f_2 L - 1/2\pi f_2 C = R$. Substituting, $f_2 = 745$ Hz.

Then BW = (745 − 681) Hz and

$$Q_0 = f_0/\text{BW} = 712/(745 - 681) = 11{\cdot}1$$

8.15. Obtain the current locus for the circuit of Fig. 8-21(*a*) with a variable inductive reactance X_L.

The **Y** locus is a semicircle with radius $r = 1/2R = 0{\cdot}1$, as shown in Fig. 8-21(*b*).

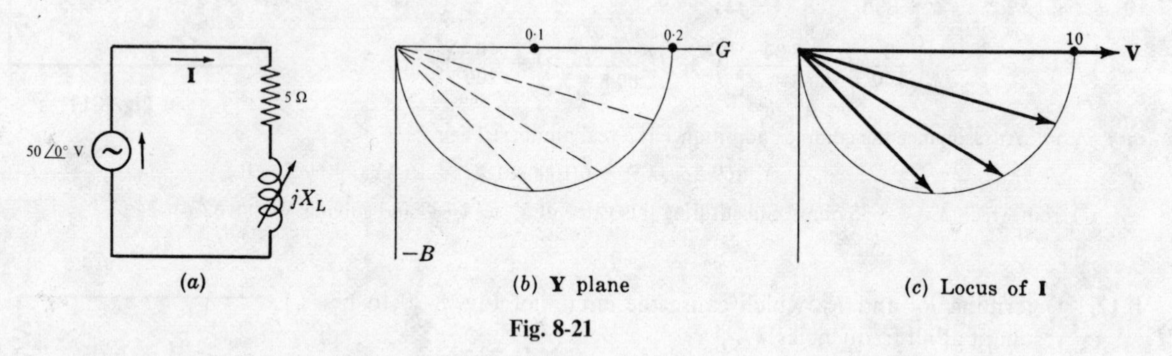

Fig. 8-21

Then the current locus is found from **I** = **VY** where **V** = 50∠0° V. Thus the current locus is similar to the **Y** locus and has a maximum value of 10 A when $X_L = 0$. See Fig. 8-21(*c*).

8.16. Obtain the current locus for the circuit of Fig. 8-22(*a*) with variable resistance R and fixed capacitive reactance.

The **Y** locus is a semicircle with radius $r = 1/2X_C = 0{\cdot}1$, as shown in Fig. 8-22(*b*).

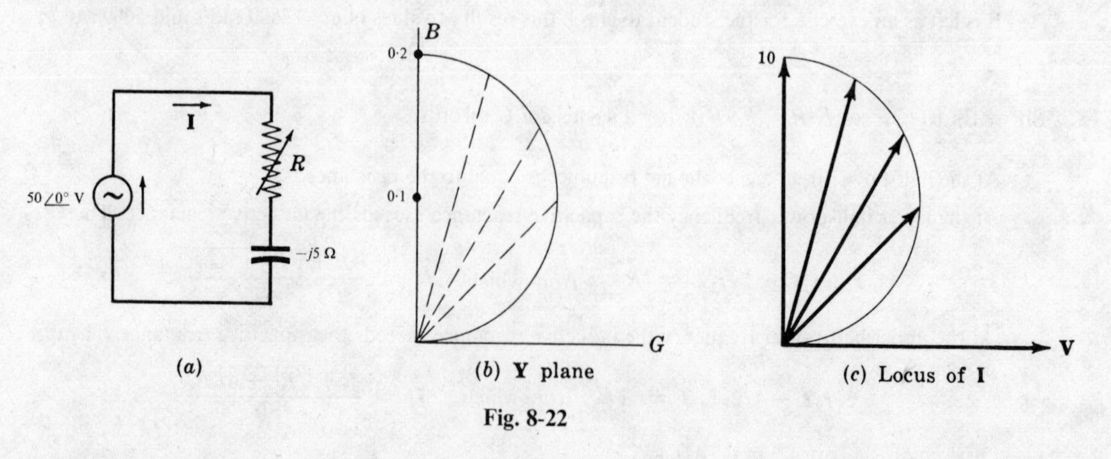

Fig. 8-22

Then the current locus is found from **I** = **VY** where **V** = 50∠0° V. Thus the current has a maximum value of 10 A when $R = 0$. See Fig. 8-22(*c*).

8.17. Find R_L which results in resonance for the circuit of Fig. 8-23(*a*). Draw the **Y** locus to explain the results.

The total admittance is

$$\mathbf{Y}_T = \frac{1}{R_L + j10} + \frac{1}{4 - j5} = \left(\frac{R_L}{R_L^2 + 100} + \frac{4}{41}\right) + j\left(\frac{5}{41} - \frac{10}{R_L^2 + 100}\right) \text{ siemens}$$

For resonance the j part of **Y** must be zero, i.e. $5/41 = 10/(R_L^2 + 100)$, from which $R_L^2 = -18$. Thus there is no value of R_L which results in resonance.

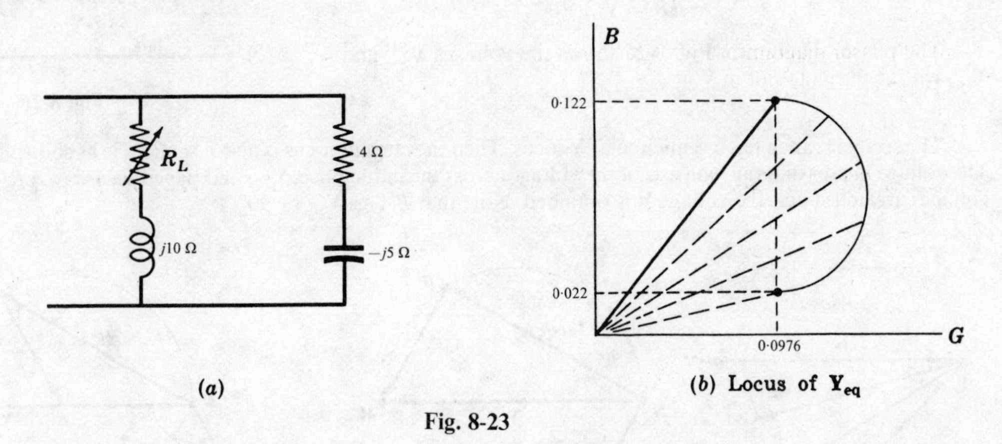

(*a*) (*b*) Locus of $\mathbf{Y}_{eq}$

Fig. 8-23

The fixed branch admittance is $1/(4 - j5) = 0{\cdot}0976 + j0{\cdot}122$ S. The semicircular locus of the adjustable branch has a radius $r = 1/2X_L = 1/20 = 0{\cdot}05$. The diameter is therefore 0·10. Since the fixed branch capacitive susceptance was 0·122 S, the locus of the variable branch does not cross the real axis and resonance is not possible.

8.18. Obtain the current locus for the circuit of Fig. 8-24(*a*) and find the value of R_C which results in a phase angle of 45° between **V** and **I**.

The fixed branch admittance is $1/R = 0{\cdot}1$ siemens. The semicircular locus of the RC branch has a radius $r = 1/2X_C = 1/8 = 0{\cdot}125$. See Fig. 8-24(*b*).

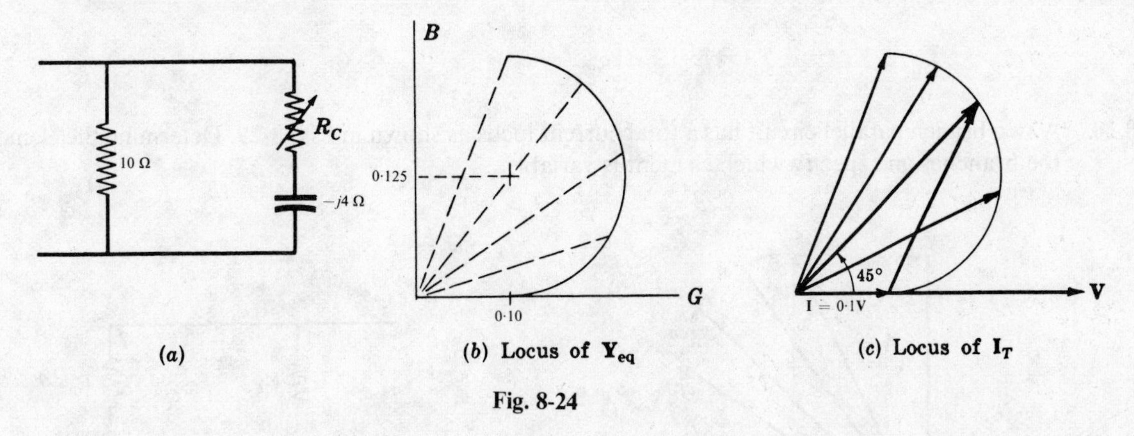

(*a*) (*b*) Locus of $\mathbf{Y}_{eq}$ (*c*) Locus of $\mathbf{I}_T$

Fig. 8-24

From Fig. 8-24(*c*) the current leads the voltage by 45° at the point shown. It follows that the real and imaginary parts of $\mathbf{Y}_T$ must be equal. If $\mathbf{Y}_T = \left(0{\cdot}1 + \frac{R_C}{R_C^2 + 16}\right) + j\left(\frac{4}{R_C^2 + 16}\right)$ siemens, then $0{\cdot}1 + \frac{R_C}{R_C^2 + 16} = \frac{4}{R_C^2 + 16}$ and $R_C = 2$ ohms

8.19. The circuit in Fig. 8-25 was examined in Prob. 6.18. It was determined that the absolute value of $\mathbf{V}_{AB}$ was constant, i.e. $V_{AB} = \frac{1}{2}V$, and the phasor V_{AB} lagged the applied voltage $\mathbf{V}$ by 2θ, where $\theta = \tan^{-1} \omega L/R$. Show these results graphically.

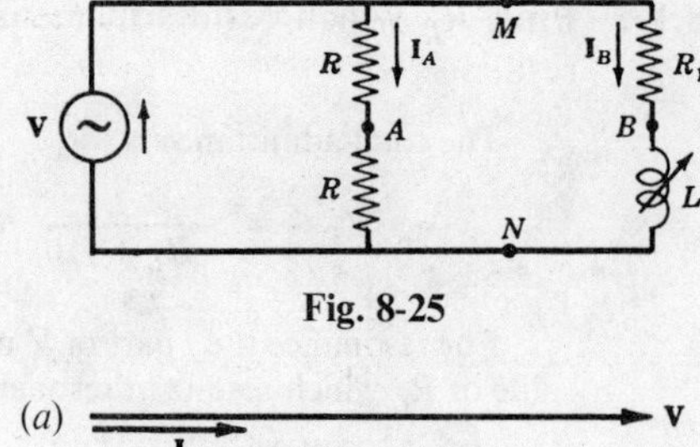

Fig. 8-25

In the first branch, $\mathbf{Z} = 2R$, $\mathbf{Y} = 1/2R$ and the current $\mathbf{I}_A = \mathbf{V}/2R$. Then the voltage across each resistor is

$$\mathbf{V}_R = \mathbf{I}R = \mathbf{V}/2$$

The phasor diagram in Fig. 8-26 shows the voltages $\mathbf{V}_{AN}$ and $\mathbf{V}_{MA}$ with A the midpoint of $\mathbf{V}$.

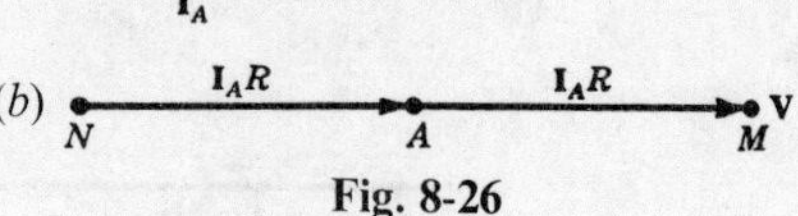

Fig. 8-26

The second branch has a semicircular $\mathbf{Y}$ locus. Then the current locus is also a semicircle as shown in Fig. 8-27(*a*). The voltage phasor diagram consists of the voltage across the inductance, $\mathbf{V}_{BN}$, and the voltage across R_1, V_{MB}. The two voltages are added and the voltage $\mathbf{V}$ is obtained. Note that I_B lags $\mathbf{V}_{BN}$ by 90°.

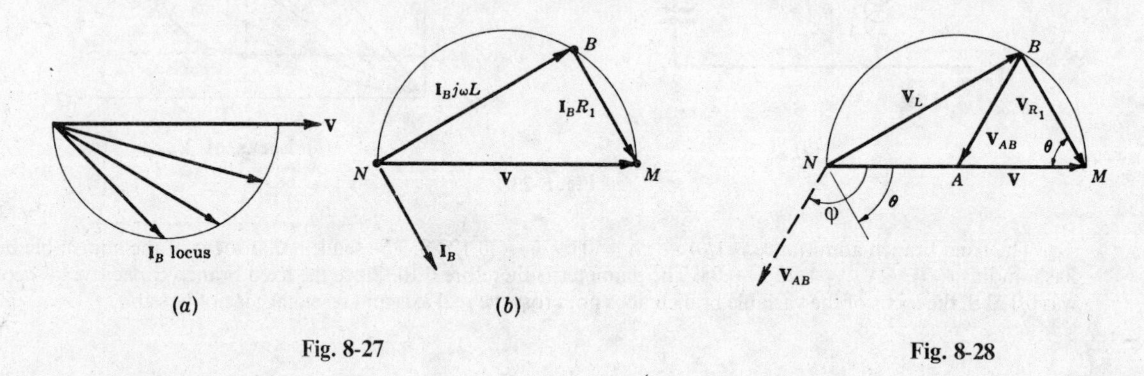

Fig. 8-27 Fig. 8-28

The voltages $\mathbf{V}_{BN}$ and $\mathbf{V}_{MB}$ are at right angles for all values of L. As L varies from 0 to ∞, B moves from M to N on the semicircular locus.

Now the two voltage phasor diagrams, Fig. 8-26(*b*) and 8-27(*b*) are superimposed in Fig. 8-28. It can be seen that $\mathbf{V}_{AB}$ is the radius $\frac{1}{2}V$ of the semicircle and therefore constant in magnitude. Furthermore, the angle φ by which $\mathbf{V}_{AB}$ lags $\mathbf{V}$ is found equal to 2θ, where $\theta = \tan^{-1} \omega L/R$.

8.20. A two branch parallel circuit has a total current locus as shown in Fig. 8-29. Determine the elements in the branches and specify which element is variable.

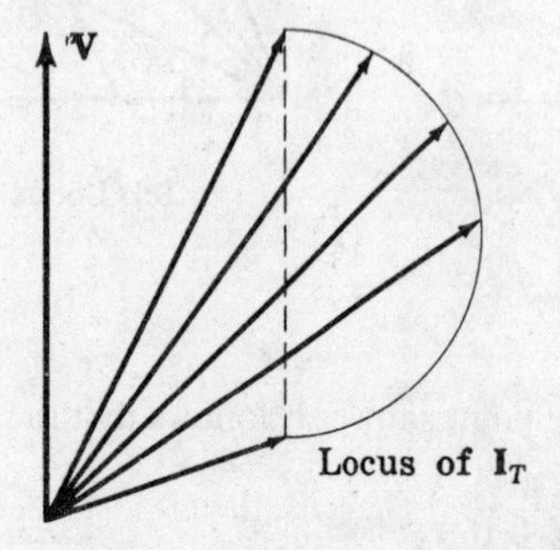

Fig. 8-29

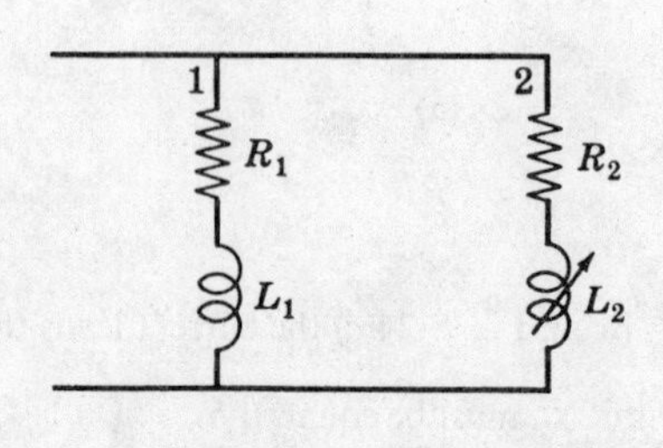

Rig. 8-30

The point at the bottom of the semicircle corresponds to the condition where the current in the variable branch is zero. Therefore the total current at this same point results entirely from the fixed branch 1. Since this current lags the voltage, the fixed branch must contain R_1 and L_1.

The semicircular locus of the current in branch 2 shows that the current is in phase with the voltage at its maximum value. At all other points on the locus, $\mathbf{I}_2$ lags $\mathbf{V}$. Therefore branch 2 consists of R_2 and L_2 with the inductance variable, as shown in Fig. 8-30 above.

Supplementary Problems

8.21. In the series *RLC* circuit of Fig. 8-31 below, the instantaneous voltage and current are $v = 70{\cdot}7 \sin(500t + 30°)$ volts and $i = 2{\cdot}83 \sin(500t + 30°)$ amperes. Find R and C.
Ans. $R = 25$ ohms, $C = 8\ \mu\text{F}$

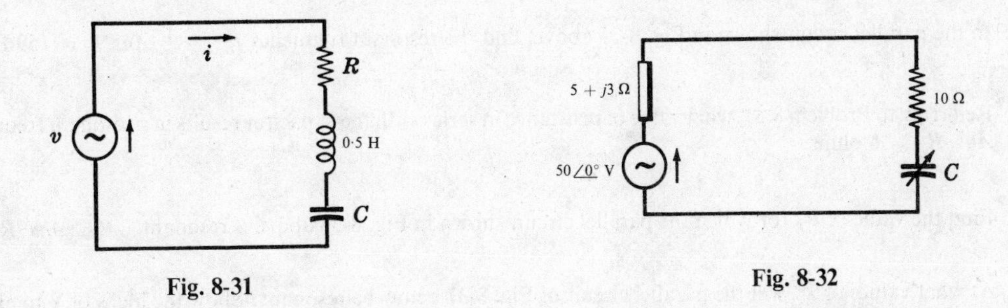

Fig. 8-31

Fig. 8-32

8.22. In the series circuit of Fig. 8-32 above, the impedance of the source is $5 + j3\ \Omega$ and the source frequency is 2000 Hz. At what value of C will the power in the 10 ohm resistor be a maximum? *Ans.* $C = 26{\cdot}6\ \mu\text{F}$, $P = 111$ W

8.23. A series *RLC* circuit with $L = 25$ mH and $C = 75\ \mu\text{F}$ has a lagging phase angle of 25° at $\omega = 2000$ rad/s. At what frequency will the phase angle be 25° leading? Find ω_0. *Ans.* $\omega = 267$ rad/s, $\omega_0 = 730$ rad/s

8.24. A series *RLC* circuit with $L = 0{\cdot}5$ H has an instantaneous voltage $v = 70{\cdot}7 \sin(500 + 30°)$ volts and an instantaneous current $i = 1{\cdot}5 \sin(500t)$ amperes. Find the values of R and C. At what frequency ω_0 will the circuit be resonant?
Ans. $R = 40{\cdot}8$ ohms, $C = 8{\cdot}83\ \mu\text{F}$, $\omega_0 = 476$ rad/s

8.25. A series circuit with $R = 10$ ohms, $L = 0{\cdot}2$ H and $C = 40\ \mu\text{F}$ has an applied voltage with a variable frequency. Find the frequencies f_1, f_0, and f_2 at which the current leads the voltage by 30°, is in phase, and lags the voltage by 30° respectively.
Ans. $f_1 = 54{\cdot}0$ Hz, $f_0 = 56{\cdot}3$ Hz, $f_2 = 58{\cdot}6$ Hz

8.26. A series *RLC* circuit with $R = 25$ ohms and $L = 0{\cdot}6$ H results in a leading phase angle of 60° at a frequency of 40 Hz. At what frequency will the circuit be resonant? *Ans.* $f_0 = 45{\cdot}4$ Hz

8.27. In the series circuit shown in Fig. 8-33, the frequency is varied until the voltage across the capacitor is a maximum. If the effective applied voltage is 100 volts, find the maximum capacitor voltage and the frequency at which it occurs. *Ans.* $\omega = 707$ rad/s, $V_C = 115{\cdot}5$ V

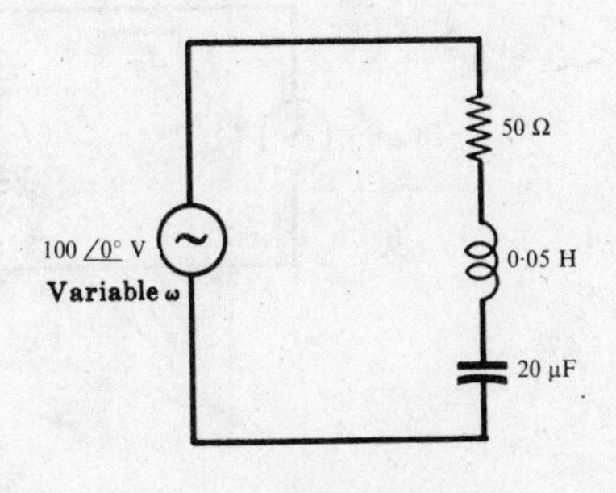

Fig. 8-33

8.28. The quality factor of the series circuit in Problem 8.27 was $Q_0 = \omega_0 L/R = 1$. Let $R = 10$ ohms which results in a Q_0 of 5 and find the frequency at which the voltage across the capacitor is maximum. Repeat for $R = 5$ ohms.
Ans. $\omega = 990$ rad/s, 998 rad/s. *Note.* With $Q_0 \geq 10$, it may be assumed that the maximum voltages across R, L and C all occur at the resonant frequency ω_0 or f_0.

8.29. In order to show the effect of Q on the current magnitude near the resonant frequency, plot the absolute value of Y vs ω for the two circuits with constants as follows. Circuit 1: $R = 5$ ohms, $L = 0{\cdot}05$ H and $C = 20\ \mu$F. Circuit 2: $R = 10$ ohms, $L = 0{\cdot}05$ H and $C = 20\ \mu$F.

8.30. In the parallel circuit of Fig. 8-34 below, $L = 0{\cdot}2$ H and $C = 30\ \mu$F. Determine the resonant frequency if $R_L = 0$ and compare it to the resonant frequency when $R = 50$ ohms. *Ans.* $\omega_0 = 408$ rad/s, $\omega_0 = 323$ rad/s

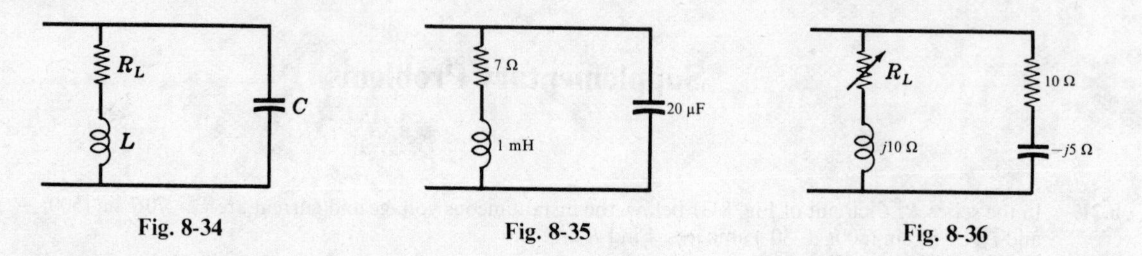

Fig. 8-34 Fig. 8-35 Fig. 8-36

8.31. In the parallel circuit shown in Fig. 8-35 above, find the resonant frequency f_0. *Ans.* $f_0 = 159$ Hz

8.32. Referring to Problem 8.31, what value of resistance in series with the capacitor results in a resonant frequency of 300 Hz? *Ans.* $R_C = 6$ ohms

8.33. Find the value of R_L for which the parallel circuit shown in Fig. 8-36 above is resonant. *Ans.* $R_L = 12{\cdot}25$ ohms

8.34. At what values of X_L will the parallel circuit of Fig. 8-37 below be resonant? Show the locus of **Y** to explain the results.

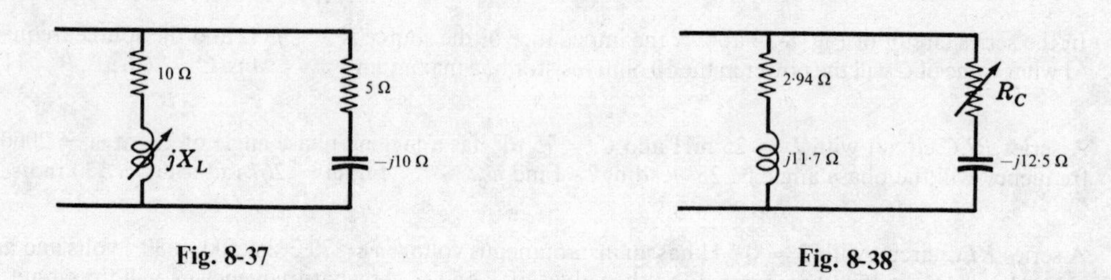

Fig. 8-37 Fig. 8-38

8.35. Find the value of R_C for which the parallel circuit of Fig. 8-38 above is resonant. Show the **Y** locus which explains this result. *Ans.* $R_C = 0$

8.36. The parallel circuit of Fig. 8-39 below is in resonance when $X_C = 9{\cdot}68\ \Omega$ and $X_C = 1{\cdot}65\ \Omega$. Find the total phasor current for each value of capacitive reactance. *Ans.* $1{\cdot}83\angle 0^\circ$ A, $3{\cdot}61\angle 0^\circ$ A

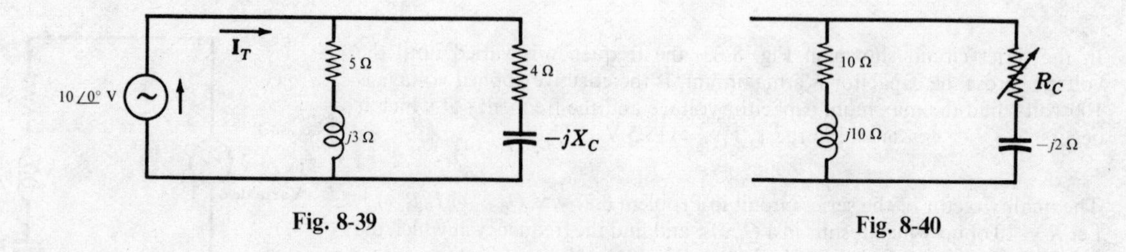

Fig. 8-39 Fig. 8-40

8.37. What value of R_C yields resonance in the parallel circuit shown in Fig. 8-40 above? *Ans.* $R_C = 6$ ohms

8.38. A voltage $\mathbf{V} = 50\angle 0^\circ$ V is applied to a series circuit consisting of a fixed inductive reactance $X_L = 5$ ohms and a variable resistance R. Sketch the admittance and current locus diagrams.

8.39. A voltage $\mathbf{V} = 50\angle 0^\circ$ V is applied to a series circuit of fixed resistance $R = 5$ ohms and variable capacitance C. Sketch the admittance and current locus diagrams.

8.40. In the parallel circuit of Fig. 8-41 below, the inductance is variable without limit. Construct the admittance locus diagram to show why it is not possible to obtain resonance.

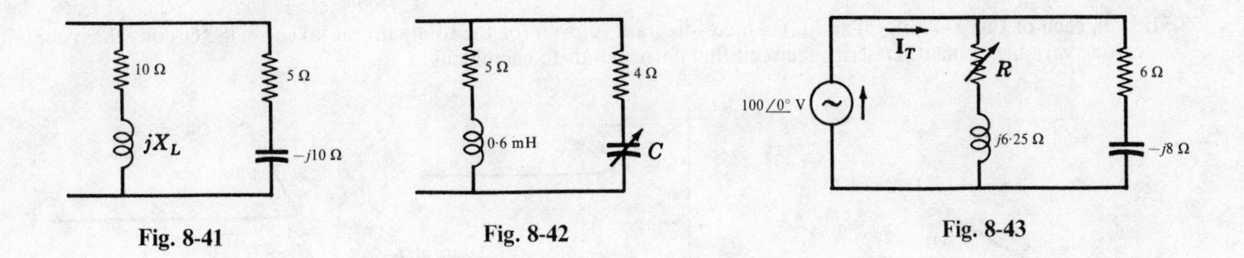

Fig. 8-41　　Fig. 8-42　　Fig. 8-43

8.41. The circuit shown in Fig. 8-42 above is resonant for two values of capacitance C when $\omega = 5000$ rad/s. Find both values of C and construct the admittance locus diagram. *Ans.* 20·6 μF, 121 μF

8.42. In the parallel circuit shown in Fig. 8-43 above, $\mathbf{I}_T$ lags the applied voltage by 53·1° when $R = 0$. Then if $R = \infty$ (open circuit), $\mathbf{I}_T$ leads the voltage by the same angle. Construct the admittance locus diagram to illustrate this condition. At what value of R is the circuit resonant? *Ans.* $R = 6{\cdot}25$ ohms

8.43. Find the value of R which makes the parallel circuit of Fig. 8-44 below resonant, and construct the admittance locus diagram to explain the result.

8.44. Referring to Problem 8.43, what change in the inductive reactance will make it possible to obtain resonance with some value of the variable resistor R? *Ans.* $X_L \leq 8{\cdot}2$ ohms

8.45. Find the value of R which results in parallel resonance for the circuit shown in Fig. 8-45 below and draw the locus diagram. *Ans.* $R = 5{\cdot}34\,\Omega$

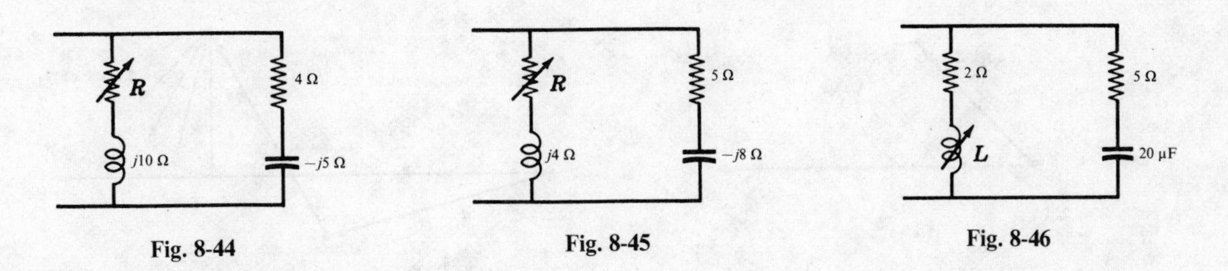

Fig. 8-44　　Fig. 8-45　　Fig. 8-46

8.46. In Problem 8.11 the parallel circuit was to be made resonant by varying the capacitance C. Use the admittance locus diagram to show why only one value of C resulted in resonance instead of the usual two.

8.47. The parallel circuit of Fig. 8-46 above is to be made resonant by varying L. Construct the admittance locus diagram and determine the values of L for resonance if $\omega = 5000$ rad/s. *Ans.* $L = 2{\cdot}43$ mH, 0·066 mH

8.48. Referring to the admittance locus of Problem 8.47, find the value of L which results in a minimum total current. What would be the magnitude of this current with an applied voltage of 100 volts effective?
Ans. $L = 2{\cdot}95$ mH, $I_T = 5{\cdot}1$ A

8.49. Referring to Problem 8·47, apply a voltage $\mathbf{V} = 150\angle 75°$ V and compute $\mathbf{I}_T$ for each value of L which caused the circuit to be resonant. *Ans.* $\mathbf{I}_T = 7{\cdot}98\angle 75°$ A, $78{\cdot}9\angle 75°$ A

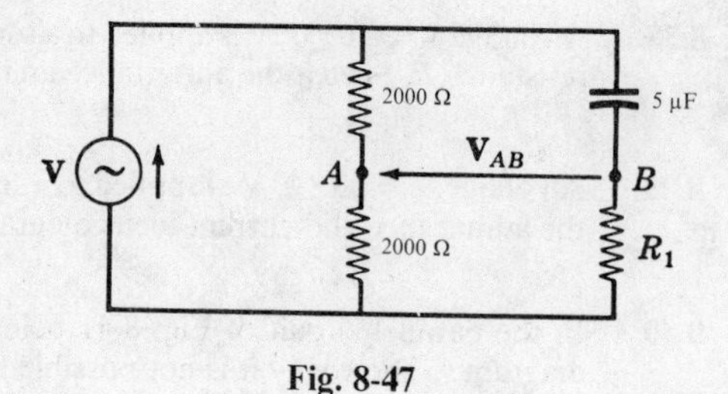

Fig. 8-47

8.50 In the phase shifting circuit shown in Fig. 8-47, the voltage $\mathbf{V}_{AB}$ is to be shifted between lag angles of 10° to 170° with respect to the applied voltage $\mathbf{V}$. At a frequency of 60 Hz, what range of R_1 is needed to satisfy the voltage shift?
Ans. 46·4 to 6080 ohms

8.51. In each of Fig. 8-48(*a*), (*b*) and (*c*), a locus diagram is given for the total current taken by the circuit which contains one variable element. Describe a circuit that corresponds to each locus.

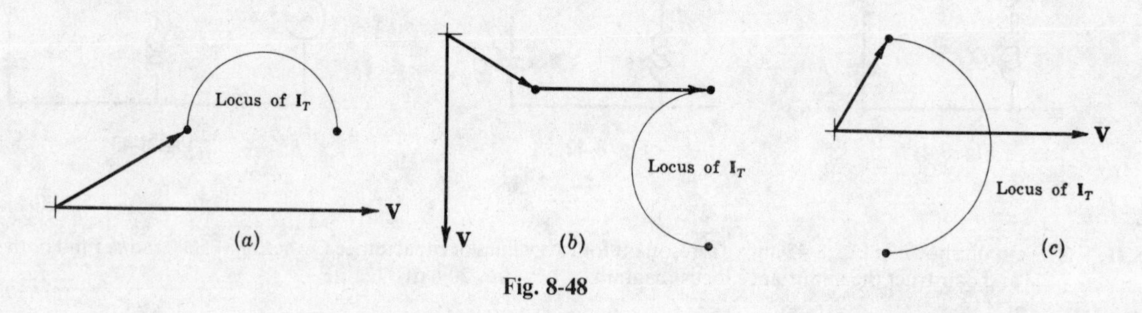

Fig. 8-48

Ans. (*a*) A two-branch parallel circuit. Branch 1: fixed R and X_C; branch 2: fixed R and variable X_C.
(*b*) A three-branch parallel circuit. Branch 1: fixed R and X_C; branch 2; fixed X_C; branch 3; fixed R and variable X_L.
(*c*) A two-branch parallel circuit. Branch 1: fixed R and X_C; branch 2: fixed X_L and variable R.

8.52. Find the circuit constants and their connection corresponding to the current locus of Fig. 8-49 if $\omega = 2000$ rad/s.
Ans. Branch 1: $R = 7{\cdot}07\ \Omega$, $L = 3{\cdot}54$ mH.
Branch 2: $R = 7{\cdot}07\ \Omega$, variable C.

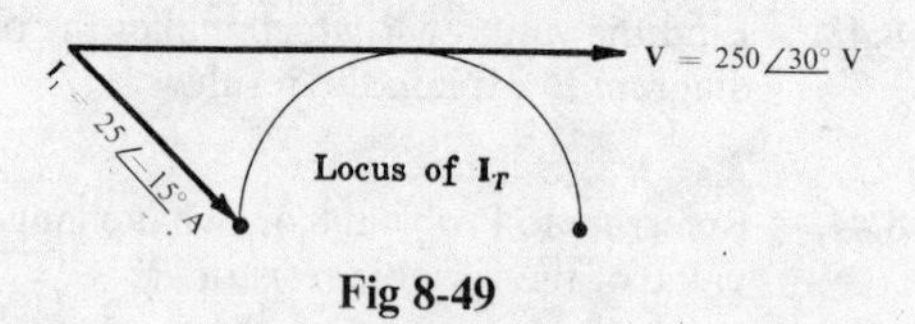

Fig 8-49

8.53. A two-branch parallel circuit has a current locus diagram as shown in Fig. 8-50. What change in the RL branch will make point A lie on the voltage phasor? *Ans.* Set X_L to 5·78 ohms.

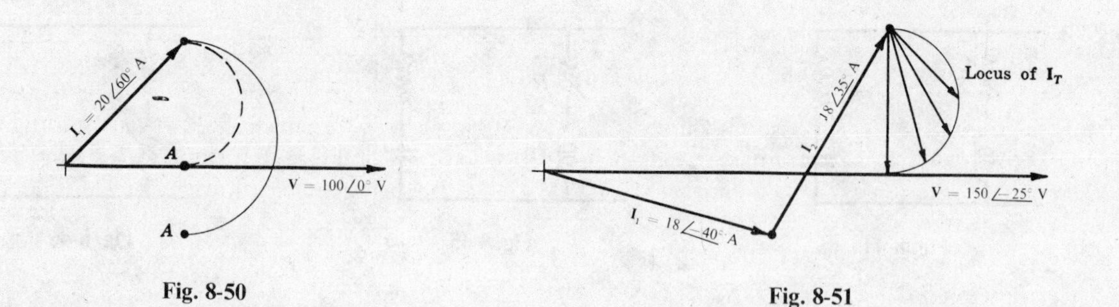

Fig. 8-50

Fig. 8-51

8.54. A three-branch parallel circuit has a current locus as given in Fig. 8-51. Determine all of the circuit constants if $\omega = 5000$ rad/s.
Ans. Branch 1: $R = 8{\cdot}05\ \Omega$, $L = 0{\cdot}423$ mH. Branch 2: $R = 4{\cdot}16\ \Omega$, $C = 27{\cdot}7\ \mu$F. Branch 3: $L = 2{\cdot}74$ mH and variable R.

CHAPTER 9

Mesh Current Network Analysis

INTRODUCTION

The voltage sources in an electric circuit or network cause currents in each of the branches and corresponding voltages across the circuit elements. The solution of the network consists in finding the currents in the branches or the voltages across the elements.

MESH CURRENTS

To apply the *mesh current method,* select closed loops of current called *mesh currents* or *loop currents* as in Fig. 9-1 below. Then write three equations in the unknowns $\mathbf{I}_1$, $\mathbf{I}_2$ and $\mathbf{I}_3$ and solve them. Now the current in any branch is given either directly by one of the mesh currents or by a combination of them.

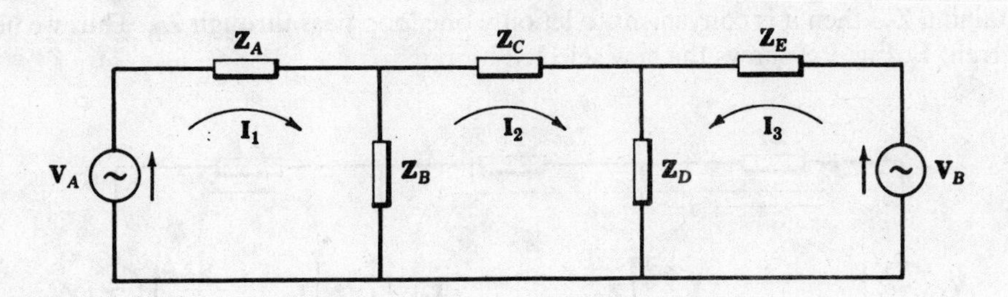

Fig. 9-1. Mesh or loop currents in a network

Thus the current in $\mathbf{Z}_A$ is $\mathbf{I}_1$ and the current in $\mathbf{Z}_B$, assuming a positive direction downward through the impedance, is $\mathbf{I}_1 - \mathbf{I}_2$. The current in any branch of the network is obtained in a similar manner. The voltage across any circuit element is then the product of the phasor current in the element and the complex impedance.

To obtain the set of three equations, apply Kirchhoff's voltage law to each current loop. The $\mathbf{I}_1$ loop is redrawn in Fig. 9-2 and the sum of the voltage drops around the loop is equated to the sum of the voltage rises.

$$\mathbf{I}_1\mathbf{Z}_A + (\mathbf{I}_1 - \mathbf{I}_2)\mathbf{Z}_B = \mathbf{V}_A \qquad (1)$$

The second loop contains no source; therefore the sum of the voltage drops is zero.

$$\mathbf{I}_2\mathbf{Z}_C + (\mathbf{I}_2 + \mathbf{I}_3)\mathbf{Z}_D + (\mathbf{I}_2 - \mathbf{I}_1)\mathbf{Z}_B = 0 \qquad (2)$$

Applying Kirchhoff's voltage law to the third loop,

$$\mathbf{I}_3\mathbf{Z}_E + (\mathbf{I}_3 + \mathbf{I}_2)\mathbf{Z}_D = \mathbf{V}_B \qquad (3)$$

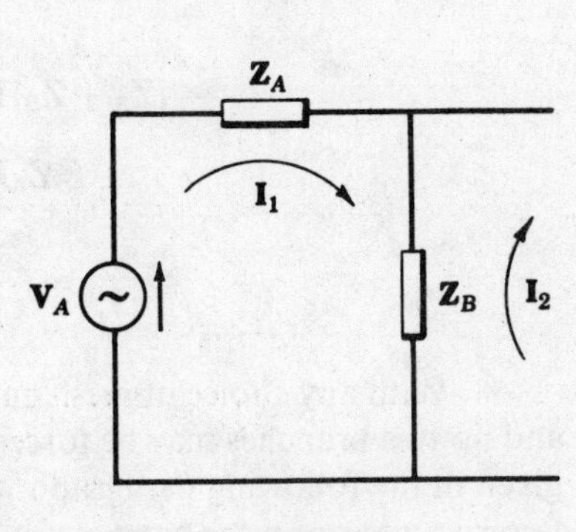

Fig. 9-2

Rearranging,

$$
\begin{aligned}
(\mathbf{Z}_A + \mathbf{Z}_B)\mathbf{I}_1 \; - \; & \mathbf{Z}_B\mathbf{I}_2 & &= \mathbf{V}_A \qquad (1')\\
-\mathbf{Z}_B\mathbf{I}_1 \; + \; & (\mathbf{Z}_B + \mathbf{Z}_C + \mathbf{Z}_D)\mathbf{I}_2 \; + & \mathbf{Z}_D\mathbf{I}_3 &= 0 \qquad (2')\\
& \mathbf{Z}_D\mathbf{I}_2 \; + & (\mathbf{Z}_D + \mathbf{Z}_E)\mathbf{I}_3 &= \mathbf{V}_B \qquad (3')
\end{aligned}
$$

We can derive the above set of equations directly. Consider loop one shown in Fig. 9-2. The direction of the current $\mathbf{I}_1$ is taken clockwise and all voltage drops in loop one caused by $\mathbf{I}_1$ are positive. Mesh current $\mathbf{I}_2$ also flows in $\mathbf{Z}_B$, but in opposite direction to $\mathbf{I}_1$. Then the voltage drop in $\mathbf{Z}_B$ caused by $\mathbf{I}_2$ is $-\mathbf{Z}_B\mathbf{I}_2$. The voltage $\mathbf{V}_A$ is positive because it has the same direction as $\mathbf{I}_1$. Now with these considerations apply Kirchhoff's law to loop one and equation (*1′*) follows. Equations (*2′*) and (*3′*) are derived similarly.

The terms *voltage rise* and *voltage drop* are carryovers from basic DC circuits where their meaning is clearer than in sinusoidal circuits where instantaneous currents and voltages have positive and negative intervals. In the sinusoidal steady state, Kirchhoff's voltage law applied to a closed loop results in a phasor equality where *the phasor sum of the voltages across the loop impedances is equal to the phasor sum of all voltage sources acting in the loop.*

CHOICE OF MESH CURRENTS

In applying the mesh current method, it is possible to simplify the solution of a given problem by the appropriate choice of the loops in the network. If in Fig. 9-1 it is required to determine only the current in the branch containing $\mathbf{Z}_B$, then it is convenient to let only one loop pass through $\mathbf{Z}_B$. Thus we need solve only for the mesh current $\mathbf{I}_1$. Fig. 9-3 shows the new selected loops.

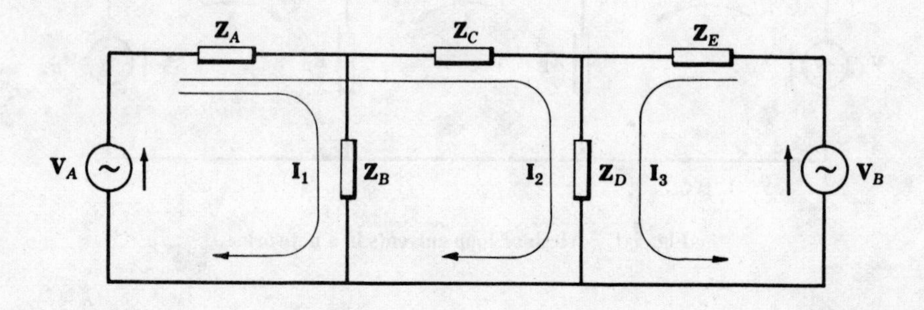

Fig. 9-3

The corresponding set of mesh current equations is

$$
\begin{aligned}
(\mathbf{Z}_A + \mathbf{Z}_B)\mathbf{I}_1 \; + \; & \mathbf{Z}_A\mathbf{I}_2 & &= \mathbf{V}_A\\
\mathbf{Z}_A\mathbf{I}_1 \; + \; & (\mathbf{Z}_A + \mathbf{Z}_C + \mathbf{Z}_D)\mathbf{I}_2 \; + & \mathbf{Z}_D\mathbf{I}_3 &= \mathbf{V}_A\\
& \mathbf{Z}_D\mathbf{I}_2 \; + & (\mathbf{Z}_D + \mathbf{Z}_E)\mathbf{I}_3 &= \mathbf{V}_B
\end{aligned}
$$

With any choice of mesh currents for the network, each circuit element must have at least one current, and no two branches may be forced to have the same current or the same combinations of currents. Rules are given in the following paragraph which indicate the required number of mesh currents to solve a network; a lesser number of mesh currents can never be a valid set.

NUMBER OF MESH CURRENTS REQUIRED

The number of mesh currents required by a simple coplanar network is apparent. More involved networks will require a method which gives the necessary number of equations.

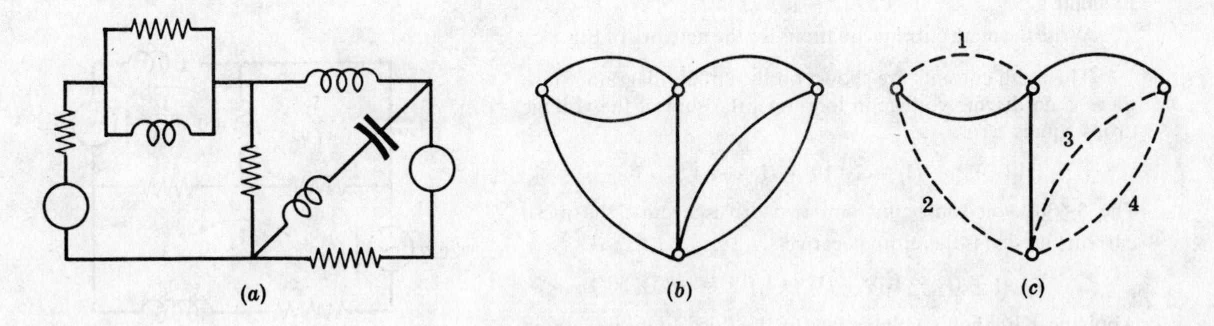

Fig. 9-4. A network, its graph and tree

A network graph is shown in Fig. 9-4(*b*) where junctions are small circles and connecting branches are replaced by lines. Next, the network tree of Fig. 9-4(*c*) is obtained by including only those branches which do not form closed paths. The network tree is not unique. The solid branches of Fig. 9-4(*c*) are called tree branches and the dotted branches link branches. Each link branch would result in a closed path. The number of mesh currents required by this network is the number of the link branches, 4. The same result is obtained by "cutting" branches of the original network such that each cut opens a closed path. When no more closed paths remain, the number of cuts is the number of mesh currents required.

A third method consists of counting the branches and junctions in the network. The necessary number of mesh currents is then given as

$$\text{number of equations} = \text{branches} - (\text{junctions} - 1)$$

In the network of Fig. 9-4(*a*) there are seven branches and four junctions. The required number of mesh currents would be $7 - (4 - 1) = 4$.

MESH EQUATIONS BY INSPECTION

The equations for a three mesh network in general notation are

$$\begin{aligned} \mathbf{Z}_{11}\mathbf{I}_1 \pm \mathbf{Z}_{12}\mathbf{I}_2 \pm \mathbf{Z}_{13}\mathbf{I}_3 &= \mathbf{V}_1 \\ \pm \mathbf{Z}_{21}\mathbf{I}_1 + \mathbf{Z}_{22}\mathbf{I}_2 \pm \mathbf{Z}_{23}\mathbf{I}_3 &= \mathbf{V}_2 \\ \pm \mathbf{Z}_{31}\mathbf{I}_1 \pm \mathbf{Z}_{32}\mathbf{I}_2 + \mathbf{Z}_{33}\mathbf{I}_3 &= \mathbf{V}_3 \end{aligned}$$

$\mathbf{Z}_{11}$ is called the self-impedance of loop one, given by the sum of all impedances through which $\mathbf{I}_1$ passes. $\mathbf{Z}_{22}$ and $\mathbf{Z}_{33}$ are the self-impedances of loops two and three, given by the sums of the impedances in their respective loops.

$\mathbf{Z}_{12}$ is the sum of all impedances common to mesh currents $\mathbf{I}_1$ and $\mathbf{I}_2$. It follows that $\mathbf{Z}_{12} = \mathbf{Z}_{21}$. The impedances $\mathbf{Z}_{13}$, $\mathbf{Z}_{31}$, $\mathbf{Z}_{23}$ and $\mathbf{Z}_{32}$ are the sums of the impedances common to the mesh currents indicated in their subscripts. The positive sign is used if both currents pass through the common impedance in the same direction, and the negative sign if they do not.

$\mathbf{V}_1$ is the sum of all voltages driving in loop one. The positive sign is used if the source drives in the direction of the mesh current, and the negative sign if it drives against the mesh current. $\mathbf{V}_2$ and $\mathbf{V}_3$ are the sums of the driving voltages in their respective loops.

Example 1.

Write the mesh current equations for the network of Fig. 9-5.

The mesh currents are shown on the circuit diagram. Since there is no driving voltage in loop one, the sum of the voltage drops equals zero.

$$\mathbf{I}_1(-j8) + (\mathbf{I}_1 - \mathbf{I}_2)\,10 + (\mathbf{I}_1 - \mathbf{I}_3)\,5 = 0$$

The $5\angle 30°$ volt source in loop two drives against the mesh current; its sign is therefore negative.

$$\mathbf{I}_2(j4) + (\mathbf{I}_2 - \mathbf{I}_3)8 + (\mathbf{I}_2 - \mathbf{I}_1)10 = -(5\angle 30°)$$

Applying Kirchhoff's voltage law to the third loop, we obtain

$$\mathbf{I}_3(3 + j4) + (\mathbf{I}_3 - \mathbf{I}_1)5 + (\mathbf{I}_3 - \mathbf{I}_2)8 = -(10\angle 0°)$$

Rearranging terms, the set of three equations is

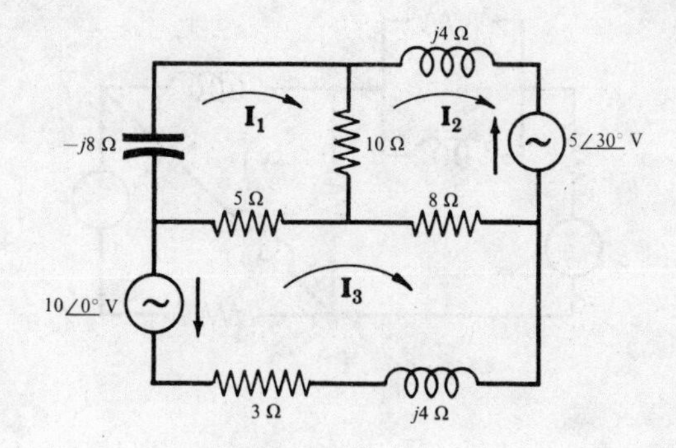

Fig. 9-5

$$\begin{aligned}
(15 - j8)\mathbf{I}_1 - 10\mathbf{I}_2 - 5\mathbf{I}_3 &= 0 \\
-10\mathbf{I}_1 + (18 + j4)\mathbf{I}_2 - 8\mathbf{I}_3 &= -(5\angle 30°) \\
-5\mathbf{I}_1 - 8\mathbf{I}_2 + (16 + j4)\mathbf{I}_3 &= -(10\angle 0°)
\end{aligned}$$

Compare the above set of equations with the set of equations of the three mesh network given in general notation. Then the self-impedance of loop one is $\mathbf{Z}_{11} = (5 + 10 - j8) = 15 - j8\ \Omega$. The impedance common to both loop one and loop two is $\mathbf{Z}_{12} = 10\,\Omega$. However, I_2 is opposite in direction to the current $\mathbf{I}_1$; hence $\mathbf{Z}_{12}$ has negative sign. Similarly, the impedance common to loops one and three is $\mathbf{Z}_{13} = -5\,\Omega$. Note that $\mathbf{Z}_{12} = \mathbf{Z}_{21}$, $\mathbf{Z}_{13} = \mathbf{Z}_{31}$, $\mathbf{Z}_{23} = \mathbf{Z}_{32}$.

The driving voltage in loop two is $5\angle 30°$ V, but it drives against the mesh current and therefore carries a negative sign. Each of the terms in the above set of equations may be checked with the general notation.

MATRICES

A matrix is a rectangular array of numbers or functions enclosed in a pair of brackets and subject to certain rules of operation. In the matrix

$$A = \begin{bmatrix} a_{11} & a_{12} & a_{13} & \dots & a_{1n} \\ a_{21} & a_{22} & a_{23} & \dots & a_{2n} \\ \dots & \dots & \dots & \dots & \dots \\ \dots & \dots & \dots & \dots & \dots \\ a_{m1} & a_{m2} & a_{m3} & \dots & a_{mn} \end{bmatrix}$$

the numbers or functions a_{ij} are called its elements. An element a_{ij} is in row i and column j. This matrix, of m rows and n columns, is of order "m by n" or "$m \times n$" and is called "the matrix A" or "the $m \times n$ matrix A" or "the $m \times n$ matrix $[a_{ij}]$".

Two matrices are equal if and only if one is the exact duplicate of the other.

ADDITION OF MATRICES

Two matrices of the same order are conformable for addition or subtraction; two matrices of different orders cannot be added or subtracted.

The sum (difference) of two $m \times n$ matrices, $A = [a_{ij}]$ and $B = [b_{ij}]$, is the $m \times n$ matrix C where each element of C is the sum (difference) of the corresponding elements of A and B. Thus $A \pm B = [a_{ij} \pm b_{ij}]$.

Example 2. If $A = \begin{bmatrix} 1 & 4 & 0 \\ 2 & 7 & 3 \end{bmatrix}$ and $B = \begin{bmatrix} 5 & 2 & 6 \\ 0 & 1 & 1 \end{bmatrix}$, then

$$A + B = \begin{bmatrix} 1+5 & 4+2 & 0+6 \\ 2+0 & 7+1 & 3+1 \end{bmatrix} = \begin{bmatrix} 6 & 6 & 6 \\ 2 & 8 & 4 \end{bmatrix} \quad \text{and} \quad A - B = \begin{bmatrix} -4 & 2 & -6 \\ 2 & 6 & 2 \end{bmatrix}$$

MULTIPLICATION OF MATRICES

The product AB, in that order, of the $1 \times m$ matrix $A = [a_{11}\, a_{12}\, a_{13} \ldots a_{1m}]$ and the

$$m \times 1 \text{ matrix } B = \begin{bmatrix} b_{11} \\ b_{21} \\ b_{31} \\ \ldots \\ b_{m1} \end{bmatrix} \text{ is the } 1 \times 1 \text{ matrix } C:$$

$$C = [a_{11}\ a_{12}\ \ldots\ a_{1m}] \cdot \begin{bmatrix} b_{11} \\ b_{21} \\ \ldots \\ b_{m1} \end{bmatrix} = [a_{11}b_{11} + a_{12}b_{21} + \cdots + a_{1m}b_{m1}] = \left[\sum_{k=1}^{m} a_{1k}\, b_{k1}\right]$$

Note that each element of the row is multiplied into each corresponding element of the column and then the products are summed.

Example 3. $[1\ 3\ 5] \begin{bmatrix} 2 \\ 4 \\ -2 \end{bmatrix} = [1(2) + 3(4) + 5(-2)] = [4]$

The product AB in that order of the $m \times s$ matrix $A = [a_{ij}]$ and the $s \times n$ matrix $B = [b_{ij}]$ is the $m \times n$ matrix $C = [c_{ij}]$ where

$$c_{ij} = \sum_{k=1}^{s} a_{ik}\, b_{kj}, \quad i = 1, 2, \ldots, m, \quad j = 1, 2, \ldots, n$$

Example 4. $\begin{bmatrix} a_{11} & a_{12} \\ a_{21} & a_{22} \\ a_{31} & a_{32} \end{bmatrix} \begin{bmatrix} b_{11} & b_{12} \\ b_{21} & b_{22} \end{bmatrix} = \begin{bmatrix} a_{11}b_{11} + a_{12}b_{21} & a_{11}b_{12} + a_{12}b_{22} \\ a_{21}b_{11} + a_{22}b_{21} & a_{21}b_{12} + a_{22}b_{22} \\ a_{31}b_{11} + a_{32}b_{21} & a_{31}b_{12} + a_{32}b_{22} \end{bmatrix}$

Example 5. $\begin{bmatrix} 3 & 5 & -8 \\ 2 & 1 & 6 \\ 4 & -6 & 7 \end{bmatrix} \begin{bmatrix} I_1 \\ I_2 \\ I_3 \end{bmatrix} = \begin{bmatrix} 3I_1 + 5I_2 - 8I_3 \\ 2I_1 + 1I_2 + 6I_3 \\ 4I_1 - 6I_2 + 7I_3 \end{bmatrix}$

Example 6. $\begin{bmatrix} 5 & -3 \\ 4 & 2 \end{bmatrix} \begin{bmatrix} 8 & -2 & 6 \\ 7 & 0 & 9 \end{bmatrix} = \begin{bmatrix} 5(8) + (-3)(7) & 5(-2) + (-3)(0) & 5(6) + (-3)(9) \\ 4(8) + 2(7) & 4(-2) + 2(0) & 4(6) + 2(9) \end{bmatrix}$

$$= \begin{bmatrix} 19 & -10 & 3 \\ 46 & -8 & 42 \end{bmatrix}$$

Matrix A is conformable to matrix B for multiplication, i.e. the product AB is defined, only when the number of columns of A is equal to the number of rows of B. Thus if A is a 3×2 matrix and B is a 2×5 matrix, then the product AB is defined but the product BA is not defined. If D is a 3×3 matrix and E is a 3×3 matrix, both products AB and BA are defined.

INVERSION

In an arrangement of positive integers, an inversion exists when a larger integer precedes a smaller integer.

For example, in 132 the integer 3 precedes 2; hence there is 1 inversion. In 321 the integer 3 precedes 2 and 1, and the integer 2 precedes 1; hence there are 3 inversions. In 4213 the integer 4 precedes 2, 1 and 3, and the integer 2 precedes 1; thus three are 4 inversions. In 3421 the integer 3 precedes 2 and 1, the integer 4 precedes 2 and 1, and the integer 2 precedes 1; thus there are 5 inversions.

DETERMINANT OF A SQUARE MATRIX

Let us take n elements of the n-square matrix

$$A = \begin{bmatrix} a_{11} & a_{12} & a_{13} & \cdots & a_{1n} \\ a_{21} & a_{22} & a_{23} & \cdots & a_{2n} \\ \cdots & \cdots & \cdots & \cdots & \cdots \\ \cdots & \cdots & \cdots & \cdots & \cdots \\ a_{n1} & a_{n2} & a_{n3} & \cdots & a_{nn} \end{bmatrix}$$

and form a product $a_{1j_1}\, a_{2j_2}\, a_{3j_3} \cdots a_{nj_n}$ such that one and only one element belongs to any row and one and only one element belongs to any column. Note that the sequence of first subscripts is, for convenience, in the order $1, 2, \ldots, n$; then the sequence $j_1, j_2, \ldots, j_n$ of second subscripts is one of the $n!$ permutations of the integers $1, 2, \ldots, n$. Associate a $+$ or $-$ sign with the product according as the number of inversions of the second subscripts is even or odd.

Then the determinant of an n-square matrix A, written $|A|$, is the sum of all $n!$ such different signed products which can be formed from the elements of A.

The determinant of a square matrix of order n is called a determinant of order n.

Example 7.

$$\begin{vmatrix} a_{11} & a_{12} \\ a_{21} & a_{22} \end{vmatrix} = a_{11}a_{22} - a_{12}a_{21}$$

Example 8.

$$\begin{vmatrix} a_{11} & a_{12} & a_{13} \\ a_{21} & a_{22} & a_{23} \\ a_{31} & a_{32} & a_{33} \end{vmatrix} = a_{11}a_{22}a_{33} - a_{11}a_{23}a_{32} - a_{12}a_{21}a_{33} + a_{12}a_{23}a_{31} + a_{13}a_{21}a_{32} - a_{13}a_{22}a_{31}$$

MINORS AND COFACTORS

The minor of an element a_{ij} of a determinant of order n is the determinant of order $(n - 1)$ obtained by deleting the row and column which contains the given element. The minor of an element a_{ij} is denoted by $|M_{ij}|$.

The signed minor, $(-1)^{i+j}|M_{ij}|$, is called the cofactor of a_{ij} and is denoted by Δ_{ij}.

Example 9.

For the third order determinant $|A| = \begin{vmatrix} a_{11} & a_{12} & a_{13} \\ a_{21} & a_{22} & a_{23} \\ a_{31} & a_{32} & a_{33} \end{vmatrix}$,

$$|M_{23}| = \begin{vmatrix} a_{11} & a_{12} \\ a_{31} & a_{32} \end{vmatrix} \quad \text{and} \quad \Delta_{23} = (-1)^{2+3}\begin{vmatrix} a_{11} & a_{12} \\ a_{31} & a_{32} \end{vmatrix} = -\begin{vmatrix} a_{11} & a_{12} \\ a_{31} & a_{32} \end{vmatrix}$$

VALUE OF A DETERMINANT

The value of a determinant $|A|$ of order n is the sum of the n products obtained by multiplying each element of any chosen row (column) of $|A|$ by its cofactor. Thus

$$|A| = \begin{vmatrix} a_{11} & a_{12} & a_{13} \\ a_{21} & a_{22} & a_{23} \\ a_{31} & a_{32} & a_{33} \end{vmatrix} = a_{12}\Delta_{12} + a_{22}\Delta_{22} + a_{32}\Delta_{32}$$

$$= -a_{12}\begin{vmatrix} a_{21} & a_{23} \\ a_{31} & a_{33} \end{vmatrix} + a_{22}\begin{vmatrix} a_{11} & a_{13} \\ a_{31} & a_{33} \end{vmatrix} - a_{32}\begin{vmatrix} a_{11} & a_{13} \\ a_{21} & a_{23} \end{vmatrix}$$

is the expansion of $|A|$ along the second column.

Example 10. $$\begin{vmatrix} 1 & 4 & 7 \\ 2 & 1 & -6 \\ 3 & 5 & 0 \end{vmatrix} = 3\begin{vmatrix} 4 & 7 \\ 1 & -6 \end{vmatrix} - 5\begin{vmatrix} 1 & 7 \\ 2 & -6 \end{vmatrix} + 0\begin{vmatrix} 1 & 4 \\ 2 & 1 \end{vmatrix}$$

$$= 3\{4(-6) - 7(1)\} - 5\{1(-6) - 7(2)\} + 0 = 7$$

Example 11. $$\begin{vmatrix} 3 & 5 & 8 \\ 1 & 0 & 2 \\ 4 & 0 & 3 \end{vmatrix} = -5\begin{vmatrix} 1 & 2 \\ 4 & 3 \end{vmatrix} = -5\{1(3) - 2(4)\} = 25$$

Example 12. $$\begin{vmatrix} 4 & 7 & -2 \\ 0 & 5 & 0 \\ 8 & 2 & -3 \end{vmatrix} = 5\begin{vmatrix} 4 & -2 \\ 8 & -3 \end{vmatrix} = 5\{4(-3) - (-2)(8)\} = 20$$

PROPERTIES OF DETERMINANTS

1. If two rows (columns) of a determinant are indentical, the value of the determinant is zero. For example,

$$\begin{vmatrix} 1 & 8 & 1 \\ -4 & 2 & -4 \\ 6 & 1 & 6 \end{vmatrix} = 0$$

2. If each element of a row (column) of a determinant is multiplied by any number k, the determinant is multiplied by k. For example,

$$2\begin{vmatrix} 3 & -4 & 2 \\ -1 & 5 & 0 \\ 2 & 6 & 7 \end{vmatrix} = \begin{vmatrix} 6 & -8 & 4 \\ -1 & 5 & 0 \\ 2 & 6 & 7 \end{vmatrix} = \begin{vmatrix} 3 & -4 & 4 \\ -1 & 5 & 0 \\ 2 & 6 & 14 \end{vmatrix}$$

3. If any two rows (columns) of a determinant are interchanged, the sign of the determinant is changed. For example,

$$\begin{vmatrix} 1 & 4 & 7 \\ -2 & 5 & 8 \\ 3 & -6 & 9 \end{vmatrix} = -\begin{vmatrix} 4 & 1 & 7 \\ 5 & -2 & 8 \\ -6 & 3 & 9 \end{vmatrix} = -\begin{vmatrix} 3 & -6 & 9 \\ -2 & 5 & 8 \\ 1 & 4 & 7 \end{vmatrix}$$

4. If each element of a row (column) of a determinant is expressed as the sum of two or more numbers, the determinant may be written as the sum of two or more determinants. For example,

$$\begin{vmatrix} 3 & -7 & 5 \\ 2 & 4 & -5 \\ 1 & 6 & 8 \end{vmatrix} = \begin{vmatrix} 3 & -9+2 & 5 \\ 2 & 4+0 & -5 \\ 1 & 8-2 & 8 \end{vmatrix} = \begin{vmatrix} 3 & -9 & 5 \\ 2 & 4 & -5 \\ 1 & 8 & 8 \end{vmatrix} + \begin{vmatrix} 3 & 2 & 5 \\ 2 & 0 & -5 \\ 1 & -2 & 8 \end{vmatrix}$$

5. If to the elements of any row (column) of a determinant there is added k times the corresponding element of any other row (column) the value of the determinant is unchanged. For example,

$$\begin{vmatrix} 1 & 9 & -3 \\ 4 & 6 & -2 \\ -3 & 1 & 5 \end{vmatrix} = \begin{vmatrix} 1 & 9+3(-3) & -3 \\ 4 & 6+3(-2) & -2 \\ -3 & 1+3(5) & 5 \end{vmatrix} = \begin{vmatrix} 1 & 0 & -3 \\ 4 & 0 & -2 \\ -3 & 16 & 5 \end{vmatrix}$$

SOLUTION OF LINEAR EQUATIONS BY DETERMINANTS. CRAMER'S RULE

The system of three linear equations in three unknowns x_1, x_2, x_3

$$\begin{aligned} a_{11}x_1 + a_{12}x_2 + a_{13}x_3 &= k_1 \\ a_{21}x_1 + a_{22}x_2 + a_{23}x_3 &= k_2 \\ a_{31}x_1 + a_{32}x_2 + a_{33}x_3 &= k_3 \end{aligned}$$

may be written in matrix form as

$$\begin{bmatrix} a_{11} & a_{12} & a_{13} \\ a_{21} & a_{22} & a_{23} \\ a_{31} & a_{32} & a_{33} \end{bmatrix} \begin{bmatrix} x_1 \\ x_2 \\ x_3 \end{bmatrix} = \begin{bmatrix} k_1 \\ k_2 \\ k_3 \end{bmatrix}$$

The numerical value of the coefficient determinant, Δ_a, is multiplied by x_1 if each element of the first column is multiplied by x_1 (Property 2).

$$\Delta_a = \begin{vmatrix} a_{11} & a_{12} & a_{13} \\ a_{21} & a_{22} & a_{23} \\ a_{31} & a_{32} & a_{33} \end{vmatrix} \quad \text{and} \quad x_1\Delta_a = \begin{vmatrix} a_{11}x_1 & a_{12} & a_{13} \\ a_{21}x_1 & a_{22} & a_{23} \\ a_{31}x_1 & a_{32} & a_{33} \end{vmatrix}$$

Now to each element of the first column of the last determinant add x_2 times the corresponding element of the second column and x_3 times the corresponding element of the third column (Property 5). Then

$$x_1\Delta_a = \begin{vmatrix} (a_{11}x_1 + a_{12}x_2 + a_{13}x_3) & a_{12} & a_{13} \\ (a_{21}x_1 + a_{22}x_2 + a_{23}x_3) & a_{22} & a_{23} \\ (a_{31}x_1 + a_{32}x_2 + a_{33}x_3) & a_{32} & a_{33} \end{vmatrix} = \begin{vmatrix} k_1 & a_{12} & a_{13} \\ k_2 & a_{22} & a_{23} \\ k_3 & a_{32} & a_{33} \end{vmatrix}$$

or

$$x_1 = \frac{\begin{vmatrix} k_1 & a_{12} & a_{13} \\ k_2 & a_{22} & a_{23} \\ k_3 & a_{32} & a_{33} \end{vmatrix}}{\Delta_a}$$

provided $\Delta_a \neq 0$. Similarly,

$$x_2 = \frac{\begin{vmatrix} a_{11} & k_1 & a_{13} \\ a_{21} & k_2 & a_{23} \\ a_{31} & k_3 & a_{33} \end{vmatrix}}{\Delta_a} \qquad x_3 = \frac{\begin{vmatrix} a_{11} & a_{12} & k_1 \\ a_{21} & a_{22} & k_2 \\ a_{31} & a_{32} & k_3 \end{vmatrix}}{\Delta_a}$$

This method of solution, called *Cramer's Rule*, can be applied to any system of n linear equations in n unknowns provided the coefficients determinant is not zero.

MATRIX METHODS AND CIRCUIT ANALYSIS

The three mesh current equations

$$\begin{aligned} \mathbf{Z}_{11}\mathbf{I}_1 \pm \mathbf{Z}_{12}\mathbf{I}_2 \pm \mathbf{Z}_{13}\mathbf{I}_3 &= \mathbf{V}_1 \\ \pm\mathbf{Z}_{21}\mathbf{I}_1 + \mathbf{Z}_{22}\mathbf{I}_2 \pm \mathbf{Z}_{23}\mathbf{I}_3 &= \mathbf{V}_2 \\ \pm\mathbf{Z}_{31}\mathbf{I}_1 \pm \mathbf{Z}_{32}\mathbf{I}_2 + \mathbf{Z}_{33}\mathbf{I}_3 &= \mathbf{V}_3 \end{aligned}$$

are now written in matrix form

$$\begin{bmatrix} \mathbf{Z}_{11} & \pm\mathbf{Z}_{12} & \pm\mathbf{Z}_{13} \\ \pm\mathbf{Z}_{21} & \mathbf{Z}_{22} & \pm\mathbf{Z}_{23} \\ \pm\mathbf{Z}_{31} & \pm\mathbf{Z}_{32} & \mathbf{Z}_{33} \end{bmatrix} \begin{bmatrix} \mathbf{I}_1 \\ \mathbf{I}_2 \\ \mathbf{I}_3 \end{bmatrix} = \begin{bmatrix} \mathbf{V}_1 \\ \mathbf{V}_2 \\ \mathbf{V}_3 \end{bmatrix}$$

or

$$[\mathbf{Z}][\mathbf{I}] = [\mathbf{V}]$$

referred to as the matrix form of Ohm's law, where $[\mathbf{Z}]$ is the impedance matrix, $[\mathbf{I}]$ the current matrix, and $[\mathbf{V}]$ the voltage matrix.

The mesh currents $\mathbf{I}_1$, $\mathbf{I}_2$ and $\mathbf{I}_3$ are found as the ratios of two determinants:

$$\mathbf{I}_1 = \frac{\begin{vmatrix} \mathbf{V}_1 & \pm\mathbf{Z}_{12} & \pm\mathbf{Z}_{13} \\ \mathbf{V}_2 & \mathbf{Z}_{22} & \pm\mathbf{Z}_{23} \\ \mathbf{V}_3 & \pm\mathbf{Z}_{32} & \mathbf{Z}_{33} \end{vmatrix}}{\Delta_z} \qquad \mathbf{I}_2 = \frac{\begin{vmatrix} \mathbf{Z}_{11} & \mathbf{V}_1 & \pm\mathbf{Z}_{13} \\ \pm\mathbf{Z}_{21} & \mathbf{V}_2 & \pm\mathbf{Z}_{23} \\ \pm\mathbf{Z}_{31} & \mathbf{V}_3 & \mathbf{Z}_{33} \end{vmatrix}}{\Delta_z} \qquad \mathbf{I}_3 = \frac{\begin{vmatrix} \mathbf{Z}_{11} & \pm\mathbf{Z}_{12} & \mathbf{V}_1 \\ \pm\mathbf{Z}_{21} & \mathbf{Z}_{22} & \mathbf{V}_2 \\ \pm\mathbf{Z}_{31} & \pm\mathbf{Z}_{32} & \mathbf{V}_3 \end{vmatrix}}{\Delta_z}$$

When the numerator determinant of each is expanded about the elements of the column containing the voltages, we obtain the following set of equations for the mesh currents:

$$\mathbf{I}_1 = \mathbf{V}_1\left(\frac{\Delta_{11}}{\Delta_z}\right) + \mathbf{V}_2\left(\frac{\Delta_{21}}{\Delta_z}\right) + \mathbf{V}_3\left(\frac{\Delta_{31}}{\Delta_z}\right) \qquad (1)$$

$$\mathbf{I}_2 = \mathbf{V}_1\left(\frac{\Delta_{12}}{\Delta_z}\right) + \mathbf{V}_2\left(\frac{\Delta_{22}}{\Delta_z}\right) + \mathbf{V}_3\left(\frac{\Delta_{32}}{\Delta_z}\right) \qquad (2)$$

$$\mathbf{I}_3 = \mathbf{V}_1\left(\frac{\Delta_{13}}{\Delta_z}\right) + \mathbf{V}_2\left(\frac{\Delta_{23}}{\Delta_z}\right) + \mathbf{V}_3\left(\frac{\Delta_{33}}{\Delta_z}\right) \qquad (3)$$

The terms on the right sides of equations (*1*), (*2*) and (*3*) are phasor components which result from the various driving voltages. Thus in (*1*) the mesh current $\mathbf{I}_1$ consists of three parts: $\mathbf{V}_1(\Delta_{11}/\Delta_z)$ due to driving voltage $\mathbf{V}_1$, $\mathbf{V}_2(\Delta_{21}/\Delta_z)$ due to driving voltage $\mathbf{V}_2$, and $\mathbf{V}_3(\Delta_{31}/\Delta_z)$ due to driving voltage $\mathbf{V}_3$.

DRIVING POINT IMPEDANCE

Consider a passive or source-free network with two external connections as shown in Fig. 9-6 below. Apply a voltage source $\mathbf{V}_1$ and call the resulting mesh current $\mathbf{I}_1$. Since there are no other sources in the network the equation for mesh current $\mathbf{I}_1$ is

$$\mathbf{I}_1 \;=\; \mathbf{V}_1\left(\frac{\Delta_{11}}{\Delta_z}\right) + (0)\left(\frac{\Delta_{21}}{\Delta_z}\right) + (0)\left(\frac{\Delta_{31}}{\Delta_z}\right) + \cdots \;=\; \mathbf{V}_1\left(\frac{\Delta_{11}}{\Delta_z}\right)$$

The *input* or *driving point impedance* is the ratio of the applied voltage $\mathbf{V}_1$ to the resulting current $\mathbf{I}_1$. Thus

$$\mathbf{Z}_{\text{input 1}} \;=\; \mathbf{V}_1/\mathbf{I}_1 \;=\; \Delta_z/\Delta_{11}$$

The input impedance of an *active network* is defined as the impedance presented by the network to the specified terminals when *all internal sources are shortened but their internal impedances retained.* Thus the ratio Δ_z/Δ_{11} is the driving point impedance of loop one regardless of whether the network is passive or active.

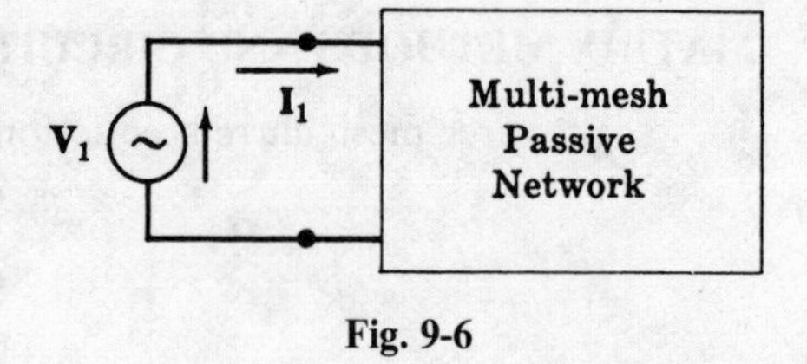

Fig. 9-6

TRANSFER IMPEDANCE

A voltage source driving in one mesh causes a current in each of the other meshes of a network. The transfer impedance is the ratio of a driving voltage in one mesh to the resulting current in another mesh, all other sources being set equal to zero.

Consider the network in Fig. 9-7 with a voltage $\mathbf{V}_r$ driving in mesh r and the resulting current $\mathbf{I}_s$ in mesh s. Then

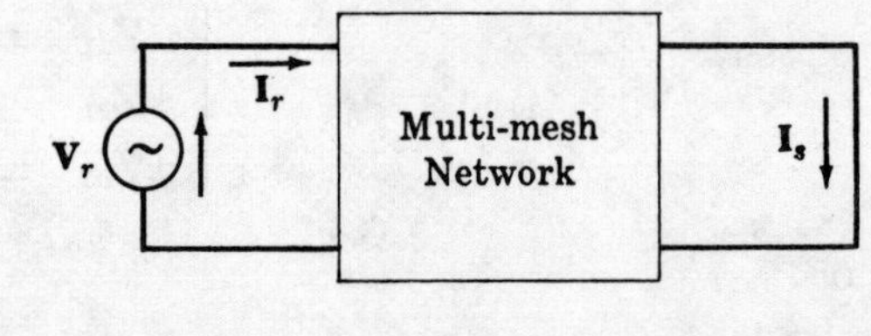

Fig. 9-7

$$\mathbf{I}_s \;=\; (0)\left(\frac{\Delta_{1s}}{\Delta_z}\right) + \cdots + \mathbf{V}_r\left(\frac{\Delta_{rs}}{\Delta_z}\right) + \cdots + (0)\left(\frac{\Delta_{ns}}{\Delta_z}\right) \;=\; \mathbf{V}_r\left(\frac{\Delta_{rs}}{\Delta_z}\right)$$

and

$$\mathbf{Z}_{\text{transfer } rs} \;=\; \mathbf{V}_r/\mathbf{I}_s \;=\; \Delta_z/\Delta_{rs}$$

The double subscripts of the transfer impedance, rs, indicate the direction of the action, i.e. the source is in mesh r and the resulting current is in mesh s. And the determinant of the denominator is the cofactor of the rs position, Δrs, with the same subscripts as the transfer impedance.

Solved Problems

9.1. Given the choice of mesh currents shown in Fig. 9-8 below, write the corresponding mesh current equations and put them in matrix form.

Applying Kirchhoff's voltage law to each of the three meshes,

$$\begin{aligned}
\mathbf{I}_1(2-j2) + (\mathbf{I}_1-\mathbf{I}_2)(j5) + (\mathbf{I}_1-\mathbf{I}_3)5 &= 10\angle 0^\circ \\
\mathbf{I}_2(10) + (\mathbf{I}_2-\mathbf{I}_3)(2-j2) + (\mathbf{I}_2-\mathbf{I}_1)(j5) &= -(5\angle 30^\circ) \\
\mathbf{I}_3(10) + (\mathbf{I}_3-\mathbf{I}_1)(5) + (\mathbf{I}_3-\mathbf{I}_2)(2-j2) &= -(10\angle 90^\circ)
\end{aligned}$$

Rearranging terms,

$$\begin{aligned}
(7+j3)\mathbf{I}_1 - (j5)\mathbf{I}_2 - (5)\mathbf{I}_3 &= 10\angle 0^\circ \\
-(j5)\mathbf{I}_1 + (12+j3)\mathbf{I}_2 - (2-j2)\mathbf{I}_3 &= -(5\angle 30^\circ) \\
-(5)\mathbf{I}_1 - (2-j2)\mathbf{I}_2 + (17-j2)\mathbf{I}_3 &= -(10\angle 90^\circ)
\end{aligned}$$

which can be expressed in the matrix form

$$\begin{bmatrix} 7+j3 & -j5 & -5 \\ -j5 & 12+j3 & -(2-j2) \\ -5 & -(2-j2) & 17-j2 \end{bmatrix} \begin{bmatrix} \mathbf{I}_1 \\ \mathbf{I}_2 \\ \mathbf{I}_3 \end{bmatrix} = \begin{bmatrix} 10\angle 0^\circ \\ -(5\angle 30^\circ) \\ -(10\angle 90^\circ) \end{bmatrix}$$

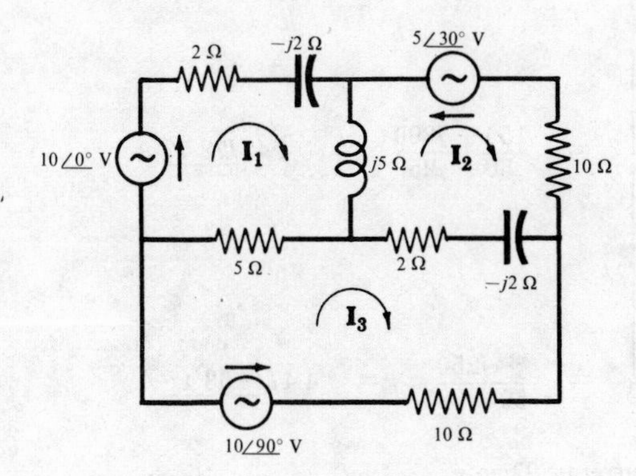

Fig. 9-8

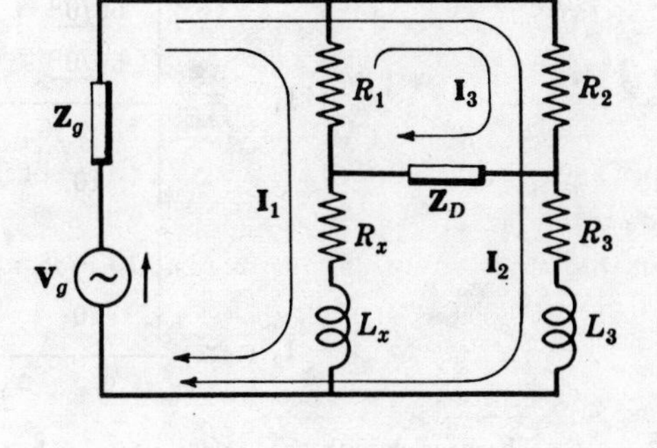

Fig. 9-9

9.2. For the network shown in Fig. 9-9 above, write the mesh current equations in matrix form by inspection.

The terms in the impedance matrix are determined by their definitions. $\mathbf{Z}_{11}$, the self-impedance of loop one, is the sum of all impedances in the loop, $(R_1 + R_x + j\omega L_x + Z_g)$ $\mathbf{Z}_{12}$, the impedance common to mesh currents one and two, is $\mathbf{Z}_g$ with positive sign since the two currents pass in the same direction. The current matrix consists simply of $\mathbf{I}_1$, $\mathbf{I}_2$ and $\mathbf{I}_3$. Then the voltage matrix is made up of the voltages driving in the loops. Thus the required matrix equation is

$$\begin{bmatrix} (R_1 + R_x + j\omega L_x + \mathbf{Z}_g) & \mathbf{Z}_g & -R_1 \\ \mathbf{Z}_g & (R_2 + R_3 + j\omega L_3 + \mathbf{Z}_g) & R_2 \\ -R_1 & R_2 & (R_1 + R_2 + \mathbf{Z}_D) \end{bmatrix} \begin{bmatrix} \mathbf{I}_1 \\ \mathbf{I}_2 \\ \mathbf{I}_3 \end{bmatrix} = \begin{bmatrix} \mathbf{V}_g \\ \mathbf{V}_g \\ 0 \end{bmatrix}$$

9.3. Find the power output of the voltage source in the circuit of Fig. 9-10. Also determine the power in the circuit resistors.

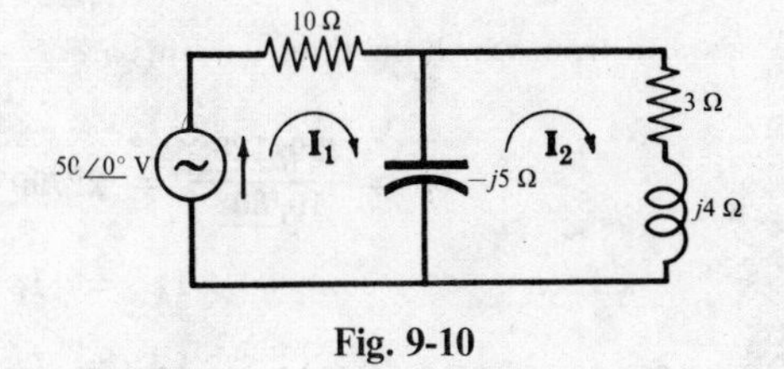

Fig. 9-10

Select the mesh currents as shown in the circuit diagram so that the source contains only one current. Then

$$\begin{bmatrix} 10-j5 & j5 \\ j5 & 3-j1 \end{bmatrix} \begin{bmatrix} \mathbf{I}_1 \\ \mathbf{I}_2 \end{bmatrix} = \begin{bmatrix} 50\angle 0^\circ \\ 0 \end{bmatrix}$$

and

$$\mathbf{I}_1 = \frac{\begin{vmatrix} 50\angle 0^\circ & j5 \\ 0 & 3-j1 \end{vmatrix}}{\begin{vmatrix} 10-j5 & j5 \\ j5 & 3-j1 \end{vmatrix}} = \frac{150-j50}{50-j25} = 2{\cdot}83\angle 8{\cdot}14^\circ \text{ A}$$

$$\mathbf{I}_2 = \frac{\begin{vmatrix} 10-j5 & 50\angle 0^\circ \\ j5 & 0 \end{vmatrix}}{\Delta_z} = \frac{-j250}{50-j25} = 4{\cdot}47\angle -63{\cdot}4^\circ \text{ A}$$

The source power $P = VI \cos\theta = 50(2{\cdot}83) \cos(8{\cdot}14^\circ) = 140$ W. The power in the 10 ohm resistor is $P_{10} = (I_1)^2 10 = (2{\cdot}83)^2 10 = 80$ W, and the power in the 3 ohm resistor is $P_3 = (I_2)^2 3 = 60$ W; their sum, $80 + 60 = 140$ W, is equal to the power output of the source.

9.4. The same circuit as in Problem 9.3 has the mesh currents selected as shown in Fig. 9-11. Find the power output of the source.

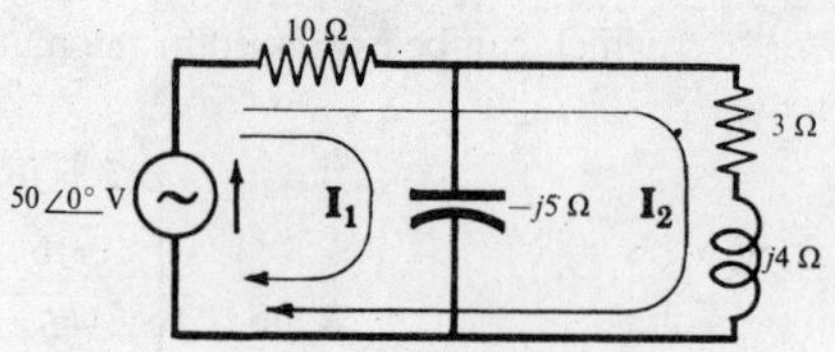

Fig. 9-11

The mesh current equations in matrix form are

$$\begin{bmatrix} 10 - j5 & 10 \\ 10 & 13 + j4 \end{bmatrix} \begin{bmatrix} \mathbf{I}_1 \\ \mathbf{I}_2 \end{bmatrix} = \begin{bmatrix} 50\underline{/0^\circ} \\ 50\underline{/0^\circ} \end{bmatrix}$$

Now

$$\mathbf{I}_1 = \frac{\begin{vmatrix} 50\underline{/0^\circ} & 10 \\ 50\underline{/0^\circ} & 13 + j4 \end{vmatrix}}{\begin{vmatrix} 10 - j5 & 10 \\ 10 & 13 + j4 \end{vmatrix}} = \frac{150 + j200}{50 - j25} = 4{\cdot}47\underline{/79{\cdot}7^\circ}\text{ A}$$

$$\mathbf{I}_2 = \frac{\begin{vmatrix} 10 - j5 & 50\underline{/0^\circ} \\ 10 & 50\underline{/0^\circ} \end{vmatrix}}{\Delta_z} = \frac{-j250}{50 - j25} = 4{\cdot}47\underline{/-63{\cdot}4^\circ}\text{ A}$$

The branch with the source contains both mesh currents. Then

$$\mathbf{I}_1 + \mathbf{I}_2 = \left(\frac{150 + j200}{50 - j25}\right) + \left(\frac{-j250}{50 - j25}\right) = 2{\cdot}83\angle{+8{\cdot}14^\circ}\text{ A}$$

and the source power $P = VI\cos\theta = 50(2{\cdot}83)\cos 8{\cdot}14^\circ = 140\text{ W}$.

9.5. The circuit shown in Fig. 9-12 has voltages given between each pair of the three lines. Find the currents $\mathbf{I}_A$, $\mathbf{I}_B$ and $\mathbf{I}_C$.

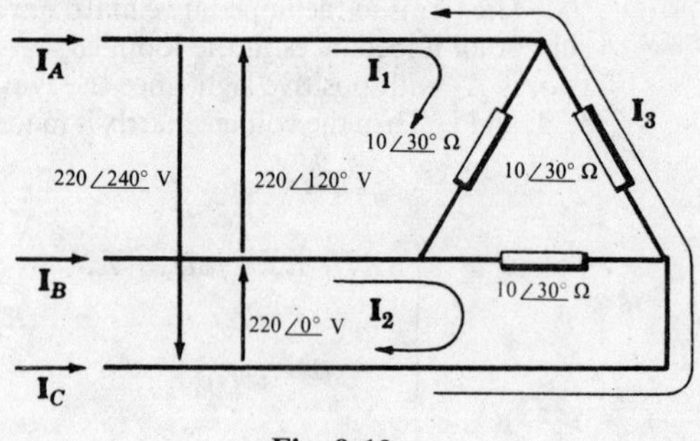

Fig. 9-12

With the mesh currents selected as shown in the diagram, the currents are independent. This is apparent when the matrix is written as

$$\begin{bmatrix} 10\underline{/30^\circ} & 0 & 0 \\ 0 & 10\underline{/30^\circ} & 0 \\ 0 & 0 & 10\underline{/30^\circ} \end{bmatrix} \begin{bmatrix} \mathbf{I}_1 \\ \mathbf{I}_2 \\ \mathbf{I}_3 \end{bmatrix} = \begin{bmatrix} 220\underline{/120^\circ} \\ 220\underline{/0^\circ} \\ 220\underline{/240^\circ} \end{bmatrix}$$

from which the three currents are

$$\mathbf{I}_1 = \frac{220\underline{/120^\circ}}{10\underline{/30^\circ}} = 22\underline{/90^\circ}\text{ A},\quad \mathbf{I}_2 = \frac{220\underline{/0^\circ}}{10\underline{/30^\circ}} = 22\underline{/-30^\circ}\text{ A},\quad \mathbf{I}_3 = \frac{220\underline{/240^\circ}}{10\underline{/30^\circ}} = 22\underline{/210^\circ}\text{ A}$$

Then

$$\mathbf{I}_A = \mathbf{I}_1 - \mathbf{I}_3 = (22\underline{/90^\circ} - 22\underline{/210^\circ}) = 38{\cdot}1\underline{/60^\circ}\text{ A}$$

$$\mathbf{I}_B = \mathbf{I}_2 - \mathbf{I}_1 = (22\underline{/-30^\circ} - 22\underline{/90^\circ}) = 38{\cdot}1\underline{/-60^\circ}\text{ A}$$

$$\mathbf{I}_C = \mathbf{I}_3 - \mathbf{I}_2 = (22\underline{/210^\circ} - 22\underline{/-30^\circ}) = 38{\cdot}1\underline{/180^\circ}\text{ A}$$

9.6. The four-mesh network of Fig. 9-13 has the mesh currents selected as shown in the diagram. In this circuit R and the two equal capacitors of C farads are adjusted until the current in $\mathbf{Z}_D$ is zero. For this condition the unknowns R_x and L_x can be expressed in terms of R, C and the source frequency ω rad/s.

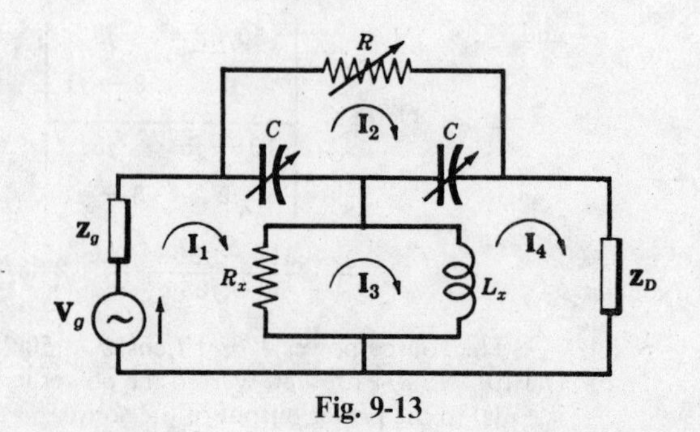

Fig. 9-13

The mesh current equations in matrix form are

$$\begin{bmatrix} \left(R_x + \frac{1}{j\omega C} + \mathbf{Z}_g\right) & -\left(\frac{1}{j\omega C}\right) & -R_x & 0 \\ -\left(\frac{1}{j\omega C}\right) & \left(R + \frac{1}{j\omega C} + \frac{1}{j\omega C}\right) & 0 & -\left(\frac{1}{j\omega C}\right) \\ -R_x & 0 & (R_x + j\omega L_x) & -(j\omega L_x) \\ 0 & -\left(\frac{1}{j\omega C}\right) & -(j\omega L_x) & \left(\frac{1}{j\omega C} + j\omega L_x + \mathbf{Z}_D\right) \end{bmatrix} \begin{bmatrix} \mathbf{I}_1 \\ \mathbf{I}_2 \\ \mathbf{I}_3 \\ \mathbf{I}_4 \end{bmatrix} = \begin{bmatrix} \mathbf{V}_g \\ 0 \\ 0 \\ 0 \end{bmatrix}$$

Express $\mathbf{I}_4$, the current in $\mathbf{Z}_D$, in determinant form and set it equal to zero.

$$\mathbf{I}_4 = \frac{\begin{vmatrix} \left(R_x + \frac{1}{j\omega C} + \mathbf{Z}_g\right) & -\left(\frac{1}{j\omega C}\right) & -R_x & \mathbf{V}_g \\ -\left(\frac{1}{j\omega C}\right) & \left(R + \frac{1}{j\omega C} + \frac{1}{j\omega C}\right) & 0 & 0 \\ -R_x & 0 & (R_x + j\omega L_x) & 0 \\ 0 & -\left(\frac{1}{j\omega C}\right) & -(j\omega L_x) & 0 \end{vmatrix}}{\Delta_z} = 0$$

Expanding the numerator about the elements of the fourth column, we obtain

$$-\mathbf{V}_g \begin{vmatrix} -\left(\frac{1}{j\omega C}\right) & \left(R + \frac{1}{j\omega C} + \frac{1}{j\omega C}\right) & 0 \\ -R_x & 0 & (R_x + j\omega L_x) \\ 0 & -\left(\frac{1}{j\omega C}\right) & -(j\omega L_x) \end{vmatrix} = 0$$

Since this determinant must have a value zero,

$$-(-R_x)(R + 1/j\omega C + 1/j\omega C)(-j\omega L_x) - (-1/j\omega C)(-1/j\omega C)(R_x + j\omega L_x) = 0$$

from which $\qquad R_x = 1/\omega^2 C^2 R \qquad$ and $\qquad L_x = 1/2\omega^2 C$

9.7. In the circuit of Fig. 9-14, determine the currents $\mathbf{I}_A$, $\mathbf{I}_B$ and $\mathbf{I}_C$.

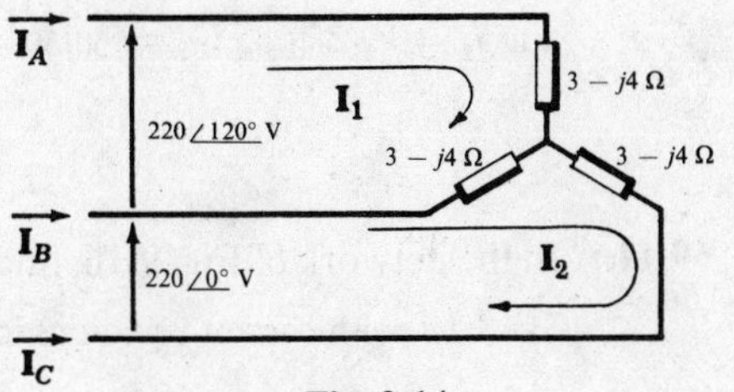

Two mesh currents are selected as shown in the diagram. Then the corresponding equations in matrix form are

$$\begin{bmatrix} 6 - j8 & -(3 - j4) \\ -(3 - j4) & 6 - j8 \end{bmatrix} \begin{bmatrix} \mathbf{I}_1 \\ \mathbf{I}_2 \end{bmatrix} = \begin{bmatrix} 220\underline{/120^\circ} \\ 220\underline{/0^\circ} \end{bmatrix}$$

Fig. 9-14

from which

$$\mathbf{I}_1 = \frac{\begin{vmatrix} 220\underline{/120^\circ} & -(3 - j4) \\ 220\underline{/0^\circ} & 6 - j8 \end{vmatrix}}{\begin{vmatrix} 6 - j8 & -(3 - j4) \\ -(3 - j4) & 6 - j8 \end{vmatrix}} = \frac{2200\underline{/66{\cdot}9^\circ} + 1100\underline{/-53{\cdot}1^\circ}}{100\underline{/-106{\cdot}2^\circ} - 25\underline{/-106{\cdot}2^\circ}} = \frac{1905\underline{/36{\cdot}9^\circ}}{75\underline{/-106{\cdot}2^\circ}} = 25{\cdot}4\underline{/143{\cdot}1^\circ}\text{ A}$$

$$\mathbf{I}_2 = \frac{\begin{vmatrix} 6 - j8 & 220\underline{/120^\circ} \\ -(3 - j4) & 220\underline{/0^\circ} \end{vmatrix}}{\Delta_z} = \frac{2200\underline{/-53{\cdot}1^\circ} + 1100\underline{/66{\cdot}9^\circ}}{75\underline{/-106{\cdot}2^\circ}} = \frac{1905\underline{/-23{\cdot}2^\circ}}{75\underline{/-106{\cdot}2^\circ}} = 25{\cdot}4\underline{/83^\circ}\text{ A}$$

and the required line currents are $\mathbf{I}_A = \mathbf{I}_1 = 25{\cdot}4\underline{/143{\cdot}1^\circ}$ A, $\mathbf{I}_B = \mathbf{I}_2 - \mathbf{I}_1 = (25{\cdot}4\underline{/83^\circ} - 25{\cdot}4\underline{/143{\cdot}1^\circ}) = 25{\cdot}4\underline{/23{\cdot}1^\circ}$ A, and $\mathbf{I}_C = -\mathbf{I}_2 = 25{\cdot}4\underline{/-97^\circ}$ A.

9.8. In the network of Fig. 9-15 use matrix methods to determine the input impedance as seen by the 50 volt source. Calculate I_1 using this impedance.

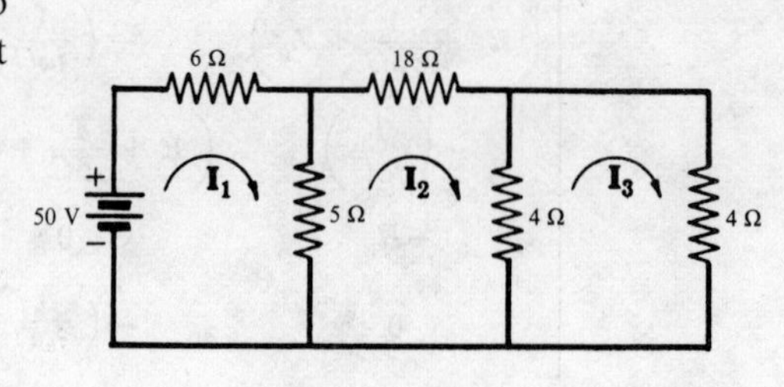

Fig. 9-15

The driving point or input impedance to loop one is

$$Z_{\text{input 1}} = \frac{\Delta_z}{\Delta_{11}} = \frac{\begin{vmatrix} 11 & -5 & 0 \\ -5 & 27 & -4 \\ 0 & -4 & 8 \end{vmatrix}}{\begin{vmatrix} 27 & -4 \\ -4 & 8 \end{vmatrix}} = \frac{2000}{200} = 10\ \Omega$$

Now $I_1 = V_1/Z_{\text{input 1}} = 50/10 = 5$ A.

9.9. Referring to the circuit of Fig. 9-15, find mesh current I_3 using the transfer impedance.

The source is in loop one and the required current in loop three. Hence the required transfer impedance is

$$Z_{\text{transfer 13}} = \frac{\Delta_z}{\Delta_{13}} = \frac{2000}{\begin{vmatrix} -5 & 27 \\ 0 & -4 \end{vmatrix}} = \frac{2000}{20} = 100\ \Omega$$

and the mesh current $I_3 = V_1/Z_{\text{transfer 13}} = 50/100 = 0{\cdot}5$ A.

9.10. Referring to the circuit of Fig. 9-15, find mesh current I_2 using the transfer impedance.

Since the source is in loop one and the required current in mesh two, the required transfer impedance is

$$Z_{\text{transfer 12}} = \frac{\Delta_z}{\Delta_{12}} = \frac{2000}{(-1)\begin{vmatrix} -5 & -4 \\ 0 & 8 \end{vmatrix}} = \frac{2000}{40} = 50\ \Omega$$

and $I_2 = V_1/Z_{\text{transfer 12}} = 50/50 = 1$ A.

9.11. In the network of Fig. 9-16, find the voltages $\mathbf{V}_{AB}$ and $\mathbf{V}_{BC}$.

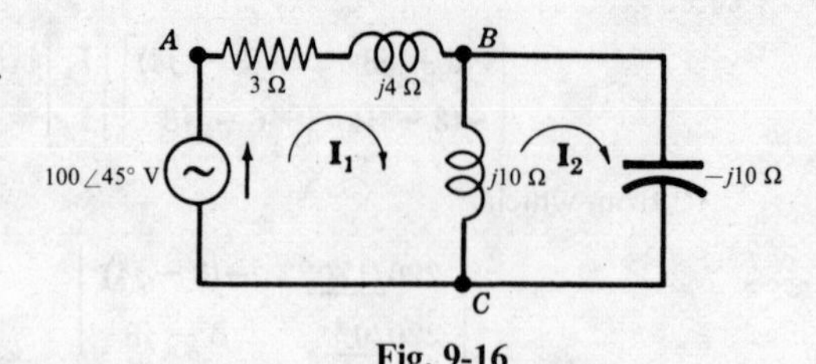

Fig. 9-16

The mesh current equations in matrix form are

$$\begin{bmatrix} 3+j14 & -j10 \\ -j10 & 0 \end{bmatrix}\begin{bmatrix} \mathbf{I}_1 \\ \mathbf{I}_2 \end{bmatrix} = \begin{bmatrix} 100\underline{/45^\circ} \\ 0 \end{bmatrix}$$

from which

$$\mathbf{I}_1 = \frac{\begin{vmatrix} 100\underline{/45^\circ} & -j10 \\ 0 & 0 \end{vmatrix}}{\begin{vmatrix} 3+j14 & -j10 \\ -j10 & 0 \end{vmatrix}} = \frac{0}{100} = 0, \qquad \mathbf{I}_2 = \frac{\begin{vmatrix} 3+j14 & 100\underline{/45^\circ} \\ -j10 & 0 \end{vmatrix}}{\Delta_z} = \frac{1000\underline{/135^\circ}}{100} = 10\underline{/135^\circ}\ \text{A}$$

Then $\mathbf{V}_{AB} = \mathbf{I}_1(3+j4) = 0$ and $\mathbf{V}_{BC} = \mathbf{I}_2(-j10) = 10\underline{/135^\circ}\,(10\underline{/-90^\circ}) = 100\underline{/45^\circ}$ V. The sum $(\mathbf{V}_{AB} + \mathbf{V}_{BC}) = 100\underline{/45^\circ}$ V, the value of the applied phasor voltage.

9.12. In the network of Fig. 9-17 obtain the three components of the power triangle for the $10\underline{/30°}$ volt source.

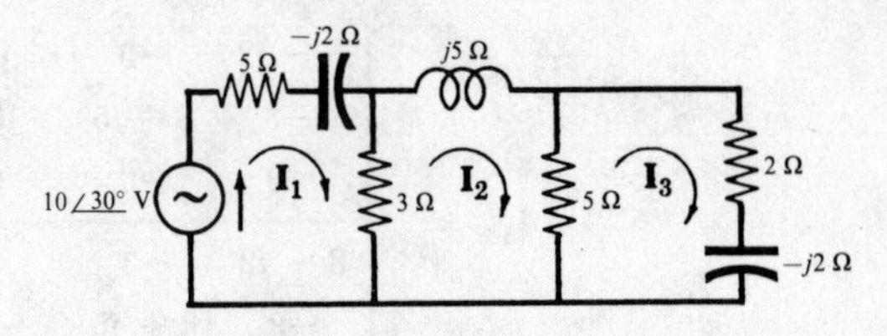

Fig. 9-17

The mesh currents are selected as shown, with only I_1 in the branch containing the source. Since there is only one source in the network, the driving point impedance can be used to find I_1.

$$Z_{\text{input 1}} = \frac{\Delta_z}{\Delta_{11}} = \frac{\begin{vmatrix} 8-j2 & -3 & 0 \\ -3 & 8+j5 & -5 \\ 0 & -5 & 7-j2 \end{vmatrix}}{\begin{vmatrix} 8+j5 & -5 \\ -5 & 7-j2 \end{vmatrix}} = \frac{315\underline{/16{\cdot}2°}}{45{\cdot}1\underline{/24{\cdot}9°}} = 6{\cdot}98\underline{/-8{\cdot}7°}\ \Omega$$

and $$I_1 = V_1/Z_{\text{input 1}} = (10\underline{/30°})/(6{\cdot}98\underline{/-8{\cdot}7°}) = 1{\cdot}43\underline{/38{\cdot}7°}\text{ A}$$

The power output of the source is $P = V_1 I_1 \cos\theta = 10(1{\cdot}43)\cos 8{\cdot}7° = 14{\cdot}1$ W. The reactive power $Q = V_1 I_1 \sin 8{\cdot}7° = 2{\cdot}16$ var leading. The apparent power $S = V_1 I_1 = 14{\cdot}3$ VA.

9.13. Referring to the network of Fig. 9-17, find mesh currents I_2 and I_3 using the transfer impedances.

Since the source is in loop one and the required current in loop two, the transfer impedance $Z_{\text{transfer 12}}$ is used.

$$Z_{\text{transfer 12}} = \frac{\Delta_z}{\Delta_{12}} = \frac{315\underline{/16{\cdot}2°}}{(-1)\begin{vmatrix} -3 & -5 \\ 0 & 7-j2 \end{vmatrix}} = \frac{315\underline{/16{\cdot}2°}}{21{\cdot}8\underline{/-16°}} = 14{\cdot}45\underline{/32{\cdot}2°}\ \Omega$$

Then $$I_2 = V_1/Z_{\text{transfer 12}} = (10\underline{/30°})/(14{\cdot}45\underline{/32{\cdot}2°}) = 0{\cdot}693\underline{/-2{\cdot}2°}\text{ A}$$

Similarly,

$$Z_{\text{transfer 13}} = \frac{\Delta_z}{\Delta_{13}} = \frac{315\underline{/16{\cdot}2°}}{\begin{vmatrix} -3 & 8+j5 \\ 0 & -5 \end{vmatrix}} = \frac{315\underline{/16{\cdot}2°}}{15} = 21\underline{/16{\cdot}2°}\ \Omega$$

and $$I_3 = V_1/Z_{\text{transfer 13}} = (10\underline{/30°})/(21\underline{/16{\cdot}2°}) = 0{\cdot}476\underline{/13{\cdot}8°}\text{ A}$$

9.14. Referring to Fig. 9-17, find the power in the network resistors and compare it with the power output of the source.

From Problems 9.12 and 9.13: $I_1 = 1{\cdot}43\underline{/38{\cdot}7°}$ A, $I_2 = 0{\cdot}693\underline{/-2{\cdot}2°}$ A, $I_3 = 0{\cdot}476\underline{/13{\cdot}8°}$ A.

The power in the first 5 ohm resistor is $P = (I_1)^2 5 = (1{\cdot}43)^2 5 = 10{\cdot}2$ W. The 3 ohm resistor has two mesh currents which are combined to give the branch current, $(I_1 - I_2) = (1{\cdot}115 + j0{\cdot}895) - (0{\cdot}693 - j0{\cdot}027) = 0{\cdot}422 + j0{\cdot}922 = 1{\cdot}01\underline{/65{\cdot}4°}$ A; then $P = (1{\cdot}01)^2 3 = 3{\cdot}06$ W. Similarly the current in the branch containing the 5 ohm resistor is $(I_2 - I_3) = (0{\cdot}693 - j0{\cdot}027) - (0{\cdot}462 + j0{\cdot}113) = (0{\cdot}231 - j0{\cdot}140) = 0{\cdot}271\underline{/-31{\cdot}2°}$ A and the power $P = (0{\cdot}271)^2 5 = 0{\cdot}367$ W. The power in the 2 ohm resistor is $P = (I_3)^2 2 = (0{\cdot}476)^2 2 = 0{\cdot}453$ W.

The total network power $P_T = 10{\cdot}2 + 3{\cdot}06 + 0{\cdot}367 + 0{\cdot}453 = 14{\cdot}1$ W, equal to the power output of Problem 9.12.

9.15. In the network of Fig. 9-18 below the source V_1 results in a voltage V_0 across the $2 - j2\ \Omega$ impedance. Find the source V_1 which corresponds to $V_0 = 5\underline{/0°}$ V.

With the given voltage V_0, mesh current $I_3 = \dfrac{V_0}{2-j2} = \dfrac{5\underline{/0°}}{2\sqrt{2}\underline{/-45°}} = 1{\cdot}76\underline{/45°}$ A. Expressed as a determinant,

$$\mathbf{I}_3 = \frac{\begin{vmatrix} 8-j2 & -3 & \mathbf{V}_1 \\ -3 & 8+j5 & 0 \\ 0 & -5 & 0 \end{vmatrix}}{\begin{vmatrix} 8-j2 & -3 & 0 \\ -3 & 8+j5 & -5 \\ 0 & -5 & 7-j2 \end{vmatrix}} = \mathbf{V}_1 \frac{\begin{vmatrix} -3 & 8+j5 \\ 0 & -5 \end{vmatrix}}{315\underline{/16{\cdot}2^\circ}} = \mathbf{V}_1\,(0{\cdot}0476\underline{/-16{\cdot}2^\circ})\text{ A}$$

Then
$$\mathbf{V}_1 = \frac{\mathbf{I}_3}{0{\cdot}0476\underline{/-16{\cdot}2^\circ}} = \frac{1{\cdot}76\underline{/45^\circ}}{0{\cdot}0476\underline{/-16{\cdot}2^\circ}} = 36{\cdot}9\underline{/61{\cdot}2^\circ}\text{ V}$$

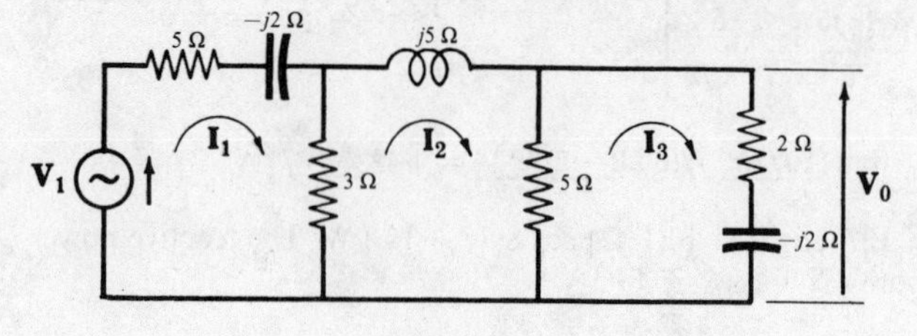

Fig. 9-18

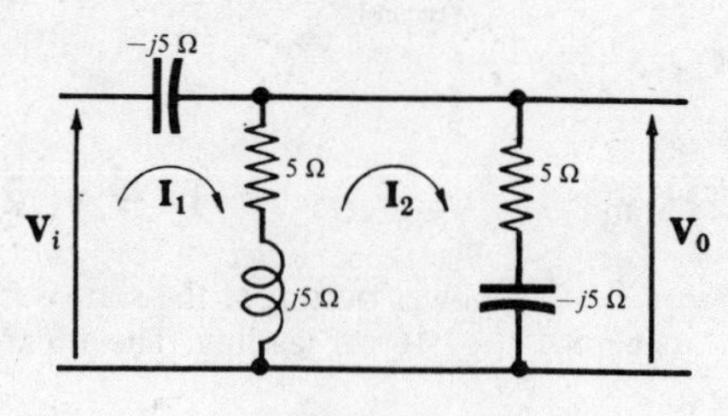

Fig. 9-19

9.16. When the network of Fig. 9-19 above is connected to a high impedance load, the output voltage $\mathbf{V}_0$ is given by the voltage drop across the $5 - j5\ \Omega$ impedance. Determine the voltage transfer function $\mathbf{V}_0/\mathbf{V}_i$ for the network.

For the two mesh currents selected as shown, the corresponding equations in matrix form are

$$\begin{bmatrix} 5 & -(5+j5) \\ -(5+j5) & 10 \end{bmatrix}\begin{bmatrix} \mathbf{I}_1 \\ \mathbf{I}_2 \end{bmatrix} = \begin{bmatrix} \mathbf{V}_i \\ 0 \end{bmatrix}$$

The output voltage $\mathbf{V}_0$ is

$$\mathbf{V}_0 = \mathbf{I}_2(5-j5) = (5-j5)\frac{\begin{vmatrix} 5 & \mathbf{V}_i \\ -(5+j5) & 0 \end{vmatrix}}{\begin{vmatrix} 5 & -(5+j5) \\ -(5+j5) & 10 \end{vmatrix}} = \frac{(5-j5)(5+j5)\mathbf{V}_i}{(50-j50)} = \frac{50\mathbf{V}_i}{50\sqrt{2}\,\underline{/-45^\circ}}$$

from which
$$\frac{\mathbf{V}_0}{\mathbf{V}_i} = \frac{50}{50\sqrt{2}\,\underline{/-45^\circ}} = 0{\cdot}707\underline{/45^\circ}$$

9.17. The network shown in Fig. 9-20 contains two voltage sources. Find the current in the $2 + j3\ \Omega$ impedance due to each of the sources.

The mesh currents are selected so that the current in the required impedance is given directly by mesh current $\mathbf{I}_2$. For the choice shown, the equations in matrix form are

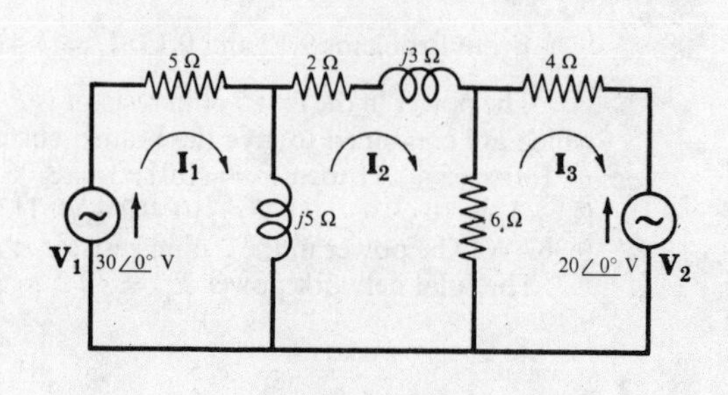

Fig. 9-20

$$\begin{bmatrix} 5+j5 & -j5 & 0 \\ -j5 & 8+j8 & -6 \\ 0 & -6 & 10 \end{bmatrix}\begin{bmatrix} \mathbf{I}_1 \\ \mathbf{I}_2 \\ \mathbf{I}_3 \end{bmatrix} = \begin{bmatrix} 30\underline{/0^\circ} \\ 0 \\ -20\underline{/0^\circ} \end{bmatrix}$$

The impedance determinant
$$\Delta_z = \begin{vmatrix} 5+j5 & -j5 & 0 \\ -j5 & 8+j8 & -6 \\ 0 & -6 & 10 \end{vmatrix} = 70 + j620 = 624\underline{/83{\cdot}55^\circ}.$$

The numerator determinant of $\mathbf{I}_2$ is expanded about the elements of the second column:

$$\mathbf{I}_2 = \frac{\begin{vmatrix} 5+j5 & 30\angle 0^\circ & 0 \\ -j5 & 0 & -6 \\ 0 & -20\angle 0^\circ & 10 \end{vmatrix}}{\Delta_z} = 30\angle 0^\circ\,(-)\frac{\begin{vmatrix} -j5 & -6 \\ 0 & 10 \end{vmatrix}}{\Delta_z} + 0 + (-20\angle 0^\circ)(-)\frac{\begin{vmatrix} 5+j5 & 0 \\ -j5 & -6 \end{vmatrix}}{\Delta_z}$$

$$= -30\angle 0^\circ\left(\frac{50\angle -90^\circ}{624\angle 83{\cdot}55^\circ}\right) + 20\angle 0^\circ\left(\frac{42{\cdot}4\angle -135^\circ}{624\angle 83{\cdot}55^\circ}\right) = 2{\cdot}41\angle 6{\cdot}45^\circ + 1{\cdot}36\angle 141{\cdot}45^\circ \text{ A}$$

The source $\mathbf{V}_1$ causes a current $2{\cdot}41\angle 6{\cdot}45^\circ$ A in the $2 + j3\ \Omega$ impedance and $\mathbf{V}_2$ causes a current $1{\cdot}36\angle 141{\cdot}45^\circ$ A. Each of these currents is a component of $\mathbf{I}_2$; hence $\mathbf{I}_2 = 2{\cdot}41\angle 6{\cdot}45^\circ + 1{\cdot}36\angle 141{\cdot}45^\circ = 1{\cdot}74\angle 40{\cdot}1^\circ$ A.

9.18. Referring to the network of Fig. 9-20, determine (*a*) the power supplied by each of the voltage sources and (*b*) the power in the network resistors.

(*a*) The current in the branch containing source $\mathbf{V}_1$ is

$$\mathbf{I}_1 = \frac{\begin{vmatrix} 30\angle 0^\circ & -j5 & 0 \\ 0 & 8+j8 & -6 \\ -20\angle 0^\circ & -6 & 10 \end{vmatrix}}{\Delta_z} = \frac{2240\angle 53{\cdot}8^\circ}{624\angle 83{\cdot}55^\circ} = 3{\cdot}59\angle -29{\cdot}75^\circ \text{ A}$$

The power output of this source $P_1 = V_1 I_1 \cos\theta = 30(3{\cdot}59)\cos 29{\cdot}75^\circ = 93{\cdot}5$ W.

The current in the branch containing $\mathbf{V}_2$ is

$$\mathbf{I}_3 = \frac{\begin{vmatrix} 5+j5 & -j5 & 30\angle 0^\circ \\ -j5 & 8+j8 & 0 \\ 0 & -6 & -20\angle 0^\circ \end{vmatrix}}{\Delta_z} = \frac{860\angle -125{\cdot}6^\circ}{624\angle 83{\cdot}55^\circ} = 1{\cdot}38\angle -209{\cdot}15^\circ \text{ A}$$

Note that $\mathbf{V}_2$ and $\mathbf{I}_3$ are not in the same direction. The output power $P_2 = V_2(I_3)\cos\theta = (-)(20)(1{\cdot}38)\cos -209{\cdot}15^\circ = 24{\cdot}1$ W.

Then the total power $P_T = P_1 + P_2 = 93{\cdot}5 + 24{\cdot}1 = 117{\cdot}6$ W.

(*b*) The power in the 5 ohm resistor is $P_5 = (I_1)^2 5 = (3{\cdot}59)^2 5 = 64{\cdot}5$ W. In the 2 ohm resistor, $P_2 = (I_2)^2 2 = (1{\cdot}74)^2 2 = 6{\cdot}05$ W. The branch current of the 6 ohm resistor is $(\mathbf{I}_2 - \mathbf{I}_3) = (1{\cdot}33 + j1{\cdot}12) - (-1{\cdot}205 + j0{\cdot}672) = 2{\cdot}535 + j0{\cdot}45 = 2{\cdot}57\angle 10{\cdot}1^\circ$ A; then $P_6 = (2{\cdot}57)^2 6 = 39{\cdot}6$ W. The power in the 4 ohm resistor is $P_4 = (I_3)^2 4 = (1{\cdot}38)^2 4 = 7{\cdot}61$ W.

Then the total power $P_T = 64{\cdot}5 + 6{\cdot}05 + 39{\cdot}6 + 7{\cdot}61 = 117{\cdot}76$ W.

9.19. The network of Fig. 9-21 contains two voltage sources, $\mathbf{V}_1$ and $\mathbf{V}_2$. With $\mathbf{V}_1 = 30\angle 0^\circ$ V, determine $\mathbf{V}_2$ such that the current in the $2 + j3\ \Omega$ impedance is zero.

Select the mesh currents as shown, with only one current in the $2 + j3\ \Omega$ impedance. The corresponding set of equations in matrix form is

$$\begin{bmatrix} 5+j5 & -j5 & 0 \\ -j5 & 8+j8 & 6 \\ 0 & 6 & 10 \end{bmatrix}\begin{bmatrix} \mathbf{I}_1 \\ \mathbf{I}_2 \\ \mathbf{I}_3 \end{bmatrix} = \begin{bmatrix} 30\angle 0^\circ \\ 0 \\ \mathbf{V}_2 \end{bmatrix}$$

from which

$$\mathbf{I}_2 = \frac{\begin{vmatrix} 5+j5 & 30\angle 0^\circ & 0 \\ -j5 & 0 & 6 \\ 0 & \mathbf{V}_2 & 10 \end{vmatrix}}{\Delta_z} = 0$$

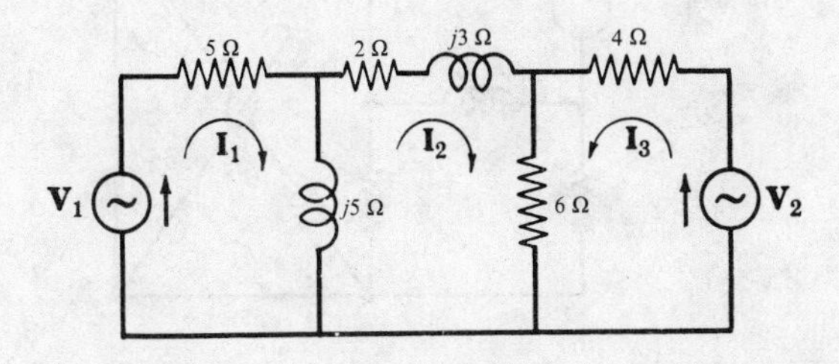

Fig. 9-21

Expanding,
$$-30\underline{/0^\circ}\begin{vmatrix} -j5 & 6 \\ 0 & 10 \end{vmatrix} - \mathbf{V}_2\begin{vmatrix} 5+j5 & 0 \\ -j5 & 6 \end{vmatrix} = 0$$

$$-30\underline{/0^\circ}\,(50\underline{/-90^\circ}) - \mathbf{V}_2\,(6)(5\sqrt{2}\,\underline{/45^\circ}) = 0$$

and
$$\mathbf{V}_2 = \frac{-30\underline{/0^\circ}\,(50\underline{/-90^\circ})}{6(5\sqrt{2}\,\underline{/45^\circ})} = 35{\cdot}4\underline{/45^\circ}\text{ V}$$

Alternate solution. If no current passes through the $2 + j3\ \Omega$ branch, $\mathbf{I}_2 = 0$ and the voltage drops across the $j5$ ohm reactance and the 6 ohm resistance must be equal, i.e.,

$$\mathbf{I}_1(j5) = \mathbf{I}_3(6)$$

Substituting $\mathbf{I}_1 = 30\underline{/0^\circ}\,/\,(5+j5)$ and $\mathbf{I}_3 = \mathbf{V}_2/10$,

$$\frac{30\underline{/0^\circ}}{5+j5}(j5) = \frac{\mathbf{V}_2}{10}(6) \qquad \text{from which} \qquad \mathbf{V}_2 = \frac{30\underline{/90^\circ}}{\sqrt{2}\,\underline{/45^\circ}}\left(\frac{10}{6}\right) = 35{\cdot}4\underline{/45^\circ}\text{ V}$$

9.20. In Fig. 9-21, source $\mathbf{V}_2 = 20\underline{/0^\circ}$ V. Determine the source voltage $\mathbf{V}_1$ which results in zero current in the branch containing $\mathbf{V}_2$.

Using the mesh currents selected in Problem 9.19, write the determinant expression for $\mathbf{I}_3$ and set it equal to zero:

$$\mathbf{I}_3 = \frac{\begin{vmatrix} 5+j5 & -j5 & \mathbf{V}_1 \\ -j5 & 8+j8 & 0 \\ 0 & 6 & 20\underline{/0^\circ} \end{vmatrix}}{\Delta_z} = 0$$

Expanding,
$$\mathbf{V}_1\begin{vmatrix} -j5 & 8+j8 \\ 0 & 6 \end{vmatrix} + 20\underline{/0^\circ}\begin{vmatrix} 5+j5 & -j5 \\ -j5 & 8+j8 \end{vmatrix} = 0$$

$$\mathbf{V}_1\,(30\underline{/-90^\circ}) + 20\underline{/0^\circ}\,(25+j80) = 0$$

and
$$\mathbf{V}_1 = \frac{-20\underline{/0^\circ}\,(25+j80)}{30\underline{/-90^\circ}} = 55{\cdot}8\underline{/-17{\cdot}4^\circ}\text{ V}$$

Supplementary Problems

9.21. Determine the number of mesh currents required to solve each of the networks shown in Fig. 9-22(*a-f*). Employ two different methods on each network. *Ans.* (*a*) 5, (*b*) 4, (*c*) 3, (*d*) 4, (*e*) 4, (*f*) 5

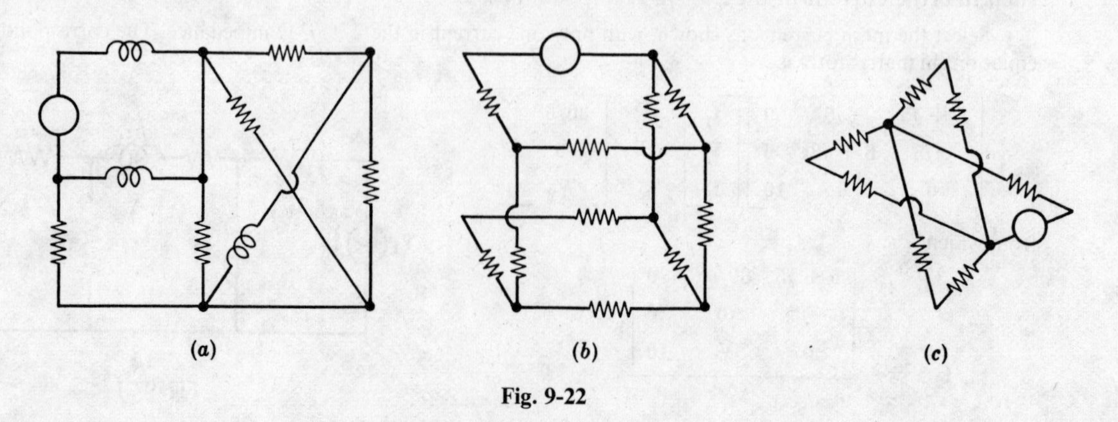

Fig. 9-22

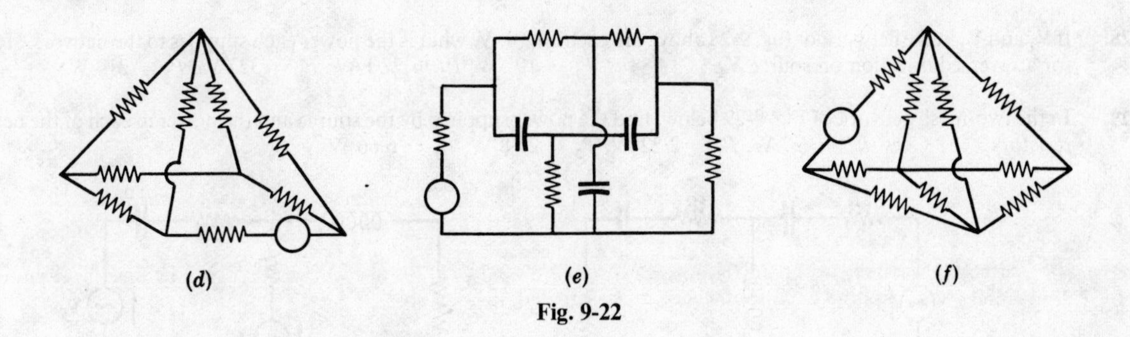

(d) (e) (f)

Fig. 9-22

9.22. In the network of Fig. 9-23 below, find the current in the 3 ohm resistor. The positive direction is shown on the diagram.
Ans. $4{\cdot}47\angle{-63{\cdot}4^\circ}$ A

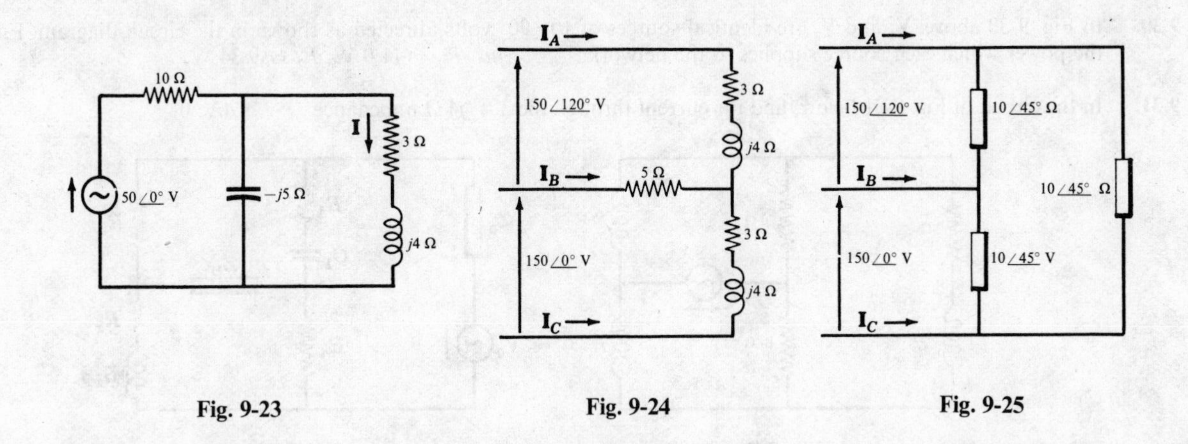

Fig. 9-23 **Fig. 9-24** **Fig. 9-25**

9.23. Find currents $\mathbf{I}_A$, $\mathbf{I}_B$ and $\mathbf{I}_C$ in the circuit shown in Fig. 9-24 above.
Ans. $\mathbf{I}_A = 12{\cdot}1\angle 46{\cdot}4^\circ$ A, $\mathbf{I}_B = 19{\cdot}1\angle{-47{\cdot}1^\circ}$ A, $\mathbf{I}_C = 22{\cdot}1\angle 166{\cdot}4^\circ$ A

9.24. In Fig. 9-25 above, find the three currents $\mathbf{I}_A$, $\mathbf{I}_B$ and $\mathbf{I}_C$. *Ans.* $26\angle 45^\circ$ A, $26\angle{-75^\circ}$ A, $26\angle{-195^\circ}$ A

9.25. In the circuit of Fig. 9-26 below, find the voltage $\mathbf{V}_{AB}$ by mesh current methods.
Ans. $\mathbf{V}_{AB} = 75{\cdot}4\angle 55{\cdot}2^\circ$ V

9.26. In Fig. 9-27 below, find the effective voltage of the source **V** which results in a power of 100 watts in the 5 ohm resistor.
Ans. 40·3 V

9.27. Select the mesh currents for the network of Fig. 9-28 below and compute Δ_z. Take a second choice of mesh currents and again compute Δ_z. *Ans.* $61 - j15\ \Omega^2$

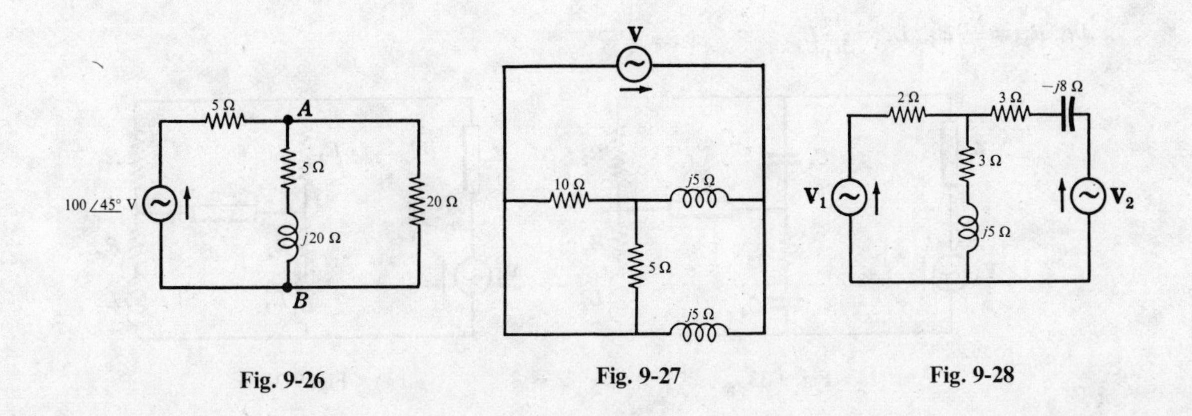

Fig. 9-26 **Fig. 9-27** **Fig. 9-28**

9.28. If $\mathbf{V}_1$ and $\mathbf{V}_2$ in the network of Fig. 9-28 above are each $50\angle 0°$ V, what is the power each supplies to the network? Repeat for a reversed direction on source $\mathbf{V}_2$. *Ans.* $P_1 = 191$ W, $P_2 = 77{\cdot}1$ W; $P_1 = 327$ W, $P_2 = 214$ W

9.29. In the two-mesh network of Fig. 9-29 below, find the power supplied by the source and the power in each of the network resistors. *Ans.* $P = 36{\cdot}7$ W, $P_1 = 2{\cdot}22$ W, $P_2 = 27{\cdot}8$ W, $P_3 = 6{\cdot}66$ W

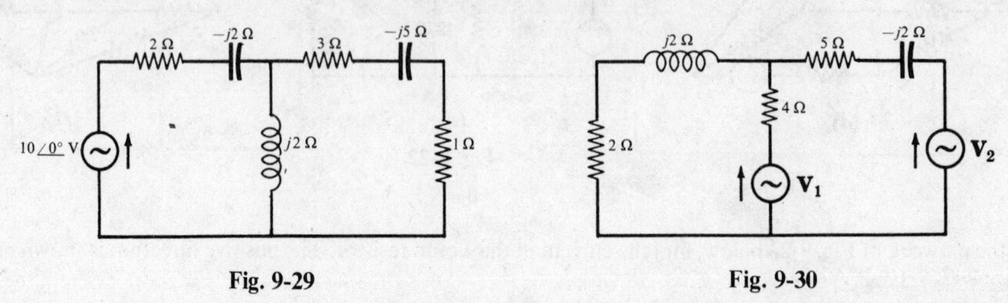

Fig. 9-29 Fig. 9-30

9.30. In Fig. 9-30 above, $\mathbf{V}_1$ and $\mathbf{V}_2$ are identical sources of $10\angle 90°$ volts, directed as shown in the circuit diagram. Find the power which each source supplies to the network. *Ans.* $P_1 = 11{\cdot}0$ W, $P_2 = 9{\cdot}34$ W

9.31. In the circuit of Fig. 9-31 below, find the current through the $3 + j4\ \Omega$ impedance. *Ans.* 0

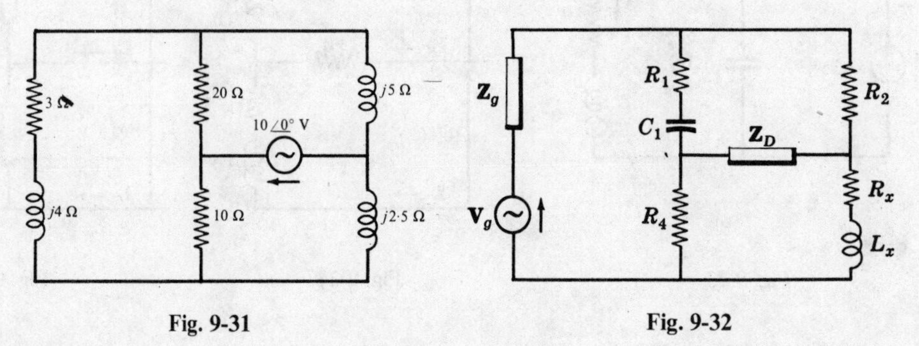

Fig. 9-31 Fig. 9-32

9.32. The circuit of Fig. 9-32 above is called the Hay bridge. Select mesh currents such that only one passes through $\mathbf{Z}_D$ and write the equations in matrix form. Then express the current through $\mathbf{Z}_D$ in determinant form and set it equal to zero. Find R_x and L_x in terms of the other bridge constants.

Ans. $R_x = \dfrac{\omega^2 C_1^2 R_1 R_2 R_4}{1 + (\omega R_1 C_1)^2}$, $L_x = \dfrac{C_1 R_2 R_4}{1 + (\omega R_1 C_1)^2}$

9.33. The circuit shown in Fig. 9-33 below is the Owen bridge. Find R_x and L_x in terms of the other bridge constants when the current through $\mathbf{Z}_D$ is zero. *Ans.* $R_x = \dfrac{C_1}{C_4} R_2$, $L_x = C_1 R_2 R_4$

9.34. The circuit shown in Fig. 9-34 below is an inductance comparison bridge. Select mesh currents and write the equations in matrix form. Find R_x and L_x when the current through Z_D is zero.

Ans. $R_x = \dfrac{R_2}{R_1} R_4$, $L_x = \dfrac{R_2}{R_1} L_4$

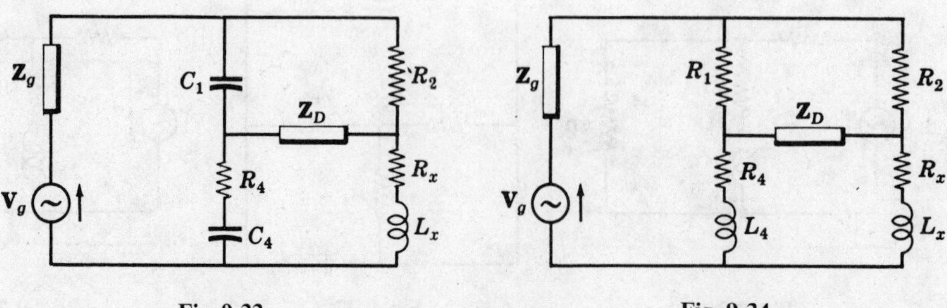

Fig. 9-33 Fig. 9-34

9.35. Find the voltage transfer function $\mathbf{V}_0/\mathbf{V}_i$ across the network of Fig. 9-35 below.
Ans. $0{\cdot}139\angle 90°$

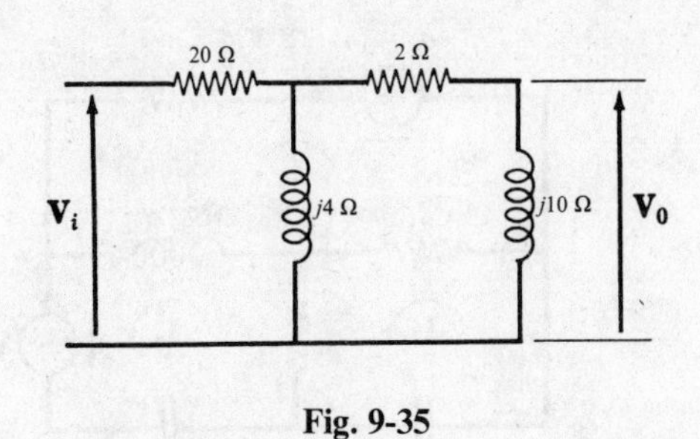

Fig. 9-35

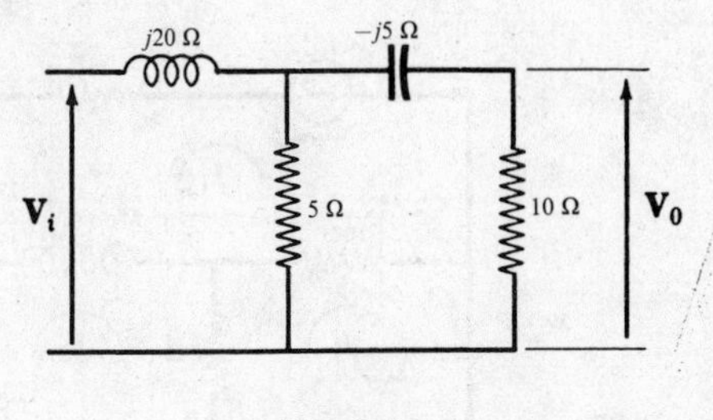

Fig. 9-36

9.36. Find the voltage transfer function $\mathbf{V}_0/\mathbf{V}_i$ across the network of Fig. 9-36 above. *Ans.* $0{\cdot}159\angle -61{\cdot}4°$

9.37. In the network of Fig. 9-37 below, find $\mathbf{V}_0$ with the polarity shown. *Ans.* $1{\cdot}56\angle 128{\cdot}7°$ V

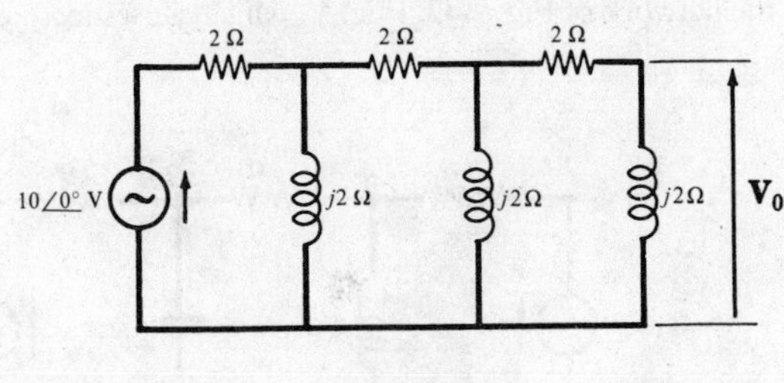

Fig. 9-37

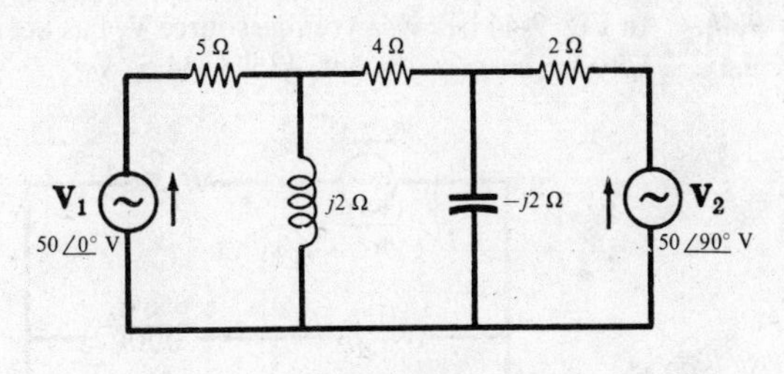

Fig. 9-38

9.38. In the network of Fig. 9-38 above, find the power in each of the three resistors. *Ans.* 471 W, 47·1 W, 471 W

9.39. Referring to the network of Fig. 9-38 above, find the power supplied by each of the voltage sources.
Ans. $P_1 = 422$ W, $P_2 = 565$ W

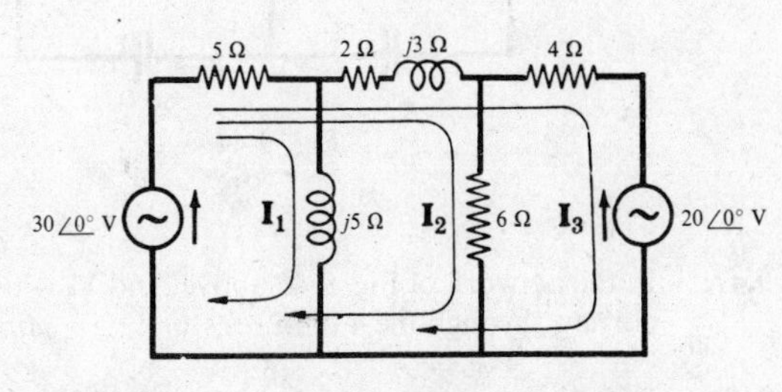

Fig. 9-39

9.40. In the network of Fig. 9-39, find mesh current $\mathbf{I}_3$ for the given choice of mesh currents. *Ans.* $1{\cdot}38\angle -209{\cdot}15°$ A

9.41. Find the current $\mathbf{I}_3$ in the network of Fig. 9-40 below. *Ans.* $11{\cdot}6\angle 113{\cdot}2°$ A

9.42. Referring to the network of Fig. 9-40 below, find the current ratio $\mathbf{I}_1/\mathbf{I}_3$. *Ans.* $-j3{\cdot}3$

9.43. In the network of Fig. 9-41 below, the three-mesh currents are shown in the primary loops. Compute $\mathbf{Z}_{\text{transfer } 13}$ and $\mathbf{Z}_{\text{transfer } 31}$. *Ans.* $4{\cdot}3\angle -68{\cdot}2°$ Ω for both parts

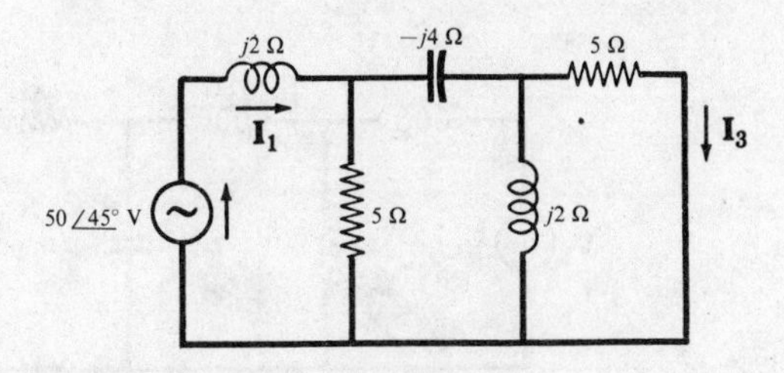

Fig. 9-40

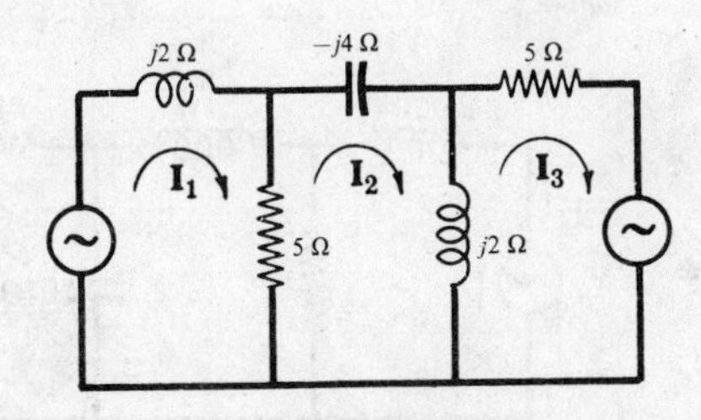

Fig. 9-41

9.44. In the three-mesh network of Fig. 9-42 below, with mesh currents as shown, find $\mathbf{Z}_{\text{input 1}}$, $\mathbf{Z}_{\text{transfer 12}}$ and $\mathbf{Z}_{\text{transfer 13}}$.
Ans. $20{\cdot}2\angle{-36{\cdot}1^\circ}\ \Omega$, $17{\cdot}4\angle{-71{\cdot}6^\circ}\ \Omega$, $6{\cdot}82\angle{-82{\cdot}9^\circ}\ \Omega$

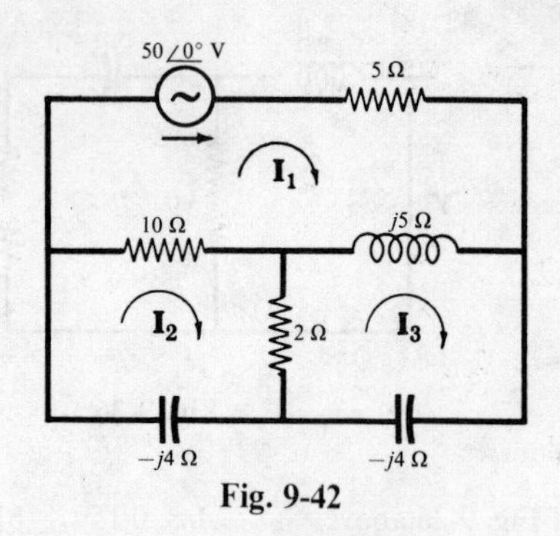

Fig. 9-42

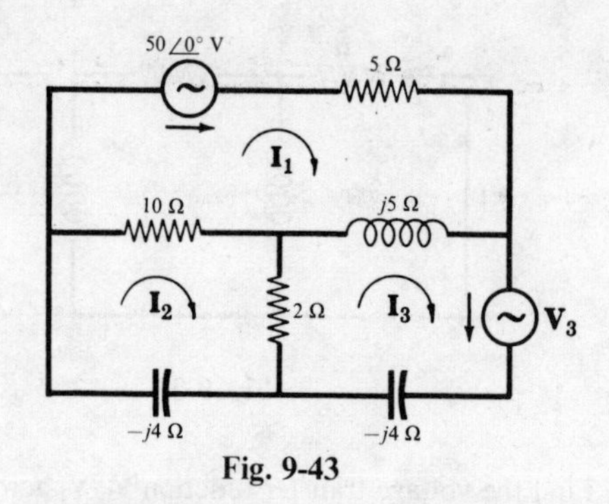

Fig. 9-43

9.45. In Fig. 9-43 above, a voltage source $\mathbf{V}_3$ is added to the network of Fig. 9-42. Find $\mathbf{V}_3$ which causes mesh current $\mathbf{I}_1$ to be zero. *Ans.* $16{\cdot}8\angle 133{\cdot}2^\circ$ V

9.46. In Fig. 9-44 below, a voltage source $\mathbf{V}_2$ has been added to the network of Fig. 9-42. Find $\mathbf{V}_2$ which causes mesh current $\mathbf{I}_1$ to be zero. *Ans.* $42{\cdot}9\angle 144{\cdot}5^\circ$ V

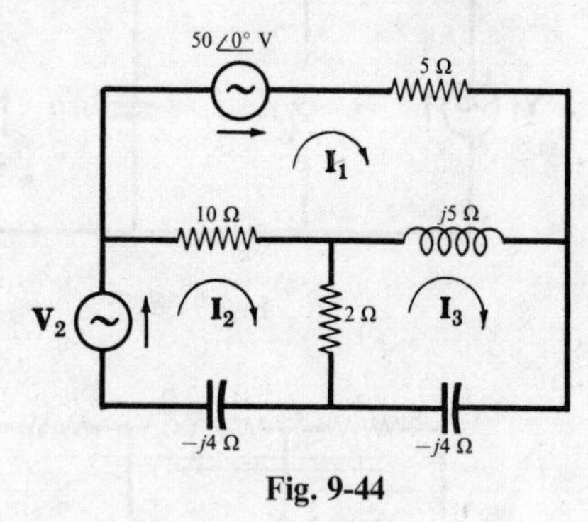

Fig. 9-44

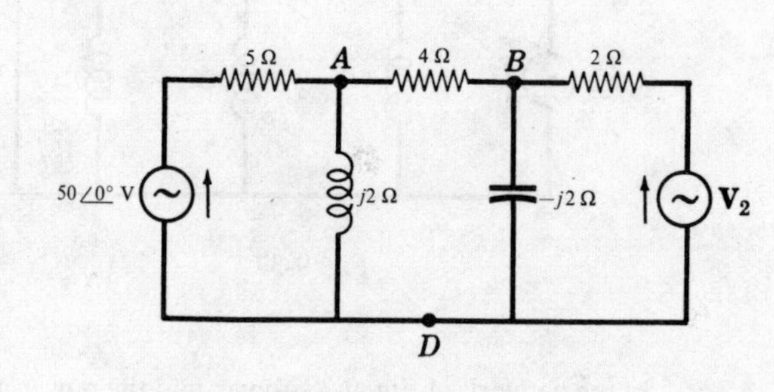

Fig. 9-45

9.47. In the network of Fig. 9-45 above, find $\mathbf{V}_2$ which results in zero current through the 4 ohm resistor. *Ans.* $26{\cdot}3\angle 113{\cdot}2^\circ$ V

9.48. Referring to the network of Fig. 9-45 above, find $\mathbf{V}_{AD}$ and $\mathbf{V}_{BD}$ when source $\mathbf{V}_2 = 26{\cdot}3\angle 113{\cdot}2^\circ$ volts.
Ans. $\mathbf{V}_{AD} = \mathbf{V}_{BD} = 18{\cdot}5\angle 68{\cdot}1^\circ$ V

9.49. Find $\mathbf{Z}_{\text{transfer 13}}$ for the choice of mesh currents shown in Fig. 9-46. Find $\mathbf{I}_3$ using this transfer impedance.
Ans. $12{\cdot}8\angle{-38{\cdot}7^\circ}\ \Omega$, $0{\cdot}782\angle 38{\cdot}7^\circ\ \Omega$

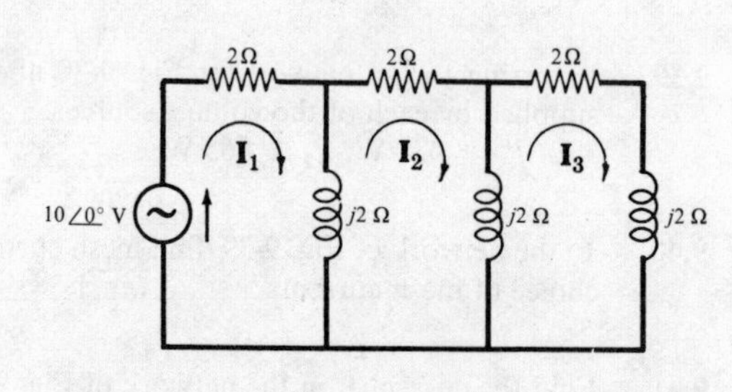

Fig. 9-46

9.50. In the network of Fig. 9-47 below, find $\mathbf{V}_2$ such that the $\mathbf{V}_2$ source current will be zero. *Ans.* $\mathbf{V}_2 = 4\angle 180^\circ$ V

9.51. Find the magnitude of the voltage source $\mathbf{V}_1$ in Fig. 9-48 below, which results in an effective voltage of 20 volts across the 5 ohm resistor. *Ans.* $69{\cdot}1$ V

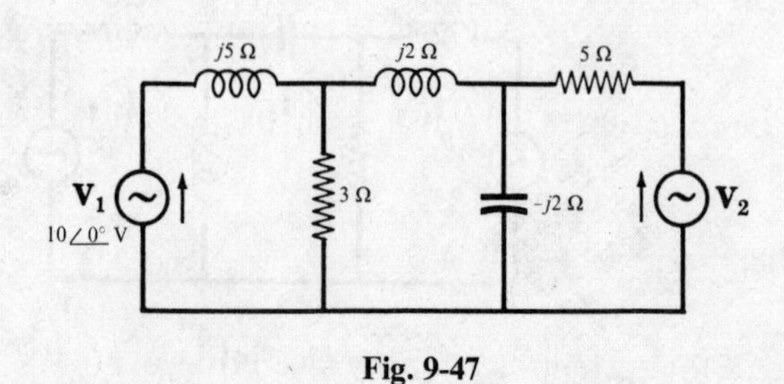

Fig. 9-47

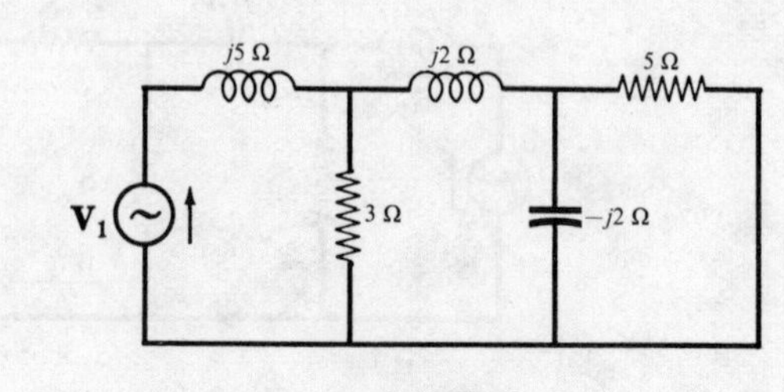

Fig. 9-48

CHAPTER 10

Node Voltage Network Analysis

INTRODUCTION

The selection of closed loops of current and the application of Kirchhoff's voltage law established the mesh current method for the solution of networks in Chapter 9. In this chapter the same solution is found by introducing sets of equations determined by the application of Kirchhoff's current law. This method is called the *node voltage method.*

NODE VOLTAGES

A *node* is a point in a network common to *two or more circuit elements*. If *three or more elements* join at a node, that node is called a *principal node* or *junction*. To each node in a circuit can be assigned a number or a letter. In Fig. 10-1, *A*, *B*, *1*, *2*, *3* are nodes and *1*, *2*, and *3* are principal nodes or junctions. A node voltage is the voltage of a given node with respect to one particular node called the *reference node*. In Fig. 10-1 select node *3* as the reference node. Then $\mathbf{V}_{13}$ is the voltage between nodes *1* and *3* and $\mathbf{V}_{23}$ is the voltage between nodes *2* and *3*. Since the node voltage is always determined with respect to a specified reference node, the notations $\mathbf{V}_1$ for $\mathbf{V}_{13}$ and $\mathbf{V}_2$ for $\mathbf{V}_{23}$ are used.

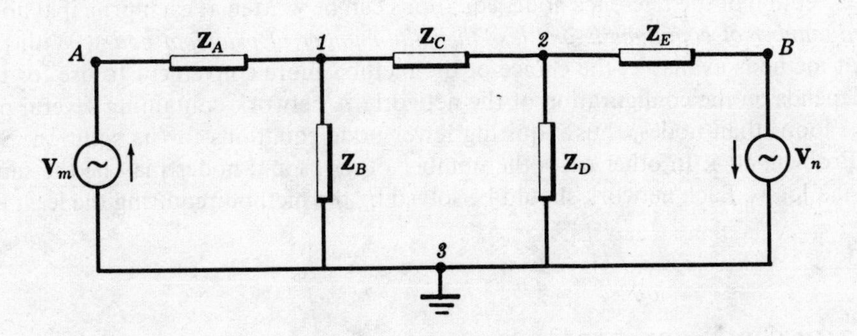

Fig. 10-1. Nodes in a Network

The node voltage method consists in finding the voltages at all the principal nodes with respect to the reference node. Kirchhoff's current law is applied to the two junctions 1 and 2, and two equations in the unknowns $\mathbf{V}_1$ and $\mathbf{V}_2$ are thus obtained. In Fig. 10-2, node *1* is redrawn with all of its connecting branches. Assume that all branch currents are leaving the node. Since the sum of currents leaving the junction is zero,

$$\frac{\mathbf{V}_1 - \mathbf{V}_m}{\mathbf{Z}_A} + \frac{\mathbf{V}_1}{\mathbf{Z}_B} + \frac{\mathbf{V}_1 - \mathbf{V}_2}{\mathbf{Z}_C} = 0 \qquad (1)$$

In formulating (*1*), the selection of the direction of the currents is arbitrary. See Problem 10.1.

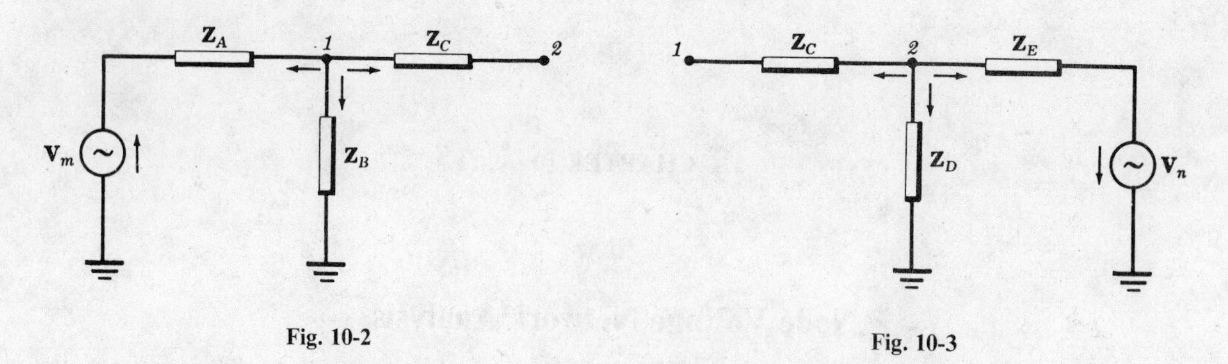

Fig. 10-2

Fig. 10-3

Repeat the same procedure for node *2* shown in Fig. 10-3. The resulting equation is

$$\frac{\mathbf{V}_2 - \mathbf{V}_1}{\mathbf{Z}_C} + \frac{\mathbf{V}_2}{\mathbf{Z}_D} + \frac{\mathbf{V}_2 + \mathbf{V}_n}{\mathbf{Z}_E} = 0 \tag{2}$$

Rearranging terms in (*1*) and (*2*), the set of the two equations is

$$\begin{aligned} \left(\frac{1}{\mathbf{Z}_A} + \frac{1}{\mathbf{Z}_B} + \frac{1}{\mathbf{Z}_C}\right)\mathbf{V}_1 - \left(\frac{1}{\mathbf{Z}_C}\right)\mathbf{V}_2 &= \left(\frac{1}{\mathbf{Z}_A}\right)\mathbf{V}_m \\ -\left(\frac{1}{\mathbf{Z}_C}\right)\mathbf{V}_1 + \left(\frac{1}{\mathbf{Z}_C} + \frac{1}{\mathbf{Z}_D} + \frac{1}{\mathbf{Z}_E}\right)\mathbf{V}_2 &= -\left(\frac{1}{\mathbf{Z}_E}\right)\mathbf{V}_n \end{aligned} \tag{3}$$

Since $1/\mathbf{Z} = \mathbf{Y}$, the set (*3*) can be rewritten in terms of admittances:

$$\begin{aligned} (\mathbf{Y}_A + \mathbf{Y}_B + \mathbf{Y}_C)\mathbf{V}_1 - \mathbf{Y}_C\mathbf{V}_2 &= \mathbf{Y}_A\mathbf{V}_m \\ -\mathbf{Y}_C\mathbf{V}_1 + (\mathbf{Y}_C + \mathbf{Y}_D + \mathbf{Y}_E)\mathbf{V}_2 &= -\mathbf{Y}_E\mathbf{V}_n \end{aligned} \tag{4}$$

NUMBER OF NODE VOLTAGE EQUATIONS

With the exception of the reference node, equations can be written at each principal node of a network. *Thus the required number of equations is one less than the number of principal nodes.* With the node voltage and mesh current methods available, the choice of the method more convenient to use for the solution of a given network depends on the configuration of the network. A network containing several parallel branches usually has more loops than nodes, thus requiring fewer node equations for its solution. See Problem 9.6, Chapter 9, and Problem 10.4. In other cases the number of loops and nodes may be the same or there may be more nodes than loops. Each network should be solved by the method requiring the least number of equations.

NODAL EQUATIONS BY INSPECTION

A network with four principal nodes requires three nodal equations for its solution. In general notation they are

$$\begin{aligned} \mathbf{Y}_{11}\mathbf{V}_1 + \mathbf{Y}_{12}\mathbf{V}_2 + \mathbf{Y}_{13}\mathbf{V}_3 &= \mathbf{I}_1 \\ \mathbf{Y}_{21}\mathbf{V}_1 + \mathbf{Y}_{22}\mathbf{V}_2 + \mathbf{Y}_{23}\mathbf{V}_3 &= \mathbf{I}_2 \\ \mathbf{Y}_{31}\mathbf{V}_1 + \mathbf{Y}_{32}\mathbf{V}_2 + \mathbf{Y}_{33}\mathbf{V}_3 &= \mathbf{I}_3 \end{aligned} \tag{5}$$

$\mathbf{Y}_{11}$ is called the self-admittance of node *1*, given by the sum of all the admittances connected to node *1*. Similarly, $\mathbf{Y}_{22}$ and $\mathbf{Y}_{33}$ are the self-admittances of nodes *2* and *3*, given by the sums of the admittances connected to the respective nodes.

$\mathbf{Y}_{12}$ is the mutual admittance between nodes *1* and *2*, given by the sum of all the admittances connecting *1* and *2*. $\mathbf{Y}_{12}$ has a negative sign as seen in the first equation of (*4*). Similarly, $\mathbf{Y}_{23}$ and $\mathbf{Y}_{13}$ are the mutual admittances of the elements connecting nodes *2* and *3* and nodes *1* and *2* respectively. All mutual admittances have negative signs. Note that $\mathbf{Y}_{13} = \mathbf{Y}_{31}$, $\mathbf{Y}_{23} = \mathbf{Y}_{32}$.

$\mathbf{I}_1$ is the sum of all the source currents at node *1*. A current which drives into the node has a positive sign, while to a current directed out of the node is assigned a negative sign. $\mathbf{I}_2$ and $\mathbf{I}_3$ are the sums of the driving currents at nodes *2* and *3* respectively.

By analogy with the matrix notation for mesh current equations (Chapter 9), the three nodal equations in (*5*) are written in matrix form:

$$\begin{bmatrix} \mathbf{Y}_{11} & \mathbf{Y}_{12} & \mathbf{Y}_{13} \\ \mathbf{Y}_{21} & \mathbf{Y}_{22} & \mathbf{Y}_{23} \\ \mathbf{Y}_{31} & \mathbf{Y}_{32} & \mathbf{Y}_{33} \end{bmatrix} \begin{bmatrix} \mathbf{V}_1 \\ \mathbf{V}_2 \\ \mathbf{V}_3 \end{bmatrix} = \begin{bmatrix} \mathbf{I}_1 \\ \mathbf{I}_2 \\ \mathbf{I}_3 \end{bmatrix} \tag{6}$$

The node voltages $\mathbf{V}_1$, $\mathbf{V}_2$ and $\mathbf{V}_3$ are given by

$$\mathbf{V}_1 = \frac{\begin{vmatrix} \mathbf{I}_1 & \mathbf{Y}_{12} & \mathbf{Y}_{13} \\ \mathbf{I}_2 & \mathbf{Y}_{22} & \mathbf{Y}_{23} \\ \mathbf{I}_3 & \mathbf{Y}_{32} & \mathbf{Y}_{33} \end{vmatrix}}{\Delta_Y}, \qquad \mathbf{V}_2 = \frac{\begin{vmatrix} \mathbf{Y}_{11} & \mathbf{I}_1 & \mathbf{Y}_{13} \\ \mathbf{Y}_{21} & \mathbf{I}_2 & \mathbf{Y}_{23} \\ \mathbf{Y}_{31} & \mathbf{I}_3 & \mathbf{Y}_{33} \end{vmatrix}}{\Delta_Y} \quad \text{and} \quad \mathbf{V}_3 = \frac{\begin{vmatrix} \mathbf{Y}_{11} & \mathbf{Y}_{12} & \mathbf{I}_1 \\ \mathbf{Y}_{21} & \mathbf{Y}_{22} & \mathbf{I}_2 \\ \mathbf{Y}_{31} & \mathbf{Y}_{32} & \mathbf{I}_3 \end{vmatrix}}{\Delta_Y}$$

When each numerator determinant is expanded about the elements of the column containing the current, the following equations for the nodal voltages are obtained.

$$\mathbf{V}_1 = \mathbf{I}_1\left(\frac{\Delta_{11}}{\Delta_Y}\right) + \mathbf{I}_2\left(\frac{\Delta_{21}}{\Delta_Y}\right) + \mathbf{I}_3\left(\frac{\Delta_{31}}{\Delta_Y}\right) \tag{7}$$

$$\mathbf{V}_2 = \mathbf{I}_1\left(\frac{\Delta_{12}}{\Delta_Y}\right) + \mathbf{I}_2\left(\frac{\Delta_{22}}{\Delta_Y}\right) + \mathbf{I}_3\left(\frac{\Delta_{32}}{\Delta_Y}\right) \tag{8}$$

$$\mathbf{V}_3 = \mathbf{I}_1\left(\frac{\Delta_{13}}{\Delta_Y}\right) + \mathbf{I}_2\left(\frac{\Delta_{23}}{\Delta_Y}\right) + \mathbf{I}_3\left(\frac{\Delta_{33}}{\Delta_Y}\right) \tag{9}$$

The terms on the right sides of (*7*), (*8*) and (*9*), are the phasor components which result from the various driving currents. Thus in (*7*) the voltage $\mathbf{V}_1$ is the sum of $\mathbf{I}_1(\Delta_{11}/\Delta_Y)$ due to the driving current $\mathbf{I}_1$, $\mathbf{I}_2(\Delta_{21}/\Delta_Y)$ due to the driving current $\mathbf{I}_2$, and $\mathbf{I}_3(\Delta_{31}/\Delta_Y)$ due to the driving current $\mathbf{I}_3$.

Example

Write the node voltage equations for the network of Fig. 10-4 and express them in matrix form.

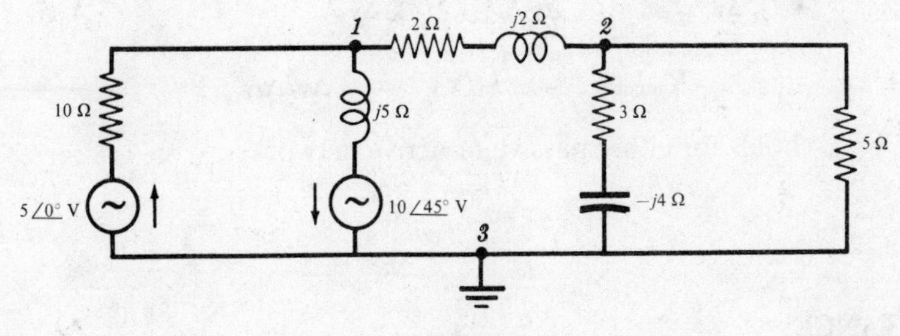

Fig. 10-4

Select the reference node *3* and number nodes *1* and *2* as shown in the circuit diagram. Assume all branch currents leave nodes *1* and *2*. Applying Kirchhoff's current law to each node, we obtain:

At node *1*: $$\frac{\mathbf{V}_1 - 5\angle 0^\circ}{10} + \frac{\mathbf{V}_1 + 10\angle 45^\circ}{j5} + \frac{\mathbf{V}_1 - \mathbf{V}_2}{2 + j2} = 0 \qquad (10)$$

At node *2*: $$\frac{\mathbf{V}_2 - \mathbf{V}_1}{2 + j2} + \frac{\mathbf{V}_2}{3 - j4} + \frac{\mathbf{V}_2}{5} = 0 \qquad (11)$$

Rearranging terms,

$$\left(\frac{1}{10} + \frac{1}{j5} + \frac{1}{2 + j2}\right)\mathbf{V}_1 - \left(\frac{1}{2 + j2}\right)\mathbf{V}_2 = \frac{5\angle 0^\circ}{10} - \frac{10\angle 45^\circ}{j5} \qquad (12)$$

$$-\left(\frac{1}{2 + j2}\right)\mathbf{V}_1 + \left(\frac{1}{2 + j2} + \frac{1}{3 - j4} + \frac{1}{5}\right)\mathbf{V}_2 = 0 \qquad (13)$$

In the square matrix containing the admittances, $\mathbf{Y}_{11} = 1/10 + 1/j5 + 1/(2 + j2)$ siemens by comparison with (*6*). This conforms to the definition of $\mathbf{Y}_{11}$ as the self-admittance of node *1*. Also, $\mathbf{Y}_{12} = \mathbf{Y}_{21} = -1/(2 + j2)$ siemens in agreement with the definition of mutual admittance.

$\mathbf{I}_1$ in the general notation was defined as the sum of all the driving currents at node *1*. According to the sign convention, the current from the left branch source directed into node *1* has a positive sign, while the current from the second branch source directed out of *1* has a negative sign. Thus $\mathbf{I}_1 = (5\angle 0^\circ)/10 - (10\angle 45^\circ)/j5$ amperes. The current $\mathbf{I}_2$ at node *2* is zero, since there is no source in the branches connected with node *2*.

DRIVING POINT ADMITTANCE

Consider a passive network with external terminals as shown in Fig. 10-5. The current source $\mathbf{I}_1$ drives into node *1* and any shunt admittances of the source are assumed to be included in the network.

Since there is no other current source within the network, the equation of $\mathbf{V}_1$ is

$$\mathbf{V}_1 = \mathbf{I}_1\left(\frac{\Delta_{11}}{\Delta_Y}\right) \qquad (14)$$

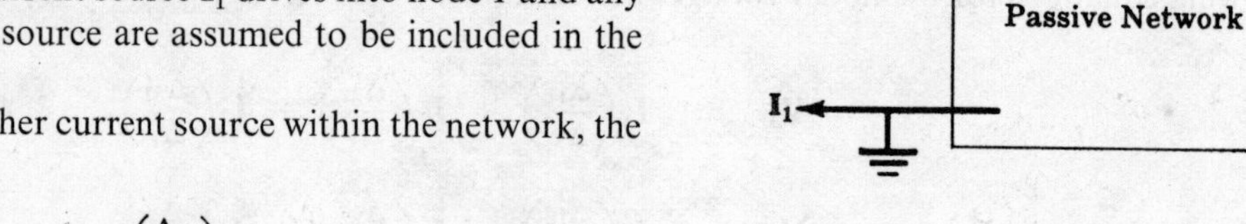

Fig. 10-5

The driving point admittance or $\mathbf{Y}_{\text{input}}$ is defined as the ratio of the driving current of the single current source existing between two nodes and the resulting voltage drop between the two nodes. Then from (*14*),

$$\mathbf{Y}_{\text{input 1}} = \frac{\mathbf{I}_1}{\mathbf{V}_1} = \frac{\Delta_Y}{\Delta_{11}}$$

In an active network the input admittance is defined as the admittance which the network presents to the specified terminals when all the internal sources are set equal to zero. Then,

$$\mathbf{V}_1 = \mathbf{I}_1\left(\frac{\Delta_{11}}{\Delta_Y}\right) + (0)\left(\frac{\Delta_{21}}{\Delta_Y}\right) + (0)\left(\frac{\Delta_{31}}{\Delta_Y}\right) + \cdots = \mathbf{I}_1\left(\frac{\Delta_{11}}{\Delta_Y}\right)$$

or

$$\mathbf{Y}_{\text{input 1}} = \mathbf{I}_1/\mathbf{V}_1 = \Delta_Y/\Delta_{11}$$

Thus the definition of $\mathbf{Y}_{\text{input}}$ holds for either passive or active networks.

TRANSFER ADMITTANCE

A current driving at one node in a network results in voltages at all nodes with respect to the reference. The transfer admittance is the ratio of a current driving in at one node to the resulting voltage at another node, all other sources being set to zero.

In the network shown in Fig. 10-6, $\mathbf{I}_r$ is the current driving into node r and the resulting voltage at node s is given by

$$\mathbf{V}_s = (0)\left(\frac{\Delta_{1s}}{\Delta_Y}\right) + \cdots + \mathbf{I}_r\left(\frac{\Delta_{rs}}{\Delta_Y}\right) + \cdots + (0)\left(\frac{\Delta_{ss}}{\Delta_Y}\right)$$

$$= \mathbf{I}_r\left(\frac{\Delta_{rs}}{\Delta_Y}\right)$$

Then $$\mathbf{Y}_{\text{transfer } rs} = \mathbf{I}_r/\mathbf{V}_s = \Delta_Y/\Delta_{rs}$$

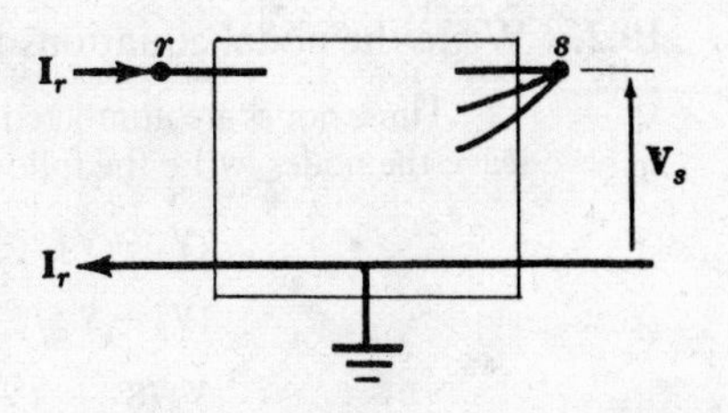

Fig. 10-6

Note that the driving current return point was selected as the reference node. This must be done, otherwise the current will appear in more than one term in the equation for $\mathbf{V}_s$ and the definition of $\mathbf{Y}_{\text{transfer}}$ will not hold.

Using the driving point and transfer admittances, we obtain the following set of equations for $\mathbf{V}_1$, $\mathbf{V}_2$ and $\mathbf{V}_3$ of a four junction network.

$$\mathbf{V}_1 = \frac{\mathbf{I}_1}{\mathbf{Y}_{\text{input 1}}} + \frac{\mathbf{I}_2}{\mathbf{Y}_{\text{transfer 21}}} + \frac{\mathbf{I}_3}{\mathbf{Y}_{\text{transfer 31}}}$$

$$\mathbf{V}_2 = \frac{\mathbf{I}_1}{\mathbf{Y}_{\text{transfer 12}}} + \frac{\mathbf{I}_2}{\mathbf{Y}_{\text{input 2}}} + \frac{\mathbf{I}_3}{\mathbf{Y}_{\text{transfer 32}}}$$

$$\mathbf{V}_3 = \frac{\mathbf{I}_1}{\mathbf{Y}_{\text{transfer 13}}} + \frac{\mathbf{I}_2}{\mathbf{Y}_{\text{transfer 23}}} + \frac{\mathbf{I}_3}{\mathbf{Y}_{\text{input 3}}}$$

When only one current source is acting on the network, with all others set equal to zero, the definitions of input and transfer admittance are apparent.

Solved Problems

10.1. Write the nodal equation for node 2 shown in Fig. 10-7(*a*) and 10-7(*b*).

Since all currents in Fig. 10-7(*a*) are directed away from node 2, write the sum of the currents leaving the node equal to zero.

$$(\mathbf{V}_2 - \mathbf{V}_1)/j2 + \mathbf{V}_2/10 + (\mathbf{V}_2 + 10\angle 0^\circ)/j5 = 0$$

Rearranging, $$-(1/j2)\mathbf{V}_1 + (1/j2 + 1/10 + 1/j5)\mathbf{V}_2 = -10\angle 0^\circ/j5$$

In Fig. 10-7(*b*) one branch current enters and the other two leave node 2. Set the current entering the node equal to the sum of the currents leaving.

$$(\mathbf{V}_1 - \mathbf{V}_2)/j2 = \mathbf{V}_2/10 + (\mathbf{V}_2 + 10/\angle 0^\circ)/j5$$

Rearranging, $$\mathbf{V}_2/10 + (\mathbf{V}_2 + 10\angle 0^\circ)/j5 + (\mathbf{V}_2 - \mathbf{V}_1)/j2 = 0$$

or $$-(1/j2)\mathbf{V}_1 + (1/j2 + 1/10 + 1/j5)\mathbf{V}_2 = -10\angle 0^\circ/j5$$

Thus any selection in the direction of the branch currents may be made in writing the nodal equations. The resulting equations will be identical.

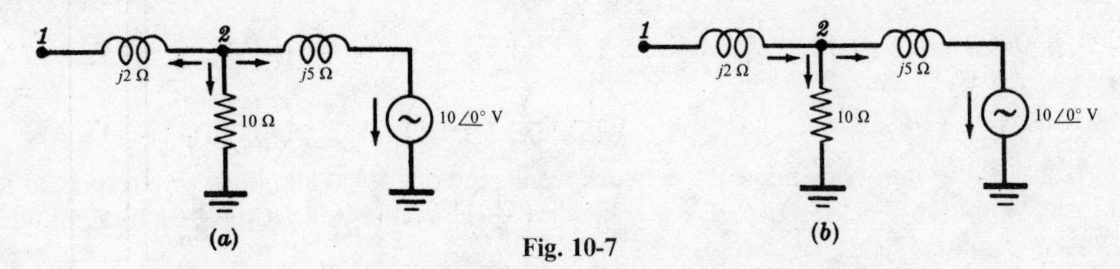

Fig. 10-7

10.2. Write the nodal equations for the network of Fig. 10-8 below and express them in matrix form.

Three nodes are numbered and the reference node selected as shown on the diagram. Assuming all branch currents leave the nodes, write the following equations at nodes *1*, *2* and *3*.

$$(\mathbf{V}_1-\mathbf{V}_2)/(-j8) \;+\; \mathbf{V}_1/5 \;+\; (\mathbf{V}_1-\mathbf{V}_3+10\angle 0^\circ)/(3+j4) \;=\; 0$$

$$(\mathbf{V}_2-\mathbf{V}_1)/(-j8) \;+\; \mathbf{V}_2/10 \;+\; (\mathbf{V}_2-\mathbf{V}_3-5\angle 0^\circ)/(j4) \;=\; 0$$

$$\mathbf{V}_3/8 \;+\; (\mathbf{V}_3-\mathbf{V}_1-10\angle 0^\circ)/(3+j4) \;+\; (\mathbf{V}_3-\mathbf{V}_2+5\angle 0^\circ)/(j4) \;=\; 0$$

Rearranging terms,

$$\left(\frac{1}{-j8}+\frac{1}{5}+\frac{1}{3+j4}\right)\mathbf{V}_1 - \left(\frac{1}{-j8}\right)\mathbf{V}_2 - \left(\frac{1}{3+j4}\right)\mathbf{V}_3 = (-10\angle 0^\circ)/(3+j4)$$

$$-\left(\frac{1}{-j8}\right)\mathbf{V}_1 + \left(\frac{1}{-j8}+\frac{1}{10}+\frac{1}{j4}\right)\mathbf{V}_2 - \left(\frac{1}{j4}\right)\mathbf{V}_3 = (5\angle 0^\circ)/(j4)$$

$$-\left(\frac{1}{3+j4}\right)\mathbf{V}_1 - \left(\frac{1}{j4}\right)\mathbf{V}_2 + \left(\frac{1}{8}+\frac{1}{j4}+\frac{1}{3+j4}\right)\mathbf{V}_3 = \left(\frac{10\angle 0^\circ}{3+j4}\right)-\left(\frac{5\angle 0^\circ}{j4}\right)$$

In matrix notation the nodal equations are

$$\begin{bmatrix} \left(\frac{1}{-j8}+\frac{1}{5}+\frac{1}{3+j4}\right) & -\left(\frac{1}{-j8}\right) & -\left(\frac{1}{3+j4}\right) \\ -\left(\frac{1}{-j8}\right) & \left(\frac{1}{-j8}+\frac{1}{10}+\frac{1}{j4}\right) & -\left(\frac{1}{j4}\right) \\ -\left(\frac{1}{3+j4}\right) & -\left(\frac{1}{j4}\right) & \left(\frac{1}{8}+\frac{1}{j4}+\frac{1}{3+j4}\right) \end{bmatrix} \begin{bmatrix} \mathbf{V}_1 \\ \mathbf{V}_2 \\ \mathbf{V}_3 \end{bmatrix} = \begin{bmatrix} -\left(\frac{10\angle 0^\circ}{3+j4}\right) \\ \left(\frac{5\angle 0^\circ}{j4}\right) \\ \left(\frac{10\angle 0^\circ}{3+j4}-\frac{5\angle 0^\circ}{j4}\right) \end{bmatrix}$$

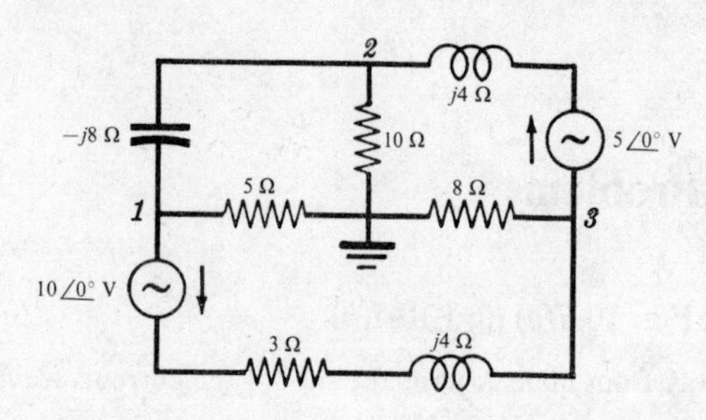

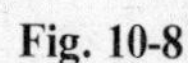

Fig. 10-8

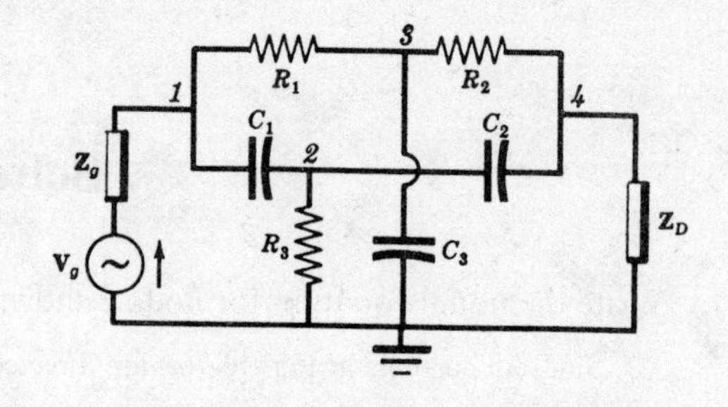

Fig. 10-9

10.3. Write the matrix form of the nodal equations for the network of Fig. 10-9 above by inspection.

The nodes are selected as shown in the diagram. In $[Y]$, $\mathbf{Y}_{11}$ is the sum of all admittances connected to node *1*, $(1/\mathbf{Z}_g+1/R_1+j\omega C_1)$. $\mathbf{Y}_{12}$ and $\mathbf{Y}_{13}$ are the negative of the sum of the admittances common to nodes *1* and *2*, and nodes *1* and *3*, i.e. $\mathbf{Y}_{12}=-(j\omega C_1)$, and $\mathbf{Y}_{13}=-(1/R_1)$ respectively. The other terms in $[Y]$ are determined in a similar manner.

There is only one driving current in the network; it drives into node *1* and therefore has a positive sign, $\mathbf{I}_1=\mathbf{V}_g/\mathbf{Z}_g$.

$$\begin{bmatrix} \left(\frac{1}{\mathbf{Z}_g}+\frac{1}{R_1}+j\omega C_1\right) & -(j\omega C_1) & -\left(\frac{1}{R_1}\right) & 0 \\ -(j\omega C_1) & \left(j\omega C_1+\frac{1}{R_3}+j\omega C_2\right) & 0 & -(j\omega C_2) \\ -\left(\frac{1}{R_1}\right) & 0 & \left(\frac{1}{R_1}+\frac{1}{R_2}+j\omega C_3\right) & -\left(\frac{1}{R_2}\right) \\ 0 & -(j\omega C_2) & -\left(\frac{1}{R_2}\right) & \left(\frac{1}{R_2}+j\omega C_2+\frac{1}{\mathbf{Z}_D}\right) \end{bmatrix} \begin{bmatrix} \mathbf{V}_1 \\ \mathbf{V}_2 \\ \mathbf{V}_3 \\ \mathbf{V}_4 \end{bmatrix} = \begin{bmatrix} \mathbf{V}_g/\mathbf{Z}_g \\ 0 \\ 0 \\ 0 \end{bmatrix}$$

10.4. In the network of Fig. 10-10 below, two equal capacitors of C farads and the resistor R are adjusted until the current through the detector impedance $\mathbf{Z}_D$ is zero. For this condition determine R_x and L_x in terms of the other circuit constants.

Nodes are selected as shown in the diagram. With the reference on one side of impedance $\mathbf{Z}_D$, a node voltage $\mathbf{V}_3$ of zero results in zero current through $\mathbf{Z}_D$. Write the nodal equations in matrix form by inspection:

$$\begin{bmatrix} \left(\dfrac{1}{\mathbf{Z}_g} + j\omega C + \dfrac{1}{R}\right) & -(j\omega C) & -\left(\dfrac{1}{R}\right) \\ -(j\omega C) & \left(j2\omega C + \dfrac{1}{R_x} + \dfrac{1}{j\omega L_x}\right) & -(j\omega C) \\ -\left(\dfrac{1}{R}\right) & -(j\omega C) & \left(j\omega C + \dfrac{1}{R} + \dfrac{1}{\mathbf{Z}_D}\right) \end{bmatrix} \begin{bmatrix} \mathbf{V}_1 \\ \mathbf{V}_2 \\ \mathbf{V}_3 \end{bmatrix} = \begin{bmatrix} \mathbf{V}_g/\mathbf{Z}_g \\ 0 \\ 0 \end{bmatrix}$$

Express $\mathbf{V}_3$ in determinant form and set it equal to zero:

$$\mathbf{V}_3 = \frac{\begin{vmatrix} \left(\dfrac{1}{\mathbf{Z}_g} + j\omega C + \dfrac{1}{R}\right) & -(j\omega C) & \mathbf{V}_g/\mathbf{Z}_g \\ -(j\omega C) & \left(j2\omega C + \dfrac{1}{R_x} + \dfrac{1}{j\omega L_x}\right) & 0 \\ -\left(\dfrac{1}{R}\right) & -(j\omega C) & 0 \end{vmatrix}}{\Delta_Y} = 0$$

The numerator determinant must be zero. Expand about the elements of the third column and obtain

$$(\mathbf{V}_g/\mathbf{Z}_g) \begin{vmatrix} -j\omega C & (j2\omega C + 1/R_x + 1/j\omega L_x) \\ -1/R & -j\omega C \end{vmatrix} = 0$$

Then
$$-\omega^2 C^2 + j2\omega C/R + 1/(RR_x) + 1/(j\omega L_x R) = 0$$

from which
$$R_x = 1/(\omega^2 C^2 R) \quad \text{and} \quad L_x = 1/(2\omega^2 C)$$

This is the same result that was obtained by the mesh current method in Problem 9.6, Chapter 9. Note that the number of equations needed in the solution is reduced from four to three by using the node voltage method.

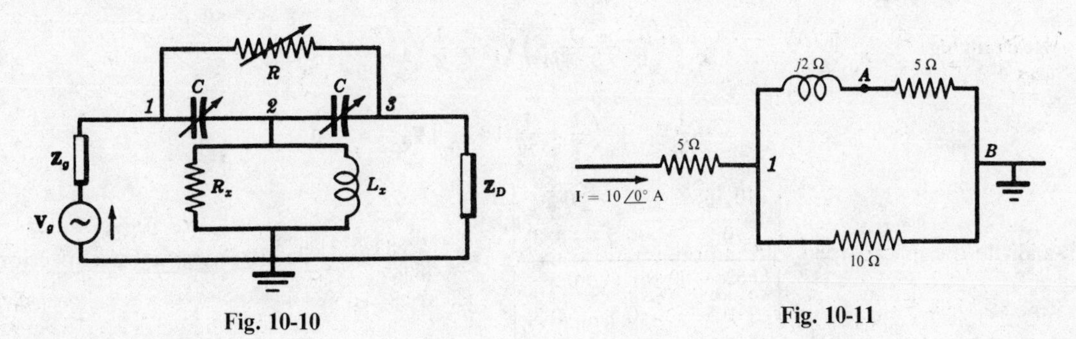

Fig. 10-10 Fig. 10-11

10.5. In the network of Fig. 10-11 above, determine the voltage $\mathbf{V}_{AB}$ by the nodal method.

With two principal nodes or junctions, only one nodal equation is required. Select the reference at B and write the equation at node 1. Applying Kirchhoff's current law the $10\angle 0°$ A current entering must equal the currents leaving:

$$10\angle 0° = \mathbf{V}_1/10 + \mathbf{V}_1/(5 + j2) \text{ and } \mathbf{V}_1 = 10\angle 0°/(0{\cdot}281\angle -14{\cdot}2°) = 35{\cdot}6\angle 14{\cdot}2° \text{ V}$$

Since the current through the $5 + j2$ Ω branch is $\mathbf{I} = \mathbf{V}_1/(5 + j2)$ A, the required voltage drop across the 5 ohm resistor is

$$\mathbf{V}_{AB} = \mathbf{I}(5) = \frac{\mathbf{V}_1}{(5 + j2)}(5) = \frac{35{\cdot}6\angle 14{\cdot}2°}{(5 + j2)}(5) = 33\angle -7{\cdot}6° \text{ V}$$

10.6. In the circuit shown in Fig. 10-12 determine the voltage $\mathbf{V}_{AB}$.

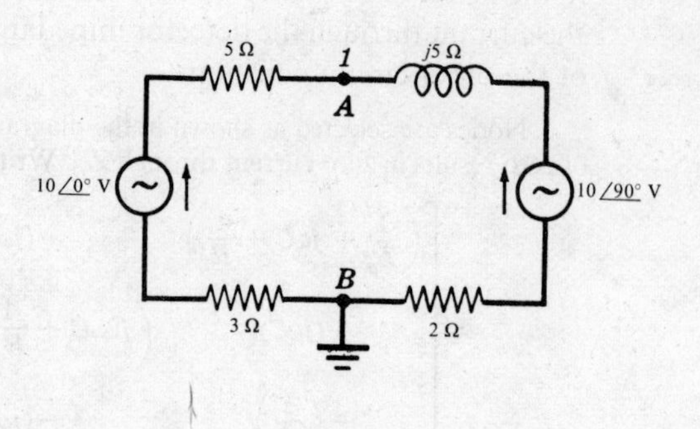

Fig. 10-12

The circuit contains no principal nodes. However, if point B is selected as the reference and point A as node 1, an equation can be written assuming that current leaves A by both branches.

$$\frac{\mathbf{V}_1 - 10\angle 0^\circ}{(5+3)} + \frac{\mathbf{V}_1 - 10\angle 90^\circ}{(2+j5)} = 0$$

Rearranging,

$$\mathbf{V}_1\left(\frac{1}{8} + \frac{1}{2+j5}\right) = \left(\frac{10\angle 0^\circ}{8} + \frac{10\angle 90^\circ}{2+j5}\right)$$

from which $V_{AB} = \mathbf{V}_1 = 11{\cdot}8\angle 55{\cdot}05^\circ$ V.

10.7. Find the voltage $\mathbf{V}_{AB}$ in the network of Fig. 10-13.

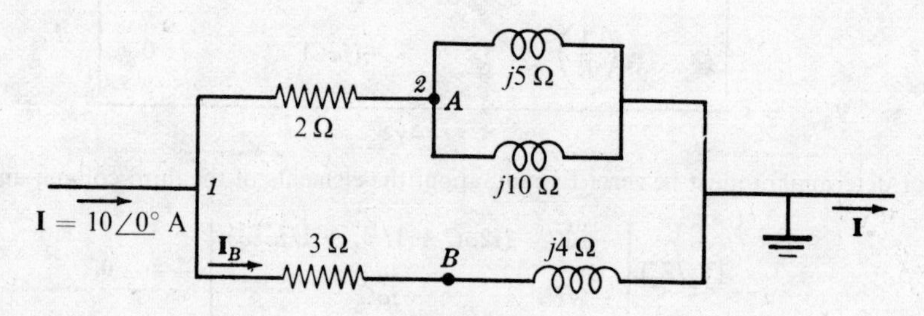

Fig. 10-13

The nodal equations are:

At node 1: $10\angle 0^\circ = (\mathbf{V}_1 - \mathbf{V}_2)/2 + \mathbf{V}_1/(3+j4)$

At node 2: $(\mathbf{V}_2 - \mathbf{V}_1)/2 + \mathbf{V}_2/j5 + \mathbf{V}_2/j10 = 0$

Rearranging,

$$\left(\frac{1}{2} + \frac{1}{3+j4}\right)\mathbf{V}_1 - \frac{1}{2}\mathbf{V}_2 = 10\angle 0^\circ$$

$$-\frac{1}{2}\mathbf{V}_1 + \left(\frac{1}{2} + \frac{1}{j5} + \frac{1}{j10}\right)\mathbf{V}_2 = 0$$

and

$$\mathbf{V}_1 = \frac{\begin{vmatrix} 10\angle 0^\circ & -0{\cdot}5 \\ 0 & (0{\cdot}5 - j0{\cdot}3) \end{vmatrix}}{\begin{vmatrix} (0{\cdot}62 - j0{\cdot}16) & -0{\cdot}5 \\ -0{\cdot}5 & (0{\cdot}5 - j0{\cdot}3) \end{vmatrix}} = \frac{5{\cdot}83\angle -31^\circ}{0{\cdot}267\angle -87{\cdot}42^\circ} = 21{\cdot}8\angle 56{\cdot}42^\circ \text{ V}$$

$$\mathbf{V}_2 = \frac{\begin{vmatrix} (0{\cdot}62 - j0{\cdot}16) & 10\angle 0^\circ \\ -0{\cdot}5 & 0 \end{vmatrix}}{\Delta_Y} = \frac{5\angle 0^\circ}{0{\cdot}267\angle -87{\cdot}42^\circ} = 18{\cdot}7\angle 87{\cdot}42^\circ \text{ V}$$

Node voltage $\mathbf{V}_2$ is the voltage of A with respect to the reference. Since $\mathbf{I}_B = \mathbf{V}_1/(3+j4)$, the voltage $\mathbf{V}_B$ with respect to the reference is

$$\mathbf{V}_B = \frac{\mathbf{V}_1}{(3+j4)}(j4) = \frac{21{\cdot}8\angle 56{\cdot}42^\circ}{(3+j4)}(j4) = 17{\cdot}45\angle 93{\cdot}32^\circ \text{ V}$$

Then the required voltage $\mathbf{V}_{AB}$ is

$$\mathbf{V}_{AB} = \mathbf{V}_A - \mathbf{V}_B = (18{\cdot}7\angle 87{\cdot}42^\circ) - (17{\cdot}45\angle 93{\cdot}32^\circ) = 2{\cdot}23\angle 34{\cdot}1^\circ \text{ V}$$

10.8. In the network shown in Fig. 10-14 find the line currents $\mathbf{I}_A$, $\mathbf{I}_B$ and $\mathbf{I}_C$.

Select node *1* and the reference point as shown in the diagram. Then solve the nodal equation

$$\frac{\mathbf{V}_1 + 100\angle 120^\circ}{20} + \frac{\mathbf{V}_1}{10} + \frac{\mathbf{V}_1 - 100\angle 0^\circ}{10} = 0$$

to obtain
$$\mathbf{V}_1 = \frac{200\angle 0^\circ - 100\angle 120^\circ}{5} = 50 - j17{\cdot}32 = 53\angle -19{\cdot}1^\circ \text{ V}$$

Now the currents in the individual branches are calculated.

$$\mathbf{I}_A = (\mathbf{V}_1 + 100\angle 120^\circ)/20 = (50 - j17{\cdot}32 - 50 + j86{\cdot}6)/20 = 3{\cdot}46\angle 90^\circ \text{ A}$$
$$\mathbf{I}_B = \mathbf{V}_1/10 = 5{\cdot}3\angle -19{\cdot}1^\circ \text{ A}$$
$$\mathbf{I}_C = (\mathbf{V}_1 - 100\angle 0^\circ)/10 = (50 - j17{\cdot}32 - 100)/10 = 5{\cdot}3\angle -160{\cdot}9^\circ \text{ A}$$

Note that the sum of the three currents entering the reference node equals zero.

$$\mathbf{I}_A + \mathbf{I}_B + \mathbf{I}_C = 3{\cdot}46\angle 90^\circ + 5{\cdot}3\angle -19{\cdot}1^\circ + 5{\cdot}3\angle -160{\cdot}9^\circ$$
$$= j3{\cdot}46 + 5{\cdot}0 - j1{\cdot}732 - 5 - j1{\cdot}732 = 0$$

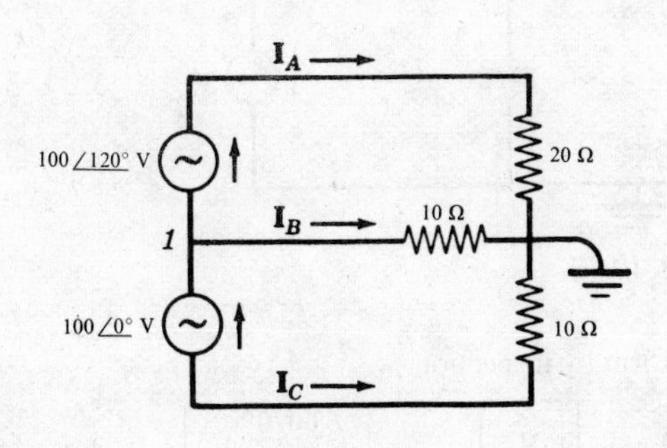

Fig. 10-14

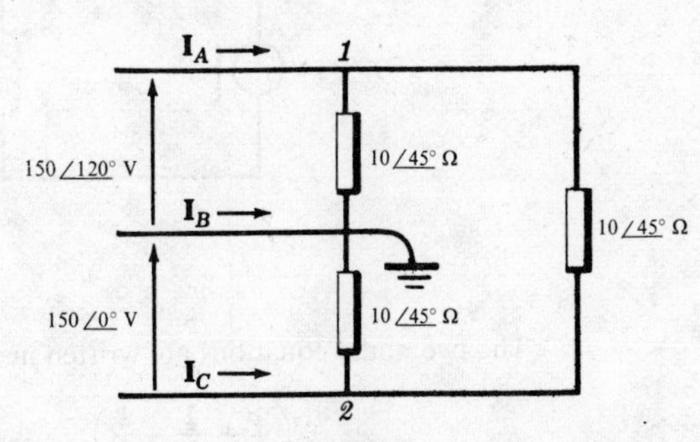

Fig. 10-15

10.9. In the circuit of Fig. 10-15 find the line currents $\mathbf{I}_A$, $\mathbf{I}_B$ and $\mathbf{I}_C$.

Nodes *1*, *2* and the reference node are identified in Fig. 10-15. The node voltages $\mathbf{V}_1$ and $\mathbf{V}_2$ can be read directly from the diagram, being equal to the given constant voltages. Thus

$$\mathbf{V}_1 = 150\angle 120^\circ \text{ V and } \mathbf{V}_2 = -150\angle 0^\circ = 150\angle 180^\circ \text{ V}$$

Applying Kirchhoff's current law to each of the three nodes, the required currents are evaluated.

At node *1*:
$$\mathbf{I}_A = \frac{\mathbf{V}_1}{10\angle 45^\circ} + \frac{\mathbf{V}_1 - \mathbf{V}_2}{10\angle 45^\circ} = \frac{300\angle 120^\circ - 150\angle 180^\circ}{10\angle 45^\circ} = 26\angle 45^\circ \text{ A}$$

At reference node:
$$\mathbf{I}_B = \frac{-\mathbf{V}_1}{10\angle 45^\circ} - \frac{\mathbf{V}_2}{10\angle 45^\circ} = \frac{150\angle -60^\circ + 150\angle 0^\circ}{10\angle 45^\circ} = 26\angle -75^\circ \text{ A}$$

At node *2*:
$$\mathbf{I}_C = \frac{\mathbf{V}_2}{10\angle 45^\circ} + \frac{\mathbf{V}_2 - \mathbf{V}_1}{10\angle 45^\circ} = \frac{300\angle 180^\circ - 150\angle 120^\circ}{10\angle 45^\circ} = 26\angle -195^\circ \text{ A}$$

10.10. In the circuit of Fig. 10-16 determine the power output of the source and the power in each of the network resistors.

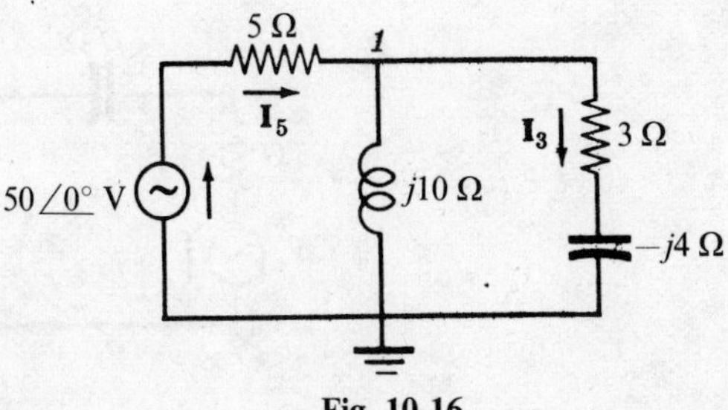

Fig. 10-16

Select the reference and node *1* as shown in the diagram. Then the nodal equation is

$$(\mathbf{V}_1 - 50\angle 0^\circ)/5 + \mathbf{V}_1/j10 + \mathbf{V}_1/(3 - j4) = 0$$

from which

$$\mathbf{V}_1 = (10\angle 0^\circ)/(0{\cdot}326\angle 10{\cdot}6^\circ) = 30{\cdot}7\angle -10{\cdot}6^\circ \text{ V}$$

Solve now for the following branch currents, assuming their directions as given in the diagram:

$$\mathbf{I}_5 = (50\angle 0^\circ - \mathbf{V}_1)/5 = (50\angle 0^\circ - 30{\cdot}7\angle -10{\cdot}6^\circ)/5 = 4{\cdot}12\angle 15{\cdot}9^\circ \text{ A}$$

$$\mathbf{I}_3 = \mathbf{V}_1/(3 - j4) = (30{\cdot}7\angle -10{\cdot}6^\circ)/(5\angle -53{\cdot}1^\circ) = 6{\cdot}14\angle 42{\cdot}5^\circ \text{ A}$$

The power output of the source is

$$P = VI_5 \cos\theta = (50)(4{\cdot}12)\cos 15{\cdot}9^\circ = 198 \text{ W}$$

From the relation $P = I^2R$, the power dissipated in each resistor is determined:

$$P_5 = (I_5)^2 5 = (4{\cdot}12)^2 5 = 85 \text{ W and } P_3 = (I_3)^2 3 = (6{\cdot}14)^2 3 = 113 \text{ W}$$

Note that the total power delivered by the source equals the sum of the powers dissipated by the two circuit resistors, i.e. $P_T = 85 + 113 = 198$ W.

10.11. In the network of Fig. 10-17 determine the voltages of nodes *1* and *2* with respect to the selected reference.

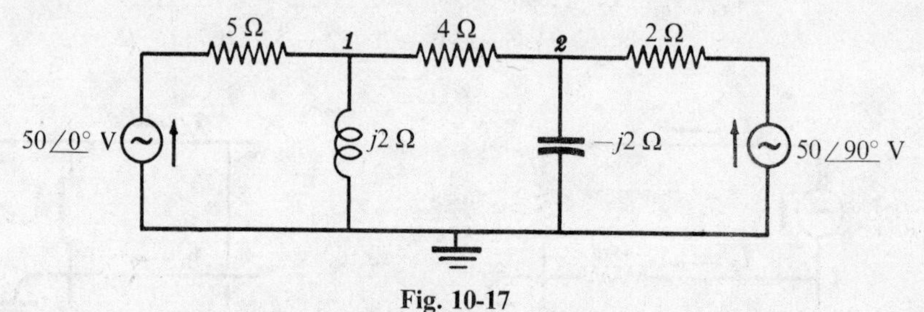

Fig. 10-17

The two nodal equations are written in matrix form by inspection.

$$\begin{bmatrix} \left(\frac{1}{5}+\frac{1}{j2}+\frac{1}{4}\right) & -\left(\frac{1}{4}\right) \\ -\left(\frac{1}{4}\right) & \left(\frac{1}{4}+\frac{1}{-j2}+\frac{1}{2}\right) \end{bmatrix} \begin{bmatrix} \mathbf{V}_1 \\ \mathbf{V}_2 \end{bmatrix} = \begin{bmatrix} \left(\frac{50\angle 0^\circ}{5}\right) \\ \left(\frac{50\angle 90^\circ}{2}\right) \end{bmatrix}$$

from which

$$\mathbf{V}_1 = \frac{\begin{vmatrix} 10 & -0{\cdot}25 \\ j25 & (0{\cdot}75 + j0{\cdot}5) \end{vmatrix}}{\begin{vmatrix} (0{\cdot}45 - j0{\cdot}5) & -0{\cdot}25 \\ -0{\cdot}25 & (0{\cdot}75 + j0{\cdot}5) \end{vmatrix}} = \frac{13{\cdot}5\angle 56{\cdot}3^\circ}{0{\cdot}546\angle -15{\cdot}95^\circ} = 24{\cdot}7\angle 72{\cdot}25^\circ \text{ V}$$

$$\mathbf{V}_2 = \frac{\begin{vmatrix} (0{\cdot}45 - j0{\cdot}5) & 10 \\ -0{\cdot}25 & j25 \end{vmatrix}}{\Delta_Y} = \frac{18{\cdot}35\angle 37{\cdot}8^\circ}{0{\cdot}546\angle -15{\cdot}95^\circ} = 33{\cdot}6\angle 53{\cdot}75^\circ \text{ V}$$

10.12. In the network of Fig. 10-18, given $\mathbf{V}_o$ as the voltage drop across the $2 - j2\ \Omega$ impedance due to the source $\mathbf{V}_i$, find the ratio $\mathbf{V}_o/\mathbf{V}_i$.

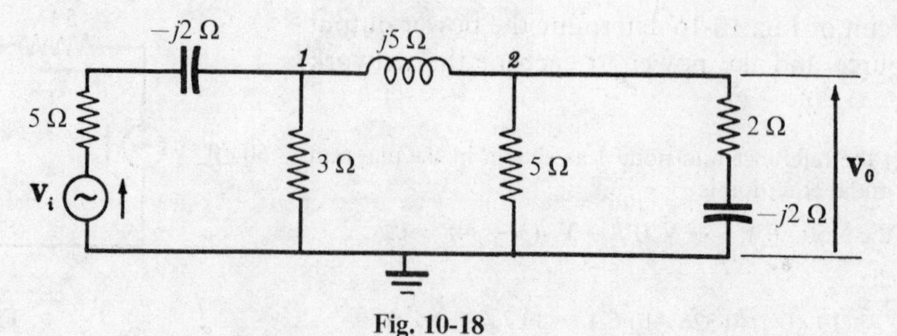

Fig. 10-18

Nodes *1*, *2* and the reference are selected as shown in the circuit diagram. With this selection $\mathbf{V}_0$ is the voltage of node *1* with respect to the reference.

Write the nodal equations in matrix form by inspection:

$$\begin{bmatrix} \left(\frac{1}{5-j2}+\frac{1}{3}+\frac{1}{j5}\right) & -\left(\frac{1}{j5}\right) \\ -\left(\frac{1}{j5}\right) & \left(\frac{1}{j5}+\frac{1}{5}+\frac{1}{2-j2}\right) \end{bmatrix} \begin{bmatrix} \mathbf{V}_1 \\ \mathbf{V}_2 \end{bmatrix} = \begin{bmatrix} \frac{\mathbf{V}_i}{5-j2} \\ 0 \end{bmatrix}$$

Solving for $\mathbf{V}_0$,

$$\mathbf{V}_0 = \mathbf{V}_2 = \frac{\begin{vmatrix} (0{\cdot}506 - j0{\cdot}131) & \mathbf{V}_i/(5-j2) \\ j0{\cdot}2 & 0 \end{vmatrix}}{\begin{vmatrix} (0{\cdot}506 - j0{\cdot}131) & j0{\cdot}2 \\ j0{\cdot}2 & (0{\cdot}45 + j0{\cdot}05) \end{vmatrix}} = \frac{(0{\cdot}2\angle{-90^\circ})\mathbf{V}_i/(5-j2)}{(0{\cdot}276\angle{-7^\circ})}$$

Then
$$\frac{\mathbf{V}_0}{\mathbf{V}_i} = \frac{0{\cdot}2\angle{-90^\circ}}{(5-j2)(0{\cdot}276\angle{-7^\circ})} = 0{\cdot}1345\angle{-61{\cdot}2^\circ}$$

This result is called a *voltage transfer function* and permits the direct computation of output voltage at the given branch for any given input voltage, i.e. $\mathbf{V}_0 = \mathbf{V}_i\,(0{\cdot}1345\angle{-61{\cdot}2^\circ})$.

10.13. Given nodes *1* and *2* in the network of Fig. 10-19, find the ratio $\mathbf{V}_1/\mathbf{V}_2$.

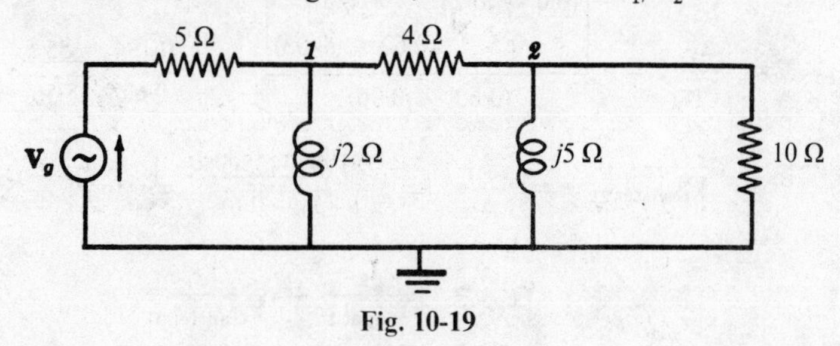

Fig. 10-19

The nodal equations are written in matrix form by inspection:

$$\begin{bmatrix} \left(\frac{1}{5}+\frac{1}{j2}+\frac{1}{4}\right) & -\left(\frac{1}{4}\right) \\ -\left(\frac{1}{4}\right) & \left(\frac{1}{4}+\frac{1}{j5}+\frac{1}{10}\right) \end{bmatrix} \begin{bmatrix} \mathbf{V}_1 \\ \mathbf{V}_2 \end{bmatrix} = \begin{bmatrix} (\mathbf{V}_g/5) \\ 0 \end{bmatrix}$$

from which
$$\mathbf{V}_1 = \frac{\begin{vmatrix} (\mathbf{V}_g/5) & -0{\cdot}25 \\ 0 & (0{\cdot}35 - j0{\cdot}2) \end{vmatrix}}{\Delta_Y} = \frac{(\mathbf{V}_g/5)(0{\cdot}403\angle{-29{\cdot}8^\circ})}{\Delta_Y}$$

$$\mathbf{V}_2 = \frac{\begin{vmatrix} (0{\cdot}45 - j0{\cdot}5) & (\mathbf{V}_g/5) \\ -0{\cdot}25 & 0 \end{vmatrix}}{\Delta_Y} = \frac{(\mathbf{V}_g/5)(0{\cdot}25)}{\Delta_Y}$$

and
$$\frac{\mathbf{V}_1}{\mathbf{V}_2} = \frac{(\mathbf{V}_g/5)(0{\cdot}403\angle{-29{\cdot}8^\circ})/\Delta_Y}{(\mathbf{V}_g/5)(0{\cdot}25)/\Delta_Y} = 1{\cdot}61\angle{-29{\cdot}8^\circ}$$

Alternate solution. Express each node voltage in terms of the cofactors. Since a single source with driving current $\mathbf{I}_1$ is acting on the circuit, $\mathbf{V}_1 = \mathbf{I}_1(\Delta_{11}/\Delta_Y)$ and $\mathbf{V}_2 = \mathbf{I}_1(\Delta_{12}/\Delta_Y)$. Then

$$\frac{\mathbf{V}_1}{\mathbf{V}_2} = \frac{\mathbf{I}_1(\Delta_{11}/\Delta_Y)}{\mathbf{I}_1(\Delta_{12}/\Delta_Y)} = \frac{\Delta_{11}}{\Delta_{12}} = \frac{0{\cdot}35 - j0{\cdot}2}{0{\cdot}25} = 1{\cdot}61\angle{-29{\cdot}8^\circ}$$

10.14. Determine the voltages of nodes *1* and *2* in the network of Fig. 10-20 using input and transfer admittances.

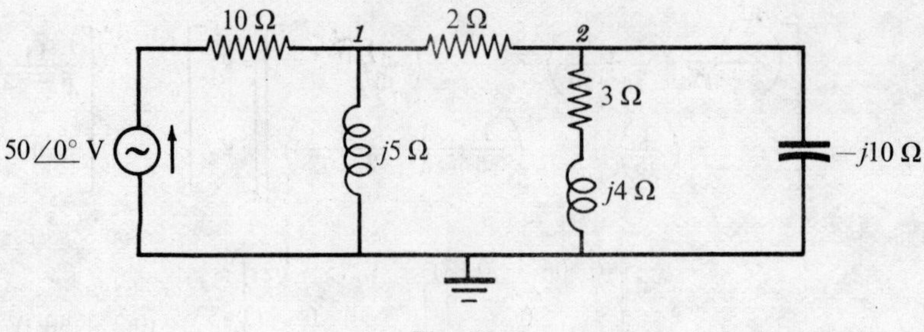

Fig. 10-20

The admittance matrix [Y] for the given selection of nodes is, by inspection,

$$[Y] = \begin{bmatrix} \left(\frac{1}{10}+\frac{1}{j5}+\frac{1}{2}\right) & -\left(\frac{1}{2}\right) \\ -\left(\frac{1}{2}\right) & \left(\frac{1}{2}+\frac{1}{3+j4}+\frac{1}{-j10}\right) \end{bmatrix} = \begin{bmatrix} (0{\cdot}6)-j0{\cdot}2) & -0{\cdot}5 \\ -0{\cdot}5 & (0{\cdot}62-j0{\cdot}06) \end{bmatrix}$$

Then

$$\mathbf{Y}_{\text{input 1}} = \frac{\Delta_Y}{\Delta_{11}} = \frac{\begin{vmatrix} (0{\cdot}6-j0{\cdot}2) & -0{\cdot}5 \\ -0{\cdot}5 & (0{\cdot}62-j0{\cdot}06) \end{vmatrix}}{(0{\cdot}62-j0{\cdot}06)} = \frac{0{\cdot}194\angle{-55{\cdot}5^\circ}}{0{\cdot}62\angle{-5{\cdot}56^\circ}} \quad 0{\cdot}313\angle{-49{\cdot}94^\circ}\text{ S}$$

and

$$\mathbf{Y}_{\text{transfer 21}} = \frac{\Delta_Y}{\Delta_{12}} = \frac{0{\cdot}194\angle{-55{\cdot}5^\circ}}{(-1)(-0{\cdot}5)} = 0{\cdot}388\angle{-55{\cdot}5^\circ}\text{ S}$$

At node *1*

$$\mathbf{V}_1 = \frac{\mathbf{I}_1}{\mathbf{Y}_{\text{input 1}}} + \frac{\mathbf{I}_2}{\mathbf{Y}_{\text{transfer 21}}}$$

Since there is no driving current at node *2*, i.e. $\mathbf{I}_2 = 0$, we have

$$\mathbf{V}_1 = \frac{\mathbf{I}_1}{\mathbf{Y}_{\text{input 1}}} \quad \frac{(50\angle{0^\circ})/10}{0{\cdot}313\angle{-49{\cdot}94^\circ}} = 15{\cdot}95\angle{49{\cdot}94^\circ}\text{ V}$$

Similarly,

$$\mathbf{V}_2 = \frac{\mathbf{I}_1}{\mathbf{Y}_{\text{transfer 12}}} + \frac{\mathbf{I}_2}{\mathbf{Y}_{\text{input 2}}} \quad \frac{(50\angle{0^\circ})/10}{0{\cdot}388\angle{-55{\cdot}5^\circ}} = 12{\cdot}9\angle{55{\cdot}5^\circ}\text{ V}$$

Supplementary Problems

10.15. Determine the number of node voltage equations required to solve each of the networks shown in Fig. 10-21(*a-f*).

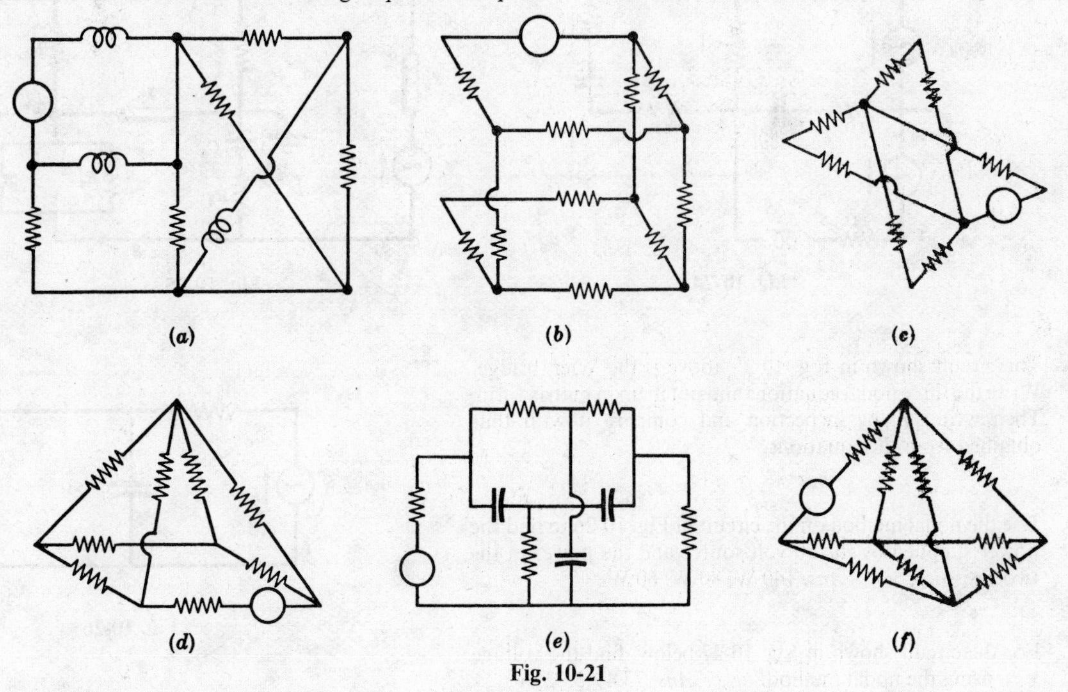

Fig. 10-21

Ans. (*a*) 3, (*b*) 5, (*c*) 1, (*d*) 4, (*e*) 4, (*f*) 4.

10.16. Write the nodal equation for the given node in the network of Fig. 10-22.

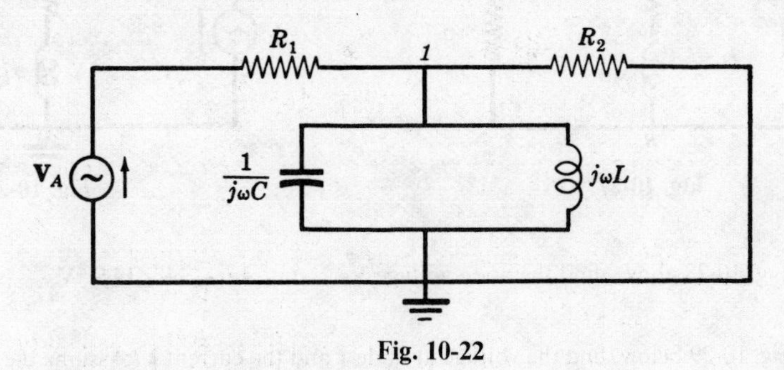

Fig. 10-22

10.17. For the network of Fig. 10-23 write the nodal equations and put them in matrix form. Then write the $[Y]$ by inspection and compare it with that obtained from the equations.

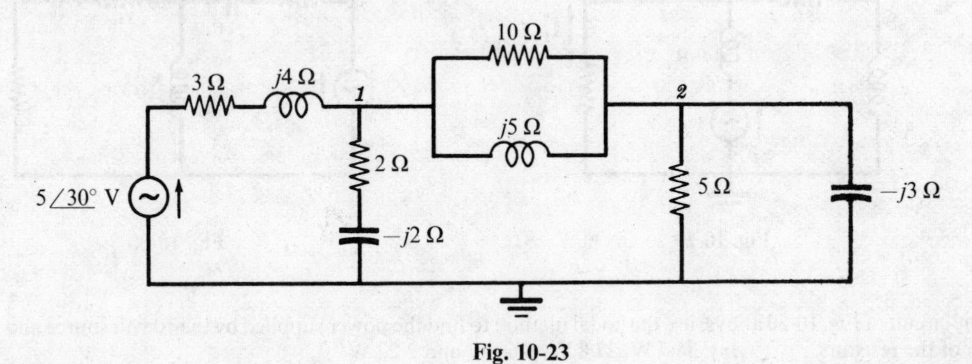

Fig. 10-23

10.18. For the given nodes of the network of Fig. 10-24 below, write the nodal equations and put them in matrix form. Then write the $[Y]$ by inspection and compare it with that obtained from the equations.

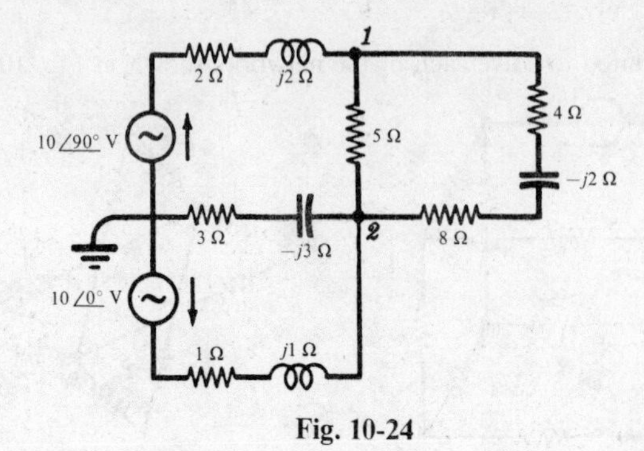

Fig. 10-24

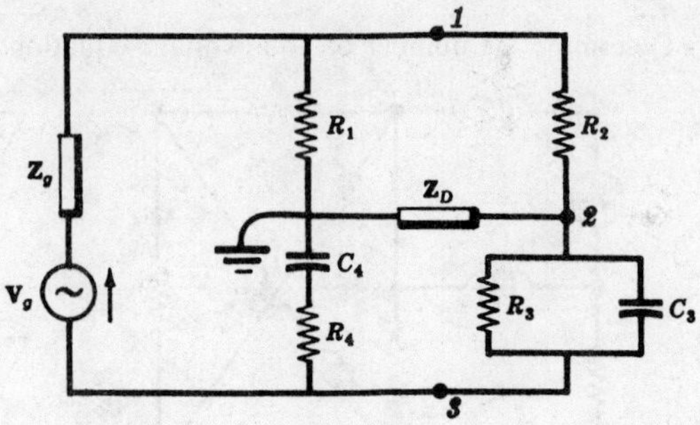

Fig. 10-25

10.19. The circuit shown in Fig. 10-25 above is the Wien bridge. Write the three nodal equations and put them in matrix form. Then write $[Y]$ by inspection and compare it with that obtained from the equations.

10.20. Use the nodal method on the circuit of Fig. 10-26 to find the power supplied by the 50 volt source and the power in the two resistors. *Ans.* 140 W, 80 W, 60 W

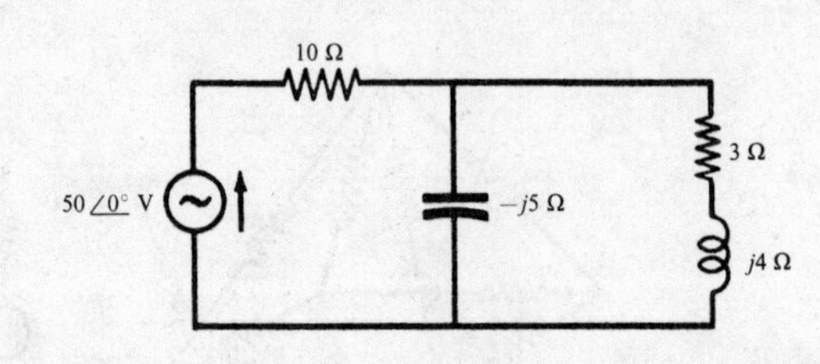

Fig. 10-26

10.21. For the circuit shown in Fig. 10-27 below, find the voltage $\mathbf{V}_{AB}$ using the nodal method. *Ans.* $75{\cdot}4\angle 55{\cdot}2^\circ$ V

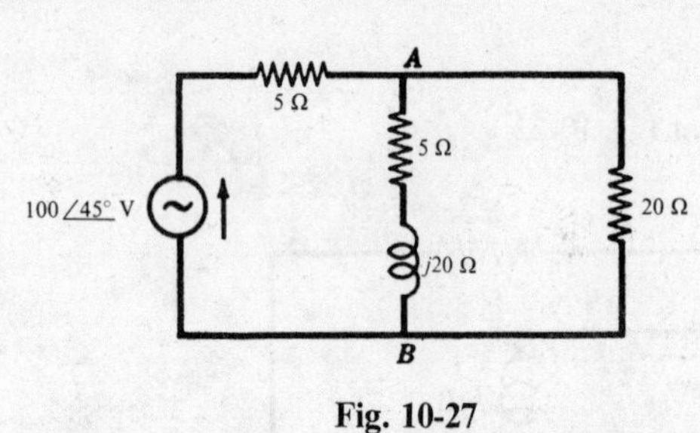

Fig. 10-27

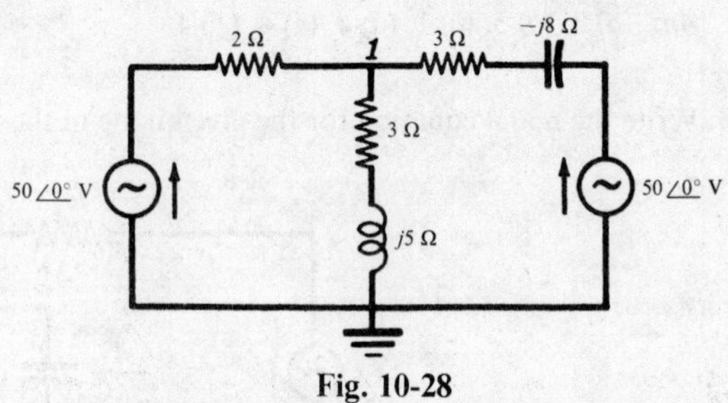

Fig. 10-28

10.22. In the circuit of Fig. 10-28 above, find the node voltage $\mathbf{V}_1$. *Ans.* $43{\cdot}9\angle 14{\cdot}9^\circ$ V

10.23. In the circuit of Fig. 10-29 below, find the voltage at node *1* and the current $\mathbf{I}_1$. Assume the direction of $\mathbf{I}_1$ as shown in the diagram. *Ans.* $17{\cdot}7\angle -45^\circ$ V, $1{\cdot}77\angle 135^\circ$ A

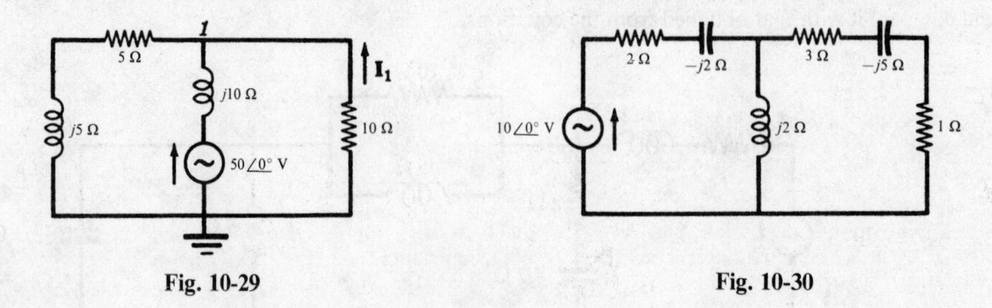

Fig. 10-29

Fig. 10-30

10.24. In the circuit of Fig. 10-30 above, use the nodal method to find the power supplied by the 10 volt source and the power in each of the resistors. *Ans.* 36·7 W, 27·8 W, 6·66 W and 2·22 W

10.25. Determine the power supplied to the circuit of Fig. 10-31 below by a source $\mathbf{V}_i = 50\angle 0^\circ$ V. Determine also the power dissipated by each resistor in the circuit.
Ans. $P = 354$ W, $P_1 = 256$ W, $P_2 = 77{\cdot}1$ W, $P_3 = 9{\cdot}12$ W, $P_4 = 11{\cdot}3$ W

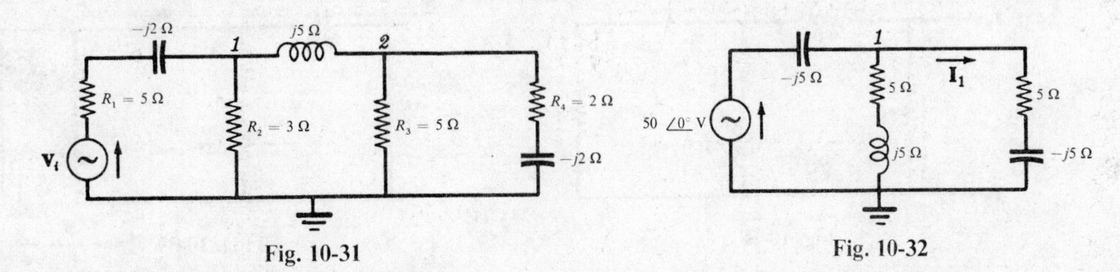

Fig. 10-31

Fig. 10-32

10.26. Use the nodal method to find $\mathbf{I}_1$ in the circuit shown in Fig. 10-32 above. *Ans.* $5\angle 90^\circ$ A

10.27. In the circuit of Fig. 10-33 below, find the effective voltage of the source **V** which results in a power of 75 watts in the 3 ohm resistor. *Ans.* 24·2 V

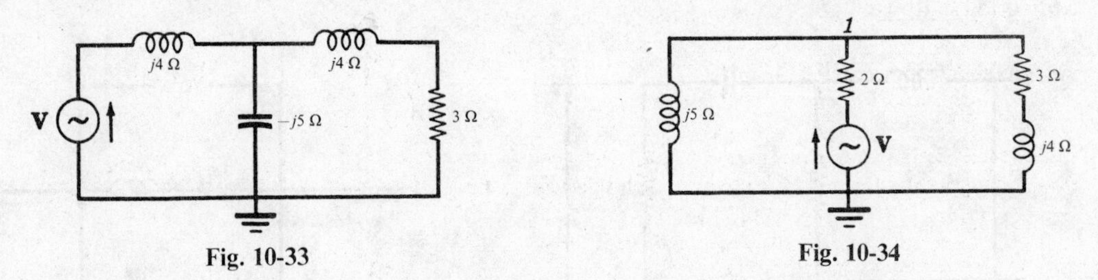

Fig. 10-33

Fig. 10-34

10.28. For the circuit of Fig. 10-34 above, find the source voltage **V** which results in a voltage at node *1* of $50\angle 0^\circ$ V.
Ans. $71{\cdot}6\angle -30{\cdot}2^\circ$ V

10.29. For the circuit shown in Fig. 10-35 below, find the voltage of node *1*. *Ans.* $179\angle 204{\cdot}8^\circ$ V

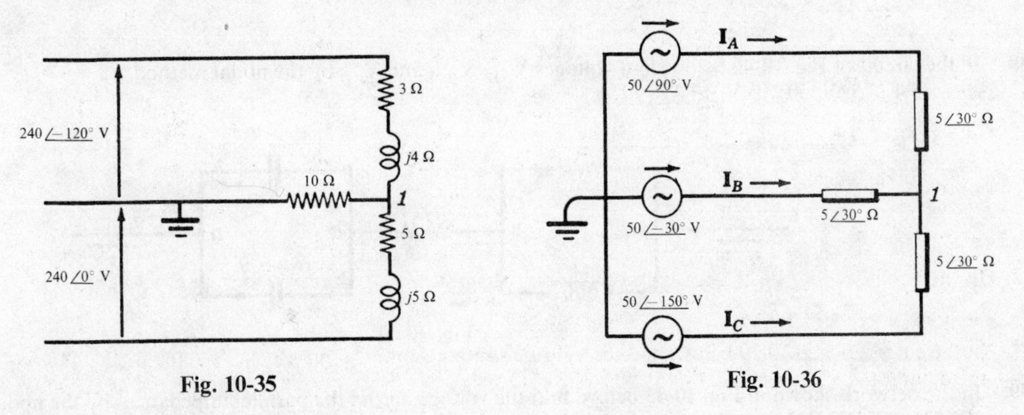

Fig. 10-35

Fig. 10-36

10.30. In the network of Fig. 10-36 above, find the three line currents $\mathbf{I}_A$, $\mathbf{I}_B$, and $\mathbf{I}_C$.
Ans. $10\angle 60^\circ$ A, $10\angle -60^\circ$ A and $10\angle 180^\circ$ A

10.31. In the circuit shown in Fig. 10-37 find the voltage source $\mathbf{V}_2$ which results in a current of zero through the $2 + j4\ \Omega$ impedance. *Ans.* $125\angle -135^\circ$ V

10.32. Referring to the circuit of Fig. 10-37, if the source $\mathbf{V}_2$ is $100\angle 30^\circ$ V what is the current in the $2 + j4\ \Omega$ impedance?
Ans. $12{\cdot}1\angle -11^\circ$ A

10.33. Referring to Problem 10.32, find the power delivered by each voltage source to the network.
Ans. $P_1 = -90{\cdot}6$ W, $P_2 = 1000$ W

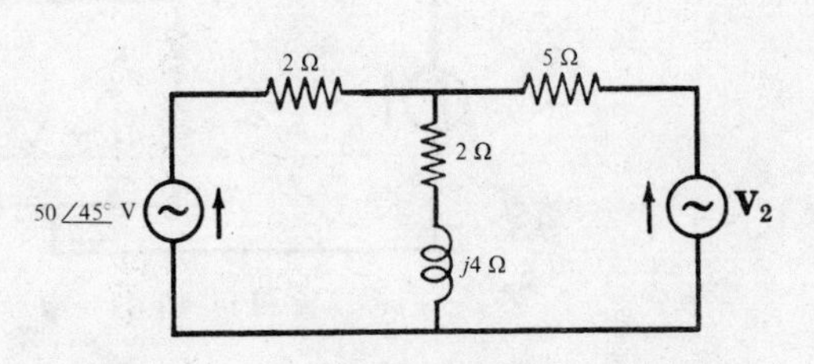

Fig. 10-37

10.34. The network of Fig. 10-38 below, has a driving current $\mathbf{I}_1$. The current in the 10 ohm resistor is $\mathbf{I}_2$. Find the ratio $\mathbf{I}_2/\mathbf{I}_1$.
Ans. $0{\cdot}151\angle 25{\cdot}8^\circ$

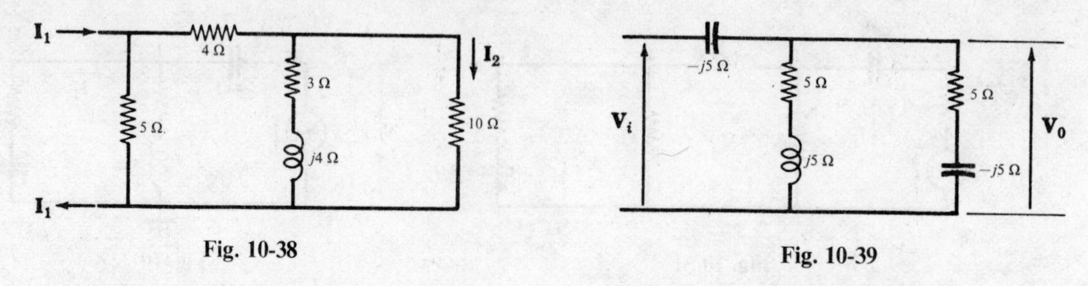

Fig. 10-38 Fig. 10-39

10.35. For the network of Fig. 10-39 above, find the voltage transfer function $\mathbf{V}_0/\mathbf{V}_i$ by the nodal method.
Ans. $0{\cdot}707\angle 45^\circ$

10.36. Find the voltage transfer function $\mathbf{V}_0/\mathbf{V}_i$ for the circuit shown in Fig. 10-40 below.
Ans. $0{\cdot}159\angle -61{\cdot}4^\circ$

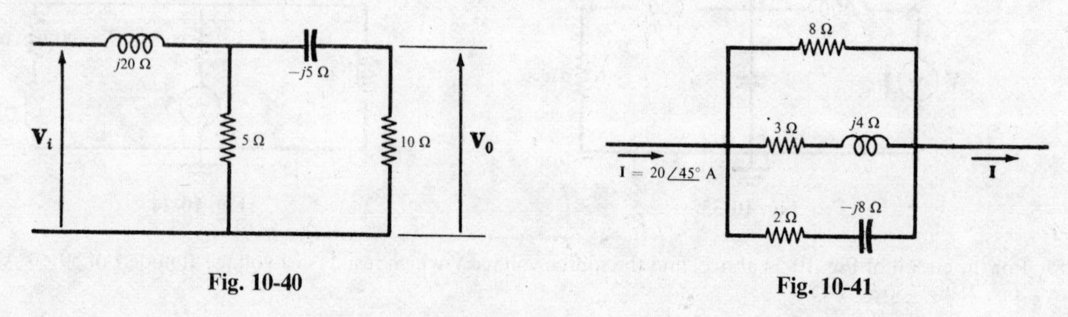

Fig. 10-40 Fig. 10-41

10.37. Use the nodal method to obtain the voltage across the parallel circuit shown in Fig. 10-41 above.
Ans. $72{\cdot}2\angle 53{\cdot}8^\circ$ V

10.38. In the circuit of Fig. 10-42 below, find voltages $\mathbf{V}_{AB}$, $\mathbf{V}_{BC}$, and $\mathbf{V}_{CD}$ by the nodal method.
Ans. $35{\cdot}4\angle 45^\circ$ V, $50\angle 0^\circ$ V, $13{\cdot}3\angle -90^\circ$ V

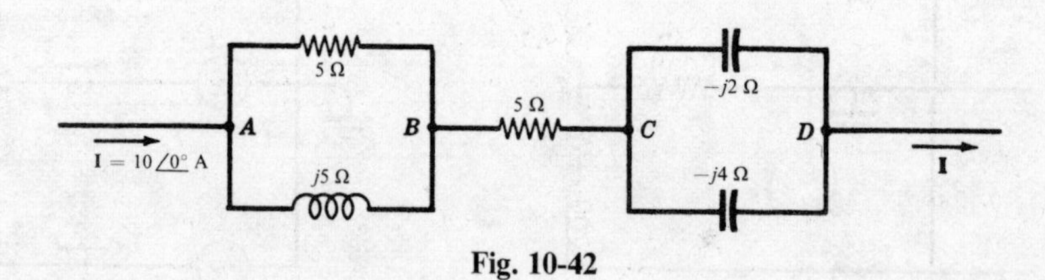

Fig. 10-42

10.39. In the network shown in Fig. 10-43 below, find the voltage across the parallel impedances by the nodal method.
Ans. $35\angle -24{\cdot}8^\circ$ V

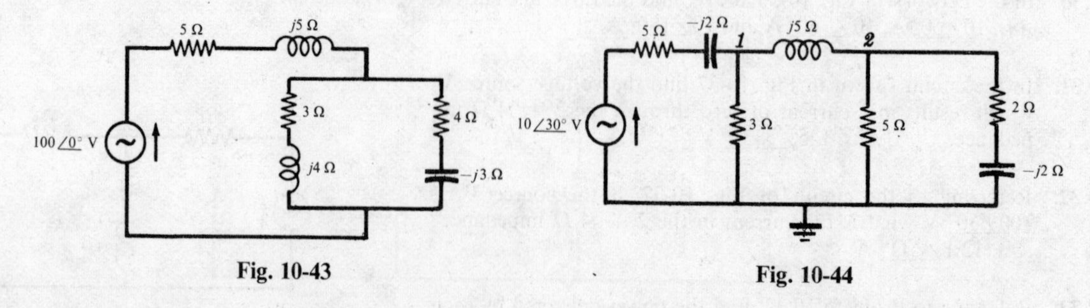

Fig. 10-43 Fig. 10-44

10.40. In the network of Fig. 10-44 above, find the node voltages $\mathbf{V}_1$ and $\mathbf{V}_2$, and the current in the $10\angle 30^\circ$ volt source.
Ans. $3{\cdot}02\angle 65{\cdot}2^\circ$ V, $1{\cdot}34\angle -31{\cdot}3^\circ$ V, $1{\cdot}44\angle 38{\cdot}8^\circ$ A

10.41. In the network of Fig. 10-45, find the power in the 6 ohm resistor by the nodal method.
Ans. 39·6 W

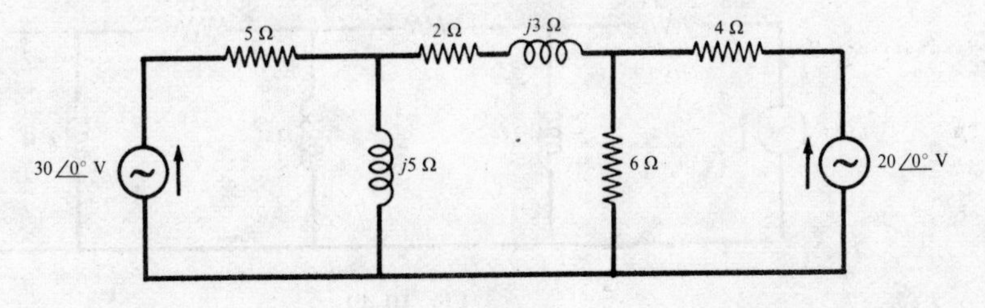

Fig. 10-45

10.42. Referring to the circuit given in Fig. 10-45 above, find the current in the 2 + $j3$ Ω impedance, with the positive direction to the right. *Ans.* 1·73 /40° A

10.43. In the network of Fig. 10-46, find the voltage $\mathbf{V}_1$ such that the current in the 4 ohm resistor is zero. Select one end of the resistor as the reference node. *Ans.* 95·4 /−23·2° V

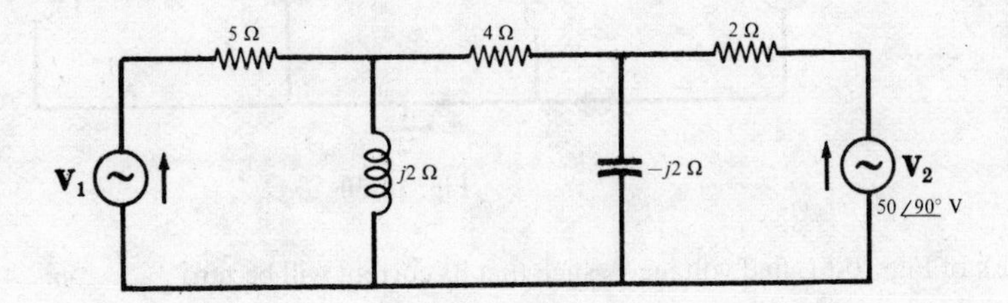

Fig. 10-46

10.44. Referring to the circuit of Fig. 10-46 above, source $\mathbf{V}_1 = 50\angle 0°$ V and $\mathbf{V}_2$ is unknown. Find $\mathbf{V}_2$ such that the current in the 4 ohm resistor is zero. *Ans.* 26·2 /113·2° V

10.45. In the circuit of Fig. 10-47, find current $\mathbf{I}_3$ with the direction given in the diagram.
Ans. 11·7 /112·9° A

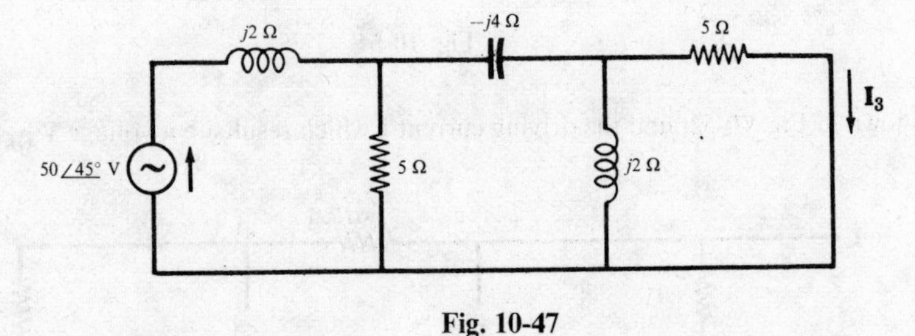

Fig. 10-47

10.46. In the network of Fig. 10-48, find the ratio of the two node voltages $\mathbf{V}_1/\mathbf{V}_2$. *Ans.* 2·26 /96·35°

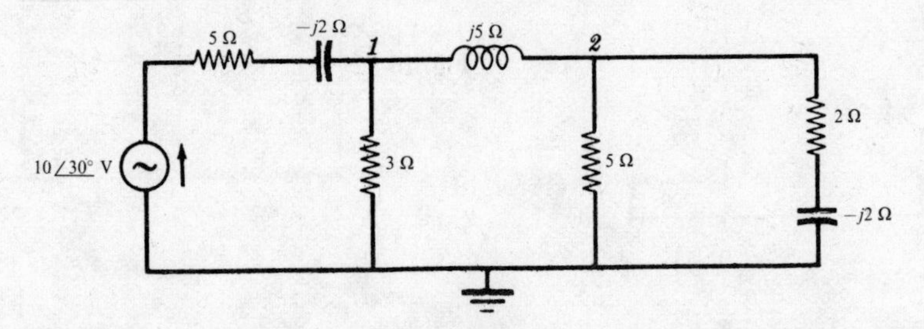

Fig. 10-48

10.47. In the network shown in Fig. 10-49, find the voltage $\mathbf{V}_0$ by the nodal method. *Ans.* $1{\cdot}56\angle 128{\cdot}7°$ V

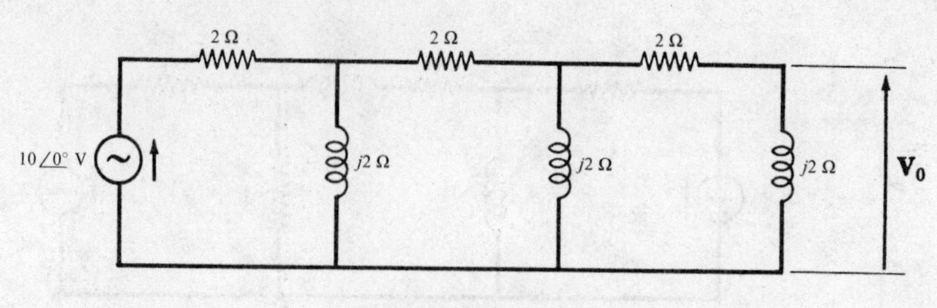

Fig. 10-49

10.48. For the network of Fig. 10-50, find node voltages $\mathbf{V}_1$ and $\mathbf{V}_2$. *Ans.* $18{\cdot}6\angle 68{\cdot}2°$ V

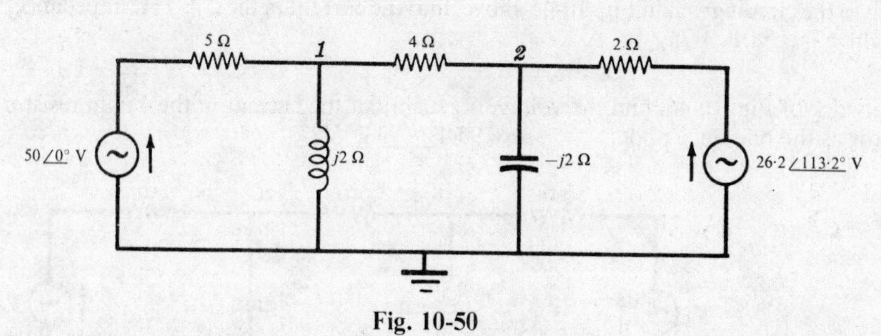

Fig. 10-50

10.49. In the network of Fig. 10-51, find voltage $\mathbf{V}_2$ such that its current will be zero. *Ans.* $4\angle 180°$ V

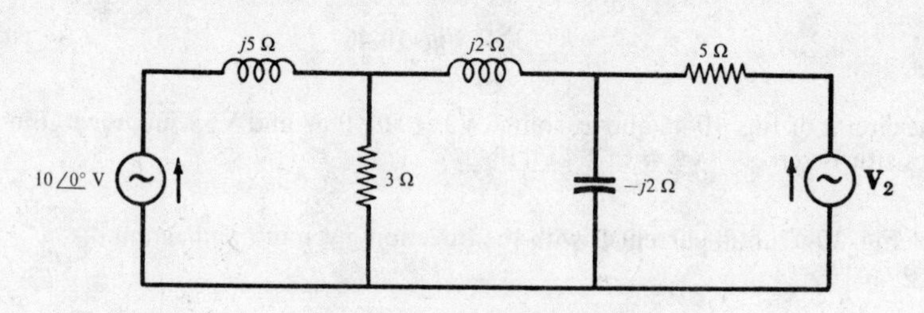

Fig. 10-51

10.50. In the network shown in Fig. 10-52, find the driving current **I** which results in a voltage $\mathbf{V}_{AB}$ of $5\angle 30°$ V.
Ans. $9{\cdot}72\angle -16°$ A

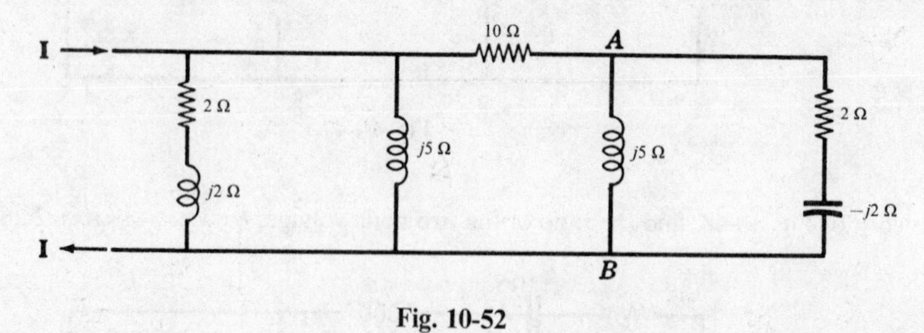

Fig. 10-52

CHAPTER 11

Thevenin's and Norton's Theorems

INTRODUCTION

A network in which all the impedances remain fixed may be solved by either the mesh current or the node voltage method. Consider now the network of Fig. 11-1. The impedances $\mathbf{Z}_1$, $\mathbf{Z}_2$ and $\mathbf{Z}_3$ are to be connected in turn to the circuit. By inserting each impedance into the circuit, a different matrix of $\mathbf{Z}$ or $\mathbf{Y}$, according to the method used, results, and consequently three different complete solutions are needed. Most of the tedious work involved is eliminated if we can replace the active network with a simple equivalent circuit. Thevenin's and Norton's theorems serve this purpose.

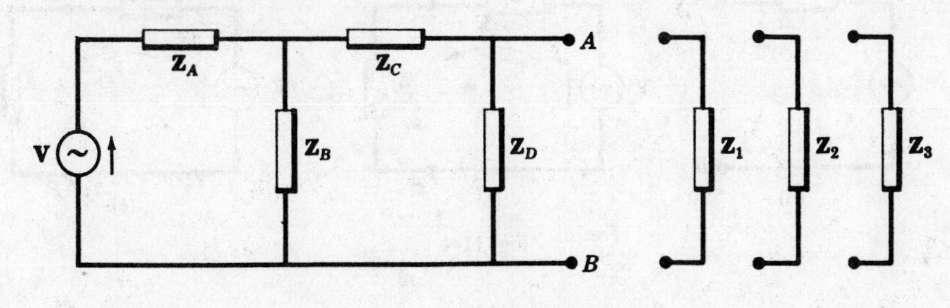

Fig. 11-1

THEVENIN'S THEOREM

Thevenin's theorem states that any linear active network with output terminals AB as shown in Fig. 11-2(*a*) can be replaced by a single voltage source $\mathbf{V}'$ in series with a single impedance $\mathbf{Z}'$ as shown in Fig. 11-2(*b*).

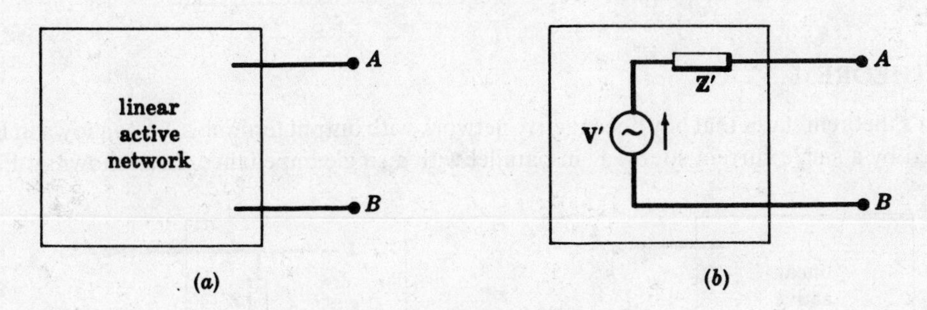

Fig. 11-2. Thevenin Equivalent Circuit

The *Thevenin equivalent voltage* $\mathbf{V}'$ is the open circuit voltage measured at the terminals AB. The *equivalent impedance* $\mathbf{Z}'$ is the driving point impedance of the network at the terminals AB when all internal sources are set equal to zero.

The polarity of the Thevenin equivalent voltage $\mathbf{V}'$ must be chosen so that the current through a connected impedance will have the same direction as would result with the impedance connected to the original active network.

Example 1.

Given the circuit of Fig. 11-3, determine the Thevenin equivalent circuit with respect to terminals AB. Use the result to find the current in two impedances, $\mathbf{Z}_1 = 5 - j5\ \Omega$ and $\mathbf{Z}_2 = 10\angle 0°\ \Omega$, connected in turn to terminals AB and determine the power delivered to them.

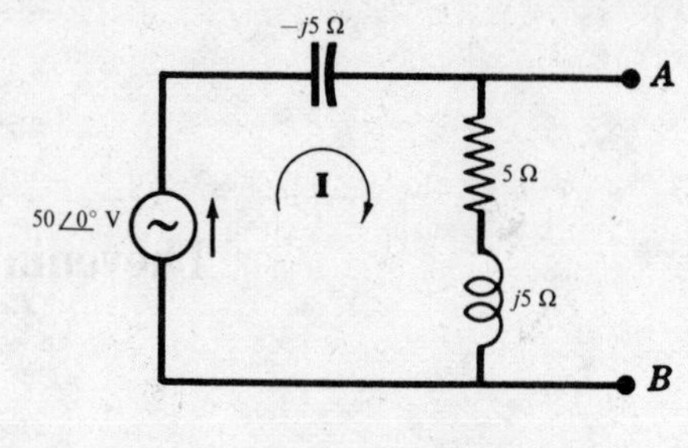

Fig. 11-3

Referring to Fig. 11-3, the current

$$\mathbf{I} = 50\angle 0°/(5 + j5 - j5) = 10\angle 0°\ \mathrm{A}$$

Then the Thevenin equivalent voltage $\mathbf{V}'$ is the voltage drop across the $5 + j5\ \Omega$ impedance. Hence

$$\mathbf{V}' = \mathbf{V}_{AB} = \mathbf{I}(5 + j5) = 70{\cdot}7\angle 45°\ \mathrm{V}$$

The driving point impedance at terminals AB is $\mathbf{Z}' = \dfrac{(5 + j5)(-j5)}{5 + j5 - j5} = 5 - j5\ \Omega$.

The Thevenin equivalent circuit is shown in Fig. 11-4(a) with the source $\mathbf{V}'$ directed to terminal A.

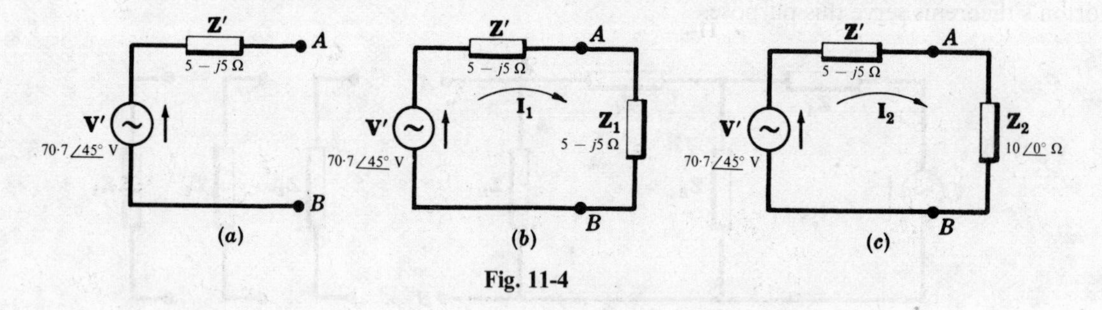

Fig. 11-4

Connect impedance $\mathbf{Z}_1$ to the terminals of the Thevenin equivalent as shown in Fig. 11-4(b). In this circuit,

$$\mathbf{I}_1 = (70{\cdot}7\angle 45°)/(5 - j5 + 5 - j5) = 5\angle 90°\ \mathrm{A} \text{ and } P_1 = (I_1)^2 5 = 125\ \mathrm{W}$$

When the impedance $\mathbf{Z}_2$ is connected to the terminals AB as shown in Fig. 11-4(c), we have

$$\mathbf{I}_2 = (70{\cdot}7\angle 45°)/(5 - j5 + 10) = 4{\cdot}47\angle 63{\cdot}43°\ \mathrm{A} \text{ and } P_2 = (I_2)^2 10 = 200\ \mathrm{W}$$

NORTON'S THEOREM

Norton's theorem states that any linear active network with output terminals AB as shown in Fig. 11-5(a) can be replaced by a single current source $\mathbf{I}'$ in parallel with a single impedance $\mathbf{Z}'$ as shown in Fig. 11-5(b).

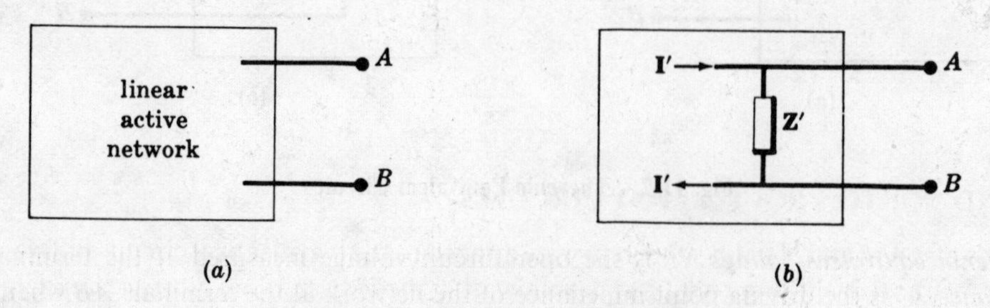

Fig. 11-5. Norton Equivalent Circuit

The *Norton equivalent current source* $\mathbf{I}'$ is the current through a short circuit applied to the terminals of the active network. The shunt impedance $\mathbf{Z}'$ is the driving point impedance of the network at the terminals AB when all internal sources are set equal to zero. Thus, given a linear active circuit, the impedances $\mathbf{Z}'$ of the Norton and Thevenin equivalent circuits are identical.

The current through an impedance connected to the terminals of the Norton equivalent circuit must have the same direction as the current through the same impedance connected to the original active network.

Example 2.

Given the circuit of Fig. 11-6, determine the Norton equivalent circuit with respect to terminals AB. Use the result to find the current in two impedances, $\mathbf{Z}_1 = 5 - j5\,\Omega$ and $\mathbf{Z}_2 = 10\angle 0°\,\Omega$, connected in turn to terminals AB and determine the power delivered to them.

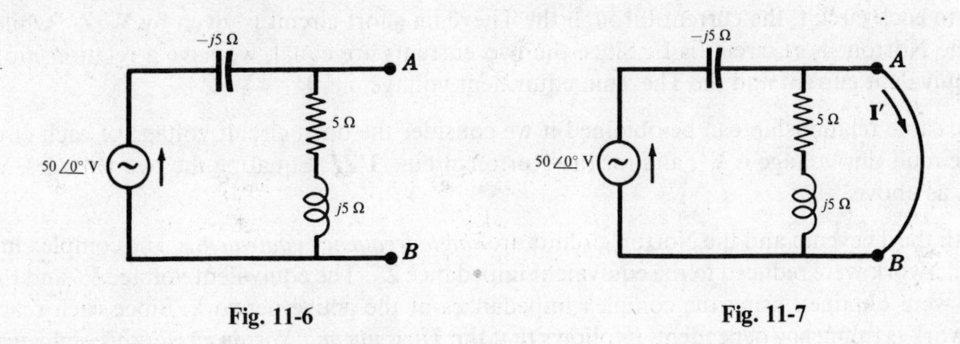

Fig. 11-6 Fig. 11-7

Referring to Fig. 11-7, when a short circuit is applied to terminals AB, $\mathbf{I}' = 50\angle 0°(-j5) = 10\angle 90°$ A. When the source is set equal to zero, $\mathbf{Z}' = \dfrac{-j5(5+j5)}{5+j5-j5} = 5 - j5\,\Omega$.

The Norton equivalent circuit is shown in Fig. 11-8(a). Note that the current is directed toward terminal A.

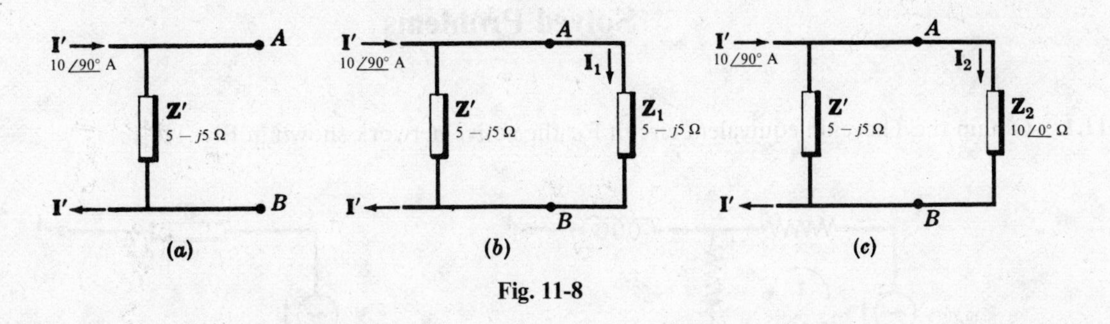

Fig. 11-8

Connect impedance $\mathbf{Z}_1$ to the terminals of the Norton equivalent as shown in Fig. 11-8(b). Then the current through $\mathbf{Z}_1$ is $\mathbf{I}_1 = \mathbf{I}'\left(\dfrac{\mathbf{Z}'}{\mathbf{Z}'+\mathbf{Z}_1}\right) = 10\angle 90°\left(\dfrac{5-j5}{10-j10}\right) = 5\angle 90°$ A. The power delivered to $\mathbf{Z}_1$ is $P_1 = (I_1)^2 5 = 125$ W.

When the impedance $\mathbf{Z}_2$ is connected to the terminals AB as shown in Fig. 11-8(c), we have

$$\mathbf{I}_2 = \mathbf{I}'(5-j5)/(15-j5) = 4{\cdot}47\angle 63{\cdot}43° \text{ A and } P_2 = (I_2)^2 10 = 200 \text{ W}$$

THEVENIN AND NORTON EQUIVALENT CIRCUITS

Thevenin's theorem and Norton's theorem were applied to the two identical circuits of Examples 1 and 2 respectively, and identical results were obtained. It follows that the Thevenin and Norton circuits are equivalent to each other.

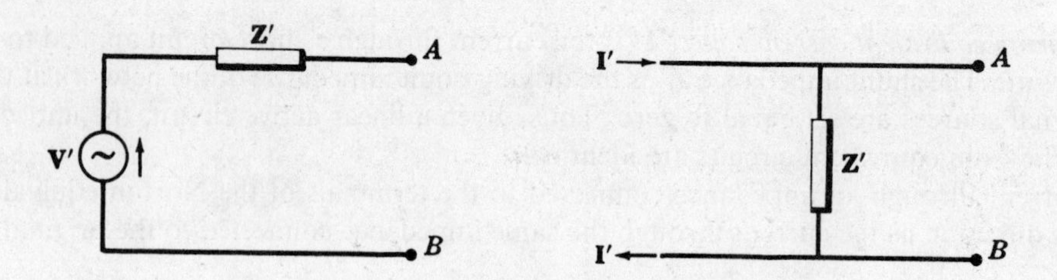

Fig. 11-9. Thevenin and Norton Circuits

In Fig. 11-9 the same impedance $\mathbf{Z}'$ is found at the left of terminals AB of both circuits. When a short is applied to each circuit, the current through the Thevenin short circuit is given by $\mathbf{V}'/\mathbf{Z}'$, while the current through the Norton short circuit is $\mathbf{I}'$. Since the two currents are equal, we have a relationship between the Norton equivalent current and the Thevenin equivalent voltage, i.e. $\mathbf{I}' = \mathbf{V}'/\mathbf{Z}'$.

The same relationship can be obtained if we consider the open circuit voltage of each circuit. For the Thevenin circuit this voltage is $\mathbf{V}'$; and for the Norton circuit, $\mathbf{I}'\mathbf{Z}'$. Equating the two voltages, $\mathbf{V}' = \mathbf{I}'\mathbf{Z}'$ or $\mathbf{I}' = \mathbf{V}'/\mathbf{Z}'$ as above.

Both the Thevenin and the Norton circuits are *single frequency equivalents*. The complex impedances of the active network were reduced to the equivalent impedance $\mathbf{Z}'$. The equivalent voltage $\mathbf{V}'$ and the equivalent current $\mathbf{I}'$ were obtained using the complex impedances of the active network. Since each reactance in the active network is frequency dependent, it follows that the *Thevenin and Norton circuits are equivalent only at the frequency for which they were computed.*

Solved Problems

11.1. Obtain the Thevenin equivalent circuit for the active network shown in Fig. 11-10.

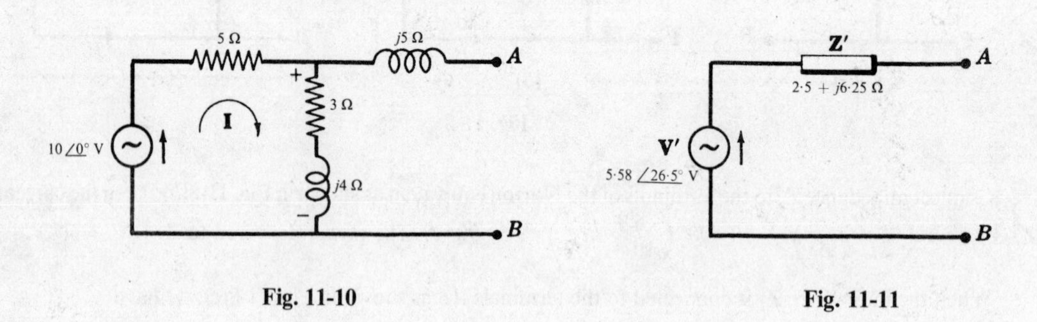

Fig. 11-10 Fig. 11-11

The equivalent impedance $\mathbf{Z}'$ of the circuit is calculated by setting the source equal to zero. Thus

$$\mathbf{Z}' = j5 + \frac{5(3+j4)}{5+3+j4} = 2{\cdot}5 + j6{\cdot}25\ \Omega$$

The current $\mathbf{I}$ shown in the open circuit of Fig. 11-10 is $\mathbf{I} = (10\angle 0^\circ)/(5+3+j4) = 1{\cdot}117\angle -26{\cdot}6^\circ$ A. Then the open circuit voltage is the drop across the $3 + j4$ ohm impedance.

$$\mathbf{V}' = \mathbf{I}(3+j4) = (1{\cdot}117\angle -26{\cdot}6^\circ)(5\angle 53{\cdot}1^\circ) = 5{\cdot}58\angle 26{\cdot}5^\circ\ \text{V}$$

The polarity of $\mathbf{V}'$ is given by the direction of the current entering the $3 + j4\ \Omega$ impedance. Thus the source $\mathbf{V}'$ drives toward terminal A in the equivalent circuit as shown in Fig. 11-11.

11.2. Obtain the Norton equivalent circuit for the active network given in Fig. 11-10.

The equivalent impedance $\mathbf{Z}' = 2{\cdot}5 + j6{\cdot}25\ \Omega$, as calculated in Problem 11.1.

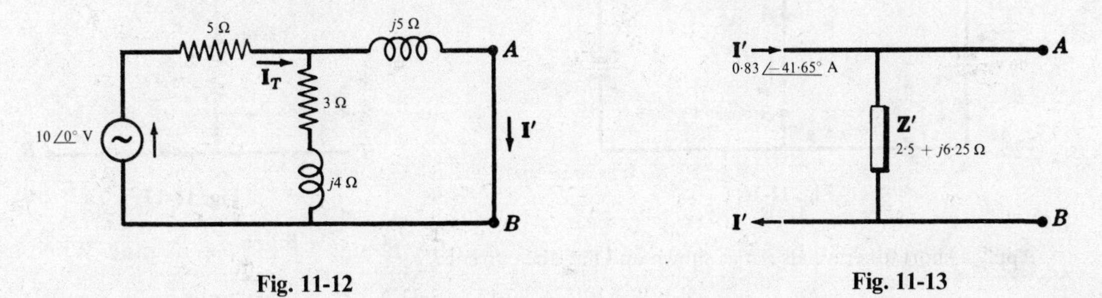

Fig. 11-12 **Fig. 11-13**

Apply a short to the terminals AB as shown in Fig. 11-12 and determine the total impedance which the circuit presents to the $10\angle 0^\circ$ V source.

$$\mathbf{Z}_T = 5 + \frac{(3 + j4)j5}{(3 + j4 + j5)} = 5{\cdot}83 + j2{\cdot}5 = 6{\cdot}35\angle 23{\cdot}2^\circ\ \Omega$$

Then the current $\mathbf{I}_T = 10\angle 0^\circ/\mathbf{Z}_T = (10\angle 0^\circ)/(6{\cdot}35\angle 23{\cdot}2^\circ) = 1{\cdot}575\angle -23{\cdot}2^\circ$ A, and

$$\mathbf{I}' = \mathbf{I}_T\left(\frac{3 + j4}{3 + j4 + j5}\right) = 1{\cdot}575\angle -23{\cdot}2^\circ \left(\frac{5\angle 53{\cdot}1^\circ}{3 + j9}\right) = 0{\cdot}83\angle -41{\cdot}65^\circ \text{ A}$$

The Norton equivalent circuit is shown in Fig. 11-13. Note that the current $\mathbf{I}'$ is directed toward A since the short circuit current entered the short at terminal A.

11.3. In the DC circuit given in Fig. 11-14 three resistors, $R_1 = 1$ ohm, $R_2 = 5$ ohms and $R_3 = 10$ ohms, are connected in turn to terminals AB. Determine the power delivered to each resistor.

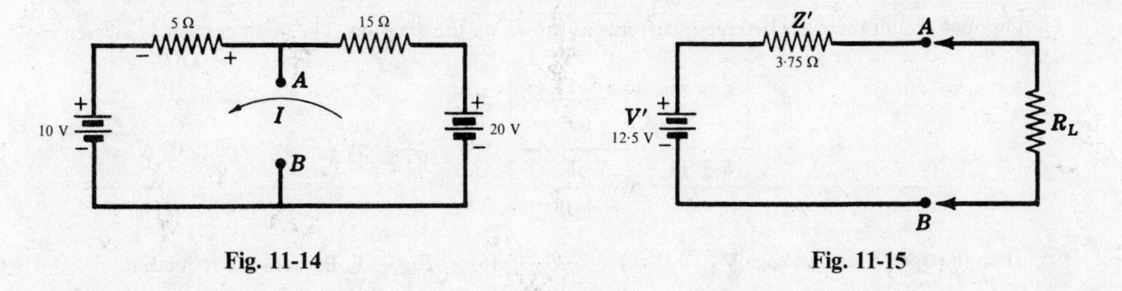

Fig. 11-14 **Fig. 11-15**

Obtain the Thevenin equivalent circuit. In Fig. 11-14 the current $\mathbf{I} = (20 - 10)/(5 + 15) = 0{\cdot}5$ A. Then the drop across the 5 ohm resistor is $\mathbf{V}_5 = I(5) = 2{\cdot}5$ volts with the polarity as shown.

Express the voltage of A with respect to B as

$$V_{AB} = V' = 10 + V_5 = 12{\cdot}5 \text{ V}$$

When the DC sources are set equal to zero, the impedance $\mathbf{Z}'$ is the parallel combination of the 5 and the 15 ohm resistors,

$$Z' = \frac{5(15)}{20} = 3{\cdot}75\ \Omega$$

The Thevenin equivalent circuit is shown in Fig. 11-15. Now connecting each of the three resistors at terminals AB, the respective powers are calculated.

With $R_1 = 1$ ohm, $\quad I_1 = 12{\cdot}5/(3{\cdot}75 + 1) = 2{\cdot}63$ A and power $P_1 = (I_1)^2(1) = (2{\cdot}63)^2(1) = 6{\cdot}91$ W.

With $R_2 = 5$ ohms, $\quad I_2 = 12{\cdot}5/(3{\cdot}75 + 5) = 1{\cdot}43$ A and power $P_2 = (I_2)^2(5) = (1{\cdot}43)^2(5) = 10{\cdot}2$ W.

With $R_3 = 10$ ohms, $I_3 = 12{\cdot}5/(3{\cdot}75 + 10) = 0{\cdot}91$ A and power $P_3 = (I_3)^2(10) = (0{\cdot}91)^2(10) = 8{\cdot}28$ W.

11.4. Obtain the Norton equivalent circuit with respect to terminals AB for the network shown in Fig. 11-16.

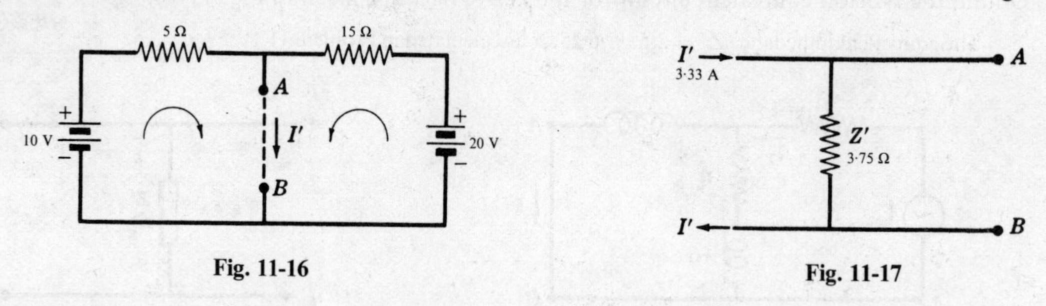

Fig. 11-16

Fig. 11-17

Apply a short to terminals AB as shown and find the current I'.

$$I' = 10/5 + 20/15 = 3{\cdot}33 \text{ A}$$

The equivalent impedance at terminals AB with the voltage sources set equal to zero is

$$Z' = 5(15)/(5 + 15) = 3{\cdot}75\ \Omega$$

The Norton equivalent circuit is shown in Fig. 11-17.

11.5. Find the Thevenin equivalent circuit for the network shown in Fig. 11-18.

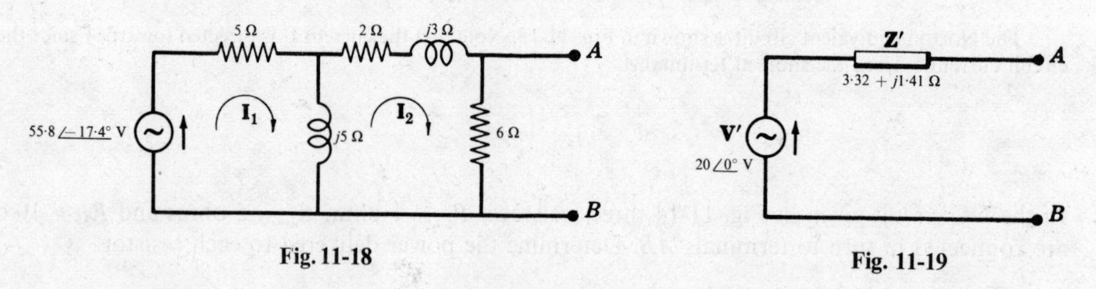

Fig. 11-18

Fig. 11-19

On open circuit there are two mesh currents as shown on the diagram. The mesh current $\mathbf{I}_2$ is given by

$$\mathbf{I}_2 = \frac{\begin{vmatrix} 5+j5 & 55{\cdot}8\angle{-17{\cdot}4^\circ} \\ -j5 & 0 \end{vmatrix}}{\begin{vmatrix} 5+j5 & -j5 \\ -j5 & 8+j8 \end{vmatrix}} = \frac{279\angle 72{\cdot}6^\circ}{83{\cdot}7\angle 72{\cdot}6^\circ} = 3{\cdot}33\angle 0^\circ \text{ A}$$

Then the open circuit voltage $\mathbf{V}_{AB} = \mathbf{I}_2(6) = 3{\cdot}33\angle 0^\circ\ (6) = 20\angle 0^\circ$ V. By network reduction,

$$Z' = \frac{6\left[\dfrac{5(j5)}{5+j5} + (2+j3)\right]}{6 + \left[\dfrac{5(j5)}{5+j5} + (2+j3)\right]} = 3{\cdot}32 + j1{\cdot}41\ \Omega$$

The Thevenin equivalent circuit is shown in Fig. 11-19 with $\mathbf{V}'$ directed to terminal A.

11.6. Obtain the Norton equivalent circuit for the network of Fig. 11-18.

Apply a short across terminals AB. Then the current $\mathbf{I}_2$ through the short is

$$\mathbf{I}_2 = \mathbf{I}' = \frac{\begin{vmatrix} 5+j5 & 55{\cdot}8\angle{-17{\cdot}4^\circ} \\ -j5 & 0 \end{vmatrix}}{\begin{vmatrix} 5+j5 & -j5 \\ -j5 & 2+j8 \end{vmatrix}} = \frac{279\angle 72{\cdot}6^\circ}{(-5+j50)} = 5{\cdot}58\angle{-23{\cdot}14^\circ} \text{ A}$$

The impedance $\mathbf{Z}' = 3{\cdot}32 + j1{\cdot}41\ \Omega$, as calculated in Problem 11.5.

As a check, the open circuit voltage of the Norton equivalent circuit shown in Fig. 11-20 can be compared to $\mathbf{V}'$ of the Thevenin in Problem 11.5.

$$\mathbf{V}_{oc} = \mathbf{I}'\mathbf{Z}' = 5{\cdot}58\angle{-23{\cdot}14^\circ}\,(3{\cdot}32 + j1{\cdot}41)$$
$$= 20{\cdot}1\angle{-0{\cdot}14^\circ}\ \mathrm{V}$$

In Problem 11.5, $\mathbf{V}' = 20\angle 0^\circ$ V.

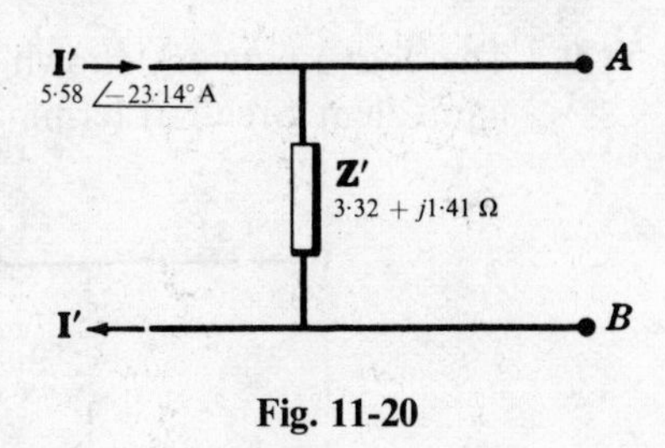

Fig. 11-20

11.7. Replace the active network shown in Fig. 11-21 with a Thevenin equivalent at the terminals *AB*.

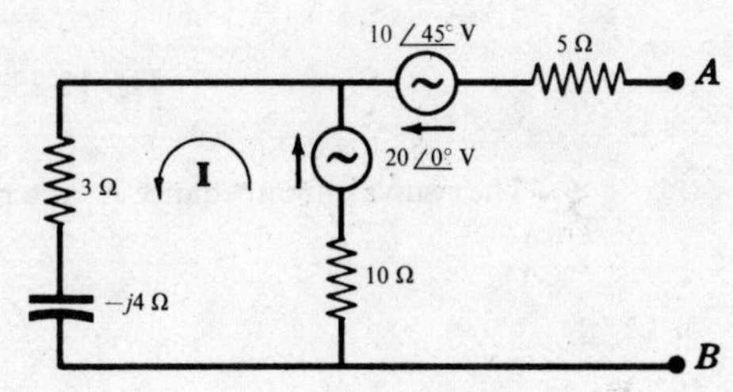

Fig. 11-21

On open circuit the current is

$$\mathbf{I} = 20\angle 0^\circ/(10 + 3 - j4) = 1{\cdot}47\angle 17{\cdot}1^\circ\ \mathrm{A}$$

Then the voltage drop across the 10 ohm resistor $\mathbf{V}_{10} = \mathbf{I}(10) = 14{\cdot}7\angle 17{\cdot}1^\circ$ V.

Now the voltage $\mathbf{V}_{AB}$ is the sum of the two source voltages and the voltage drop across the 10 ohm resistor, with polarities as shown in Fig. 11-22. Then

$$\mathbf{V}' = \mathbf{V}_{AB} = 20\angle 0^\circ - 10\angle 45^\circ - 14{\cdot}7\angle 17{\cdot}1^\circ = 11{\cdot}39\angle 264{\cdot}4^\circ\ \mathrm{V}$$

The impedance $\mathbf{Z}' = 5 + \dfrac{10(3 - j4)}{10 + 3 - j4} = 7{\cdot}97 - j2{\cdot}16\ \Omega$.

The Thevenin equivalent circuit is shown in Fig. 11-23.

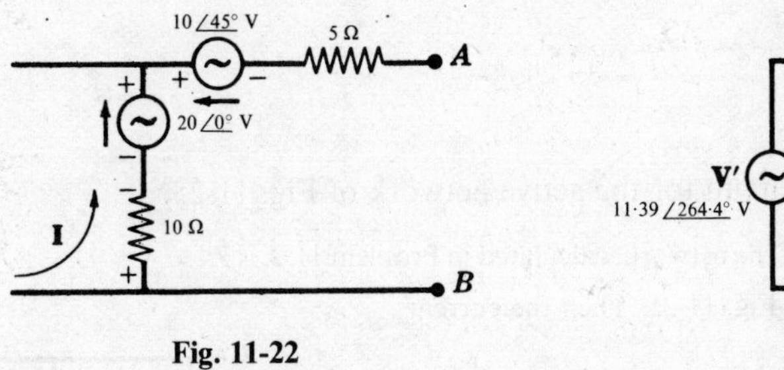

Fig. 11-22

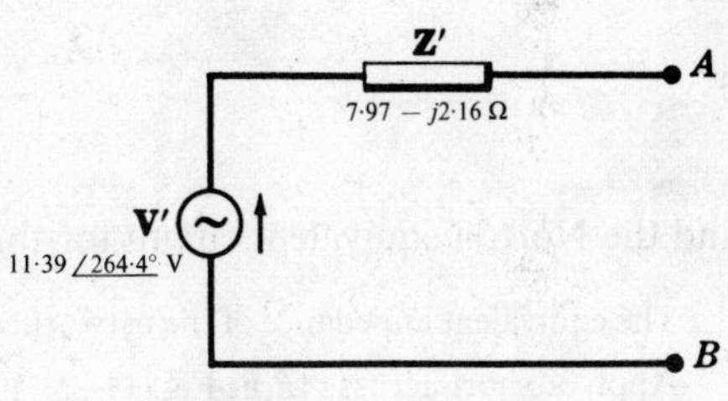

Fig. 11-23

11.8. Obtain the Norton equivalent circuit for the network given in Fig. 11-21.

$\mathbf{Z}' = 7{\cdot}97 - j2{\cdot}16\ \Omega$, as calculated in Problem 11.7.

Apply a short at terminals *AB* and select clockwise primary loops for the mesh currents. Then

$$\mathbf{I}' = \mathbf{I}_2 = \frac{\begin{vmatrix} 13 - j4 & -20 \\ -10 & (20 - 10\angle 45^\circ) \end{vmatrix}}{\begin{vmatrix} 13 - j4 & -10 \\ -10 & 15 \end{vmatrix}} = \frac{156\angle 247{\cdot}4^\circ}{112{\cdot}3\angle{-32{\cdot}3^\circ}} = 1{\cdot}39\angle 279{\cdot}7^\circ\ \mathrm{A}$$

The Norton current source $\mathbf{I}'$ drives into terminal *A* as shown in Fig. 11-24.

Comparing the open circuit voltage $\mathbf{V}_{oc}$ of this circuit with the Thevenin equivalent voltage $\mathbf{V}'$ of Problem 11.7,

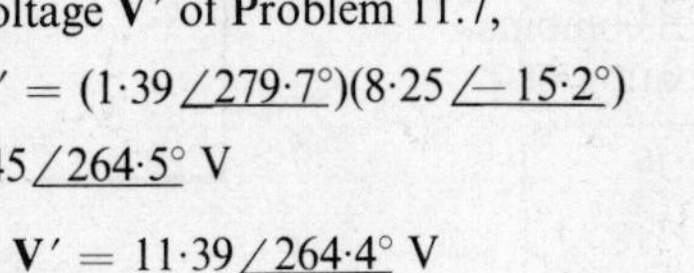

$$\mathbf{V}_{oc} = \mathbf{I}'\mathbf{Z}' = (1{\cdot}39\angle 279{\cdot}7^\circ)(8{\cdot}25\angle{-15{\cdot}2^\circ})$$
$$= 11{\cdot}45\angle 264{\cdot}5^\circ\ \mathrm{V}$$

and $\qquad \mathbf{V}' = 11{\cdot}39\angle 264{\cdot}4^\circ$ V

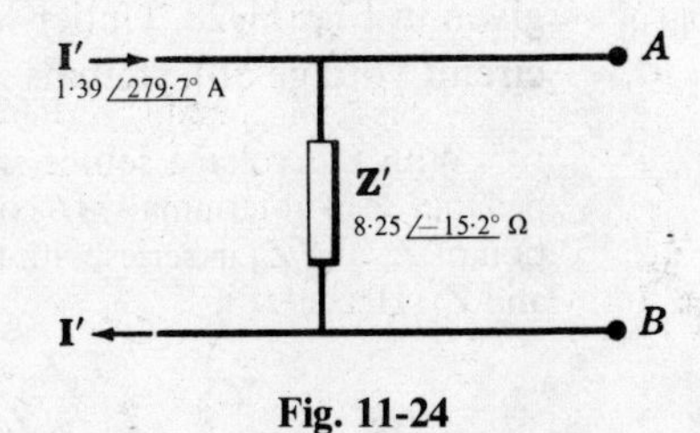

Fig. 11-24

11.9. The active network shown in Fig. 11-25 contains a current source $\mathbf{I} = 5\angle 30^\circ$ A. Find the Thevenin equivalent circuit at terminals AB.

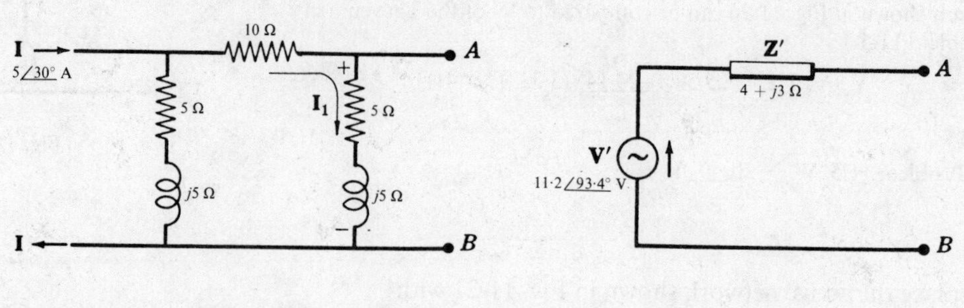

Fig. 11-25 Fig. 11-26

The equivalent impedance $\mathbf{Z}'$ at terminals AB with the source reduced to zero consists of two branches in parallel. Then

$$\mathbf{Z}' = \frac{(5 + j5)(15 + j5)}{(5 + j5 + 15 + j5)} = 4 + j3\ \Omega$$

On open circuit the current $\mathbf{I}$ divides between the two branches. Solving for $\mathbf{I}_1$ shown in the diagram, we have

$$\mathbf{I}_1 = 5\angle 30^\circ \left(\frac{5 + j5}{20 + j10}\right) = 1{\cdot}585\angle 48{\cdot}4^\circ\ \Omega$$

Since the voltage $\mathbf{V}_{AB} = \mathbf{V}'$ is the voltage drop across the $5 + j5$ impedance,

$$\mathbf{V}' = \mathbf{I}_1(5 + j5) = (1{\cdot}585\angle 48{\cdot}4^\circ)(7{\cdot}07\angle 45^\circ) = 11{\cdot}2\angle 93{\cdot}4^\circ\ \text{V}$$

The Thevenin equivalent circuit is shown in Fig. 11-26.

11.10. Find the Norton equivalent circuit for the active network of Fig. 11-25.

The equivalent impedance of the network, calculated in Problem 11.9, is $\mathbf{Z}' = 4 + j3 = 5\angle 36{\cdot}9^\circ\ \Omega$.

Apply a short across AB in Fig. 11-25. Then the current through the short circuit is

$$\mathbf{I}' = 5\angle 30^\circ \left(\frac{5 + j5}{5 + j5 + 10}\right) = 2{\cdot}24\angle 56{\cdot}6^\circ\ \text{A}$$

The Norton equivalent circuit is shown in Fig. 11-27.

On open circuit the Norton equivalent has a voltage $\mathbf{V}_{oc} = (2{\cdot}24\angle 56{\cdot}6^\circ)(5\angle 36{\cdot}9^\circ) = 11{\cdot}2\angle 93{\cdot}5^\circ$ V. In Problem 11.9, the Thevenin equivalent voltage $\mathbf{V}' = 11{\cdot}2\angle 93{\cdot}4^\circ$ V.

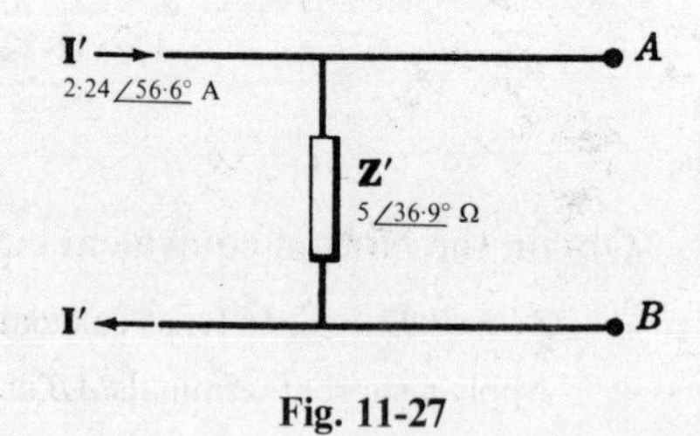

Fig. 11-27

11.11. Obtain the Thevenin equivalent for the bridge circuit given in Fig. 11-28. Under what conditions is the open circuit voltage at terminals AB zero?

With the voltage source set to zero, the equivalent impedance seen at terminals AB consists of the parallel combination of $\mathbf{Z}_1$ and $\mathbf{Z}_4$ in series with the parallel combination of $\mathbf{Z}_2$ and $\mathbf{Z}_3$. Hence

$$\mathbf{Z}' = \frac{\mathbf{Z}_1\mathbf{Z}_4}{\mathbf{Z}_1 + \mathbf{Z}_4} + \frac{\mathbf{Z}_2\mathbf{Z}_3}{\mathbf{Z}_2 + \mathbf{Z}_3}$$

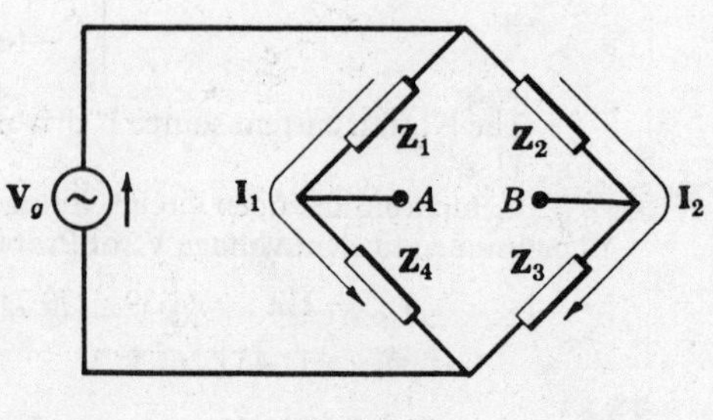

Fig. 11-28

On the open circuit the source $\mathbf{V}_g$ results in currents $\mathbf{I}_1$ and $\mathbf{I}_2$ as shown in the diagram.

$$\mathbf{I}_1 = \mathbf{V}_g/(\mathbf{Z}_1 + \mathbf{Z}_4) \text{ and } \mathbf{I}_2 = \mathbf{V}_g/(\mathbf{Z}_2 + \mathbf{Z}_3)$$

Assuming the potential of A greater than B, we have

$$\mathbf{V}' = \mathbf{V}_{AB} = \mathbf{I}_1\mathbf{Z}_4 - \mathbf{I}_2\mathbf{Z}_3$$

$$= \frac{\mathbf{V}_g\,\mathbf{Z}_4}{\mathbf{Z}_1 + \mathbf{Z}_4} - \frac{\mathbf{V}_g\,\mathbf{Z}_3}{\mathbf{Z}_2 + \mathbf{Z}_3}$$

$$= \mathbf{V}_g\left[\frac{\mathbf{Z}_2\mathbf{Z}_4 - \mathbf{Z}_1\mathbf{Z}_3}{(\mathbf{Z}_1 + \mathbf{Z}_4)(\mathbf{Z}_2 + \mathbf{Z}_3)}\right]$$

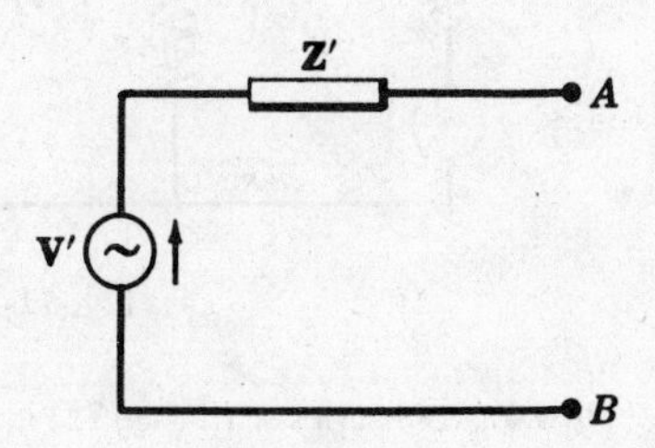

Fig. 11-29

The Thevenin equivalent voltage $\mathbf{V}'$ is proportional to the difference $\mathbf{Z}_2\mathbf{Z}_4 - \mathbf{Z}_1\mathbf{Z}_3$. When $\mathbf{Z}_2\mathbf{Z}_4 = \mathbf{Z}_1\mathbf{Z}_3$, the voltage $\mathbf{V}' = 0$.

11.12. Obtain the Thevenin equivalent for the bridge circuit shown in Fig. 11-30.

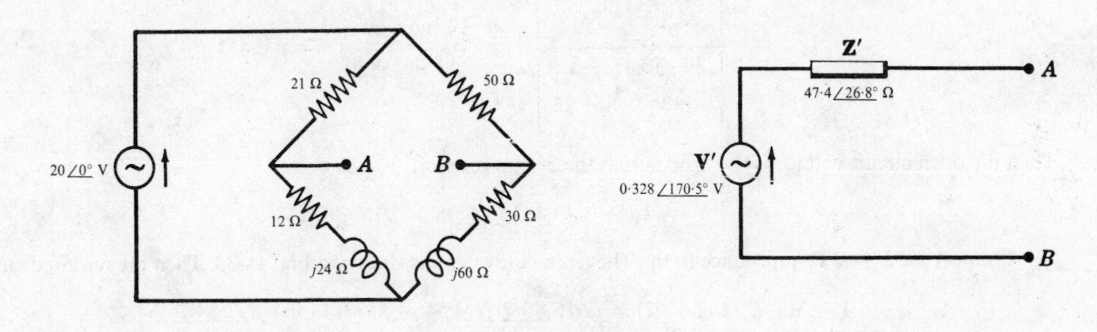

Fig. 11-30

Fig. 11-31

The equivalent impedance at terminals AB with the source set equal to zero is

$$\mathbf{Z}' = \frac{21(12 + j24)}{33 + j24} + \frac{50(30 + j60)}{80 + j60} = 47{\cdot}4\,\angle 26{\cdot}8^\circ\ \Omega$$

On open circuit, the current on the left side of the bridge is $\mathbf{I}_1 = (20\angle 0^\circ)/(33 + j24)$ A. On the right side of the bridge, $\mathbf{I}_2 = (20\angle 0^\circ)/(80 + j60)$ A.

Assuming point A at a higher potential than B, we have

$$\mathbf{V}' = \mathbf{V}_{AB} = \frac{(20\angle 0^\circ)(12 + j24)}{33 + j24} - \frac{(20\angle 0^\circ)(30 + j60)}{80 + j60}$$

$$= (20\angle 0^\circ)(1 + j2)\left[\frac{12}{33 + j24} - \frac{30}{80 + j60}\right] = 0{\cdot}328\,\angle 170{\cdot}5^\circ\ \text{V}$$

11.13. For the network shown in Fig. 11-32 below, replace the circuit to the left of terminals AB with a Thevenin equivalent. Then determine the current in the $2 - j2\ \Omega$ impedance connected to the equivalent circuit.

The impedance $\mathbf{Z}'$ can be found by network reduction. The $5 - j2\ \Omega$ impedance is in parallel with the 3 ohm resistance. The equivalent impedance is

$$\mathbf{Z}_1 = \frac{(5 - j2)3}{8 - j2} = 1{\cdot}94 - j0{\cdot}265\ \Omega$$

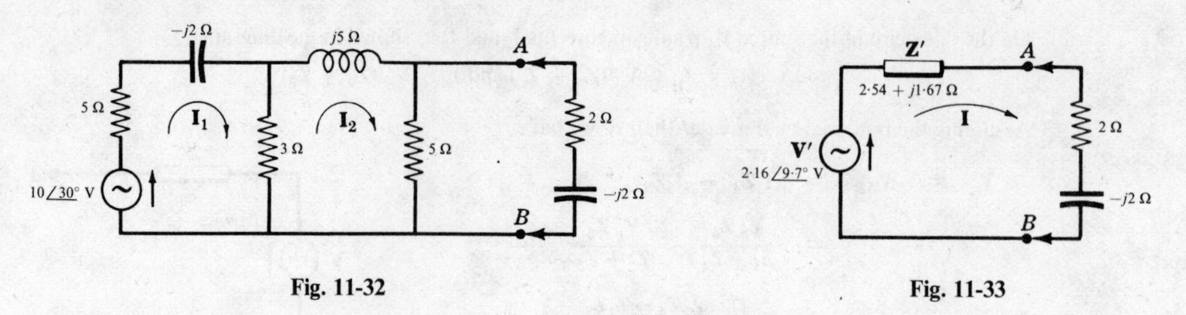

Fig. 11-32

Fig. 11-33

Now $\mathbf{Z}_1$ is in series with the $j5\ \Omega$ impedance. Add the two impedances to obtain

$$\mathbf{Z}_2 = 1{\cdot}94 - j0{\cdot}265 + j5 = 1{\cdot}94 + j4{\cdot}735\ \Omega$$

$\mathbf{Z}'$ is now found by combining $\mathbf{Z}_2$ and the 5 ohm resistor. Thus

$$\mathbf{Z}' \quad = \quad \frac{(1{\cdot}94 + j4{\cdot}735)5}{6{\cdot}94 + j4{\cdot}735} \quad = \quad 3{\cdot}04\angle 33{\cdot}4^\circ \quad = \quad 2{\cdot}54 + j1{\cdot}67\ \Omega$$

Consider the open circuit and solve for $\mathbf{I}_2$ using the mesh current method.

$$\mathbf{I}_2 \quad = \quad \frac{\begin{vmatrix} 8 - j2 & 10\angle 30^\circ \\ -3 & 0 \end{vmatrix}}{\begin{vmatrix} 8 - j2 & -3 \\ -3 & 8 + j5 \end{vmatrix}} \quad = \quad \frac{30\angle 30^\circ}{69{\cdot}25\angle 20{\cdot}3^\circ} \quad = \quad 0{\cdot}433\angle 9{\cdot}7^\circ\ \Omega$$

Then the open circuit voltage is the drop across the 5 ohm resistor.

$$\mathbf{V}' = \mathbf{I}_2(5) = (0{\cdot}433\angle 9{\cdot}7^\circ)5 = 2{\cdot}16\angle 9{\cdot}7^\circ\ \text{V}$$

Connect the $2 - j2\ \Omega$ impedance to the Thevenin equivalent as shown in Fig. 11-33. Then the required current is

$$\mathbf{I} = \mathbf{V}'/(\mathbf{Z}' + 2 - j2) = (2{\cdot}16\angle 9{\cdot}7^\circ)/(4{\cdot}54 - j0{\cdot}33) = 0{\cdot}476\angle 13{\cdot}87^\circ\ \text{A}$$

11.14. In the network shown in Fig. 11-34, find $\mathbf{V}_2$ such that the current through the $2 + j3\ \Omega$ impedance is zero.

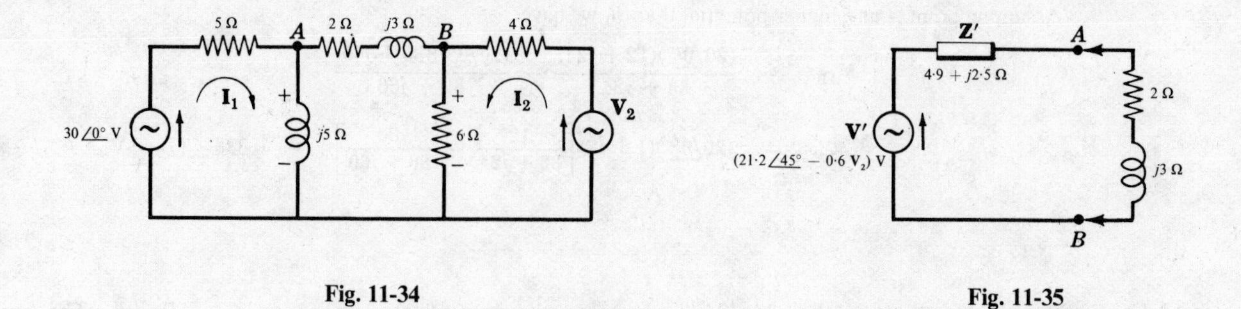

Fig. 11-34

Fig. 11-35

Apply the Thevenin theorem to the given circuit and obtain the equivalent voltage measured at terminals AB. On open circuit the two loop currents are $\mathbf{I}_1 = (30\angle 0^\circ)/(5 + j5)$ A and $\mathbf{I}_2 = \mathbf{V}_2/10$ amperes.

Assuming point A with a potential greater than B, we have

$$\mathbf{V}' = \mathbf{V}_{AB} = \mathbf{I}_1(j5) - \mathbf{I}_2(6) = 30\angle 0^\circ(j5)/(5 + j5) - \mathbf{V}_2(6)/10 = 21{\cdot}2\angle 45^\circ - 0{\cdot}6\mathbf{V}_2\ \text{volts.}$$

The current in the Thevenin equivalent circuit of Fig. 11-35 above is zero if $\mathbf{V}' = 0$. Hence

$$0 = 21{\cdot}2\angle 45^\circ - 0{\cdot}6\mathbf{V}_2 \text{ and } \mathbf{V}_2 = 35{\cdot}4\angle 45^\circ \text{ V}$$

Note. The value of the impedance $\mathbf{Z}'$ shown in Fig. 11-35 is not needed in this problem but its calculation is left as an exercise for the reader.

11.15. In the network of Fig. 11-36, find the source voltage $\mathbf{V}_1$ for which the current in the $20\angle 0^\circ$ V source is zero.

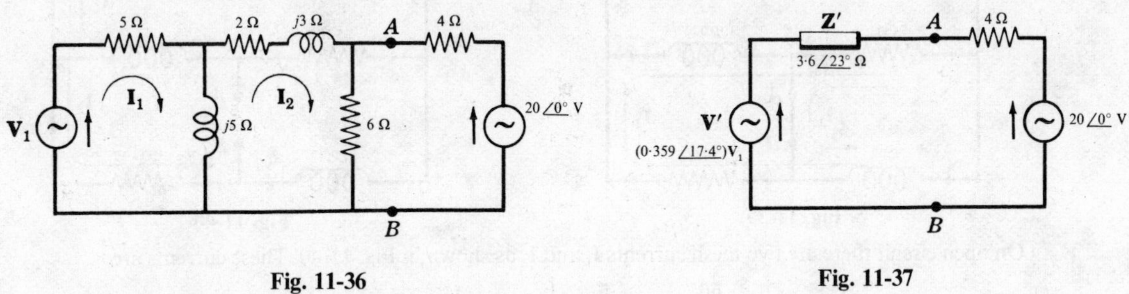

Fig. 11-36 Fig. 11-37

Obtain the Thevenin equivalent for the active network to the left of terminals AB. On open circuit there are two mesh currents $\mathbf{I}_1$ and $\mathbf{I}_2$ as shown. Solving for $\mathbf{I}_2$,

$$\mathbf{I}_2 = \frac{\begin{vmatrix} 5 + j5 & \mathbf{V}_1 \\ -j5 & 0 \end{vmatrix}}{\begin{vmatrix} 5 + j5 & -j5 \\ -j5 & 8 + j8 \end{vmatrix}} = \frac{\mathbf{V}_1\, 5\angle 90^\circ}{83{\cdot}6\angle 72{\cdot}6^\circ} \text{ amperes.}$$

Now the open circuit voltage is the drop across the 6 ohm resistor, $\mathbf{I}_2(6)$.

$$\mathbf{V}' = \frac{\mathbf{V}_1\, 5\angle 90^\circ}{83{\cdot}6\angle 72{\cdot}6^\circ}(6) = (0{\cdot}359\angle 17{\cdot}4^\circ)\mathbf{V}_1 \text{ volts.}$$

When the Thevenin equivalent circuit is connected to the terminals AB as shown in Fig. 11-37, it is apparent that for zero current $\mathbf{V}'$ must equal the other source, i.e. $\mathbf{V}' = 20\angle 0^\circ$ V. Thus $(0{\cdot}359\angle 17{\cdot}4^\circ)\mathbf{V}_1 = 20\angle 0^\circ$ V, from which $\mathbf{V}_1 = 55{\cdot}7\angle -17{\cdot}4^\circ$ V.

The note at the end of Problem 11.14 applies also to this problem.

11.16. The active network shown in Fig. 11-38 has three impedances $\mathbf{Z}_1 = 10\angle 30^\circ\,\Omega$, $\mathbf{Z}_2 = 20\angle 0^\circ\,\Omega$ and $\mathbf{Z}_3 = 5 - j5\,\Omega$ which are to be connected in turn at terminals AB. Find the power in each of the three impedances.

Replace the network with a Thevenin equivalent at terminals AB and connect the impedances to the equivalent circuit.

To compute the input impedance select three mesh currents as if a source were driving between AB as shown in Fig. 11-39. Then the input impedance $\mathbf{Z}_{\text{input }1}$ is $\mathbf{Z}'$ of the Thevenin circuit. From the definition of $\mathbf{Z}_{\text{input}}$ we have $\mathbf{Z}_{\text{input }1} = \Delta_z/\Delta_{11}$, where

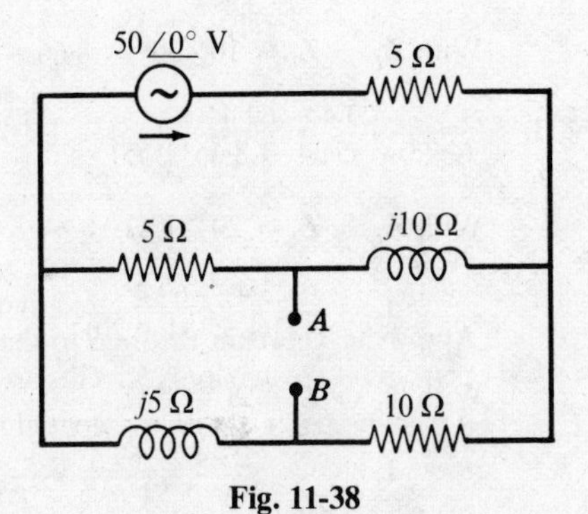

Fig. 11-38

$$\Delta_z = \begin{vmatrix} (5+j5) & -5 & (5+j5) \\ -5 & (10+j10) & (-5-j10) \\ (5+j5) & (-5-j10) & (15+j15) \end{vmatrix} = 1455\underline{/121^\circ}$$

and
$$\Delta_{11} = \begin{vmatrix} (10+j10) & (-5-j10) \\ (-5-j10) & (15+j15) \end{vmatrix} = 213{\cdot}5\underline{/69{\cdot}4^\circ}$$

Substituting,
$$\mathbf{Z}' = \mathbf{Z}_{\text{input 1}} = \Delta_z/\Delta_{11} = 1455\underline{/121^\circ}\,/213{\cdot}5\underline{/69{\cdot}4^\circ} = 6{\cdot}82\underline{/51{\cdot}6^\circ} = 4{\cdot}23 + j5{\cdot}34\ \Omega$$

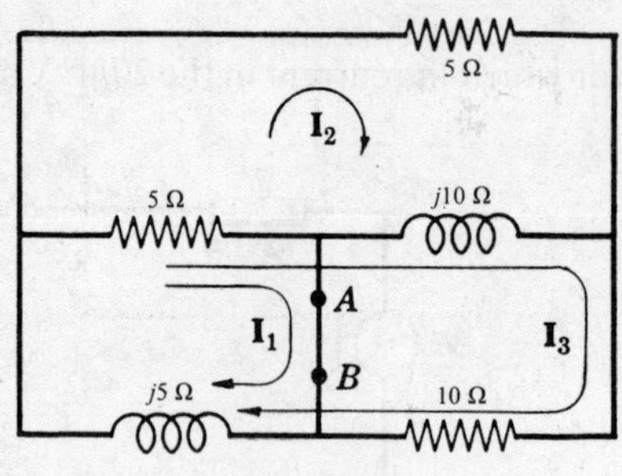

Fig. 11-39

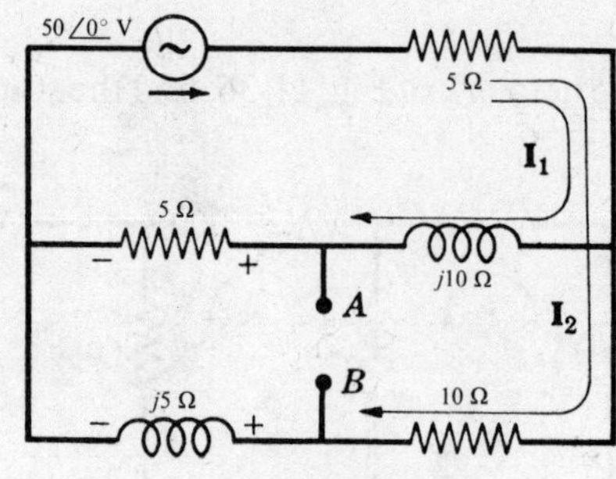

Fig. 11-40

On open circuit there are two mesh currents $\mathbf{I}_1$ and $\mathbf{I}_2$ as shown in Fig. 11-40. These currents are

$$\mathbf{I}_1 = \frac{\begin{vmatrix} 50 & 5 \\ 50 & 15+j5 \end{vmatrix}}{\begin{vmatrix} 10+j10 & 5 \\ 5 & 15+j5 \end{vmatrix}} = \frac{558\underline{/26{\cdot}6^\circ}}{213{\cdot}5\underline{/69{\cdot}4^\circ}} = 2{\cdot}62\underline{/-42{\cdot}8^\circ}\ \text{A}$$

and
$$\mathbf{I}_2 = \frac{\begin{vmatrix} 10+j10 & 50 \\ 5 & 50 \end{vmatrix}}{\Delta_z} = \frac{558\underline{/63{\cdot}4^\circ}}{213{\cdot}5\underline{/69{\cdot}4^\circ}} = 2{\cdot}62\underline{/-6^\circ}\ \text{A}$$

Now the Thevenin equivalent voltage $\mathbf{V}'$ is the open circuit voltage $\mathbf{V}_{AB}$, assuming point A at a higher potential than B. In Fig. 11-40 the voltage drops across the 5 ohm resistor in the centre branch and the $j5\ \Omega$ reactance in the lower branch are marked with instantaneous polarities. Then

$$\mathbf{V}' = \mathbf{V}_{AB} = \mathbf{I}_1(5) - \mathbf{I}_2(j5)$$
$$= (2{\cdot}62\underline{/-42{\cdot}8^\circ})(5) - (2{\cdot}62\underline{/-6^\circ})(5\underline{/90^\circ})$$
$$= 23{\cdot}4\underline{/-69{\cdot}4^\circ}\ \text{V}$$

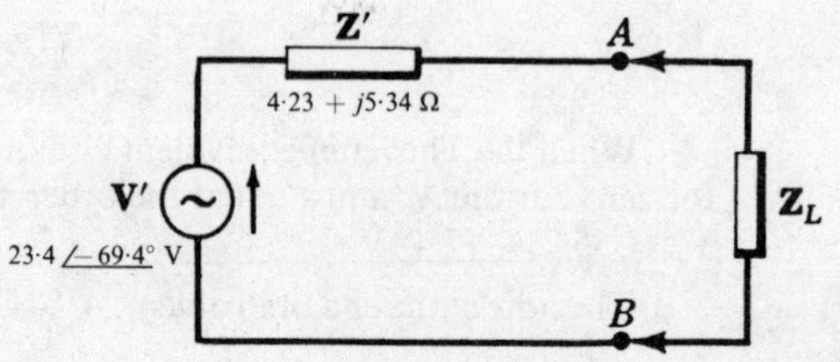

Z′ A 4·23 + j5·34 Ω V′ 23·4 /−69·4° V Z_L B

Fig. 11-41

The Thevenin equivalent circuit is shown in Fig. 11-41 with the load impedance $\mathbf{Z}_L$ shown connected to terminals AB.

Substituting the given values of $\mathbf{Z}_L$ in $\mathbf{I} = \mathbf{V}'/(\mathbf{Z}' + \mathbf{Z}_L)$, the currents and the required powers are obtained. Hence:

With $\mathbf{Z}_L = \mathbf{Z}_1 = 10\underline{/30^\circ} = 8{\cdot}66 + j5\ \Omega$,

$$\mathbf{I}_1 = \frac{23{\cdot}4\underline{/-69{\cdot}4^\circ}}{(4{\cdot}23 + j5{\cdot}34 + 8{\cdot}66 + j5)} = 1{\cdot}414\underline{/-108{\cdot}2^\circ}\ \text{A and } P_1 = (I_1)^2\,\text{Re}\mathbf{Z}_1 = (1{\cdot}414)^2(8{\cdot}66) = 17{\cdot}32\ \text{W}$$

With $\mathbf{Z}_L = \mathbf{Z}_2 = 20\underline{/0^\circ}\ \Omega$,

$$\mathbf{I}_2 = \frac{23{\cdot}4\underline{/-69{\cdot}4^\circ}}{4{\cdot}23 + j5{\cdot}34 + 20)} = 0{\cdot}940\underline{/-81{\cdot}8^\circ}\ \text{A and } P_2 = (0{\cdot}940)^2(20) = 17{\cdot}65\ \text{W}$$

With $\mathbf{Z}_L = \mathbf{Z}_3 = 5 - j5\ \Omega$,

$$\mathbf{I}_3 = \frac{23{\cdot}4\underline{/-69{\cdot}4^\circ}}{(4{\cdot}23 + j5{\cdot}34 + 5 - j5)} = 2{\cdot}54\underline{/-71{\cdot}5^\circ}\ \text{A and } P_3 = (2{\cdot}54)^2(5) = 32{\cdot}3\ \text{W}$$

Supplementary Problems

11.17. Obtain the Thevenin equivalent circuit at terminals AB of the active network given in Fig. 11-42.
Ans. $Z' = 9{\cdot}43$ ohms, $V' = 6{\cdot}29$ V ($B+$)

11.18. Obtain the Norton equivalent circuit for the network of Fig. 11-42. *Ans.* $Z' = 9{\cdot}43$ ohms, $I' = 0{\cdot}667$ A

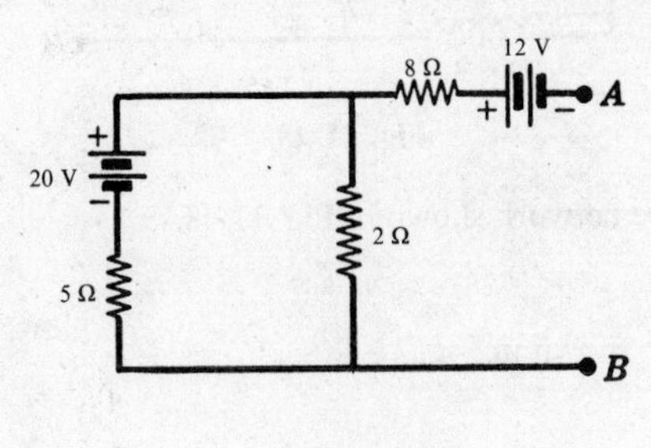

Fig. 11-42

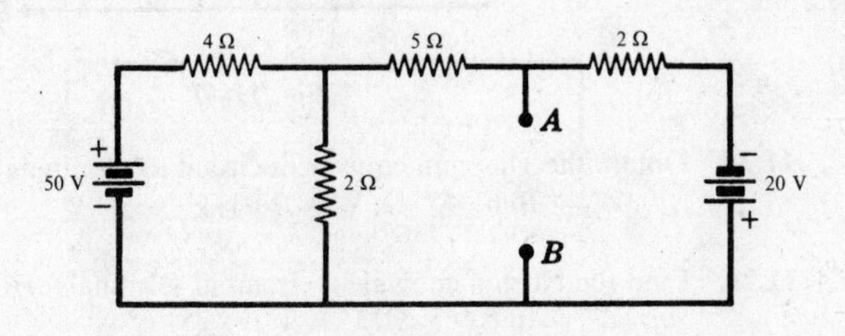

Fig. 11-43

11.19. Obtain the Thevenin equivalent circuit at terminals AB of the active network given in Fig. 11-43.
Ans. $Z' = 1{\cdot}52$ ohms, $V' = 11{\cdot}18$ V ($B+$)

11.20. Obtain the Norton equivalent circuit for the network of Fig. 11-43.
Ans. $Z' = 1{\cdot}52$ ohms, $I' = 7{\cdot}35$ A

11.21. Find the Thevenin equivalent at terminals AB of the bridge circuit shown in Fig. 11-44. *Ans.* $Z' = 55{\cdot}5$ ohms, $V' = 0$

11.22. In the bridge circuit of Fig. 11-44 the 500 ohm resistor is changed to 475 ohms. Find the Thevenin equivalent circuit.
Ans. $Z' = 55{\cdot}4$ ohms, $V' = 0{\cdot}0863$ V ($A+$)

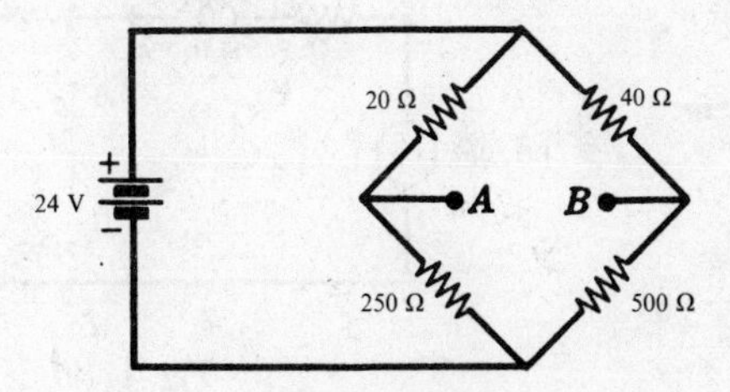

Fig. 11-44

11.23. Use the Thevenin theorem on the bridge circuit of Fig. 11-45 to find the deflection of a galvanometer connected to AB with a resistance of 100 ohms and a sensitivity of 0·5 μA/mm. *Ans.* $D = 0{\cdot}195$ m

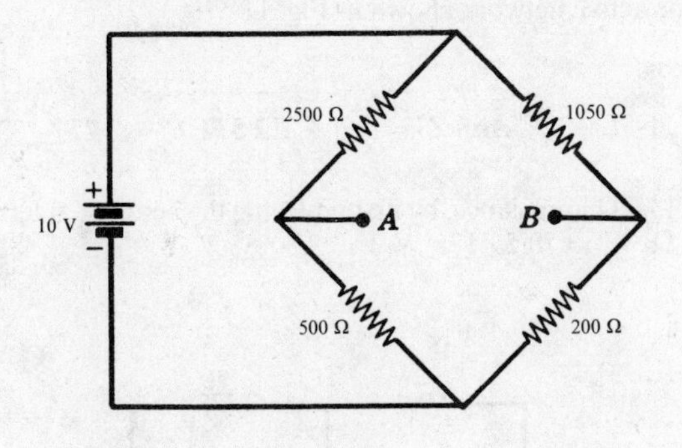

Fig. 11-45

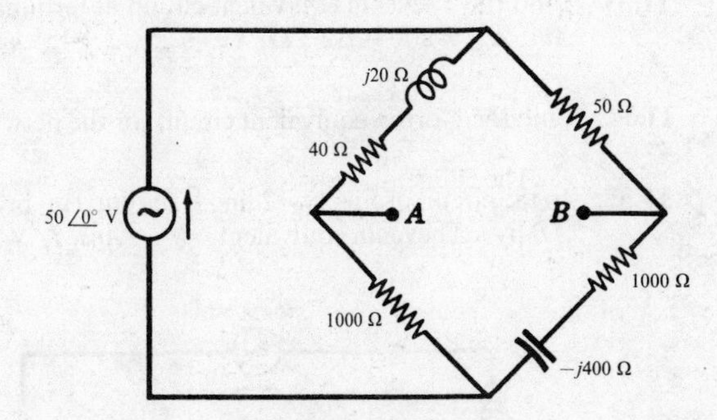

Fig. 11-46

11.24. Find the Thevenin equivalent circuit at terminals AB of the AC bridge shown in Fig. 11-46.
Ans. $Z' = 88{\cdot}7\angle 11{\cdot}55^\circ$ Ω, $V' = 0{\cdot}192\angle -43{\cdot}4^\circ$ V

11.25. Use the Thevenin theorem to find the power in a 1 ohm resistor connected to terminals AB of the network shown in Fig. 11-47 below. *Ans.* 2·22 W

11.26. Repeat Problem 11.25 using the Norton equivalent circuit.

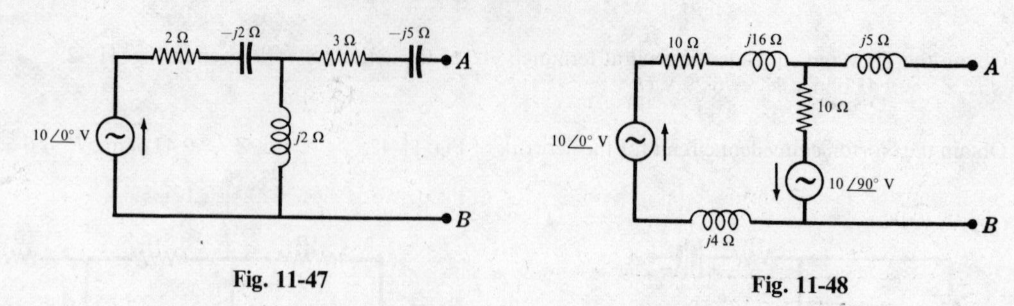

Fig. 11-47 Fig. 11-48

11.27. Obtain the Thevenin equivalent circuit at terminals AB of the active network shown in Fig. 11-48.
Ans. $\mathbf{Z}' = 10{\cdot}6\angle 45^\circ\ \Omega$, $\mathbf{V}' = 11{\cdot}17\angle -63{\cdot}4^\circ$ V

11.28. Find the Norton equivalent circuit at terminals AB of the network shown in Fig. 11-48.
Ans. $\mathbf{Z}' = 10{\cdot}6\angle 45^\circ\ \Omega$, $\mathbf{I}' = 1{\cdot}05\angle 251{\cdot}6^\circ$ A

11.29. Use the Thevenin theorem to find the power in an impedance of $2 + j4\ \Omega$ connected to terminals AB of the active network shown in Fig. 11-49. *Ans.* 475 W

11.30. Repeat Problem 11.29 using the Norton theorem.

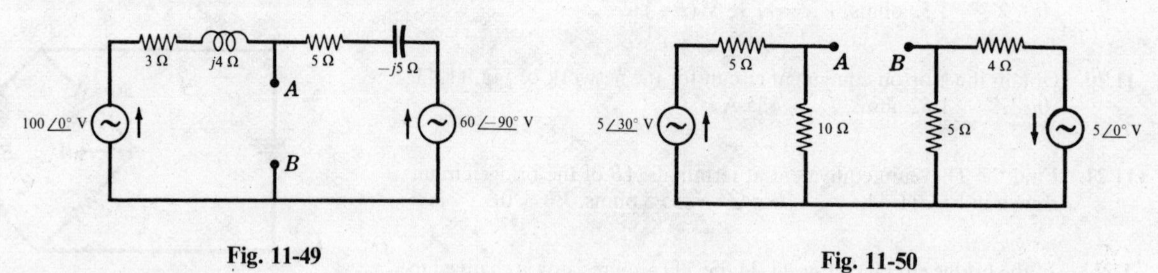

Fig. 11-49 Fig. 11-50

11.31. Find the Thevenin equivalent circuit at terminals AB of the active network of Fig. 11-50.
Ans. $\mathbf{Z}' = 5{\cdot}55\angle 0^\circ\ \Omega$, $\mathbf{V}' = 5{\cdot}9\angle 16{\cdot}4^\circ$ V

11.32. Find the Norton equivalent circuit for the network of Fig. 11-50. *Ans.* $\mathbf{Z}' = 5{\cdot}55\angle 0^\circ\ \Omega$, $\mathbf{I}' = 1{\cdot}06\angle 16{\cdot}4^\circ$ A

11.33. Find the Thevenin equivalent circuit at terminals AB of the active network shown in Fig. 11-51.
Ans. $\mathbf{Z}' = 2{\cdot}5 + j12{\cdot}5\ \Omega$, $\mathbf{V}' = 25\sqrt{2}\angle 45^\circ$ V

11.34. Find the Norton equivalent circuit for the network in Fig. 11-51. *Ans.* $\mathbf{Z}' = 2{\cdot}5 + j12{\cdot}5\ \Omega$, $\mathbf{I}' = 2{\cdot}77\angle -33{\cdot}7^\circ$ A

11.35. In the circuit of Fig. 11-52 find $\mathbf{I}$, the current hrough the $3 + j4\ \Omega$ impedance, by first replacing the network at terminals AB by a Thevenin equivalent. *Ans.* $\mathbf{Z}' = 3{\cdot}53\angle 45^\circ\ \Omega$, $\mathbf{V}' = 70{\cdot}7\angle 135^\circ$ V, $\mathbf{I} = 8{\cdot}3\angle 85{\cdot}2^\circ$ A

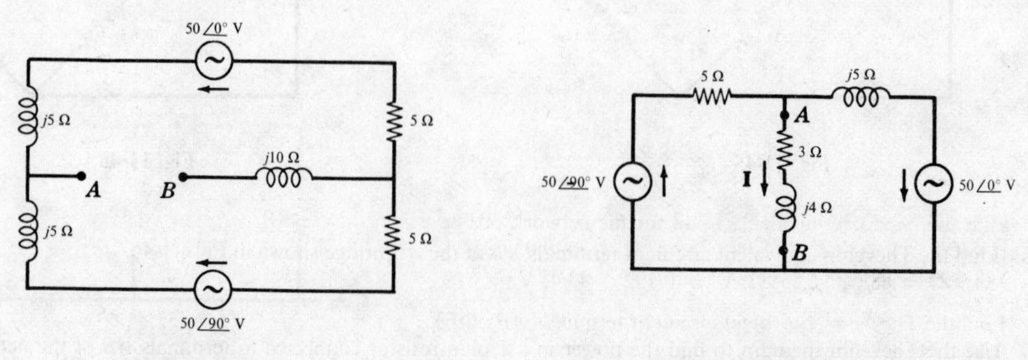

Fig. 11-51 Fig. 11-52

11.36. Repeat Problem 11.35 using a Norton equivalent circuit at terminals AB.
Ans. $\mathbf{Z}' = 3{\cdot}53\angle 45^\circ\ \Omega$, $\mathbf{I}' = 20\angle 90^\circ$ A, $\mathbf{I} = 8{\cdot}3\angle 85{\cdot}2^\circ$ A

11.37. In the network of Fig. 11-53 a current source of $15\angle 45^\circ$ A drives at the terminals shown on the diagram. Replace the network at AB with a Thevenin equivalent circuit.
Ans. $\mathbf{Z}' = 11{\cdot}48 + j1{\cdot}19\ \Omega$, $\mathbf{V}' = 28{\cdot}6\angle 83{\cdot}8^\circ$ V

11.38. Obtain the Norton equivalent circuit at terminals AB of the network shown in Fig. 11-53.
Ans. $\mathbf{Z}' = 11{\cdot}48 + j1{\cdot}19\ \Omega$, $\mathbf{I}' = 2{\cdot}47\angle 77{\cdot}9^\circ$ A

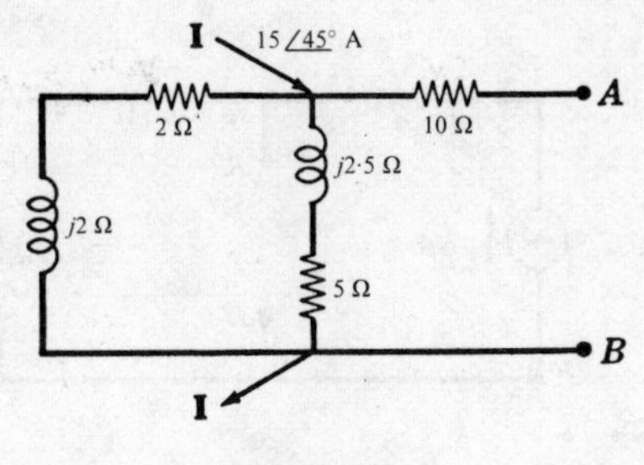

Fig. 11-53

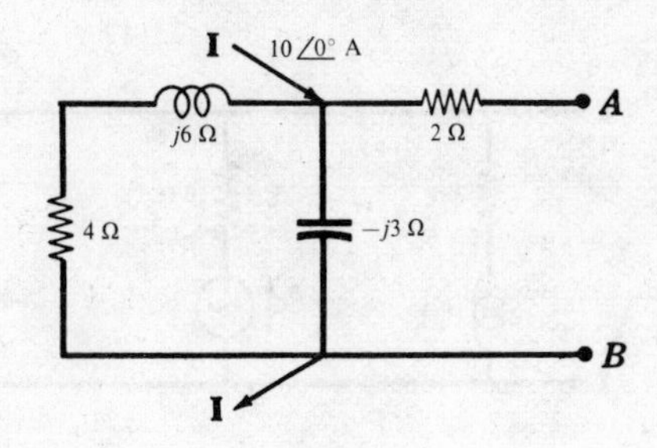

Fig. 11-54

11.39. Obtain a Thevenin equivalent circuit at terminals AB of the network shown in Fig. 11-54.
Ans. $\mathbf{Z}' = 5{\cdot}34\angle -49{\cdot}8^\circ\ \Omega$, $\mathbf{V}' = 43{\cdot}3\angle -70{\cdot}6^\circ$ V

11.40. Find the Norton equivalent circuit for the network of Fig. 11-54.
Ans. $\mathbf{Z}' = 5{\cdot}34\angle -49{\cdot}8^\circ\ \Omega$, $\mathbf{I}' = 8{\cdot}1\angle -20{\cdot}8^\circ$ A

11.41. Use the Thevenin theorem to find the power in an impedance $\mathbf{Z} = 10\angle 60^\circ\ \Omega$ connected to terminals AB of the network shown in Fig. 11-55. *Ans.* 23 W

11.42. Repeat Problem 11.41 using a Norton equivalent circuit.

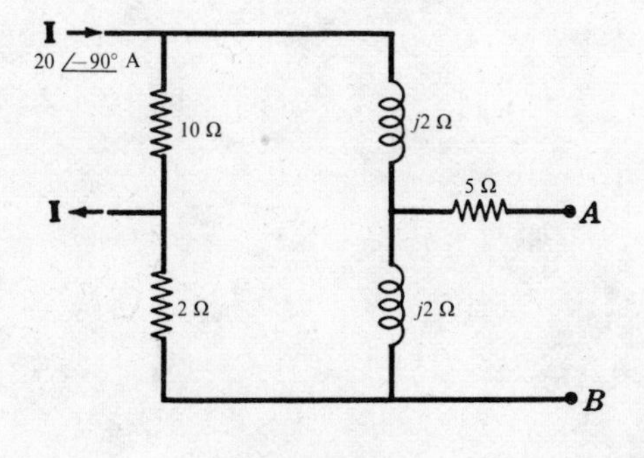

Fig. 11-55

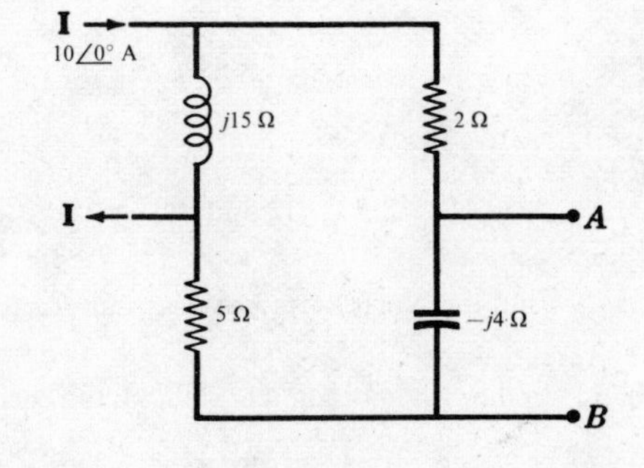

Fig. 11-56

11.43. Obtain the Thevenin equivalent circuit for the active network shown in Fig. 11-56.
Ans. $\mathbf{Z}' = 5{\cdot}09\angle -82{\cdot}5^\circ\ \Omega$, $\mathbf{V}' = 46{\cdot}2\angle -57{\cdot}5^\circ$ V

11.44. Find the Norton equivalent circuit for the network of Fig. 11-56. *Ans.* $\mathbf{Z}' = 5{\cdot}09\angle -82{\cdot}5^\circ\ \Omega$, $\mathbf{I}' = 9{\cdot}05\angle 25^\circ$ A

11.45. Find the Thevenin equivalent circuit at terminals AB of the active network shown in Fig. 11-57.
Ans. $\mathbf{Z}' = 6{\cdot}2\angle 51{\cdot}8^\circ\ \Omega$, $\mathbf{V}' = 62{\cdot}6\angle 44{\cdot}17^\circ$ V

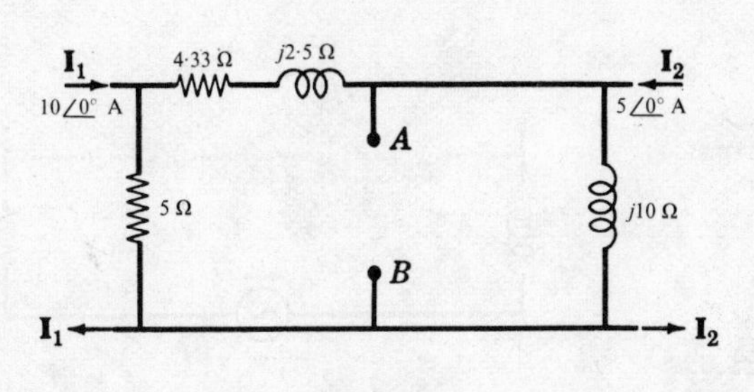

Fig. 11-57

11.46. Obtain the Norton equivalent circuit for the network of Fig. 11-57.
Ans. $\mathbf{Z}' = 6{\cdot}2\underline{/51{\cdot}8^\circ}\ \Omega$, $\mathbf{I}' = 10{\cdot}1\underline{/-7{\cdot}63^\circ}$ A

11.47. The active network of Fig. 11-58 contains a current source of $4\underline{/45^\circ}$ A and a voltage source of $25\underline{/90^\circ}$ V. Find the Thevenin equivalent circuit at terminals *AB*.
Ans. $\mathbf{Z}' = 3{\cdot}68\ \underline{/36^\circ}\ \Omega$, $\mathbf{V}' = 22{\cdot}2\ \underline{/98^\circ}$ V

11.48. Obtain the Norton equivalent circuit for the network of Fig. 11-58.
Ans. $\mathbf{Z}' = 3{\cdot}68\underline{/36^\circ}\ \Omega$, $\mathbf{I}' = 6{\cdot}03\underline{/62^\circ}$ A

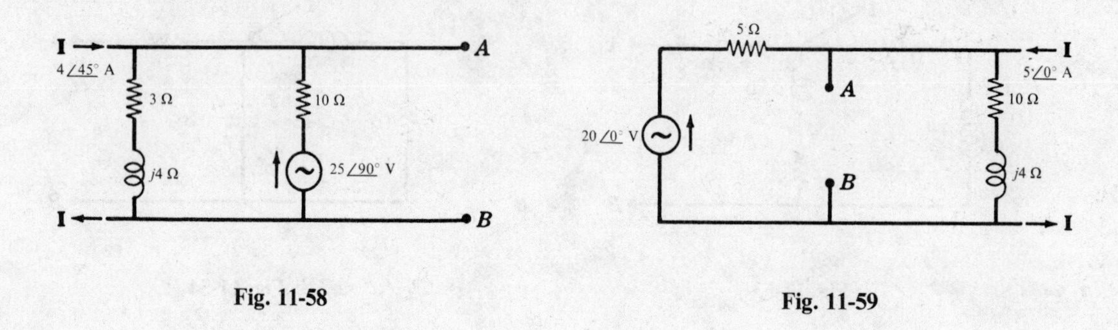

Fig. 11-58

Fig. 11-59

11.49. Find the Thevenin equivalent circuit at terminals *AB* of the active network shown in Fig. 11-59.
Ans. $\mathbf{Z}' = 3{\cdot}47\underline{/6{\cdot}85^\circ}\ \Omega$, $\mathbf{V}' = 31{\cdot}2\underline{/6{\cdot}89^\circ}$ V

11.50. Obtain the Norton equivalent circuit for the network of Fig. 11-59. *Ans.* $\mathbf{Z}' = 3{\cdot}47\underline{/6{\cdot}85^\circ}\ \Omega$, $\mathbf{I}' = 9{\cdot}0\underline{/0^\circ}$ A

CHAPTER 12

Network Theorems

INTRODUCTION

The mesh current and node voltage methods make it possible to solve nearly all circuit problems. The introduction of Thevenin's and Norton's theorems in Chapter 11 proved to be effective in reducing numerical computations when several impedances were to be connected singly at a pair of terminals. Similarly, the theorems presented in this chapter achieve the same purpose of simplifying the solution of particular types of circuit problems. For this reason this chapter can be considered an extension of Chapter 11.

STAR-DELTA TRANSFORMATION

The passive three-terminal network consisting of three impedances Z_A, $\mathbf{Z}_B$ and $\mathbf{Z}_C$, as shown in Fig. 12-1(*a*), is said to form a delta or π connection. The passive three-terminal network consisting of three impedances $\mathbf{Z}_1$, $\mathbf{Z}_2$ and $\mathbf{Z}_3$, as shown in Fig. 12-1(*b*), is said to form a star or T connection. The two circuits are equivalent if their respective input, output and transfer impedances are equal.

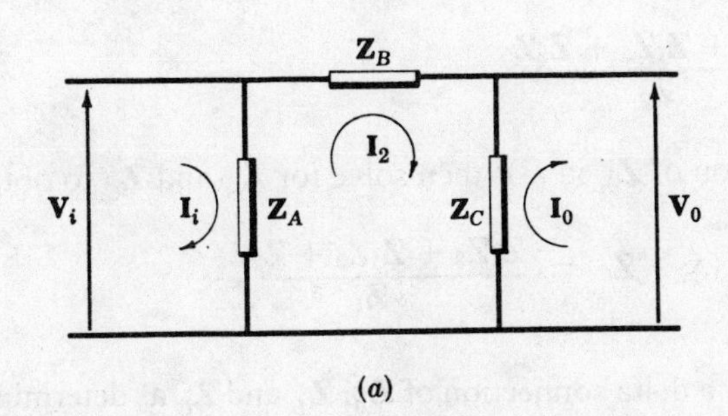

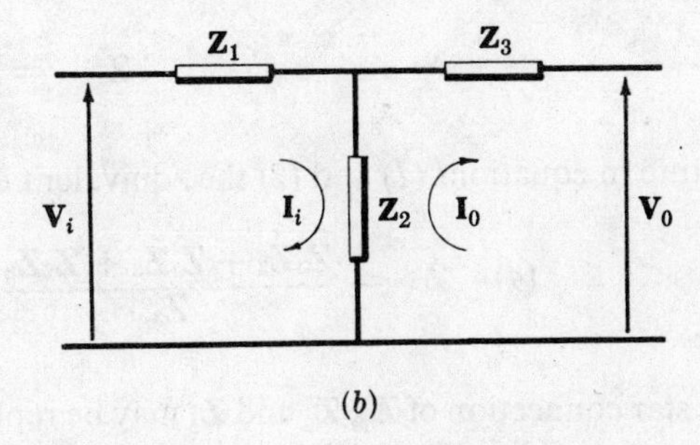

Fig. 12-1

Let $\mathbf{V}_i$ be the input voltage and $\mathbf{V}_0$ the corresponding output voltage in each circuit. Select the input current $\mathbf{I}_i$ and output current $\mathbf{I}_0$ with the same clockwise direction in each circuit. Let the centre mesh current of the delta-connected circuit be $\mathbf{I}_2$ with direction as shown.

The matrix form of the mesh current equations of the delta-connected circuit is

$$\begin{bmatrix} \mathbf{Z}_A & -\mathbf{Z}_A & 0 \\ -\mathbf{Z}_A & \mathbf{Z}_A + \mathbf{Z}_B + \mathbf{Z}_C & -\mathbf{Z}_C \\ 0 & -\mathbf{Z}_C & \mathbf{Z}_C \end{bmatrix} \begin{bmatrix} \mathbf{I}_i \\ \mathbf{I}_2 \\ \mathbf{I}_0 \end{bmatrix} = \begin{bmatrix} \mathbf{V}_i \\ 0 \\ -\mathbf{V}_0 \end{bmatrix}$$

The corresponding input, output and transfer impedances are

$$\mathbf{Z}_{\text{input}} = \frac{\Delta_z}{\Delta_{11}} = \frac{\mathbf{Z}_A\mathbf{Z}_B}{\mathbf{Z}_A + \mathbf{Z}_B}$$

$$\mathbf{Z}_{\text{output}} = \frac{\Delta_z}{\Delta_{33}} = \frac{\mathbf{Z}_B\mathbf{Z}_C}{\mathbf{Z}_B + \mathbf{Z}_C}$$

$$\mathbf{Z}_{\text{transfer i0}} = \frac{\Delta_z}{\Delta_{13}} = \mathbf{Z}_B$$

The mesh current equations of the star connected circuit of Fig. 12-1(*b*) are

$$\begin{bmatrix} \mathbf{Z}_1 + \mathbf{Z}_2 & -\mathbf{Z}_2 \\ -\mathbf{Z}_2 & \mathbf{Z}_2 + \mathbf{Z}_3 \end{bmatrix}\begin{bmatrix} \mathbf{I}_i \\ \mathbf{I}_0 \end{bmatrix} = \begin{bmatrix} \mathbf{V}_i \\ -\mathbf{V}_0 \end{bmatrix}$$

The corresponding input, output and transfer impedances are

$$\mathbf{Z}_{\text{input}} = \frac{\Delta_z}{\Delta_{11}} = \frac{\mathbf{Z}_1\mathbf{Z}_2 + \mathbf{Z}_1\mathbf{Z}_3 + \mathbf{Z}_2\mathbf{Z}_3}{\mathbf{Z}_2 + \mathbf{Z}_3}$$

$$\mathbf{Z}_{\text{output}} = \frac{\Delta_z}{\Delta_{22}} = \frac{\mathbf{Z}_1\mathbf{Z}_2 + \mathbf{Z}_1\mathbf{Z}_3 + \mathbf{Z}_2\mathbf{Z}_3}{\mathbf{Z}_1 + \mathbf{Z}_2}$$

$$\mathbf{Z}_{\text{transfer i0}} = \frac{\Delta_z}{\Delta_{12}} = \frac{\mathbf{Z}_1\mathbf{Z}_2 + \mathbf{Z}_1\mathbf{Z}_3 + \mathbf{Z}_2\mathbf{Z}_3}{\mathbf{Z}_2}$$

Now equate the impedances of the delta and star circuits.

$$\frac{\mathbf{Z}_A\mathbf{Z}_B}{\mathbf{Z}_A + \mathbf{Z}_B} = \frac{\mathbf{Z}_1\mathbf{Z}_2 + \mathbf{Z}_1\mathbf{Z}_3 + \mathbf{Z}_2\mathbf{Z}_3}{\mathbf{Z}_2 + \mathbf{Z}_3} \qquad (1)$$

$$\frac{\mathbf{Z}_B\mathbf{Z}_C}{\mathbf{Z}_B + \mathbf{Z}_C} = \frac{\mathbf{Z}_1\mathbf{Z}_2 + \mathbf{Z}_1\mathbf{Z}_3 + \mathbf{Z}_2\mathbf{Z}_3}{\mathbf{Z}_1 + \mathbf{Z}_2} \qquad (2)$$

$$\mathbf{Z}_B = \frac{\mathbf{Z}_1\mathbf{Z}_2 + \mathbf{Z}_1\mathbf{Z}_3 + \mathbf{Z}_2\mathbf{Z}_3}{\mathbf{Z}_2} \qquad (3)$$

Substitute in equations (*1*) and (*2*) the equivalent expression of $\mathbf{Z}_B$ in (*3*); then solve for $\mathbf{Z}_A$ and $\mathbf{Z}_C$ to obtain

$$(4)\quad \mathbf{Z}_A = \frac{\mathbf{Z}_1\mathbf{Z}_2 + \mathbf{Z}_1\mathbf{Z}_3 + \mathbf{Z}_2\mathbf{Z}_3}{\mathbf{Z}_3}, \qquad (5)\quad \mathbf{Z}_C = \frac{\mathbf{Z}_1\mathbf{Z}_2 + \mathbf{Z}_1\mathbf{Z}_3 + \mathbf{Z}_2\mathbf{Z}_3}{\mathbf{Z}_1}$$

Thus a star connection of $\mathbf{Z}_1$, $\mathbf{Z}_2$ and $\mathbf{Z}_3$ may be replaced by a delta connection of $\mathbf{Z}_A$, $\mathbf{Z}_B$ and $\mathbf{Z}_C$ as determined from equations (*3*), (*4*) and (*5*).

To obtain the delta to star transformation add the three equations (*3*), (*4*) and (*5*) and invert the sum. Then

$$\frac{1}{\mathbf{Z}_A + \mathbf{Z}_B + \mathbf{Z}_C} = \frac{\mathbf{Z}_1\mathbf{Z}_2\mathbf{Z}_3}{(\mathbf{Z}_1\mathbf{Z}_2 + \mathbf{Z}_1\mathbf{Z}_3 + \mathbf{Z}_2\mathbf{Z}_3)^2} \qquad (6)$$

Now multiply the left side of (*6*) by $\mathbf{Z}_A\mathbf{Z}_B$ and the right side of (*6*) by the expressions for $\mathbf{Z}_A$ in (*4*) and $\mathbf{Z}_B$ in (*3*).

$$\left(\frac{1}{\mathbf{Z}_A + \mathbf{Z}_B + \mathbf{Z}_C}\right)\mathbf{Z}_A\mathbf{Z}_B = \frac{\mathbf{Z}_1\mathbf{Z}_2\mathbf{Z}_3}{(\mathbf{Z}_1\mathbf{Z}_2 + \mathbf{Z}_1\mathbf{Z}_3 + \mathbf{Z}_2\mathbf{Z}_3)^2}\left(\frac{\mathbf{Z}_1\mathbf{Z}_2 + \mathbf{Z}_1\mathbf{Z}_3 + \mathbf{Z}_2\mathbf{Z}_3}{\mathbf{Z}_3}\right)\left(\frac{\mathbf{Z}_1\mathbf{Z}_2 + \mathbf{Z}_1\mathbf{Z}_3 + \mathbf{Z}_2\mathbf{Z}_3}{\mathbf{Z}_2}\right)$$

from which

$$\mathbf{Z}_1 = \frac{\mathbf{Z}_A\mathbf{Z}_B}{\mathbf{Z}_A + \mathbf{Z}_B + \mathbf{Z}_C}$$

Using a similar procedure, expressions for $\mathbf{Z}_2$ and $\mathbf{Z}_3$ in terms of $\mathbf{Z}_A$, $\mathbf{Z}_B$ and $\mathbf{Z}_C$ are obtained. For convenience, the complete results of the star-delta transformations are tabulated below.

star to delta Transformation	delta to star Transformation
$\mathbf{Z}_A = \dfrac{\mathbf{Z}_1\mathbf{Z}_2 + \mathbf{Z}_1\mathbf{Z}_3 + \mathbf{Z}_2\mathbf{Z}_3}{\mathbf{Z}_3}$	$\mathbf{Z}_1 = \dfrac{\mathbf{Z}_A\mathbf{Z}_B}{\mathbf{Z}_A + \mathbf{Z}_B + \mathbf{Z}_C}$
$\mathbf{Z}_B = \dfrac{\mathbf{Z}_1\mathbf{Z}_2 + \mathbf{Z}_1\mathbf{Z}_3 + \mathbf{Z}_2\mathbf{Z}_3}{\mathbf{Z}_2}$	$\mathbf{Z}_2 = \dfrac{\mathbf{Z}_A\mathbf{Z}_C}{\mathbf{Z}_A + \mathbf{Z}_B + \mathbf{Z}_C}$
$\mathbf{Z}_C = \dfrac{\mathbf{Z}_1\mathbf{Z}_2 + \mathbf{Z}_1\mathbf{Z}_3 + \mathbf{Z}_2\mathbf{Z}_3}{\mathbf{Z}_1}$	$\mathbf{Z}_3 = \dfrac{\mathbf{Z}_B\mathbf{Z}_C}{\mathbf{Z}_A + \mathbf{Z}_B + \mathbf{Z}_C}$

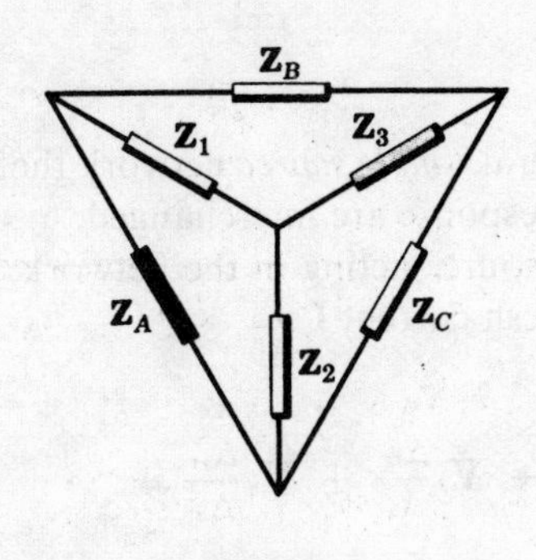

Fig. 12-2

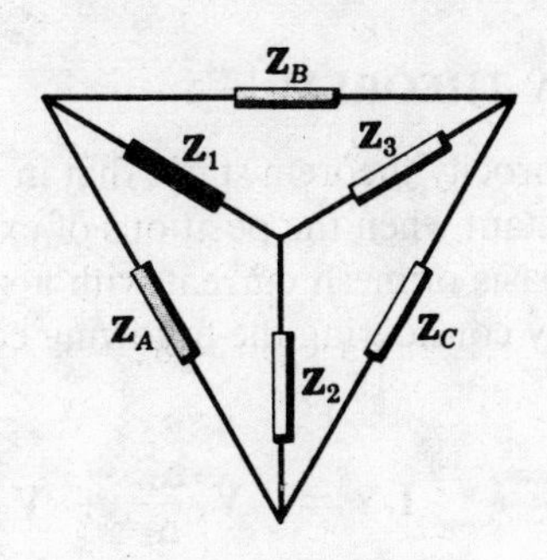

Fig. 12-3

Two mnemonic rules to determine the above relationships are given below.

1. Star to delta transformation

   Any impedance of the delta circuit is equal to the sum of the products of all possible pairs of the star impedances divided by the opposite impedance of the star circuit.

   Referring to Fig. 12-2, $\mathbf{Z}_A$ is given by the sum of the three products divided by $\mathbf{Z}_3$, the opposite impedance of the star circuit.

2. Delta to star transformation

   Any impedance of the star circuit is equal to the product of the two adjacent delta impedances divided by the sum of the three delta impedances.

   Referring to Fig. 12-3, $\mathbf{Z}_1$ is given by the product $\mathbf{Z}_A\mathbf{Z}_B$, the adjacent delta circuit impedances, divided by the sum of the three delta impedances.

SUPERPOSITION THEOREM

The superposition theorem states that the response in any element of a linear bilateral network containing two or more sources is the sum of the responses obtained by each source acting separately and with all the other sources set equal to zero.

In both the mesh current and node voltage methods of analysis the superposition principle was implicit. The mesh currents and the node voltages were found as ratios of two determinants (see Chapters 9 and 10). The expansions of the numerator determinants about the elements of the column containing the sources resulted in equations of the following type.

$$\mathbf{I}_1 = \mathbf{V}_1\frac{\Delta_{11}}{\Delta_\mathbf{Z}} + \mathbf{V}_2\frac{\Delta_{21}}{\Delta_\mathbf{Z}} + \mathbf{V}_3\frac{\Delta_{31}}{\Delta_\mathbf{Z}} + \cdots \tag{7}$$

and

$$\mathbf{V}_1 = \mathbf{I}_1\frac{\Delta_{11}}{\Delta_\mathbf{Y}} + \mathbf{I}_2\frac{\Delta_{21}}{\Delta_\mathbf{Y}} + \mathbf{I}_3\frac{\Delta_{31}}{\Delta_\mathbf{Y}} + \cdots \tag{8}$$

The terms in (*7*) are component currents of the mesh current $\mathbf{I}_1$ due to the driving voltages $\mathbf{V}_1$, $\mathbf{V}_2$, etc. In (*8*) the terms are components of the node voltage $\mathbf{V}_1$ due to the driving currents $\mathbf{I}_1$, $\mathbf{I}_2$, etc.

If the mesh currents are selected such that the sources are all in uncoupled branches, then the terms of (*7*) will be identical to the currents which result if the sources are permitted to act separately. Similarly, if the current sources of a network to be solved by the nodal method all have the same return point, then with this point as reference the terms of (*8*) will be identical to the node voltages which result when each source acts separately.

The superposition principle is applied to determine currents and node voltages which are linearly related to the sources acting on the network. Power cannot be determined by superposition since the relationship between power and current or voltage is quadratic.

RECIPROCITY THEOREM

The reciprocity theorem states that in a linear, bilateral, *single source* network the ratio of excitation to response is constant when the positions of excitation and response are interchanged.

On the basis of mesh current with a single voltage source acting in the network, the theorem may be demonstrated by considering the following equation for mesh current $\mathbf{I}_r$.

$$\mathbf{I}_r = \mathbf{V}_1\frac{\Delta_{1r}}{\Delta_z} + \mathbf{V}_2\frac{\Delta_{2r}}{\Delta_z} + \cdots + \mathbf{V}_r\frac{\Delta_{rr}}{\Delta_z} + \mathbf{V}_s\frac{\Delta_{sr}}{\Delta_z} + \cdots$$

Let the only source in the network be $\mathbf{V}_s$. Then

$$\mathbf{I}_r = \mathbf{V}_s\frac{\Delta_{sr}}{\Delta_z}$$

The ratio of excitation to response is

$$\frac{\mathbf{V}_s}{\mathbf{I}_r} = \frac{\Delta_z}{\Delta_{sr}} = \mathbf{Z}_{\text{transfer } sr} \qquad (9)$$

Now when the positions of the excitation and response are interchanged the source becomes $\mathbf{V}_r$ and the current $\mathbf{I}_s$.

$$\mathbf{I}_s = \mathbf{V}_r\frac{\Delta_{rs}}{\Delta_z}$$

Then the ratio of excitation to response is

$$\frac{\mathbf{V}_r}{\mathbf{I}_s} = \frac{\Delta_z}{\Delta_{rs}} = \mathbf{Z}_{\text{transfer } rs} \qquad (10)$$

The two transfer impedances in (*9*) and (*10*) are equal in any linear, bilateral network since in such networks the impedance matrix [**Z**] is symmetrical with respect to the principal diagonal, and the cofactors Δ_{sr} and Δ_{rs} are equal. Thus the current in mesh r which results from a voltage source in mesh s is the same as the current in mesh s when the voltage source is moved to mesh r. It must be noted that *currents in other parts of the network will not remain the same.*

The reciprocity theorem also applies to networks containing a single current source. Here the theorem states that the voltage which results at a pair of terminals $m\,n$ due to a current source acting at terminals $a\,b$ is the same as the voltage at terminals $a\,b$ when the current source is moved to terminals $m\,n$. It should be noted that *voltages at other points in the network will not remain the same.* See Problem 12.9.

COMPENSATION THEOREM

A network impedance **Z** which contains a current **I** has a voltage drop given by **IZ**. Then according to the compensation theorem this impedance may be replaced by a *compensation* emf where the magnitude and phase of this source are equal to **IZ**. Similarly, if the voltage across an element or branch of a network containing an impedance **Z** is **V**, the element or branch may be replaced by a current source $\mathbf{I} = \mathbf{V}/\mathbf{Z}$. Currents and voltages in all other parts of the network remain unchanged after the substitution of the compensation source. The compensation theorem is also called the substitution theorem.

In Fig. 12-4(*a*), a branch of a network contains impedances $\mathbf{Z}_A$ and $\mathbf{Z}_B$. If the current in this branch is $\mathbf{I}_1$, the voltage drop across $\mathbf{Z}_A$ is $\mathbf{I}_1\mathbf{Z}_A$ with polarity as shown. Fig. 12-4(*b*) shows the compensation source $\mathbf{V}_c = \mathbf{I}_1\mathbf{Z}_A$ which replaces $\mathbf{Z}_A$. $\mathbf{V}_c$ must be polarized as shown since the conventional arrow points to the positive terminal.

If any change which would affect $\mathbf{I}_1$ occurs in the network, then the compensation source must be changed accordingly. For this reason the *compensation source* $\mathbf{V}_c$ *is called a dependent source*.

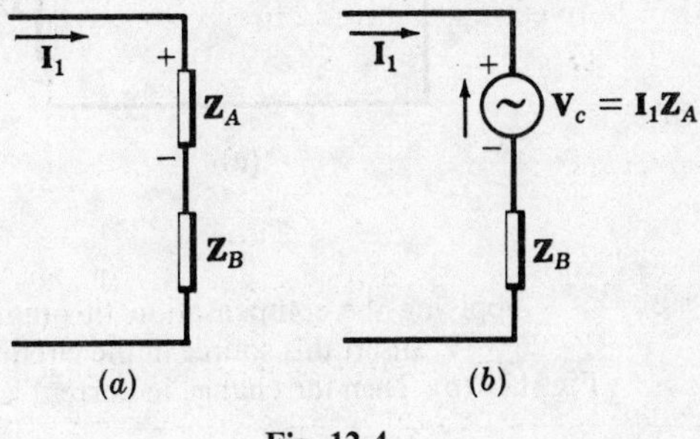

Fig. 12-4

The compensation theorem is useful in determining current and voltage changes in a circuit element when the value of its impedance is changed. This application occurs in bridge and potentiometer circuits where a slight change in one impedance results in a shift from the null conditions.

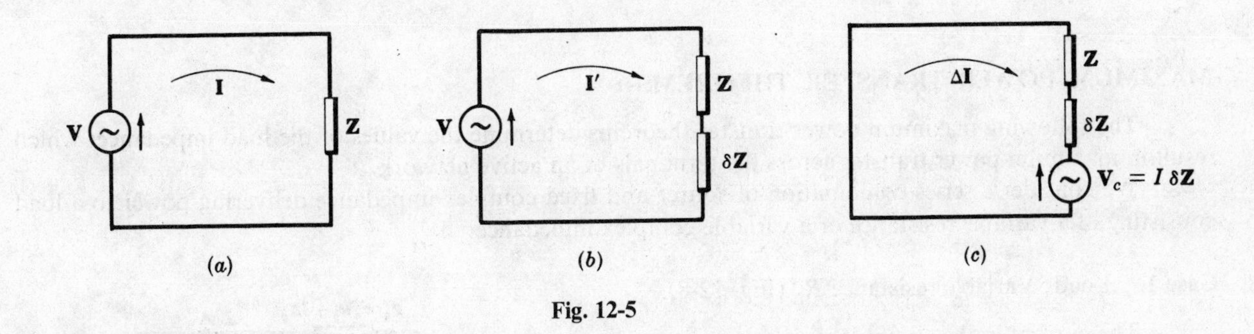

Fig. 12-5

In Fig. 12-5(*a*) the source **V** applied to the circuit results in a current $\mathbf{I} = \mathbf{V}/\mathbf{Z}$. In Fig. 12-5(*b*) the total circuit impedance is changed to $(\mathbf{Z} + \delta\mathbf{Z})$. Then the current in the circuit is $\mathbf{I}' = \mathbf{V}/(\mathbf{Z} + \delta\mathbf{Z})$. Now a compensation voltage source $\mathbf{V}_a = \mathbf{I}\delta\mathbf{Z}$ acting in the circuit containing **Z** and $\delta\mathbf{Z}$ *with the original source set to zero*, results in the current $\Delta\mathbf{I}$ as shown in Fig. 12-5(*c*). $\Delta\mathbf{I}$ is the change in current which results from the change $\delta\mathbf{Z}$ in circuit impedance. By the superposition theorem, $\mathbf{I} + \Delta\mathbf{I} = \mathbf{I}'$ or $\Delta\mathbf{I} = \mathbf{I}' - \mathbf{I}$.

Example.

In the circuit shown in Fig. 12-6, the $3 + j4\ \Omega$ impedance is changed to $5 + j5\ \Omega$, i.e. $\delta\mathbf{Z} = 2 + j1\ \Omega$. Find the change in current by direct calculation and verify the result by applying the compensation theorem.

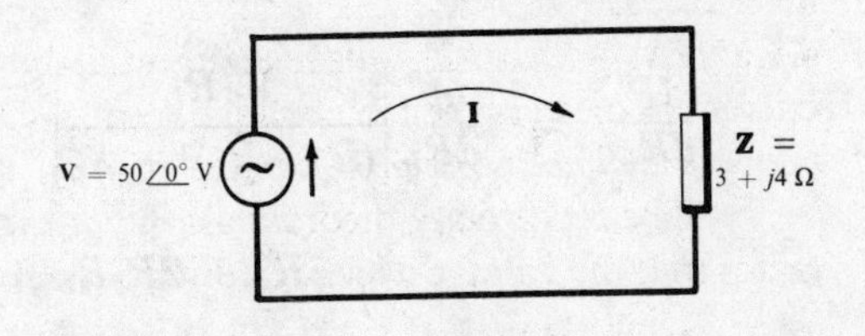

Fig. 12-6

Before the change, $\mathbf{I} = \mathbf{V}/\mathbf{Z} = (50\angle 0°)/(5\angle 53{\cdot}1°) = 10\angle -53{\cdot}1°$ A. When $\delta\mathbf{Z}$ is added to the circuit as shown in Fig. 12-7(*a*) below, we have

$$\mathbf{I}' = \mathbf{V}/(\mathbf{Z} + \delta\mathbf{Z}) = (50\angle 0°)/(5 + j5) = 7{\cdot}07\angle -45°\ \text{A}$$

The change in current is

$$\Delta\mathbf{I} = \mathbf{I}' - \mathbf{I} = (5 - j5) - (6 - j8) = -1 + j3 = 3{\cdot}16\underline{/108{\cdot}45^\circ}\ \text{A}$$

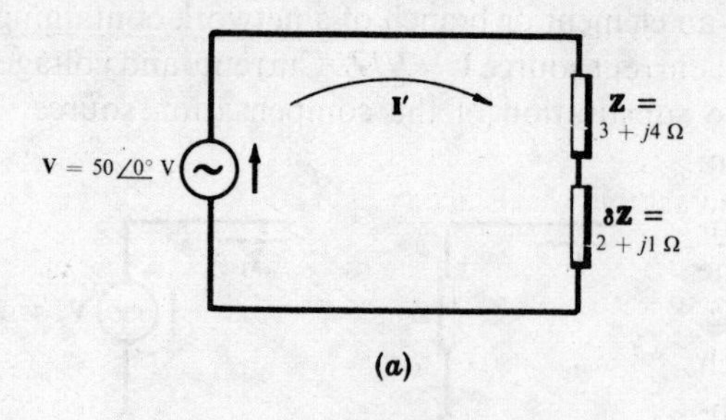

(a)

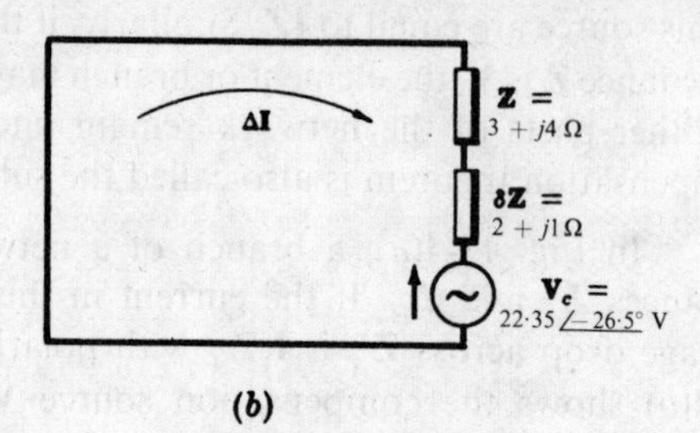

(b)

Fig. 12-7

Applying the compensation theorem, the compensation source $\mathbf{V}_c = \mathbf{I}\delta\mathbf{Z} = (10\underline{/-53{\cdot}1^\circ})(2 + j1) = 22{\cdot}35\underline{/-26{\cdot}5^\circ}$ V. Insert this source in the circuit containing **Z** and δ**Z**, and set the $50\underline{/0^\circ}$ source equal to zero as shown in Fig. 12-7(*b*). Then the change in current is

$$\Delta\mathbf{I} = -\frac{\mathbf{V}_c}{\mathbf{Z} + \delta\mathbf{Z}} = -\frac{22{\cdot}35\underline{/-26{\cdot}5^\circ}}{5 + j5} = 3\ 16\underline{/108{\cdot}45^\circ}\ \text{A}$$

Thus when an impedance is changed in a circuit and the corresponding change in current $\Delta\mathbf{I}$ is required, $\Delta\mathbf{I}$ is determined by letting the compensation source $\mathbf{V}_c$ act in the network with all other sources set equal to zero.

MAXIMUM POWER TRANSFER THEOREMS

The following maximum power transfer theorems determine the values of the load impedances which result in maximum power transfer across the terminals of an active network.

We consider a series combination of source and fixed complex impedance delivering power to a load consisting of a variable resistance or a variable complex impedance.

Case 1. Load: Variable resistance R_L (Fig. 12-8).

The current in the circuit is

$$\mathbf{I} = \frac{\mathbf{V}_g}{(R_g + R_L) + jX_g}$$

$$I = |\mathbf{I}| = \frac{V_g}{\sqrt{(R_g + R_L)^2 + X_g^2}}$$

$\mathbf{Z}_g = R_g + jX_g$, A, I, $\mathbf{V}_g$, R_L, B

Fig. 12-8

Then the power delivered to R_L is

$$P = I^2 R_L = \frac{V_g^2 R_L}{(R_g + R_L)^2 + X_g^2}$$

To determine the value of R_L for maximum power transferred to the load, set the first derivative dP/dR_L equal to zero.

$$\frac{dP}{dR_L} = \frac{d}{dR_L}\left[\frac{V_g^2 R_L}{(R_g + R_L)^2 + X_g^2}\right] = V_g^2\left\{\frac{[(R_g + R_L)^2 + X_g^2] - R_L(2)(R_g + R_L)}{[(R_g + R_L)^2 + X_g^2]^2}\right\} = 0$$

or

$$R_g^2 + 2R_gR_L + R_L^2 + X_g^2 - 2R_LR_g - 2R_L^2 = 0$$

and

$$R_g^2 + X_g^2 = R_L^2$$

Hence $$R_L = \sqrt{R_g^2 + X_g^2} = |\mathbf{Z}_g|$$

With a variable pure resistance load the maximum power is delivered across the terminals of the active network if the load resistance is made equal to the absolute value of the active network impedance.

If the reactive component of the impedance in series with the source is zero, i.e. $X_g = 0$, then the maximum power is transferred to the load when the load and source resistances are equal, $R_L = R_g$.

Case 2. Load: Impedance $\mathbf{Z}_L$ with variable resistance and variable reactance (Fig. 12-9).

The circuit current is

$$\mathbf{I} = \frac{\mathbf{V}_g}{(R_g + R_L) + j(X_g + X_L)}$$

$$I = |\mathbf{I}| = \frac{V_g}{\sqrt{(R_g + R_L)^2 + (X_g + X_L)^2}}$$

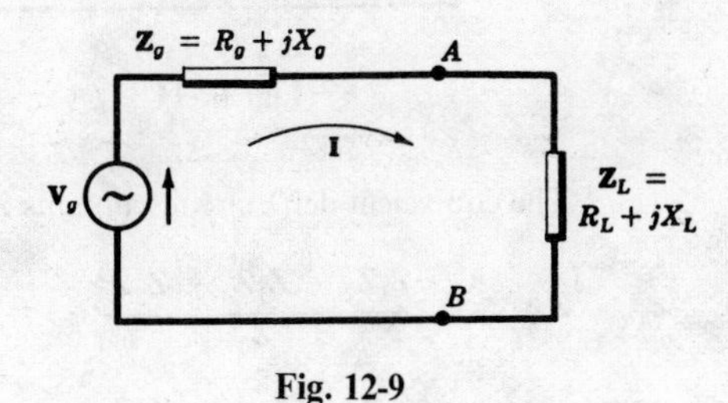

Fig. 12-9

The power delivered by the source is

$$P = I^2 R_L = \frac{V_g^2 R_L}{(R_g + R_L)^2 + (X_g + X_L)^2} \tag{11}$$

If R_L in (*11*) is held fixed, the value of P is maximum when $X_g = -X_L$. Then equation (*11*) becomes

$$P = \frac{V_g^2 R_L}{(R_g + R_L)^2}$$

Consider now R_L to be variable. As shown in case 1, the maximum power is delivered to the load when $R_L = R_g$. If $R_L = R_g$ and $X_L = -X_g$, $\mathbf{Z}_L = \mathbf{Z}_g^*$.

With the load impedance consisting of variable resistance and variable reactance, maximum power transfer across the terminals of the active network occurs when the load impedance $\mathbf{Z}_L$ is equal to the complex conjugate of the network impedance $\mathbf{Z}_g$.

Case 3. Load: Impedance $\mathbf{Z}_L$ with variable resistance and fixed reactance (Fig. 12-10).

We obtain the same equations for current I and power P as in case 2 with the condition that X_L be kept constant.

When the first derivative of P with respect to R_L is set equal to zero, it is found that

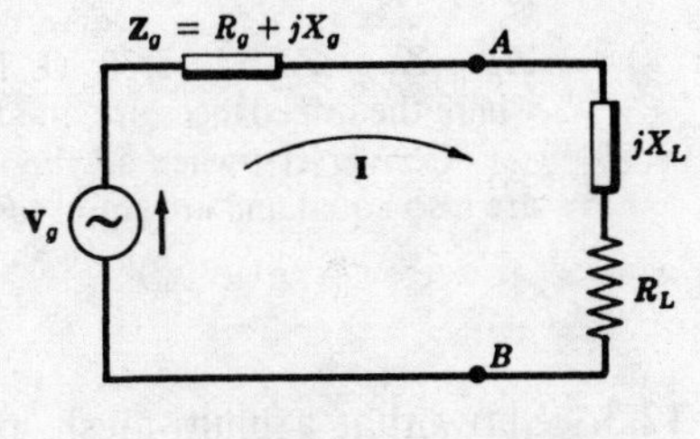

Fig. 12-10

$$R_L^2 = R_g^2 + (X_g + X_L)^2$$

and $$R_L = |\mathbf{Z}_g + jX_L|$$

Since $\mathbf{Z}_g$ and X_L are both fixed quantities, they could be combined into a single impedance. Then, with R_L variable, case 3 is reduced to case 1 and the maximum power results when R_L is equal to the absolute value of the network impedance.

Solved Problems

12.1. Determine the delta-connected equivalent circuit for the star-connected impedances shown in Fig. 12-11.

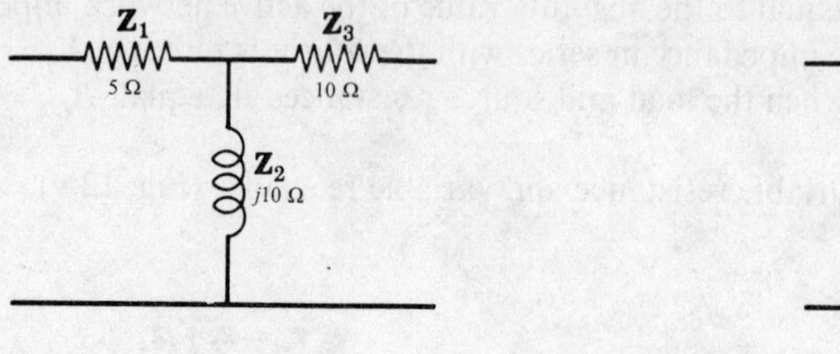

Fig. 12-11

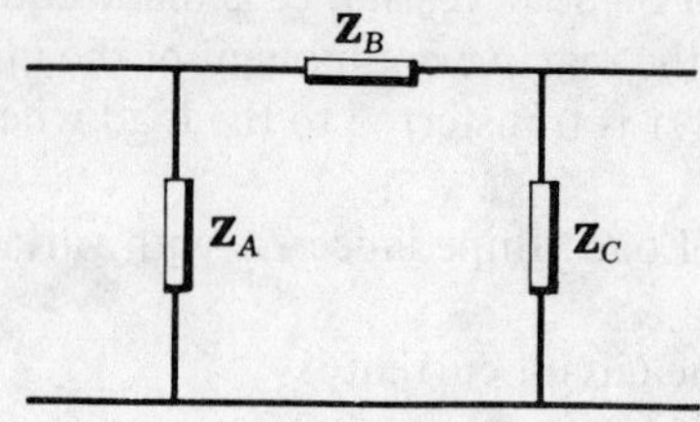

Fig. 12-12

The equivalent delta circuit contains $\mathbf{Z}_A$, $\mathbf{Z}_B$ and $\mathbf{Z}_C$ as shown in Fig. 12-12. Then

$$\mathbf{Z}_A = \frac{\mathbf{Z}_1\mathbf{Z}_2 + \mathbf{Z}_1\mathbf{Z}_3 + \mathbf{Z}_2\mathbf{Z}_3}{\mathbf{Z}_3} = \frac{5(j10) + 5(10) + 10(j10)}{10} = \frac{50 + j150}{10} = 5 + j15\ \Omega$$

$$\mathbf{Z}_B = \frac{\mathbf{Z}_1\mathbf{Z}_2 + \mathbf{Z}_1\mathbf{Z}_3 + \mathbf{Z}_2\mathbf{Z}_3}{\mathbf{Z}_2} = \frac{50 + j150}{j10} = 15 - j5\ \Omega$$

$$\mathbf{Z}_C = \frac{\mathbf{Z}_1\mathbf{Z}_2 + \mathbf{Z}_1\mathbf{Z}_3 + \mathbf{Z}_2\mathbf{Z}_3}{\mathbf{Z}_1} = \frac{50 + j150}{5} = 10 + j30\ \Omega$$

As a check convert the delta circuit impedances of Fig. 12-12 back to the star circuit. Thus

$$\mathbf{Z}_1 = \frac{\mathbf{Z}_A\mathbf{Z}_B}{\mathbf{Z}_A + \mathbf{Z}_B + \mathbf{Z}_C} = \frac{(5 + j15)(15 - j5)}{5 + j15 + 15 - j5 + 10 + j30} = \frac{150 + j200}{30 + j40} = 5\ \Omega$$

$$\mathbf{Z}_2 = \frac{\mathbf{Z}_A\mathbf{Z}_C}{\mathbf{Z}_A + \mathbf{Z}_B + \mathbf{Z}_C} = \frac{(5 + j15)(10 + j30)}{30 + j40} = j10\ \Omega$$

$$\mathbf{Z}_3 = \frac{\mathbf{Z}_B\mathbf{Z}_C}{\mathbf{Z}_A + \mathbf{Z}_B + \mathbf{Z}_C} = \frac{(15 - j5)(10 + j30)}{30 + j40} = 10\ \Omega$$

12.2. A delta connection contains three equal impedances $\mathbf{Z}\Delta = 15\underline{/30°}\ \Omega$. Find the impedances of the equivalent star connection.

$\mathbf{Z}_1 = \dfrac{\mathbf{Z}_A\mathbf{Z}_B}{\mathbf{Z}_A + \mathbf{Z}_B + \mathbf{Z}_C}$, where $\mathbf{Z}_A = \mathbf{Z}_B = \mathbf{Z}_C = \mathbf{Z}_\Delta$. Then $\mathbf{Z}_1 = \mathbf{Z}_\Delta/3 = (15\angle 30°)/3 = 5\angle 30°\ \Omega$. Similarly, $\mathbf{Z}_2 = \mathbf{Z}_3 = \mathbf{Z}_\Delta/3 = 5/30°\ \Omega$. Thus any delta circuit with three identical impedances has a star-connected equivalent where the impedances are one-third those of the delta circuit.

Conversely, when all the impedances in a star circuit are equal, the impedances in the equivalent delta circuit are also equal and are equal to three times the impedances of the star circuit.

12.3. Show that a multi-mesh, passive, three-terminal network may be replaced with a delta connection of three impedances.

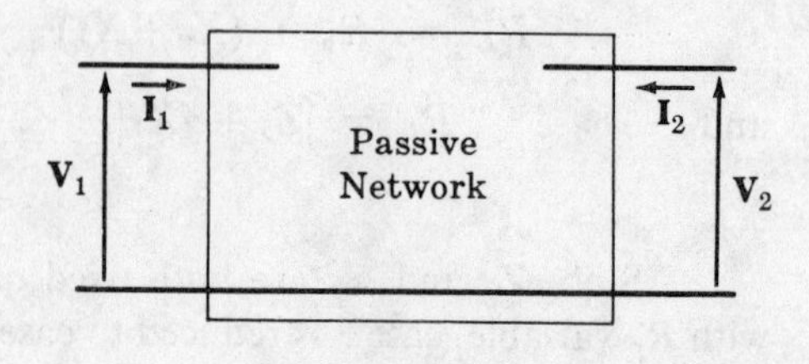

Fig. 12-13

Apply a voltage $\mathbf{V}_1$ to the left terminals as shown in Fig. 12-13 and label the current $\mathbf{I}_1$ which enters the network. Also, label $\mathbf{V}_2$ and $\mathbf{I}_2$ at the right terminals as shown. Since the network is passive, all other driving voltages are zero.

The mesh current equations in matrix form are

$$\begin{bmatrix} \mathbf{Z}_{11} & \mathbf{Z}_{12} & \cdots & \mathbf{Z}_{1n} \\ \mathbf{Z}_{21} & \mathbf{Z}_{22} & \cdots & \\ \cdots & \cdots & \cdots & \cdots \\ \cdots & \cdots & \cdots & \cdots \\ \mathbf{Z}_{n1} & \cdots & & \mathbf{Z}_{nn} \end{bmatrix} \begin{bmatrix} \mathbf{I}_1 \\ \mathbf{I}_2 \\ \cdot \\ \cdot \\ \mathbf{I}_n \end{bmatrix} = \begin{bmatrix} \mathbf{V}_1 \\ \mathbf{V}_2 \\ 0 \\ 0 \\ 0 \end{bmatrix}$$

from which $\mathbf{I}_1 = \mathbf{V}_1 \dfrac{\Delta_{11}}{\Delta_Z} + \mathbf{V}_2 \dfrac{\Delta_{21}}{\Delta_Z}$ and $\mathbf{I}_2 = \mathbf{V}_1 \dfrac{\Delta_{12}}{\Delta_Z} + \mathbf{V}_2 \dfrac{\Delta_{22}}{\Delta_Z}$

Now express these two simultaneous equations in matrix form.

$$\begin{bmatrix} \dfrac{\Delta_{11}}{\Delta_Z} & \dfrac{\Delta_{21}}{\Delta_Z} \\ \dfrac{\Delta_{12}}{\Delta_Z} & \dfrac{\Delta_{22}}{\Delta_Z} \end{bmatrix} \begin{bmatrix} \mathbf{V}_1 \\ \mathbf{V}_2 \end{bmatrix} = \begin{bmatrix} \mathbf{I}_1 \\ \mathbf{I}_2 \end{bmatrix}$$

This matrix equation is similar to that which results from a network of three nodes where one has been selected as the reference. Such a network is shown in Fig. 12-14 with $\mathbf{Z}_A$, $\mathbf{Z}_B$ and $\mathbf{Z}_C$ in a delta connection. Insert $\mathbf{V}_1$, $\mathbf{I}_1$, $\mathbf{V}_2$ and $\mathbf{I}_2$ with the same directions as in Fig. 12-13 and write the corresponding matrix equations applying the node voltage method of analysis.

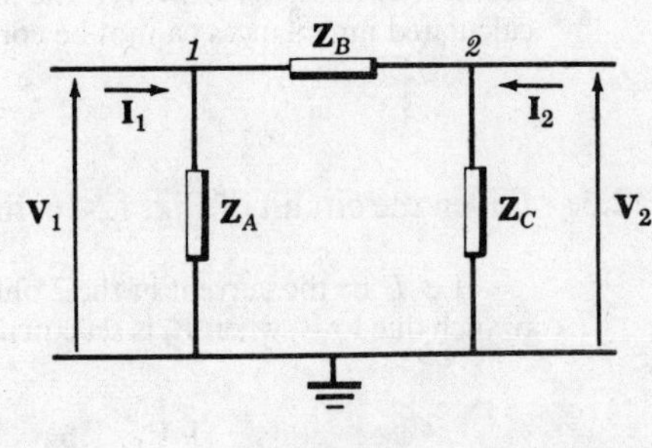

Fig. 12-14

$$\begin{bmatrix} \left(\dfrac{1}{\mathbf{Z}_A} + \dfrac{1}{\mathbf{Z}_B}\right) & -\dfrac{1}{\mathbf{Z}_B} \\ -\dfrac{1}{\mathbf{Z}_B} & \left(\dfrac{1}{\mathbf{Z}_B} + \dfrac{1}{\mathbf{Z}_C}\right) \end{bmatrix} \begin{bmatrix} \mathbf{V}_1 \\ \mathbf{V}_2 \end{bmatrix} = \begin{bmatrix} \mathbf{I}_1 \\ \mathbf{I}_2 \end{bmatrix}$$

Equate the corresponding elements of the two coefficient matrices. Then

$$(1)\quad \left(\frac{1}{\mathbf{Z}_A} + \frac{1}{\mathbf{Z}_B}\right) = \frac{\Delta_{11}}{\Delta_Z}, \qquad (2)\quad \left(\frac{1}{\mathbf{Z}_B} + \frac{1}{\mathbf{Z}_C}\right) = \frac{\Delta_{22}}{\Delta_Z}, \qquad (3)\quad -\frac{1}{\mathbf{Z}_B} = \frac{\Delta_{21}}{\Delta_Z}$$

Substituting *(3)* in *(1)* and *(2)*, we obtain

$$\mathbf{Z}_A = \frac{\Delta_Z}{\Delta_{11} + \Delta_{21}}, \qquad \mathbf{Z}_B = -\frac{\Delta_Z}{\Delta_{21}}, \qquad \mathbf{Z}_C = \frac{\Delta_Z}{\Delta_{22} + \Delta_{21}}$$

A transformation of a three-terminal network to an equivalent star or delta circuit is always possible mathematically as shown above. However, the elements of the equivalent circuit may not be physically realizable. See Problem 12-4.

12.4. Apply the results of Problem 12.3 to the network shown in Fig. 12-15 to obtain the delta connected equivalent circuit.

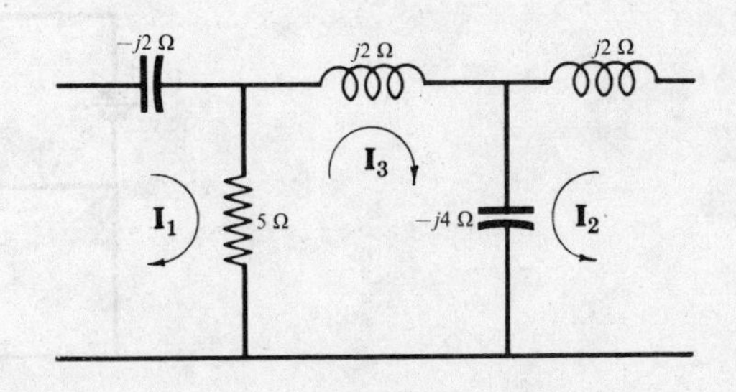

Fig. 12-15

Select the mesh currents as shown in the diagram. Then

$$\Delta_Z = \begin{vmatrix} 5 - j2 & 0 & -5 \\ 0 & -j2 & -j4 \\ -5 & -j4 & 5 - j2 \end{vmatrix} = 40 - j24 = 46{\cdot}6\underline{/-31^\circ}$$

and

$$\Delta_{11} = \begin{vmatrix} -j2 & -j4 \\ -j4 & 5-j2 \end{vmatrix} = 12 - j10, \qquad \Delta_{22} = \begin{vmatrix} 5-j2 & -5 \\ -5 & 5-j2 \end{vmatrix} = -4 - j20,$$

$$\Delta_{21} = (-)\begin{vmatrix} 0 & -5 \\ -j4 & 5-j2 \end{vmatrix} = j20$$

Using the expressions from Problem 12.3,

$$\mathbf{Z}_A = \frac{\Delta_Z}{\Delta_{11}+\Delta_{21}} = \frac{46{\cdot}6\angle{-31^\circ}}{12 - j10 + j20} = 2{\cdot}98\angle{-70{\cdot}8^\circ}\ \Omega$$

$$\mathbf{Z}_B = -\frac{\Delta_Z}{\Delta_{21}} = -\frac{46{\cdot}6\angle{-31^\circ}}{j20} = 2{\cdot}33\angle{59^\circ}\ \Omega$$

$$\mathbf{Z}_C = \frac{\Delta_Z}{\Delta_{22}+\Delta_{21}} = \frac{46{\cdot}6\angle{-31^\circ}}{-4 - j20 + j20} = 11{\cdot}65\angle{149^\circ}\ \Omega$$

Note that the impedance $\mathbf{Z}_A$ can be realised as a resistance and capacitance in series and $\mathbf{Z}_B$ as a resistance and inductance in series. However, the impedance $\mathbf{Z}_C$ would require a negative resistance. Hence a circuit with the three calculated impedances cannot be constructed.

12.5. **Given the circuit of Fig. 12-16, find the current in the 2 ohm resistor by using the superposition principle.**

Let I' be the current in the 2 ohm resistor due to V_1 when V_2 is set equal to zero, and I'' the current in the same branch due to V_2 when V_1 is set equal to zero. Then selecting mesh currents as shown in Fig. 12-16, solve for I' and I''.

$$I' = \frac{\begin{vmatrix} V_1 & 5 & 0 \\ V_1 & 12 & -4 \\ 0 & -4 & 6 \end{vmatrix}}{\begin{vmatrix} 7 & 5 & 0 \\ 5 & 12 & -4 \\ 0 & -4 & 6 \end{vmatrix}} = \frac{10\begin{vmatrix} 12 & -4 \\ -4 & 6 \end{vmatrix} - 10\begin{vmatrix} 5 & 0 \\ -4 & 6 \end{vmatrix}}{242} = 1{\cdot}075\ \text{A}$$

$$I'' = \frac{\begin{vmatrix} 0 & 5 & 0 \\ -V_2 & 12 & -4 \\ 0 & -4 & 6 \end{vmatrix}}{242} = \frac{-(-20)\begin{vmatrix} 5 & 0 \\ -4 & 6 \end{vmatrix}}{242} = 2{\cdot}48\ \text{A}$$

Applying the superposition theorem, the current I_1 due to the two sources acting simultaneously is

$$I_1 = I' + I'' = 1{\cdot}075 + 2{\cdot}48 = 3{\cdot}555\ \text{A}$$

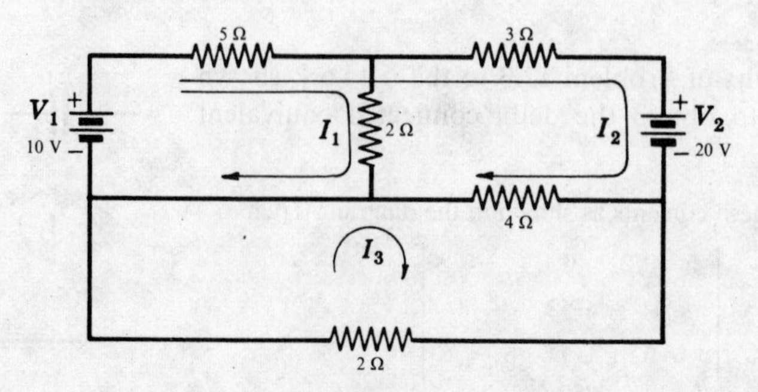

Fig. 12-16

12.6. Apply the superposition theorem to the network shown in Fig. 12-17 and obtain the current in the $3 + j4$ ohm impedance.

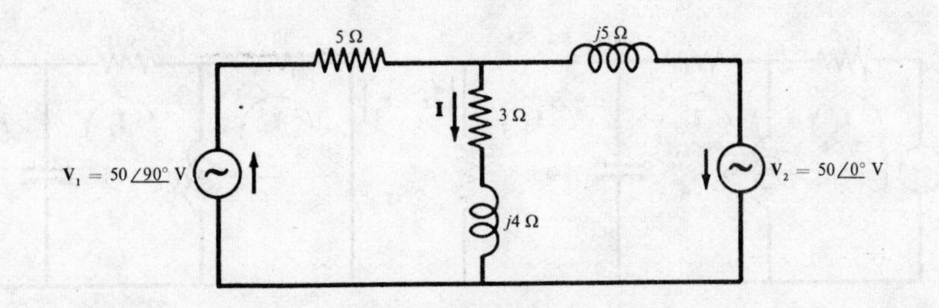

Fig. 12-17

Set $\mathbf{V}_2 = 0$ and let $\mathbf{V}_1$ be the only source present in the circuit. Then

$$\mathbf{Z}_{T_1} = 5 + \frac{(3+j4)j5}{3+j9} = 5{\cdot}83 + j2{\cdot}5 = 6{\cdot}35\underline{/23{\cdot}2^\circ}\ \Omega$$

and

$$\mathbf{I}_{T_1} = \frac{\mathbf{V}_1}{\mathbf{Z}_{T_1}} = \frac{50\underline{/90^\circ}}{6{\cdot}35\underline{/23{\cdot}2^\circ}} = 7{\cdot}87\underline{/66{\cdot}8^\circ}\ \text{A}$$

The current in the $3 + j4$ ohm branch due to $\mathbf{V}_1$ alone is

$$\mathbf{I}_1 = \mathbf{I}_{T_1}\left(\frac{j5}{3+j9}\right) = 7{\cdot}87\underline{/66{\cdot}8^\circ}\left(\frac{j5}{3+j9}\right) = 4{\cdot}15\underline{/85{\cdot}3^\circ}\ \text{A}$$

Now set $\mathbf{V}_1 = 0$ and let $\mathbf{V}_2$ be the only source in the circuit. Then

$$\mathbf{Z}_{T_2} = j5 + \frac{5(3+j4)}{8+j4} = 2{\cdot}5 + j6{\cdot}25 = 6{\cdot}74\underline{/68{\cdot}2^\circ}\ \Omega$$

and

$$\mathbf{I}_{T_2} = \frac{\mathbf{V}_2}{\mathbf{Z}_{T_2}} = \frac{50\underline{/0^\circ}}{6{\cdot}74\underline{/68{\cdot}2^\circ}} = 7{\cdot}42\underline{/-68{\cdot}2^\circ}\ \text{A}$$

The current in the $3 + j4$ ohm branch due to $\mathbf{V}_2$ alone is

$$\mathbf{I}_2 = -(7{\cdot}42\underline{/-68{\cdot}2^\circ})\left(\frac{5}{8+j4}\right) = 4{\cdot}15\underline{/85{\cdot}3^\circ}\ \text{A}$$

where the minus sign gives $\mathbf{I}_2$ the same direction as the branch current $\mathbf{I}$ shown in the diagram.

The total current in the $3 + j4$ ohm branch is

$$\mathbf{I} = \mathbf{I}_1 + \mathbf{I}_2 = 4{\cdot}15\underline{/85{\cdot}3^\circ} + 4{\cdot}15\underline{/85{\cdot}3^\circ} = 8{\cdot}30\underline{/85{\cdot}3^\circ}\ \text{A}$$

12.7. Apply the superposition theorem to the network of Fig. 12-18 to find the voltage V_{AB}.

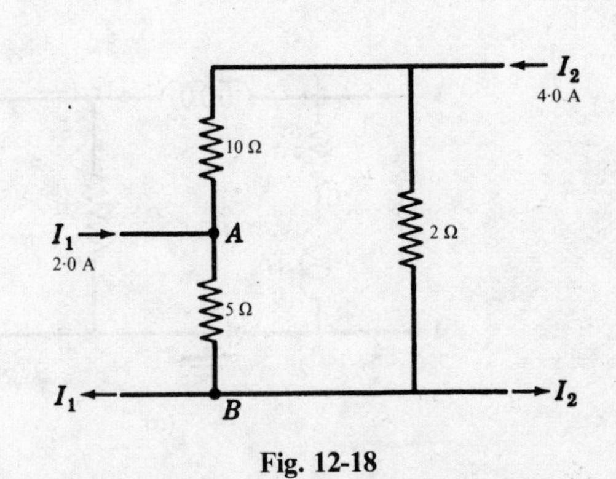

Fig. 12-18

Let the source $I_1 = 2$ A act on the network and set the source $I_2 = 0$. Then $V'_{AB} = 2\frac{5(12)}{17} = 7{\cdot}06$ V.

Now set $I_1 = 0$ and let $I_2 = 4$ A act on the network. The current through the 5 ohm resistor is $I_5 = 4(2/17) = 8/17$ A. Then the voltage $V''_{AB} = (8/17)5 = 2{\cdot}35$ volts.

The voltage V_{AB} with both sources acting is $V_{AB} = V'_{AB} + V''_{AB} = 7{\cdot}06 + 2{\cdot}35 = 9{\cdot}41$ V

12.8. In the single source network shown in Fig. 12-19(a) the voltage source $100\angle 45^\circ$ V causes a current $\mathbf{I}_x$ in the 5 ohm branch. Find $\mathbf{I}_x$ and then verify the reciprocity theorem for this circuit.

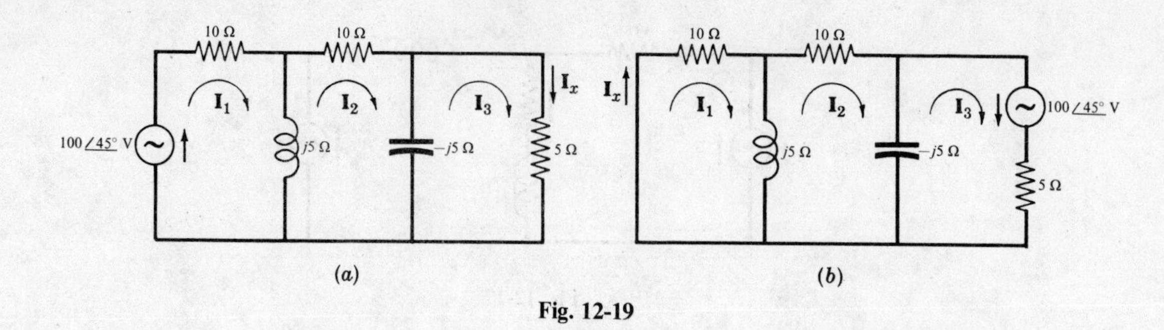

Fig. 12-19

Mesh currents $\mathbf{I}_1$, $\mathbf{I}_2$ and $\mathbf{I}_3$ are shown in Fig. 12-19(a). The required current $\mathbf{I}_x$ is mesh current $\mathbf{I}_3$.

$$\mathbf{I}_x = \mathbf{I}_3 = \frac{\begin{vmatrix} 10+j5 & -j5 & 100\angle 45^\circ \\ -j5 & 10 & 0 \\ 0 & j5 & 0 \end{vmatrix}}{\begin{vmatrix} 10+j5 & -j5 & 0 \\ -j5 & 10 & j5 \\ 0 & j5 & 5-j5 \end{vmatrix}} = 100\angle 45^\circ \left(\frac{25}{1155\angle -12{\cdot}5^\circ}\right) = 2{\cdot}16\angle 57{\cdot}5^\circ \text{ A} \qquad (1)$$

Now apply the reciprocity theorem by interchanging the positions of the excitation and the response as shown in Fig. 12-19(b). Again use clockwise primary loop currents as shown and note that $\mathbf{I}_x = \mathbf{I}_1$.

$$\mathbf{I}_x = \mathbf{I}_1 = \frac{\begin{vmatrix} 0 & -j5 & 0 \\ 0 & 10 & j5 \\ 100\angle 45^\circ & j5 & 5-j5 \end{vmatrix}}{\Delta_Z} = 100\angle 45^\circ \left(\frac{25}{1155\angle -12{\cdot}5^\circ}\right) = 2{\cdot}16\angle 57{\cdot}5^\circ \text{ A} \qquad (2)$$

Compare the results in (1) and (2); $\mathbf{I}_x$ is the same in both equations and the reciprocity theorem is thus verified.

12.9. The network of Fig. 12-20(a) contains a single current source $\mathbf{I} = 12\angle 90^\circ$ A. Determine the voltage $\mathbf{V}_2$ at node 2. Apply the reciprocity theorem and compare the results.

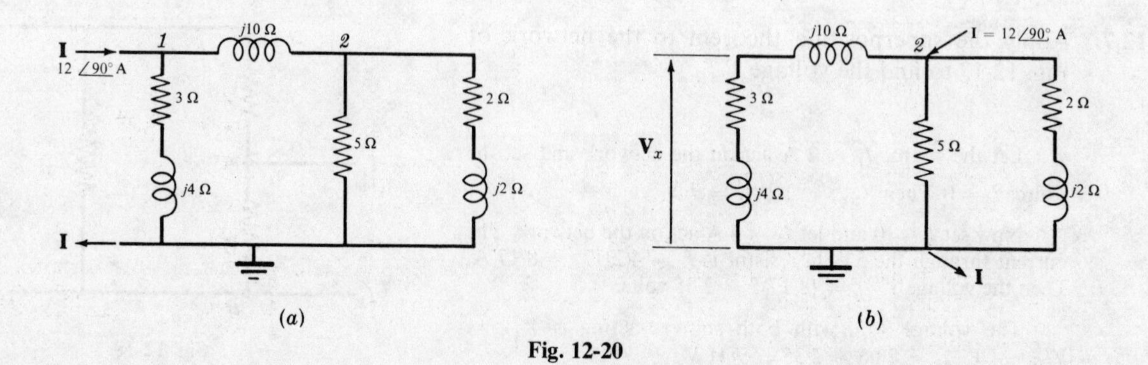

Fig. 12-20

The two nodal equations for the network in Fig. 12-20(a) are written in matrix form.

$$\begin{bmatrix} \left(\frac{1}{3+j4}+\frac{1}{j10}\right) & -\frac{1}{j10} \\ -\frac{1}{j10} & \left(\frac{1}{j10}+\frac{1}{5}+\frac{1}{2+j2}\right) \end{bmatrix} \begin{bmatrix} \mathbf{V}_1 \\ \mathbf{V}_2 \end{bmatrix} = \begin{bmatrix} 12\angle 90^\circ \\ 0 \end{bmatrix}$$

from which

$$\mathbf{V}_2 = \frac{\begin{vmatrix} 0{\cdot}12 - j0{\cdot}26 & 12\angle 90^\circ \\ j0{\cdot}1 & 0 \end{vmatrix}}{\begin{vmatrix} 0{\cdot}12 - j0{\cdot}26 & j0{\cdot}1 \\ j0{\cdot}1 & 0{\cdot}45 - j0{\cdot}35 \end{vmatrix}} = 12\angle 90^\circ \left(\frac{-j0{\cdot}1}{0{\cdot}161\angle 260{\cdot}35^\circ}\right) = 7{\cdot}45\angle 99{\cdot}65^\circ \text{ V}$$

Using the reciprocity theorem, apply the current **I** between node 2 and the reference node in the circuit of Fig. 12-20(b). Then calculate the voltage at the terminals where the current source was driving before. Since there are only two nodes in this network, only one nodal equation is required.

$$\left(\frac{1}{3+j14}+\frac{1}{5}+\frac{1}{2+j2}\right)\mathbf{V}_2 = 12\angle 90^\circ \text{ from which } \mathbf{V}_2 = \frac{12\angle 90^\circ}{0{\cdot}563\angle -34{\cdot}4^\circ} = 21{\cdot}3\angle 124{\cdot}4^\circ \text{ V}$$

Then the voltage $\mathbf{V}_x$ is

$$\mathbf{V}_x = \mathbf{V}_2\left(\frac{3+j4}{3+j4+j10}\right) = 21{\cdot}3\angle 124{\cdot}4^\circ \left(\frac{3+j4}{3+j14}\right) = 7{\cdot}45\angle 99{\cdot}6^\circ \text{ V}$$

Comparing the calculated values of $\mathbf{V}_2$ in the network of Fig. 12-20(a) and $\mathbf{V}_x$ in the network of Fig. 12-20(b) we find that they are equal, thus proving the reciprocity theorem. Note also that $\mathbf{V}_2$ does not remain the same after after the interchange of the positions of excitation and response.

12.10. In the single current source circuit shown in Fig. 12-21(a), find the voltage $\mathbf{V}_x$. Interchange the current source and the resulting voltage $\mathbf{V}_x$. Is the reciprocity theorem verified?

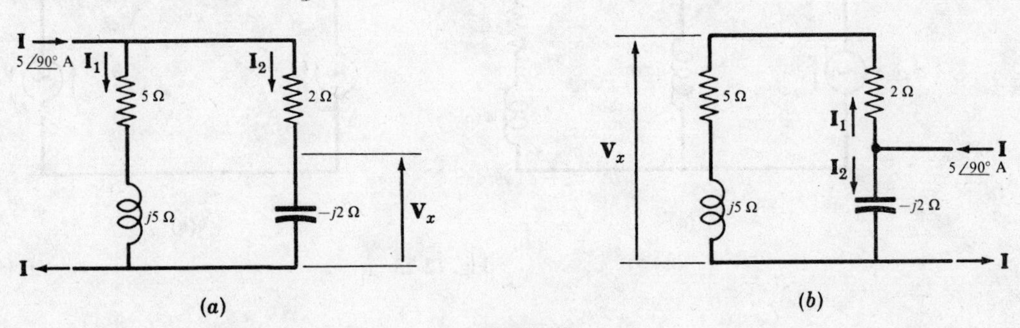

Fig. 12-21

In Fig. 12-21(a) the current $\mathbf{I}_2 = \mathbf{I}\left(\frac{5+j5}{7+j3}\right) = 5\angle 90^\circ\left(\frac{5+j5}{7+j3}\right) = 4{\cdot}64\angle 111{\cdot}8^\circ$ A. Then the voltage $\mathbf{V}_x = \mathbf{I}_2(-j2) = 4{\cdot}64\angle 111{\cdot}8^\circ\,(2\angle -90^\circ) = 9{\cdot}28\angle 21{\cdot}8^\circ$ V.

The current source **I** and the terminals at which the voltage $\mathbf{V}_x$ is measured are interchanged as shown in Fig. 12-21(b). Then the current $\mathbf{I}_1 = \mathbf{I}\left(\frac{-j2}{7+j3}\right) = 5\angle 90^\circ\left(\frac{-j2}{7+j3}\right) = 1{\cdot}31\angle -23{\cdot}2^\circ$ A. Since $\mathbf{V}_x = 1{\cdot}31\angle -23{\cdot}2^\circ\,(5+j5) = 9{\cdot}27\angle 21{\cdot}8$ V as before, the reciprocity theorem is verified.

12.11. In the network of Fig. 12-22(*a*) replace the $j4\ \Omega$ reactance with a compensation emf.

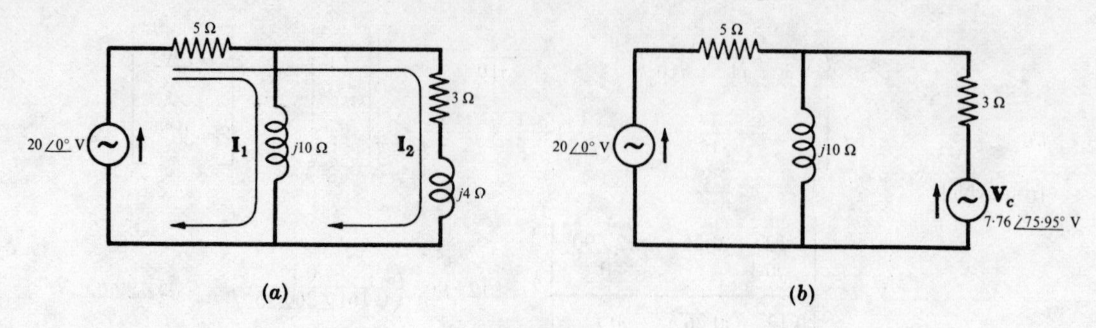

Fig. 12-22

Select mesh currents $\mathbf{I}_1$ and $\mathbf{I}_2$ as shown in the diagram and solve for $\mathbf{I}_2$, the current in the $j4\ \Omega$ reactance.

$$\mathbf{I}_2 = \frac{\begin{vmatrix} 5+j10 & 20 \\ 5 & 20 \end{vmatrix}}{\begin{vmatrix} 5+j10 & 5 \\ 5 & 8+j4 \end{vmatrix}} = \frac{20(j10)}{103\angle 104{\cdot}05^\circ} = 1{\cdot}94\angle -14{\cdot}05^\circ \text{ A}$$

The compensation source $\mathbf{V}_c = \mathbf{I}_2(j4) \angle -14{\cdot}05^\circ\ (j4) = 7{\cdot}76\angle 75{\cdot}95^\circ$ V. The circuit is shown in Fig. 12-22(*b*) with the compensation source in place of the $j4\ \Omega$ reactance. To show that the two circuits are equivalent, find one branch current in each circuit and compare results.

12.12. In the network shown in Fig. 12-23(*a*) replace the parallel combination of the $j10\ \Omega$ and $3 + j4\ \Omega$ impedances with a compensation source.

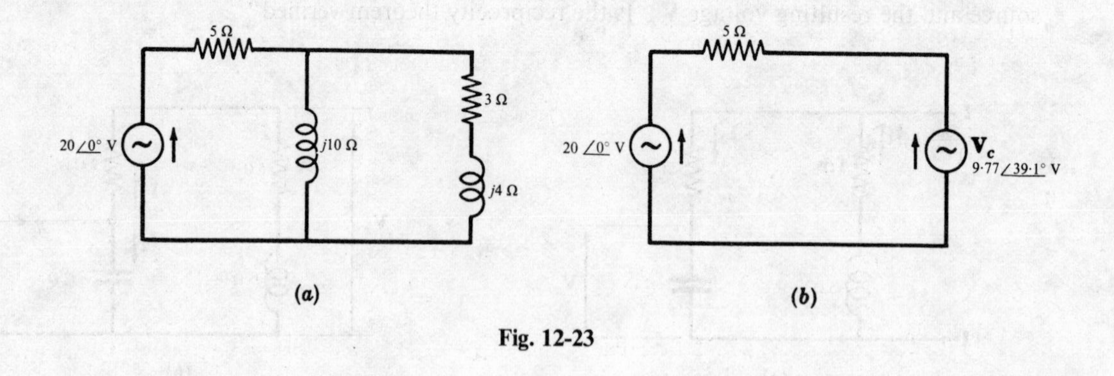

Fig. 12-23

The equivalent impedance of the parallel impedances is $\mathbf{Z}_{eq} = \dfrac{j10(3+j4)}{3+j14} = 1{\cdot}46 + j3{\cdot}17 = 3{\cdot}50\angle 65{\cdot}3^\circ\ \Omega$.

Then

$$\mathbf{Z}_T = 5 + 1{\cdot}46 + j3{\cdot}17 = 7{\cdot}18\angle 26{\cdot}2^\circ\ \Omega \text{ and } \mathbf{I}_T = \frac{\mathbf{V}}{\mathbf{Z}_T} = \frac{20\angle 0^\circ}{7{\cdot}18\angle 26{\cdot}2^\circ} = 2{\cdot}79\angle -26{\cdot}2^\circ \text{ A}$$

The compensation source is

$$\mathbf{V}_c = \mathbf{I}_T\mathbf{Z}_{eq} = 2{\cdot}79\angle -26{\cdot}2^\circ\ (3{\cdot}50\angle 65{\cdot}3^\circ) = 9{\cdot}77\angle 39{\cdot}1^\circ \text{ V}$$

The circuit with the properly polarised compensation voltage source is shown in Fig. 12-23(*b*).

12.13. In the network of Fig. 12-24(a), the $3 + j4$ ohm impedance is changed to $4 + j4$ ohms, Fig. 12-24(b). Find the current in the 10 ohm resistor before and after the change. Then apply the compensation theorem to determine the difference in the two currents in the 10 ohm resistor.

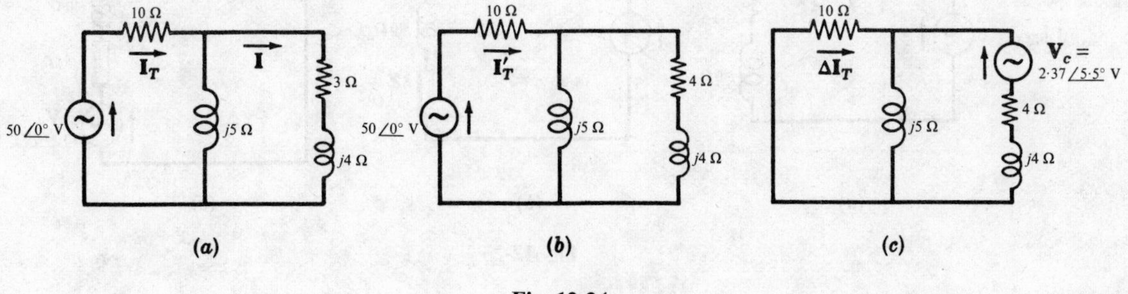

Fig. 12-24

Before the change in the $3 + j4\ \Omega$ impedance [Fig. 12-24(a)]

$$\mathbf{Z}_T = 10 + \frac{j5(3+j4)}{3+j9} = 11{\cdot}1\angle 13^\circ\ \Omega \text{ and } \mathbf{I}_T = \frac{\mathbf{V}}{\mathbf{Z}_T} = \frac{50\angle 0^\circ}{11{\cdot}1\angle 13^\circ} = 4{\cdot}50\angle{-13^\circ}\ \text{A}$$

After the change [Fig. 12-24(b)],

$$\mathbf{Z}'_T = 10 + \frac{j5(4+j4)}{4+j9} = 11{\cdot}03 + j2{\cdot}68 = 11{\cdot}35\angle 13{\cdot}65^\circ\ \Omega \text{ and } \mathbf{I}'_T = \frac{\mathbf{V}}{\mathbf{Z}'_T} = 4{\cdot}41\angle{-13{\cdot}65^\circ}\ \text{A}$$

The compensation voltage source $\mathbf{V}_c = \mathbf{I}(\delta\mathbf{Z})$ where $\mathbf{I}$, the initial current in the $3 + j4\ \Omega$ branch, is

$$\mathbf{I} = \mathbf{I}_T\left(\frac{j5}{3+j9}\right) = 4{\cdot}5\angle{-13^\circ}\left(\frac{j5}{3+j9}\right) = 2{\cdot}37\angle 5{\cdot}5^\circ\ \text{A}$$

and $\delta\mathbf{Z} = (4 + j4) - (3 + j4) = 1\ \Omega$. Then $\mathbf{V}_c = 2{\cdot}37\angle 5{\cdot}5^\circ(1) = 2{\cdot}37\angle 5{\cdot}5^\circ\ \Omega$ with direction opposite to the current $\mathbf{I}$. The change in current $\Delta\mathbf{I}_T$ is found by setting the initial voltage source equal to zero and leaving $\mathbf{V}_c$ as the only source acting in the circuit as shown in Fig. 12-24(c). Then for this circuit, $\mathbf{Z}_T'' = 4 + j4 + \dfrac{j5(10)}{10+j5} = 10\angle 53{\cdot}1^\circ\ \Omega$ and

$$\Delta\mathbf{I}_T = -\left(\frac{\mathbf{V}_c}{\mathbf{Z}_T}\right)\left(\frac{j5}{10+j5}\right) = -\left(\frac{2{\cdot}37\angle 5{\cdot}5^\circ}{10\angle 53{\cdot}1^\circ}\right)\left(\frac{j5}{10+j5}\right) = 0{\cdot}1055\angle 195{\cdot}8^\circ\ \text{A}$$

Comparing ΔI_T with the difference between I'_T and I_T,

$$\mathbf{I}' - \mathbf{I}_T = (4{\cdot}41\angle{-13{\cdot}65^\circ}) - (4{\cdot}50\angle{-13^\circ}) = -0{\cdot}10 - j0{\cdot}03 = 0{\cdot}1045\angle 196{\cdot}7^\circ\ \text{A}$$

Note that the two values of $\Delta\mathbf{I}_T$ are not exactly the same. The value of $\Delta\mathbf{I}_T$ computed using the compensation voltage $\mathbf{V}_c$ is more accurate than the value of $\Delta\mathbf{I}_T$ obtained by subtracting the initial current $\mathbf{I}_T$ from $\mathbf{I}'_T$. This is particularly true when the impedance change is small. As seen above, this results in a small current change, thus introducing an error when computing the difference of two quantities which are very nearly equal.

12.14. Calculate the change in current in the series circuit of Fig. 12-25(a) below when the value of the reactance is reduced to $j35\ \Omega$.

Let $\mathbf{I}$ and $\mathbf{I}'$ be the respective currents in the circuit before and after the reactance change, as shown in Fig. 12-25(a) and (b) below. Then

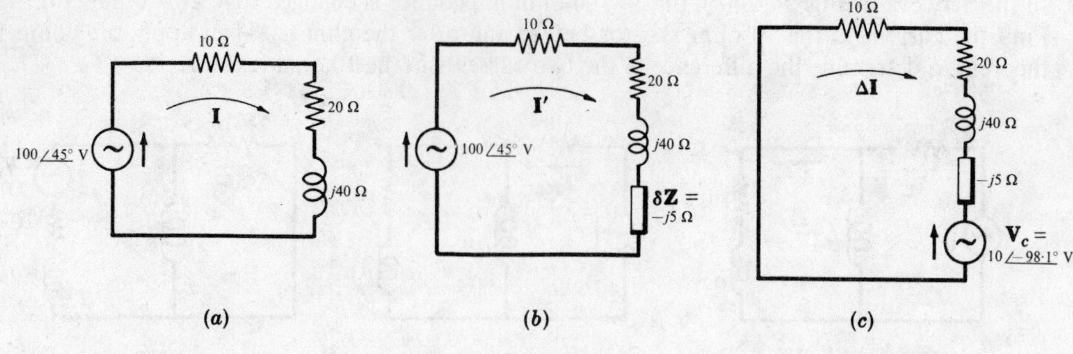

Fig. 12-25

$$\mathbf{I} = \frac{\mathbf{V}}{\mathbf{Z}} = \frac{100\angle 45^\circ}{50\angle 53{\cdot}1^\circ} = 2{\cdot}0\angle{-8{\cdot}1^\circ}\ \text{A};\ \mathbf{I}' = \frac{\mathbf{V}}{\mathbf{Z}+\delta\mathbf{Z}} = \frac{100\angle 45^\circ}{30+j35} = 2{\cdot}17\angle{-4{\cdot}4^\circ}\ \text{A}$$

and
$$\Delta\mathbf{I} = \mathbf{I}' - \mathbf{I} = 2{\cdot}17\angle{-4{\cdot}4^\circ} - 2{\cdot}0\angle{-8{\cdot}1^\circ} = 0{\cdot}223\angle 31{\cdot}6^\circ\ \text{A}$$

If we calculate ΔI by applying the compensation theorem, we have $\mathbf{V}_c = \mathbf{I}(\delta\mathbf{Z}) = 2{\cdot}0\angle{-8{\cdot}1^\circ}(-j5) = 10\angle{-98{\cdot}1^\circ}$ V directed as shown in Fig. 12-25(*c*). The current change is

$$\Delta\mathbf{I} = -\mathbf{V}_c/(\mathbf{Z}+\delta\mathbf{Z}) = -(10\angle{-98{\cdot}1^\circ})/(30+j35) = (10\angle 81{\cdot}9^\circ)/(46{\cdot}1\angle 49{\cdot}4^\circ) = 0{\cdot}217\angle 32{\cdot}5^\circ\ \text{A}$$

12.15. In the circuit of Fig. 12-26 the load $\mathbf{Z}_L$ consists of a pure resistance R_L. Find the value of R_L for which the source delivers maximum power to the load. Determine the value of the maximum power P.

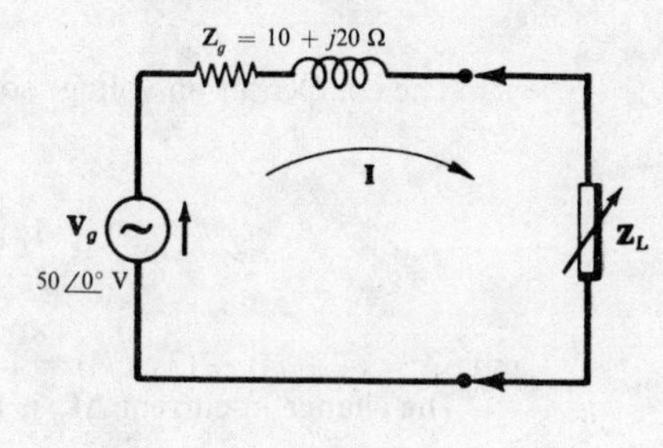

Fig. 12-26

Maximum power transfer occurs when

$$R_L = |\mathbf{Z}_g| = |10 + j20| = 22{\cdot}4 \text{ ohms}$$

Now $\mathbf{I} = \mathbf{V}/(\mathbf{Z}_g + R) = (50\angle 0^\circ)/(10 + j20 + 22{\cdot}4) = 1{\cdot}31\angle{-31{\cdot}7^\circ}$ A and the maximum power delivered to the load is $P = I^2R_L = (1{\cdot}31)^2 22{\cdot}4 = 38{\cdot}5$ watts.

12.16. If the load in the circuit of Fig. 12-26 consists of a complex impedance $\mathbf{Z}_L$ which is variable in both R_L and X_L, determine the value of $\mathbf{Z}_L$ which results in maximum power transfer. Calculate the value of the maximum power.

Maximum power transfer occurs when $\mathbf{Z}_L = \mathbf{Z}_g^*$. Since $\mathbf{Z}_g = 10 + j20\ \Omega$, $\mathbf{Z}_L = 10 - j20\ \Omega$. The total impedance of the circuit is $\mathbf{Z}_T = (10 + j20) + (10 - j20) = 20\ \Omega$. Then $\mathbf{I} = \mathbf{V}/\mathbf{Z}_T = (50\angle 0^\circ)/20 = 2{\cdot}5\angle 0^\circ$ A and $P = I^2R_L = (2{\cdot}5)^2 10 = 62{\cdot}5$ watts.

12.17. In the network shown in Fig. 12-27 the load connected across terminals *AB* consists of a variable resistance R_L and a capacitive reactance X_C which is variable between 2 and 8 ohms. Determine the values of R_L and X_C which result in maximum power transfer. Calculate the maximum power P delivered to the load.

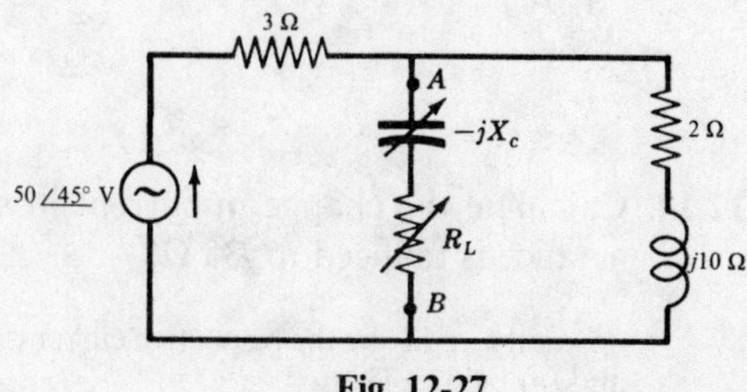

Fig. 12-27

The Thevenin equivalent voltage at terminals AB is $\mathbf{V}' = \frac{50\angle 45^\circ}{5 + j10}(2 + j10) = 45{\cdot}6\angle 60{\cdot}3^\circ$ V. The impedance of the active network connected to terminals AB is $\mathbf{Z}' = 3(2 + j10)/(5 + j10) = 2{\cdot}64 + j0{\cdot}72\ \Omega$.

In the given circuit the maximum power transfer occurs with an impedance $\mathbf{Z}_L = \mathbf{Z}'^* = 2{\cdot}64 - j0{\cdot}72\ \Omega$. Under the conditions of the problem, X_C is adjustable between 2 and 8 ohms. Hence the closest value of X_C is 2 ohms and

$$R_L = |\mathbf{Z}_g - jX_C| = |2{\cdot}64 + j0{\cdot}72 - j2| = |2{\cdot}64 - j1{\cdot}28| = 2{\cdot}93 \text{ ohms}$$

Now
$$\mathbf{Z}_T = \mathbf{Z}' + \mathbf{Z}_L = (2{\cdot}64 + 2{\cdot}93) + j(0{\cdot}72 - 2) = 5{\cdot}57 - j1{\cdot}28 = 5{\cdot}70\angle -13^\circ\ \Omega$$

$$\mathbf{I} = \frac{\mathbf{V}'}{\mathbf{Z}_T} = \frac{45{\cdot}6\angle 60{\cdot}3^\circ}{5{\cdot}70\angle -13^\circ} = 8{\cdot}0\angle 73{\cdot}3^\circ \text{ A and } P = I^2R_L = (8{\cdot}0)^2 2{\cdot}93 = 187{\cdot}5 \text{ W}$$

12.18. In the circuit shown in Fig. 12-28, the resistance R_g is variable between 2 and 55 ohms. What value of R_g results in maximum power transfer across the terminals AB?

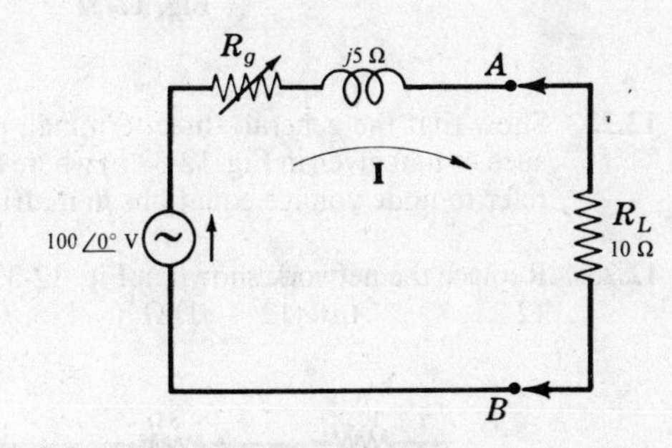

Fig. 12-28

In the given circuit the load resistance R_L is fixed. Thus the *maximum power transfer theorems do not apply*. Obviously the greatest current results when R_g is a minimum.

Set R_g equal to 2 ohms. Then

$$\mathbf{Z}_T = (2 + j5 + 10) = 13\angle 22{\cdot}6^\circ\ \Omega$$

and

$$\mathbf{I} = \mathbf{V}/\mathbf{Z}_T = 100\angle 0^\circ/(13\angle 22{\cdot}6^\circ) = 7{\cdot}7\angle -22{\cdot}6^\circ \text{ A}$$

The maximum power $P = (7{\cdot}7)^2 10 = 593$ W.

Supplementary Problems

12.19. Obtain the star-connected equivalent of the delta-connected circuit shown in Fig. 12-29.
Ans. $(0{\cdot}5 - j0{\cdot}5)\ \Omega$, $(3 - j1)\ \Omega$, $(1 + j3)\ \Omega$

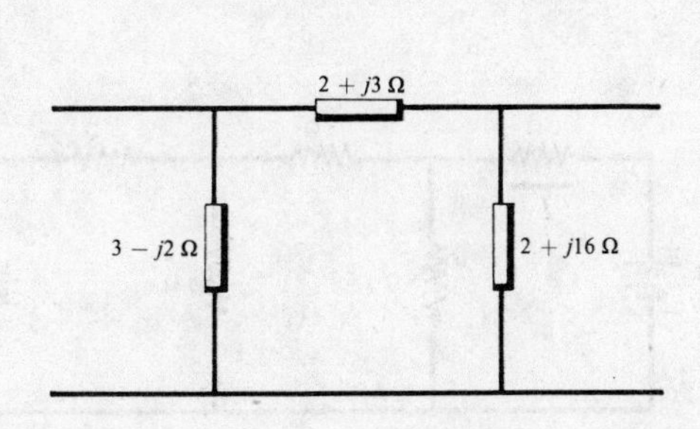

Fig. 12-29

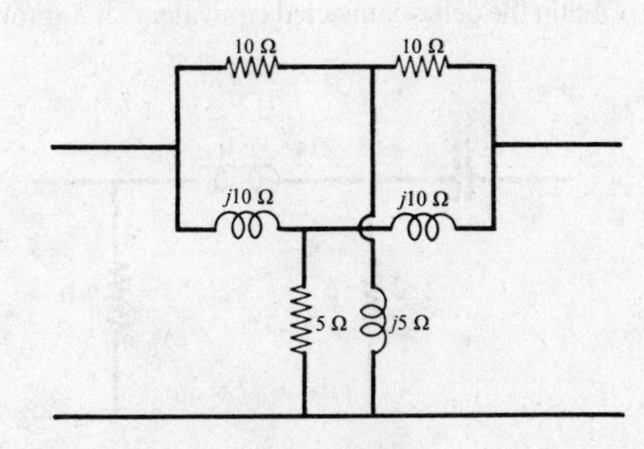

Fig. 12-30

12.20. The network shown in Fig. 12-30 consists of two star-connected circuits in parallel. Obtain the single delta-connected equivalent. *Ans.* $(5 + j5)\ \Omega$, ∞, $(5 + j5)\ \Omega$

12.21. In Fig. 12-31 a balanced delta-connected circuit with $\mathbf{Z} = 10\angle 30^\circ\ \Omega$ is in parallel with a balanced star-connected circuit with $\mathbf{Z} = 4\angle -45^\circ\ \Omega$. Obtain the star-connected equivalent. *Ans.* $\mathbf{Z} = 2{\cdot}29\angle -3{\cdot}5^\circ\ \Omega$

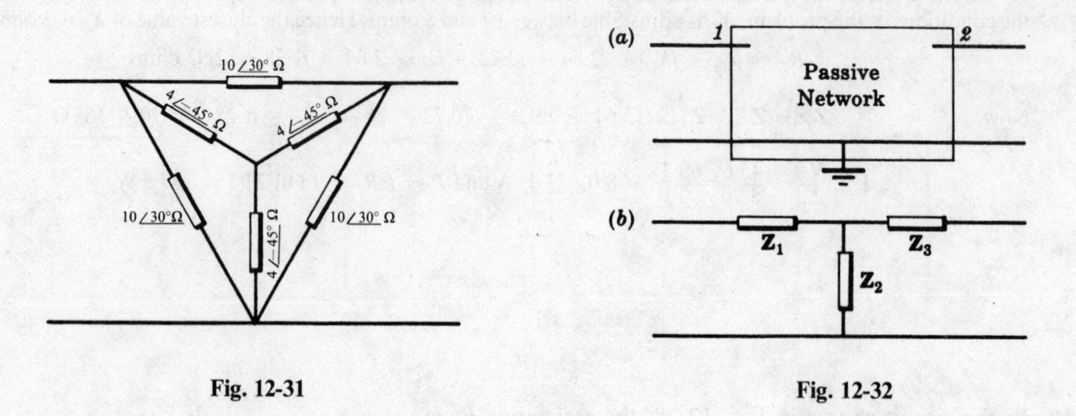

Fig. 12-31

Fig. 12-32

12.22. Show that the general, three terminal, passive network of Fig. 12-32(*a*) can be replaced by a star-connected circuit such as that given in Fig. 12-32(*b*) where $\mathbf{Z}_1 = (\Delta_{11} - \Delta_{12})/\Delta_Y$, $\mathbf{Z}_2 = \Delta_{12}/\Delta_Y$ and $\mathbf{Z}_3 = (\Delta_{22} - \Delta_{12})/\Delta_Y$. ($\Delta_Y$ and cofactors refer to node voltage equations in matrix form.)

12.23. Replace the network shown in Fig. 12-33 with an equivalent star connection using the methods developed in Problem 12.22. *Ans.* $(12 + j1)\ \Omega$, $(-1 + j2)\ \Omega$, $(4 + j1)\ \Omega$

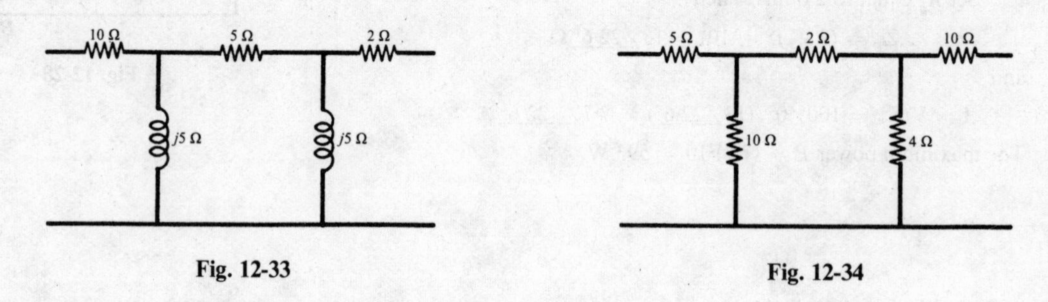

Fig. 12-33

Fig. 12-34

12.24. Obtain the star connection of three impedances equivalent to the network shown in Fig. 12-34.
Ans. $6{\cdot}25\ \Omega$, $2{\cdot}5\ \Omega$, $10{\cdot}5\ \Omega$

12.25. Referring to the network of Fig. 12-34 obtain the delta-connected equivalent circuit. *Ans.* $10{\cdot}25\ \Omega$, $43\ \Omega$, $17{\cdot}2\ \Omega$

12.26. Obtain the delta-connected equivalent circuit for the network of Fig. 12-35. *Ans.* $(3 - j2)\ \Omega$, $(2 + j3)\ \Omega$, $(2 + j16)\ \Omega$

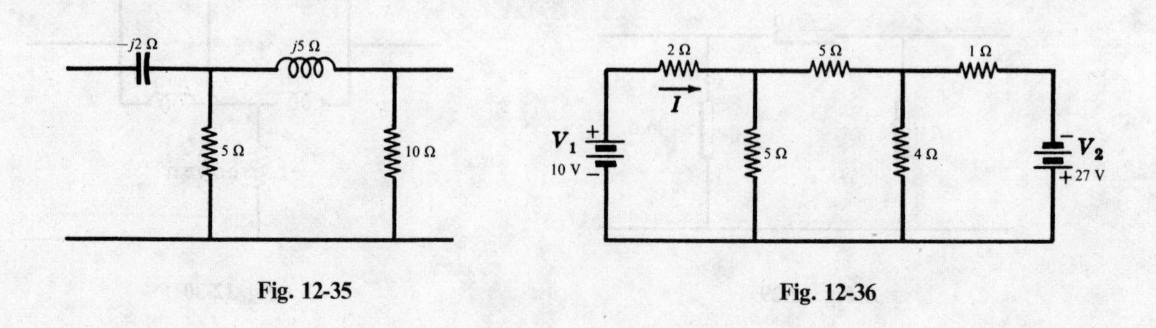

Fig. 12-35

Fig. 12-36

12.27. Find the current in the 2 ohm resistor in the circuit of Fig. 12-36, using the superposition theorem. *Ans.* $I = 4{\cdot}27$ A

12.28. Referring to the network of Fig. 12-36 above, the voltage source V_2 is changed to 8·93 volts, positive at the upper terminal. Obtain the current in the 2 ohm resistor by using the superposition theorem. *Ans.* $I = 1{\cdot}43$ A

12.29. In the network shown in Fig. 12-37 obtain the current in the 5 ohm resistor due to each of the voltage sources.
Ans. 2·27 A, 3·41 A

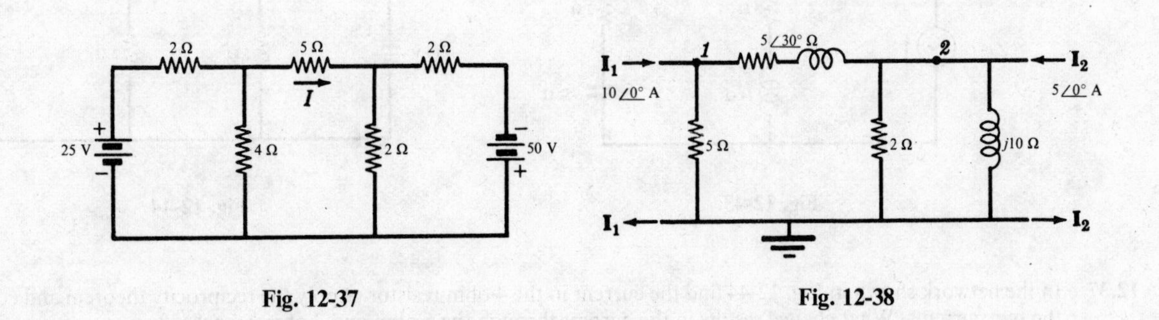

Fig. 12-37 Fig. 12-38

12.30. In the network of Fig. 12-38 determine the components of node voltage $\mathbf{V}_2$ due to each current source.
Ans. $8{\cdot}48\angle{-2{\cdot}8^\circ}$ V, $8{\cdot}20\angle{12{\cdot}2^\circ}$ V

12.31. In the network shown in Fig. 12-39 find the current in the 4 ohm resistor due to each of the voltage sources.
Ans. $3{\cdot}24\angle{60{\cdot}95^\circ}$ A, $6{\cdot}16\angle{-142{\cdot}2^\circ}$ A

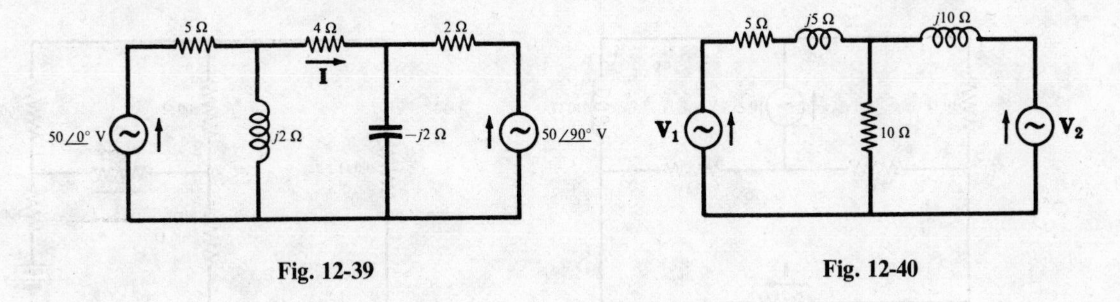

Fig. 12-39 Fig. 12-40

12.32. In the network shown in Fig. 12-40 let the voltage sources act separately on the circuit. If the corresponding currents in the 10 ohm resistor are equal, what is the value of the ratio $\mathbf{V}_1/\mathbf{V}_2$? *Ans.* $0{\cdot}707\angle{45^\circ}$

12.33. In the network shown in Fig. 12-41 obtain the components of node voltage $\mathbf{V}_2$ due to each current source $\mathbf{I}_1$ and $\mathbf{I}_2$.
Ans. $5{\cdot}82\angle{-5{\cdot}5^\circ}$ V, $9{\cdot}22\angle{72{\cdot}9^\circ}$ V

12.34. Referring to the network of Fig. 12-41, current source $\mathbf{I}_2$ is changed to $3{\cdot}16\angle{191{\cdot}6^\circ}$ A. Determine the node voltage $\mathbf{V}_2$ by means of the superposition theorem.

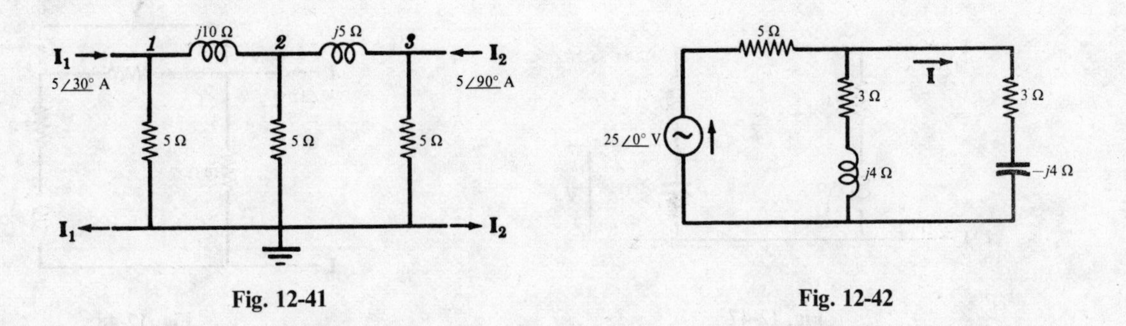

Fig. 12-41 Fig. 12-42

12.35. In the circuit shown in Fig. 12-42 find the current **I** in the $3 - j4$ ohm impedance. Apply the reciprocity theorem and compare the two currents. *Ans.* $2{\cdot}27\angle{53{\cdot}2^\circ}$ A

12.36. In the circuit shown in Fig. 12-43 find the current **I** in the $2 - j2$ ohm impedance. Apply the reciprocity theorem and compare the two currents. *Ans.* $10{\cdot}1\angle 129{\cdot}1^\circ$ A

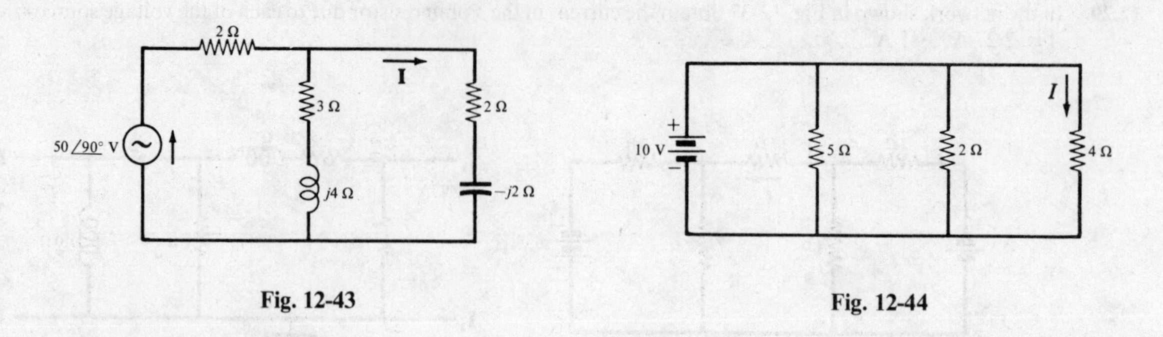

Fig. 12-43

Fig. 12-44

12.37. In the network shown in Fig. 12-44 find the current in the 4 ohm resistor. Apply the reciprocity theorem and compare the two currents. What change results in the current through the 5 ohm and 2 ohm branches?
Ans. 2·5 A. After the reciprocity theorem is applied, the 5 ohm and 2 ohm branch currents are zero. They previously had currents of 2 and 5 A respectively.

12.38. In the network shown in Fig. 12-45 determine the current in the 5 ohm resistor. Apply the reciprocity theorem and compare the two currents. *Ans.* $0{\cdot}270\angle 53{\cdot}75^\circ$ A

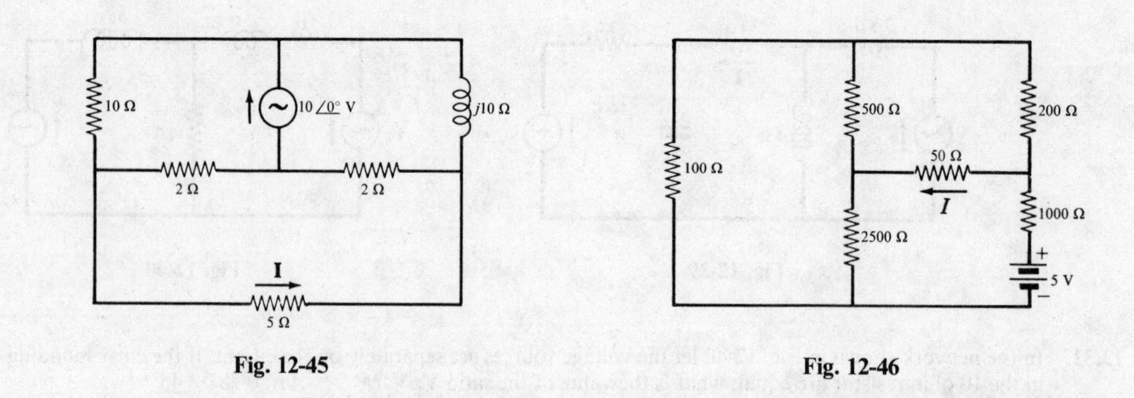

Fig. 12-45

Fig. 12-46

12.39. In the network of Fig. 12-46 calculate the current I in the 50 ohm resistor. Verify the reciprocity theorem by interchanging the voltage source and the resulting current I. *Ans.* 1·32 mA

12.40. In the circuit shown in Fig. 12-47 determine the voltage $\mathbf{V}_x$. Then apply the reciprocity theorem and compare the two voltages. *Ans.* $35\angle -12{\cdot}1^\circ$ V

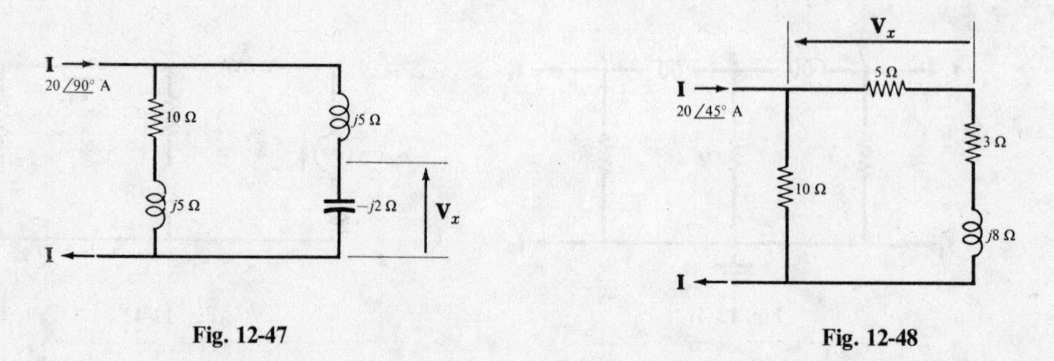

Fig. 12-47

Fig. 12-48

12.41. Find $\mathbf{V}_x$ in the circuit of Fig. 12-48. Then verify the reciprocity theorem. *Ans.* $50{\cdot}8\angle 21^\circ$ V

12.42. In the network shown in Fig. 12-49 find the voltage $\mathbf{V}_x$. Interchange the position of the current source and the voltage $\mathbf{V}_x$ and verify the reciprocity theorem. *Ans.* $2{\cdot}53\,\underline{/-162{\cdot}3^\circ}$ V

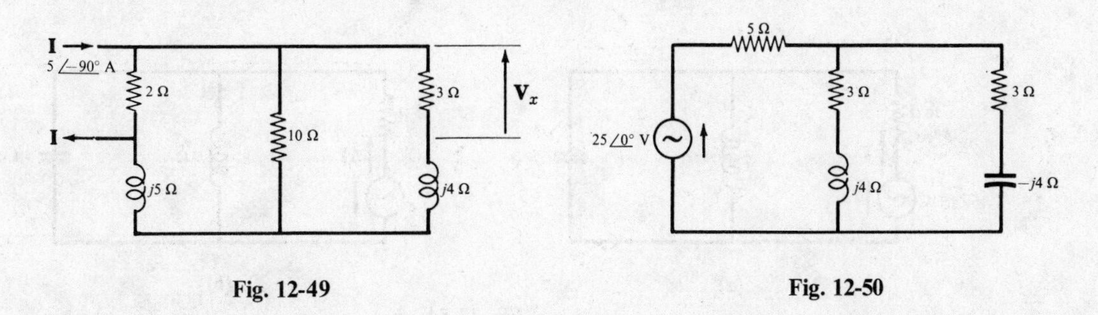

Fig. 12-49

Fig. 12-50

12.43. In the network of Fig. 12-50 replace the parallel impedances $3 + j4\,\Omega$ and $3 - j4\,\Omega$ with a compensation voltage source. As a check find the current in the 5 ohm resistor before and after the substitution.
Ans. $\mathbf{V}_c = 11{\cdot}35\,\underline{/0^\circ}$ V, $\mathbf{I} = 2{\cdot}73\,\underline{/0^\circ}$ A

12.44. Referring to the network of Fig. 12-50 replace the 5 ohm resistor with a compensation voltage source and find the total current from the $25\,\underline{/0^\circ}$ V source before and after the substitution. *Ans.* $\mathbf{V}_C = 13{\cdot}65\,\underline{/0^\circ}$ V, $\mathbf{I} = 2{\cdot}73\,\underline{/0^\circ}$ A

12.45. In the network shown in Fig. 12-51 replace each of the parallel combinations of resistors with a compensation voltage source and calculate the total current output of the 50 volt source. *Ans.* 11·35 V, 4·55 V, 3·41 A

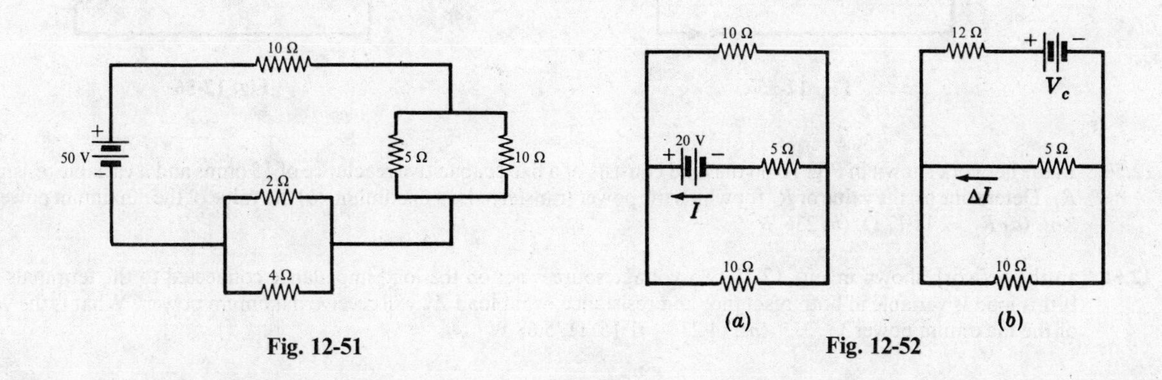

Fig. 12-51

Fig. 12-52

12.46. In the network of Fig. 12-52(*a*) the 20 volt source contains a current I as shown. When the upper 10 ohm resistor is changed to 12 ohms the current output of the source changes to I'. Find the change in current $\Delta I = (I' - I)$ by use of the compensation voltage source as shown in Fig. 12-52(*b*). *Ans.* $\Delta I = -0{\cdot}087$ A

12.47. In the network shown in Fig. 12-53(*a*) the 5 ohm resistor is changed to 8 ohms. Determine the resulting change in current $\Delta\mathbf{I}$ through the $3 + j4$ ohm impedance. *Ans.* $0{\cdot}271\,\underline{/159{\cdot}5^\circ}$ A

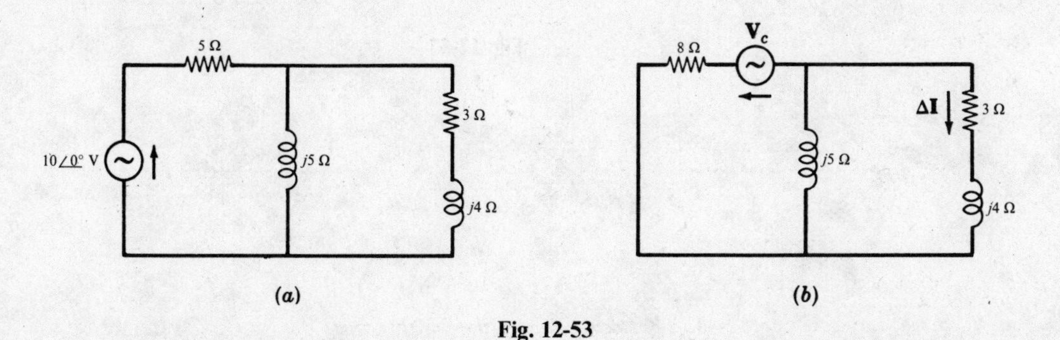

Fig. 12-53

12.48. In the network shown in Fig. 12-54(a) the $50\angle 45°$ V source contains a current **I**. The 10 ohm resistor is changed to 5 ohms. Use the compensation theorem to find $\mathbf{V}_c$ and $\Delta\mathbf{I}$ shown in Fig. 12-54(b). *Ans.* $21{\cdot}45\angle -166°$ V, $2{\cdot}74\angle -36°$ A

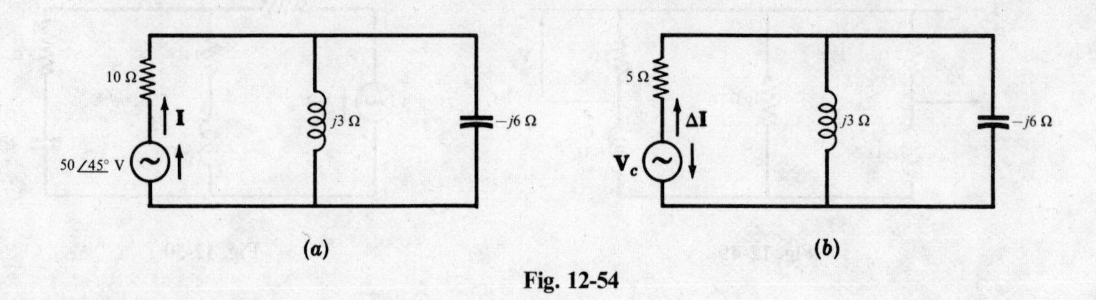

Fig. 12-54

12.49. In the circuit shown in Fig. 12-55 find the value of R_L which results in maximum power transfer. Calculate the value of the maximum power. *Ans.* 11·17 ohms, 309 W

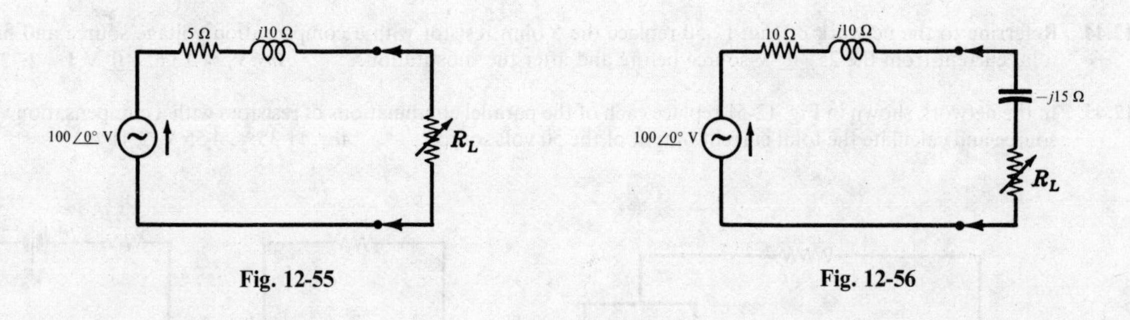

Fig. 12-55 Fig. 12-56

12.50. In the network shown in Fig. 12-56 the load consists of a fixed capacitive reactance of 15 ohms and a variable resistance R_L. Determine (a) the value of R_L for which the power transferred is a maximum, (b) the value of the maximum power. *Ans.* (a) $R_L = 11{\cdot}17\ \Omega$, ($b$) 236 W

12.51. In the network shown in Fig. 12-57 two voltage sources act on the load impedance connected to the terminals AB. If this load is variable in both reactance and resistance, what load $\mathbf{Z}_L$ will receive maximum power? What is the value of the maximum power? *Ans.* $(4{\cdot}23 + j1{\cdot}15)\ \Omega$, 5·68 W

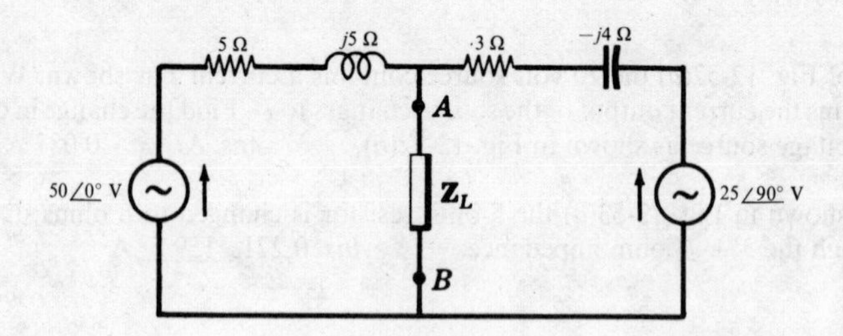

Fig. 12-57

CHAPTER 13

Mutual Inductance

INTRODUCTION

The networks studied in the previous chapters consisted of loops or meshes and nodes. Since two loops have a common branch and two nodes are joined by either passive or active elements, meshes and nodes are said to be conductively coupled. Methods were developed for the solution of these networks.

In this chapter we analyze another type of coupling, namely, the magnetic coupling. When the interaction between two loops takes place through a magnetic field instead of through common elements, the loops are said to be inductively or magnetically coupled.

SELF-INDUCTANCE

When a current is changing in a circuit, the magnetic flux linking the same circuit changes and an emf is induced in the circuit. Assuming constant permeability, the induced emf is proportional to the rate of change of the current, i.e.,

$$v_L = L\frac{di}{dt} \tag{1}$$

where the constant of proportionality L is called the self-inductance of the circuit. The unit of self-inductance is the henry(H).

In a coil of N turns, the induced emf is given by

$$v_L = N\frac{d\varphi}{dt} \tag{2}$$

where $N\,d\varphi$ is the flux linkage of the circuit. Combining equations (*1*) and (*2*) we have

$$L\frac{di}{dt} = N\frac{d}{dt}$$

from which

$$L = N\frac{d\varphi}{di}$$

MUTUAL INDUCTANCE

Consider a time-varying current i_1 in coil 1 of Fig. 13-1. The changing current i_1 establishes a magnetic flux φ_1. Part of this flux links only coil 1 and is called leakage flux φ_{11}. The remaining flux φ_{12} links also coil 2 as shown. The induced voltage in coil 2 is given by Faraday's law,

$$v_2 = N_2\frac{d\varphi_{12}}{dt} \tag{3}$$

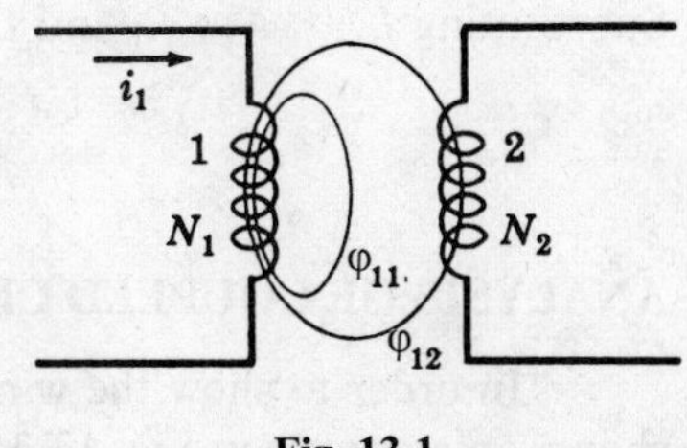

Fig. 13-1

Since φ_{12} is related to the current i_1, v_2 is proportional to the rate of change of i_1, or

$$v_2 = M\frac{di_1}{dt} \tag{4}$$

where the constant of proportionality M is called the mutual inductance between the two coils. The unit of mutual inductance is the same as the unit of self-inductance (the henry).

Combining equations (3) and (4), we have

$$v_2 = N_2\frac{d\varphi_{12}}{dt} = M\frac{di_1}{dt}$$

and

$$M = N_2\frac{d\varphi_{12}}{di_1} \tag{5}$$

With a set of coils wound on the same iron core, the flux and current are not linearly related and the mutual inductance is given by equation (5). If the coils are linked with air as the medium, the flux and current are linearly related and the mutual inductance is

$$M = \frac{N_2\varphi_{12}}{i_1} \tag{6}$$

Mutual coupling is bilateral and analogous results are obtained if a time-varying current i_2 is introduced in coil 2 of Fig. 13-1. Then the linking fluxes are φ_2, φ_{21} and φ_{22}, the induced voltage in coil 1 is $v_1 = M(di_2/dt)$ and equations (5) and (6) become respectively

$$(7)\quad M = \frac{N_1\,d\varphi_{21}}{di_2} \qquad\text{and}\qquad (8)\quad M = \frac{N_1\varphi_{21}}{i_2} \tag{7}$$

COUPLING COEFFICIENT k

In Fig. 13-1 the linkage flux depends on spacing and orientation of the axes of the coils and on the permeability of the medium. The fraction of total flux which links the coils is called the coefficient of coupling k. Then

$$k = \frac{\varphi_{12}}{\varphi_1} = \frac{\varphi_{21}}{\varphi_2} \tag{8}$$

Since $\varphi_{12} \leqslant \varphi_1$ and $\varphi_{21} \leqslant \varphi_2$, the maximum value of k is unity.

An expression for M in terms of self-inductances L_1 and L_2 is obtained as follows. Multiply equation (6) by (8) and obtain

$$M^2 = \left(\frac{N_2\varphi_{12}}{i_1}\right)\left(\frac{N_1\varphi_{21}}{i_2}\right) = \left(\frac{N_2k\varphi_1}{i_1}\right)\left(\frac{N_1k\varphi_2}{i_2}\right) = k^2\left(\frac{N_1\varphi_1}{i_1}\right)\left(\frac{N_2\varphi_2}{i_2}\right) \tag{9}$$

Substituting $L_1 = N_1\varphi_1/i_1$ and $L_2 = N_2\varphi_2/i_2$ in (9),

$$M^2 = k^2L_1L_2 \text{ and } M = k\sqrt{L_1L_2}$$

ANALYSIS OF COUPLED CIRCUITS

In order to show the winding sense and its effects on the voltages of mutual inductance, the coils are shown on a core as in Fig. 13-2 on page 179.

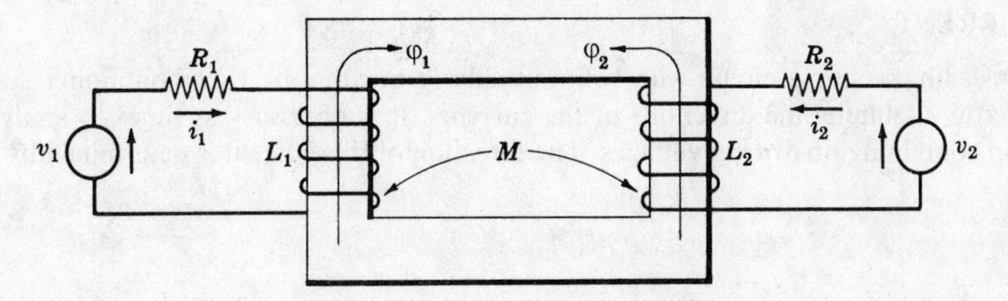

Fig. 13-2 .

Since each circuit contains a voltage source, select mesh currents i_1 and i_2 in the same direction as the sources and write the two mesh equations according to Kirchhoff's voltage law.

$$\begin{aligned} R_1 i_1 + L_1 \frac{di_1}{dt} \pm M \frac{di_2}{dt} &= v_1 \\ R_2 i_2 + L_2 \frac{di_2}{dt} \pm M \frac{di_1}{dt} &= v_2 \end{aligned} \tag{10}$$

The voltages of mutual inductance may be of either polarity depending on the winding sense. To determine the correct signs in (*10*) apply the right hand rule to each coil, allowing the fingers to wrap around in the direction of the assumed current. Then the right thumb points in the direction of the flux. Thus the positive directions of φ_1 and φ_2 are as shown in the figure. *If fluxes φ_1 and φ_2 due to the assumed positive current directions aid one another, then the signs on the voltages of mutual inductance are the same as the signs on the voltages of self-inductance.* Referring to Fig. 13-2 we note that φ_1 and φ_2 oppose each other. Then rewriting (*10*) with the correct signs, we have

$$\begin{aligned} R_1 i_1 + L_1 \frac{di_1}{dt} - M \frac{di_2}{dt} &= v_1 \\ R_2 i_2 + L_2 \frac{di_2}{dt} - M \frac{di_1}{dt} &= v_2 \end{aligned} \tag{11}$$

Assuming sinusoidal voltage sources, the set (*11*) in the sinusoidal steady state becomes

$$\begin{aligned} (R_1 + j\omega L_1)\mathbf{I}_1 - j\omega M\mathbf{I}_2 &= \mathbf{V}_1 \\ -j\omega M\mathbf{I}_1 + (R_2 + j\omega L_2)\mathbf{I}_2 &= \mathbf{V}_2 \end{aligned} \tag{12}$$

Recalling the general set of two simultaneous mesh current equations (Chapter 9), we have

$$\begin{aligned} \mathbf{Z}_{11}\mathbf{I}_1 \pm \mathbf{Z}_{12}\mathbf{I}_2 = \mathbf{V}_1 \\ \pm\mathbf{Z}_{21}\mathbf{I}_1 + \mathbf{Z}_{22}\mathbf{I}_2 = \mathbf{V}_2 \end{aligned} \tag{13}$$

We found that $\mathbf{Z}_{12} = \mathbf{Z}_{21}$ were the impedances common to the two mesh currents $\mathbf{I}_1$ and $\mathbf{I}_2$. The meshes were conductively coupled since the currents passed through a common branch. Now in the circuit of Fig. 13-2 we have a similar set of equations where $j\omega M$ corresponds to $\mathbf{Z}_{12}$ and $\mathbf{Z}_{21}$ of equations (*13*). The meshes are not conductively coupled since the two currents do not have any common impedances. However, the equations indicate that coupling does exist. In such cases the coupling is called *mutual* or *magnetic coupling*.

NATURAL CURRENT

In the preceding section a circuit with two mutually coupled loops, each containing a voltage source, were examined after assuming the directions of the currents. It is necessary at times to analyze the natural current in a loop containing no driving voltages. The direction of this current is determined by application of Lenz's law.

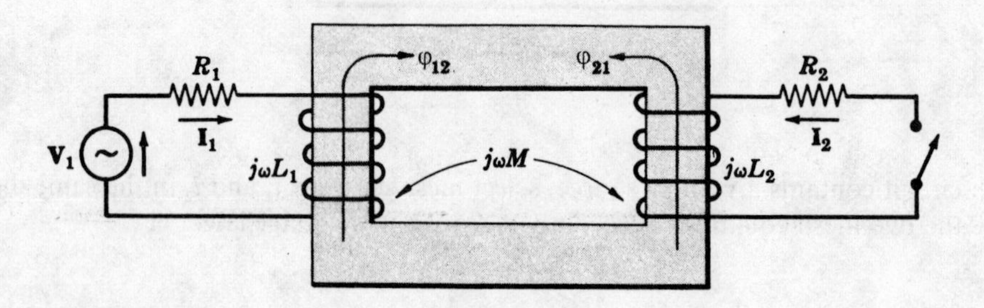

Fig. 13-3

Consider the circuit shown in Fig. 13-3 where only mesh 1 contains a driving voltage. Select current $\mathbf{I}_1$ in agreement with the source $\mathbf{V}_1$ and apply the right hand rule to determine the direction of the flux φ_{12}. Now Lenz's law states that the polarity of the induced voltage is such that if the circuit is completed, a current will pass through the coil in a direction which creates a flux opposing the main flux set up by current $\mathbf{I}_1$. Therefore when the switch is closed in the circuit of Fig. 13-3, the direction of flux φ_{21} according to Lenz's law is as shown. Now apply the right hand rule with the thumb pointing in the direction of φ_{21}; the fingers will wrap around coil 2 in the direction of the *natural current*. Then the mesh current equations are

$$\begin{aligned} (R_1 + j\omega L_1)\mathbf{I}_1 \; - \; j\omega M\mathbf{I}_2 \; &= \; \mathbf{V}_1 \\ -j\omega M\mathbf{I}_1 \; + \; (R_2 + j\omega L_2)\mathbf{I}_2 \; &= \; 0 \end{aligned} \tag{14}$$

Since mesh 2 has no driving voltage, it follows that the natural current $\mathbf{I}_2$ resulted from the voltage of mutual inductance, $(R_2 + j\omega L_2)\,\mathbf{I}_2 = (j\omega M\mathbf{I}_1)$. In Fig. 13-4 this voltage is shown as a source. The direction of the source must be as indicated by the arrow for the positive direction of $\mathbf{I}_2$. Hence, *the instantaneous polarity of the voltage of mutual inductance at coil two is positive at the terminal where the natural current leaves the winding.*

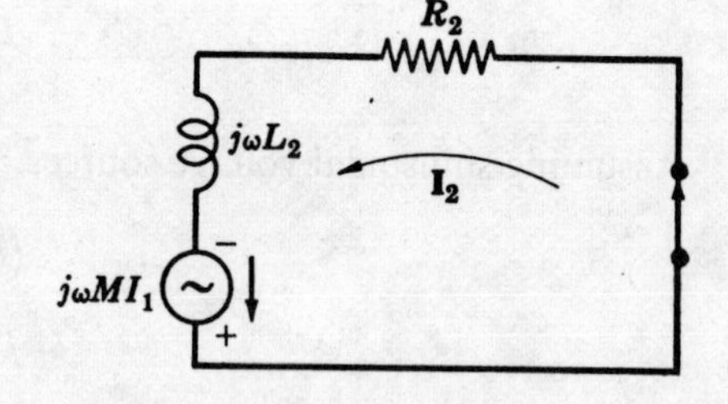

Fig. 13-4

DOT RULE FOR COUPLED COILS

While the relative polarity for voltages of mutual inductance can be determined from sketches of the core which show the winding sense, the method is not practical. To simplify the diagrammatic representation of coupled circuits, the coils are marked with dots as shown in Fig. 13-5(*c*). On each coil, a dot is placed at the terminals which are instantaneously of the same polarity on the basis of the mutual inductance alone. Then to apply the dot notation we must know at which terminal of the coils the dots are assigned. Moreover, we must determine the sign associated with the voltage of mutual inductance when we write the mesh current equations.

To assign the dots on a pair of coupled coils, select a current direction in one coil of the pair and place a dot at the terminal where this current enters the winding. The dotted terminal is instantaneously positive with respect to the other terminal of the coil. Apply the right hand rule to find the corresponding flux as shown in Fig. 13-5(*a*). Now in the second coil the flux must oppose the original flux, according to Lenz's law. See Fig. 13-5(*b*).

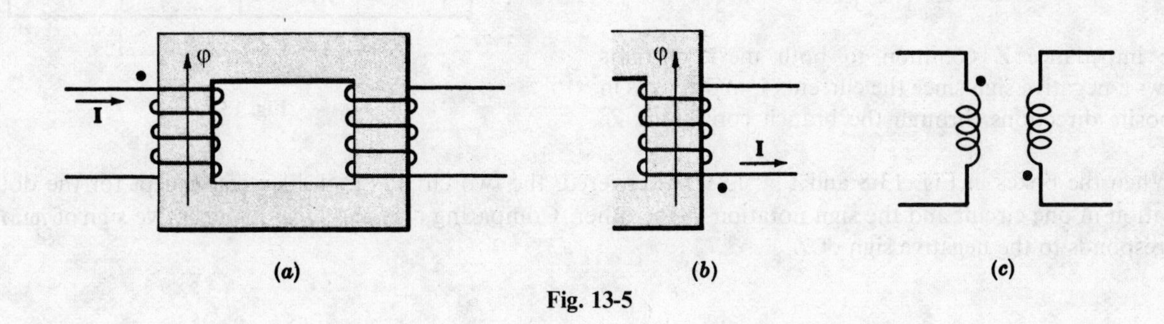

Fig. 13-5

Use the right hand rule to find the direction of the natural current, and since the voltage of mutual inductance is positive at the terminal where this natural current leaves the winding, place a dot at this terminal as shown in Fig. 13-5(*b*). With the instantaneous polarity of the coils given by the dots, the core is no longer needed in the diagram and the coupled coils may be illustrated as in Fig. 13-5(*c*).

To determine the sign of the voltage of mutual inductance in the mesh current equations, we use the dot rule which states: (1) *When both assumed currents enter or leave a pair of coupled coils at the dotted terminals, the signs of the M terms will be the same as the signs of the L terms*; (2) *If one current enters at a dotted terminal and one leaves by a dotted terminal, the signs of the M terms are opposite to the signs of the L terms.*

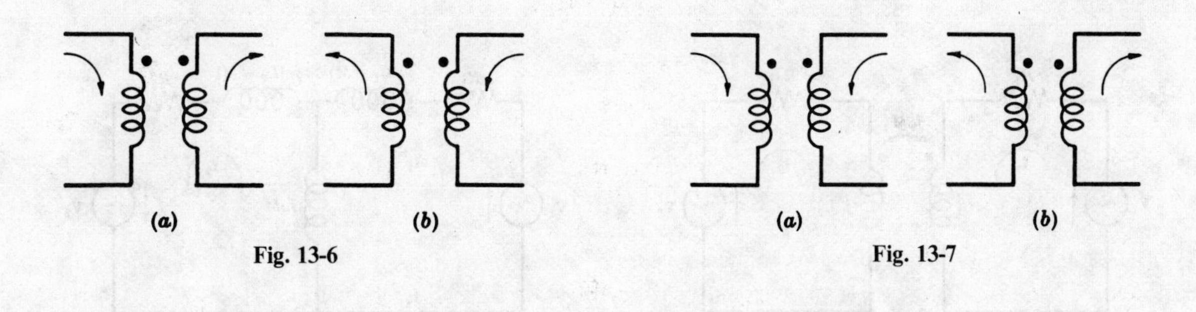

Fig. 13-6 Fig. 13-7

Fig. 13-6 above shows when the signs of the M and L terms are opposite. Fig. 13-7 above shows two cases in which the signs of M and L are the same.

As a further illustration of the relative polarities in connection with mutual coupled circuits consider the circuit of Fig. 13-8 where the dots are marked and the currents $\mathbf{I}_1$ and $\mathbf{I}_2$ selected as shown. Since one current enters at a dotted terminal and the other leaves by a dotted terminal, the sign on the M terms is opposite to the sign on the L terms. For this circuit the mesh current equations in matrix form are

$$\begin{bmatrix} \mathbf{Z}_{11} & -j\omega M \\ -j\omega M & \mathbf{Z}_{22} \end{bmatrix} \begin{bmatrix} \mathbf{I}_1 \\ \mathbf{I}_2 \end{bmatrix} = \begin{bmatrix} \mathbf{V}_1 \\ 0 \end{bmatrix} \quad (15)$$

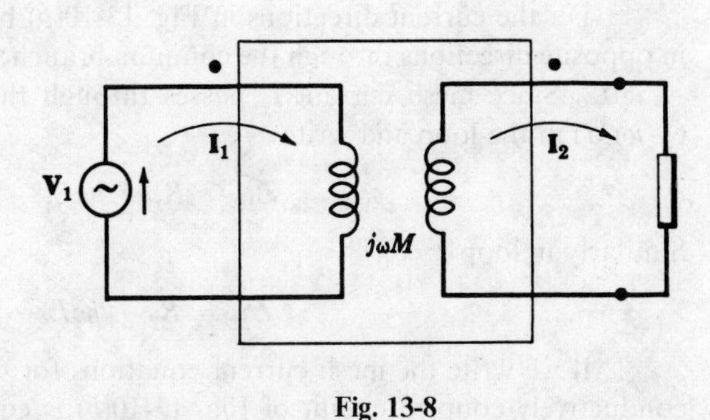

Fig. 13-8

Now a simple conductive coupled two mesh network is given in Fig. 13-9 and the positive terminals are indicated. The mesh current equations in matrix form are

$$\begin{bmatrix} \mathbf{Z}_{11} & -\mathbf{Z} \\ -\mathbf{Z} & \mathbf{Z}_{22} \end{bmatrix} \begin{bmatrix} \mathbf{I}_1 \\ \mathbf{I}_2 \end{bmatrix} = \begin{bmatrix} \mathbf{V}_1 \\ 0 \end{bmatrix} \quad (16)$$

The impedance $\mathbf{Z}$ common to both mesh currents shows a negative sign since the currents $\mathbf{I}_1$ and $\mathbf{I}_2$ pass in opposite directions through the branch containing $\mathbf{Z}$.

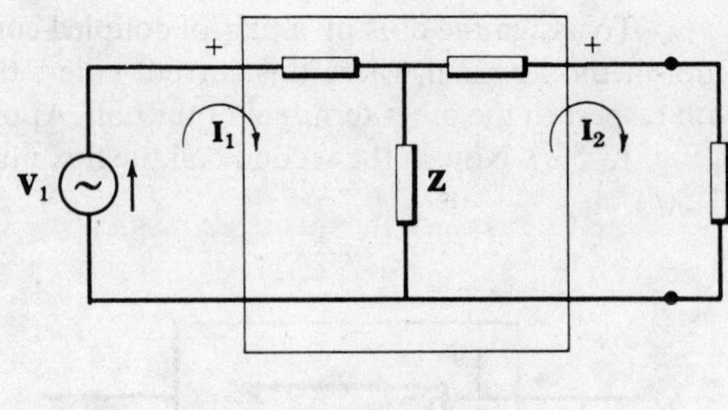

Fig. 13-9

When the boxes in Fig. 13-8 and Fig. 13-9 are covered, the two circuits look identical except for the dot notation in one circuit and the sign notation in the other. Comparing (*15*) and (*16*), the negative sign of $j\omega M$ corresponds to the negative sign of $\mathbf{Z}$.

CONDUCTIVELY COUPLED EQUIVALENT CIRCUITS

It is possible in analysis to replace a mutually coupled circuit with a conductively coupled equivalent circuit. Consider the circuit of Fig. 13-10(*a*). Slected the directions of currents $\mathbf{I}_1$ and $\mathbf{I}_2$ as shown. Then the mesh current equations in matrix form are

$$\begin{bmatrix} R_1 + j\omega L_1 & -j\omega M \\ -j\omega M & R_2 + j\omega L_2 \end{bmatrix} \begin{bmatrix} \mathbf{I}_1 \\ \mathbf{I}_2 \end{bmatrix} = \begin{bmatrix} \mathbf{V}_1 \\ \mathbf{V}_2 \end{bmatrix} \quad (17)$$

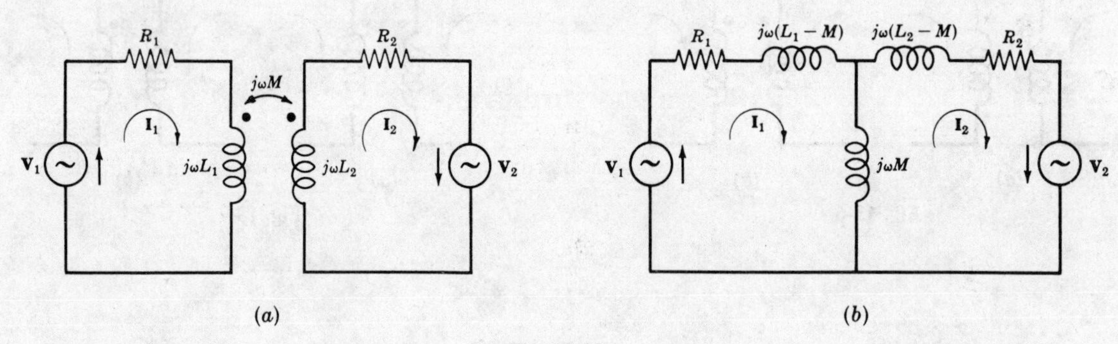

Fig. 13-10

Let the current directions in Fig. 13-10(*b*) be the same as in Fig. 13-10(*a*). The currents $\mathbf{I}_1$ and $\mathbf{I}_2$ pass in opposite directions through the common branch; then the required impedance here is $j\omega M$. In (*17*), $\mathbf{Z}_{11} = R_1 + j\omega L_1$. Since mesh current $\mathbf{I}_1$ passes through the common branch with impedance $j\omega M$, we must insert $(-j\omega M)$ in the loop and write

$$\mathbf{Z}_{11} = R_1 + j\omega L_1 - j\omega M + j\omega M = R_1 + j\omega L_1$$

Similarly in loop 2,

$$\mathbf{Z}_{22} = R_2 + j\omega L - j\omega M + j\omega M = R_2 + j\omega L_2$$

If we write the mesh current equations for the circuit in Fig. 13-10(*b*) we obtain the set (*17*). Thus the conductively coupled circuit of Fig. 13-10(*b*) is equivalent to the mutually coupled circuit of Fig. 13-10(*a*).

The above method of analysis does not always lead to a physically realizable equivalent circuit. This is true when $M > L_1$ or $M > L_2$.

To replace the series connection of the mutually coupled coils shown in Fig. 13-11(*a*), proceed in the following manner. First apply the methods described above and obtain the dotted equivalent shown in Fig. 13-11(*b*). Then replace the dotted equivalent by the conductive equivalent shown in Fig. 13-11(*c*).

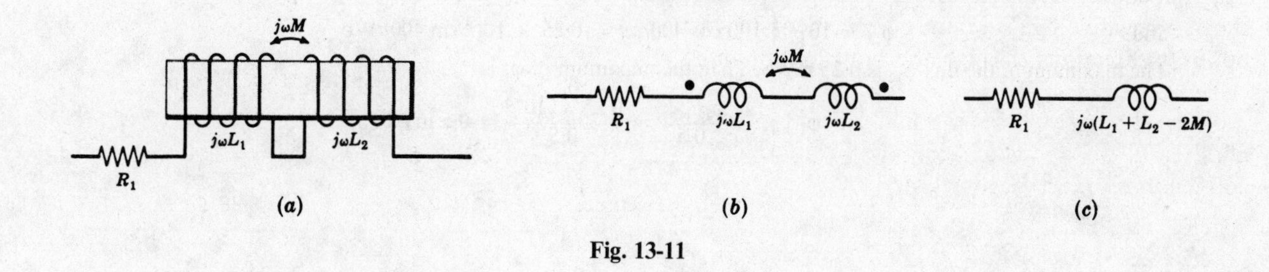

Fig. 13-11

Analysis of the circuit shown in Fig. 13-11(*a*) requires consideration of magnetic fluxes to determine the signs of the voltages of mutual inductance. With the circuit of Fig. 13-11(*b*), no fluxes need be considered but the dot rule is required. Then with the circuit of Fig. 13-11(*c*) the necessary equations can be written in the usual manner with no special attention to either flux, dots or mutual inductance. All three circuits have the same complex impedance $\mathbf{Z} = R_1 + j\omega(L_1 + L_2 - 2M)$.

Solved Problems

13.1. Coil 1 of a pair of coupled coils has a continuous current of 5 A, and the corresponding fluxes φ_{11} and φ_{12} are 0·2 and 0·4 mWb respectively. If the turns are $N_1 = 500$ and $N_2 = 1500$, find L_1, L_2, M and k.

The total flux is $\varphi_1 = \varphi_{11} + \varphi_{12} = 6 \times 10^{-4}$ webers. Then the self-inductance of coil 1 is $L_1 = N_1\varphi_1/I_1 = 500(6 \times 10^{-4})/5 = 0{\cdot}06$ H.

The coupling coefficient $k = \varphi_{12}/\varphi_1 = 0{\cdot}4/0{\cdot}6 = 0{\cdot}667$.

The mutual inductance $M = N_2\varphi_{12}/I_1 = 1500(4 \times 10^{-4})/5 = 0{\cdot}12$ H.

Since $M = k\sqrt{L_1L_2}$, $0{\cdot}12 = 0{\cdot}667\sqrt{0{\cdot}06L_2}$ and $L_2 = 0{\cdot}539$ H.

13.2. Two coupled coils of $L_1 = 0{\cdot}8$ H and $L_2 = 0{\cdot}2$ H have a coupling coefficient $k = 0{\cdot}9$. Find the mutual inductance M and the turns ratio N_1/N_2.

The mutual inductance is $M = k\sqrt{L_1L_2} = 0{\cdot}9\sqrt{0{\cdot}8(0{\cdot}2)} = 0{\cdot}36$ H.

Using $M = N_2\varphi_{12}/i_1$, substitute $k\varphi_1$ for φ_{12} and multiply by N_1/N_1 to obtain

$$M = k\frac{N_2}{N_1}\left(\frac{N_1\varphi_1}{i_1}\right) = k\frac{N_2}{N_1}L_1 \text{ and } N_1/N_2 = kL_1/M = 0{\cdot}9(0{\cdot}8)/0{\cdot}36 = 2$$

13.3. Two coupled coils with respective self-inductances $L_1 = 0{\cdot}5$ H and $L_2 = 0{\cdot}20$ H have a coupling coefficient $k = 0{\cdot}5$. Coil 2 has 1000 turns. If the current in coil 1 is $i_1 = 5 \sin 400t$ amperes, determine the voltage at coil 2 and the maximum flux set up by coil 1.

The mutual inductance is $M = k\sqrt{L_1L_2} = 0{\cdot}5\sqrt{0{\cdot}05(0{\cdot}20)} = 0{\cdot}05$ H. Then the voltage at coil 2 is given by $v_2 = M(di_1/dt) = 0{\cdot}05 \frac{d}{dt}(5 \sin 400t) = 100 \cos 400t$. Since the voltage at coil 2 is also given by $v_2 = N_2(d\varphi_{12}/dt)$, we obtain

$$100 \cos 400t = 1000(d\varphi_{12}/dt)$$

and

$$\varphi_{12} = 10^{-3} \int 100 \cos 400t \, dt = 0{\cdot}25 \times 10^{-3} \sin 400t \text{ Wb}$$

The maximum of the flux φ_{12} is 0·25 mWb. Then the maximum of φ_1 is

$$\varphi_{1\,max} = \frac{\varphi_{12\,max}}{0{\cdot}5} = \frac{0{\cdot}25 \times 10^{-3}}{0{\cdot}5} = 0{\cdot}5 \text{ mWb}$$

13.4. Apply Kirchhoff's voltage law to the coupled circuit shown in Fig. 13-12 and write the equation in instantaneous form.

Examination of the winding sense of the coils shows that the signs on the M terms are opposite to the signs on the L terms. Also note that the voltage of mutual inductance appears at each coil due to the current i in the other coil of the pair.

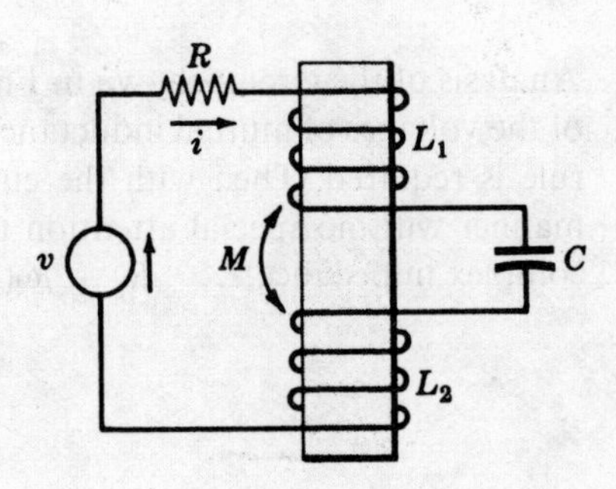

Fig. 13-12

$$Ri + L_1\frac{di}{dt} - M\frac{di}{dt} + \frac{1}{C}\int i\,dt + L_2\frac{di}{dt} - M\frac{di}{dt} = v$$

or

$$Ri + (L_1 + L_2 - 2M)\frac{di}{dt} + \frac{1}{C}\int i\,dt = v$$

13.5. Write the mesh current equations in instantaneous form for the coupled circuit of Fig. 13-13 below.

Select mesh currents i_1 and i_2 as shown in the diagram and apply the right hand rule to each winding. Since the fluxes aid, the sign of the M terms are the same as the sign of the L terms. Then

$$R_1i_1 + L_1\frac{di_1}{dt} + M\frac{di_2}{dt} = v$$

$$R_2i_2 + L_2\frac{di_2}{dt} + M\frac{di_1}{dt} = v$$

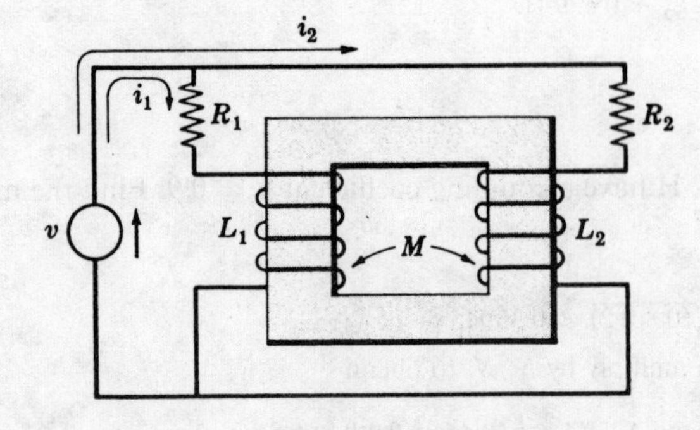

Fig. 13-13

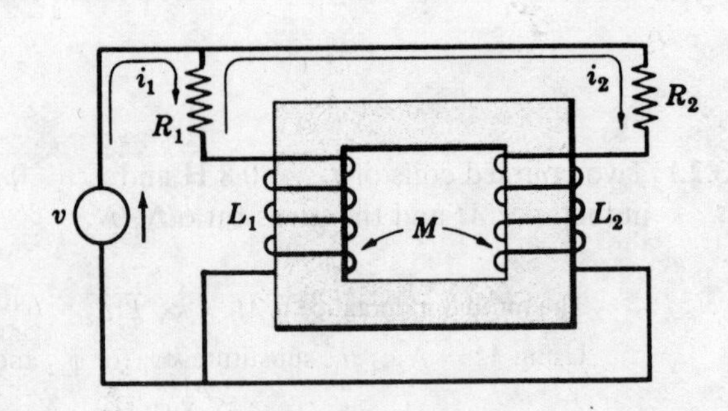

Fig. 13-14

13.6. Repeat Problem 13.5 with mesh current i_2 as shown in Fig. 13-14 above.

In applying Kirchhoff's voltage law to the loop of current i_2, the voltages of mutual inductance are negative. Thus

$$R_1(i_1 - i_2) + L_1\frac{d}{dt}(i_1 - i_2) + M\frac{di_2}{dt} = v$$

$$R_1(i_2 - i_1) + R_2 i_2 + L_2\frac{di_2}{dt} - M\frac{d}{dt}(i_2 - i_1) + L_1\frac{d}{dt}(i_2 - i_1) - M\frac{di_2}{dt} = 0$$

13.7. Two coils connected in series have an equivalent inductance L_A if the connection is aiding and an equivalent inductance L_B if the connection is opposing. Find the mutual inductance M in terms of L_A and L_B.

When the connection is aiding, the equivalent inductance is given by

$$L_A = L_1 + L_2 + 2M \qquad (1)$$

When the connection is opposing, we have

$$L_B = L_1 + L_2 - 2M \qquad (2)$$

Subtracing (2) from (1),

$$L_A - L_B = 4M \text{ and } M = \tfrac{1}{4}(L_A - L_B)$$

This solution points out an experimental method for the determination of M. Connect the coils both ways and obtain the equivalent inductances on an *AC* bridge. The resulting inductance is one-fourth of the difference between the two equivalent inductances.

13.8. Obtain the dotted equivalent circuit for the coupled circuit shown in Fig. 13-15. Find the voltage across the $-j10\ \Omega$ reactance using the equivalent circuit.

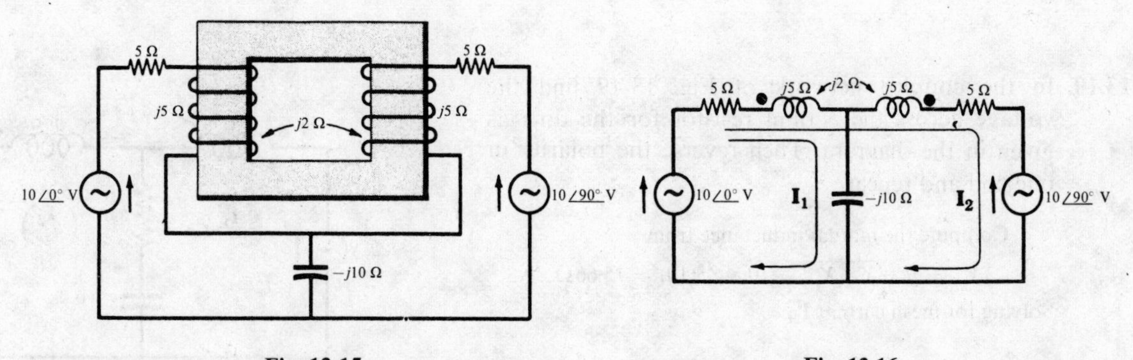

Fig. 13-15 Fig. 13-16

To place the dots on the circuit, consider only the coils and their winding sense. Drive a current into the top of the left coil and place a dot at this terminal. The corresponding flux direction is upward on the left side of the core. By Lenz's law the flux at the right coil must also be upward. Then the right hand rule gives the direction of the natural current. This current leaves the winding by the upper terminal, which should then be marked with a dot as shown in Fig. 13-16.

For i_1 and i_2 selected as shown, the mesh current equations in matrix form are

$$\begin{bmatrix} 5 - j5 & 5 + j3 \\ 5 + j3 & 10 + j6 \end{bmatrix}\begin{bmatrix} \mathbf{I}_1 \\ \mathbf{I}_2 \end{bmatrix} = \begin{bmatrix} 10 \\ 10 - j10 \end{bmatrix}$$

from which

$$\mathbf{I}_1 = \frac{\begin{vmatrix} 10 & 5+j3 \\ 10-j10 & 10+j6 \end{vmatrix}}{\Delta_z} = 1{\cdot}015\,\underline{/113{\cdot}95^\circ}\text{ A}$$

Then the voltage across the $-j10\ \Omega$ reactance is

$$\mathbf{V} = \mathbf{I}_1(-j10) = 10{\cdot}15\,\underline{/23{\cdot}95^\circ}\text{ V}$$

13.9. Obtain the dotted equivalent for the coupled coils shown in Fig. 13-17 and write the corresponding equation.

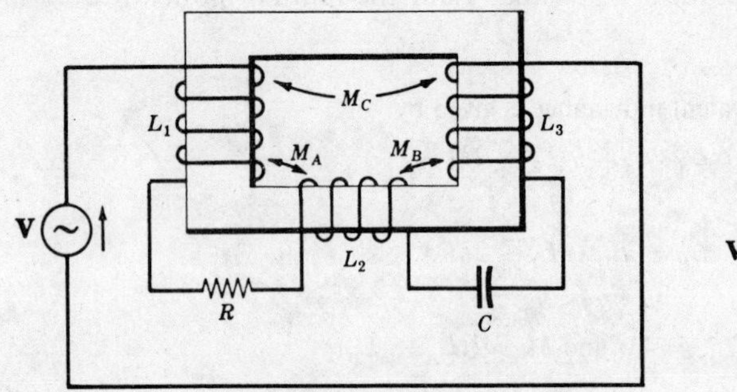

Fig. 13-17

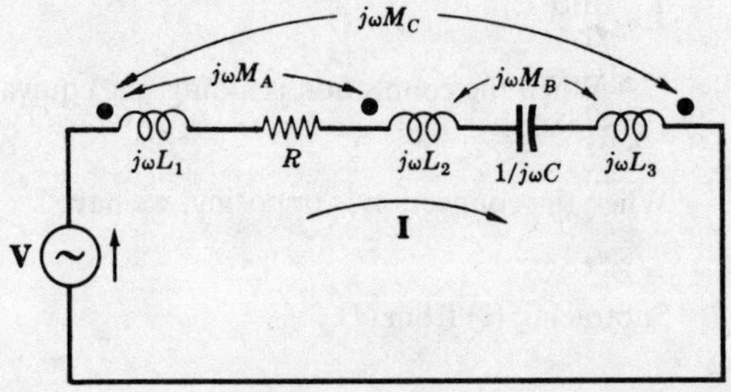

Fig. 13-18

Assign the dots using the methods of Problem 13.8 to obtain the circuit shown in Fig. 13-18. Applying Kirchhoff's voltage law to the single loop,

$$\left[R + \frac{1}{j\omega C} + j\omega(L_1 + L_2 + L_3 + 2M_A - 2M_B - 2M_C)\right]\mathbf{I} = \mathbf{V}$$

13.10. In the coupled network of Fig. 13-19 find the voltage across the 5 ohm resistor for the dots as given in the diagram. Then reverse the polarity in one coil and repeat.

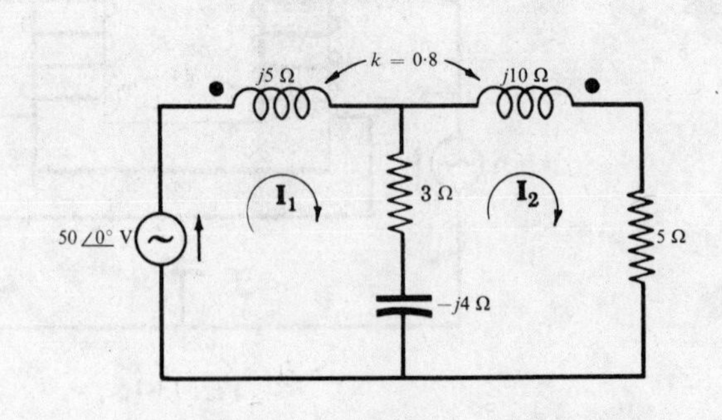

Fig. 13-19

Compute the mutual inductance from

$$jX_m = jk\sqrt{X_{L1}X_{L2}} = j0{\cdot}8\sqrt{5(10)} = j5{\cdot}66\,\Omega$$

Solving for mesh current $\mathbf{I}_2$,

$$\mathbf{I}_2 = \frac{\begin{vmatrix} 3+j1 & 50 \\ -3-j1{\cdot}66 & 0 \end{vmatrix}}{\begin{vmatrix} 3+j1 & -3-j1{\cdot}66 \\ -3-j1{\cdot}66 & 8+j6 \end{vmatrix}} = \frac{171\,\underline{/29^\circ}}{19{\cdot}9\,\underline{/53{\cdot}8^\circ}} = 8{\cdot}60\,\underline{/-24{\cdot}8^\circ}\text{ A}$$

Then the voltage across the 5 ohm resistor is $\mathbf{V}_5 = \mathbf{I}_2(5) = 43\,\underline{/-24{\cdot}8^\circ}$ V.

With a change in polarity in one coil the impedance matrix changes, resulting in a new value of the mesh current $\mathbf{I}_2$.

$$\mathbf{I}_2 = \frac{\begin{vmatrix} 3+j1 & 50 \\ -3+j9{\cdot}66 & 0 \end{vmatrix}}{\begin{vmatrix} 3+j1 & -3+j9{\cdot}66 \\ -3+j9{\cdot}66 & 8+j6 \end{vmatrix}} = \frac{505\angle -72{\cdot}7^\circ}{132\angle 39{\cdot}4^\circ} = 3{\cdot}83\angle -112{\cdot}1^\circ \text{ A}$$

The voltage across the 5 ohm resistor is $\mathbf{V}_5 = \mathbf{I}_2(5) = 19{\cdot}15\angle -112{\cdot}1^\circ$ V.

13.11. Obtain the equivalent inductance of the parallel connection of L_1 and L_2 shown in Fig. 13-20(*a*).

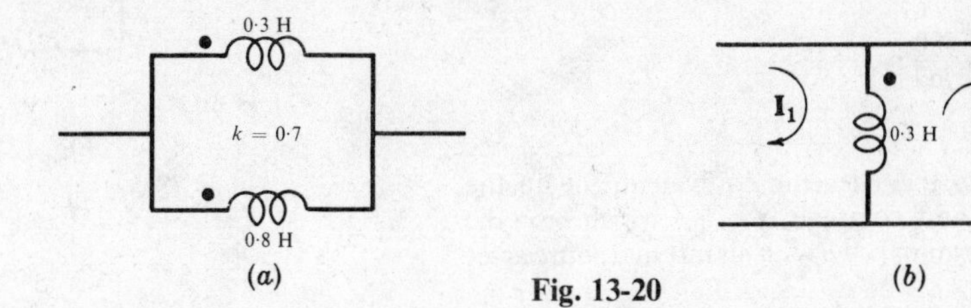

Fig. 13-20

The mutual inductance $M = k\sqrt{L_1L_2} = 0{\cdot}7\sqrt{0{\cdot}3(0{\cdot}8)} = 0{\cdot}343$ H. Set up the circuit as in Fig. 13-20(*b*) and insert the mesh currents. Now

$$[Z] = \begin{bmatrix} j\omega 0{\cdot}3 & j\omega 0{\cdot}043 \\ j\omega 0{\cdot}043 & j\omega 0{\cdot}414 \end{bmatrix}$$

$$\mathbf{Z}_{\text{input 1}} = \frac{\Delta_z}{\Delta_{11}} = \frac{j\omega 0{\cdot}3(j\omega 0{\cdot}414) - (j\omega 0{\cdot}043)^2}{j\omega 0{\cdot}414} = j\omega 0{\cdot}296\ \Omega$$

Thus the equivalent inductance of the coupled coils is 0·296 H.

13.12. The Heaviside bridge circuit in Fig. 13-21 is used to determine the mutual inductance between a pair of coils. Find M in terms of the other bridge constants when detector current $\mathbf{I}_D$ equals zero.

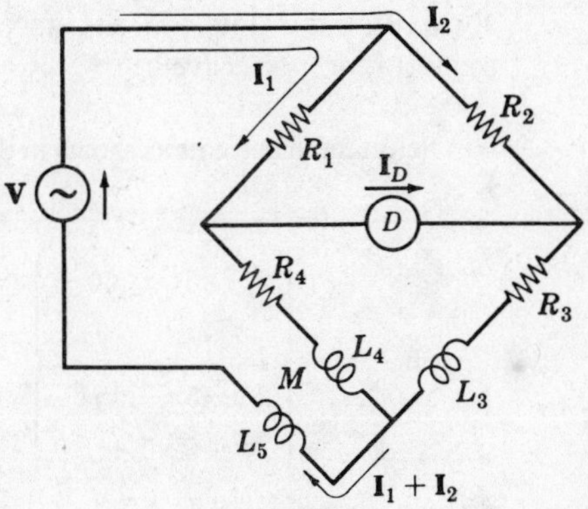

Select two mesh currents $\mathbf{I}_1$ and $\mathbf{I}_2$ as shown in the diagram. If $\mathbf{I}_D = 0$ the voltage drops across R_1 and R_2 must be equal:

$$\mathbf{I}_1R_1 = \mathbf{I}_2R_2 \tag{1}$$

Similarly the drops across $(R_4 + j\omega L_4)$ and $(R_3 + j\omega L_3)$ must be equal. However, at L_4 there appears a voltage of mutual induction and the current in the other coil of the pair, L_5, is the sum $\mathbf{I}_1 + \mathbf{I}_2$.

$$\mathbf{I}_1(R_4 + j\omega L_4) + j\omega M(\mathbf{I}_1 + \mathbf{I}_2) = \mathbf{I}_2(R_3 + j\omega L_3) \tag{2}$$

Fig. 13-21

Substituting $\mathbf{I}_2 = (R_1/R_2)\mathbf{I}_1$ into (*2*),

$$\mathbf{I}_1(R_4 + j\omega L_4 + j\omega M) + (R_1/R_2)\mathbf{I}_1(j\omega M) = (R_1/R_2)\mathbf{I}_1(R_3 + j\omega L_3) \tag{3}$$

Equating real and imaginary parts of (*3*),

$$R_4R_2 = R_1R_3 \text{ and } j\omega\left(L_4 + M + \frac{R_1}{R_2}M\right) = j\omega\frac{R_1}{R_2}L_3 \text{ from which } M = \frac{R_1L_3 - R_2L_4}{R_1 + R_2}$$

13.13. Replace the coupled network shown in Fig. 13-22 with a Thevenin equivalent at the terminals AB.

The Thevenin equivalent voltage $\mathbf{V}'$ is the open circuit voltage at the terminals AB. Select mesh currents $\mathbf{I}_1$ and $\mathbf{I}_2$ as shown and solve for $\mathbf{I}_2$.

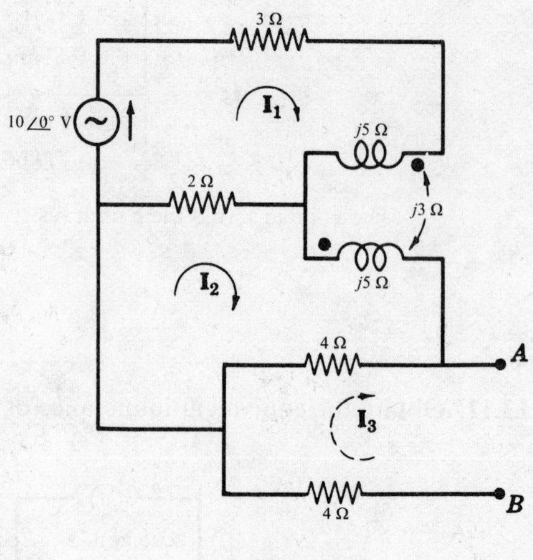

Fig. 13-22

$$\mathbf{I}_2 = \frac{\begin{vmatrix} 5+j5 & 10 \\ -2+j3 & 0 \end{vmatrix}}{\begin{vmatrix} 5+j5 & -2+j3 \\ -2+j3 & 6+j5 \end{vmatrix}} = \frac{20-j30}{10+j67} = 0{\cdot}533\angle{-137{\cdot}8^\circ}\ \text{A}$$

Now $\mathbf{V}' = \mathbf{V}_{AB} = \mathbf{I}_2(4) = 2{\cdot}13\angle{-137{\cdot}8^\circ}$ V

To determine $\mathbf{Z}'$ of the Thevenin equivalent, set up the third mesh current $\mathbf{I}_3$ and compute $\mathbf{Z}_{\text{input 3}}$ which is the impedance seen at the terminals AB with all internal sources set to zero.

$$\mathbf{Z}' = \mathbf{Z}_{\text{input 3}} = \frac{\Delta_z}{\Delta_{33}} = \frac{\begin{vmatrix} 5+j5 & -2+j3 & 0 \\ -2+j3 & 6+j5 & -4 \\ 0 & -4 & 8 \end{vmatrix}}{\begin{vmatrix} 5+j5 & -2+j3 \\ -2+j3 & 6+j5 \end{vmatrix}} = \frac{j456}{10+j67} = 6{\cdot}74\angle{8{\cdot}5^\circ}\ \Omega$$

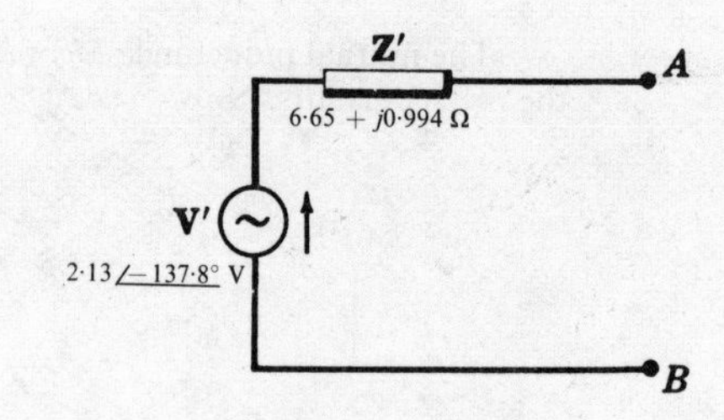

Fig. 13-23

The Thevenin equivalent circuit is shown in Fig. 13-23.

13.14. In the two loop coupled circuit shown in Fig. 13-24, show that the dots are not necessary so long as the second loop is passive.

Select mesh currents as shown in the diagram and solve for $\mathbf{I}_2$.

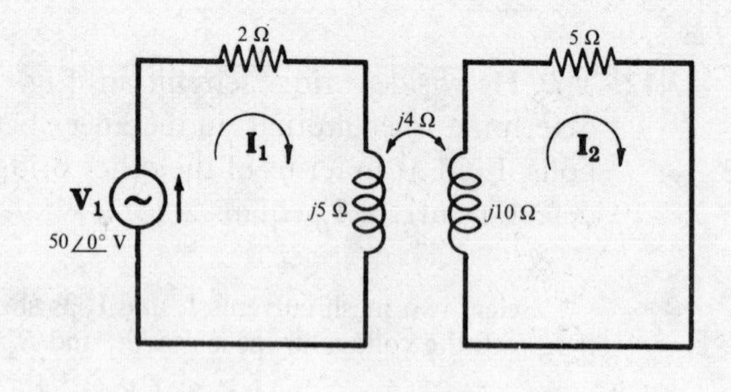

Fig. 13-24

$$\mathbf{I}_2 = \frac{\begin{vmatrix} 2+j5 & 50 \\ \pm j4 & 0 \end{vmatrix}}{\begin{vmatrix} 2+j5 & \pm j4 \\ \pm j4 & 5+j10 \end{vmatrix}} = \frac{-50(\pm j4)}{-24+j45} = 3{\cdot}92\angle{61{\cdot}9^\circ \pm 90^\circ}\ \text{A}$$

The value of Δ_z is unaffected by the sign on M, and the current $\mathbf{I}_2$ will have a phase angle of either $151{\cdot}9^\circ$ or $-28{\cdot}1^\circ$. Since there is no voltage source in the loop it is unnecessary that the polarity of the voltage of mutual inductance be known. Voltage drops across the loop impedances would be equal in magnitude and differ in phase by 180°. Power in an impedance would be unaffected. It is also true that $\mathbf{I}_1$ is identical for either sign on the mutual.

13.15. In the circuit shown in Fig. 13-25, find R_L which results in maximum power transfer after selecting the best connection for the coils and finding k.

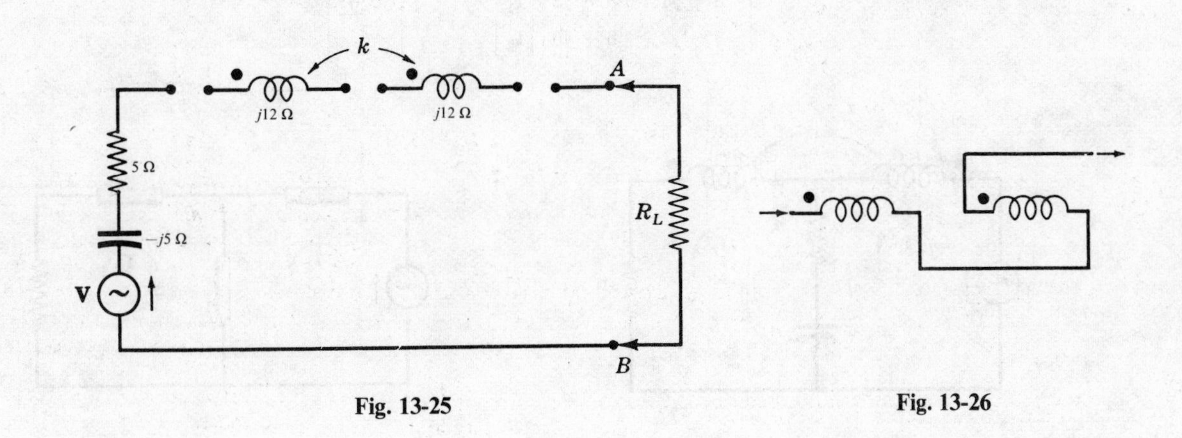

Fig. 13-25

Fig. 13-26

The circuit impedance to the left of AB must be a minimum. Express the impedance of this part of the circuit.

$$\mathbf{Z} = 5 - j5 + j12 + j12 \pm j2X_M = 5 + j19 \pm j2\sqrt{12(12)}\,\Omega$$

For minimum impedance the reactance should be zero; hence, the proper sign for the mutual is negative.

$$19 - 2k\sqrt{12(12)} = 0 \text{ and } k = 19/24 = 0{\cdot}792$$

The connection shown in Fig. 13-26 results in a negative sign on the voltages of mutual inductance as required. Then the circuit impedance to the left of AB is 5 ohms pure resistance, and the maximum power results when $R_L = R_g = 5$ ohms.

13.16. The circuit of Fig. 13-25 has a load resistance $R_L = 10$ ohms and a source $\mathbf{V} = 50/0°$ V. With both connections of the coils possible and k variable from 0 to 1, find the range of power that can be delivered to the load resistor.

With the coupling shown in Fig. 13-26 the sign on the mutual is negative and the total circuit impedance including the load is $\mathbf{Z}_T = 5 - j5 + j12 + j12 - j24k + 10\,\Omega$. Set $k = 1$; then

$$\mathbf{Z}_T = 15 - j5 = 15{\cdot}8\angle{-18{\cdot}45°}\,\Omega,\ \mathbf{I} = \frac{\mathbf{V}}{\mathbf{Z}_T} = \frac{50\angle 0°}{15{\cdot}8\angle{-18{\cdot}45°}} = 3{\cdot}16\angle 18{\cdot}45°\text{ A}$$

The power in the 10 ohm resistor is $P = I^2R = (3{\cdot}16)^2(10) = 100$ W.

Now set $k = 0$; then

$$\mathbf{Z}_T = 15 + j19 = 24{\cdot}2\angle 51{\cdot}7°\,\Omega,\ \mathbf{I} = 50\angle 0°/(24{\cdot}2\angle 51{\cdot}7°) = 2{\cdot}06\angle{-51{\cdot}7°}\text{ A}$$

The power in the 10 ohm resistor is $(2{\cdot}06)^2(10) = 42{\cdot}4$ watts. With $k = 0{\cdot}792$, $P_{max} = 111$ W.

Change the connection of the coils to result in a positive sign of the mutual inductance. Then the impedance becomes $\mathbf{Z}_T = 15 + j19 + jk24\,\Omega$.

Set $k = 1$; then

$$\mathbf{Z}_T = 15 + j43 = 45{\cdot}6\angle 70{\cdot}8°\,\Omega,\ \mathbf{I} = 50\angle 0°/(45{\cdot}6\angle 70{\cdot}8°) = 1{\cdot}095\angle{-70{\cdot}8°}\text{ A}$$

The corresponding power $P = I^2R = (1{\cdot}095)^2(10) = 12$ W.

Thus the 10 ohm resistor can expect a power within the range 12 to 100 watts.

13.17. Obtain a conductively coupled equivalent circuit for the mutually coupled circuit shown in Fig. 13-27.

Select mesh currents $\mathbf{I}_1$ and $\mathbf{I}_2$ as shown and write the matrix equation.

$$\begin{bmatrix} 3 + j1 & -3 - j2 \\ -3 - j2 & 8 + j6 \end{bmatrix} \begin{bmatrix} \mathbf{I}_1 \\ \mathbf{I}_2 \end{bmatrix} = \begin{bmatrix} 50\angle 0^\circ \\ 0 \end{bmatrix}$$

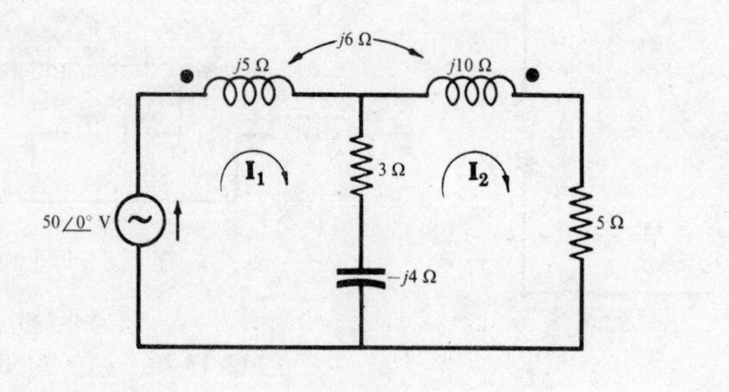

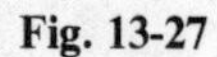
Fig. 13-27

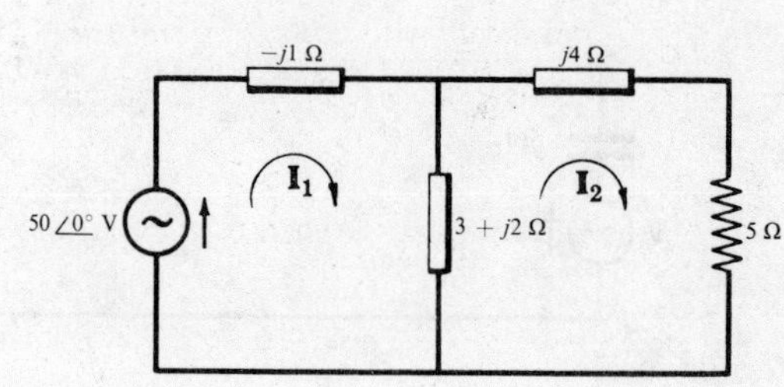

Fig. 13-28

Select the mesh currents in the conductively coupled circuit with the same directions as in the mutually coupled circuit. From the impedance matrix, $\mathbf{Z}_{12} = -3 - j2\ \Omega$. Since the currents pass through the common branch in opposite directions, the required branch impedance is $3 + j2\ \Omega$. Now the self-impedance of loop 1 is $\mathbf{Z}_{11} = 3 + j1\ \Omega$. Then a $-j1$ Ω impedance is required in the loop. Similarly, since $\mathbf{Z}_{22} = 8 + j6\ \Omega$, the loop requires a $5 + j4\ \Omega$ impedance in addition to the elements of the common branch as shown in Fig. 13-28.

13.18. Obtain the conductively coupled equivalent circuit for the mutually coupled network shown in Fig. 13-29.

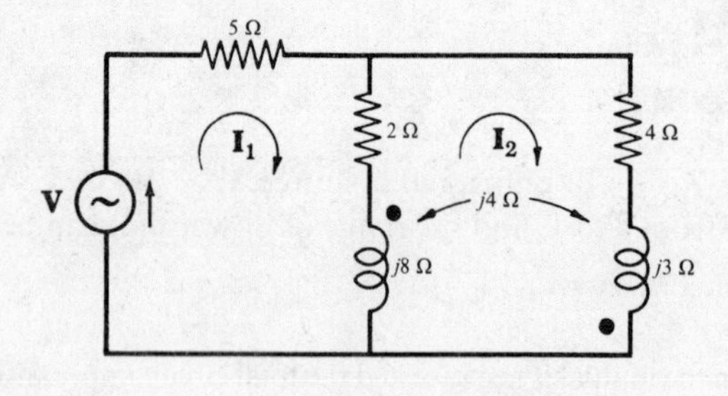

Fig. 13-29

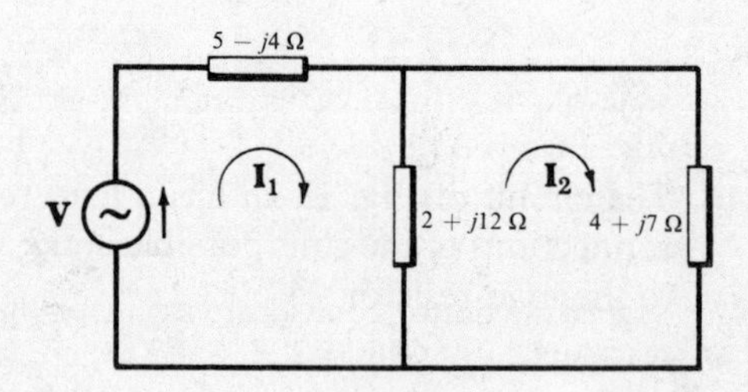

Fig. 13-30

Select the mesh currents $\mathbf{I}_1$ and $\mathbf{I}_2$ and write the mesh current equations in matrix form.

$$\begin{bmatrix} 7 + j8 & -2 - j12 \\ -2 - j12 & 6 + j19 \end{bmatrix} \begin{bmatrix} \mathbf{I}_1 \\ \mathbf{I}_2 \end{bmatrix} = \begin{bmatrix} \mathbf{V} \\ 0 \end{bmatrix}$$

The mesh currents in the conductively coupled circuit pass through the common branch in opposite directions. Since $\mathbf{Z}_{12}$ in the impedance matrix is $-2 - j12\ \Omega$, the impedance of this branch must be $2 + j12\ \Omega$. Also, from the impedance matrix, $\mathbf{Z}_{11} = 7 + j8\ \Omega$ and $\mathbf{Z}_{22} = 6 + j19\ \Omega$. Then the remaining impedances in loops 1 and 2 of the equivalent circuit are respectively

$$\mathbf{Z}_1 = (7 + j8) - (2 + j12) = 5 - j4\ \Omega \quad \text{and} \quad \mathbf{Z}_2 = (6 + j19) - (2 + j12) = 4 + j7\ \Omega$$

The conductively coupled equivalent circuit is shown in Fig. 13-30.

Supplementary Problems

13.19. Two coils have a coupling coefficient $k = 0{\cdot}85$ and coil 1 has 250 turns. With a current $i_1 = 2$ A in coil 1, the total flux φ_1 is 0·3 mWb. When i_1 is reduced linearly to zero in two milliseconds the voltage induced in coil 2 is 63·75 volts. Find L_1, L_2, M and N_2. *Ans.* 37·5 mH, 150 mH, 63·8 mH, 500

13.20. Two coupled coils with turns $N_1 = 100$ and $N_2 = 800$ have a coupling coefficient of 0·85. With coil 1 open and a current of 5 amperes in coil 2, the flux φ_2 is 0·35 mWb. Find L_1, L_2 and M. *Ans.* 0·875 mH, 56 mH, 5·95 mH

13.21. If two identical coils have an equivalent inductance of 0·080 H in series aiding and 0·035 H in series opposing, what are the values of L_1, L_2, M and k? *Ans.* $L_1 = 28{\cdot}8$ mH, $L_2 = 28{\cdot}8$ mH, $M = 11{\cdot}25$ mH, 0·392

13.22. Two coupled coils with $L_1 = 0{\cdot}02$ H, $L_2 = 0{\cdot}01$ H and $k = 0{\cdot}5$ are connected in four different ways; series aiding, series opposing, and parallel with both arrangements of the winding sense. What are the four equivalent inductances? *Ans.* 15·9 mH, 44·1 mH, 9·47 mH, 3·39 mH

13.23. Two identical coils with $L = 0{\cdot}02$ H have a coupling coefficient $k = 0{\cdot}8$. Find M and the two equivalent inductances with the coils connected in series aiding and series opposing. *Ans.* 16 mH, 72 mH, 8 mH

13.24. Two coils with inductances in the ratio of four to one have a coupling coefficient $k = 0{\cdot}6$. When these coils are connected in series aiding the equivalent inductance is 44·4 mH. Find L_1, L_2 and M. *Ans.* 6 mH, 24 mH, 7·2 mH

13.25. Two coils with inductances $L_1 = 6{\cdot}8$ mH and $L_2 = 4{\cdot}5$ mH are connected in series aiding and series opposing. The equivalent inductances of these connections are 19·6 and 3 mH respectively. Find M and k. *Ans.* 4·15 mH, 0·75

13.26. Select mesh currents for the coupled circuit of Fig. 13-31 and write the equations in instantaneous form. Obtain the dotted equivalent circuit, write the equations and compare the results.

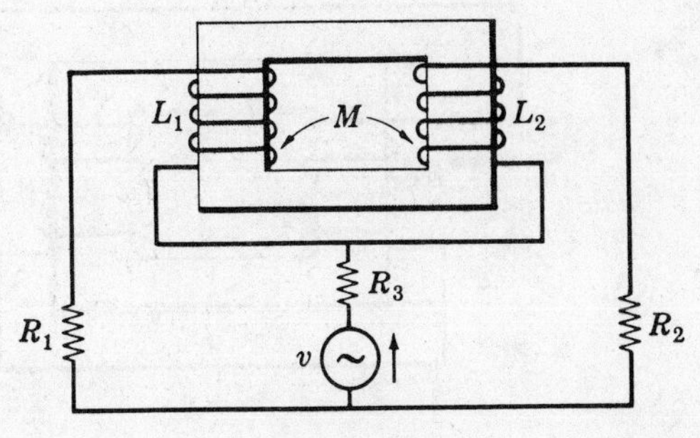

Fig. 13-31

13.27. Sketch the dotted equivalent circuit for the coupled coils shown in Fig. 13-32 and find the equivalent inductance reactance. *Ans.* $j12\ \Omega$

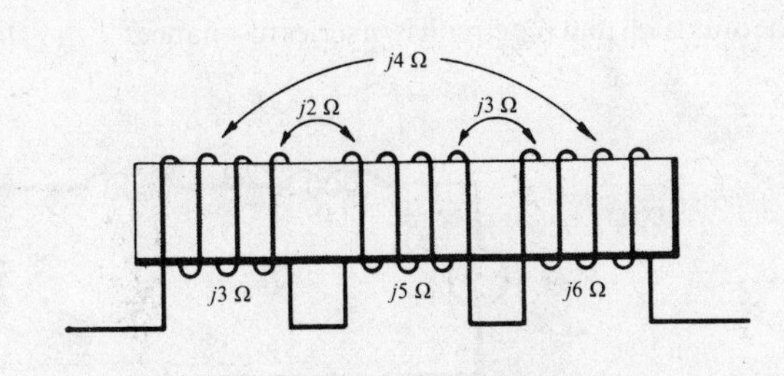

Fig. 13-32

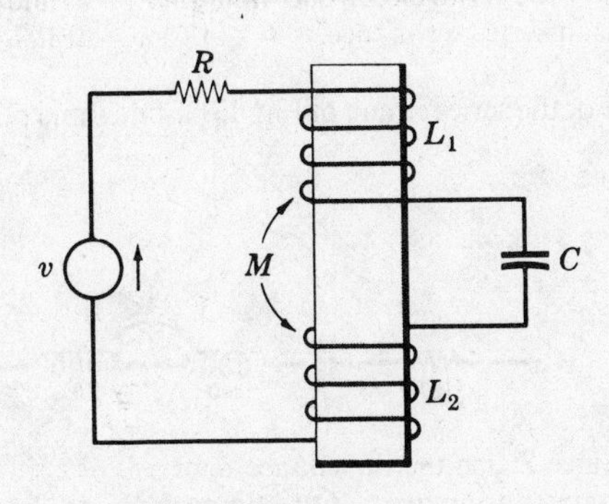

Fig. 13-33

13.28. Obtain the dotted equivalent circuit for the coupled coils of Fig. 13-33 and write the equation in instantaneous form.

13.29. Sketch the dotted equivalent circuit for the coupled coils shown in Fig. 13-34 and find the current **I**.
Ans. $4{\cdot}47\,\angle 26{\cdot}7^\circ$ A

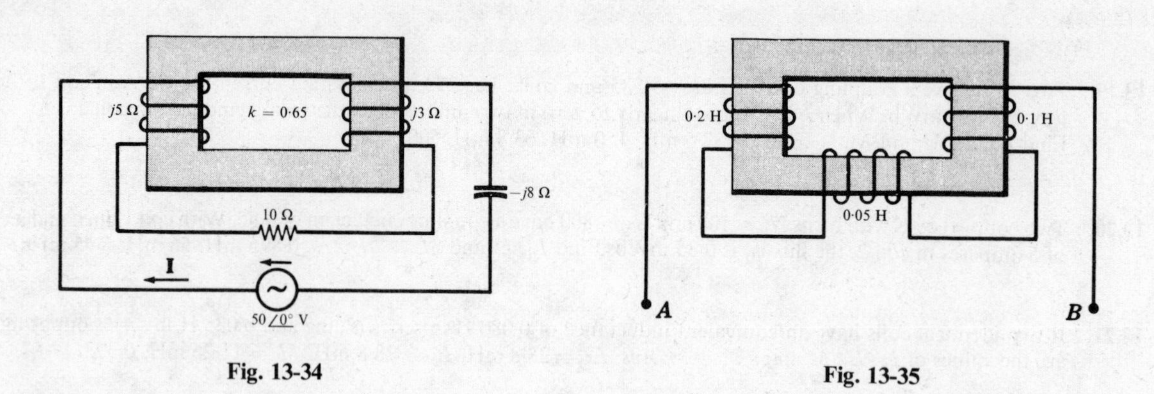

Fig. 13-34 **Fig. 13-35**

13.30. Obtain the dotted equivalent circuit for the three coupled coils shown in Fig. 13-35 and find the equivalent inductance at the terminals *AB*. All coupling coefficients are 0·5. *Ans.* 0·239 H

13.31. Obtain the dotted equivalent circuit for the coupled circuit of Fig. 13-36 and find the equivalent impedance at the terminals *AB*. *Ans.* $2{\cdot}54 + j2{\cdot}26$ Ω

13.32. Referring to the coupled circuit of Fig. 13-36, reverse the winding of one coil and find the equivalent impedance.
Ans. $2{\cdot}53 + j0{\cdot}238$ Ω

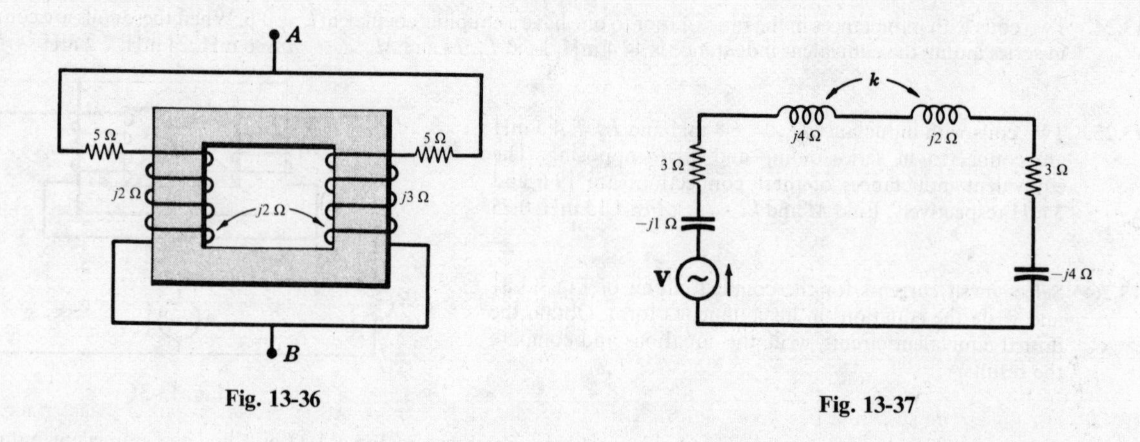

Fig. 13-36 **Fig. 13-37**

13.33. For the series circuit shown in Fig. 13-37 find the value of *k* and place the dots on the coupled coils such that the circuit is in series resonance. *Ans.* $k = 0{\cdot}177$

13.34. For the series circuit of Fig. 13-38 find *k* and place the dots such that the circuit is in series resonance. *Ans.* $k = 0{\cdot}112$

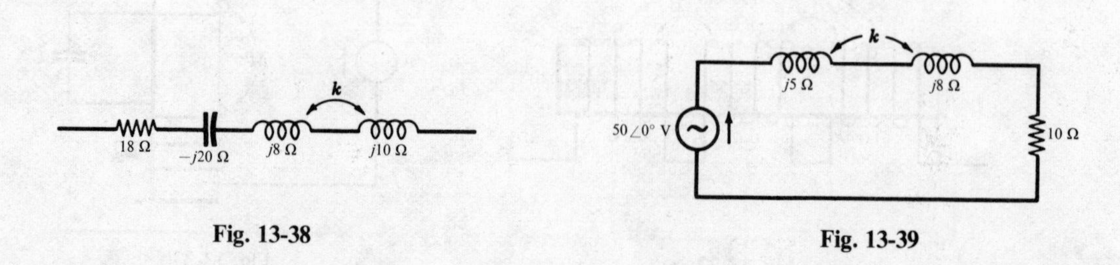

Fig. 13-38 **Fig. 13-39**

13.35. For the circuit shown in Fig. 13-39 find *k* and place the dots so that the power output of the $50\angle 0^\circ$ volts source is 168 watts. *Ans.* $k = 0{\cdot}475$

13.36. Referring to Problem 13.35, find the power output of the source when the dots are reversed. Use the value of k found in Problem 13.35. *Ans.* 54·2 W

13.37. For the coupled circuit shown in Fig. 13-40, find the voltage ratio $\mathbf{V}_2/\mathbf{V}_1$ which results in zero current $\mathbf{I}_1$. Repeat for a zero current $\mathbf{I}_2$. *Ans.* $1{\cdot}414\angle{-45^\circ}$, $0{\cdot}212\angle{32^\circ}$

13.38. Referring to Problem 13.37 what voltage appears across the $j8\ \Omega$ reactance when $\mathbf{V}_1$ is $100\angle{0^\circ}$ V and $\mathbf{I}_1 = 0$? *Ans.* $100\angle{0^\circ}$ V (+ at dot)

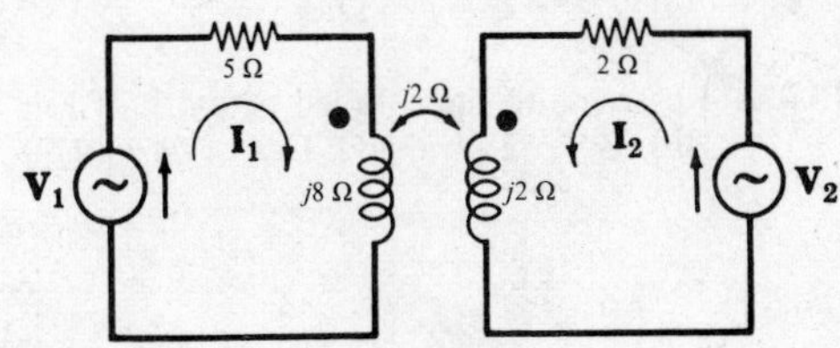

Fig. 13-40

13.39. In the coupled circuit of Fig. 13-41 find the mutual inductive reactance $j\omega M$ if the power in the 5 ohm resistor is 45·2 watts. *Ans.* $j4\ \Omega$

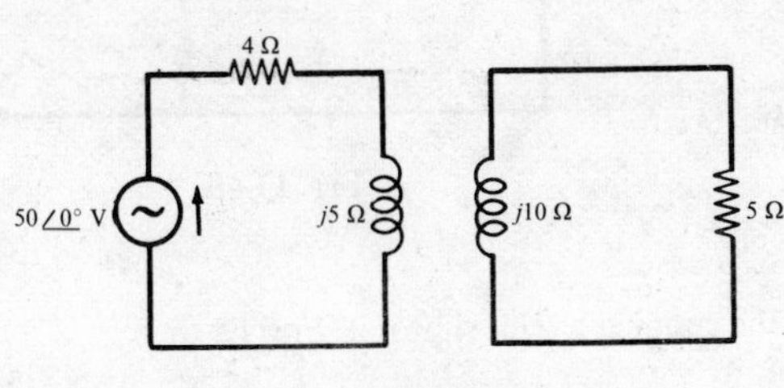

Fig. 13-41

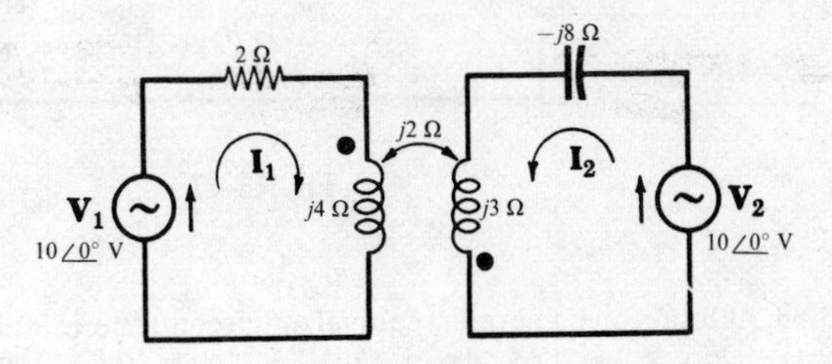

Fig. 13-42

13.40. For the coupled circuit of Fig. 13-42 find the components of current $\mathbf{I}_2$ caused by each source $\mathbf{V}_1$ and $\mathbf{V}_2$. *Ans.* $0{\cdot}77\angle{112{\cdot}6^\circ}$ A, $1{\cdot}72\angle{86^\circ}$ A

13.41. Determine the value of k in the coupled circuit of Fig. 13-43 if the power in the 10 ohm resistor is 32 watts. *Ans.* 0·791

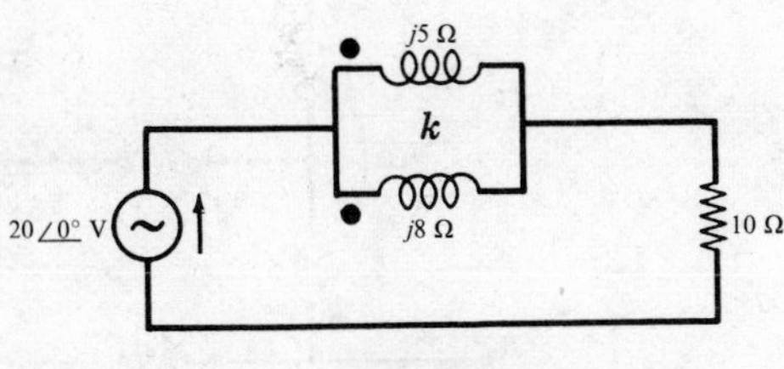

Fig. 13-43

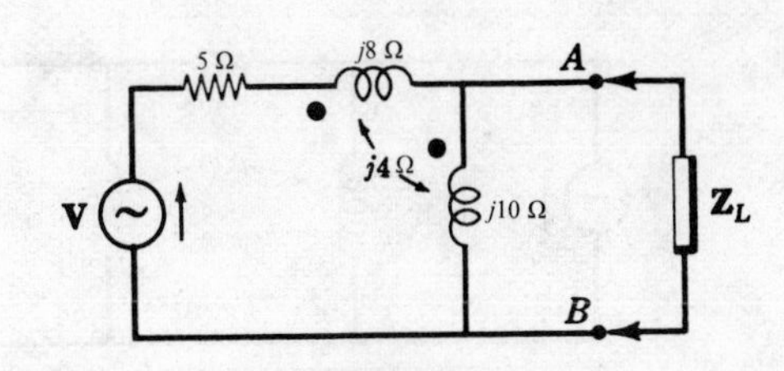

Fig. 13-44

13.42. For the circuit shown in Fig. 13-44 find the load impedance $\mathbf{Z}_L$ which results in maximum power transfer across the terminals AB. *Ans.* $1{\cdot}4 - j2{\cdot}74\ \Omega$

13.43. For the coupled circuit shown in Fig. 13-45 find the input impedance at the terminals of the source. *Ans.* $3 + j36{\cdot}3\ \Omega$

13.44. Referring to the circuit of Fig. 13-45, find the voltage across the $j5\ \Omega$ reactance if the source $\mathbf{V} = 50\angle{45^\circ}$ V. *Ans.* $25{\cdot}2\angle{49{\cdot}74^\circ}$ V

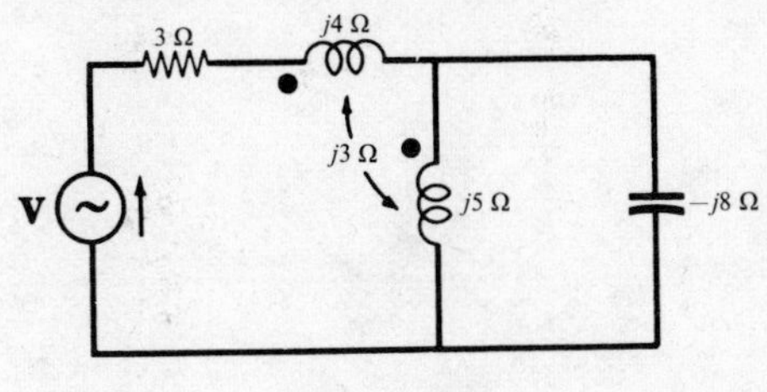

Fig. 13-45

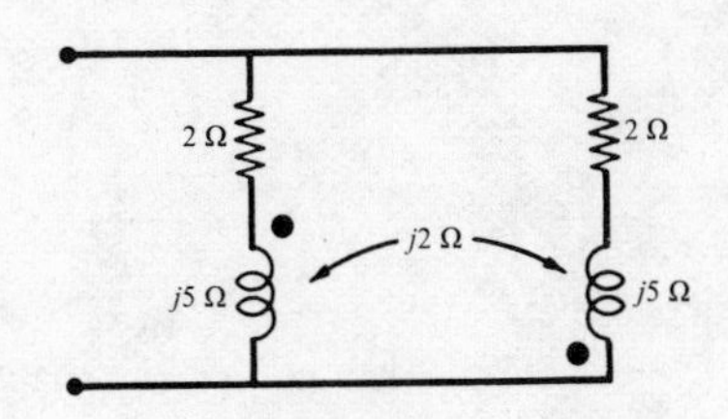

Fig. 13-46

13.45. Find the equivalent impedance of the coupled circuit shown in Fig. 13-46. *Ans.* $1 + j1{\cdot}5\ \Omega$

13.46. Obtain the Thevenin equivalent circuit at terminals AB of the coupled circuit given in Fig. 13-47.
Ans. $\mathbf{Z}' = 2 + j6{\cdot}5\ \Omega$, $\mathbf{V}' = 5 + j5$ V

13.47. Referring to the coupled network of Fig. 13-47, obtain the Norton equivalent circuit at terminals AB.
Ans. $\mathbf{Z}' = 2 + j6{\cdot}5\ \Omega$, $\mathbf{I}' = 1{\cdot}04\angle{-27{\cdot}9^\circ}$ A

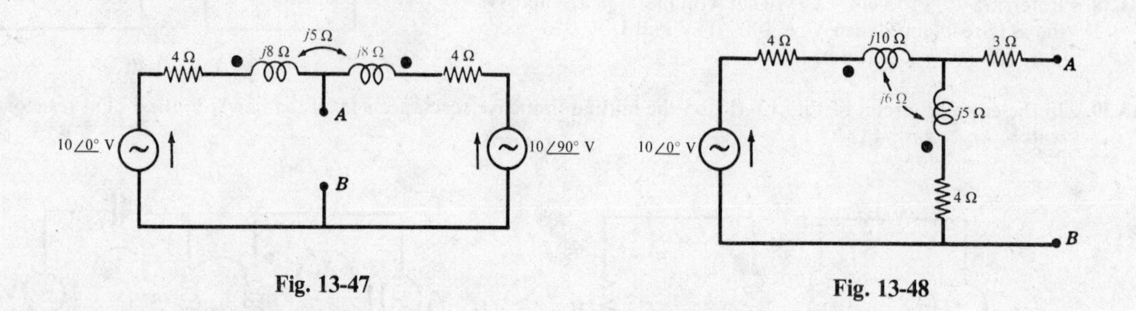

Fig. 13-47 Fig. 13-48

13.48. Obtain the Thevenin equivalent circuit at terminals AB of the coupled circuit shown in Fig. 13-48.
Ans. $\mathbf{Z}' = 8{\cdot}63\angle 48{\cdot}75^\circ\ \Omega$, $\mathbf{V}' = 4{\cdot}84\angle{-34{\cdot}7^\circ}$ V

13.49. Find the Norton equivalent circuit for the coupled network of Fig. 13-48.
Ans. $\mathbf{Z}' = 8{\cdot}63\angle 48{\cdot}75^\circ\ \Omega$, $\mathbf{I}' = 0{\cdot}560\angle{-83{\cdot}4^\circ}$ A

13.50. For the coupled circuit shown in Fig. 13-49 find the input impedance at the terminals of the voltage source **V**.
Ans. $7{\cdot}06 + j3{\cdot}22\ \Omega$

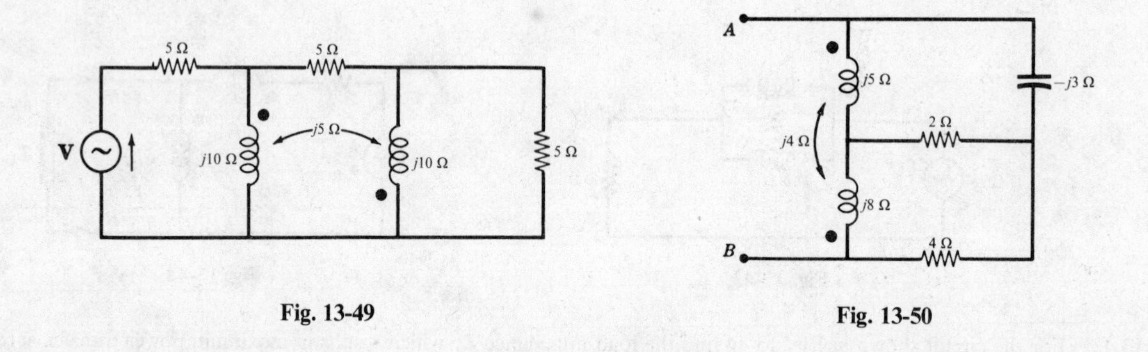

Fig. 13-49 Fig. 13-50

13.51. Find the equivalent impedance at terminals AB of the coupled network of Fig. 13-50. *Ans.* $6{\cdot}22 + j4{\cdot}65\ \Omega$

CHAPTER 14

Polyphase Systems

INTRODUCTION

A polyphase system consists of two or more equal voltages with fixed phase differences which supply power to loads connected to the lines. In the two-phase system two equal voltages differ in phase by 90°, while in the three-phase system the voltages have a phase difference of 120°. Systems of six or more phases are sometimes used in polyphase rectifiers to obtain a rectified voltage with low ripple, but three-phase is the system commonly used for generation and transmission of electric power.

TWO-PHASE SYSTEM

Rotation of the pair of perpendicular coils in Fig. 14-1(*a*) in the constant magnetic field results in induced voltages with a fixed 90° phase difference. With equal number of turns of the coils, the phasor and instantaneous voltages have equal magnitudes as shown in their respective diagrams in Figures 14-1(*b*) and (*c*).

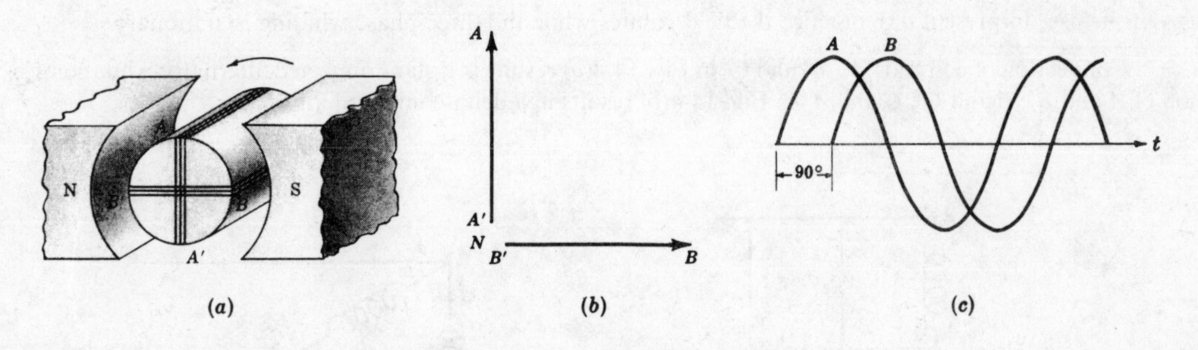

Fig. 14-1. Two-phase System

The voltage phasor diagram in Fig. 14-1(*b*) has as a reference $\mathbf{V}_{BN} = V_{\text{coil}}\angle 0^\circ$ and the voltage $\mathbf{V}_{AN} = V_{\text{coil}}\angle 90^\circ$. If the coil ends A' and B' are joined as line N, the two-phase system is contained on the three lines A, B and N. The potential between lines A and B exceeds the line to neutral voltages by the factor $\sqrt{2}$ and is obtained from the sum, $\mathbf{V}_{AB} = \mathbf{V}_{AN} + \mathbf{V}_{NB} = V_{\text{coil}}\angle 90^\circ + V_{\text{coil}}\angle 180^\circ = \sqrt{2}\,V_{\text{coil}}\angle 135^\circ$.

THREE-PHASE SYSTEM

The induced voltages in the three equally spaced coils in Fig. 14-2(*a*) below have a phase difference of 120°. The voltage in coil A reaches a maximum first, followed by B and then C for sequence ABC. This sequence is evident from the phasor diagram with its positive rotation counterclockwise where the phasors would pass a fixed point in the order A-B-C-A-B-$C\cdots$, and also from the instantaneous voltage plot of Fig. 14-2(*c*) below where the maxima occur in the same order.

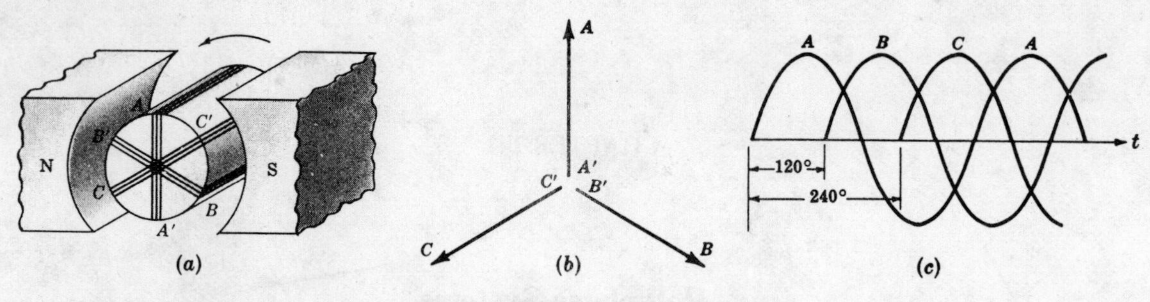

Fig. 14-2. Three-phase System

Rotation of the coils in the opposite direction results in the sequence CBA shown in Fig. 14-3.

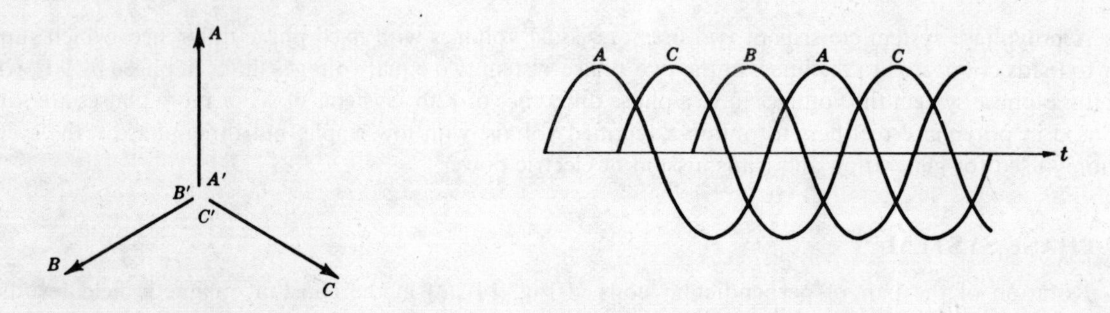

Fig. 14-3. Sequence CBA

Although the machine sketched in Fig. 14-2(a) is sound in theory, a number of practical limitations prevent its use. In present day practice the field rotates while the three-phase winding is stationary.

Connection of coil ends A', B' and C' in Fig. 14-4(a) results in a star-connected alternator while connection of A and B', B and C', C and A' in Fig. 14-4(b) results in a delta-connected alternator.

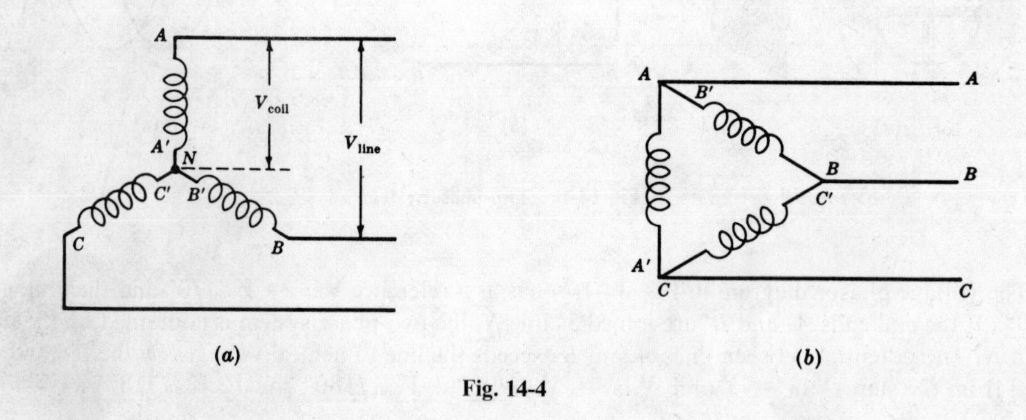

Fig. 14-4

In the star-connection the coil and line currents are equal and line to line voltage is $\sqrt{3}$ times the coil voltage. The delta-connection results in equal line and coil voltages but coil currents are $1/\sqrt{3}$ times the line currents. See Problem 14.2.

For either connection the lines A, B and C provide a three-phase system of voltages. The neutral point in the star-connection is the fourth conductor of the three-phase, four-wire system.

THREE-PHASE SYSTEM VOLTAGES

Selection of one voltage as the reference with a phase angle of zero determines the phase angles of all other voltages in the system. In this chapter $\mathbf{V}_{BC}$ is chosen as reference. The triangles in Figures 14-5(*a*) and (*b*) show all the voltages for sequences *ABC* and *CBA* respectively.

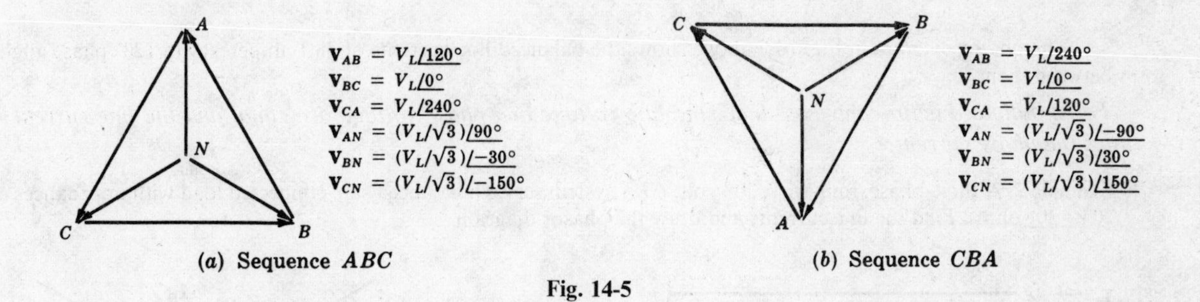

(*a*) Sequence *ABC* (*b*) Sequence *CBA*

Fig. 14-5

The *system voltage* is the voltage between any pair of lines, *A* and *B*, *B* and *C* or *C* and *A*. And in the four-wire system the magnitude of the line to neutral voltage is $1/\sqrt{3}$ times line voltage. For example, a three-phase, four-wire, 208 volt. *CBA* system has line voltages of 208 volts and line to neutral voltages of $208/\sqrt{3}$ or 120 volts. Referring to Fig. 14-5(*b*), the phase angles of the voltages are determined. Thus, $\mathbf{V}_{BC} = 208\angle 0°$ V, $\mathbf{V}_{AB} = 208\angle 240°$ V, $\mathbf{V}_{CA} = 208\angle 120°$ V, $\mathbf{V}_{AN} = 120\angle -90°$ V, $\mathbf{V}_{BN} = 120\angle 30°$ V and $\mathbf{V}_{CN} = 120\angle 150°$ V.

BALANCED THREE-PHASE LOADS

Example 1. A three-phase, three wire, 110 volt, *ABC* system supplies a delta connection of three equal impedances of $5\angle 45°$ ohms. Determine the line currents $\mathbf{I}_A$, $\mathbf{I}_B$, and $\mathbf{I}_C$ and draw the phasor diagram.

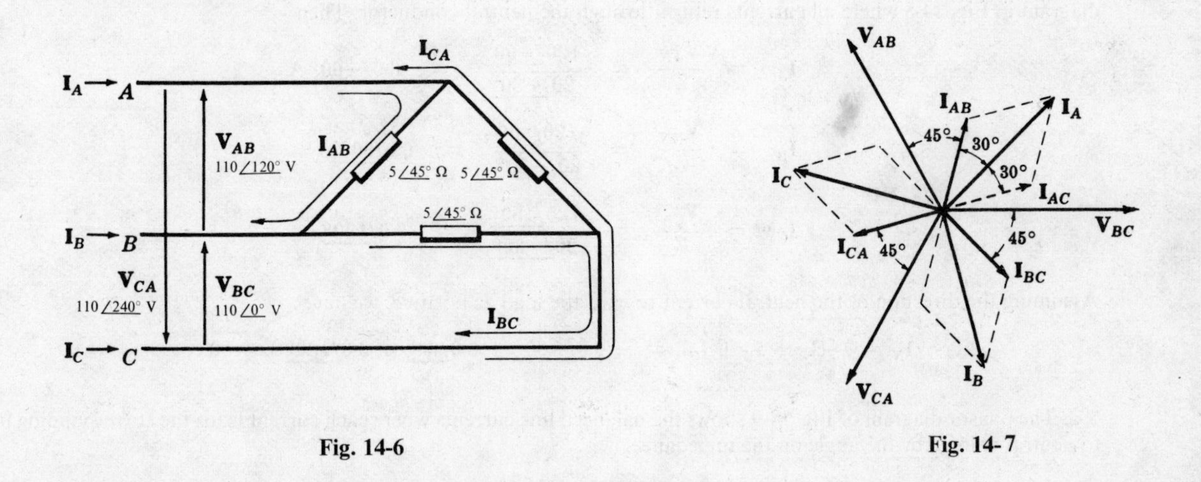

Fig. 14-6 Fig. 14-7

Sketch the circuit and apply the voltages as shown in Fig. 14-6. The positive directions of the line and phase currents are given on the circuit diagram. Then

$$\mathbf{I}_{AB} = \frac{\mathbf{V}_{AB}}{\mathbf{Z}} = \frac{110\angle 120°}{5\angle 45°} = 22\angle 75° = 5{\cdot}7 + j21{\cdot}2 \text{ A}$$

$$\mathbf{I}_{BC} = \frac{\mathbf{V}_{BC}}{\mathbf{Z}} = \frac{110\angle 0°}{5\angle 45°} = 22\angle -45° = 15{\cdot}55 - j15{\cdot}55 \text{ A}$$

$$\mathbf{I}_{CA} = \frac{\mathbf{V}_{CA}}{\mathbf{Z}} = \frac{110\angle 240°}{5\angle 45°} = 22\angle 195° = -21{\cdot}2 - j5{\cdot}7 \text{ A}$$

Apply Kirchhoff's current law at each corner of the load and write,

$$\mathbf{I}_A = \mathbf{I}_{AB} + \mathbf{I}_{AC} = 22\angle 75^\circ - 22\angle 195^\circ = 38{\cdot}1\angle 45^\circ \text{ A}$$

$$\mathbf{I}_B = \mathbf{I}_{BA} + \mathbf{I}_{BC} = -22\angle 75^\circ + 22\angle -45^\circ = 38{\cdot}1\angle -75^\circ \text{ A}$$

$$\mathbf{I}_C = \mathbf{I}_{CA} + \mathbf{I}_{CB} = 22\angle 195^\circ - 22\angle -45^\circ = 38{\cdot}1\angle 165^\circ \text{ A}$$

The phasor diagram in Fig. 14-7 above shows the balanced line currents of 38.1 amperes with 120° phase angles between them.

For a balanced delta-connected load, the line voltage and phase voltage are equal and the line current is $\sqrt{3}$ times the phase current.

Example 2. A three-phase, four-wire, 208 volt, *CBA* system serves a balanced star-connected load with impedances of $20\angle -30^\circ$ ohms. Find the line currents and draw the phasor diagram.

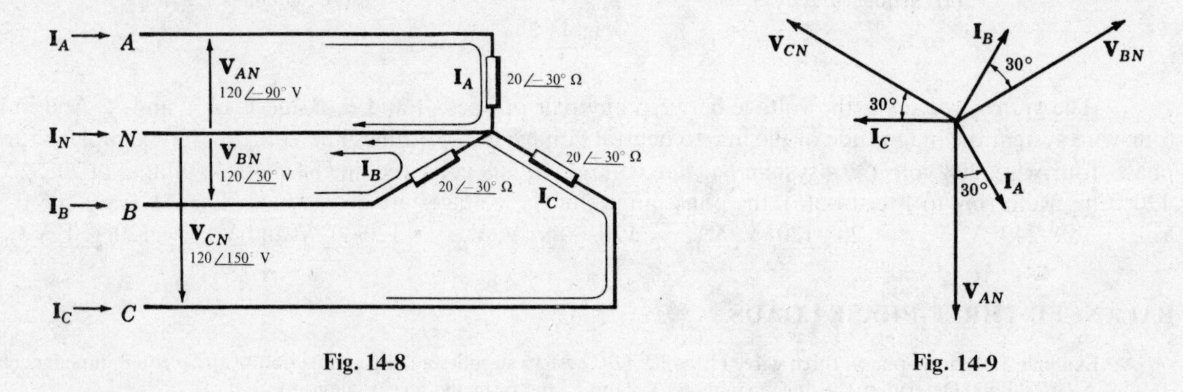

Fig. 14-8

Fig. 14-9

Sketch the circuit and apply the line to neutral voltages using Fig. 14-5(*b*). Select line currents as shown on the diagram in Fig. 14-8 where all currents return through the neutral conductor. Then

$$\mathbf{I}_A = \frac{\mathbf{V}_{AN}}{\mathbf{Z}} = \frac{120\angle -90^\circ}{20\angle -30^\circ} = 6{\cdot}0\angle -60^\circ \text{ A}$$

$$\mathbf{I}_B = \frac{\mathbf{V}_{BN}}{\mathbf{Z}} = \frac{120\angle 30^\circ}{20\angle -30^\circ} = 6{\cdot}0\angle 60^\circ \text{ A}$$

$$\mathbf{I}_C = \frac{\mathbf{V}_{CN}}{\mathbf{Z}} = \frac{120\angle 150^\circ}{20\angle -30^\circ} = 6{\cdot}0\angle 180^\circ \text{ A}$$

Assuming the direction of the neutral current toward the load as positive, we have

$$\mathbf{I}_N = -(\mathbf{I}_A + \mathbf{I}_B + \mathbf{I}_C) = -(6{\cdot}0\angle -60^\circ + 6{\cdot}0\angle 60^\circ + 6{\cdot}0\angle 180^\circ) = 0$$

The phasor diagram of Fig. 14-9 shows the balanced line currents where each current leads the corresponding line to neutral voltage by the angle on the impedance.

In a balanced star-connected load the line currents and phase currents are equal, the neutral current is zero, and the line voltage is $\sqrt{3}$ times the phase voltage, i.e., $V_L = \sqrt{3}\, V_P$.

ONE-LINE EQUIVALENT CIRCUIT FOR BALANCED LOADS

According to the Y-Δ transformations of Chapter 12 a set of three equal impedances $\mathbf{Z}_\Delta$ in a delta connection is equivalent to a set of three equal star-connected impedances $\mathbf{Z}_Y$, where $\mathbf{Z}_Y = (1/3)\mathbf{Z}_\Delta$. Then a more direct computation of the star circuit is possible for balanced three-phase loads of either type.

The one-line equivalent circuit is one phase of the three-phase, four-wire, star-connected circuit in Fig. 14-10, except that a voltage is used which has the line to neutral magnitude and a phase angle of zero. The line current calculated for this circuit has a phase angle with respect to the phase angle of zero on the voltage. Then the actual line currents $\mathbf{I}_A$, $\mathbf{I}_B$ and $\mathbf{I}_C$ will lead or lag their respective line to neutral voltages by this same phase angle.

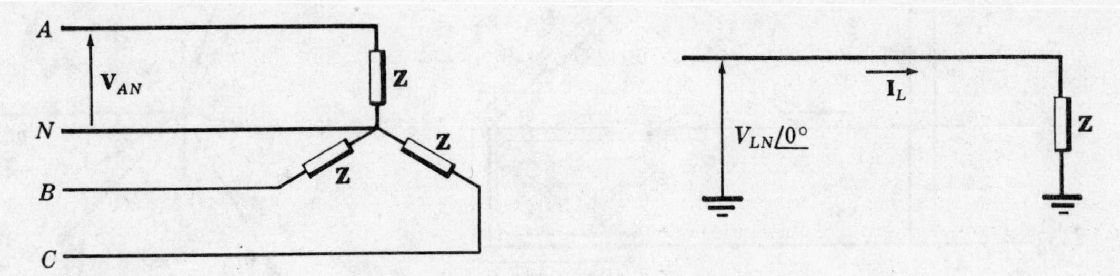

Fig. 14-10. One-line Equivalent Circuit

Example 3. Calculate the line currents of Example 1 by the one-line equivalent method.

Draw the one-line circuit and mark a Δ at the load to show that the actual impedances were in a delta-connection. The impedance of the star-connected equivalent is

$$\mathbf{Z}_Y = \mathbf{Z}_\Delta/3 = (5/3)\underline{/45°}\,\Omega$$

and the line to neutral voltage is

$$V_{LN} = V_L/\sqrt{3} = 110/\sqrt{3} = 63{\cdot}5\ \text{V}$$

Then the line current is

$$\mathbf{I}_L = \frac{\mathbf{V}_{LN}}{\mathbf{Z}} = \frac{63{\cdot}5\underline{/0°}}{(5/3)\underline{/45°}} = 38{\cdot}1\underline{/-45°}\,\text{A}$$

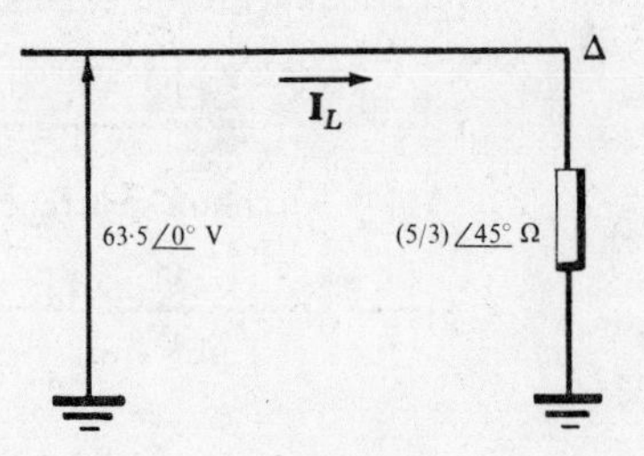

Fig. 14-11

Since this current lags the voltage by 45°, the line currents $\mathbf{I}_A$, $\mathbf{I}_B$ and $\mathbf{I}_C$ lag their respective voltages $\mathbf{V}_{AN}$, $\mathbf{V}_{BN}$ and $\mathbf{V}_{CN}$ by 45°. The angles on these voltages are obtained from the *ABC* triangle of Fig. 14-5(*a*). The line to neutral voltages and the corresponding line currents are tabulated below.

$$\mathbf{V}_{AN} = 63{\cdot}5\underline{/90°}\ \text{V} \qquad \mathbf{I}_A = 38{\cdot}1\underline{/90° - 45°} = 38{\cdot}1\underline{/45°}\ \text{A}$$

$$\mathbf{V}_{BN} = 63{\cdot}5\underline{/-30°}\ \text{V} \qquad \mathbf{I}_B = 38{\cdot}1\underline{/-30° - 45°} = 38{\cdot}1\underline{/-75°}\ \text{A}$$

$$\mathbf{V}_{CN} = 63{\cdot}5\underline{/-150°}\ \text{V} \qquad \mathbf{I}_C = 38{\cdot}1\underline{/-150° - 45°} = 38{\cdot}1\underline{/-195°}\ \text{A}$$

These currents are identical to those obtained in Example 1. If the phase currents in the delta-connected impedances are required, they may be found from $I_P = I_L/\sqrt{3} = 38{\cdot}1/\sqrt{3} = 22$ A. The phase angles on these currents are obtained by first setting the phase angles on the line to line voltages and then determining the currents such that they lag by 45°. Hence,

$$\mathbf{V}_{AB} = 110\underline{/120°}\ \text{V} \qquad \mathbf{I}_{AB} = 22\underline{/120° - 45°} = 22\underline{/75°}\ \text{A}$$

$$\mathbf{V}_{BC} = 110\underline{/0°}\ \text{V} \qquad \mathbf{I}_{BC} = 22\underline{/0° - 45°} = 22\underline{/-45°}\ \text{A}$$

$$\mathbf{V}_{CA} = 110\underline{/240°}\ \text{V} \qquad \mathbf{I}_{CA} = 22\underline{/240° - 45°} = 22\underline{/195°}\ \text{A}$$

UNBALANCED DELTA-CONNECTED LOAD

The solution of the unbalanced delta-connected load consists of computing the phase currents and then applying Kirchhoff's current law to the junctions to obtain the three line currents. The line currents will not be equal nor will they have a 120° phase difference as was the case with balanced loads.

Example 4.

A three-phase, three-wire, 240 volt, ABC system has a delta-connected load with $\mathbf{Z}_{AB} = 10\angle 0°\ \Omega$, $\mathbf{Z}_{BC} = 10\angle 30°\ \Omega$ and $\mathbf{Z}_{CA} = 15\angle -30°\ \Omega$. Obtain the three line currents and draw the phasor diagram.

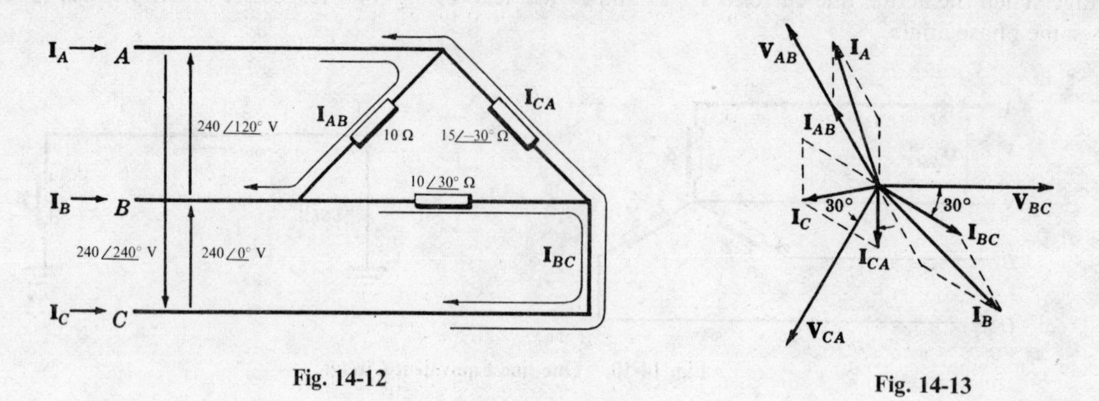

Fig. 14-12

Fig. 14-13

Construct the circuit diagram as in Fig. 14-12, and apply the phasor voltages. Then the phase currents as shown on the diagram are independent and given by

$$\mathbf{I}_{AB} = \frac{\mathbf{V}_{AB}}{\mathbf{Z}_{AB}} = \frac{240\angle 120°}{10\angle 0°} = 24\angle 120°\ \text{A},\ \mathbf{I}_{BC} = \frac{\mathbf{V}_{BC}}{\mathbf{Z}_{BC}} = 24\angle -30°\ \text{A},\ \mathbf{I}_{CA} = \frac{\mathbf{V}_{CA}}{\mathbf{Z}_{CA}} = 16\angle 270°\ \text{A}$$

Apply Kirchhoff's current law to the junctions of the load and write

$$\mathbf{I}_A = \mathbf{I}_{AB} + \mathbf{I}_{AC} = 24\angle 120° - 16\angle 270° = 38{\cdot}7\angle 108{\cdot}1°\ \text{A}$$

$$\mathbf{I}_B = \mathbf{I}_{BA} + \mathbf{I}_{BC} = -24\angle 120° + 24\angle -30° = 46{\cdot}4\angle -45°\ \text{A}$$

$$\mathbf{I}_C = \mathbf{I}_{CA} + \mathbf{I}_{CB} = 16\angle 270° - 24\angle -30° = 21{\cdot}2\angle 190{\cdot}9°\ \text{A}$$

The corresponding phasor diagram is shown in Fig. 14-13.

UNBALANCED FOUR-WIRE, STAR-CONNECTED LOAD

On a four-wire system the neutral conductor will carry a current when the load is unbalanced and the voltage across each of the load impedances remains fixed with the same magnitude as the line to neutral voltage. The line currents are unequal and do not have a 120° phase difference.

Example 5.

A three-phase, four-wire, 208 volt, CBA system has a star-connected load with $\mathbf{Z}_A = 6\angle 0°\ \Omega$, $\mathbf{Z}_B = 6\angle 30°\ \Omega$ and $\mathbf{Z}_C = 5\angle 45°\ \Omega$. Obtain the three line currents and the neutral current. Draw the phasor diagram.

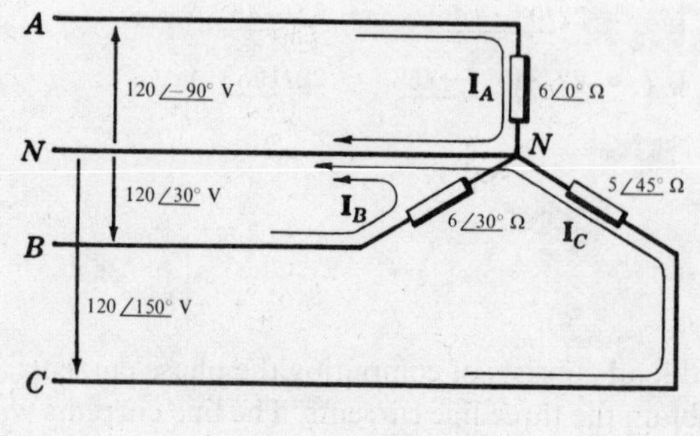

Fig. 14-14

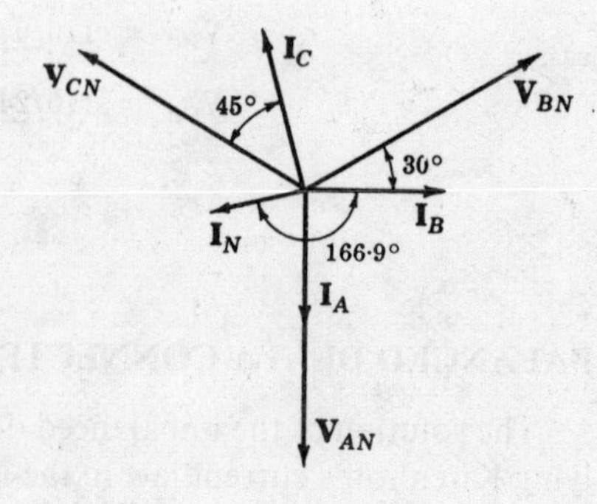

Fig. 14-15

Construct the circuit diagram as in Fig. 14-14 above. Apply the phasor voltages and select the line currents as shown. The currents are independent and given by

$$\mathbf{I}_A = \frac{\mathbf{V}_{AN}}{\mathbf{Z}_A} = \frac{120\angle -90^\circ}{6\angle 0^\circ} = 20\angle -90^\circ \text{ A}, \quad \mathbf{I}_B = \frac{\mathbf{V}_{BN}}{\mathbf{Z}_B} = 20\angle 0^\circ \text{ A}, \quad \mathbf{I}_C = \frac{\mathbf{V}_{CN}}{\mathbf{Z}_C} = 24\angle 105^\circ \text{ A}$$

The neutral conductor contains the sum of the line currents $\mathbf{I}_A$, $\mathbf{I}_B$ and $\mathbf{I}_C$. Then assuming a positive direction of $\mathbf{I}_N$ toward the load,

$$\mathbf{I}_N = -(\mathbf{I}_A + \mathbf{I}_B + \mathbf{I}_C) = -(20\angle -90^\circ + 20\angle 0^\circ + 24\angle 105^\circ) = 14{\cdot}1\angle -166{\cdot}9^\circ \text{ A}$$

The diagram phasor is shown in Fig. 14-15 above.

UNBALANCED THREE-WIRE, STAR-CONNECTED LOAD

With only the three lines A, B and C connected to an unbalanced star load the common point of the three load impedances is not at the potential of the neutral and is marked "O" instead of N. The voltages across the three impedances can vary considerably from line to neutral magnitude, as shown by the voltage triangle which relates all of the voltages in the circuit. Of particular interest is the displacement of "O" from N, the *displacement neutral voltage*.

Example 6.

A three-phase, three-wire, 208 volt, CBA system has a star-connected load with $\mathbf{Z}_A = 6\angle 0^\circ\,\Omega$, $\mathbf{Z}_B = 6\angle 30^\circ\,\Omega$ and $\mathbf{Z}_C = 5\angle 45^\circ\,\Omega$. Obtain the line currents and the phasor voltage across each impedance. Construct the voltage triangle and determine the displacement neutral voltage, $\mathbf{V}_{ON}$.

Draw the circuit diagram and select mesh currents $\mathbf{I}_1$ and $\mathbf{I}_2$ as shown in Fig. 14-16. Write the corresponding matrix equations of $\mathbf{I}_1$ and $\mathbf{I}_2$ as follows.

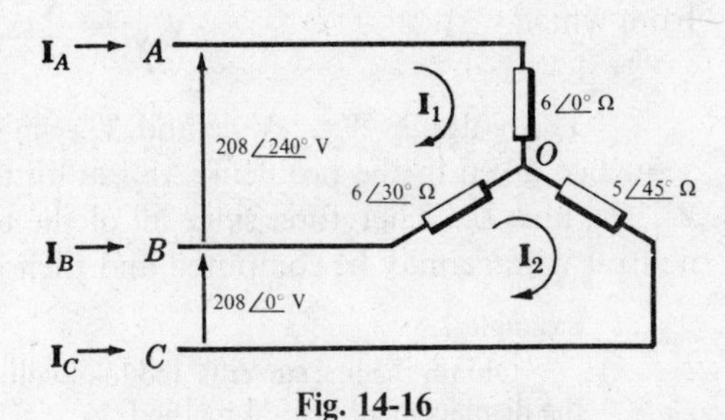

Fig. 14-16

$$\begin{bmatrix} 6\angle 0^\circ + 6\angle 30^\circ & -6\angle 30^\circ \\ -6\angle 30^\circ & 6\angle 30^\circ + 5\angle 45^\circ \end{bmatrix} \begin{bmatrix} \mathbf{I}_1 \\ \mathbf{I}_2 \end{bmatrix} = \begin{bmatrix} 208\angle 240^\circ \\ 208\angle 0^\circ \end{bmatrix}$$

from which $\mathbf{I}_1 = 23{\cdot}3\angle 261{\cdot}1^\circ$ A and $\mathbf{I}_2 = 26{\cdot}5\angle -63{\cdot}4^\circ$ A. Then line currents $\mathbf{I}_A$, $\mathbf{I}_B$ and $\mathbf{I}_R$ directed as shown on the diagram are

$$\mathbf{I}_A = \mathbf{I}_1 = 23{\cdot}3\angle 261{\cdot}1^\circ \text{ A}$$

$$\mathbf{I}_B = \mathbf{I}_2 - \mathbf{I}_1 = 26{\cdot}5\angle -63{\cdot}4^\circ - 23{\cdot}3\angle 261{\cdot}1^\circ = 15{\cdot}45\angle -2{\cdot}5^\circ \text{ A}$$

$$\mathbf{I}_C = -\mathbf{I}_2 = 26{\cdot}5\angle 116{\cdot}6^\circ \text{ A}$$

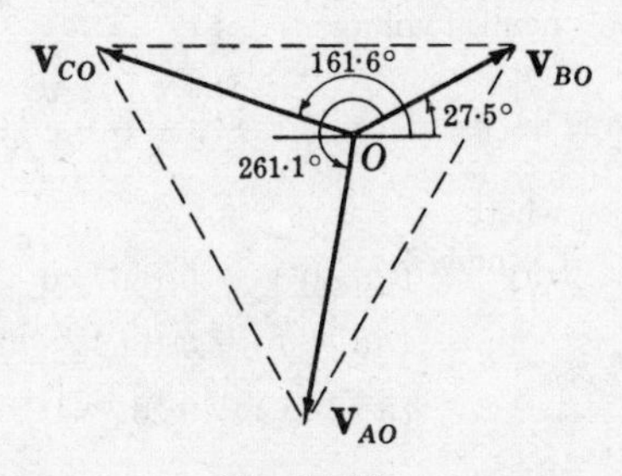

Fig. 14-17

Now the voltages across the three impedances are given by the products of the line currents and the corresponding impedances.

$$\mathbf{V}_{AO} = \mathbf{I}_A\mathbf{Z}_A = 23{\cdot}3\angle 261{\cdot}1^\circ\,(6\angle 0^\circ) = 139{\cdot}8\angle 261{\cdot}1^\circ \text{ V}$$

$$\mathbf{V}_{BO} = \mathbf{I}_B\mathbf{Z}_B = 15{\cdot}45\angle -2{\cdot}5^\circ\,(6\angle 30^\circ) = 92{\cdot}7\angle 27{\cdot}5^\circ \text{ V}$$

$$\mathbf{V}_{CO} = \mathbf{I}_C\mathbf{Z}_C = 26{\cdot}5\angle 116{\cdot}6^\circ\,(5\angle 45^\circ) = 132{\cdot}5\angle 161{\cdot}6^\circ \text{ V}$$

The phasor diagram of these three voltages shown in Fig. 14-17 forms an equilateral triangle. In Fig. 14-18 this triangle is redrawn and the neutral is added, thus showing the displacement neutral voltage $\mathbf{V}_{ON}$. This voltage may be computed using any of the three points A, B or C and following the conventional double subscript notation. Using point A, we obtain

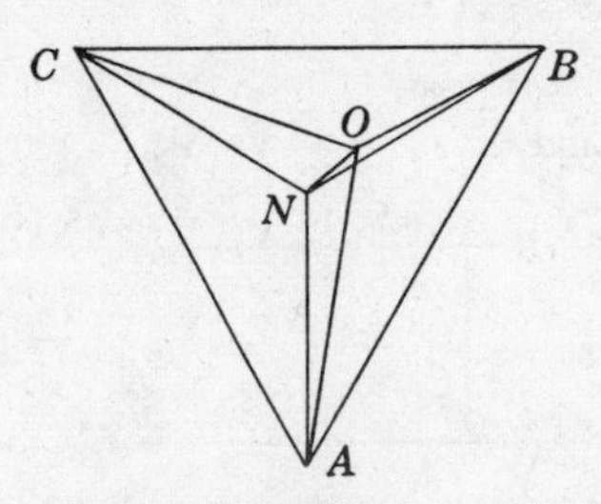

Fig. 14-18

$$\mathbf{V}_{ON} = \mathbf{V}_{OA} + \mathbf{V}_{AN} = -139{\cdot}8\angle 261{\cdot}1^\circ + 120\angle -90^\circ = 28{\cdot}1\angle 39{\cdot}8^\circ \text{ V}$$

DISPLACEMENT NEUTRAL METHOD, UNBALANCED THREE-WIRE STAR LOAD

In Example 6 the displacement neutral voltage $\mathbf{V}_{ON}$ was obtained in terms of load voltages. If we determine a relation for $\mathbf{V}_{ON}$ independent of the load voltages, then the required currents and voltages of Example 6 are obtained more directly as shown in Example 7.

To obtain the displacement neutral voltage write the line currents in terms of the load voltages and load admittances.

$$\mathbf{I}_A = \mathbf{V}_{AO}\mathbf{Y}_A,\ \mathbf{I}_B = \mathbf{V}_{BO}\mathbf{Y}_B,\ \mathbf{I}_C = \mathbf{V}_{CO}\mathbf{Y}_C \qquad (1)$$

Now apply Kirchhoff's current law at point O in Fig. 14-19 and write

$$\mathbf{I}_A + \mathbf{I}_B + \mathbf{I}_C = 0 \qquad (2)$$

or

$$\mathbf{V}_{AO}\mathbf{Y}_A + \mathbf{V}_{BO}\mathbf{Y}_B + \mathbf{V}_{CO}\mathbf{Y}_C = 0 \qquad (3)$$

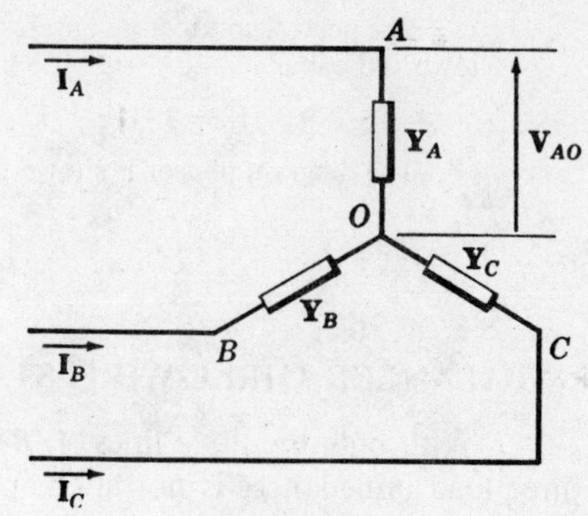

Fig. 14-19

Referring to the diagram of Fig. 14-18, express the voltages $\mathbf{V}_{AO}$, $\mathbf{V}_{BO}$ and $\mathbf{V}_{CO}$ in terms of their two component voltages, i.e.,

$$\mathbf{V}_{AO} = \mathbf{V}_{AN} + \mathbf{V}_{NO} \qquad \mathbf{V}_{BO} = \mathbf{V}_{BN} + \mathbf{V}_{NO} \qquad \mathbf{V}_{CO} = \mathbf{V}_{CN} + \mathbf{V}_{NO} \qquad (4)$$

Substituting the expressions of *(4)* into *(3)* we obtain

$$(\mathbf{V}_{AN} + \mathbf{V}_{NO})\mathbf{Y}_A + (\mathbf{V}_{BN} + \mathbf{V}_{NO})\mathbf{Y}_B + (\mathbf{V}_{CN} + \mathbf{V}_{NO})\mathbf{Y}_C = 0 \qquad (5)$$

from which

$$\mathbf{V}_{ON} = \frac{\mathbf{V}_{AN}\,\mathbf{Y}_A + \mathbf{V}_{BN}\,\mathbf{Y}_B + \mathbf{V}_{CN}\,\mathbf{Y}_C}{\mathbf{Y}_A + \mathbf{Y}_B + \mathbf{Y}_C} \qquad (6)$$

The voltages $\mathbf{V}_{AN}$, $\mathbf{V}_{BN}$ and $\mathbf{V}_{CN}$ in equation *(6)* are obtained from the triangle of Fig. 14-5 for the sequence given in the problem. And admittances $\mathbf{Y}_A$, $\mathbf{Y}_B$ and $\mathbf{Y}_C$ are the reciprocals of the load impedances $\mathbf{Z}_A$, $\mathbf{Z}_B$ and $\mathbf{Z}_C$. Therefore, since all of the terms in *(6)* are either given or readily obtained, the displacement neutral voltage may be computed and then used to determine the line currents.

Example 7.

Obtain the line currents and load voltages in Example 6 by the displacement neutral method.

Referring to Fig. 14-20, the equation for the displacement neutral voltage is

$$\mathbf{V}_{ON} = \frac{\mathbf{V}_{AN}\mathbf{Y}_A + \mathbf{V}_{BN}\mathbf{Y}_B + \mathbf{V}_{CN}\mathbf{Y}_C}{\mathbf{Y}_A + \mathbf{Y}_B + \mathbf{Y}_C}$$

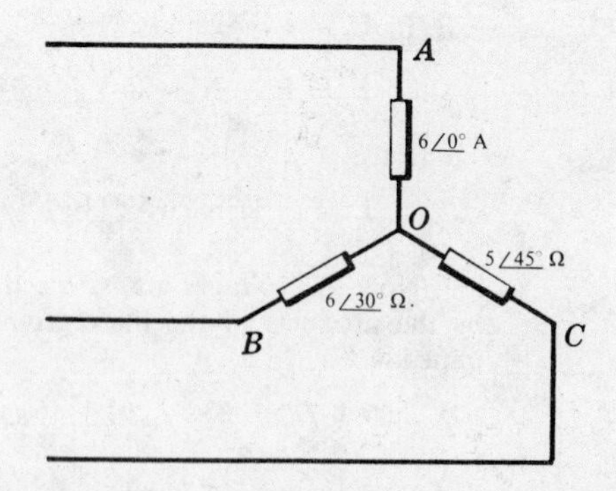

Fig. 14-20

where

$$\mathbf{Y}_A = 1/(6\angle 0^\circ) = 0{\cdot}1667\angle 0^\circ = 0{\cdot}1667\ \text{S}$$
$$\mathbf{Y}_B = 1/(6\angle 30^\circ) = 0{\cdot}1667\angle -30^\circ = 0{\cdot}1443 - j0{\cdot}0833\ \text{S}$$
$$\mathbf{Y}_C = 1/(5\angle 45^\circ) = 0{\cdot}20\angle -45^\circ = 0{\cdot}1414 - j0{\cdot}1414\ \text{S}$$

$$\mathbf{Y}_A + \mathbf{Y}_B + \mathbf{Y}_C = 0{\cdot}4524 - j0{\cdot}2247\ \text{S}$$
$$= 0{\cdot}504\angle -26{\cdot}5^\circ\ \text{S}$$

and

$$\mathbf{V}_{AN}\mathbf{Y}_A = 120\angle -90^\circ\,(0{\cdot}1667\angle 0^\circ) = 20\angle -90^\circ = -j20\ \text{A}$$
$$\mathbf{V}_{BN}\mathbf{Y}_B = 120\angle 30^\circ\,(0{\cdot}1667\angle -30^\circ) = 20\angle 0^\circ = 20\ \text{A}$$
$$\mathbf{V}_{CN}\mathbf{Y}_C = 120\angle 150^\circ\,(0{\cdot}20\angle -45^\circ) = 24\angle 105^\circ = -6{\cdot}2 + j23{\cdot}2\ \text{A}$$

$$\mathbf{V}_{AN}\mathbf{Y}_A + \mathbf{V}_{BN}\mathbf{Y}_B + \mathbf{V}_{CN}\mathbf{Y}_C = 13{\cdot}8 + j3{\cdot}2 = 14{\cdot}1\angle 13{\cdot}1^\circ\ \text{A}$$

Then

$$\mathbf{V}_{ON} = 14{\cdot}1\angle 13{\cdot}1^\circ/0{\cdot}504\angle -26{\cdot}5^\circ = 28{\cdot}0\angle 39{\cdot}6^\circ\ \text{V}$$

Voltages $\mathbf{V}_{AO}$, $\mathbf{V}_{BO}$ and $\mathbf{V}_{CO}$ are obtained using $\mathbf{V}_{NO}$ and the appropriate line to neutral voltage.

$$\mathbf{V}_{AO} = \mathbf{V}_{AN} + \mathbf{V}_{NO} = 120\angle{-90^\circ} - 28{\cdot}0\angle{39{\cdot}6^\circ} = 139{\cdot}5\angle{261{\cdot}1}\ \text{V}$$
$$\mathbf{V}_{BO} = \mathbf{V}_{BN} + \mathbf{V}_{NO} = 120\angle{30^\circ} - 28{\cdot}0\angle{39{\cdot}6^\circ} = 92{\cdot}5\angle{27{\cdot}1^\circ}\ \text{V}$$
$$\mathbf{V}_{CO} = \mathbf{V}_{CN} + \mathbf{V}_{NO} = 120\angle{150^\circ} - 28{\cdot}0\angle{39{\cdot}6^\circ} = 132{\cdot}5\angle{161{\cdot}45^\circ}\ \text{V}$$

The line currents are readily obtained from the voltages and the corresponding load admittances:

$$\mathbf{I}_A = \mathbf{V}_{AO}\mathbf{Y}_A = 139{\cdot}5\angle{261{\cdot}1^\circ}\,(0{\cdot}1667\angle{0^\circ}) = 23{\cdot}2\angle{261{\cdot}1^\circ}\ \text{A}$$
$$\mathbf{I}_B = \mathbf{V}_{BO}\mathbf{Y}_B = 92{\cdot}5\angle{27{\cdot}1^\circ}\,(0{\cdot}1667\angle{-30^\circ}) = 15{\cdot}4\angle{-2{\cdot}9^\circ}\ \text{A}$$
$$\mathbf{I}_C = \mathbf{V}_{CO}\mathbf{Y}_C = 132{\cdot}5\angle{161{\cdot}45^\circ}\,(0{\cdot}20\angle{-45^\circ}) = 26{\cdot}5\angle{116{\cdot}45^\circ}\ \text{A}$$

The above currents and voltages compare favourably with the results of Example 6.

POWER IN BALANCED THREE-PHASE LOADS

Since the phase impedances of balanced star or delta loads contain equal currents, the phase power is one-third of the total power. The voltage across impedance $\mathbf{Z}_\Delta$ in Fig. 14-21(*a*) is *line voltage* and the current is *phase current*. The angle between the voltage and current is the angle on the impedance. Then the phase power is

$$P_P = V_L I_P \cos\theta \qquad (7)$$

and the total power is

$$P_T = 3\,V_L I_P \cos\theta \qquad (8)$$

Since $I_L = \sqrt{3}\,I_P$ in balanced delta-connected loads,

$$P_T = \sqrt{3}\,V_L I_L \cos\theta \qquad (9)$$

The star-connected impedances of Fig. 14-21(*b*) contain the *line currents*, and the voltage across $\mathbf{Z}_Y$ is a *phase voltage*. The angle between them is the angle on the impedance. Then the phase power is

$$P_P = V_P I_L \cos\theta \qquad (10)$$

and the total power is

$$P_T = 3\,V_P I_L \cos\theta \qquad (11)$$

Since $V_L = \sqrt{3}\,V_P$,

$$P_T = \sqrt{3}\,V_L I_L \cos\theta \qquad (12)$$

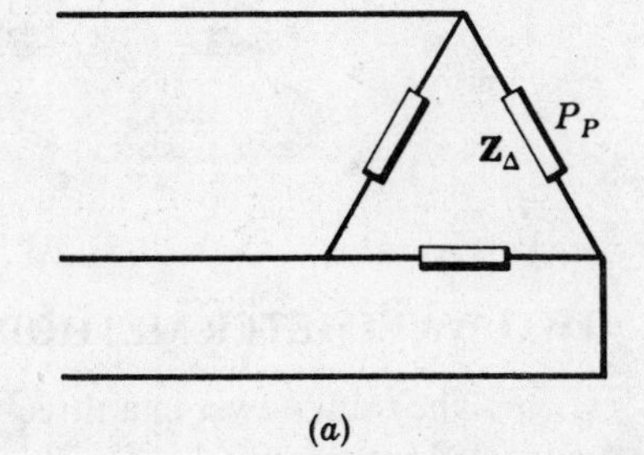

(*a*)

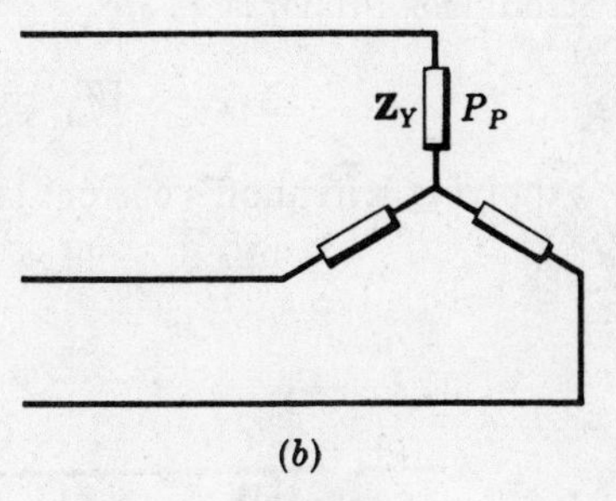

(*b*)

Fig. 14-21

Since equations (*9*) and (*12*) are identical, the total power in any balanced three-phase load is given by $\sqrt{3}\,V_L I_L \cos\theta$ where θ is the angle on the load impedance or the angle on an equivalent impedance in the case where several balanced loads are served from the same system.

The total volt-amperes S_T and the total reactive power Q_T are related to P_T in Chapter 7. Therefore a balanced three-phase load has the power, apparent power, and reactive power given by

$$P_T = \sqrt{3}\,V_L I_L \cos\theta \qquad S_T = \sqrt{3}\,V_L I_L \qquad Q_T = \sqrt{3}\,V_L I_L \sin\theta \qquad (13)$$

WATTMETERS AND FOUR-WIRE STAR LOADS

A wattmeter is an instrument with a potential coil and a current coil so arranged that its deflection is proportional to $VI\cos\theta$ where θ is the angle between the voltage and current. A four-wire, star-connected load requires three wattmeters with one meter in each line as shown in Fig. 14-22(*a*).

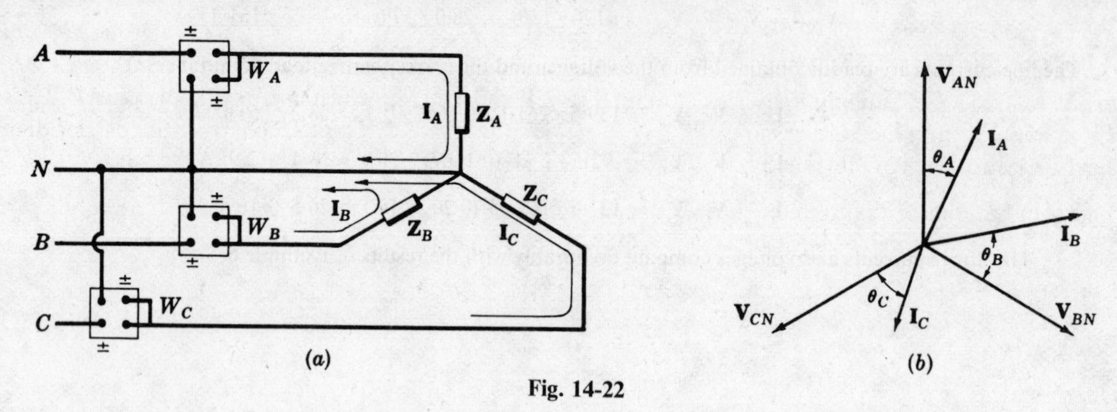

Fig. 14-22

The phasor diagram in Fig. 14-22(*b*) assumes a lagging current in phase *A* and leading currents in phases *B* and *C* with phase angles θ_A, θ_B and θ_C respectively. Then the wattmeter readings are

$$W_A = V_{AN} I_A \cos \measuredangle_A^{AN}, \qquad W_B = V_{BN} I_B \cos \measuredangle_B^{BN}, \qquad W_C = V_{CN} I_C \cos \measuredangle_C^{CN} \tag{14}$$

where $\measuredangle_A^{AN}$ indicates the angle between V_{AN} and $\mathbf{I}_A$. Wattmeter W_A reads the power in phase *A* and wattmeters W_B and W_C the power in phases *B* and *C* respectively. The total power is

$$P_T = W_A + W_B + W_C \tag{15}$$

TWO-WATTMETER METHOD

The total power in a three-phase, three-wire load is given by the sum of the readings on two wattmeters connected in any two lines with their potential coils connected to the third line as shown in Fig. 14-23. The readings of the meters are

$$W_A = V_{AB} I_A \cos \measuredangle_A^{AB} \quad \text{and} \quad W_C = V_{CB} I_C \cos \measuredangle_C^{CB} \tag{16}$$

Applying Kirchhoff's current law to junctions *A* and *C* of the delta load, we obtain

$$\mathbf{I}_A = \mathbf{I}_{AB} + \mathbf{I}_{AC} \quad \text{and} \quad \mathbf{I}_C = \mathbf{I}_{CA} + \mathbf{I}_{CB} \tag{17}$$

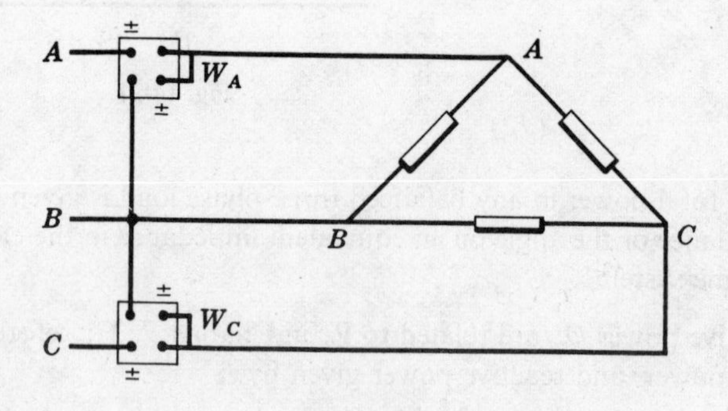

Fig. 14-23

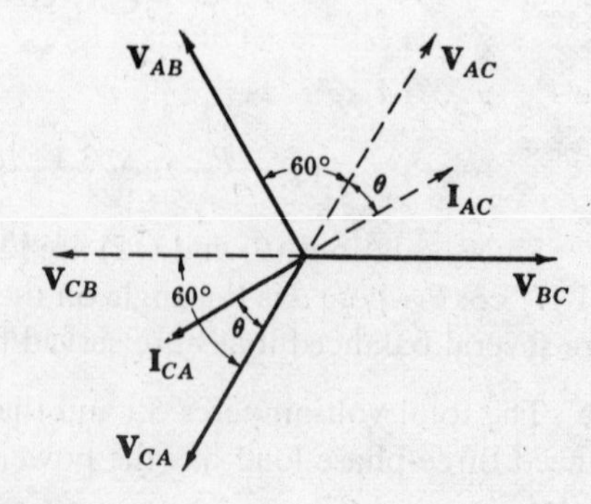

Fig. 14-24

Substituting the expressions for $\mathbf{I}_A$ and $\mathbf{I}_C$ given in (*17*) into the wattmeter equations (*16*), we obtain

$$\begin{aligned} W_A &= V_{AB} I_{AB} \cos \measuredangle_{AB}^{AB} + V_{AB} I_{AC} \cos \measuredangle_{AC}^{AB} \\ W_C &= V_{CB} I_{CA} \cos \measuredangle_{CA}^{CB} + V_{CB} I_{CB} \cos \measuredangle_{CB}^{CB} \end{aligned} \tag{18}$$

The terms $V_{AB} I_{AB} \cos \measuredangle_{AB}^{AB}$ and $V_{CB} I_{CB} \cos \measuredangle_{CB}^{CB}$ are immediately recognized as the power in phases *AB* and *CB* of the load. The two remaining terms contain $V_{AB} I_{AC}$ and $V_{CB} I_{CA}$ which can now be written $V_L I_{AC}$ since both V_{AB} and V_{CB} are line voltages and $I_{AC} = I_{CA}$. To identify these two terms, construct the phasor diagram of Fig. 14-24 above where current $\mathbf{I}_{AC}$ is assumed to lag $\mathbf{V}_{AC}$ by θ.

From the phasor diagram,

$$\measuredangle_{AC}^{AB} = 60° + \theta \quad \text{and} \quad \measuredangle_{CA}^{CB} = 60° - \theta \tag{19}$$

Now add the two remaining wattmeter terms from (*18*) and substitute $(60° + \theta)$ and $(60° - \theta)$ for $\measuredangle_{AC}^{AB}$ and $\measuredangle_{CA}^{CB}$ respectively.

$$V_L I_{AC} \cos(60° + \theta) + V_L I_{AC} \cos(60° - \theta) \tag{20}$$

Since $\cos(x \pm y) = \cos x \cos y \pm \sin x \sin y$, we write

$$V_L I_{AC} (\cos 60° \cos\theta - \sin 60° \sin\theta + \cos 60° \cos\theta + \sin 60° \sin\theta) \tag{21}$$

or

$$V_L I_{AC} \cos\theta \tag{22}$$

which is the power in the remaining phase of the load, phase *AC*. Thus we find that two wattmeters indicate the total power in a delta-connected load. The two-wattmeter method for a star-connected load is left as an exercise for the reader.

TWO-WATTMETER METHOD APPLIED TO BALANCED LOADS

To show the application of the two-wattmeter method to balanced loads, consider the star-connection of the three equal impedances shown in Fig. 14-25(*a*). The phasor diagram is drawn in Fig. 14-25(*b*) for the *ABC* sequence with the assumption of a lagging current with phase angle θ.

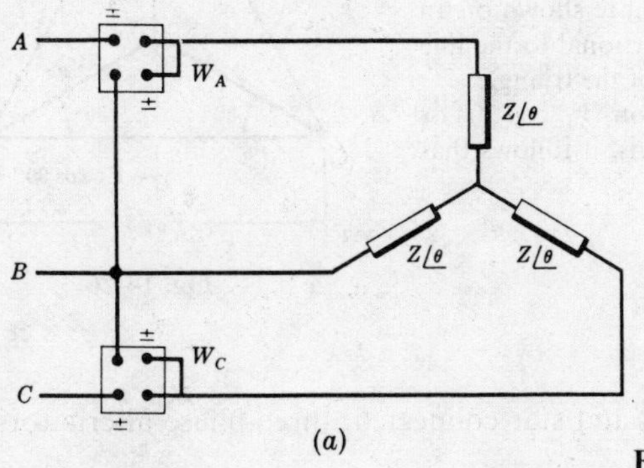

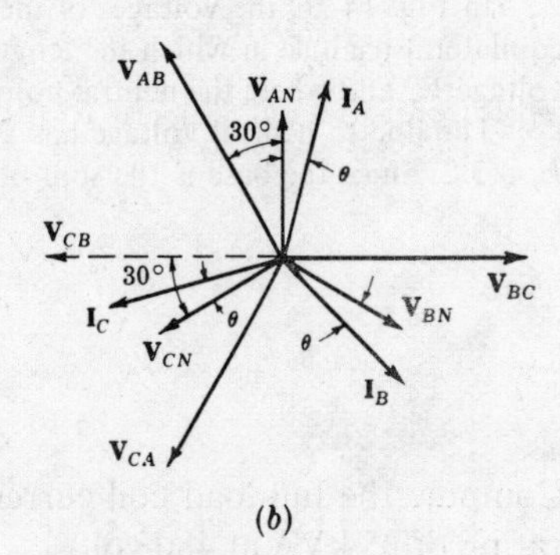

Fig. 14-25

Now with the wattmeters in lies *A* and *C*, their readings are

$$W_A = V_{AB} I_A \cos \measuredangle_A^{AB} \quad \text{and} \quad W_C = V_{CB} I_C \cos \measuredangle_C^{CB} \tag{23}$$

From the phasor diagram,

$$\measuredangle_A^{AB} = 30° + \theta \quad \text{and} \quad \measuredangle_C^{CB} = 30° - \theta \tag{24}$$

Substituting (*24*) in (*23*), we have

$$W_A = V_{AB}I_A \cos(30° + \theta) \text{ and } W_C = V_{CB}I_C \cos(30° - \theta) \tag{25}$$

When the two-wattmeter method is used on a balanced load, the wattmeter readings are $V_L I_L (30° + \theta)$ and $V_L I_L \cos(30° - \theta)$ where θ is the angle on the impedance. The two readings can be used to determine the angle θ.

Writing the expression for W_1 and using the cosine of the sum of two angles, we obtain

$$W_1 = V_L I_L (\cos 30° \cos\theta - \sin 30° \sin\theta) \tag{26}$$

Similarly,

$$W_2 = V_L I_L (\cos 30° \cos\theta + \sin 30° \sin\theta) \tag{27}$$

Then the sum $W_1 + W_2 = \sqrt{3}\, V_L I_L \cos\theta$ and the difference $W_2 - W_1 = V_L I_L \sin 0$, from which

$$\tan\theta = \sqrt{3}\left(\frac{W_2 - W_1}{W_1 + W_2}\right) \tag{28}$$

Thus the tangent of the angle on **Z** is $\sqrt{3}$ times the ratio of the difference between the two wattmeter readings and their sum. With no knowledge of the lines on which the meters are located nor of the system sequence, it is not possible to distinguish between $+\theta$ and $-\theta$. However, when both the sequence and meter location are known, the sign can be fixed by the following expressions. For sequence *ABC*,

$$\tan\theta = \sqrt{3}\,\frac{W_A - W_B}{W_A + W_B} = \sqrt{3}\,\frac{W_B - W_C}{W_B + W_C} = \sqrt{3}\,\frac{W_C - W_A}{W_C + W_A} \tag{29}$$

and for *CBA*,

$$\tan\theta = \sqrt{3}\,\frac{W_B - W_A}{W_B + W_A} = \sqrt{3}\,\frac{W_C - W_B}{W_C + W_B} = \sqrt{3}\,\frac{W_A - W_C}{W_A + W_C} \tag{30}$$

Solved Problems

14.1. Show that the line voltage V_L in the three-phase system is $\sqrt{3}$ times the line to neutral voltage V_P.

In Fig. 14-26, the voltages of the three-phase system are shown on an equilateral triangle in which the length of a side is proportional to the line voltage V_L and where the neutral point N is at the centre of the triangle.

The line to neutral voltage has a horizontal projection, $V_P \cos 30°$ or $V_P\sqrt{3}/2$. Since the base is the sum of two such projections, it follows that

$$V_L = 2(V_P\sqrt{3}/2) = \sqrt{3}V_P$$

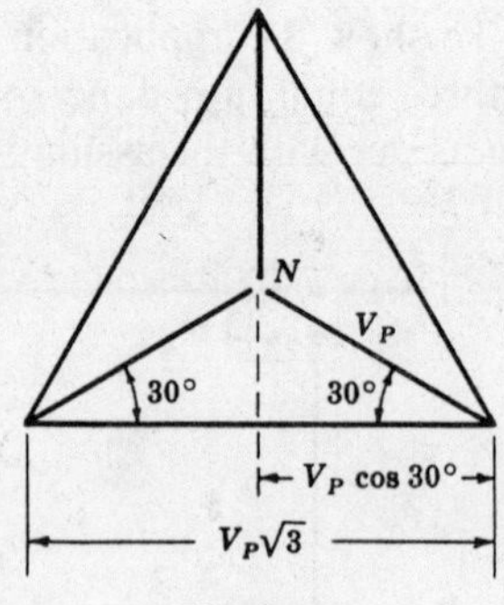

Fig. 14-26

14.2. Compute the full-load coil currents for both delta and star connected, three-phase alternators with a rating of 25 kVA at 480 volts.

With the star-connection the line current and coil current have the same magnitude. For a balanced three-phase system,

$$\text{kVA} = \sqrt{3}V_L I_L \times 10^{-3} \text{ and } I_L = \frac{\text{kVA}}{\sqrt{3}V_L \times 10^{-3}} = \frac{25}{\sqrt{3}(480 \times 10^{-3})} = 30{\cdot}1 \text{ A}$$

The delta-connected alternator with the same kVA rating also has full-load line currents of 30·1 A. The coil currents are $I_L/\sqrt{3}$. Then $I_{coil} = 30{\cdot}1/\sqrt{3} = 17{\cdot}35$ A.

14.3. A two-phase system with a line to neutral voltage of 150 volts supplies a balanced delta-connected load with impedances of $10\underline{/53{\cdot}1^\circ}$ ohms. Find the line currents and the total power.

In a two-phase system the two line to neutral voltages have a 90° phase difference. Then, if $\mathbf{V}_{BN}$ is the reference, $\mathbf{V}_{AN}$ is at 90° as shown in Fig. 14-27. The line to line voltage $\sqrt{2}$ times the line to neutral voltages. Thus $V_{AB} = \sqrt{2}(150) = 212$ V. The phase currents are

$$\mathbf{I}_{AB} = \frac{\mathbf{V}_{AB}}{\mathbf{Z}} = \frac{212\underline{/135^\circ}}{10\underline{/53{\cdot}1^\circ}} = 21{\cdot}2\underline{/81{\cdot}9^\circ}\ \text{A}$$

$$\mathbf{I}_{AN} = \frac{\mathbf{V}_{AN}}{\mathbf{Z}} = \frac{150\underline{/90^\circ}}{10\underline{/53{\cdot}1^\circ}} = 15{\cdot}0\underline{/36{\cdot}9^\circ}\ \text{A}$$

$$\mathbf{I}_{BN} = \frac{\mathbf{V}_{BN}}{\mathbf{Z}} = \frac{150\underline{/0^\circ}}{10\underline{/53{\cdot}1^\circ}} = 15{\cdot}0\underline{/-53{\cdot}1^\circ}\ \text{A}$$

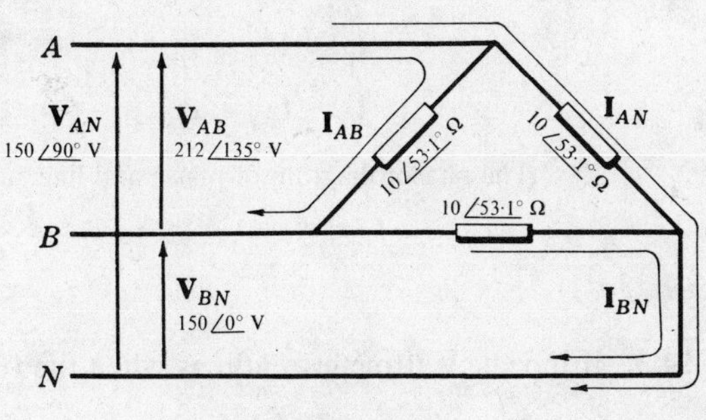

Fig. 14-27

The line currents are given in terms of the phase currents by applying Kirchhoff's current law at the junctions of the delta-connected load. If we assume the positive direction of these currents to be directed toward the load, then

$$\mathbf{I}_A = \mathbf{I}_{AN} + \mathbf{I}_{AB} = \quad 15{\cdot}0\underline{/36{\cdot}9^\circ} + 21{\cdot}2\underline{/81{\cdot}9^\circ} \quad = 33{\cdot}5\underline{/63{\cdot}4^\circ}\ \text{A}$$
$$\mathbf{I}_B = \mathbf{I}_{BN} + \mathbf{I}_{BA} = \quad 15{\cdot}0\underline{/-53{\cdot}1^\circ} - 21{\cdot}2\underline{/81{\cdot}9^\circ} = 33{\cdot}6\underline{/-79{\cdot}7^\circ}\ \text{A}$$
$$\mathbf{I}_N = \mathbf{I}_{NA} + \mathbf{I}_{NB} = -15{\cdot}0\underline{/36{\cdot}9^\circ} - 15{\cdot}0\underline{/-53{\cdot}1^\circ} = 21{\cdot}2\underline{/171{\cdot}86^\circ}\ \text{A}$$

The total power is obtained by using the effective current in the load impedances. Thus,

$$P_{AB} = I_{AB}^2 R = (21{\cdot}2)^2 6 = 2700\ \text{W}$$
$$P_{AN} = I_{AN}^2 R = (15{\cdot}0)^2 6 = 1350\ \text{W}$$
$$P_{BN} = I_{BN}^2 R = (15{\cdot}0)^2 6 = 1350\ \text{W}$$
$$\text{Total power} = 5400\ \text{W}$$

14.4. A three-phase, three-wire, 100 volt, *ABC* system supplies a balanced delta-connected load with impedances of $20\underline{/45^\circ}$ ohms. Determine the line currents and draw the phasor diagram.

Apply the line voltages of the *ABC* sequence to the circuit given in Fig. 14-28. Then the selected phase currents are

$$\mathbf{I}_{AB} = \frac{\mathbf{V}_{AB}}{\mathbf{Z}} = \frac{100\underline{/120^\circ}}{20\underline{/45^\circ}} = 5{\cdot}0\underline{/75^\circ}\ \text{A},\ \mathbf{I}_{BC} = \frac{\mathbf{V}_{BC}}{\mathbf{Z}} = 5{\cdot}0\underline{/-45^\circ}\ \text{A},\ \mathbf{I}_{CA} = \frac{\mathbf{V}_{CA}}{\mathbf{Z}} = 5{\cdot}0\underline{/195^\circ}\ \text{A}$$

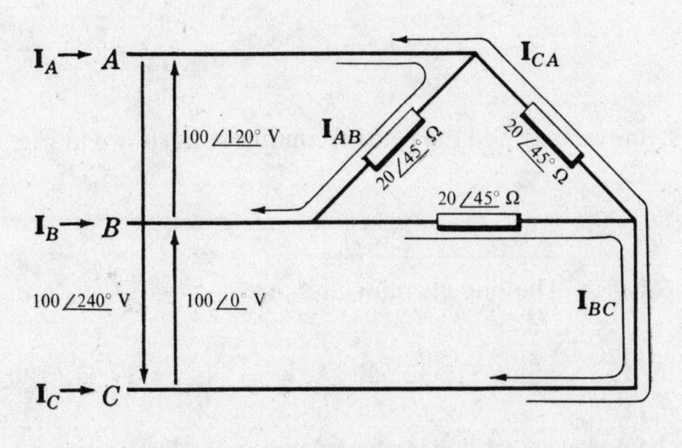

Fig. 14-28

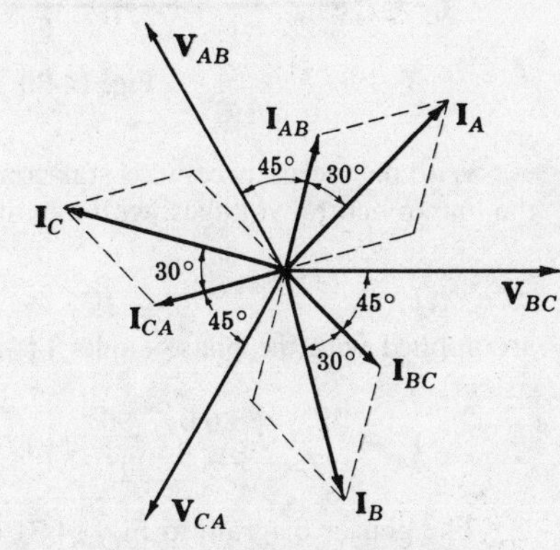

Fig. 14-29

To obtain the line currents as shown in the circuit diagram, we apply Kirchhoff's current law at each junction of the load. Thus

$$\mathbf{I}_A = \mathbf{I}_{AB} + \mathbf{I}_{AC} = 5{\cdot}0\angle 75^\circ - 5{\cdot}0\angle 195^\circ = 8{\cdot}66\angle 45^\circ \text{ A}$$

$$\mathbf{I}_B = \mathbf{I}_{BA} + \mathbf{I}_{BC} = -5{\cdot}0\angle 75^\circ + 5{\cdot}0\angle -45^\circ = 8{\cdot}66\angle -75^\circ \text{ A}$$

$$\mathbf{I}_C = \mathbf{I}_{CA} + \mathbf{I}_{CB} = 5{\cdot}0\angle 195^\circ - 5{\cdot}0\angle -45^\circ = 8{\cdot}66\angle 165^\circ \text{ A}$$

The phasor diagram of phase and line currents is shown in Fig. 14-29 above.

14.5. Find the wattmeter readings when the two-wattmeter method is applied to the circuit of Problem 14.4.

With a three-phase, three-wire balanced load, the wattmeter readings are

$$W_1 = V_L I_L \cos(30^\circ + \theta) \text{ and } W_2 = V_L I_L \cos(30^\circ - \theta) \qquad (1)$$

where θ is the angle on the load impedance. From Problem 14.4, $V_L = 100$, $I_L = 8{\cdot}66$ and the angle on the load impedance is 45°. Substituting these values in (1), we obtain

$$W_1 = 100(8{\cdot}66)\cos(30^\circ + 45^\circ) = 866 \cos 75^\circ = 224 \text{ W}$$

$$W_2 = 100(8{\cdot}66)\cos(30^\circ - 45^\circ) = 866 \cos(-15^\circ) = 836 \text{ W}$$

Then the total power is $P_T = W_1 + W_2 = 1060$ W.

As a check we can compute the total power in any balanced, three-phase load from

$$P = \sqrt{3} V_L I_L \cos\theta = \sqrt{3}\, 100(8{\cdot}66) \cos 45^\circ = 1060 \text{ W}$$

14.6. Three identical impedances of $5\angle -30^\circ$ ohms are connected in star to a three-phase, three-wire, 150 volt, *CBA* system. Find the line currents and draw the phasor diagram.

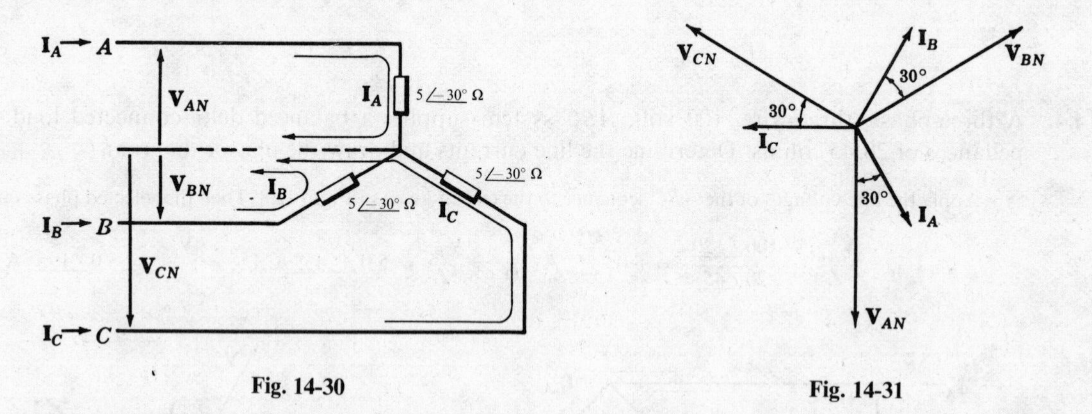

Fig. 14-30 Fig. 14-31

With balanced, three-wire, star-connected systems we may add the neutral conductor as shown in Fig. 14-30. Then the line to neutral voltages, with magnitudes

$$V_{LN} = V_L/\sqrt{3} = 150/\sqrt{3} = 86{\cdot}6 \text{ V}$$

are applied with the phase angles of the *CBA* sequence. The line currents are

$$\mathbf{I}_A = \frac{\mathbf{V}_{AN}}{\mathbf{Z}} = \frac{86{\cdot}6\angle -90^\circ}{5\angle -30^\circ} = 17{\cdot}32\angle -60^\circ \text{ A},\ \mathbf{I}_B = \frac{\mathbf{V}_{BN}}{\mathbf{Z}} = 17{\cdot}32\angle 60^\circ \text{ A},\ \mathbf{I}_C = \frac{\mathbf{V}_{CN}}{\mathbf{Z}} = 17{\cdot}32\angle 180^\circ \text{ A}$$

The phasor diagram in Fig. 14-31 shows the balanced set of line currents leading the line to neutral voltages by 30°, the angle on the load impedance.

14.7. Find the wattmeter readings when the two-wattmeter method is applied to the circuit of Problem 14.6.

With a balanced three-phase load,

$$W_1 = V_L I_L \cos(30^\circ + \theta) = 150(17{\cdot}32)\cos(30^\circ + 30^\circ) = 1300 \text{ W}$$

$$W_2 = V_L I_L \cos(30^\circ - \theta) = 150(17{\cdot}32)\cos(30^\circ - 30^\circ) = 2600 \text{ W}$$

The total power $P_T = W_1 + W_2 = 3900$ W.

As a check we can compute the phase power $P_P = I_L^2 R = (17{\cdot}32)^2 4{\cdot}33 = 1300$ W and then the total power is

$$P_T = 3P_P = 3(1300) = 3900 \text{ W}$$

Or with balanced three-phase loads, the total power is

$$P = \sqrt{3} V_L I_L \cos\theta = \sqrt{3}(150)(17{\cdot}32)\cos(-30^\circ) = 3900 \text{ W}$$

14.8. Three identical impedances of $15\angle 30^\circ$ ohms are connected in delta to a three-phase, three-wire, 200 volt *ABC* system. Find the line currents using the one-line equivalent method.

Since the load is delta-connected, we first obtain the equivalent impedances of the star-connected load:

$$\mathbf{Z}_Y = \mathbf{Z}_\Delta/3 = 15\angle 30^\circ/3 = 5\angle 30^\circ\ \Omega$$

The magnitude of the line to neutral voltage

$$V_{LN} = V_L/\sqrt{3} = 200/\sqrt{3} = 115{\cdot}5 \text{ V}$$

Now in the one-line equivalent circuit of Fig. 14-32 the applied voltage is $115{\cdot}5\angle 0^\circ$ V, and the resulting current

$$\mathbf{I}_L = \frac{\mathbf{V}_{LN}}{\mathbf{Z}} = \frac{115{\cdot}5\angle 0^\circ}{5\angle 30^\circ} = 23{\cdot}1\angle -30^\circ \text{ A}$$

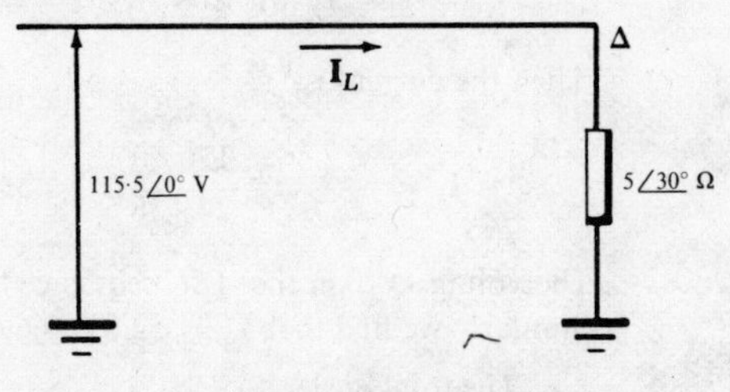

Fig. 14-32

To obtain the line currents $\mathbf{I}_A$, $\mathbf{I}_B$ and $\mathbf{I}_C$ we first determine the phase angle on the corresponding line to neutral voltages in the *ABC* sequence. Since V_{AN} has a phase angle of 90°, $I_A = 23{\cdot}1\angle 90^\circ - 30^\circ = 23{\cdot}1\angle 60^\circ$ A. In the same manner we find that $\mathbf{I}_B = 23{\cdot}1\angle -60^\circ$ A and $\mathbf{I}_C = 23{\cdot}1\angle 180^\circ$ A.

The currents in the delta-connected impedances are related to the line currents by $I_L = \sqrt{3} I_P$ from which $I_P = 23{\cdot}1\sqrt{3} = 13{\cdot}3$ A.

The angle on $\mathbf{V}_{AB}$ in the *ABC* sequence is 120° and thus $\mathbf{I}_{AB} = 13{\cdot}3\angle 120^\circ - 30^\circ = 13{\cdot}3\angle 90^\circ$ A. By the same method we find that $\mathbf{I}_{BC} = 13{\cdot}3\angle -30^\circ$ A and $\mathbf{I}_{CA} = 13{\cdot}3\angle 210^\circ$ A.

14.9. Three identical impedances of $10\angle 30^\circ$ ohms in a star connection and three identical impedances of $15\angle 0$ ohms also in a star connection are both on the same three-phase, three-wire 250 volt system. Find the total power.

Since both loads are star-connected, their phase impedances can be put directly on the one-line equivalent circuit as shown in Fig. 14-33. The voltage required in the one-line equivalent circuit is

$$V_{LN} = V_L/\sqrt{3} = 250/\sqrt{3} = 144{\cdot}5 \text{ V}$$

Then the current is

$$\mathbf{I}_L = \frac{144{\cdot}5\angle 0^\circ}{10\angle 30^\circ} + \frac{144{\cdot}5\angle 0^\circ}{15\angle 0^\circ}$$

$$= 14{\cdot}45\angle -30^\circ + 9{\cdot}62\angle 0^\circ = 23{\cdot}2\angle -18{\cdot}1^\circ \text{ A}$$

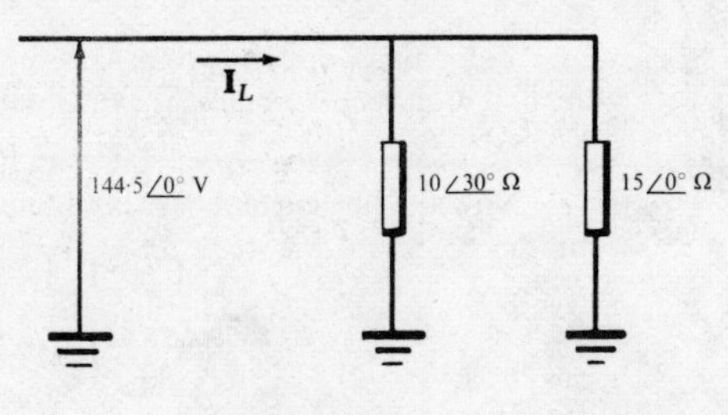

Fig. 14-33

In the power formula $P = \sqrt{3}V_L I_L \cos\theta$, θ is the angle of the load impedance when there is a single load. With several loads on the same system, θ is the angle of the equivalent load impedance. In computing the current $\mathbf{I}_L$, both loads were considered and the current was found to lag the voltage by 18·1°. Therefore we know that the equivalent impedance is inductive and has an angle of 18·1°. Then

$$P = \sqrt{3}V_L I_L \cos\theta = \sqrt{3}\,250(23{\cdot}2)\cos 18{\cdot}1° = 9530 \text{ W}$$

14.10. Three identical impedances of $12\angle 30°$ ohms in a delta connection and three identical impedances of $5\angle 45°$ ohms in a star connection are on the same three-phase, three-wire, 208 volt, *ABC* system. Find the line currents and the total power.

Since the first of the loads is delta-connected we obtain the star-connected equivalent.

$$\mathbf{Z}_Y = \mathbf{Z}_\Delta/3 = 12\angle 30°\ /3 = 4\angle 30°\ \Omega$$

With a line voltage of 208, the line to neutral voltage is $208/\sqrt{3}$ or 120 volts.

The one-line equivalent circuit is shown in Fig. 14-34 with the two load impedances $4\angle 30°$ and $5\angle 45°$ ohms. These impedances can be replaced with an equivalent where

$$\mathbf{Z}_{eq} = \frac{4\angle 30°\,(5\angle 45°)}{4\angle 30° + 5\angle 45°} = 2{\cdot}24\angle 36{\cdot}6°\ \Omega$$

Then the current is

$$\mathbf{I}_L = \frac{\mathbf{V}_{LN}}{\mathbf{Z}_{eq}} = \frac{120\angle 0°}{2{\cdot}24\angle 36{\cdot}6°} = 53{\cdot}6\angle -36{\cdot}6°\ \text{A}$$

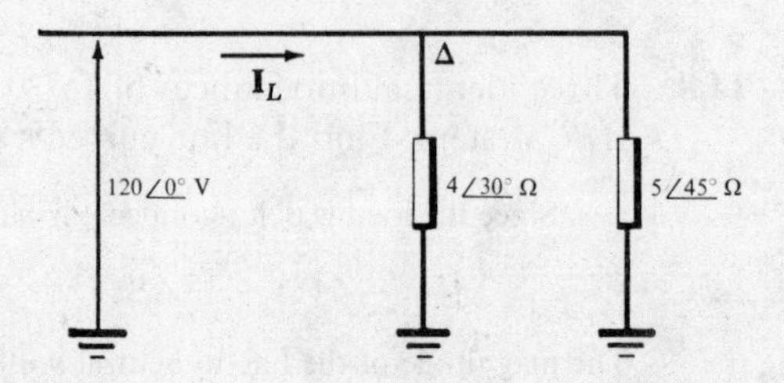

Fig. 14-34

The voltage $\mathbf{V}_{AN}$ in the *ABC* sequence has a phase angle of 90° and thus $\mathbf{I}_A = 53{\cdot}6\angle(90° - 36{\cdot}6°) = 53{\cdot}6\angle 53{\cdot}4°$ A. Similarly we find that $\mathbf{I}_B = 53{\cdot}6\angle -66{\cdot}6°$ A and $\mathbf{I}_C = 53{\cdot}6\angle -186{\cdot}6°$ A.

The total power

$$P = \sqrt{3}V_L I_L \cos\theta = \sqrt{3}\,208(53{\cdot}6)\cos 36{\cdot}6° = 15\,500 \text{ W}$$

14.11. A three-phase, three-wire, 240 volt, *CBA* system supplies a delta-connected load in which $\mathbf{Z}_{AB} = 25\angle 90°$ ohms, $\mathbf{Z}_{BC} = 15\angle 30°$ ohms and $\mathbf{Z}_{CA} = 20\angle 0°$ ohms. Find the line currents and the total power.

Apply the line voltages of the *CBA* sequence to the delta-connected load in Fig. 14-35, and select the phase currents as shown on the diagram. Then

$$\mathbf{I}_{AB} = \frac{\mathbf{V}_{AB}}{\mathbf{Z}_{AB}} = \frac{240\angle 240°}{25\angle 90°} = 9{\cdot}6\angle 150°\ \text{A}$$

$$\mathbf{I}_{BC} = \frac{\mathbf{V}_{BC}}{\mathbf{Z}_{BC}} = \frac{240\angle 0°}{15\angle 30°} = 16{\cdot}0\angle -30°\ \text{A}$$

$$\mathbf{I}_{CA} = \frac{\mathbf{V}_{CA}}{\mathbf{Z}_{CA}} = \frac{240\angle 120°}{20\angle 0°} = 12{\cdot}0\angle 120°\ \text{A}$$

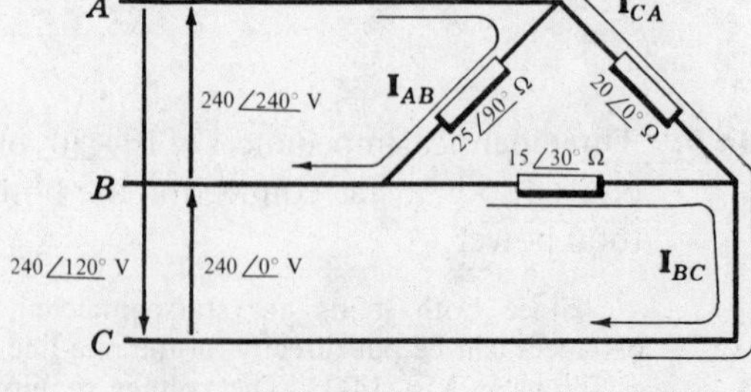

Fig. 14-35

Now the line currents are computed in terms of the phase currents.

$$\mathbf{I}_A = \mathbf{I}_{AB} + \mathbf{I}_{AC} = 9{\cdot}6\angle 150° - 12\angle 120° = 6{\cdot}06\angle 247°\ \text{A}$$

$$\mathbf{I}_B = \mathbf{I}_{BA} + \mathbf{I}_{BC} = -9{\cdot}6\angle 150° + 16\angle -30° = 25{\cdot}6\angle -30°\ \text{A}$$

$$\mathbf{I}_C = \mathbf{I}_{CA} + \mathbf{I}_{CB} = 12\angle 120° - 16\angle -30° = 27{\cdot}1\angle 137{\cdot}2°\ \text{A}$$

As expected with an unbalanced load, the line currents are not equal.

The power in each phase is calculated as follows.

Impedance $\mathbf{Z}_{AB} = 25\angle 90^\circ = 0 + j25$ ohms, $R_{AB} = 0$ and $I_{AB} = 9{\cdot}6$ A. Then

$$P_{AB} = I_{AB}^2 R_{AB} = (9{\cdot}6)^2(0) = 0$$

Impedance $\mathbf{Z}_{BC} = 15\angle 30^\circ = 13 + j7{\cdot}5$ ohms, $R_{BC} = 13$ ohms and $I_{BC} = 16$ A. Then

$$P_{BC} = I_{BC}^2 R_{BC} = (16)^2(13) = 3330 \text{ W}$$

Impedance $\mathbf{Z}_{CA} = 20\angle 0^\circ = 20 + j0$ ohms, $R_{CA} = 20$ ohms and $I_{CA} = 12$ A. Then

$$P_{CA} = I_{CA}^2 R_{CA} = (12)^2(20) = 2880 \text{ W}$$

The total power is the sum of the power in the phases,

$$P_T = P_{AB} + P_{BC} + P_{CA} = 0 + 3330 + 2880 = 6210 \text{ W}$$

14.12. Find the wattmeter readings when the two-wattmeter method is used on the circuit of Problem 14.11 with meters in lines (*a*) *A* and *B*, (*b*) *A* and *C*.

(*a*) With the wattmeters in lines *A* and *B*,

$$(1)\quad W_A = V_{AC} I_A \cos \measuredangle_A^{AC} \qquad (2)\quad W_B = V_{BC} I_B \cos \measuredangle_B^{BC}$$

From Problem 14.11, $V_{AC} = 240\angle -60^\circ$ V and $\mathbf{I}_A = 6{\cdot}06\angle 247{\cdot}7^\circ$ A. Then the angle $\measuredangle_A^{AC}$ is the angle between $247{\cdot}7^\circ$ and -60° or $52{\cdot}3^\circ$. Substituting in (*1*),

$$W_A = 240(6{\cdot}06) \cos 52{\cdot}3^\circ = 890 \text{ W}$$

Also from Problem 14.11, $V_{BC} = 240\angle 0^\circ$ V and $\mathbf{I}_B = 25{\cdot}6\angle -30^\circ$ A. Then $\measuredangle_B^{BC} = 30^\circ$. Substituting in (*2*),

$$W_B = 240(25{\cdot}6) \cos 30^\circ = 5320 \text{ W}$$

The total power $P_T = W_A + W_B = 890 + 5320 = 6210$ W.

(*b*) With the wattmeters in lines *A* and *C*,

$$(3)\quad W_A = V_{AB} I_A \cos \measuredangle_A^{AB} \qquad (4)\quad W_C = V_{CB} I_C \cos \measuredangle_C^{CB}$$

From Problem 14.11, $V_{AB} = 240\angle 240^\circ$ V. Since $\mathbf{I}_A = 6{\cdot}06\angle 247{\cdot}7^\circ$ A, $\measuredangle_A^{AB} = 7{\cdot}7^\circ$ Substituting in (*3*),

$$W_A = 240(6{\cdot}06) \cos 7{\cdot}7^\circ = 1440 \text{ W}$$

Also $V_{CB} = 240\angle 180^\circ$ V and $\mathbf{I}_C = 27{\cdot}1\angle 132{\cdot}2^\circ$ A from which $\measuredangle_C^{CB} = 42{\cdot}8^\circ$. Substituting in (*4*),

$$W_C = 240(27{\cdot}1) \cos 42{\cdot}8^\circ = 4770 \text{ W}$$

The total power $P_T = W_A + W_C = 1440 + 4770 = 6210$ W.

14.13. A three-phase, four-wire, 208 volt, *ABC* system supplies a star-connected load in which $\mathbf{Z}_A = 10\angle 0^\circ$ ohms, $\mathbf{Z}_B = 15\angle 30^\circ$ ohms and $\mathbf{Z}_C = 10\angle -30^\circ$ ohms. Find the line currents, the neutral current and the total power.

Apply the line to neutral voltages of the *ABC* sequence to the circuit as shown in Fig. 14-36 and compute the line currents assuming the positive direction toward the load.

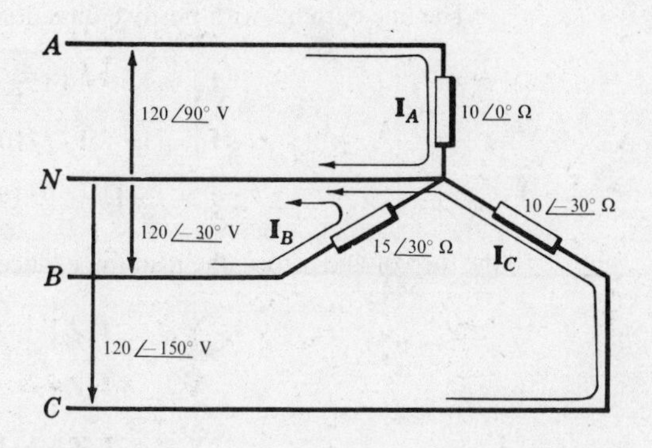

Fig. 14-39

$$\mathbf{I}_A = \mathbf{V}_{AN}/\mathbf{Z}_A = (120\angle 90^\circ)/(10\angle 0^\circ) = 12\angle 90^\circ \text{ A}$$

$$\mathbf{I}_B = \mathbf{V}_{BN}/\mathbf{Z}_B = (120\angle -30^\circ)/(15\angle 30^\circ) = 8\angle -60^\circ \text{ A}$$

$$\mathbf{I}_C = \mathbf{V}_{CN}/\mathbf{Z}_C = (120\angle -150^\circ)/(10\angle -30^\circ) = 12\angle -120^\circ \text{ A}$$

The neutral conductor contains the phasor sum of the line currents and if the positive direction is toward the load,

$$\mathbf{I}_N = -(\mathbf{I}_A + \mathbf{I}_B + \mathbf{I}_C) = -(12\angle 90^\circ + 8\angle -60^\circ + 12\angle -120^\circ) = 5{\cdot}69\angle 69{\cdot}4^\circ \text{ A}$$

The impedance $\mathbf{Z}_A = 10 + j0$ ohms passes current $\mathbf{I}_A = 12\angle 90^\circ$ A, and the power in this phase of the load is $P_A = (12)^2(10) = 1440$ W. Impedance $\mathbf{Z}_B = 15\angle 30^\circ = 13 + j7{\cdot}5\ \Omega$ contains the current $\mathbf{I}_B = 8\angle -60^\circ$ A, and the phase power is $P_B = (8)^2 13 = 832$ W. Similarly $\mathbf{Z}_C = 10\angle -30^\circ = 8{\cdot}66 - j5$ ohms contains $\mathbf{I}_C = 12\angle -120^\circ$ A and $P_C = (12)^2 8{\cdot}66 = 1247$ W.

The total power is $P_T = P_A + P_B + P_C = 1440 + 832 + 1247 = 3519$ W.

14.14. The load impedances of Problem 14.13 are connected to a three-phase, three-wire, 208 volt, *ABC* system. Find the line currents and the voltages across the load impedances.

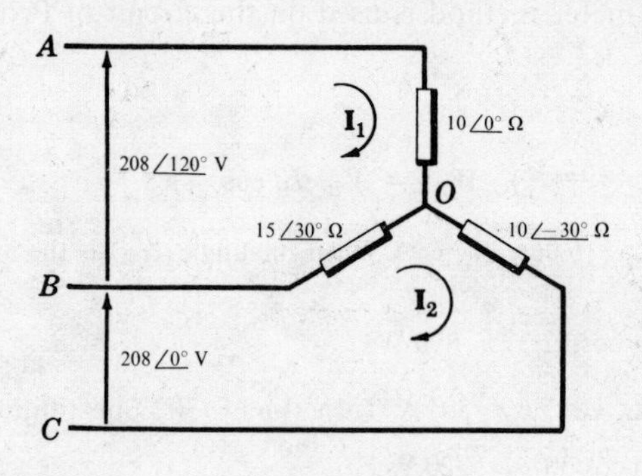

Fig. 14-37

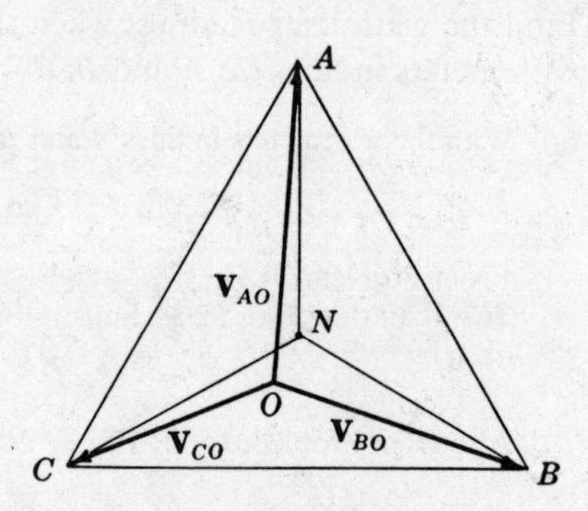

Fig. 14-38

The circuit of Fig. 14-37 shows the two line voltages $\mathbf{V}_{AB}$ and $\mathbf{V}_{BC}$. With the mesh currents $\mathbf{I}_1$ and $\mathbf{I}_2$ selected as shown, the matrix form of the mesh current equations is

$$\begin{bmatrix} 10\angle 0^\circ + 15\angle 30^\circ & -15\angle 30^\circ \\ -15\angle 30^\circ & 15\angle 30^\circ + 10\angle -30^\circ \end{bmatrix} \begin{bmatrix} \mathbf{I}_1 \\ \mathbf{I}_2 \end{bmatrix} = \begin{bmatrix} 208\angle 120^\circ \\ 208\angle 0^\circ \end{bmatrix}$$

from which
$$\mathbf{I}_1 = \frac{5210\angle 90^\circ}{367{\cdot}5\angle 3{\cdot}9^\circ} = 14{\cdot}15\angle 86{\cdot}1^\circ \text{ A}$$

$$\mathbf{I}_2 = \frac{3730\angle 56{\cdot}6^\circ}{367{\cdot}5\angle 3{\cdot}9^\circ} = 10{\cdot}15\angle 52{\cdot}7^\circ \text{ A}$$

The line currents with positive directions toward the load are given in terms of $\mathbf{I}_1$ and $\mathbf{I}_2$ as

$$\mathbf{I}_A = \mathbf{I}_1 = 14{\cdot}15\angle 86{\cdot}1^\circ \text{ A}$$

$$\mathbf{I}_B = \mathbf{I}_2 - \mathbf{I}_1 = 10{\cdot}15\angle 52{\cdot}7^\circ - 14{\cdot}15\angle 86{\cdot}1^\circ = 8{\cdot}0\angle -49{\cdot}5^\circ \text{ A}$$

$$\mathbf{I}_C = -\mathbf{I}_2 = 10{\cdot}15\angle (52{\cdot}7^\circ - 180^\circ) = 10{\cdot}15\angle -127{\cdot}3^\circ \text{ A}$$

Now the voltages across the load impedances are

$$\mathbf{V}_{AO} = \mathbf{I}_A\mathbf{Z}_A = 14{\cdot}15\angle 86{\cdot}1^\circ (10\angle 0^\circ) = 141{\cdot}5\angle 86{\cdot}1^\circ \text{ V}$$

$$\mathbf{V}_{BO} = \mathbf{I}_B\mathbf{Z}_B = 8{\cdot}0\angle -49{\cdot}5^\circ (15\angle 30^\circ) = 120\angle -19{\cdot}5^\circ \text{ V}$$

$$\mathbf{V}_{CO} = \mathbf{I}_C\mathbf{Z}_C = 10{\cdot}15\angle -127{\cdot}3^\circ (10\angle -30^\circ) = 101{\cdot}5\angle -157{\cdot}3^\circ \text{ V}$$

A plot of the three voltages $\mathbf{V}_{AO}$, $\mathbf{V}_{BO}$ and $\mathbf{V}_{CO}$ shows the triangle of the *ABC* sequence when the ends of the phasors are joined by straight lines. Then point *N* can be added as shown in Fig. 14-38 above.

14.15. Repeat the solution of Problem 14.14 using the displacement neutral method.

In the displacement neutral method the voltage $\mathbf{V}_{ON}$ is computed from the formula

$$\mathbf{V}_{ON} = \frac{\mathbf{V}_{AN}\,\mathbf{Y}_A + \mathbf{V}_{BN}\,\mathbf{Y}_B + \mathbf{V}_{CN}\,\mathbf{Y}_C}{\mathbf{Y}_A + \mathbf{Y}_B + \mathbf{Y}_C}$$

From Problem 14.14 $\mathbf{Y}_A = 1/10 = 0{\cdot}1$ S, $\mathbf{Y}_B = 1/(15\angle 30^\circ) = 0{\cdot}0577 - j0{\cdot}033$ S and $\mathbf{Y}_C = 1/(10\angle -30^\circ) = 0{\cdot}0866 + j0{\cdot}050$ S. Then

$$\mathbf{Y}_A + \mathbf{Y}_B + \mathbf{Y}_C = 0{\cdot}244 + j0{\cdot}0167 = 0{\cdot}244\angle 3{\cdot}93^\circ \text{ S}$$

and

$$\mathbf{V}_{AN}\mathbf{Y}_A = 120\angle 90^\circ\,(0{\cdot}1) = 12\angle 90^\circ = j12 \text{ A}$$

$$\mathbf{V}_{BN}\mathbf{Y}_B = 120\angle -30^\circ\,(0{\cdot}0667\angle -30^\circ) = 8{\cdot}0\angle -60^\circ = 4{\cdot}0 - j6{\cdot}93 \text{ A}$$

$$\mathbf{V}_{CN}\mathbf{Y}_C = 120\angle -150^\circ\,(0{\cdot}1\angle 30^\circ) = 12\angle -120^\circ = -6{\cdot}0 - j10{\cdot}4 \text{ A}$$

$$\mathbf{V}_{AN}\mathbf{Y}_A + \mathbf{V}_{BN}\mathbf{Y}_B + \mathbf{V}_{CN}\mathbf{Y}_C = -2{\cdot}0 - j5{\cdot}33 = 5{\cdot}69\angle 249{\cdot}4^\circ \text{ A}$$

Thus

$$\mathbf{V}_{ON} = (5{\cdot}69\angle 249{\cdot}4^\circ)/(0{\cdot}244\angle 3{\cdot}93^\circ) = 23{\cdot}3\angle 245{\cdot}5^\circ = -9{\cdot}66 - j21{\cdot}2 \text{ V}$$

The voltages across the load impedances can be expressed in terms of the corresponding line to neutral voltage and the displacement neutral voltage as follows.

$$\mathbf{V}_{AO} = \mathbf{V}_{AN} + \mathbf{V}_{NO} = 120\angle 90^\circ + (9{\cdot}66 + j21{\cdot}2) = 141{\cdot}2\angle 86{\cdot}08^\circ \text{ V}$$

$$\mathbf{V}_{BO} = \mathbf{V}_{BN} + \mathbf{V}_{NO} = 120\angle -30^\circ + (9{\cdot}66 + j21{\cdot}2) = 120\angle -18{\cdot}9^\circ \text{ V}$$

$$\mathbf{V}_{CO} = \mathbf{V}_{CN} + \mathbf{V}_{NO} = 120\angle -150^\circ + (9{\cdot}66 + j21{\cdot}2) = 102\angle 202{\cdot}4^\circ \text{ V}$$

To obtain the line currents we take the products of these voltages and the corresponding admittances.

$$\mathbf{I}_A = \mathbf{V}_{AO}\mathbf{Y}_A = 141{\cdot}2\angle 86{\cdot}08^\circ\,(0{\cdot}1\angle 0^\circ) = 14{\cdot}12\angle 86{\cdot}08^\circ \text{ A}$$

$$\mathbf{I}_B = \mathbf{V}_{BO}\mathbf{Y}_B = 120\angle -18{\cdot}9^\circ\,(0{\cdot}0667\angle -30^\circ) = 8{\cdot}0\angle -48{\cdot}9^\circ \text{ A}$$

$$\mathbf{I}_C = \mathbf{V}_{CO}\mathbf{Y}_C = 102\angle 202{\cdot}4^\circ\,(0{\cdot}1\angle 30^\circ) = 10{\cdot}2\angle 232{\cdot}4^\circ \text{ or } 10{\cdot}2\angle -127{\cdot}6^\circ \text{ A}$$

The above results check with those of Problem 14.14 within slide rule accuracy.

14.16. Readings of 1154 and 577 watts are obtained when the two-wattmeter method is used on a balanced load. Find the delta-connected load impedances if the system voltage is 100 volts.

For balanced three-phase loads,

$$\tan\theta = \sqrt{3}\frac{W_1 - W_2}{W_1 + W_2} = \pm\sqrt{3}\frac{1154 - 577}{1154 + 577} = \pm 0{\cdot}577$$

from which $\theta = \pm 30^\circ$. (We write $\pm$ since without knowing both the sequence and the location of the meters the sign cannot be determined.)

The total power $P = \sqrt{3}V_L I_L \cos\theta$ and

$$I_L = \frac{P}{\sqrt{3}V_L\cos\theta} = \frac{1731}{\sqrt{3}(100)(0{\cdot}866)} = 11{\cdot}55 \text{ A}$$

We sketch the one-line equivalent circuit and apply the voltage $(100/\sqrt{3})\angle 0^\circ = 57{\cdot}7\angle 0^\circ$ V as shown in Fig. 14-39. Then the star-connected impedance

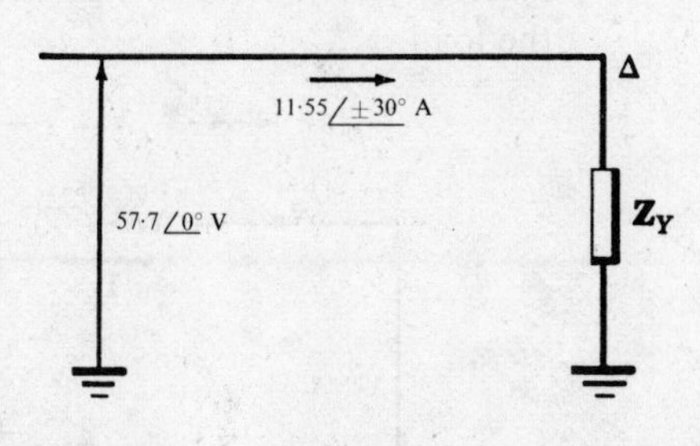

Fig. 14-39

$$\mathbf{Z}_Y = \frac{\mathbf{V}}{\mathbf{I}} = \frac{57{\cdot}7\angle 0^\circ}{11{\cdot}55\angle \pm 30^\circ} = 5{\cdot}0\angle \mp 30^\circ\,\Omega$$

and

$$\mathbf{Z}_\Delta = 3\mathbf{Z}_Y = 15\angle \mp 30^\circ\,\Omega$$

14.17. The two-wattmeter method is applied to a three-phase, three-wire, 100 volt ABC system with the meters in lines B and C, $W_B = 836$ W and $W_C = 224$ W. Find the impedance of the balanced delta-connected load.

Since the sequence and the meter locations are known, the sign of θ is established. Thus

$$\tan\theta \;=\; \sqrt{3}\,\frac{W_B - W_C}{W_B + W_C} \;=\; \sqrt{3}\,\frac{836 - 224}{836 + 224} \;=\; 1 \quad \text{or} \quad \theta = 45°$$

Since $P = \sqrt{3}V_L I_L \cos\theta$, $I_L = \dfrac{P}{\sqrt{3}V_L\cos\theta} = \dfrac{1060}{\sqrt{3}(100)(0{\cdot}707)} = 8{\cdot}66$ A. Then the one-line equivalent circuit has a voltage of $57{\cdot}7\angle 0°$ V and the star-connected impedance is $\mathbf{Z}_Y = \mathbf{V}/\mathbf{I} = (57{\cdot}7\angle 0°)/(8{\cdot}66\angle -45°) = 6{\cdot}67\angle 45°\ \Omega$. The required delta-connected load impedance is $\mathbf{Z}_\Delta = 3\mathbf{Z}_Y = 20\angle 45°\ \Omega$.

14.18. A three-phase, 1500 watt, unity power factor heating unit and a 5 hp induction motor with a full load efficiency of 80% and a power factor of 0·85 are served from the same three-phase, three-wire, 208 volt system. Find the magnitude of the line current for rated output from the 5 hp motor.

Since 746 W = 1 hp, the motor output is (5 hp)(746 W/hp) = 3730 W. Then the input to the motor is 3730/0·80 = 4662 W.

The motor is a balanced three-phase load. Then

$$P = \sqrt{3}V_L I_L \cos\theta,\ 4662 = \sqrt{3}(208 I_L)(0{\cdot}85),\ I_L = 15{\cdot}25\ \text{A}$$

On the one-line equivalent circuit the phasor current lags the voltage by θ and $\theta = \cos^{-1} 0{\cdot}85 = 31{\cdot}7°$. Thus the motor line current is $\mathbf{I}_L = 15{\cdot}25\angle -31{\cdot}7°$ A.

Now for the heating load, $P = \sqrt{3}V_L I_L \cos\theta$ where $\theta = 0°$. Substituting, $1500 = \sqrt{3}(208)I_L$, $I_L = 4{\cdot}16$ A, $\mathbf{I}_L = 4{\cdot}16\angle 0°$ A.

The total line current is the phasor sum of the motor and heating load currents:

$$\mathbf{I}_L = 15{\cdot}25\angle -31{\cdot}7 + 4{\cdot}17\angle 0 = 18{\cdot}9\angle -25{\cdot}1°\ \text{A}$$

Thus the current in each line will be 18·9 A for rated output of 5 hp from the induction motor.

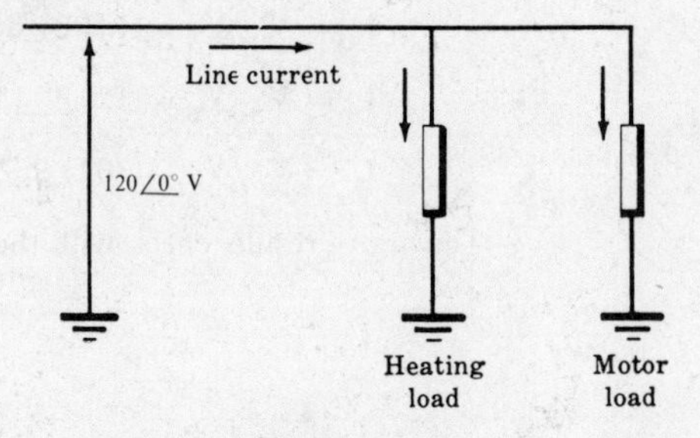

Fig. 14-40

14.19. Three identical impedances of $30\angle 30°$ ohms are connected in delta to a three-phase, three-wire, 208 volt system by conductors which have impedances of $0{\cdot}8 + j\,0{\cdot}6\ \Omega$. Find the magnitude of the line voltage at the load.

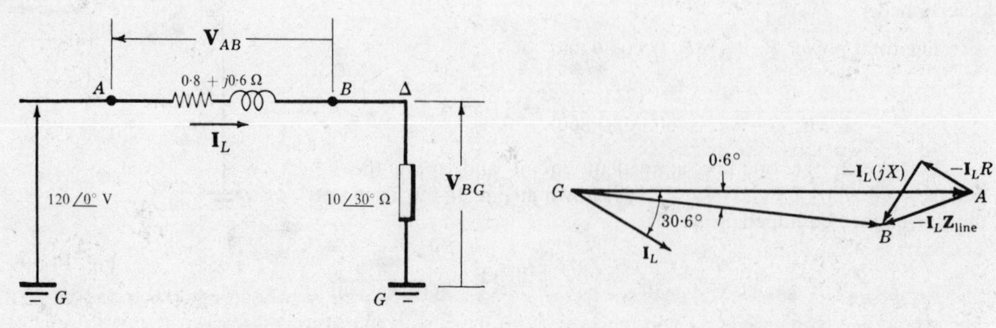

Fig. 14-41

Fig. 14-42

The circuit is shown in Fig. 14-41 above with the star equivalent impedance of $\frac{1}{3}\mathbf{Z}_\Delta$ or $10\angle 30^\circ$ ohms. The line impedance is in series with the load and

$$\mathbf{Z}_{eq} = \mathbf{Z}_{line} + \mathbf{Z}_{load} = 0{\cdot}8 + j0{\cdot}6 + 8{\cdot}66 + j5{\cdot}0 = 9{\cdot}46 + j5{\cdot}6 = 11{\cdot}0\angle 30{\cdot}6^\circ\ \Omega$$

Then
$$\mathbf{I}_L = \frac{\mathbf{V}}{\mathbf{Z}_{eq}} = \frac{120\angle 0^\circ}{11{\cdot}0\angle 30{\cdot}6^\circ} = 10{\cdot}9\angle -30{\cdot}6^\circ\ \text{A}$$

The load voltage is $\mathbf{V}_{BG} = \mathbf{I}_L\mathbf{Z}_{load} = 10{\cdot}9\angle -30{\cdot}6^\circ\ (10\angle 30^\circ) = 109\angle -0{\cdot}6^\circ$ V.

The required line voltage is

$$V_L = \sqrt{3}(109) = 189\ \text{V}$$

Thus the system voltage of 208 has dropped to 189 volts at the load due to the impedance in the lines.

The phasor diagram is shown in Fig. 14-42 on page 214 with the line drop $\mathbf{V}_{AB} = \mathbf{I}_L\mathbf{Z}_{line} = (10{\cdot}9\angle -30{\cdot}6^\circ)(0{\cdot}8 + j0{\cdot}6) = 10{\cdot}9\angle 6{\cdot}3^\circ$ V and $\mathbf{V}_{AG} = \mathbf{V}_{AB} + \mathbf{V}_{BG}$.

14.20. In Problem 14.19, find the line voltage at the load when a set of capacitors with reactance of $-j60\ \Omega$ is connected in parallel with the load.

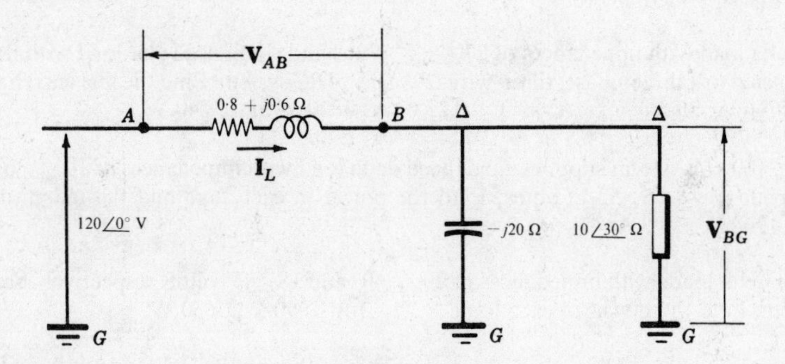

Fig. 14-43

In the one-line equivalent circuit of Fig. 14-43 the $-j20\ \Omega$ and the $10\angle 30^\circ\ \Omega$ are in parallel and

$$\mathbf{Z}_P = \frac{10\angle 30^\circ(-j20)}{(8{\cdot}66 + j5) - j20} = 11{\cdot}55\angle 0^\circ\ \Omega$$

Then $\mathbf{Z}_P$ is in series with the line impedances and

$$\mathbf{Z}_{eq} = \mathbf{Z}_{line} + \mathbf{Z}_P = (0{\cdot}8 + j0{\cdot}6) + (11{\cdot}55\angle 0^\circ) = 12{\cdot}35\angle 2{\cdot}78^\circ\ \Omega$$

Now the line current is

$$\mathbf{I}_L = \frac{\mathbf{V}}{\mathbf{Z}_{eq}} = \frac{120\angle 0^\circ}{12{\cdot}35\angle 2{\cdot}78^\circ} = 9{\cdot}73\angle -2{\cdot}78^\circ\ \text{A}$$

and the voltage across the load is

$$\mathbf{V}_{BG} = \mathbf{I}_L\mathbf{Z}_P = (9{\cdot}73\angle -2{\cdot}78^\circ)(11{\cdot}55\angle 0^\circ) = 112\angle -2{\cdot}78^\circ\ \text{V}$$

The corresponding line voltage $V_L = \sqrt{3}(112) = 194$ volts.

As shown in Chapter 7, the power factor is improved when the capacitors are connected in parallel to the load. This results in a reduction in the voltage drop in the line impedance. Thus in this problem the system voltage of 208 volts has dropped to 194 volts instead of 189 volts as in Problem 14.19.

Supplementary Problems

14.21. Three identical impedances of $10\angle 53{\cdot}1^\circ$ ohms are connected in delta to a three-phase, three-wire, 240 volt, *CBA* system. Find the line currents. *Ans.* $41{\cdot}6\angle -143{\cdot}1^\circ$ A, $41{\cdot}6\angle -23{\cdot}1^\circ$ A and $41{\cdot}6\angle 96{\cdot}9^\circ$ A

14.22. Three impedances of $15{\cdot}9\angle 70^\circ$ ohms are connected in delta to a three-phase, three-wire, 100 volt, *CBA* system. Find the line currents and the total power. *Ans.* $10{\cdot}9\angle -160^\circ$ A, $10{\cdot}9\angle -40^\circ$ A, $10{\cdot}9\angle 80^\circ$ A; 646 W

14.23. Three impedances of $42\angle -35^\circ$ ohms are connected in delta to a three-phase, three-wire, 350 volt, *ABC* system. Find the line currents and the total power. *Ans.* $14{\cdot}4\angle 125^\circ$ A, $14{\cdot}4\angle 5^\circ$ A, $14{\cdot}4\angle -115^\circ$ A, 7130 W

14.24. A balanced star load with impedances of $6\angle 45^\circ$ ohms is connected to a three-phase, four-wire, 208 volt, *CBA* system. Find the line currents including the neutral. *Ans.* $20\angle -135^\circ$ A, $20\angle -15^\circ$ A, $20\angle 105^\circ$ A and 0

14.25. A balanced star load with impedances of $65\angle -20^\circ$ ohms is connected to a three-phase, three-wire, 480 volt, *CBA* system. Find the line currents and the total power. *Ans.* $4{\cdot}26\angle -70^\circ$ A, $4{\cdot}26\angle 50^\circ$ A, $4{\cdot}26\angle 170^\circ$ A; 3320 W

14.26. A 50 hp induction motor with a full load efficiency of 85% and a power factor of 0·80 is connected to a three-phase, 480 volt system. Find the equivalent star connected impedance which can replace this motor. *Ans.* $4{\cdot}2\angle 36{\cdot}9^\circ\ \Omega$

14.27. A 25 hp three-phase, induction motor with a full load efficiency of 82% and a power factor of 0·75 is connected to a 208 volt system. Find the equivalent delta-connected impedance which can replace this motor and find the readings obtained by the two-wattmeter method. *Ans.* $4{\cdot}28\angle 41{\cdot}4^\circ\ \Omega$; 5·58 kW, 17·15 kW

14.28. Three identical impedances of $9\angle -30^\circ$ ohms in delta and three impedances of $5\angle 45^\circ$ ohms in star are both connected to the same three-phase, three-wire, 480 volt, *ABC* system. Find the magnitude of the line current and the total power. *Ans.* 119·2 A, 99 kW

14.29. A balanced delta load with impedances of $27\angle -25^\circ$ ohms and a balanced star load with impedances of $10\angle -30^\circ$ ohms are both connected to a three-phase, three-wire, 208 volt, *ABC* system. Find the line currents and the power in each load. *Ans.* $25{\cdot}3\angle 117{\cdot}4^\circ$ A, $25{\cdot}3\angle -2{\cdot}6^\circ$ A, $25{\cdot}3\angle -122{\cdot}6^\circ$ A; 4340 W and 3740 W

14.30. A three-phase, 100 volt system supplies a balanced delta load with impedances of $10\angle -36{\cdot}9^\circ$ ohms and a balanced star load with impedances of $5\angle 53{\cdot}1^\circ$ ohms. Find the power in each load and the magnitude of the total line current. *Ans.* 2400 W, 1200 W; 20·8 A

14.31. Two balanced delta loads with impedances of $20\angle -60^\circ$ and $18\angle 45^\circ$ ohms respectively are connected to a three-phase, 150 volt system. Find the power in each load. *Ans.* 1690 W, 2650 W

14.32. A three-phase, three-wire, 173·2 volt, *CBA* system serves three balanced loads with connections and impedances as follows: star-connected with impedances of $10\angle 0^\circ$ ohms, delta-connected with impedances of $24\angle 90^\circ$ ohms, and the third delta-connected with an unknown impedance. Find this impedance if the current in line *A* with a positive direction toward the load is given as $32{\cdot}7\angle -138{\cdot}1^\circ$ A. *Ans.* $18\angle 45^\circ\ \Omega$

14.33. The wattmeters in lines *A* and *B* of a 120 volt, *CBA* system read 1500 and 500 watts respectively. Find the impedance of the balanced delta-connected load. *Ans.* $16{\cdot}3\angle -41^\circ\ \Omega$

14.34. The wattmeters in lines *A* and *B* of a 173·2 volt, *ABC* system read −301 and +1327 watts respectively. Find the impedances of the balanced star-connected load. *Ans.* $10\angle -70^\circ\ \Omega$

14.35. Find the readings of the two wattmeters used on a three-wire, 240 volt system with a balanced delta-connected load of $20\angle 80^\circ$ ohms. *Ans.* −1710 W, 3210 W

14.36. Wattmeters are in lines *B* and *C* of a three-wire, 173·2 volt, *CBA* system which supplies a balanced load. Find the wattmeter readings if line current $\mathbf{I}_A = 32{\cdot}7\angle -41{\cdot}9^\circ$ A. *Ans.* 1170 W, 5370 W

14.37. A 100 volt, *CBA* system supplies a balanced load and has wattmeters in lines *A* and *B*. If $\mathbf{I}_B = 10{\cdot}9\angle -40^\circ$ A is the current in line *B*, find the two wattmeter readings. *Ans.* −189 W, 835 W

14.38. A delta load with $\mathbf{Z}_{AB} = 10\angle 30^\circ$ ohms, $\mathbf{Z}_{BC} = 25\angle 0^\circ$ ohms and $\mathbf{Z}_{CA} = 20\angle -30^\circ$ ohms is connected to a three-phase, three-wire, 500 volt, *ABC* system. Find the line currents and the total power. *Ans.* $75\angle 90^\circ$ A, $53{\cdot}9\angle -68{\cdot}2^\circ$ A, $32\angle 231{\cdot}3^\circ$ A; 42·4 kW

14.39. A three-phase, three-wire, 208 volt, *ABC* system supplies a delta-connected load where $\mathbf{Z}_{AB} = 5\angle 0^\circ$ ohms, $\mathbf{Z}_{BC} = 4\angle 30^\circ$ ohms and $\mathbf{Z}_{CA} = 6\angle -15^\circ$ ohms. Find the line currents and the readings of wattmeters in lines *A* and *C*. *Ans.* $70{\cdot}5\angle 99{\cdot}65^\circ$ A, $90{\cdot}5\angle -43{\cdot}3^\circ$ A, $54{\cdot}6\angle 187{\cdot}9^\circ$ A; 13·7 kW, 11·25 kW

14.40. A star load with $\mathbf{Z}_A = 3 + j0$ ohms, $\mathbf{Z}_B = 2 + j3$ ohms and $\mathbf{Z}_C = 2 - j1$ ohms is connected to a three-phase, four-wire, 100 volt, *CBA* system. Find the line currents including the neutral assuming the positive direction is toward the load.
Ans. $19{\cdot}25\angle{-90^\circ}$ A, $16\angle{-26{\cdot}3^\circ}$ A, $25{\cdot}8\angle{176{\cdot}6^\circ}$ A, $27{\cdot}3\angle{65{\cdot}3^\circ}$ A

14.41. A star load with $\mathbf{Z}_A = 12\angle{45^\circ}$ ohms, $\mathbf{Z}_B = 10\angle{30^\circ}$ ohms and $\mathbf{Z}_C = 8\angle{0^\circ}$ ohms is connected to a four-wire, 208 volt system. Find the total power. *Ans.* 3898 W

14.42. The line currents in a three-phase, three-wire, 220 volt, *ABC* system are $\mathbf{I}_A = 43{\cdot}5\angle{116{\cdot}6^\circ}$ A, $\mathbf{I}_B = 43{\cdot}3\angle{-48^\circ}$ A and $\mathbf{I}_C = 11{\cdot}39\angle{218^\circ}$ A. Find the readings of wattmeters in lines (*a*) *A* and *B*, (*b*) *B* and *C*, and (*c*) *A* and *C*.
Ans. (*a*) 5270 W, 6370 W; (*b*) 9310 W, 2330 W; (*c*) 9550 W, 1980 W

14.43. The line currents of a three-phase, three-wire, 440 volt, *ABC* system are $\mathbf{I}_A = 19{\cdot}72\angle{90^\circ}$ A, $\mathbf{I}_B = 57{\cdot}3\angle{-9{\cdot}9^\circ}$ A and $\mathbf{I}_C = 57{\cdot}3\angle{189{\cdot}9^\circ}$ A. Find the readings of the wattmeters in lines (*a*) *A* and *B*, (*b*) *B* and *C*.
Ans. 7·52 kW, 24·8 kW; (*b*) 16·15 kW, 16·15 kW

14.44. The phasor diagram in Fig. 14-44 shows the line currents and line to line voltages of a three-phase, three-wire, 346 volt, *ABC* system. If the line current magnitude is 10 A find the impedance of the star-connected load. *Ans.* $20\angle{90^\circ}$ Ω

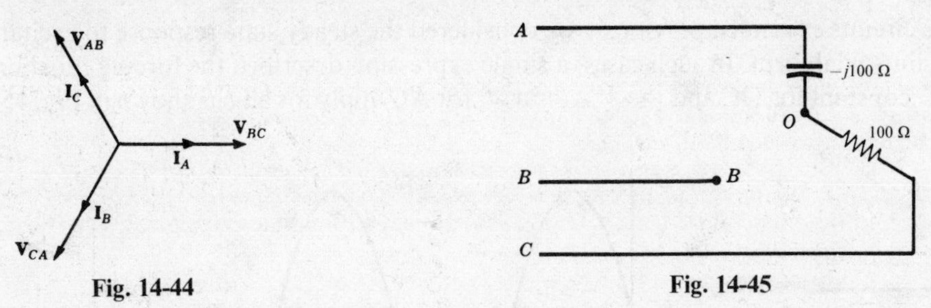

Fig. 14-44

Fig. 14-45

14.45. The circuit of Fig. 14-45 shows an infinite impedance (open circuit) in phase *B* of the star-connected load. Find the phasor voltage $\mathbf{V}_{OB}$ if the system is 208 volt, sequence *ABC*. *Ans.* $284\angle{150^\circ}$ V

14.46. A star-connected, three-phase, 400 volt alternator has a limit of 35 A per coil (*a*) What is the kVA rating of the machine? (*b*) If the alternator supplies a line current of 20 A at a power factor of 0·65, what is the kVA per phase of the machine? *Ans.* 26·6 kVA, 5·08 kVA

14.47. The balanced set of line currents in the phasor diagram in Fig. 14-46 have current magnitudes of 10 A and the line voltage is 120 volts. Find the corresponding total power and total volt-amperes.
Ans. 1·47 kW, 2·08 kVA

14.48. A star load with $\mathbf{Z}_A = 10\angle{0^\circ}$ ohms, $\mathbf{Z}_B = 10\angle{60^\circ}$ ohms and $\mathbf{Z}_C = 10\angle{-60^\circ}$ ohms is connected to a three-phase, three-wire, 200 volt, *ABC* system. Find the voltages across the load impedances, $\mathbf{V}_{AO}$, $\mathbf{V}_{BO}$ and $\mathbf{V}_{CO}$. *Ans.* $173\angle{90^\circ}$ V, $100\angle{0^\circ}$ V, $100\angle{180^\circ}$ V

14.49. A star load with $\mathbf{Z}_A = 10\angle{-60^\circ}$ ohms, $\mathbf{Z}_B = 10\angle{0^\circ}$ ohms and $\mathbf{Z}_C = 10\angle{60^\circ}$ ohms is connected to a three-phase, three-wire, 208 volt, *CBA* system. Find the voltages across the load impedances.
Ans. $208\angle{-120^\circ}$ V, 0, $208\angle{180^\circ}$ V

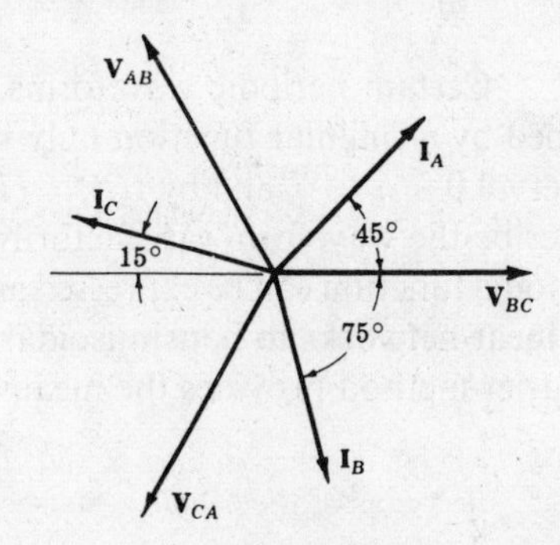

Fig. 14-46

14.50. A three-wire, 480 volt, *ABC* system supplies a star-connected load in which $\mathbf{Z}_A = 10\angle{0^\circ}$ Ω, $\mathbf{Z}_B = 5\angle{-30^\circ}$ Ω and $\mathbf{Z}_C = 5\angle{30^\circ}$ Ω. Find the readings of the wattmeters in lines *A* and *B*. *Ans.* 8·92 kW, 29·6 kW

14.51. A three-wire, 100 volt, *CBA* system supplies a star-connected load in which $\mathbf{Z}_A = 3 + j0$ ohms, $\mathbf{Z}_B = 2 + j3$ ohms and $\mathbf{Z}_C = 2 - j1$ ohms. Find the voltages across the impedances of the load.
Ans. $31{\cdot}6\angle{-67{\cdot}9^\circ}$ V, $84{\cdot}3\angle{42{\cdot}7^\circ}$ V, $68{\cdot}6\angle{123{\cdot}8^\circ}$ V

14.52. Three identical impedances of $15\angle{60^\circ}$ ohms are connected in star to a three-phase, three-wire, 240 volt system. The lines between the supply and the load have impedances of $2 + j1$ ohms. Find the line voltage magnitude at the load.
Ans. 213 V

14.53. Repeat Problem 14.52 for identical star-connected impedances of $15\angle{-60^\circ}$ ohms and compare the results by drawing the voltage phasor diagrams. *Ans.* 235 V

CHAPTER 15

Fourier Method of Waveform Analysis

INTRODUCTION

In the circuits examined previously we considered the steady state response to excitations having either constant or sinusoidal form. In such cases, a single expression described the forcing functions for all values of time, e.g. $v =$ constant for DC and $v = V_{max} \sin \omega t$ for AC apply for all t as shown in Fig. 15-1(*a*) and (*b*).

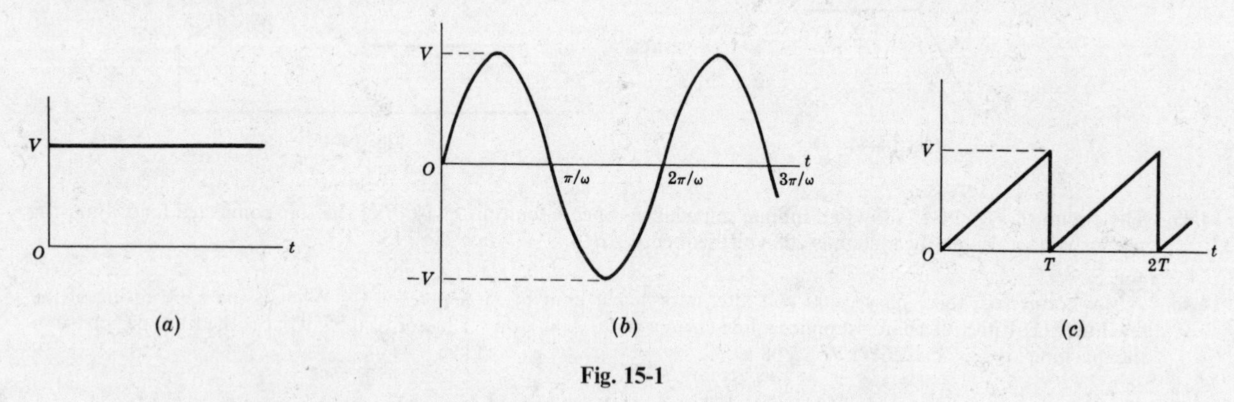

Fig. 15-1

Certain periodic waveforms, of which the sawtooth in Fig. 15-1(*c*) above is an example, can be described by a singular function only within an interval. Thus, the sawtooth is expressed by $f(t) = (V/T)t$ in the interval $0 < t < T$ and by $f(t) = (V/T)(t - T)$ in the interval $T < t < 2T$. While such piecemeal expressions describe the waveform satisfactorily, they do not permit the determination of the circuit response. Now, if a periodic function can be expressed as the sum of a finite or infinite number of sinusoidal functions, the responses of linear networks to nonsinusoidal exitations can be determined by applying the superposition theorem. The Fourier method provides the means for solving this type of problem.

TRIGONOMETRIC FOURIER SERIES

Any periodic waveform, i.e. one for which $f(t) = f(t + T)$, can be expressed by a Fourier series provided that

(1) if it is discontinuous there are a finite number of discontinuities in the period T,

(2) it has a finite average value for the period T,

(3) it has a finite number of positive and negative maxima.

When these conditions, called the *Dirichlet conditions*, are satisfied the Fourier series exists and can be written in trigonometric form:

$$f(t) = \tfrac{1}{2}a_0 + a_1 \cos \omega t + a_2 \cos 2\omega t + a_3 \cos 3\omega t + \cdots + b_1 \sin \omega t + b_2 \sin 2\omega t + b_3 \sin 3\omega t + \cdots \qquad (1)$$

The Fourier coefficients, a's and b's, are determined for a given waveform by the evaluation integrals. We obtain the cosine coefficient evaluation integral by multiplying both sides of (*1*) by $\cos n\omega t$ and integrating over a full period. The period of the fundamental, $2\pi/\omega$, is the period of the series since each term in the series has a frequency which is an integral multiple of the fundamental.

$$\int_0^{2\pi/\omega} f(t)\cos n\omega t\,dt = \int_0^{2\pi/\omega} \tfrac{1}{2}a_0\cos n\omega t\,dt + \int_0^{2\pi/\omega} a_1\cos\omega t\cos n\omega t\,dt + \cdots$$
$$+ \int_0^{2\pi/\omega} a_n\cos^2 n\omega t\,dt + \cdots + \int_0^{2\pi/\omega} b_1\sin\omega t\cos n\omega t\,dt$$
$$+ \int_0^{2\pi/\omega} b_2\sin 2\omega t\cos n\omega t\,dt + \cdots \qquad (2)$$

The definite integrals on the right side of (*2*) are all zero except $\int_0^{2\pi/\omega} a_n\cos^2 n\omega t\,dt$ which has a value $\frac{\pi}{\omega}a_n$. Then

$$a_n = \frac{\omega}{\pi}\int_0^{2\pi/\omega} f(t)\cos n\omega t\,dt = \frac{2}{T}\int_0^T f(t)\cos n\omega t\,dt \qquad (3)$$

Multiplying (*1*) by $\sin n\omega t$ and integrating as above results in the sine coefficient evaluation integral.

$$b_n = \frac{\omega}{\pi}\int_0^{2\pi/\omega} f(t)\sin n\omega t\,dt = \frac{2}{T}\int_0^T f(t)\sin n\omega t\,dt \qquad (4)$$

An alternate form of the evaluation integrals with the variable ωt and the corresponding period of 2π radians is

$$a_n = \frac{1}{\pi}\int_0^{2\pi} f(t)\cos n\omega t\,d(\omega t) \qquad (5)$$

$$b_n = \frac{1}{\pi}\int_0^{2\pi} f(t)\sin n\omega t\,d(\omega t) \qquad (6)$$

The limits of integration must include one full period but need not be from 0 to T or 0 to 2π. Instead, the integration can be carried out from $-T/2$ to $T/2$, $-\pi$ to $+\pi$, or any other full period which simplifies the integration. The constant a_0 is obtained from (*3*) or (*5*) with $n = 0$; however, since $\frac{1}{2}a_0$ is the average value of the function, it can frequently be determined by inspection of the waveform. The series with coefficients obtained from the above evaluation integrals converges uniformly to the function at all continuous points and converges to the mean value at points of discontinuity.

Example 1.

Find the Fourier series for the waveform shown in Fig. 15-2. The waveform is continuous for $0 < \omega t < 2\pi$ and given by $f(t) = (10/2\pi)\omega t$, with discontinuities at $\omega t = n2\pi$ where $n = 0, 1, 2, \ldots$. The Dirichlet conditions are satisfied and the Fourier coefficients are evaluated using (*5*) and (*6*). The average value of the function is 5 by inspection and thus $\frac{1}{2}a_0 = 5$. Now, using equation (*5*),

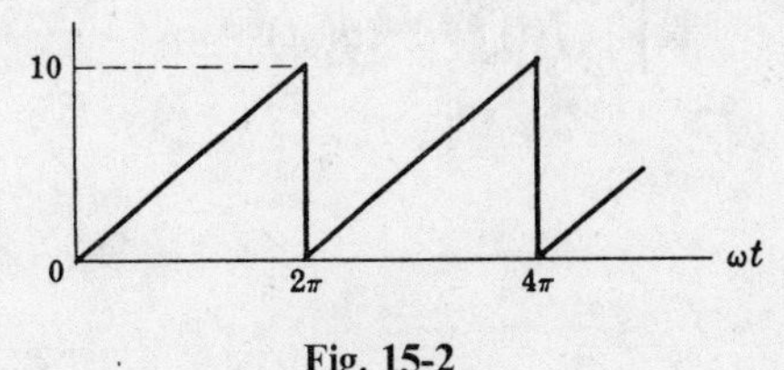

Fig. 15-2

$$a_n = \frac{1}{\pi}\int_0^{2\pi}\left(\frac{10}{2\pi}\right)\omega t\cos n\omega t\,d(\omega t) = \frac{10}{2\pi^2}\left[\frac{\omega t}{n}\sin n\omega t + \frac{1}{n^2}\cos n\omega t\right]_0^{2\pi}$$

$$= \frac{10}{2\pi^2 n^2}(\cos n2\pi - \cos 0) = 0 \qquad \text{for all integer values of } n$$

Thus the series contains no cosine terms. Using equation (*6*) we obtain

$$b_n = \frac{1}{\pi}\int_0^{2\pi}\left(\frac{10}{2\pi}\right)\omega t\sin n\omega t\,d(\omega t) = \frac{10}{2\pi^2}\left[-\frac{\omega t}{n}\cos n\omega t + \frac{1}{n^2}\sin n\omega t\right]_0^{2\pi} = -\frac{10}{\pi n}$$

Using these sine term coefficients and the average term, the series is

$$f(t) \;=\; 5 - \frac{10}{\pi}\sin\omega t - \frac{10}{2\pi}\sin 2\omega t - \frac{10}{3\pi}\sin 3\omega t - \cdots \;=\; 5 - \frac{10}{\pi}\sum_{n=1}^{\infty}\frac{\sin n\omega t}{n}$$

The sine and cosine terms of like frequency can be combined as a single sine or cosine term with a phase angle. Two alternate forms of the trigonometric series result.

$$f(t) \;=\; \tfrac{1}{2}a_0 + \sum c_n \cos(n\omega t - \theta_n) \tag{7}$$

and

$$f(t) \;=\; \tfrac{1}{2}a_0 + \sum c_n \sin(n\omega t + \varphi_n) \tag{8}$$

where $c_n = \sqrt{a_n^2 + b_n^2}$, $\theta_n = \tan^{-1}(b_n/a_n)$ and $\varphi_n = \tan^{-1}(a_n/b_n)$. c_n in (7) and (8) is the harmonic amplitude, and the harmonic phase angles are θ_n or φ_n.

EXPONENTIAL FOURIER SERIES

If we express each of the sine and cosine terms in the trigonometric series by its exponential equivalent the result is a series of exponential terms:

$$\begin{aligned} f(t) \;=\; & \frac{a_0}{2} + a_1\left(\frac{e^{j\omega t}+e^{-j\omega t}}{2}\right) + a_2\left(\frac{e^{j2\omega t}+e^{-j2\omega t}}{2}\right) + \cdots \\ & + b_1\left(\frac{e^{j\omega t}-e^{-j\omega t}}{2j}\right) + b_2\left(\frac{e^{j2\omega t}-e^{-j2\omega t}}{2j}\right) + \cdots \end{aligned} \tag{9}$$

Rearranging,

$$\begin{aligned} f(t) \;=\; & \cdots + \left(\frac{a_2}{2} - \frac{b_2}{2j}\right)e^{-j2\omega t} + \left(\frac{a_1}{2} - \frac{b_1}{2j}\right)e^{-j\omega t} \\ & + \frac{a_0}{2} + \left(\frac{a_1}{2} + \frac{b_1}{2j}\right)e^{j\omega t} + \left(\frac{a_2}{2} + \frac{b_2}{2j}\right)e^{j2\omega t} + \cdots \end{aligned} \tag{10}$$

We now define a new complex constant **A** such that

$$A_0 = \tfrac{1}{2}a_0, \qquad \mathbf{A}_n = \tfrac{1}{2}(a_n - jb_n), \qquad \mathbf{A}_{-n} = \tfrac{1}{2}(a_n + jb_n) \tag{11}$$

and rewrite (10) as

$$f(t) \;=\; \{\cdots + \mathbf{A}_{-2}e^{-j2\omega t} + \mathbf{A}_{-1}e^{-j\omega t} + A_0 + \mathbf{A}_1 e^{j\omega t} + \mathbf{A}_2 e^{j2\omega t} + \cdots\} \tag{12}$$

To obtain the evaluation integral for the $\mathbf{A}_n$ coefficients, we multiply (12) on both sides by $e^{-jn\omega t}$ and integrate over the full period:

$$\begin{aligned} \int_0^{2\pi} f(t)\,e^{-jn\omega t}\,d(\omega t) \;=\; & \cdots + \int_0^{2\pi}\mathbf{A}_{-2}\,e^{-j2\omega t}\,e^{-jn\omega t}\,d(\omega t) + \int_0^{2\pi}\mathbf{A}_{-1}\,e^{-j\omega t}\,e^{-jn\omega t}\,d(\omega t) \\ & + \int_0^{2\pi} A_0\,e^{-jn\omega t}\,d(\omega t) + \int_0^{2\pi}\mathbf{A}_1\,e^{j\omega t}\,e^{-jn\omega t}\,d(\omega t) + \cdots \\ & + \int_0^{2\pi}\mathbf{A}_n\,e^{jn\omega t}\,e^{-jn\omega t}\,d(\omega t) + \cdots \end{aligned} \tag{13}$$

The definite integrals on the right side of (13) are all zero except $\int_0^{2\pi}\mathbf{A}_n\,d(\omega t)$ which has the value $2\pi\mathbf{A}_n$.

Then

$$\mathbf{A}_n \;=\; \frac{1}{2\pi}\int_0^{2\pi} f(t)\,e^{-jn\omega t}\,d(\omega t)$$

or with t as the variable,

$$\mathbf{A}_n \;=\; \frac{1}{T}\int_0^{T} f(t)\,e^{-jn\omega t}\,dt \tag{14}$$

Just as with the a_n and b_n evaluation integrals, the limits of integration in (14) need cover any convenient full period and not necessarily 0 to 2π or 0 to T.

The trigonometric series coefficients are derived from the exponential series coefficients as follows: first add and then subtract the expressions for $\mathbf{A}_n$ and $\mathbf{A}_{-n}$ in *(11)*. Thus

$$\mathbf{A}_n + \mathbf{A}_{-n} = \tfrac{1}{2}(a_n - jb_n + a_n + jb_n)$$

from which

$$a_n = \mathbf{A}_n + \mathbf{A}_{-n} \tag{15}$$

and

$$\mathbf{A}_n - \mathbf{A}_{-n} = \tfrac{1}{2}(a_n - jb_n - a_n - jb_n)$$

or

$$b_n = j(\mathbf{A}_n - \mathbf{A}_{-n}) \tag{16}$$

Example 2.

Find the exponential Fourier series for the waveform shown in Fig. 15-3. Using the coefficients of this exponential series obtain a_n and b_n of the trigonometric series and compare with Example 1.

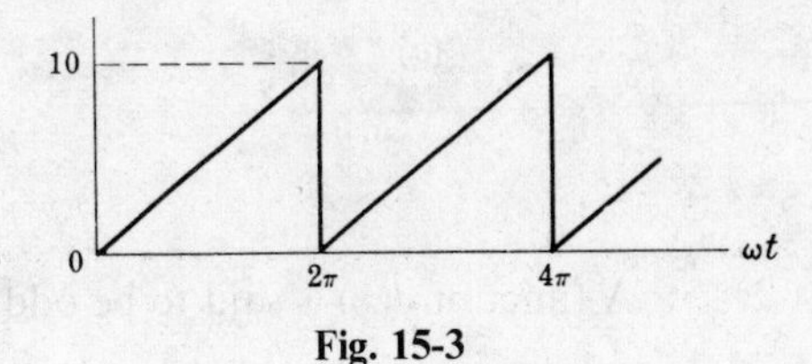

Fig. 15-3

In the interval $0 < \omega t < 2\pi$ the function is given by $f(t) = (10/2\pi)\omega t$. By inspection we note the average value of the function is 5. Substituting $f(t)$ in *(14)* we obtain the coefficients $\mathbf{A}_n$.

$$\mathbf{A}_n = \frac{1}{2\pi}\int_0^{2\pi}\left(\frac{10}{2\pi}\right)\omega t\, e^{-jn\omega t}\, d(\omega t) = \frac{10}{(2\pi)^2}\left[\frac{e^{-jn\omega t}}{(-jn)^2}(-jn\omega t - 1)\right]_0^{2\pi} = j\frac{10}{2\pi n}$$

Inserting the coefficients $\mathbf{A}_n$ in *(12)*, the exponential form of the Fourier series for the given waveform is

$$f(t) = \cdots - j\frac{10}{4\pi}e^{-j2\omega t} - j\frac{10}{2\pi}e^{-j\omega t} + 5 + j\frac{10}{2\pi}e^{j\omega t} + j\frac{10}{4\pi}e^{j2\omega t} + \cdots \tag{17}$$

The trigonometric series cosine term coefficient is

$$a_n = \mathbf{A}_n + \mathbf{A}_{-n} = j\frac{10}{2\pi n} + j\frac{10}{2\pi(-n)} = 0$$

and the sine term coefficient is

$$b_n = j(\mathbf{A}_n - \mathbf{A}_{-n}) = j\left(j\frac{10}{2\pi n} - j\frac{10}{2\pi(-n)}\right) = -\frac{10}{\pi n}$$

Thus the trigonometric series has no cosine terms since $a_n = 0$ for all n, and the sine term coefficients are $-10/(\pi n)$. The average value is 5 and the series is

$$f(t) = 5 - \frac{10}{\pi}\sin\omega t - \frac{10}{2\pi}\sin 2\omega t - \frac{10}{3\pi}\sin 3\omega t - \cdots$$

which is the same as in Example 1.

WAVEFORM SYMMETRY

The series obtained in Example 1 contained only sine terms in addition to a constant term. Other waveforms will have only cosine terms, and sometimes only odd harmonics are present in the series whether the series contains sine, cosine or both types of terms. This is the result of certain types of symmetry associated with the waveform. Knowledge of such symmetry results in reduced calculations in determining the series. For this purpose the following definitions are important.

1. A function $f(x)$ is said to be even if $f(x) = f(-x)$.

The function $f(x) = 2 + x^2 + x^4$ is an example of even functions since the functional values for x and $-x$ are equal. The cosine is an even function since it can be expressed in series form as

$$\cos x = 1 - \frac{x^2}{2!} + \frac{x^4}{4!} - \frac{x^6}{6!} + \frac{x^8}{8!} - \cdots$$

The sum of two or more even functions is an even function, and with the addition of a constant the even nature of the function is still preserved.

In Fig. 15-4 the waveforms shown represent even functions. They are symmetrical with respect to the vertical axis.

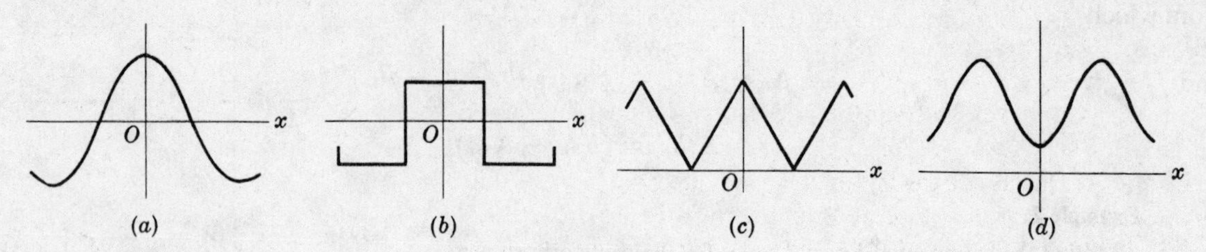

Fig. 15-4

2. A function $f(x)$ is said to be odd if $f(x) = -f(-x)$.

The function $f(x) = x + x^3 + x^5$ is an example of odd functions since the values of the function for x and $-x$ are of opposite sign. The sine is an odd function since it can be expressed in series form as

$$\sin x \quad = \quad x - \frac{x^3}{3!} + \frac{x^5}{5!} - \frac{x^7}{7!} + \frac{x^9}{9!} - \cdots$$

The sum of two or more odd functions is an odd function, but the addition of a constant removes the odd nature of the function since $f(x)$ is no longer equal to $-f(-x)$. The product of two odd functions is an even function.

The waveforms shown in Fig. 15-5 represent odd functions.

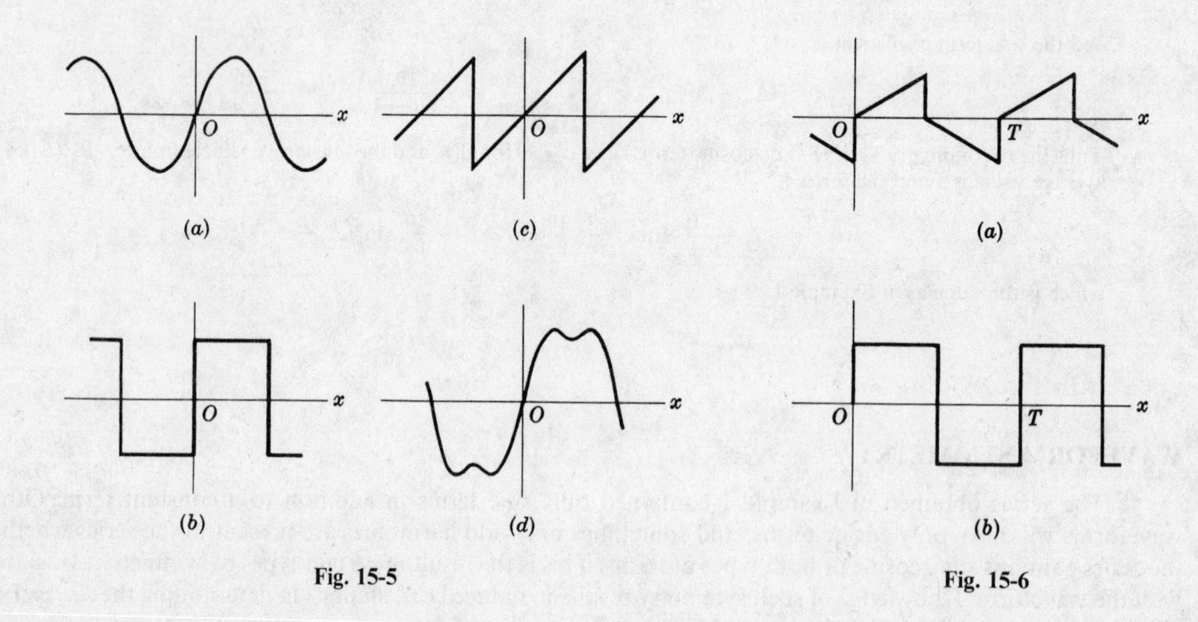

Fig. 15-5 **Fig. 15-6**

3. A periodic function $f(x)$ is said to have half-wave symmetry if $f(x) = -f(x + T/2)$ where T is the period. Two waveforms with half-wave symmetry are shown in Fig. 15-6.

When the type of symmetry of a waveform is established, the following conclusions are reached. If the waveform is even, all terms of the corresponding series are cosine terms and possibly a constant if the waveform has an average value. Hence there is no need of evaluating the integral for the coefficients b_n since no sine terms can be present. If odd, the series contains only sine terms. The wave may be odd only after the constant is removed, in which case its Fourier representation will simply contain that constant and a series of sine terms. If the waveform has half-wave symmetry, only odd harmonics are present in the series. This series will contain both sine and cosine terms unless the function is also odd or even. In any case, a_n and b_n are equal to zero for $n = 2, 4, 6, \ldots$ for any waveform with half-wave symmetry.

Certain waveforms can be odd or even, depending upon the location of the vertical axis. The square wave of Fig. 15-7(*a*) meets the condition of an even function, i.e. $f(x) = f(-x)$. A shift of the vertical axis to the position shown in Fig. 15-7(*b*) results in an odd function where $f(x) = -f(-x)$. With the vertical axis placed at any points other than those shown in Fig. 15-7, the square wave is neither even nor odd and its series contains both sine and cosine terms. Thus in the analysis of periodic functions the vertical axis should be conveniently chosen to result in either an even or odd function provided that the type of waveform makes this possible.

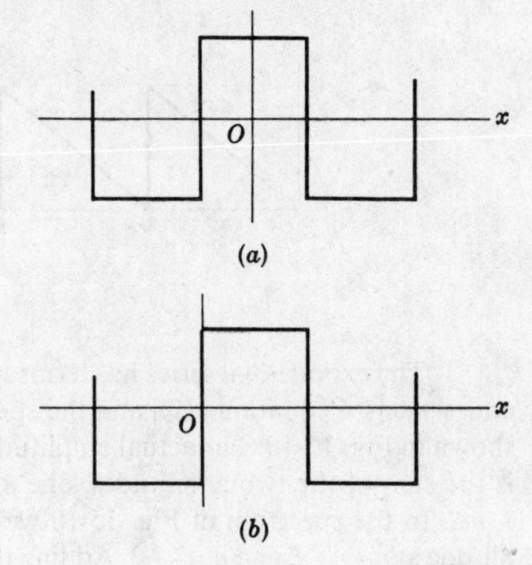

Fig. 15-7

The shifting of the horizontal axis may simplify the series representation of the function. As an example, the waveform of Fig. 15-8(*a*) does not meet the requirements of an odd function until the average value is removed as shown in Fig. 15-8(*b*). Thus its series will contain a constant term and all sine terms.

Since the exponential equivalent of the sine is pure imaginary and the exponential of a cosine pure real, the above symmetry considerations can be used to check the coefficients of the exponential series. An even waveform contains only cosine terms in its trigonometric series and therefore the exponential Fourier coefficients must be pure real numbers. Similarly, an odd function whose trigonometric series consists of sine terms has pure imaginary coefficients in its exponential series.

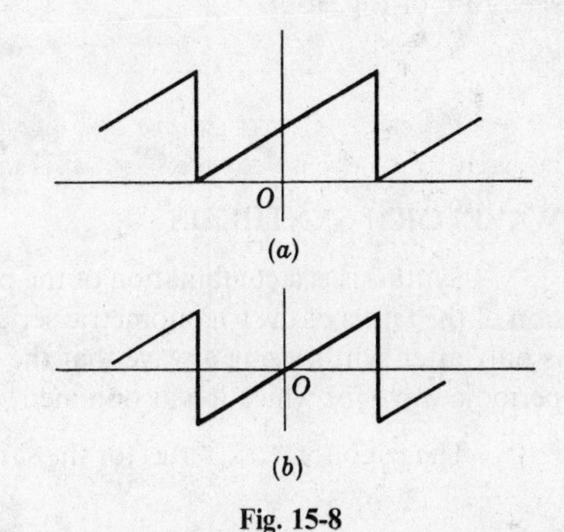

Fig. 15-8

LINE SPECTRUM

A plot showing each of the harmonic amplitudes in the wave is called the *line spectrum.* The lines decrease rapidly for waves with rapidly convergent series. Waves with discontinuities such as the sawtooth and square wave have spectra with slowly decreasing amplitudes since their series have strong high harmonics. Their 10th harmonics will often have amplitudes of significant value compared to the fundamental. In contrast the series of waveforms without discontinuities and with a generally smooth appearance will converge rapidly to the function and only a few terms are required to generate the wave. Such rapid convergence will be evident from the line spectrum where the harmonic amplitudes decrease rapidly, so that any above the 5th or 6th are insignificant.

The harmonic content and the line spectrum of a wave are part of the very nature of that wave and never change regardless of the method of analysis. Shifting the zero axis gives the trigonometric series a completely different appearance, and the exponential series coefficients also change greatly with a shift in the zero axis, but the same harmonics always appear in the series and their amplitude given by $c_n = \sqrt{a_n^2 + b_n^2}$ or $c_n = |\mathbf{A}_n| + |\mathbf{A}_{-n}|$ remains constant.

In Fig. 15-9 the sawtooth wave of Example 1 and its spectrum are shown. Since there were only sine terms in the series, the harmonic amplitudes c_n are given directly by b_n.

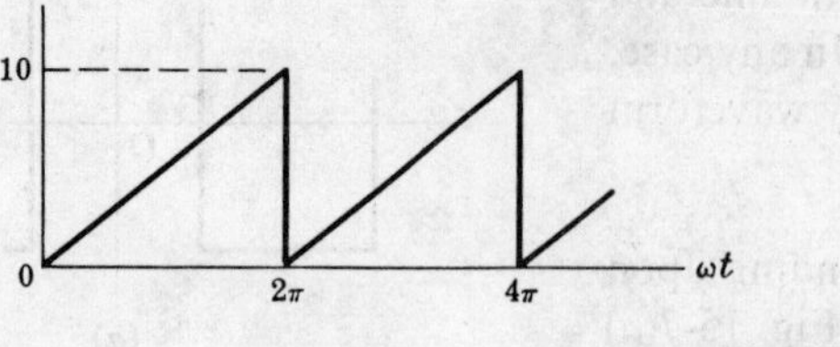

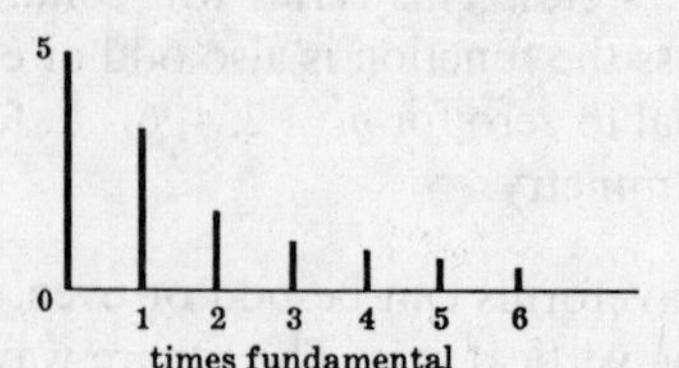

Fig. 15-9

The exponential series has terms with frequencies of $+n\omega$ and $-n\omega$ [see equation (*17*)] and the spectrum is constructed as shown in Fig. 15-10. The actual amplitude of a specific harmonic is the sum of the two amplitudes, one at $+n\omega$ and the other at $-n\omega$. In the spectrum of Fig. 15-10 we find lines of $10/4\pi$ amplitude at $n = -2$ and $n = +2$. Adding these we obtain $10/2\pi$ for the actual amplitude of this harmonic, which agrees with the spectrum of Fig. 15-9.

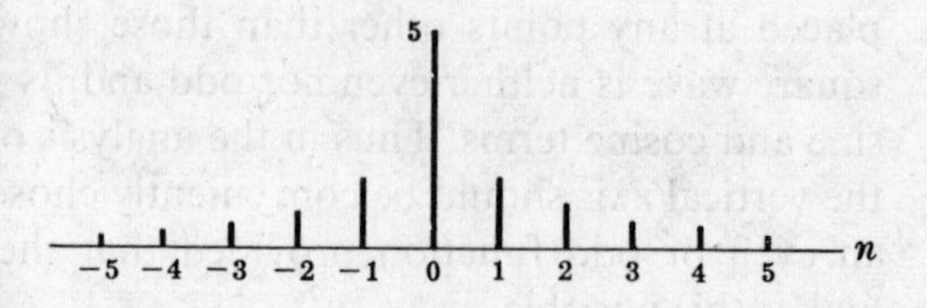

Fig. 15-10

WAVEFORM SYNTHESIS

Synthesis is a combination of the parts so as to form the whole. In Fourier analysis it is the recombination of the terms of the trigonometric series, usually the first four or five, to produce the original wave. Often it is only after synthesizing a wave that the student is convinced that the Fourier series does in fact express the periodic wave for which it was obtained.

The trigonometric series for the sawtooth wave of Example 1 with a peak amplitude of 10 is

$$f(t) \quad = \quad 5 - \frac{10}{\pi}\sin\omega t - \frac{10}{2\pi}\sin 2\omega t - \frac{10}{3\pi}\sin 3\omega t - \cdots$$

These four terms are plotted and added in Fig. 15-11 and although the result is not a perfect sawtooth wave it appears that with more terms included the sketch will more nearly resemble a sawtooth. Since this wave has discontinuities, its series is not rapidly convergent and consequently the synthesis using only four terms does not produce a very good result. The next term at the frequency 4ω has amplitude $10/4\pi$ which is certainly significant compared to the fundamental with amplitude $10/\pi$. As each term is added in synthesizing the waveform, the irregularities of the resultant are reduced and the approximation to the original wave is improved. This is what was meant when we said earlier that *the series converges to the function at all continuous points and to the mean value at points of discontinuity*. In Fig. 15-11 at 0 and 2π it is clear that a value of 5 will remain since all sine terms are zero at these points. These are the points of discontinuity; and the value of the function when they are approached from the left is 10, and from the right 0, with the mean value 5.

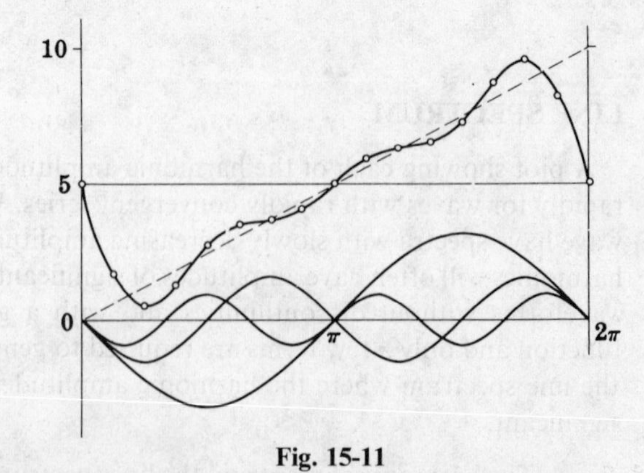

Fig. 15-11

EFFECTIVE VALUE AND POWER

A periodic, non-sinusoidal wave of current passing through a resistor results in a power which is determined by the effective or rms value of the wave. In Chapter 2 the effective value of a function such as

$$f(t) \quad = \quad \tfrac{1}{2}a_0 + a_1 \cos \omega t + a_2 \cos 2\omega t + \cdots + b_1 \sin \omega t + b_2 \sin 2\omega t + \cdots$$

was found to be

$$F_{\text{rms}} \quad = \quad \sqrt{(\tfrac{1}{2}a_0)^2 + \tfrac{1}{2}a_1^2 + \tfrac{1}{2}a_2^2 + \cdots + \tfrac{1}{2}b_1^2 + \tfrac{1}{2}b_2^2 + \cdots} \qquad (18)$$

Expressing the harmonic amplitude by $c_n = \sqrt{a_n^2 + b_n^2}$ and writing c_0 for the average value, from equation (18) we have

$$F_{\text{rms}} \quad = \quad \sqrt{c_0^2 + \tfrac{1}{2}c_1^2 + \tfrac{1}{2}c_2^2 + \tfrac{1}{2}c_3^2 + \cdots}$$

Considering a linear network with an applied voltage which is periodic, we would expect that the resulting current would contain the same harmonic terms as the voltage but with harmonic amplitudes of different relative mangitude since the impedance varies with $n\omega$. It is possible that some harmonics would not appear in the current since parallel resonance results in an infinite impedance. In general we could write

$$v \; = \; V_0 + \sum V_n \sin (n\omega t + \varphi_n) \qquad \text{and} \qquad i \; = \; I_0 + \sum I_n \sin (n\omega t + \psi_n) \qquad (19)$$

with corresponding effective values of

$$V_{\text{rms}} \; = \; \sqrt{V_0^2 + \tfrac{1}{2}V_1^2 + \tfrac{1}{2}V_2^2 + \cdots} \qquad \text{and} \qquad I_{\text{rms}} \; = \; \sqrt{I_0^2 + \tfrac{1}{2}I_1^2 + \tfrac{1}{2}I_2^2 + \cdots} \qquad (20)$$

The average power P follows from integration of the instantaneous power given by the product of v and i,

$$p \; = \; vi \; = \; [V_0 + \sum V_n \sin (n\omega t + \varphi_n)][I_0 + \sum I_n \sin (n\omega t + \psi_n)] \qquad (21)$$

Since v and i both have periods of T sec, their product must have an integral number of its periods in T. (Recall that for a single sine wave of applied voltage the product vi has a period of half that of the voltage wave). The average power

$$P \; = \; \frac{1}{T}\int_0^T [V_0 + \sum V_n \sin (n\omega t + \varphi_n)][I_0 + \sum I_n \sin (n\omega t + \psi_n)]\, dt \qquad (22)$$

Examination of the possible terms in the product of the two infinite series shows them to be of the following types: the product of two constants, the product of a constant and a sine function, the product of two sine functions of different frequencies, and sine functions squared. After integration, the product of the two constants is still $V_0 I_0$ and the sine functions squared with the limits applied appear as $(V_n I_n/2) \cos (\varphi_n - \psi_n)$ while all other products upon integration over the period T are zero. Then the average power

$$P \; = \; V_0 I_0 + \tfrac{1}{2}V_1 I_1 \cos \theta_1 + \tfrac{1}{2}V_2 I_2 \cos \theta_2 + \tfrac{1}{2}V_3 I_3 \cos \theta_3 + \cdots \qquad (23)$$

where $\theta_n = (\varphi_n - \psi_n)$ is the angle on the equivalent impedance of the network at the frequency $n\omega$ rad/sec, and V_n and I_n are the maximum values of the respective sine functions. In the single frequency AC circuits, we found that the average power $P = VI \cos \theta$ which is included in (23) since V is an effective voltage, $V = V_{\max}/\sqrt{2}$ and $I = I_{\max}/\sqrt{2}$ so that $P = \tfrac{1}{2}V_{\max} I_{\max} \cos \theta$. In simple DC circuits the power is VI, included in (23) as $V_0 I_0$. Therefore, the power equation (23) is perfectly general, including DC, single frequency AC and also periodic non-sinusoidal waves. We note also in (23) that there is no contribution to the average power from voltage and current of different frequencies. In regard to power then, each harmonic acts independently.

APPLICATIONS IN CIRCUIT ANALYSIS

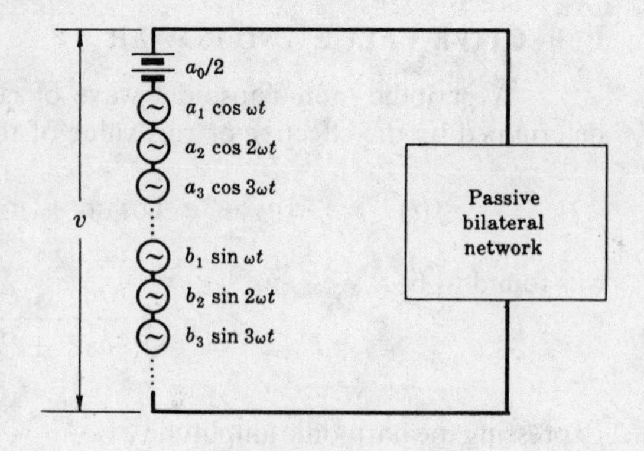

Fig. 15-12

It has already been suggested above that we could apply the terms of a voltage series to a linear network and obtain the corresponding harmonic terms of the current series. This result is obtained by superposition. Thus we consider each term of the Fourier series representing the voltage as a single source as shown in Fig. 15-12. Now the equivalent impedance of the network at each harmonic frequency $n\omega$ is used to compute the current at that harmonic. And the sum of these individual responses is the total response i in series form due to the applied voltage.

Example 3.

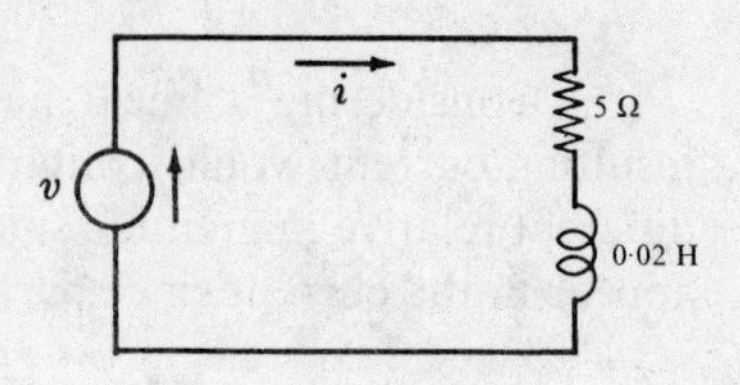

Fig. 15-13

As series RL circuit in which $R = 5$ ohms and $L = 0{\cdot}02$ H has an applied voltage $v = 100 + 50 \sin \omega t + 25 \sin 3\omega t$ volts where $\omega = 500$ rad/s. Find the current and the average power.

Compute the equivalent impedance of the circuit at each frequency. Then obtain the respective currents.

At $\omega = 0$, $\mathbf{Z} = 5\ \Omega$ and

$$I_0 = V_0/R = 100/5 = 20 \text{ A}$$

At $\omega = 500$ rad/s, $\mathbf{Z}_1 = 5 + j(0{\cdot}02)(500) = 5 + j10\ \Omega$ and

$$i_1 = \frac{V_{1\,\max}}{|\mathbf{Z}_1|} \sin (\omega t - \theta_1) = \frac{50}{11{\cdot}15} \sin (\omega t - 63{\cdot}4^\circ) = 4{\cdot}48 \sin (\omega t - 63{\cdot}4^\circ) \text{ amperes}$$

At $3\omega = 1500$ rad/s, $\mathbf{Z}_3 = 5 + j30\ \Omega$ and

$$i_3 = \frac{V_{3\,\max}}{|\mathbf{Z}_3|} \sin (3\omega t - \theta_3) = \frac{25}{30{\cdot}4} \sin (3\omega t - 80{\cdot}54^\circ) = 0{\cdot}823 \sin (3\omega t - 80{\cdot}54^\circ) \text{ amperes}$$

The sum of the harmonic currents is the required total response.

$$i = 20 + 4{\cdot}48 \sin (\omega t - 63{\cdot}4^\circ) + 0{\cdot}823 \sin (3\omega t - 80{\cdot}54^\circ) \text{ amperes}$$

This current has an effective value

$$I_{\text{rms}} = \sqrt{20^2 + 4{\cdot}48^2/2 + 0{\cdot}823^2/2} = \sqrt{410{\cdot}6} = 20{\cdot}25 \text{ A}$$

which results in a power in the 5 ohm resistor

$$P = I_{\text{rms}}^2 R = (410{\cdot}6)5 = 2053 \text{ W}$$

As a check we compute the total average power by calculating first the power contributed by each harmonic and then adding the results.

At $\omega = 0$, $\quad P = V_0 I_0 = 100(20) = 2000$ W

At $\omega = 500$ rad/s, $\quad P = \frac{1}{2} V_1 I_1 \cos \theta_1 = \frac{1}{2}(50)(4{\cdot}48) \cos 63{\cdot}4^\circ = 50{\cdot}1$ W

At $3\omega = 1500$ rad/s, $\quad P = \frac{1}{2} V_3 I_3 \cos \theta_3 = \frac{1}{2}(25)(0{\cdot}823) \cos 80{\cdot}54^\circ = 1{\cdot}69$ W

Then $\quad P_T = 2000 + 50{\cdot}1 + 1{\cdot}69 = 2052$ W

Alternate method.

The series expression for the voltage across the resistor is

$$v_R = Ri = 100 + 22{\cdot}4 \sin(\omega t - 63{\cdot}4^\circ) + 4{\cdot}11 \sin(3\omega t - 80{\cdot}54^\circ) \text{ volts}$$

and

$$V_R = \sqrt{100^2 + \tfrac{1}{2}(22{\cdot}4)^2 + \tfrac{1}{2}(4{\cdot}11)^2} = \sqrt{10{,}259} = 101{\cdot}3 \text{ V}$$

Then the power delivered by the source is $P = V_R^2/R = (101{\cdot}3)^2/5 = 2052$ W.

The exponential Fourier series is used in the same way except that frequently the circuit impedance can be expressed in terms of $n\omega$ and the coefficients of the current series $\mathbf{I}_n$ can be computed from the ratio $\mathbf{V}_n/\mathbf{Z}_n$ as shown in Example 4 below.

Example 4.

A voltage represented by the triangular wave shown in Fig. 15-14 is applied to a pure capacitor of C farads. Determine the resulting current.

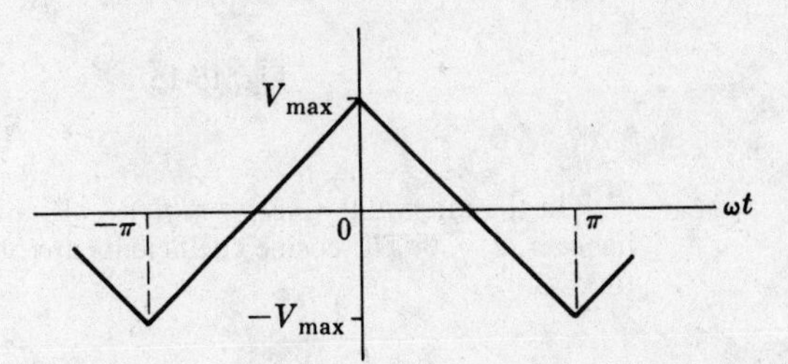

Fig. 15-14

In the interval $-\pi < \omega t < 0$ the voltage function is $v = V_{max} + (2V_{max}/\pi)\omega t$; and for $0 < \omega t < \pi$, $v = V_{max} - (2V_{max}/\pi)\omega t$. Then the coefficients of the exponential series are determined by the evaluation integral

$$\mathbf{A}_n = \frac{1}{2\pi}\int_{-\pi}^{0} [V_{max} + (2V_{max}/\pi)\omega t]e^{-jn\omega t}\, d(\omega t)$$

$$+ \frac{1}{2\pi}\int_{0}^{\pi} [V_{max} - (2V_{max}/\pi)\omega t]e^{-jn\omega t}\, d(\omega t)$$

from which $\mathbf{A}_n = \dfrac{4V_{max}}{\pi^2 n^2}$ for odd n, and $\mathbf{A}_n = 0$ for even n.

The circuit impedance $\mathbf{Z} = 1/j\omega C$ can be expressed as a function of n, i.e. $\mathbf{Z}_n = 1/jn\omega C$. Now

$$\mathbf{I}_n = \frac{\mathbf{V}_n}{\mathbf{Z}_n} = \frac{4V_{max}}{\pi^2 n^2}(jn\omega C) = j\left(\frac{4V_{max}\,\omega C}{\pi^2 n}\right)$$

and the current series is

$$i = j\left(\frac{4V_{max}\,\omega C}{\pi^2}\right)\sum \frac{e^{jn\omega t}}{n} \quad \text{for odd } n \text{ only}$$

The series could be converted to the trigonometric form and then synthesized to show the current waveform. However, this series is of the same form as the result in Problem 15.8 where the coefficient $\mathbf{A}_n = -j(2V/n\pi)$ for odd n only. The sign here is negative, indicating that our current wave is the negative of the square wave of Problem 15.8, and with a peak value $(2V_{max}\omega C)/\pi$.

Solved Problems

15.1. Find the trigonometric Fourier series for the square wave shown in Fig. 15-15 and plot the line spectrum.

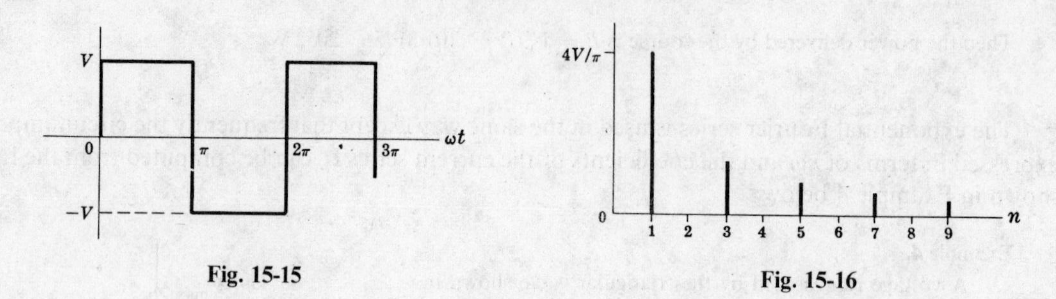

Fig. 15-15

Fig. 15-16

In the interval $0 < \omega t < \pi$, $f(t) = V$; and for $\pi < \omega t < 2\pi$, $f(t) = -V$. The average value of the wave is zero; hence $a_0/2 = 0$. The cosine coefficients are obtained by writing the evaluation integral with the functions inserted as follows.

$$a_n = \frac{1}{\pi}\left\{\int_0^{\pi} V\cos n\omega t\, d(\omega t) + \int_{\pi}^{2\pi} (-V)\cos n\omega t\, d(\omega t)\right\} = \frac{V}{\pi}\left\{\left[\frac{1}{n}\sin n\omega t\right]_0^{\pi} - \left[\frac{1}{n}\sin n\omega t\right]_{\pi}^{2\pi}\right\}$$
$$= 0 \quad \text{for all } n$$

Thus the series contains no cosine terms. Proceeding with the evaluation integral for the sine terms,

$$b_n = \frac{1}{\pi}\left\{\int_0^{\pi} V\sin n\omega t\, d(\omega t) + \int_{\pi}^{2\pi} (-V)\sin n\omega t\, d(\omega t)\right\}$$
$$= \frac{V}{\pi}\left\{\left[-\frac{1}{n}\cos n\omega t\right]_0^{\pi} + \left[\frac{1}{n}\cos n\omega t\right]_{\pi}^{2\pi}\right\}$$
$$= \frac{V}{\pi n}(-\cos n\pi + \cos 0 + \cos n2\pi - \cos n\pi) = \frac{2V}{\pi n}(1 - \cos n\pi)$$

Then $b_n = 4V/\pi n$ for $n = 1, 3, 5, \ldots$, and $b_n = 0$ for $n = 2, 4, 6, \ldots$. The series for the square wave is

$$f(t) = \frac{4V}{\pi}\sin \omega t + \frac{4V}{3\pi}\sin 3\omega t + \frac{4V}{5\pi}\sin 5\omega t + \cdots$$

The line spectrum for this series is shown in Fig. 15-16 above. The series contains only odd harmonic sine terms which could have been anticipated by examination of the wave-form for symmetry. Since the wave in Fig. 15-15 is odd, its series contains only sine terms; and since it also has half-wave symmetry, only odd harmonics are present.

15.2. Find the trigonometric Fourier series for the triangular wave shown in Fig. 15-17 and plot the spectrum.

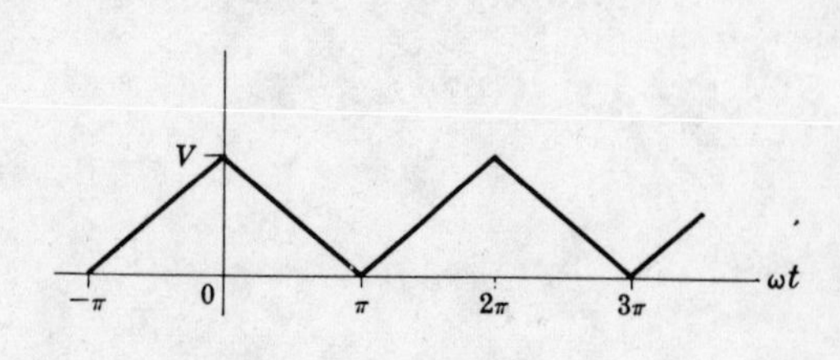

Fig. 15-17

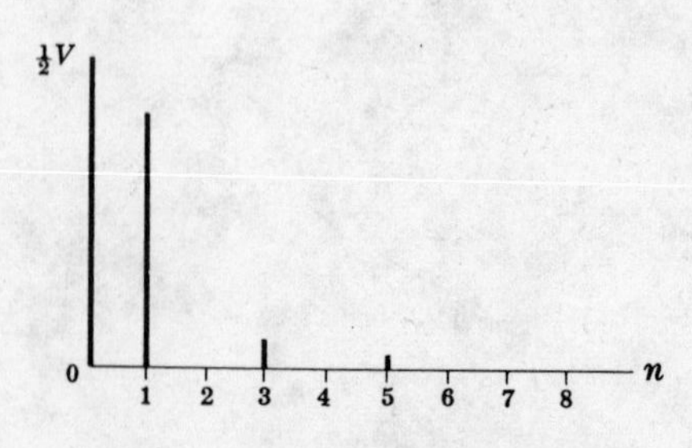

Fig. 15-18

The wave is an even function since $f(t) = f(-t)$, and if the average value $V/2$ is removed it also has half-wave symmetry, i.e. $f(t) = -f(t + T/2)$. In the interval $-\pi < \omega t < 0$, $f(t) = V + (V/\pi)\omega t$; and for $0 < \omega t < \pi$, $f(t) = V - (V/\pi)\omega t$. Since even waveforms have only cosine terms, $b_n = 0$ for all integer n.

$$a_n = \frac{1}{\pi}\int_{-\pi}^{0}[V + (V/\pi)\omega t]\cos n\omega t\, d(\omega t) + \frac{1}{\pi}\int_{0}^{\pi}[V - (V/\pi)\omega t]\cos n\omega t\, d(\omega t)$$

$$= \frac{V}{\pi}\left\{\int_{-\pi}^{\pi}\cos n\omega t\, d(\omega t) + \int_{-\pi}^{0}\frac{\omega t}{\pi}\cos n\omega t\, d(\omega t) - \int_{0}^{\pi}\frac{\omega t}{\pi}\cos n\omega t\, d(\omega t)\right\}$$

$$= \frac{V}{\pi^2}\left\{\left[\frac{1}{n^2}\cos n\omega t + \frac{\omega t}{n}\sin n\omega t\right]_{-\pi}^{0} - \left[\frac{1}{n^2}\cos n\omega t + \frac{\omega t}{n}\sin n\omega t\right]_{0}^{\pi}\right\}$$

$$= \frac{V}{\pi^2 n^2}\{\cos 0 - \cos(-n\pi) - \cos n\pi + \cos 0\} = \frac{2V}{\pi^2 n^2}(1 - \cos n\pi)$$

As the half-wave symmetry predicted, the series contains only odd terms since $a_n = 0$ for $n = 2, 4, 6, \ldots$. For $n = 1, 3, 5, \ldots$, $a_n = 4V/\pi^2 n^2$. Then the required Fourier series is

$$f(t) = \frac{V}{2} + \frac{4V}{\pi^2}\cos\omega t + \frac{4V}{(3\pi)^2}\cos 3\omega t + \frac{4V}{(5\pi)^2}\cos 5\omega t + \cdots$$

The coefficients decrease as $1/n^2$, and thus the series converges more rapidly than that of Problem 15.1. This fact is evident from the line spectrum shown in Fig. 15-18.

15.3. Find the trigonometric Fourier series for the sawtooth wave shown in Fig. 15-19 and plot the spectrum.

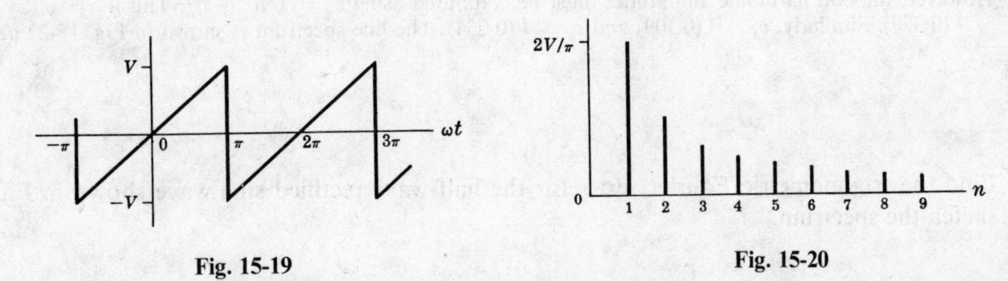

Fig. 15-19

Fig. 15-20

By inspection we note that the average value of the wave is zero and that the wave is odd. Consequently the series will contain only sine terms. A singular expression, $f(t) = (V/\pi)\omega t$, describes the wave over the period from $-\pi$ to $+\pi$ and we will use these limits on our evaluation integral for b_n.

$$b_n = \frac{1}{\pi}\int_{-\pi}^{\pi}(V/\pi)\omega t\sin n\omega t\, d(\omega t) = \frac{V}{\pi^2}\left[\frac{1}{n^2}\sin n\omega t - \frac{\omega t}{n}\cos n\omega t\right]_{-\pi}^{\pi} = -\frac{2V}{n\pi}(\cos n\pi)$$

Cos $n\pi$ is positive for even n and negative for odd n, and thus the signs of the coefficients alternate. The required series is

$$f(t) = \frac{2V}{\pi}\{\sin\omega t - \tfrac{1}{2}\sin 2\omega t + \tfrac{1}{3}\sin 3\omega t - \tfrac{1}{4}\sin 4\omega t + \cdots\}$$

The coefficients decrease as $1/n$, and thus the series converges slowly as shown by the spectrum in Fig. 15-20 above. Except for the shift in the zero axis and the average term, this waveform is the same as in Example 1. Compare the line spectrum of Fig. 15-9 with that of Fig. 15-20 and note the similarity.

15.4. Find the trigonometric Fourier series for the waveform shown in Fig. 15-21 below and sketch the spectrum.

In the interval $0 < \omega t < \pi$, $f(t) = (V/\pi)\omega t$; and for $\pi < \omega t < 2\pi$, $f(t) = 0$. By inspection the average value of the wave is $V/4$. Since the wave is neither even nor odd, the series will contain both sine and cosine terms. In the interval 0 to π, we have

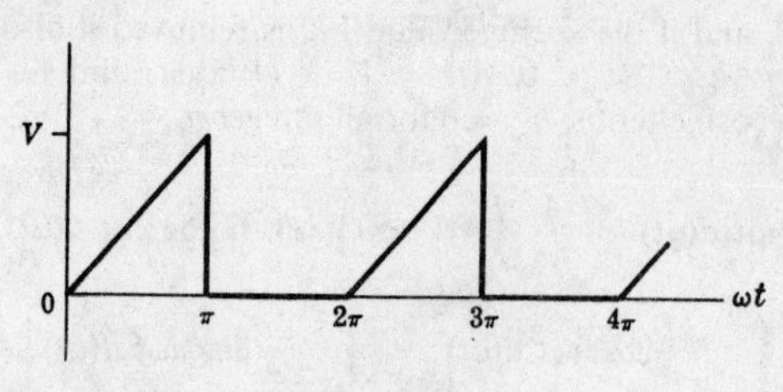

Fig. 15-21

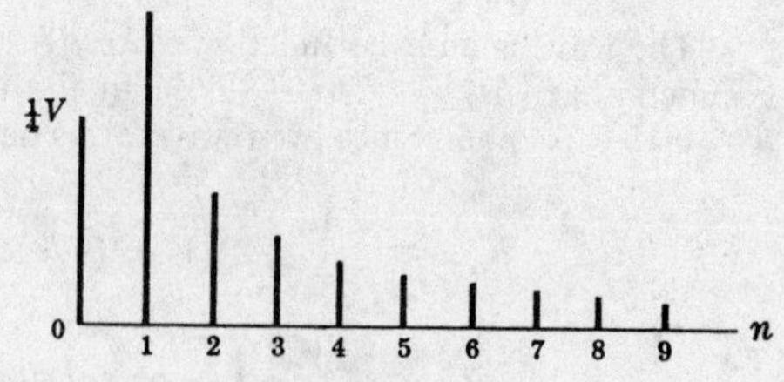

Fig. 15-22

$$a_n = \frac{1}{\pi}\int_0^{\pi}(V/\pi)\omega t\cos n\omega t\,d(\omega t) = \frac{V}{\pi^2}\left[\frac{1}{n^2}\cos n\omega t + \frac{\omega t}{n}\sin n\omega t\right]_0^{\pi} = \frac{V}{\pi^2 n^2}(\cos n\pi - 1)$$

When n is even, $(\cos n\pi - 1) = 0$ and $a_n = 0$. When n is odd, $a_n = -2V/(\pi^2 n^2)$. The b_n coefficients are

$$b_n = \frac{1}{\pi}\int_0^{\pi}(V/\pi)\omega t\sin n\omega t\,d(\omega t) = \frac{V}{\pi^2}\left[\frac{1}{n^2}\sin n\omega t - \frac{\omega t}{n}\cos n\omega t\right]_0^{\pi} = -\frac{V}{\pi n}(\cos n\pi)$$

The sign alternates with $b_n = -V/\pi n$ for even n, and $b_n = +V/\pi n$ for odd n. Then the required Fourier series is

$$f(t) = \frac{V}{4} - \frac{2V}{\pi^2}\cos\omega t - \frac{2V}{(3\pi)^2}\cos 3\omega t - \frac{2V}{(5\pi)^2}\cos 5\omega t - \cdots$$

$$+ \frac{V}{\pi}\sin\omega t - \frac{V}{2\pi}\sin 2\omega t + \frac{V}{3\pi}\sin 3\omega t - \cdots$$

The even harmonic amplitudes are given directly by the b_n coefficients since there are no even cosine terms. However, the odd harmonic amplitudes must be computed using $c_n = \sqrt{a_n^2 + b_n^2}$. Thus $c_1 = \sqrt{(2V/\pi^2)^2 + (V/\pi)^2} = V(0{\cdot}377)$. Similarly, $c_3 = V(0{\cdot}109)$ and $c_5 = V(0{\cdot}064)$. The line spectrum is shown in Fig. 15-22 above.

15.5. Find the trigonometric Fourier series for the half-wave rectified sine wave shown in Fig. 15-23 and sketch the spectrum.

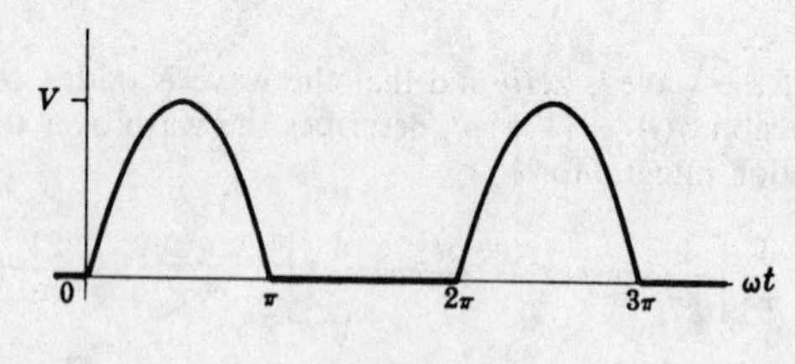

Fig. 15-23

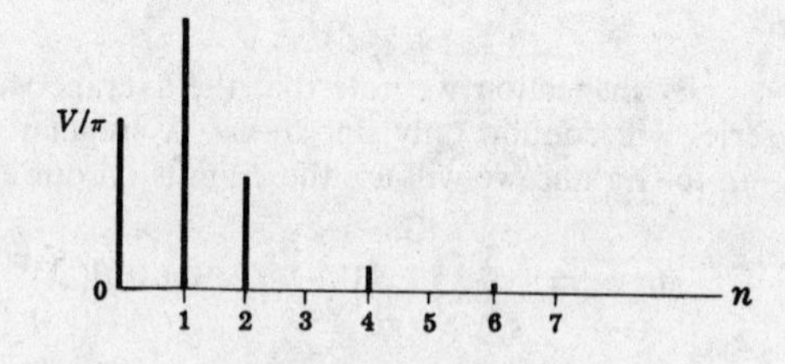

Fig. 15-24

The wave shows no symmetry and we therefore expect the series to contain both sine and cosine terms. Since the average value is not obtainable by inspection, we evaluate a_0 for the term $a_0/2$ in the series.

$$a_0 = \frac{1}{\pi}\int_0^{\pi}V\sin\omega t\,d(\omega t) = \frac{V}{\pi}\Big[-\cos\omega t\Big]_0^{\pi} = \frac{2V}{\pi}$$

Next we determine a_n:

$$a_n = \frac{1}{\pi}\int_0^{\pi}V\sin\omega t\cos n\omega t\,d(\omega t)$$

$$= \frac{V}{\pi}\left[\frac{-n\sin\omega t\sin n\omega t - \cos n\omega t\cos\omega t}{-n^2+1}\right]_0^{\pi} = \frac{V}{\pi(1-n^2)}(\cos n\pi + 1)$$

With n even, $a_n = 2V/\pi(1-n^2)$; and with n odd, $a_n = 0$. However, this expression is indeterminate for $n = 1$ and therefore we must integrate separately for a_1.

$$a_1 = \frac{1}{\pi}\int_0^\pi V \sin \omega t \cos \omega t \, d(\omega t) = \frac{V}{\pi}\int_0^\pi \tfrac{1}{2}\sin 2\omega t \, d(\omega t) = 0$$

Now we evaluate b_n:

$$b_n = \frac{1}{\pi}\int_0^\pi V \sin \omega t \sin n\omega t \, d(\omega t) = \frac{V}{\pi}\left[\frac{n \sin \omega t \cos n\omega t - \sin n\omega t \cos \omega t}{-n^2+1}\right]_0^\pi = 0$$

Here again the expression is indeterminate for $n = 1$, and b_1 is evaluated separately.

$$b_1 = \frac{1}{\pi}\int_0^\pi V \sin^2 \omega t \, d(\omega t) = \frac{V}{\pi}\left[\frac{\omega t}{2} - \frac{\sin 2\omega t}{4}\right]_0^\pi = \frac{V}{2}$$

Then the required Fourier series is

$$f(t) = \frac{V}{\pi}\left\{1 + \frac{\pi}{2}\sin \omega t - \frac{2}{3}\cos 2\omega t - \frac{2}{15}\cos 4\omega t - \frac{2}{35}\cos 6\omega t - \cdots\right\}$$

The spectrum in Fig. 15-24 above shows the strong fundamental term in the series and the rapidly decreasing amplitudes of the higher harmonics.

15.6. Find the trigonometric Fourier series for the half-wave rectified sine wave shown in Fig. 15-25 where the vertical axis is shifted from its position in Problem 15.5.

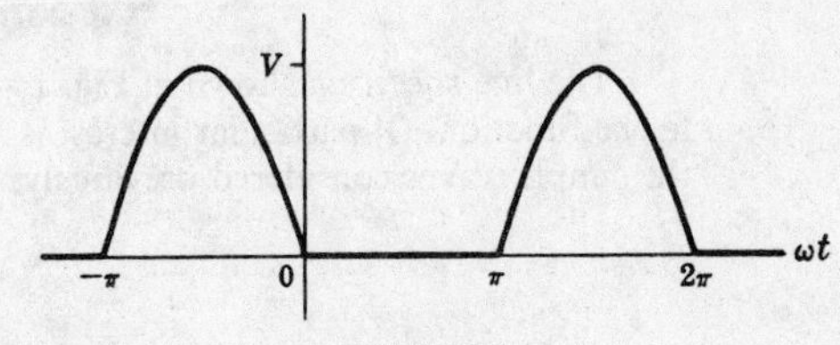

Fig. 15-25

The function is described in the interval $-\pi < \omega t < 0$ by $f(t) = -V \sin \omega t$. The average value is the same as that of Problem 15.5, i.e. $a_0 = 2V/\pi$. For the coefficients a_n, we have

$$a_n = \frac{1}{\pi}\int_{-\pi}^0 (-V \sin \omega t) \cos n\omega t \, d(\omega t) = \frac{V}{\pi(1-n^2)}(1 + \cos n\pi)$$

For n even, $a_n = 2V/\pi(1-n^2)$; and for n odd, $a_n = 0$ except that $n = 1$ must be examined separately.

$$a_1 = \frac{1}{\pi}\int_{-\pi}^0 (-V \sin \omega t) \cos \omega t \, d(\omega t) = 0$$

For the coefficients b_n, we obtain

$$b_n = \frac{1}{\pi}\int_{-\pi}^0 (-V \sin \omega t) \sin n\omega t \, d(\omega t) = 0$$

But again this expression is indeterminate for $n = 1$, so we evaluate b_1 separately.

$$b_1 = \frac{1}{\pi}\int_{-\pi}^0 (-V) \sin^2 \omega t \, d(\omega t) = -\frac{V}{2}$$

Thus the series is

$$f(t) = \frac{V}{\pi}\left\{1 - \frac{\pi}{2}\sin \omega t - \frac{2}{3}\cos 2\omega t - \frac{2}{15}\cos 4\omega t - \frac{2}{35}\cos 6\omega t - \cdots\right\}$$

This series is identical to that of Problem 15.5 except for the fundamental term which is negative in this series. The spectrum would obviously be identical to that in Fig. 15.24.

15.7. Find the trigonometric Fourier series for the rectangular pulse shown in Fig. 15-26 below and plot the spectrum.

With the zero axis positioned as shown, the wave is even and the series will contain only cosine terms and a constant term. The period from $-\pi$ to $+\pi$ is used for the evaluation integrals and the function is zero except for the interval from $-\pi/6$ to $+\pi/6$.

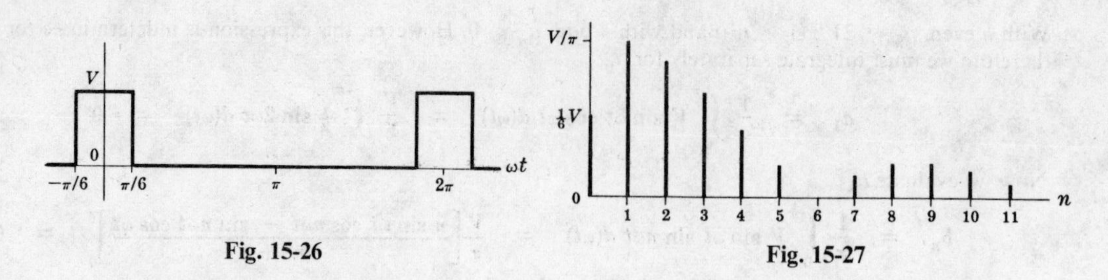

Fig. 15-26

Fig. 15-27

$$a_0 = \frac{1}{\pi}\int_{-\pi/6}^{\pi/6} V\, d(\omega t) = \frac{V}{3}, \qquad a_n = \frac{1}{\pi}\int_{-\pi/6}^{\pi/6} V \cos n\omega t\, d(\omega t) = \frac{2V}{n\pi}\sin\frac{n\pi}{6}$$

Since $\sin n\pi/6 = 1/2, \sqrt{3}/2, 1, \sqrt{3}/2, 1/2, 0, -1/2, \ldots$ for $n = 1, 2, 3, 4, 5, 6, 7, \ldots$ respectively, the series is

$$f(t) = \frac{V}{6} + \frac{2V}{\pi}\left\{\frac{1}{2}\cos\omega t + \frac{\sqrt{3}}{2}\left(\frac{1}{2}\right)\cos 2\omega t + 1\left(\frac{1}{3}\right)\cos 3\omega t + \frac{\sqrt{3}}{2}\left(\frac{1}{4}\right)\cos 4\omega t \right.$$
$$\left. + \frac{1}{2}\left(\frac{1}{5}\right)\cos 5\omega t - \frac{1}{2}\left(\frac{1}{7}\right)\cos 7\omega t - \cdots\right\}$$

or

$$f(t) = \frac{V}{6} + \frac{2V}{\pi}\sum_{n=1}^{\infty}\frac{1}{n}\sin(n\pi/6)\cos n\omega t$$

The line spectrum shown in Fig. 15-27 decreases very slowly for this wave, since the series converges very slowly to the function. Of particular interest is the fact that the 8th, 9th and 10th harmonic amplitudes exceed the 7th. With the simple waves considered previously, the higher harmonic amplitudes were progressively lower.

15.8. Find the exponential Fourier series for the square wave shown in Fig. 15-28 and sketch the line spectrum. Obtain the trigonometric series coefficients and compare them with Problem 15.1.

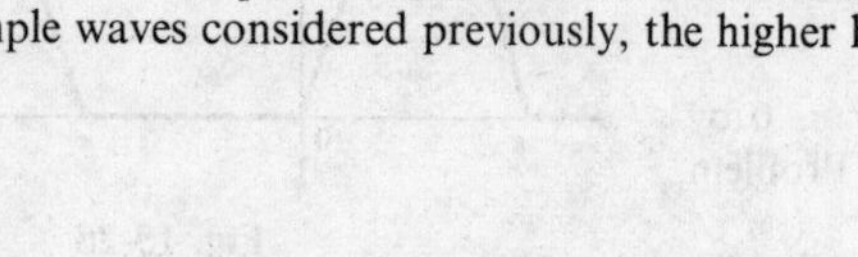

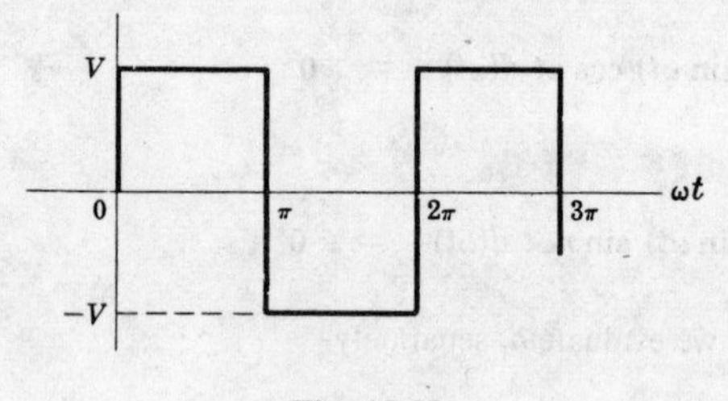

Fig. 15-28

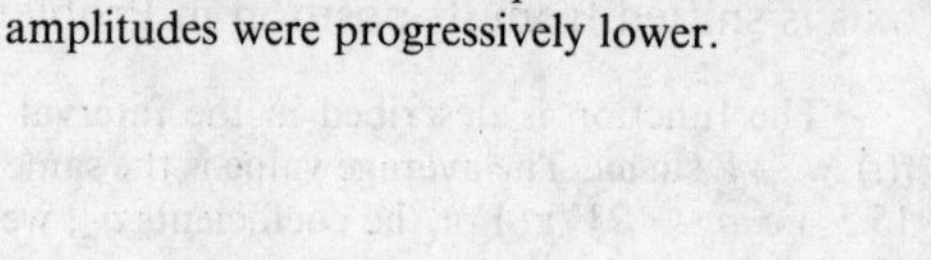

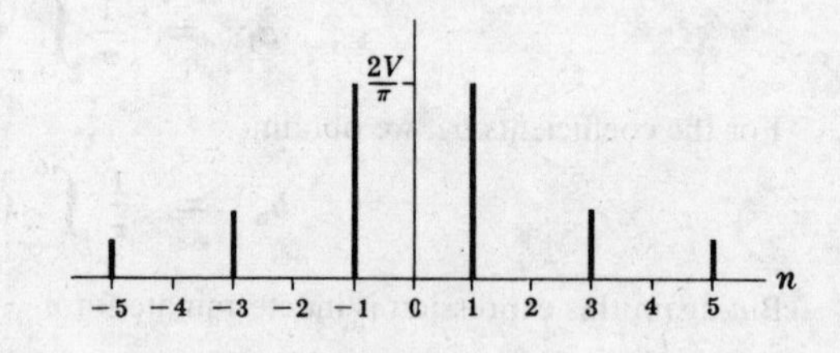

Fig. 15-29

In the interval $-\pi < \omega t < 0$, $f(t) = -V$; and for $0 < \omega t < \pi$, $f(t) = V$. The average value of the wave is zero. The wave is odd, therefore the $\mathbf{A}_n$ coefficients will be pure imaginaries.

$$\mathbf{A}_n = \frac{1}{2\pi}\left\{\int_{-\pi}^{0}(-V)e^{-jn\omega t}\,d(\omega t) + \int_0^{\pi} Ve^{-jn\omega t}\,d(\omega t)\right\}$$

$$= \frac{V}{2\pi}\left\{-\left[\frac{1}{(-jn)}e^{-jn\omega t}\right]_{-\pi}^{0} + \left[\frac{1}{(-jn)}e^{-jn\omega t}\right]_0^{\pi}\right\}$$

$$= \frac{V}{(-j2\pi n)}(-e^0 + e^{jn\pi} + e^{-jn\pi} - e^0) = j\frac{V}{n\pi}(e^{jn\pi} - 1)$$

For n even, $e^{jn\pi} = +1$ and $\mathbf{A}_n = 0$; for n odd, $e^{jn\pi} = -1$ and $\mathbf{A}_n = -j(2V/n\pi)$. The required Fourier series is

$$f(t) = \cdots + j\frac{2V}{3\pi}e^{-j3\omega t} + j\frac{2V}{\pi}e^{-j\omega t} - j\frac{2V}{\pi}e^{j\omega t} - j\frac{2V}{3\pi}e^{j3\omega t} - \cdots$$

The spectrum in Fig. 15-29 above shows amplitudes for both positive and negative frequencies. Combining the values at $+n$ and $-n$ yields the amplitude plotted for the trigonometric series in Fig. 15-16.

The trigonometric series cosine coefficients are

$$a_n = \mathbf{A}_n + \mathbf{A}_{-n} = -j\frac{2V}{n\pi} + \left(-j\frac{2V}{(-n\pi)}\right) = 0$$

and
$$b_n = j[\mathbf{A}_n - \mathbf{A}_{-n}] = j\left[-j\frac{2V}{n\pi} + j\frac{2V}{(-n\pi)}\right] = \frac{4V}{n\pi} \quad \text{for odd } n \text{ only}$$

This agrees with the trigonometric series coefficients obtained in Problem 15.1.

15.9. Find the exponential Fourier series for the triangular wave shown in Fig. 15-30 and sketch the spectrum.

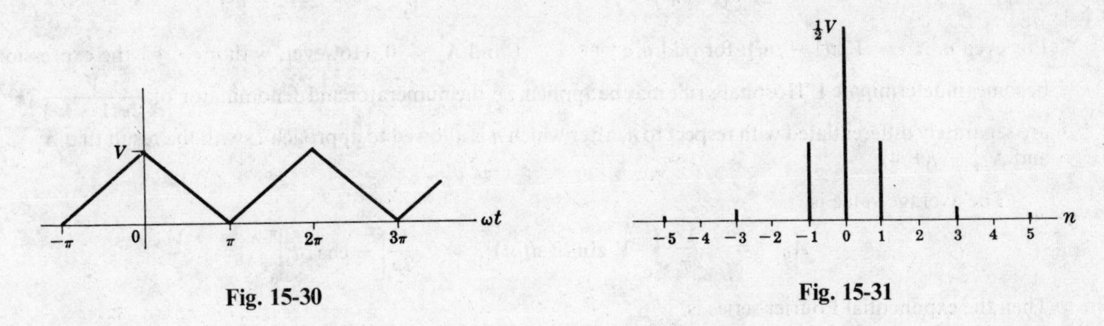

Fig. 15-30 **Fig. 15-31**

In the interval $-\pi < \omega t < 0$, $f(t) = V + (V/\pi)\omega t$; and for $0 < \omega t < \pi$, $f(t) = V - (V/\pi)\omega t$. The wave is even and therefore the $\mathbf{A}_n$ coefficients will be pure real. By inspection the average value is $V/2$.

$$\mathbf{A}_n = \frac{1}{2\pi}\left\{\int_{-\pi}^{0}[V + (V/\pi)\omega t]e^{-jn\omega t}\,d(\omega t) + \int_{0}^{\pi}[V - (V/\pi)\omega t]e^{-jn\omega t}\,d(\omega t)\right\}$$

$$= \frac{V}{2\pi^2}\left\{\int_{-\pi}^{0}\omega t\, e^{-jn\omega t}\,d(\omega t) + \int_{0}^{\pi}(-\omega t)e^{-jn\omega t}\,d(\omega t) + \int_{-\pi}^{\pi}\pi\, e^{-jn\omega t}\,d(\omega t)\right\}$$

$$= \frac{V}{2\pi^2}\left\{\left[\frac{e^{-jn\omega t}}{(-jn)^2}(-jn\omega t - 1)\right]_{-\pi}^{0} - \left[\frac{e^{-jn\omega t}}{(-jn)^2}(-jn\omega t - 1)\right]_{0}^{\pi}\right\} = \frac{V}{\pi^2 n^2}[1 - e^{jn\pi}]$$

For even n, $e^{jn\pi} = +1$ and $\mathbf{A}_n = 0$; for odd n, $\mathbf{A}_n = 2V/\pi^2 n^2$. Thus the series is

$$f(t) = \cdots + \frac{2V}{(-3\pi)^2}e^{-j3\omega t} + \frac{2V}{(-\pi)^2}e^{-j\omega t} + \frac{V}{2} + \frac{2V}{(\pi)^2}e^{j\omega t} + \frac{2V}{(3\pi)^2}e^{j3\omega t} + \cdots$$

The spectrum is shown in Fig. 15-31 with lines at $-n$ and $+n$ which when added are the same as the amplitudes on the spectrum in Fig. 15-18.

The trigonometric series coefficients are

$$a_n = \mathbf{A}_n + \mathbf{A}_{-n} = \frac{2V}{\pi^2 n^2} + \frac{2V}{\pi^2(-n)^2} = \frac{4V}{\pi^2 n^2} \quad \text{for odd } n \text{ only}$$

and
$$b_n = j[\mathbf{A}_n - \mathbf{A}_{-n}] = j\left[\frac{2V}{\pi^2 n^2} - \frac{2V}{\pi^2(-n)^2}\right] = 0$$

These coefficients agree with the results of Problem 15.2.

15.10. Find the exponential Fourier series for the half-wave rectified sine wave shown in Fig. 15-32 below.

In the interval $0 < \omega t < \pi$, $f(t) = V \sin \omega t$; and from π to 2π, $f(t) = 0$. Then

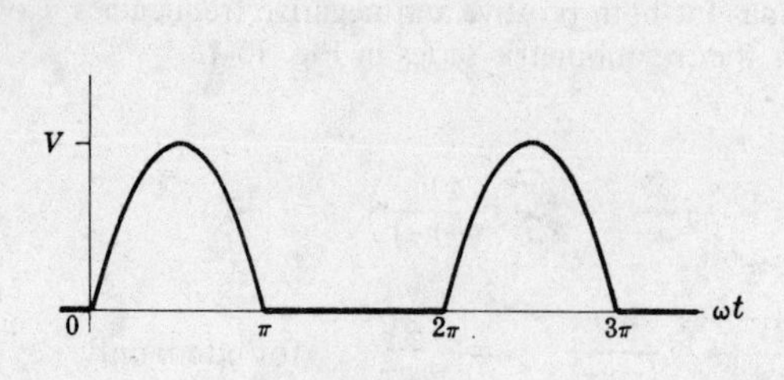

Fig. 15-32

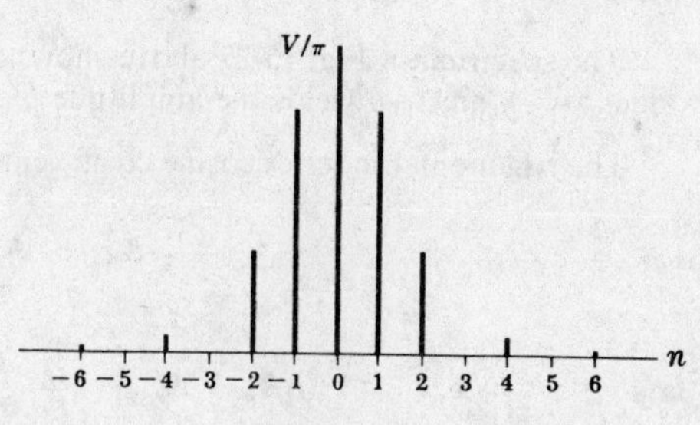

Fig. 15-33

$$\mathbf{A}_n = \frac{1}{2\pi}\int_0^{\pi} V \sin \omega t\, e^{-jn\omega t}\, d(\omega t)$$

$$= \frac{V}{2\pi}\left[\frac{e^{-jn\omega t}}{(1-n^2)}(-jn \sin \omega t - \cos \omega t)\right]_0^{\pi} = \frac{V}{2\pi(1-n^2)}(e^{-jn\pi}+1)$$

For even n, $\mathbf{A}_n = V/\pi(1-n^2)$; for odd $n, e^{-jn\pi} = -1$ and $\mathbf{A}_n = 0$. However, with $n = \pm 1$ the expression for $\mathbf{A}_n$ becomes indeterminate. L'Hospital's rule may be applied, i.e. the numerator and denominator of $\frac{V}{2\pi(1-n^2)}(e^{-jn\pi}+1)$ are separately differentiated with respect to n, after which n is allowed to approach 1, with the result that $\mathbf{A}_1 = -j(V/4)$ and $\mathbf{A}_{-1} = j(V/4)$.

The average value is

$$A_0 = \frac{1}{2\pi}\int_0^{\pi} V \sin \omega t\, d(\omega t) = \frac{V}{2\pi}\Big[-\cos \omega t\Big]_0^{\pi} = \frac{V}{\pi}$$

Then the exponential Fourier series is

$$f(t) = \cdots - \frac{V}{15\pi}e^{-j4\omega t} - \frac{V}{3\pi}e^{-j2\omega t} + j\frac{V}{4}e^{-j\omega t} + \frac{V}{\pi} - j\frac{V}{4}e^{j\omega t} - \frac{V}{3\pi}e^{j2\omega t} - \frac{V}{15\pi}e^{j4\omega t} - \cdots$$

It is interesting to note that there are only two imaginary coefficients in the series at $n = \pm 1$ and that the single sine term in the trigonometric series of Problem 15.6 has the coefficient $b_1 = j[\mathbf{A}_1 - \mathbf{A}_{-1}] = j[-j(V/4)] = \frac{1}{2}V$.

The line spectrum in Fig. 15-33 shows the harmonic amplitudes of the wave and should be compared with that of Fig. 15-24 above.

15.11. Find the average power in a resistance $R = 10$ ohms if the current in series form is $i = 10 \sin \omega t + 5 \sin 3\omega t + 2 \sin 5\omega t$ amperes.

The current has an effective value $I = \sqrt{\frac{1}{2}(10)^2 + \frac{1}{2}(5)^2 + \frac{1}{2}(2)^2} = \sqrt{64 \cdot 5} = 8 \cdot 03$ A. Then the average power is $P = I^2R = (64\cdot5)10 = 654$ W.

Another method:

The total power is the sum of the harmonic powers which are given by $\frac{1}{2}V_{max}I_{max} \cos \theta$. But the voltage across the resistor and the current are in phase for all harmonics and $\theta_n = 0$. Then

$$v_R = Ri = 100 \sin \omega t + 50 \sin 3\omega t + 20 \sin 5\omega t \text{ volts}$$

and $P = \frac{1}{2}(10)(100) + \frac{1}{2}(5)(50) + \frac{1}{2}(2)(20) = 645$ W.

15.12. Find the average power supplied to a network if the applied voltage and resulting current are

$$v = 50 + 50 \sin 5 \times 10^3 t + 30 \sin 10^4 t + 20 \sin 2 \times 10^4 t \text{ volts}$$

$$i = 11\cdot2 \sin(5 \times 10^3 t + 63\cdot4°) + 10\cdot6 \sin(10^4 t + 45°) + 8\cdot97 \sin(2 \times 10^4 t + 26\cdot6°) \text{ amperes}$$

The total average power is the sum of the harmonic power:

$$P = \tfrac{1}{2}(50)(11\cdot2) \cos 63\cdot4° + \tfrac{1}{2}(30)(10\cdot6) \cos 45° + \tfrac{1}{2}(20)(8\cdot97) \cos 26\cdot6° = 317\cdot7 \text{ W}$$

15.13. Obtain the constants of the two-element series circuit with the applied voltage and resultant current given in Problem 15.12.

The voltage series contains a constant term 50 but there is no corresponding term in the current series, thus indicating that one of the elements is a capacitor. Since power is delivered to the circuit, the other element must be a resistor.

The effective current is $I = \sqrt{\frac{1}{2}(11{\cdot}2)^2 + \frac{1}{2}(10{\cdot}6)^2 + \frac{1}{2}(8{\cdot}97)^2} = 12{\cdot}6$ A.

The average power $P = I^2R$, from which $R = P/I^2 = 317{\cdot}7/159{\cdot}2 = 2$ ohms.

At $\omega = 5 \times 10^3$ rad/s, $|\mathbf{Z}| = V_{max}/I_{max} = 50/11{\cdot}2 = 4{\cdot}47\ \Omega$. Since $|\mathbf{Z}| = \sqrt{R^2 + X_C^2}$, $X_C = \sqrt{(4{\cdot}47)^2 - 4} = 4\ \Omega$. Then $X_C = 1/(\omega C)$ and $C = 1/(\omega X_C) = 1/4(4 \times 5 \times 10^3) = 50\ \mu$F.

Therefore the two-element series circuit consists of a resistor of 2 ohms and a capacitor of 50 µF.

15.14. The voltage wave shown in Fig. 15-34 is applied to a series circuit of $R = 2000$ ohms and $L = 10$ H. Use the trigonometric Fourier series and obtain the voltage across the resistor. Plot the line spectrum of the applied voltage and v_R to show the effect of the inductance on the harmonics. $\omega = 377$ rad/s.

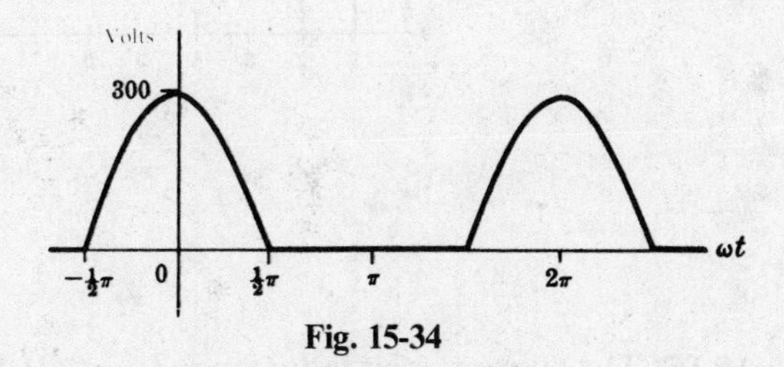

Fig. 15-34

The applied voltage has an average value of V_{max}/π, the same as in Problem 15.5. The wave function is even and hence the series contains only cosine terms with coefficients obtained by the following evaluation integral.

$$a_n = \frac{1}{\pi}\int_{-\pi/2}^{\pi/2} 300 \cos \omega t \cos n\omega t\, d(\omega t) = \frac{600}{\pi(1-n^2)} \cos n\pi/2$$

Cos $n\pi/2$ has a value of -1 for $n = 2, 6, 10, \ldots$ and $+1$ for $n = 4, 8, 12, \ldots$. For n odd, $\cos n\pi/2 = 0$. However, for $n = 1$ the expression is indeterminate and must be evaluated separately.

$$a_1 = \frac{1}{\pi}\int_{-\pi/2}^{\pi/2} 300 \cos^2 \omega t\, d(\omega t) = \frac{300}{\pi}\left[\frac{\omega t}{2} + \frac{\sin 2\omega t}{4}\right]_{-\pi/2}^{\pi/2} = \frac{300}{2}$$

Then the series form of the voltage is

$$v = \frac{300}{\pi}\left\{1 + \frac{\pi}{2}\cos \omega t + \frac{2}{3}\cos 2\omega t - \frac{2}{15}\cos 4\omega t + \frac{2}{35}\cos 6\omega t - \cdots\right\} \text{ volts}$$

The total impedance of the series circuit, $\mathbf{Z} = R + jn\omega L$, is computed for each harmonic in the voltage expression. The results appear in the adjacent table.

The terms of the current series have coefficients which are the voltage series coefficients divided by Z, and the corresponding current term lags by the phase angle θ.

n	$n\omega$	R	$n\omega L$	$\lvert Z\rvert$	θ
0	0	2 k	0	2 k	0°
1	377	2 k	3·77 k	4·26 k	62°
2	754	2 k	7·54 k	7·78 k	75·1°
4	1508	2 k	15·08 k	15·2 k	82·45°
6	2262	2 k	22·62 k	22·6 k	84·92°

$$n = 0,\ I_0 = \frac{300/\pi}{2\text{ k}} \text{ amperes;}$$

$$n = 1,\ i_1 = \frac{300/2}{4{\cdot}26\text{ k}} \cos(\omega t - 62°) \text{ amperes;}$$

$$n = 2,\ i_2 = \frac{600/3\pi}{7{\cdot}78\text{ k}} \cos(2\omega t - 75{\cdot}1°) \text{ amperes; etc.}$$

Then the current series is

$$i = \frac{300}{2\text{ k }\pi} + \frac{300}{(2)4{\cdot}26\text{ k}}\cos(\omega t - 62°) + \frac{600}{3\pi(7{\cdot}78\text{ k})}\cos(2\omega t - 75{\cdot}1°)$$
$$- \frac{600}{15\pi(15{\cdot}2\text{ k})}\cos(4\omega t - 82{\cdot}45°) + \frac{600}{35\pi(22{\cdot}6\text{ k})}\cos(6\omega t - 84{\cdot}92°) - \cdots \text{ amperes}$$

The voltage across the 2 k resistor is simply $i(2\text{ k})$ or

$$v_R = 95{\cdot}5 + 70{\cdot}4\cos(\omega t - 62°) + 16{\cdot}4\cos(2\omega t - 75{\cdot}1°)$$
$$- 1{\cdot}67\cos(4\omega t - 82{\cdot}45°) + 0{\cdot}483\cos(6\omega t - 84{\cdot}92°) - \cdots \text{ volts}$$

In Fig. 15-35 the spectra of the applied voltage and v_R show clearly how the harmonic amplitudes have been reduced by the 10 H series inductance.

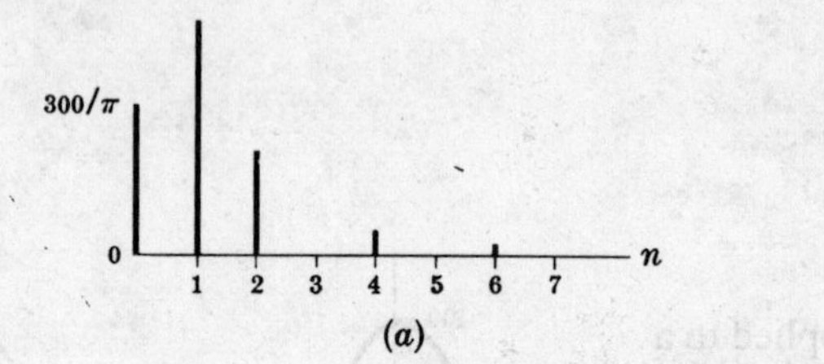

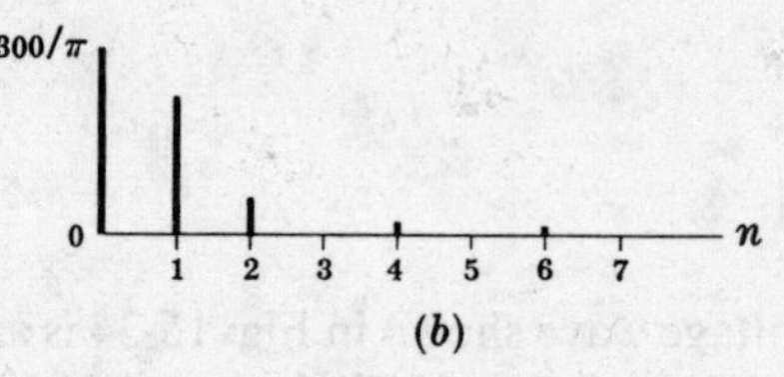

Fig. 15-35

15.15. The current in an inductance $L = 0{\cdot}01$ H has a waveform as shown in Fig. 15-36. Obtain the trigonometric series for v_L, the voltage across the inductance. $\omega = 500$ rad/s.

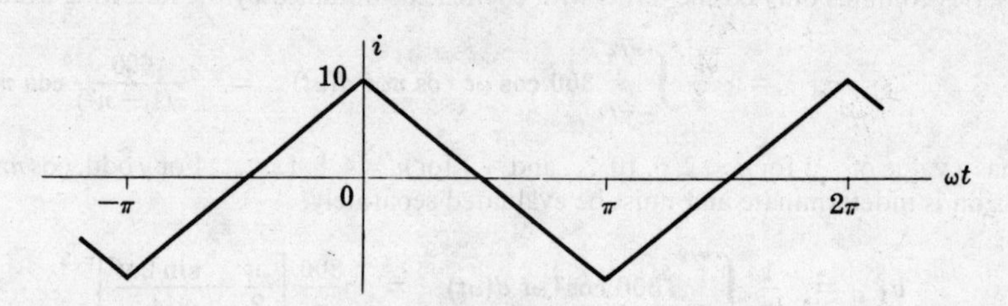

Fig. 15-36

The average value of the current is zero and the waveform is even. Hence the series will contain only cosine terms. In the interval $-\pi < \omega t < 0$, $i = 10 + (20/\pi)\omega t$; and for $0 < \omega t < \pi$, $i = 10 - (20/\pi)\omega t$.

$$a_n = \frac{1}{\pi}\left\{\int_{-\pi}^{0}[10 + (20/\pi)\omega t]\cos n\omega t\, d(\omega t) + \int_{0}^{\pi}[10 - (20/\pi)\omega t]\cos n\omega t\, d(\omega t)\right\}$$

$$= \frac{40}{\pi^2 n^2}(1 - \cos n\pi) = \frac{80}{\pi^2 n^2} \quad \text{for odd } n \text{ only}$$

Then the current series is

$$i = \frac{80}{\pi^2}\left\{\cos\omega t + \frac{1}{9}\cos 3\omega t + \frac{1}{25}\cos 5\omega t + \frac{1}{49}\cos 7\omega t + \cdots\right\}$$

The voltage across the inductance is

$$v_L = L\frac{di}{dt} = 0{\cdot}01\left(\frac{80}{\pi^2}\right)\frac{d}{dt}\{\cos\omega t + \tfrac{1}{9}\cos 3\omega t + \tfrac{1}{25}\cos 5\omega t + \cdots\}$$

$$= \frac{400}{\pi^2}\{-\sin\omega t - \tfrac{1}{3}\sin 3\omega t - \tfrac{1}{5}\sin 5\omega t - \tfrac{1}{7}\sin 7\omega t - \cdots\} \text{ volts}$$

The waveform could be obtained by synthesis, but this series differs from that of Problem 15.1 by a minus sign. Thus v_L is a square wave, the negative of the waveform shown in Fig. 15-15.

Supplementary Problems

15.16. Synthesize the waveform for which the trigonometric Fourier series is

$$f(t) \quad = \quad \frac{8V}{\pi^2}\{\sin \omega t - \tfrac{1}{9}\sin 3\omega t + \tfrac{1}{25}\sin 5\omega t - \tfrac{1}{49}\sin 7\omega t + \cdots\}$$

15.17. Synthesize the waveform if its Fourier series is

$$f(t) \quad = \quad 5 - \frac{40}{\pi^2}(\cos \omega t + \tfrac{1}{9}\cos 3\omega t + \tfrac{1}{25}\cos 5\omega t + \cdots)$$
$$+ \frac{20}{\pi}(\sin \omega t - \tfrac{1}{2}\sin 2\omega t + \tfrac{1}{3}\sin 3\omega t - \tfrac{1}{4}\sin 4\omega t + \cdots)$$

15.18. Synthesize the waveform for the given Fourier series.

$$f(t) \quad = \quad V\left\{\frac{1}{2\pi} - \frac{1}{\pi}\cos \omega t - \frac{1}{3\pi}\cos 2\omega t + \frac{1}{2\pi}\cos 3\omega t - \frac{1}{15\pi}\cos 4\omega t - \frac{1}{6\pi}\cos 6\omega t \right.$$
$$\left. + \cdots + \frac{1}{4}\sin \omega t - \frac{2}{3\pi}\sin 2\omega t + \frac{4}{15\pi}\sin 4\omega t - \cdots \right\}$$

15.19. Find the trigonometric Fourier series for the sawtooth wave shown in Fig. 15-37 and plot the line spectrum. Compare with Example 1.

Ans. $f(t) \quad = \quad \frac{V}{2} + \frac{V}{\pi}\{\sin \omega t + \tfrac{1}{2}\sin 2\omega t + \tfrac{1}{3}\sin 3\omega t + \cdots\}$

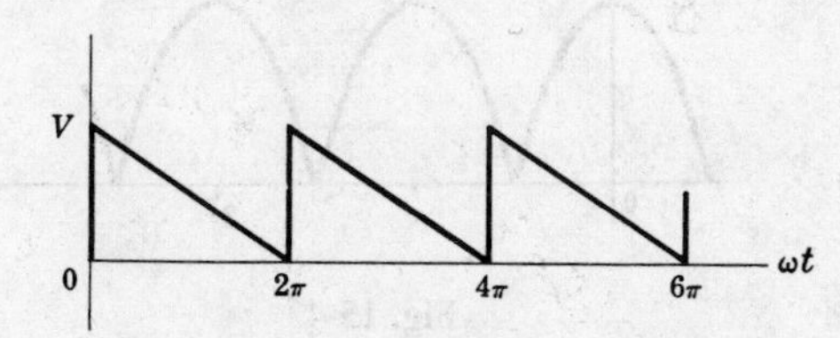

Fig. 15-37

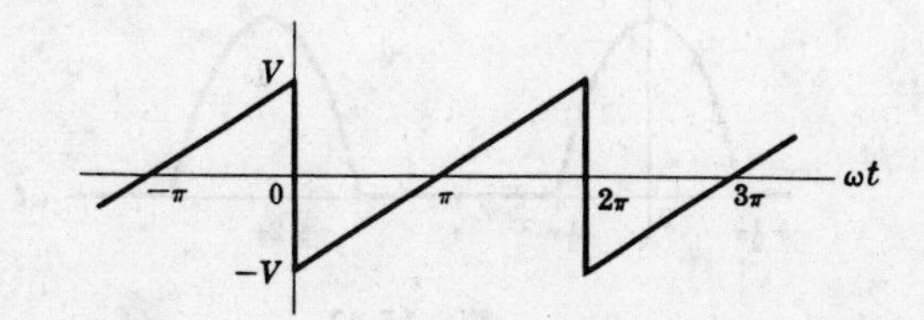

Fig. 15-38

15.20. Find the trigonometric Fourier series for the sawtooth wave shown in Fig. 15-38 and plot the spectrum. Compare with the result of Problem 15.3.

Ans. $f(t) \quad = \quad \frac{-2V}{\pi}\{\sin \omega t + \tfrac{1}{2}\sin 2\omega t + \tfrac{1}{3}\sin 3\omega t + \tfrac{1}{4}\sin 4\omega t + \cdots\}$

15.21. Find the trigonometric Fourier series for the waveform shown in Fig. 15-39 and plot the line spectrum.

Ans. $f(t) \quad = \quad \frac{4V}{\pi^2}\{\cos \omega t + \tfrac{1}{9}\cos 3\omega t + \tfrac{1}{25}\cos 5\omega t + \cdots\}$
$$- \frac{2V}{\pi}\{\sin \omega t + \tfrac{1}{3}\sin 3\omega t + \tfrac{1}{5}\sin 5\omega t + \cdots\}$$

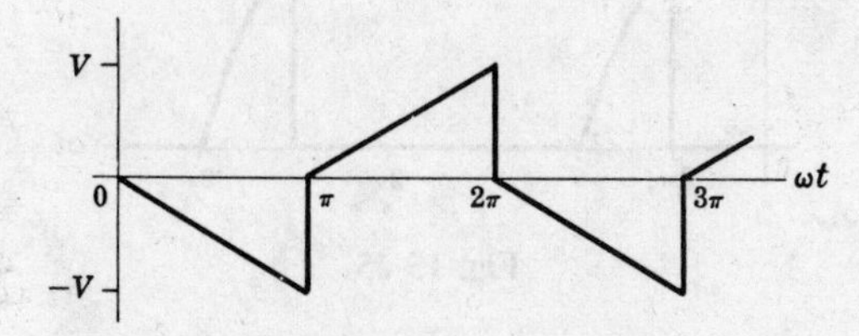

Fig. 15-39

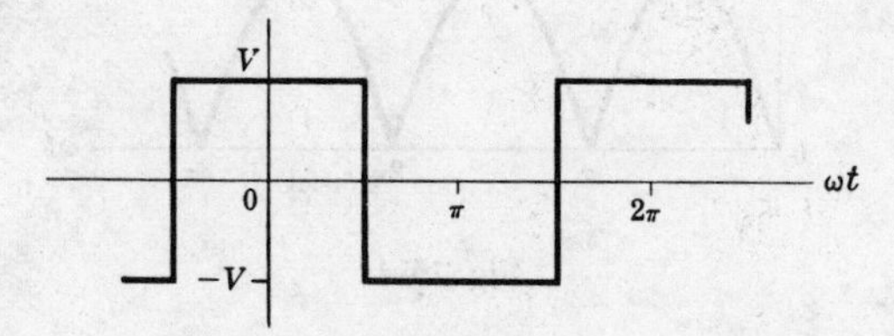

Fig. 15-40

15.22. Find the trigonometric Fourier series of the square wave shown in Fig. 15-40 and plot the line spectrum. Compare with the result of Problem 15.1.

Ans. $f(t) \quad = \quad \frac{4V}{\pi}\{\cos \omega t - \tfrac{1}{3}\cos 3\omega t + \tfrac{1}{5}\cos 5\omega t - \tfrac{1}{7}\cos 7\omega t + \cdots\}$

15.23. Find the trigonometric Fourier series for the waveforms shown in Fig. 15-41(*a*) and (*b*). Plot the line spectrum of each and compare.

Ans.
$$f_1(t) = \frac{5}{12} + \sum_{n=1}^{\infty}\left\{\frac{10}{n\pi}\left(\sin\frac{n\pi}{12}\right)\cos n\omega t + \frac{10}{n\pi}\left(1 - \cos\frac{n\pi}{12}\right)\sin n\omega t\right\}$$

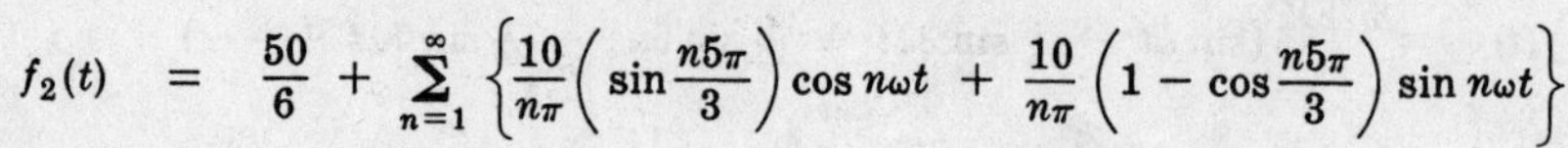

$$f_2(t) = \frac{50}{6} + \sum_{n=1}^{\infty}\left\{\frac{10}{n\pi}\left(\sin\frac{n5\pi}{3}\right)\cos n\omega t + \frac{10}{n\pi}\left(1 - \cos\frac{n5\pi}{3}\right)\sin n\omega t\right\}$$

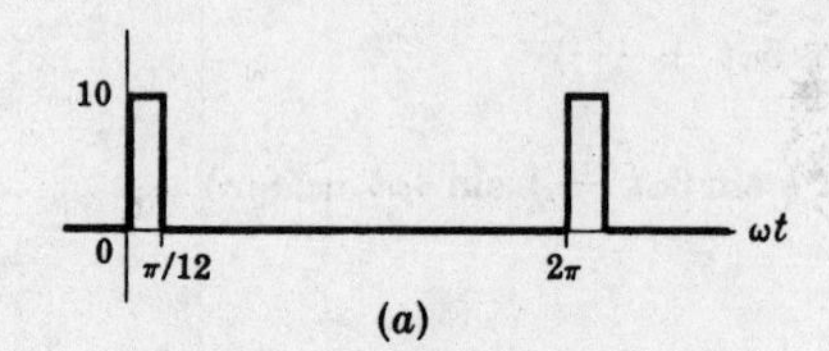

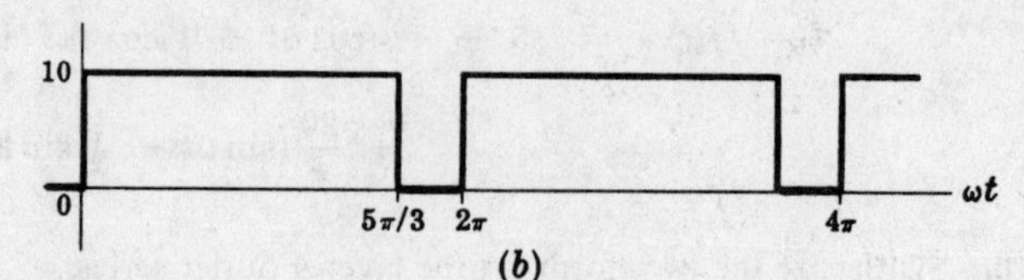

Fig. 15-41

15.24. Find the trigonometric Fourier series for the half-wave rectified sine wave shown in Fig. 15-42 and plot the line spectrum. Compare the answer with the results of Problems 15.5 and 15.6.

Ans.
$$f(t) = \frac{V}{\pi}\left\{1 + \frac{\pi}{2}\cos\omega t + \frac{2}{3}\cos 2\omega t - \frac{2}{15}\cos 4\omega t + \frac{2}{35}\cos 6\omega t - \cdots\right\}$$

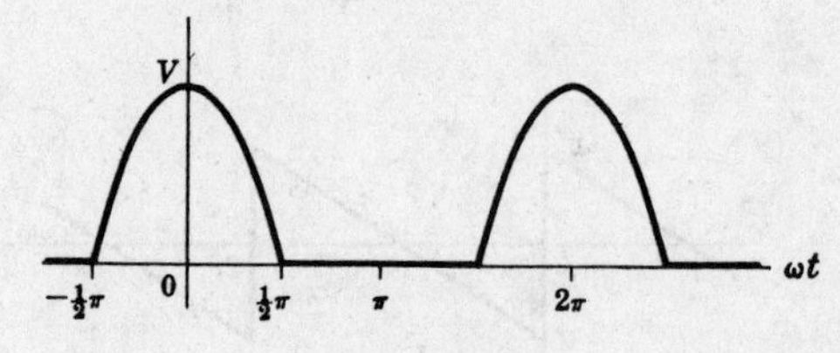

Fig. 15-42

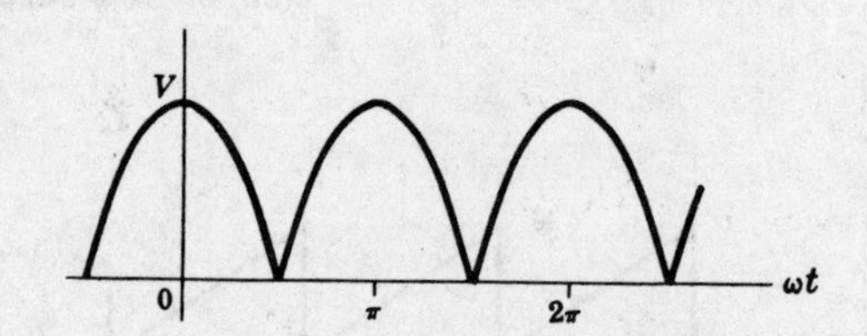

Fig. 15-43

15.25. Find the trigonometric Fourier series for the full-wave rectified sine wave shown in Fig. 15-43 and plot the spectrum.

Ans.
$$f(t) = \frac{2V}{\pi}\{1 + \tfrac{2}{3}\cos 2\omega t - \tfrac{2}{15}\cos 4\omega t + \tfrac{2}{35}\cos 6\omega t - \cdots\}$$

15.26. The waveform in Fig. 15-44 is similar to that of Problem 15.25 above but with the position of the zero axis changed. Find the Fourier series and compare the two results.

Ans.
$$f(t) = \frac{2V}{\pi}\{1 - \tfrac{2}{3}\cos 2\omega t - \tfrac{2}{15}\cos 4\omega t - \tfrac{2}{35}\cos 6\omega t - \cdots\}$$

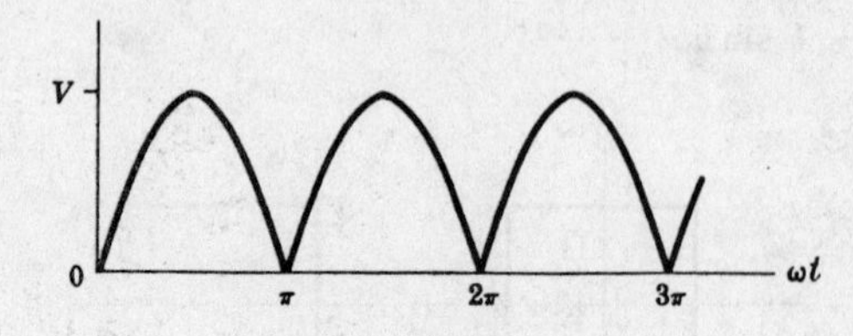

Fig. 15-44

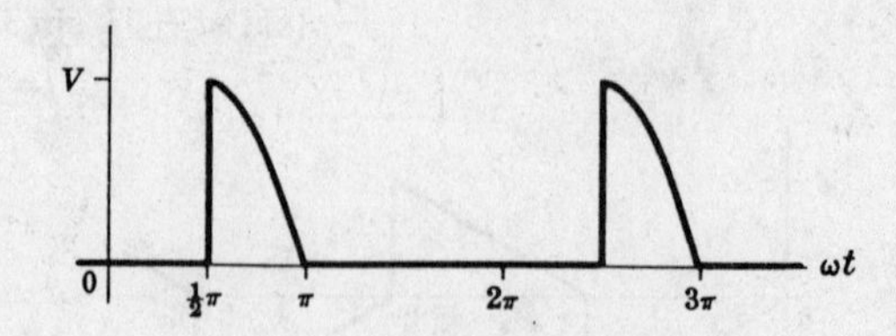

Fig. 15-45

15.27. Find the trigonometric Fourier series for the waveform shown in Fig. 15-45.

Ans.
$$f(t) = \frac{V}{2\pi} - \frac{V}{2\pi}\cos\omega t + \sum_{n=2}^{\infty}\frac{V}{\pi(1-n^2)}(\cos n\pi + n\sin n\pi/2)\cos n\omega t$$
$$+ \frac{V}{4}\sin\omega t + \sum_{n=2}^{\infty}\left[\frac{-nV\cos n\pi/2}{\pi(1-n^2)}\right]\sin n\omega t$$

15.28. Find the trigonometric Fourier series for the waveform shown in Fig. 15-46. Add this series with that of Problem 15.27 and compare the sum with the series obtained in Problem 15.5.

Ans.
$$f(t) = \frac{V}{2\pi} + \frac{V}{2\pi}\cos\omega t + \sum_{n=2}^{\infty}\frac{V[n\sin n\pi/2 - 1]}{\pi(n^2-1)}\cos n\omega t + \frac{V}{4}\sin\omega t + \sum_{n=2}^{\infty}\frac{Vn\cos n\pi/2}{\pi(1-n^2)}\sin n\omega t$$

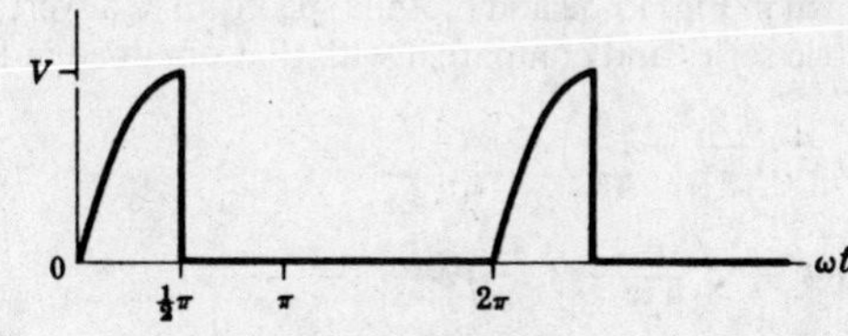

Fig. 15-46

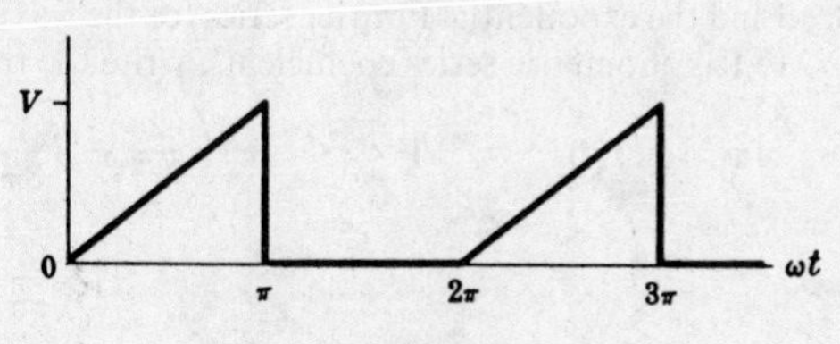

Fig. 15-47

15.29. Find the exponential Fourier series for the waveform shown in Fig. 15-47 and plot the line spectrum. Convert the coefficients obtained here into the trigonometric series coefficients, write the trigonometric series and compare it with the result of Problem 15.4.

Ans.
$$f(t) = V\left\{\cdots - \left(\frac{1}{9\pi^2} - j\frac{1}{6\pi}\right)e^{-j3\omega t} - j\frac{1}{4\pi}e^{-j2\omega t} - \left(\frac{1}{\pi^2} - j\frac{1}{2\pi}\right)e^{-j\omega t} + \frac{1}{4} - \left(\frac{1}{\pi^2} + j\frac{1}{2\pi}\right)e^{j\omega t} + j\frac{1}{4\pi}e^{j2\omega t} - \left(\frac{1}{9\pi^2} + j\frac{1}{6\pi}\right)e^{j3\omega t} - \cdots\right\}$$

15.30. Find the exponential Fourier series for the waveform shown in Fig. 15-48 and plot the line spectrum.

Ans.
$$f(t) = V\left\{\cdots + \left(\frac{1}{9\pi^2} + j\frac{1}{6\pi}\right)e^{-j3\omega t} + j\frac{1}{4\pi}e^{-j2\omega t} + \left(\frac{1}{\pi^2} + j\frac{1}{2\pi}\right)e^{-j\omega t} + \frac{1}{4} + \left(\frac{1}{\pi^2} - j\frac{1}{2\pi}\right)e^{j\omega t} - j\frac{1}{4\pi}e^{j2\omega t} + \left(\frac{1}{9\pi^2} - j\frac{1}{6\pi}\right)e^{j3\omega t} + \cdots\right\}$$

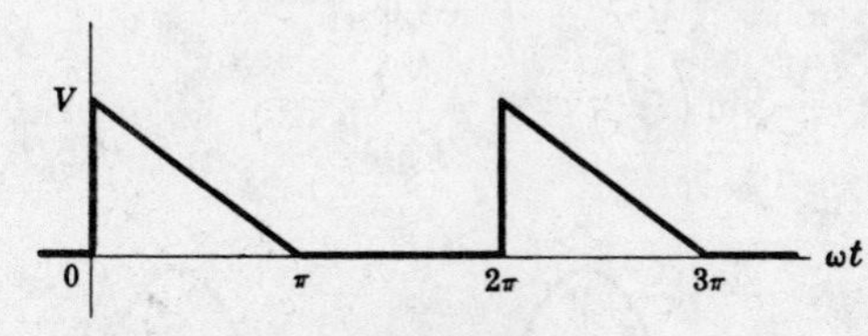

Fig. 15-48

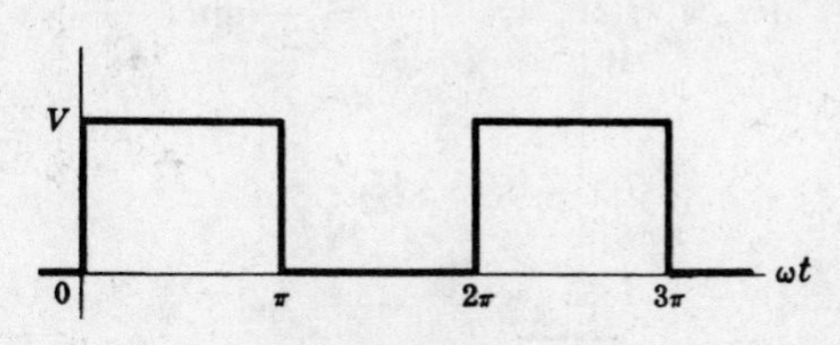

Fig. 15-49

15.31. Find the exponential Fourier series for the square wave shown in Fig. 15-49 and plot the line spectrum. Add the exponential series of Problems 15.29 and 15.30 and compare the sum to the series obtained here.

Ans.
$$f(t) = V\left\{\cdots + j\frac{1}{3\pi}e^{-j3\omega t} + j\frac{1}{\pi}e^{-j\omega t} + \frac{1}{2} - j\frac{1}{\pi}e^{j\omega t} - j\frac{1}{3\pi}e^{j3\omega t} - \cdots\right\}$$

15.32. Find the exponential Fourier series for the sawtooth waveform shown in Fig. 15-50 and plot the spectrum. Convert the coefficients obtained here into the trigonometric series coefficients, write the trigonometric series and compare the result with the series obtained in Problem 15.19.

Ans.
$$f(t) = V\left\{\cdots + j\frac{1}{4\pi}e^{-j2\omega t} + j\frac{1}{2\pi}e^{-j\omega t} + \frac{1}{2} - j\frac{1}{2\pi}e^{j\omega t} - j\frac{1}{4\pi}e^{j2\omega t} - \cdots\right\}$$

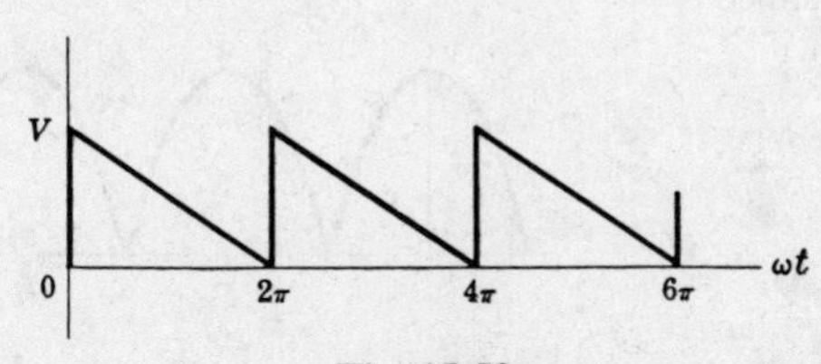

Fig. 15-50

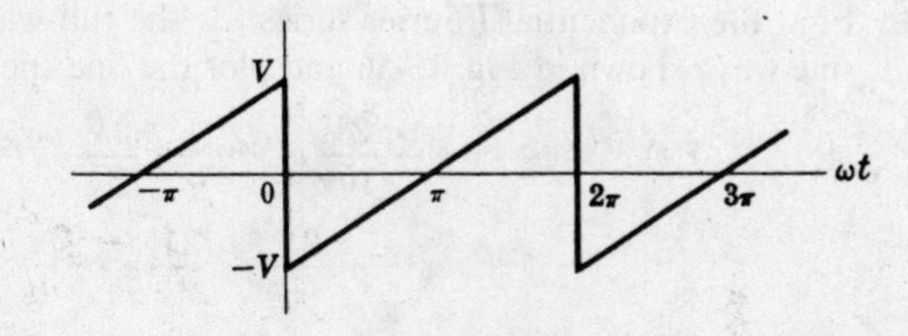

Fig. 15-51

15.33. Find the exponential Fourier series for the waveform shown in Fig. 15-51 above and plot the spectrum. Convert the trigonometric series coefficients found in Problem 15.20 into exponential series coefficients and compare them with the coefficients of the series obtained here.

Ans. $$f(t) = V\left\{\cdots - j\frac{1}{2\pi}e^{-j2\omega t} - j\frac{1}{\pi}e^{-j\omega t} + j\frac{1}{\pi}e^{j\omega t} + j\frac{1}{2\pi}e^{j2\omega t} + \cdots\right\}$$

15.34. Find the exponential Fourier series for the waveform shown in Fig. 15-52 and plot the spectrum. Convert the coefficients to trigonometric series coefficients, write the trigonometric series and compare it with that obtained in Problem 15.21.

Ans. $$f(t) = V\left\{\cdots + \left(\frac{2}{9\pi^2} - j\frac{1}{3\pi}\right)e^{-j3\omega t} + \left(\frac{2}{\pi^2} - j\frac{1}{\pi}\right)e^{-j\omega t} + \left(\frac{2}{\pi^2} + j\frac{1}{\pi}\right)e^{j\omega t} + \left(\frac{2}{9\pi^2} + j\frac{1}{3\pi}\right)e^{j3\omega t} + \cdots\right\}$$

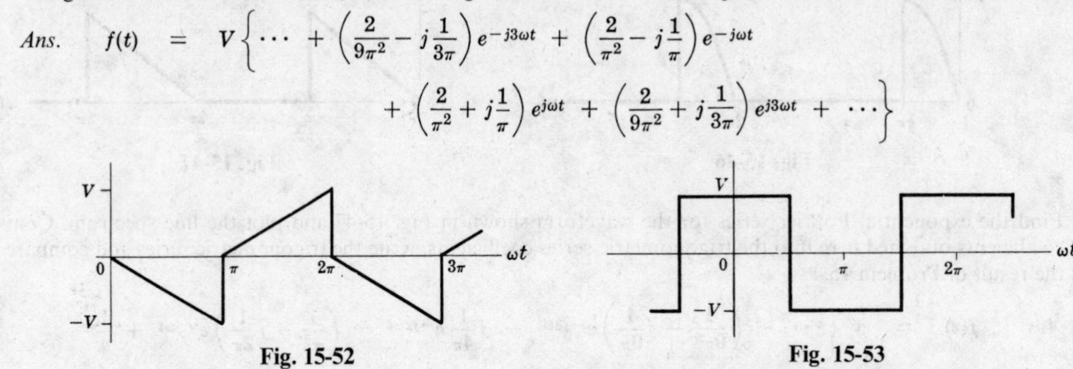

Fig. 15-52 Fig. 15-53

15.35. Find the exponential Fourier series for the square wave shown in Fig. 15-53 and plot the line spectrum. Convert the trigonometric series coefficients of Problem 15.22 into exponential series coefficients and compare with the coefficients in the result obtained here.

Ans. $$f(t) = \frac{2V}{\pi}\{\cdots + \tfrac{1}{5}e^{-j5\omega t} - \tfrac{1}{3}e^{-j3\omega t} + e^{-j\omega t} + e^{j\omega t} - \tfrac{1}{3}e^{j3\omega t} + \tfrac{1}{5}e^{j5\omega t} - \cdots\}$$

15.36. Find the exponential Fourier series for the waveform shown in Fig. 15-54 and plot the line spectrum.

Ans. $$f(t) = \cdots - \frac{V}{2\pi}\sin\left(\frac{-2\pi}{6}\right)e^{-j2\omega t} - \frac{V}{\pi}\sin\left(\frac{-\pi}{6}\right)e^{-j\omega t} + \frac{V}{6} + \frac{V}{\pi}\sin\left(\frac{\pi}{6}\right)e^{j\omega t} + \frac{V}{2\pi}\sin\left(\frac{2\pi}{6}\right)e^{j2\omega t} + \cdots$$

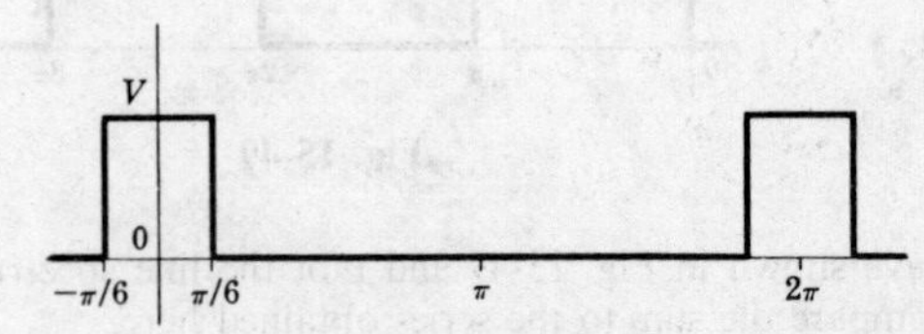

Fig. 15-54 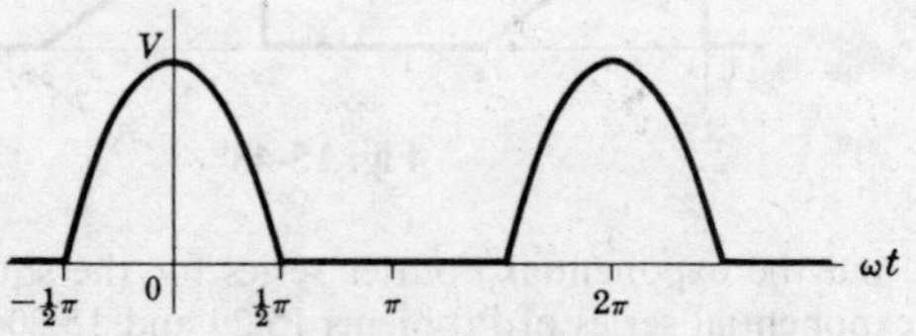 Fig. 15-55

15.37. Find the exponential Fourier series for the half-wave rectified sine wave shown in Fig. 15-55. Convert these coefficients into the trigonometric series coefficients, write the trigonometric series and compare it with the result of Problem 15.24.

Ans. $$f(t) = \cdots - \frac{V}{15\pi}e^{-j4\omega t} + \frac{V}{3\pi}e^{-j2\omega t} + \frac{V}{4}e^{-j\omega t} + \frac{V}{\pi} + \frac{V}{4}e^{j\omega t} + \frac{V}{3\pi}e^{j2\omega t} - \frac{V}{15\pi}e^{j4\omega t} + \cdots$$

15.38. Find the exponential Fourier series for the full-wave rectified sine wave shown in Fig. 15-56 and plot the line spectrum.

Ans. $$f(t) = \cdots - \frac{2V}{15\pi}e^{-j4\omega t} + \frac{2V}{3\pi}e^{-j2\omega t} + \frac{2V}{\pi} + \frac{2V}{3\pi}e^{j2\omega t} - \frac{2V}{15\pi}e^{j4\omega t} + \cdots$$

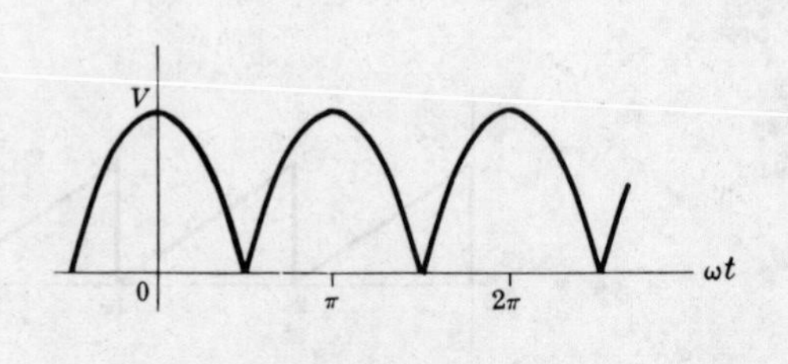

Fig. 15-56

15.39. Find the effective voltage, effective current and average power supplied to a passive network if the applied voltage $v = 200 + 100 \cos (500t + 30°) + 75 \cos (1500 + 60°)$ volts and the resulting current $i = 3{\cdot}53 \cos (500t + 75°) + 3{\cdot}55 \cos (1500 + 78{\cdot}45°)$ amperes. *Ans.* 218·5 V, 3·54 A, 250·8 W

15.40. A voltage $v = 50 + 25 \sin 500t + 10 \sin 1500t + 5 \sin 2500t$ volts is applied to the terminals of a passive network and the resulting current is

$$i = 5 + 2{\cdot}23 \sin (500t - 26{\cdot}6°) + 0{\cdot}566 \sin (1500t - 56{\cdot}3°) + 0{\cdot}186 \sin (2500t - 68{\cdot}2°) \text{ amperes}$$

Find the effective voltage, effective current and the average power.
Ans. 53·6 V, 5·25 A, 276·5 W

15.41. A three element series circuit with $R = 5$ ohms, $L = 5$ mH and $C = 50$ μF has an applied voltage $v = 150 \sin 1000t + 100 \sin 2000t + 75 \sin 3000t$ volts. Find the effective current and the average power for the circuit. Sketch the line spectrum of the voltage and the current and note the effect of series resonance. *Ans.* 16·58 A, 1374 W

15.42. A two-element series circuit with $R = 10$ ohms and $L = 0{\cdot}02$ H contains a current $i = 5 \sin 100t + 3 \sin 300t + 2 \sin 500t$ amperes. Find the effective applied voltage and the average power. *Ans.* 48 V, 190 W

15.43. A pure inductance of $L = 0{\cdot}01$ H contains the triangular current wave shown in Fig. 15-57 where $\omega = 500$ rad/s. Obtain the exponential Fourier series for the current and find the series expression for the voltage across the inductance v_L. Compare the answer with the result of Problem 15.8.

Ans. $$v_L = \frac{200}{\pi^2}\{\cdots - j\tfrac{1}{3}e^{-j3\omega t} - je^{-j\omega t} + je^{j\omega t} + j\tfrac{1}{3}e^{j3\omega t} + \cdots\} \text{ volts}$$

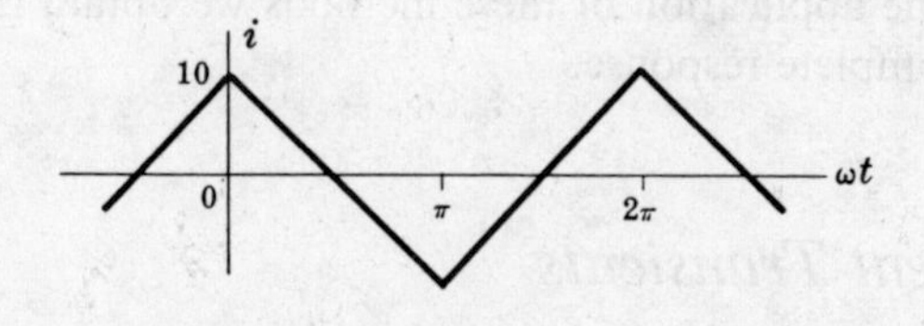

Fig. 15-57

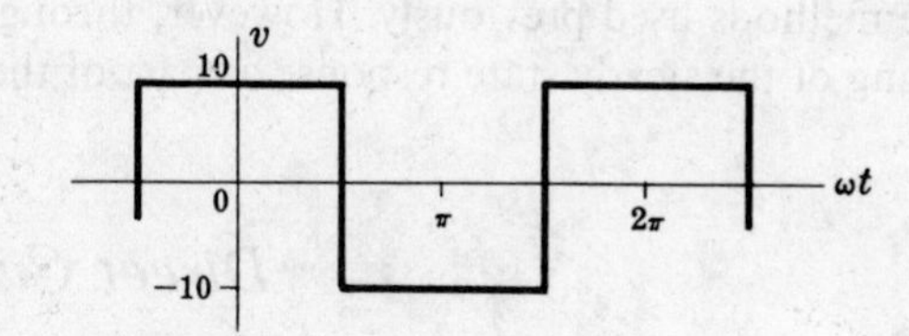

Fig. 15-58

15.44. A pure inductance of $L = 0{\cdot}01$ H has an applied voltage with a waveform shown in Fig. 15-58 where $\omega = 200$ rad/s. Obtain the current series in trigonometric form and identify the current waveform.

Ans. $$i = \frac{20}{\pi}\{\sin \omega t - \tfrac{1}{9} \sin 3\omega t + \tfrac{1}{25} \sin 5\omega t - \tfrac{1}{49} \sin 7\omega t + \cdots\} \text{ amperes}$$

15.45. Fig. 15-59 shows a full-wave rectified sine wave representing a voltage applied to the terminals of an LC circuit. The maximum value of the voltage is 170 V and $\omega = 377$ rad/s. Use the trigonometric Fourier series and find the voltage across the inductor and the capacitor. Plot the line spectrum of each.

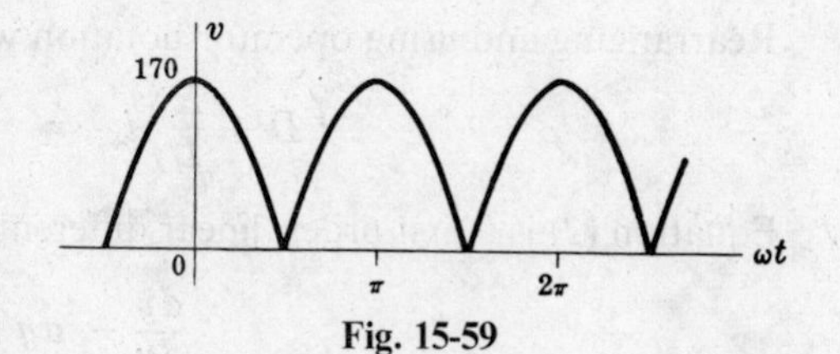

Fig. 15-59

15.46. A three element circuit consists of $R = 5$ ohms in series with a parallel combination of L and C. At $\omega = 500$ rad/s the corresponding reactances are $j2\ \Omega$ and $-j8\ \Omega$. Find the total current if the applied voltage is given by $v = 50 + 20 \sin 500t + 10 \sin 1000t$ volts. *Ans.* $i = 10 + 3{\cdot}53 \sin (500t - 28{,}1°)$ amperes

CHAPTER 16

Circuit Transients

INTRODUCTION

When a circuit is switched from one condition to another either by a change in the applied voltage or a change in one of the circuit elements, there is a transitional period during which the branch currents and voltage drops change from their former values to new ones. After this transition interval called the *transient*, the circuit is said to be in the steady state.

The application of Kirchhoff's voltage law to a circuit containing energy storage elements results in a differential equation which is solved by any of several available methods. This solution consists of two parts, the *complementary function* and the *particular solution.* For equations in circuit analysis the complementary function always goes to zero in a relatively short time and is the transient part of the solution. The particular solution is the steady state response which was the subject of our attention in the previous chapters. The methods by which the particular solution is obtained in this chapter are generally long and involved and never as direct as the methods used previously. However, through the application of these methods we obtain the physical meaning of the steady state response as part of the complete response.

Direct Current Transients

RL TRANSIENT

The series *RL* circuit shown in Fig. 16-1 has a constant voltage V applied when the switch is closed. Kirchhoff's voltage law results in the following differential equation

$$Ri + L\frac{di}{dt} = V \qquad (1)$$

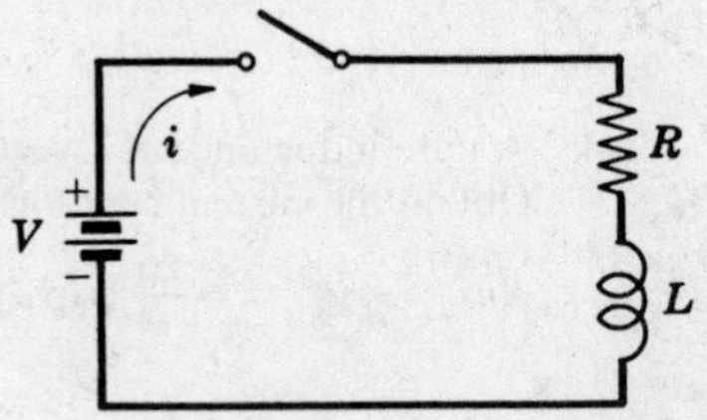

Rearranging and using operator notation where $D = d/dt$,

$$\left(D + \frac{R}{L}\right)i = \frac{V}{L} \qquad (2)$$

Fig. 16-1

Equation (2) is a first order, linear differential equation of the type

$$\frac{dy}{dx} - ay = \mathcal{R} \quad \text{or} \quad (D-a)y = \mathcal{R} \qquad (3)$$

where $D = d/dx$, a is a constant, and R may be a function of x but not of y. The complete solution of (3), consisting of the complementary function and the particular solution, is

$$y = y_c + y_p = ce^{ax} + e^{ax}\int e^{-ax}\mathcal{R}\,dx \qquad (4)$$

where c is an arbitrary constant determined by known initial conditions. By (4), the solution of (2) is

$$i = ce^{-(R/L)t} + e^{-(R/L)t}\int e^{(R/L)t}\left(\frac{V}{L}\right)dt = ce^{-(R/L)t} + \frac{V}{R} \qquad (5)$$

To determine c, we set $t = 0$ in (5) and substitute the initial current i_0 for i. This initial current is the current just after the switch is closed. The inductance has the voltage-current relationships $v = L\frac{di}{dt}$ and $i = \frac{1}{L}\int v\,dt$. The second expression assures us that whatever the applied voltage, the current through an inductor must be a continuous function. Then since the current was zero at $t = 0-$, it must also be zero at $t = 0+$. Substituting in (5) we obtain

$$i_0 = 0 = c(1) + V/R \text{ or } c = -V/R \qquad (6)$$

Substituting this value of c in (5) results in

$$i = -\frac{V}{R}e^{-(R/L)t} + \frac{V}{R} = \frac{V}{R}(1 - e^{-(R/L)t}) \qquad (7)$$

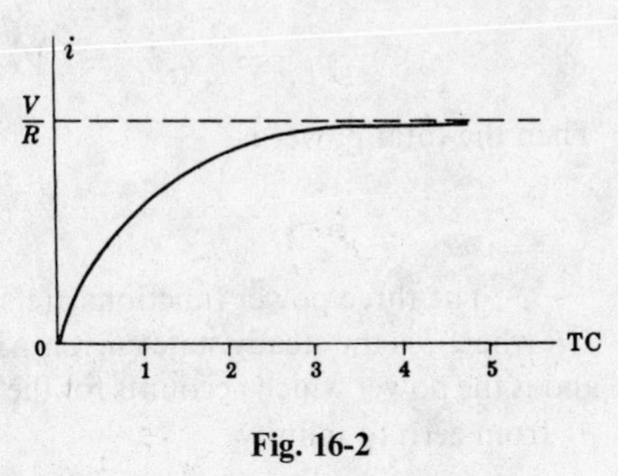

Fig. 16-2

This type of equation is known as an exponential rise, as shown in Fig. 16-2. The plot shows the transition period during which the current adjusts from its initial value of zero to the final value V/R, the steady state.

The time constant TC of a function such as (7) is the time at which the exponent of e is unity. Thus for the RL transient the time constant $TC = L/R$ seconds. At 1 TC the quantity within the parentheses in (7) has the value $(1 - e^{-1}) = (1 - 0{\cdot}368) = 0{\cdot}632$. At this time the current is 63·2% of its final value. Similarly at 2 TC, $(1 - e^{-2}) = (1 - 0{\cdot}135) = 0{\cdot}865$ and the current is 86·5% of its final value. After 5 TC the transient is generally regarded as terminated. For convenience, the time constant is the unit used to plot the current of equation (7).

As another example, for the exponential decay shown in Fig. 16-3 with the following equation

$$f(t) = Ae^{-\alpha t} \qquad (8)$$

the time constant is again the time at which the exponent on e is unity, i.e. $\text{TC} = 1/\alpha$. The value at 1 TC is $e^{-1} = 0{\cdot}368$ and the function has decayed to 36·8% of its initial value A. Then at 2 TC, $e^{-2} = 0{\cdot}135$ and the function is 13·5% of A. After 5 TC the transient is considered terminated.

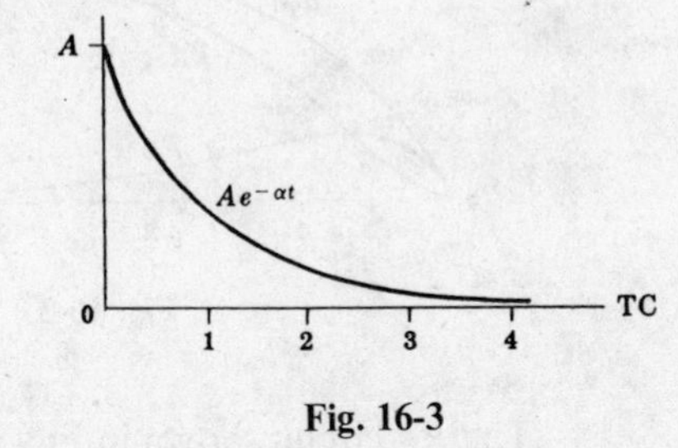

Fig. 16-3

The transient voltages across the elements of the RL circuit are obtained from the current. Accordingly, the voltage across the resistor is

$$v_R = Ri = V(1 - e^{-(R/L)t}) \qquad (9)$$

and the voltage across the inductance is

$$v_L = L\frac{di}{dt} = L\frac{d}{dt}\left\{\frac{V}{R}(1 - e^{-(R/L)t})\right\} = Ve^{-(R/L)t} \qquad (10)$$

The resistor voltage transient is an exponential rise with the same time constant as the current, while the voltage across the inductance is an exponential decay but with the same time constant. The sum of v_R and v_L satisfies Kirchhoff's law throughout the transient period. See Fig. 16–4.

$$v_R + v_L = V(1 - e^{-(R/L)t}) + Ve^{-(R/L)t} = V \qquad (11)$$

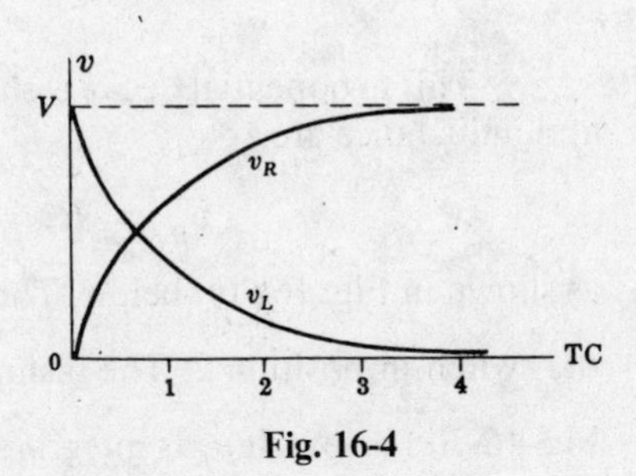

Fig. 16-4

The instantaneous power in any circuit element is given by the product of the voltage and the current. Thus the power in the resistor is

$$p_R = v_R i = V(1 - e^{-(R/L)t})\frac{V}{R}(1 - e^{-(R/L)t}) = \frac{V^2}{R}(1 - 2e^{-(R/L)t} + e^{-2(R/L)t}) \qquad (12)$$

and the power in the inductance is

$$p_L = v_L i = Ve^{-(R/L)t}\frac{V}{R}(1 - e^{-(R/L)t}) = \frac{V^2}{R}(e^{-(R/L)t} - e^{-2(R/L)t}) \qquad (13)$$

Then the total power is

$$p_T = p_R + p_L = \frac{V^2}{R}(1 - e^{-(R/L)t}) \qquad (14)$$

The three power functions are shown in Fig. 16-5 where p_R and p_T have the steady state value V^2/R or I^2R where I is the steady state current. The transient power in the inductance has initial and final values of zero and is the power which accounts for the energy stored in the magnetic field of the coil. To show this we integrate p_L from zero to infinity.

$$W = \int_0^\infty \frac{V^2}{R}(e^{-(R/L)t} - e^{-2(R/L)t})\,dt = \frac{V^2}{R}\left[-\frac{L}{R}e^{-(R/L)t} + \frac{L}{2R}e^{-2(R/L)t}\right]_0^\infty$$
$$= \frac{1}{2}\frac{V^2}{R}\left(\frac{L}{R}\right) = \frac{1}{2}LI^2 \text{ joules} \qquad (15)$$

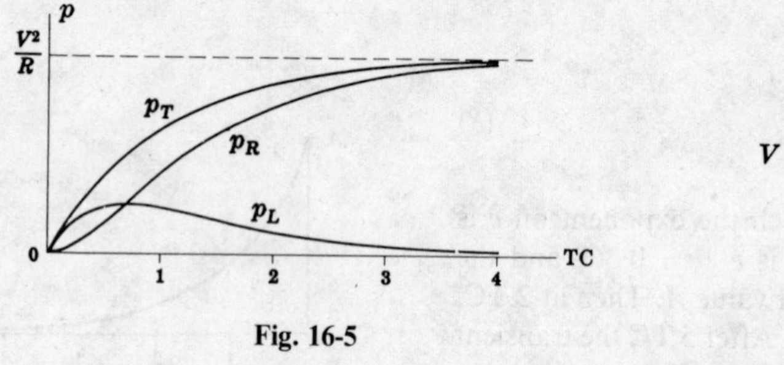

Fig. 16-5

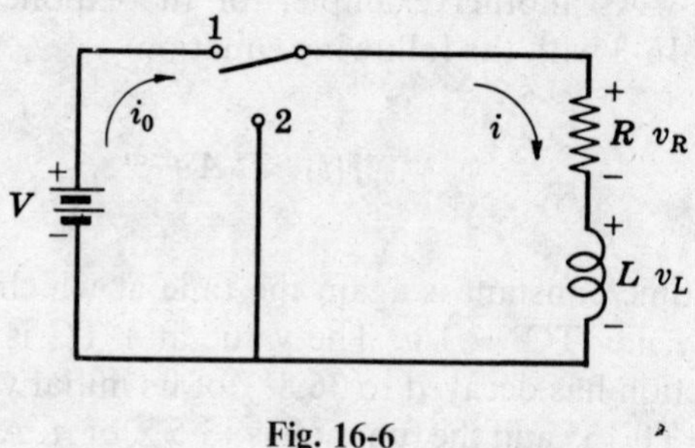

Fig. 16-6

The RL circuit shown in Fig. 16-6 contains an initial current $i_0 = V/R$. At $t = 0$ the switch is turned to position 2 which removes the source and at the same time puts a short circuit across the series RL branch. The application of Kirchhoff's voltage law to the source-free circuit results in the equation

$$L\frac{di}{dt} + Ri = 0 \quad \text{or} \quad \left(D + \frac{R}{L}\right)i = 0 \qquad (16)$$

whose solution is

$$i = ce^{-(R/L)t} \qquad (17)$$

At $t = 0$, the initial current is $i_0 = V/R$. Substituting in (17), $c = V/R$ and the current equation is

$$i = \frac{V}{R}e^{-(R/L)t} \qquad (18)$$

This exponential decay is shown in Fig. 16-7(*a*) below. The corresponding voltages across the resistance and inductance are

$$v_R = Ri = Ve^{-(R/L)t} \quad \text{and} \quad v_L = L\frac{di}{dt} = -Ve^{-(R/L)t} \qquad (19)$$

as shown in Fig. 16-7(*b*) below. The sum $v_R + v_L$ satisfies Kirchhoff's law since the applied voltage is zero with the switch in position 2. The instantaneous powers $p_R = \frac{V^2}{R}e^{-2(R/L)t}$ and $p_L = -\frac{V^2}{R}e^{-2(R/L)t}$ are shown in Fig. 16-7(*c*) below. If p_L is integrated from zero to infinity we find that the energy released is exactly that which

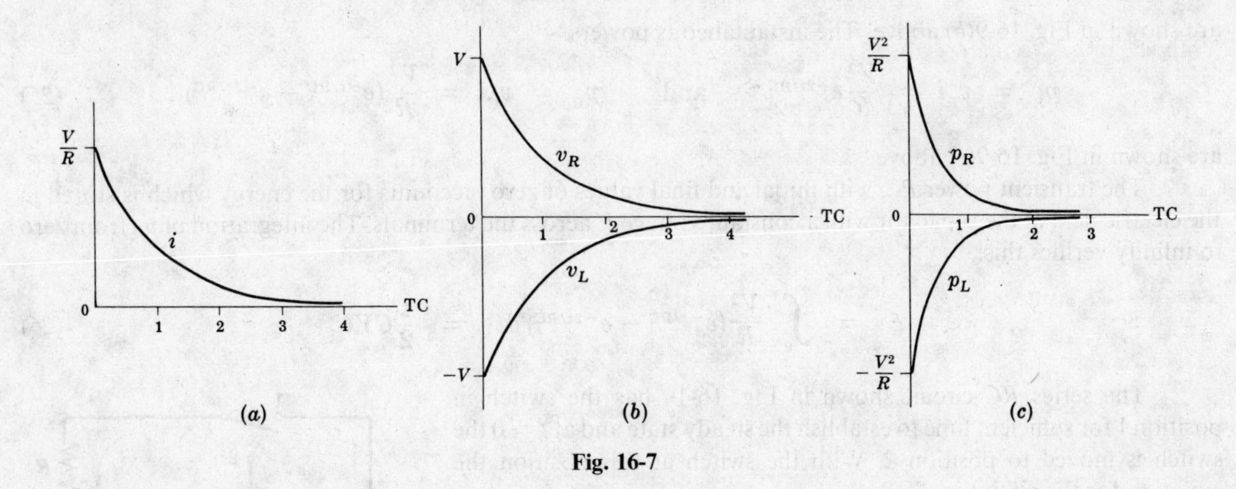

Fig. 16-7

was stored in the magnetic field during the previous transient, $\frac{1}{2}LI^2$. During the decay transient this energy is transferred to the resistor.

RC TRANSIENT

The application of Kirchhoff's voltage law to the series RC circuit shown in Fig. 16-8 results in the following differential equation

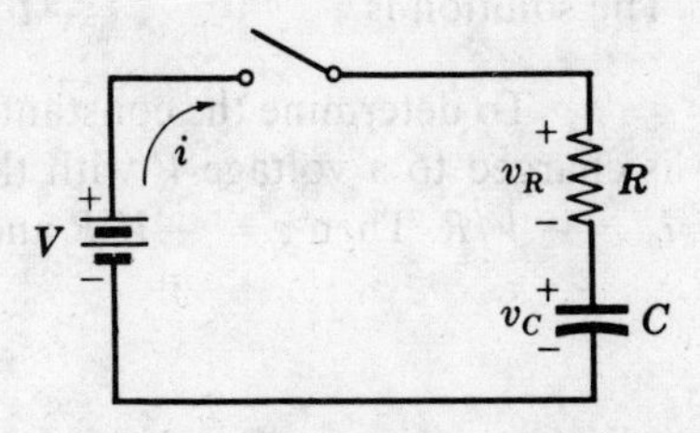

Fig. 16-8

$$\frac{1}{C}\int i\,dt + Ri = V \tag{20}$$

and after differentiating,

$$\frac{i}{C} + R\frac{di}{dt} = 0 \quad \text{or} \quad \left(D + \frac{1}{RC}\right)i = 0 \tag{21}$$

The solution to this homogeneous equation consists of only the complementary function since the particular solution is zero. Thus

$$i = ce^{-t/RC} \tag{22}$$

To determine the constant c we note that equation (*20*) at $t = 0$ is $Ri_0 = V$ or $i_0 = V/R$. Now substituting the value of i_0 into (*22*), we obtain $c = V/R$ at $t = 0$. Then

$$i = \frac{V}{R}e^{-t/RC} \tag{23}$$

Equation (*23*) has the form of an exponential decay as shown in Fig. 16-9(*a*).

The corresponding transient voltages

$$v_R = Ri = Ve^{-t/RC} \quad \text{and} \quad v_C = \frac{1}{C}\int i\,dt = V(1 - e^{-t/RC}) \tag{24}$$

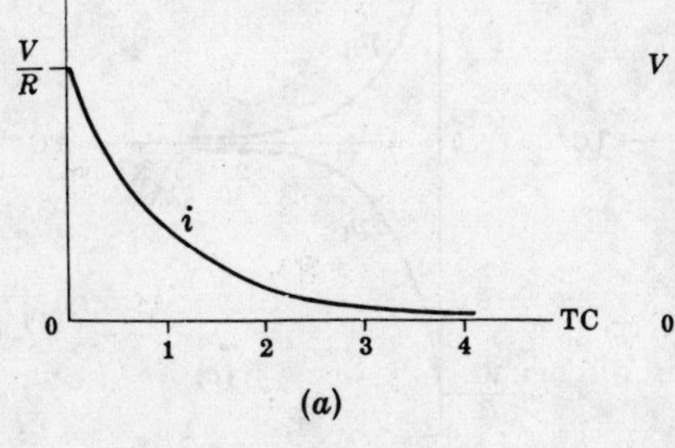

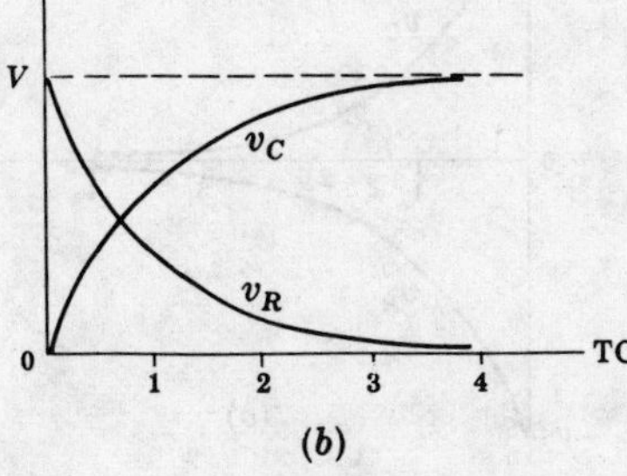

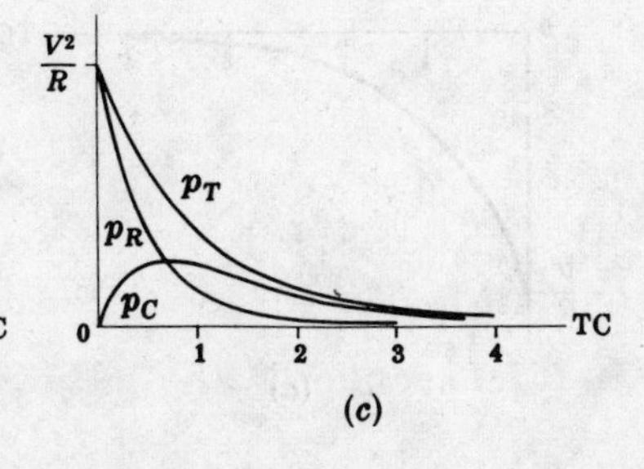

Fig. 16-9

are shown in Fig. 16-9(*b*) above. The instantaneous powers

$$p_R = v_R i = \frac{V^2}{R} e^{-2t/RC} \qquad \text{and} \qquad p_C = v_C i = \frac{V^2}{R}(e^{-t/RC} - e^{-2t/RC}) \tag{25}$$

are shown in Fig. 16-9(*c*) above.

The transient power P_C, with initial and final values of zero, accounts for the energy which is stored in the electric field of the capacitor with a constant voltage V across the terminals. The integration of p_R from zero to infinity verifies this.

$$\mathcal{E} = \int_0^\infty \frac{V^2}{R}(e^{-t/RC} - e^{-2t/RC})\,dt = \frac{1}{2}CV^2 \tag{26}$$

The series RC circuit shown in Fig. 16-10 has the switch in position 1 for sufficient time to establish the steady state and at $t = 0$ the switch is moved to position 2. With the switch at this position the equation for the circuit is

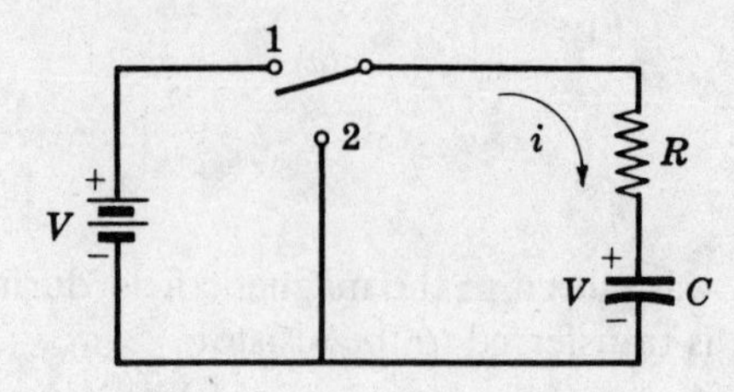

Fig. 16-10

$$\frac{1}{C}\int i\,dt + Ri = 0 \qquad \text{or} \qquad \left(D + \frac{1}{RC}\right)i = 0 \tag{27}$$

The solution is

$$i = ce^{-t/RC} \tag{28}$$

To determine the constant c we set $t = 0$ in (*28*) and substitute the initial current i_0. Since the capacitor is charged to a voltage V with the polarity shown in the diagram, the initial current is opposite to i; hence $i_0 = -V/R$. Then $c = -V/R$ and the current is

$$i = -\frac{V}{R}e^{-t/RC} \tag{29}$$

This decay transient is plotted in Fig. 16-11(*a*). The corresponding transient voltages for the circuit elements,

$$v_R = Ri = -Ve^{-t/RC} \qquad \text{and} \qquad v_C = \frac{1}{C}\int i\,dt = Ve^{-t/RC} \tag{30}$$

are shown in Fig. 16-11(*b*). Note that $v_R + v_C = 0$, satisfying Kirchhoff's law since there is no applied voltage while the switch is in position 2. The transient powers

$$p_R = v_R i = \frac{V^2}{R}e^{-2t/RC} \qquad \text{and} \qquad p_C = v_C i = -\frac{V^2}{R}e^{-2t/RC} \tag{31}$$

are shown in Fig. 16-11(*c*). There is no source to account for p_R but it is apparent that the energy stored in the capacitor is transferred to the resistor during this transient. The integration of p_C with the limits of zero and infinity resulting in $-\frac{1}{2}CV^2$ is left to the reader.

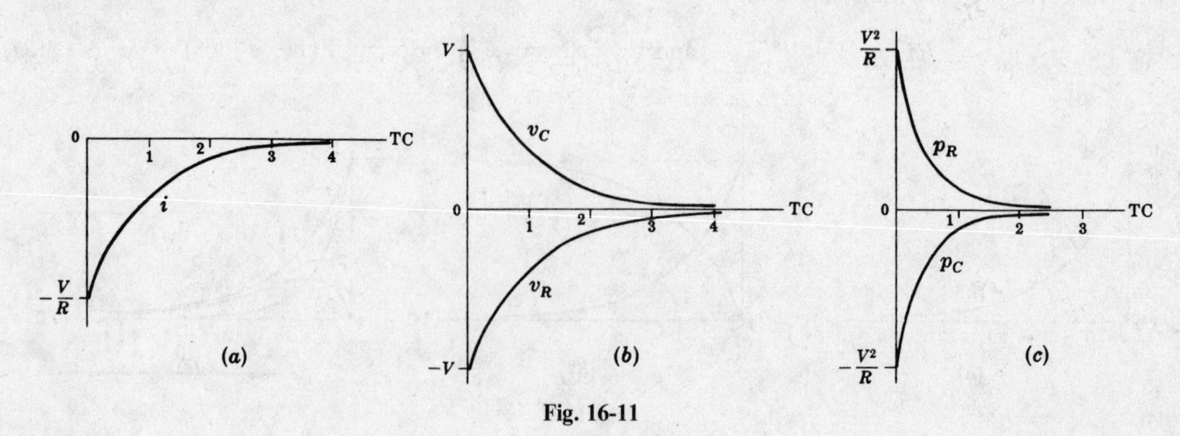

Fig. 16-11

RC TRANSIENT CHARGE BASIS

In a series *RC* circuit it is convenient sometimes to know the equation representing the transient charge q. Then, since current and charge are related by $i = dq/dt$, the current, if needed, may be obtained by differentiation.

In Fig. 16-12 the capacitor is charged with polarity on the plates as shown, since q has the same direction as that of i in Fig. 16-8. The current basis equation

$$\frac{1}{C}\int i\,dt + Ri = V \tag{32}$$

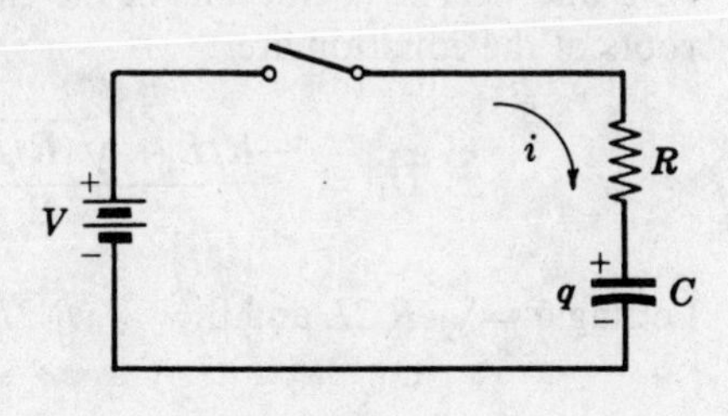

Fig. 16-12

is written on the charge basis by substituting dq/dt for i. Thus

$$\frac{q}{C} + R\frac{dq}{dt} = V \quad \text{or} \quad \left(D + \frac{1}{RC}\right)q = \frac{V}{R} \tag{33}$$

Using the method outlined in deriving equation (5), the solution is

$$q = ce^{-t/RC} + CV \tag{34}$$

At $t = 0$, the initial charge on the capacitor is $q_0 = 0$ and

$$q_0 = 0 = c(1) + CV \text{ or } c = -CV \tag{35}$$

Substituting this value of c in (34), we obtain

$$q = CV(1 - e^{-t/RC}) \tag{36}$$

The charge transient is an exponential rise to a final value of CV. Then if a decay circuit such as that in Fig. 16-10 is analyzed on the charge basis, the result is a charge decay from the value CV represented by the equation

$$q = CVe^{-t/RC} \tag{37}$$

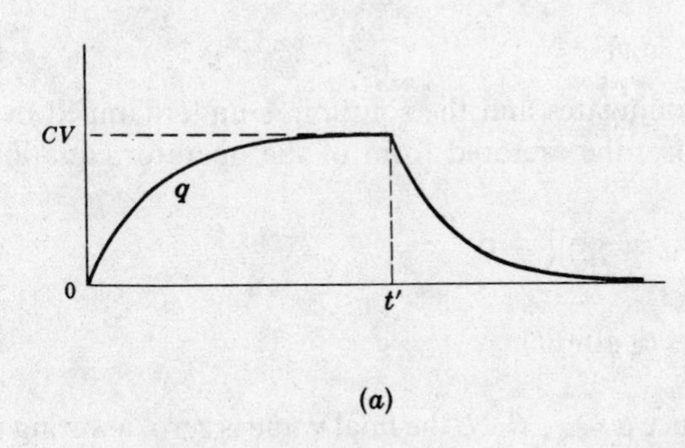

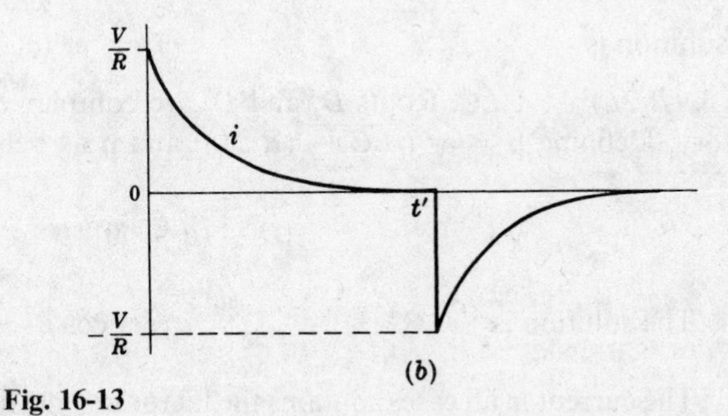

Fig. 16-13

The charge functions for buildup and decay are shown in Fig. 16-13(*a*), and the corresponding current functions in Fig. 16-13(*b*). Since charge must be a continuous function, $q = CV$ at $t'(-)$ and $t'(+)$ while i at $t'(-)$ is zero and at $t'(+)$ has the value $-V/R$.

RLC TRANSIENT

The application of Kirchhoff's voltage law to the series *RLC* circuit in Fig. 16-14 results in the following integro-differential equation

$$Ri + L\frac{di}{dt} + \frac{1}{C}\int i\,dt = V \tag{38}$$

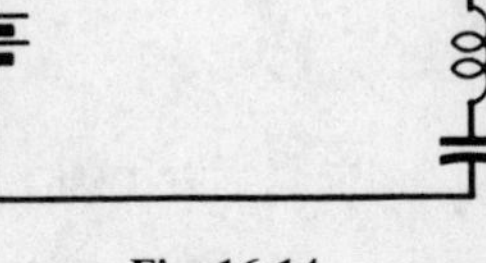

Fig. 16-14

Differentiating, we obtain

$$L\frac{d^2i}{dt^2} + R\frac{di}{dt} + \frac{i}{C} = 0 \quad \text{or} \quad \left(D^2 + \frac{R}{L}D + \frac{1}{LC}\right)i = 0 \tag{39}$$

This second order, linear differential equation is of the homogeneous type with a particular solution of zero. The complementary function can be one of three different types depending upon the relative magnitudes of R, L and C. The coefficients in the characteristic equation $D^2 + (R/L)D + 1/LC = 0$ are constants and the roots of the equation are

$$D_1 = \frac{-R/L + \sqrt{(R/L)^2 - 4/LC}}{2} \quad \text{and} \quad D_2 = \frac{-R/L - \sqrt{(R/L)^2 - 4/LC}}{2} \tag{40}$$

Letting $\alpha = -R/2L$ and $\beta = \sqrt{(R/2L)^2 - 1/LC}$,

$$D_1 = \alpha + \beta \text{ and } D_2 = \alpha - \beta \tag{41}$$

The radicand of β can be positive, zero or negative and the solution is then overdamped, critically damped or underdamped (oscillatory).

Case 1. $(R/2L)^2 > 1/LC$. Roots D_1 and D_2 are real and unequal resulting in the overdamped case. Then in factored form equation (*39*) is written

$$[D - (\alpha + \beta)][D - (\alpha - \beta)]i = 0 \tag{42}$$

and the current is

$$i = c_1e^{(\alpha+\beta)t} + c_2e^{(\alpha-\beta)t} \quad \text{or} \quad i = e^{\alpha t}(c_1e^{\beta t} + c_2e^{-\beta t}) \tag{43}$$

Case 2. $(R/2L)^2 = 1/LC$. Roots D_1 and D_2 are equal and the solution is the critically damped case. In factored form equation (*39*) becomes

$$(D - \alpha)(D - \alpha)i = 0 \tag{44}$$

The solution is

$$i = e^{\alpha t}(c_1 + c_2t) \tag{45}$$

Case 3. $(R/2L)^2 < 1/LC$. Roots D_1 and D_2 are complex conjugates and the solution is underdamped or oscillatory. Defining $\beta = \sqrt{1/LC - (R/2L)^2}$ and a as before, the factored form of the operator equation is

$$[D - (\alpha + j\beta)][D - (\alpha - j\beta)]i = 0 \tag{46}$$

The solution is

$$i = e^{\alpha t}(c_1 \cos \beta t + c_2 \sin \beta t) \tag{47}$$

The current in all cases contains the factor $e^{\alpha t}$ and since $\alpha = -R/2L$ the final value is zero, assuring that the complementary function decays in a relatively short time. The three cases are sketched in Fig. 16-15 when the initial value is zero and the initial slope is positive.

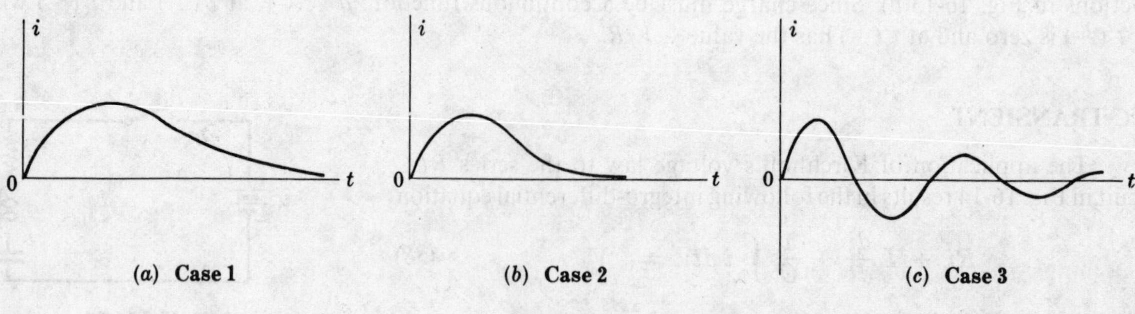

Fig. 16-15

Alternating Current Transients

RL SINUSOIDAL TRANSIENT

The *RL* circuit shown in Fig. 16-16 has a sinusoidal voltage applied when the switch is closed. The voltage function could be at any point in the period at the instant of closing the switch, and therefore the phase angle φ can take on values from 0 to 2π rad/sec. Application of Kirchhoff's voltage law results in the following equation.

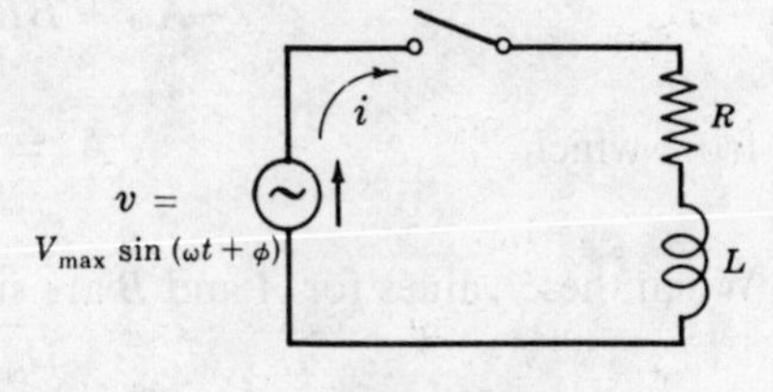

Fig. 16-16

$$Ri + L\frac{di}{dt} = V_{max}\sin(\omega t+\varphi) \quad \text{or} \quad \left(D+\frac{R}{L}\right)i = \frac{V_{max}}{L}\sin(\omega t+\varphi) \tag{48}$$

The complementary function is $i_c = ce^{-(R/L)t}$ and the particular solution is

$$i_p = e^{-(R/L)t}\int e^{(R/L)t}\frac{V_{max}}{L}\sin(\omega t+\varphi)\,dt = \frac{V_{max}}{\sqrt{R^2+\omega^2L^2}}\sin(\omega t+\varphi-\tan^{-1}\omega L/R)$$

The complete solution is

$$i = i_c + i_p = ce^{-(R/L)t} + \frac{V_{max}}{\sqrt{R^2+\omega^2L^2}}\sin(\omega t+\varphi-\tan^{-1}\omega L/R) \tag{49}$$

The inductance prevents any sudden change in the current and since before the switch was closed the current was zero, it follows that $i_0 = 0$. Then at $t = 0$,

$$i_0 = 0 = c(1) + \frac{V_{max}}{\sqrt{R^2+\omega^2L^2}}\sin(\varphi-\tan^{-1}\omega L/R) \quad \text{and} \quad c = \frac{-V_{max}}{\sqrt{R^2+\omega^2L^2}}\sin(\varphi-\tan^{-1}\omega L/R)$$

Substituting in *(49)*, the current is

$$i = e^{-(R/L)t}\left[\frac{-V_{max}}{\sqrt{R^2+\omega^2L^2}}\sin(\varphi-\tan^{-1}\omega L/R)\right] + \frac{V_{max}}{\sqrt{R^2+\omega^2L^2}}\sin(\omega t+\varphi-\tan^{-1}\omega L/R) \tag{50}$$

The first part of *(50)* contains the factor $e^{-(R/L)t}$ which has a value of zero in a relatively short time. The expression within the brackets is simply a rather involved constant. The magnitude of this constant is dependent upon the time in the cycle φ at which the switch is closed. If $(\varphi - \tan^{-1}\omega L/R) = n\pi$ where $n =$ 0, 1, 2, 3, . . ., then the constant is zero and the current goes directly into the steady state. And if $(\varphi - \tan^{-1}\omega L/R) = (1 + 2n)\pi/2$, the transient will have the maximum possible amplitude.

The second part of *(50)* is the steady state current which lags the applied voltage by $\tan^{-1}\omega L/R$. This particular solution, obtained above by integration, can be found by the method of undetermined coefficients. The method is applicable when the forcing function is a sine, cosine or exponential function, since with these functions successive differentiations repeat the same set of functions. To apply the method to equation *(48)* where the right hand side is $V_{max}\sin(\omega t + \varphi)$, we assume a particular current

$$i_p = A\cos(\omega t + \varphi) + B\sin(\omega t + \varphi) \tag{51}$$

where *A* and *B* are constants. Then the first derivative is

$$i_p' = -A\omega\sin(\omega t + \varphi) + B\omega\cos(\omega t + \varphi) \tag{52}$$

Substituting these expressions for i_p and i_p' in *(48)*, we obtain

$$\{-A\omega\sin(\omega t+\varphi) + B\omega\cos(\omega t+\varphi)\} + \frac{R}{L}\{A\cos(\omega t+\varphi) + B\sin(\omega t+\varphi)\} = \frac{V_{max}}{L}\sin(\omega t+\varphi) \tag{53}$$

Combining coefficients of like terms,

$$(-A\omega + BR/L)\sin(\omega t+\varphi) + (B\omega + AR/L)\cos(\omega t+\varphi) = \frac{V_{max}}{L}\sin(\omega t+\varphi) \tag{54}$$

Now equating coefficients of like terms results in two equations in A and B,

$$-A\omega + BR/L = V_{max}/L \quad \text{and} \quad B\omega + AR/L = 0 \tag{55}$$

from which

$$A = \frac{-\omega L V_{max}}{R^2 + \omega^2 L^2} \quad \text{and} \quad B = \frac{R V_{max}}{R^2 + \omega^2 L^2} \tag{56}$$

When these values for A and B are substituted into equation (*51*), the particular current is

$$i_p = \frac{-\omega L V_{max}}{R^2 + \omega^2 L^2} \cos(\omega t + \varphi) + \frac{R V_{max}}{R^2 + \omega^2 L^2} \sin(\omega t + \varphi) \tag{57}$$

or

$$i_p = \frac{V_{max}}{\sqrt{R^2 + \omega^2 L^2}} \sin(\omega t + \varphi - \tan^{-1} \omega L/R) \tag{58}$$

which is the same as the particular solution obtained above by integration.

RC SINUSOIDAL TRANSIENT

The *RC* circuit shown in Fig. 16-17 has a sinusoidal voltage applied at the time the switch is closed. The application of Kirchhoff's voltage law to the circuit results in the following equation

$$Ri + \frac{1}{C}\int i\,dt = V_{max} \sin(\omega t + \varphi) \tag{59}$$

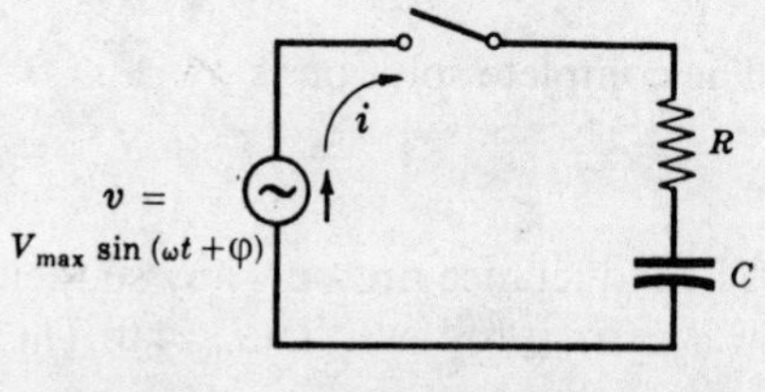

$v = V_{max} \sin(\omega t + \varphi)$, i, R, C

Fig. 16-17

Differentiating and using operator notation, we obtain

$$\left(D + \frac{1}{RC}\right) i = \frac{\omega V_{max}}{R} \cos(\omega t + \varphi) \tag{60}$$

The complementary function is $\qquad i_c = ce^{-t/RC} \qquad (61)$

and the particular solution, obtained either by integration or undetermined coefficients, is

$$i_p = \frac{V_{max}}{\sqrt{R^2 + (1/\omega C)^2}} \sin(\omega t + \varphi + \tan^{-1} 1/\omega CR) \tag{62}$$

Then the complete solution is

$$i = ce^{-t/RC} + \frac{V_{max}}{\sqrt{R^2 + (1/\omega C)^2}} \sin(\omega t + \varphi + \tan^{-1} 1/\omega CR) \tag{63}$$

To determine the constant c, let $t = 0$ in equation (*59*); then the initial current $i_0 = \frac{V_{max}}{R} \sin \varphi$. Substituting this into (*63*) and setting $t = 0$, we obtain

$$\frac{V_{max}}{R} \sin \varphi = c(1) + \frac{V_{max}}{\sqrt{R^2 + (1/\omega C)^2}} \sin(\varphi + \tan^{-1} 1/\omega CR) \tag{64}$$

or

$$c = \frac{V_{max}}{R} \sin \varphi - \frac{V_{max}}{\sqrt{R^2 + (1/\omega C)^2}} \sin(\varphi + \tan^{-1} 1/\omega CR) \tag{65}$$

Substitution of c from (*65*) into (*63*) results in the complete current

$$i = e^{-t/RC}\left[\frac{V_{max}}{R} \sin \varphi - \frac{V_{max}}{\sqrt{R^2 + (1/\omega C)^2}} \sin(\varphi + \tan^{-1} 1/\omega CR)\right] + \frac{V_{max}}{\sqrt{R^2 + (1/\omega C)^2}} \sin(\omega t + \varphi + \tan^{-1} 1/\omega CR) \tag{66}$$

The first term is the transient with the decay factor $e^{-t/RC}$. The entire quantity within the brackets is simply a constant. The second term is the steady state current which leads the applied voltage by $\tan^{-1} 1/\omega CR$.

RLC SINUSOIDAL TRANSIENT

The series *RLC* circuit shown in Fig. 16-18 has a sinusoidal voltage applied when the switch is closed. The resulting equation is

$$Ri + L\frac{di}{dt} + \frac{1}{C}\int i\,dt = V_{max}\sin(\omega t + \varphi) \qquad (67)$$

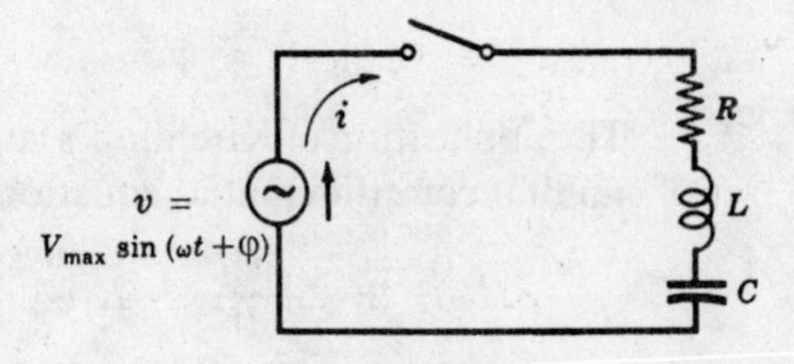

Fig. 16-18

Differentiating and using operator notation, we obtain

$$\left(D^2 + \frac{R}{L}D + \frac{1}{LC}\right)i = \frac{\omega V_{max}}{L}\cos(\omega t + \varphi) \qquad (68)$$

The particular solution is obtained by the method of undetermined coefficients as follows. First let $i_p = A\cos(\omega t + \varphi) + B\sin(\omega t + \varphi)$. Then evaluate i_p' and i_p' and substitute the results in equation (*67*). The values of *A* and *B* are then found by equating coefficients of like terms as done before in the case of the *RL* sinusoidal transient. Expressing the result as a single sine function, the particular solution is

$$i_p = \frac{V_{max}}{\sqrt{R^2 + (1/\omega C - \omega L)^2}}\sin\left(\omega t + \varphi + \tan^{-1}\frac{(1/\omega C - \omega L)}{R}\right) \qquad (69)$$

The complementary function is identical to that of the *DC* series *RLC* circuit examined previously where the result was overdamped, critically damped or oscillatory, depending upon *R*, *L* and *C*.

Case 1. $(R/2L)^2 > 1/LC$. The roots are real and unequal resulting in the overdamped case. $D_1 = \alpha + \beta$ and $D_2 = \alpha - \beta$, where $\alpha = -R/2L$ and $\beta = \sqrt{(R/2L)^2 - 1/LC}$. The complete solution is

$$i = e^{\alpha t}(c_1 e^{\beta t} + c_2 e^{-\beta t}) + \frac{V_{max}}{\sqrt{R^2 + (1/\omega C - \omega L)^2}}\sin\left(\omega t + \varphi + \tan^{-1}\frac{(1/\omega C - \omega L)}{R}\right) \qquad (70)$$

Case 2. $(R/2L)^2 = 1/LC$. The roots are real and equal, resulting in the critically damped case and the complete current is

$$i = e^{\alpha t}(c_1 + c_2 t) + \frac{V_{max}}{\sqrt{R^2 + (1/\omega C - \omega L)^2}}\sin\left(\omega t + \varphi + \tan^{-1}\frac{(1/\omega C - \omega L)}{R}\right) \qquad (71)$$

Case 3. $(R/2L)^2 < 1/LC$. The roots are complex conjugates resulting in the oscillatory case and the complete current is

$$i = e^{\alpha t}(c_1\cos\beta t + c_2\sin\beta t) + \frac{V_{max}}{\sqrt{R^2 + (1/\omega C - \omega L)^2}}\sin\left(\omega t + \varphi + \tan^{-1}\frac{(1/\omega C - \omega L)}{R}\right) \qquad (72)$$

where $\beta = \sqrt{1/LC - (R/2L)^2}$.

The particular solutions of equations (*70*), (*71*) and (*72*) are identical while the transient current given by the complementary function differs in each case. For example, in Case 3 the transient contains a set of sinusoidal functions of frequency β rad/sec, a frequency which is in general different from ω of the particular solution. Consequently the appearance of the current during the transient period is impossible to predict, often having a very irregular shape. Once the decay factor brings the transient to zero, the current then leads or lags the applied voltage depending upon the relative magnitudes of the reactances $1/\omega C$ and ωL in $\tan^{-1}(1/\omega C - \omega L)/R$.

Two Mesh Transients

The application of Kirchhoff's voltage law to the two-mesh network in Fig. 16-19 results in the following set of simultaneous differential equations:

$$\begin{aligned} R_1 i_1 + L_1\frac{di_1}{dt} + R_1 i_2 &= V \\ R_1 i_1 + (R_1+R_2)i_2 + L_2\frac{di_2}{dt} &= V \end{aligned} \qquad (73)$$

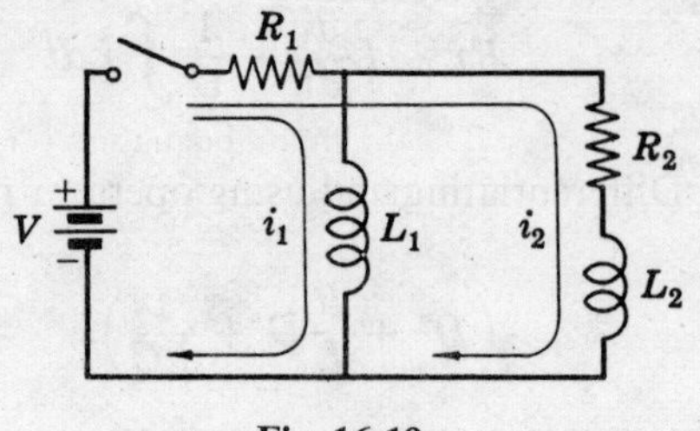

Fig. 16-19

Using operator notation and rearranging, we obtain

$$(D + R_1/L_1)i_1 + (R_1/L_1)i_2 = V/L_1$$

$$(R_1/L_2)i_1 + \left(D + \frac{R_1+R_2}{L_2}\right)i_2 = V/L_2$$

or

$$\begin{bmatrix} D + R_1/L_1 & R_1/L_1 \\ R_1/L_2 & D + \dfrac{R_1+R_2}{L_2} \end{bmatrix} \begin{bmatrix} i_1 \\ i_2 \end{bmatrix} = \begin{bmatrix} V/L_1 \\ V/L_2 \end{bmatrix} \qquad (74)$$

In order to obtain an equation for i_1 independent of i_2, we use determinants and write

$$\begin{vmatrix} D + R_1/L_1 & R_1/L_1 \\ R_1/L_2 & D + \dfrac{R_1+R_2}{L_2} \end{vmatrix} i_1 = \begin{vmatrix} V/L_1 & R_1/L_1 \\ V/L_2 & D + \dfrac{R_1+R_2}{L_2} \end{vmatrix} \qquad (75)$$

The determinant on the left is expanded and rearranged in order of descending powers of D. In the expansion of the determinant on the right a term $D(V/L_1)$ appears; but since $D = d/dt$ and V/L_1 is a constant, this term is zero.

$$\left[D^2 + \left(\frac{R_1L_1 + R_2L_1 + R_1L_2}{L_1L_2}\right)D + \frac{R_1R_2}{L_1L_2}\right]i_1 = VR_2/L_1L_2 \qquad (76)$$

The characteristic equation is of the form $D^2 + AD + B = 0$, but since in this case $A^2 - 4B > 0$ for all values of circuit constants (except that neither L_1 nor $L_2 = 0$) the complementary function is of the form given in equation (*43*). Since the forcing function is a constant, a particular solution is the constant which satisfies the equation

$$\left(\frac{R_1R_2}{L_1L_2}\right)i_{1p} = VR_2/L_1L_2 \quad \text{or} \quad i_{1p} = V/R_1 \qquad (77)$$

Now with the same methods applied to i_2, we obtain

$$\begin{vmatrix} D + R_1/L_1 & R_1/L_1 \\ R_1/L_2 & D + \dfrac{R_1+R_2}{L_2} \end{vmatrix} i_2 = \begin{vmatrix} D + R_1/L_1 & V/L_1 \\ R_1/L_2 & V/L_2 \end{vmatrix} \qquad (78)$$

After expanding the two determinants, we obtain

$$\left[D^2 + \left(\frac{R_1L_1 + R_2L_1 + R_1L_2}{L_1L_2}\right)D + \frac{R_1R_2}{L_1L_2}\right]i_2 = 0$$

The characteristic equation is the same as that of equation (*76*), and consequently the complementary functions are identical. However, the particular solution for i_2 is zero since the equation is of the homogeneous type.

Examination of the circuit shows this to be perfectly reasonable, since in the steady state L_1 appears as a short circuit across the R_2L_2 branch thereby shunting current away from this branch. The R_1 is the only limiting impedance in the steady state and it follows that the current $i_1 = V/R_1$ as given above in equation (*77*).

Solved Problems

16.1. A series RL circuit with $R = 50$ ohms and $L = 10$ H has a constant voltage $V = 100$ V applied at $t = 0$ by the closing of a switch. Find (*a*) the equations for i, v_R and v_L, (*b*) the current at $t = 0{\cdot}5$ seconds and (*c*) the time at which $v_R = v_L$.

(*a*) The differential equation for the given circuit is

$$50i + 10\frac{di}{dt} = 100 \text{ or } (D+5)i = 10 \qquad (1)$$

and the complete solution is

$$i = i_c + i_p = ce^{-5t} + 2 \qquad (2)$$

At $t = 0$, $i_0 = 0$ and $0 = c(1) + 2$ or $c = -2$. Then

$$i = 2(1 - e^{-5t}) \text{ amperes} \qquad (3)$$

shown in Fig. 16-20(*a*).

The corresponding voltages across the circuit elements

$$v_R = Ri = 100(1 - e^{-5t}) \text{ volts and } v_L = L\frac{di}{dt} = 100e^{-5t} \text{ volts} \qquad (4)$$

are shown in Fig. 16-20(*b*).

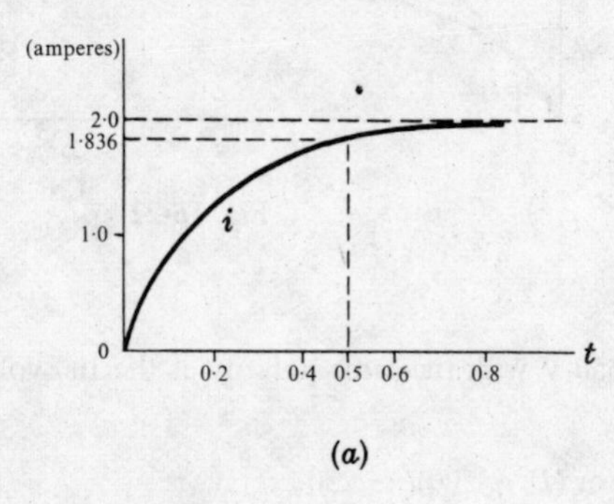

(*a*) (*b*)

Fig. 16-20

(*b*) Put $t = 0{\cdot}5$ sec in (3) and obtain $i = 2(1 - e^{-5(0\cdot5)}) = 2(1 - 0{\cdot}082) = 1{\cdot}836$ A.

(*c*) When $v_R = v_L$, each must be 50 volts; and since the applied voltage is 100, we set either v_R or v_L to 50 and solve for t. From (4), $v_L = 50 = 100e^{-5t}$ volts. Then $e^{-5t} = 0{\cdot}5$ or $5t = 0{\cdot}693$, and $t = 0{\cdot}1386$ s.

16.2. Referring to Problem 16.1, find the equations for p_R and p_L and show that the power in the inductance accounts for the steady state energy stored in the magnetic field.

Using the current and the voltages obtained in Problem 16.1, the instantaneous powers are

$$p_R = v_R i = 100(1 - e^{-5t})\,2(1 - e^{-5t}) = 200(1 - 2e^{-5t} + e^{-10t}) \text{ watts}$$

$$p_L = v_L i = 100e^{-5t}\,2(1 - e^{-5t}) = 200(e^{-5t} - e^{-10t}) \text{ watts}$$

$$p_T = p_R + p_L = 200(1 - e^{-5t}) \text{ watts}$$

The steady state energy stored in the magnetic field is $W = \frac{1}{2}LI^2 = \frac{1}{2}(10)(2)^2 = 20$ joules.

The integral of p_L from $t = 0$ to $t = \infty$ is $W = \int_0^\infty 200(e^{-5t} - e^{-10t})\,dt = 20$ joules.

16.3. In the series circuit shown in Fig. 16-21 the switch is closed on position 1 at $t = 0$ thereby applying the 100 volt source to the RL branch, and at $t = 500$ μs the switch is moved to position 2. Obtain the equations for the current in both intervals and sketch the transient.

With the switch in position 1 the equation is

$$100i + 0{\cdot}2\frac{di}{dt} = 100 \text{ or } (D + 500)i = 500 \qquad (1)$$

and the complete current is $\qquad i = c_1 e^{-500t} + 1{\cdot}0$ amperes $\qquad (2)$

At $t = 0$, $i = 0$. Using the initial condition in (2), $0 = c_1(1) + 1{\cdot}0$ or $c_1 = -1{\cdot}0$. Then the current is

$$i = 1{\cdot}0(1 - e^{-500t}) \text{ amperes} \qquad (3)$$

Now at 500 μ sec this transient is interrupted and the current is

$$i = 1{\cdot}0(1 - e^{-500(500 \times 10^{-6})}) = 1{\cdot}0(1 - 0{\cdot}779) = 0{\cdot}221 \text{ A} \qquad (4)$$

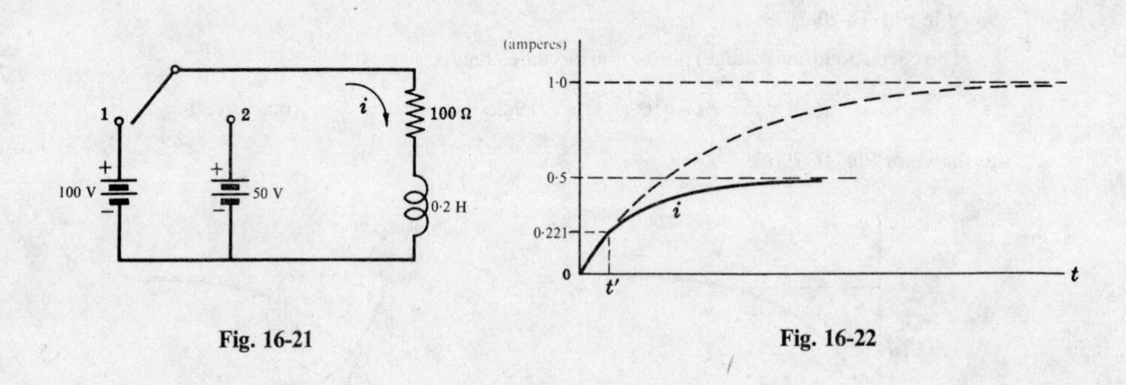

Fig. 16-21 Fig. 16-22

With the switch in position 2 the applied voltage is 50 V with the same polarity as the 100 volt source and the equation is

$$100i + 0{\cdot}2\frac{di}{dt} = 50 \text{ or } (D + 500)i = 250 \qquad (5)$$

which gives $\qquad i = c_2 e^{-500(t-t')} + 0{\cdot}5$ amperes $\qquad (6)$

where $t' = 500$μs. Now when $t = t'$ in equation (6), the current is 0·221 A as found in (4).

$$i = 0{\cdot}221 = c_2(1) + 0{\cdot}5 \text{ and } c_2 = -0{\cdot}279$$

Then for $t < t'$, $\qquad i = -0{\cdot}279e^{-500(t-t')} + 0{\cdot}5$ amperes $\qquad (7)$

Equation (3) applies for $0 < t < t'$ and the transient shown dotted in Fig. 16-22 is heading for a steady state value of 1·0. Then at t' when the current is 0·221 A the switch is moved to position 2 and equation (7) applies for $t < t'$ with a final value of 0·5 A as shown.

16.4. Repeat Problem 16.3 with the polarity of the 50 volt source reversed.

The first part of the transient with the switch in position 1 is the same as obtained in Problem 16.3: $i = 1{\cdot}0(1 - e^{-500t})$ with $i = 0{\cdot}221$ A at $t = 500$ μs.

The reversed polarity on the 50 volt source results in the following equation

$$100i + 0{\cdot}2\frac{di}{dt} = -50 \text{ or } (D + 500)i = -250 \qquad (1)$$

whose solution is $\qquad i = c\, e^{-500(t-t')} - 0{\cdot}5$ amperes $\qquad (2)$

Now at $t = t'$ the current is 0·221 amps. Substituting into equation (2), $0·221 = c(1) - 0·5$ or $c = 0·721$. Then the current equation for $t > t'$ is

$$i = 0·721\, e^{-500(t-t')} - 0·5 \text{ amperes}$$

The current transient is shown in Fig. 16-23 and the final value is $-0·5$ since its direction with the 50 volt source applied is opposite to the assumed positive direction for i.

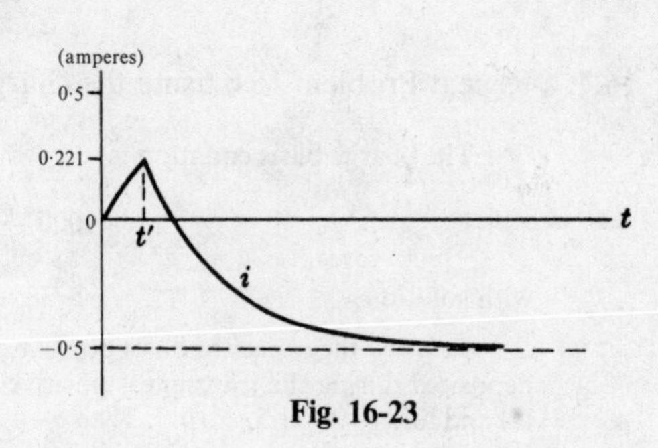

Fig. 16-23

16.5. A series RC circuit with $R = 5000$ ohms and $C = 20\ \mu$F has a constant voltage $V = 100$ V applied at $t = 0$ and the capacitor has no initial charge. Find the equations of i, v_R and v_C.

When the switch is closed the equation is

$$5000i + \frac{1}{20 \times 10^{-6}} \int i\, dt = 100 \qquad (1)$$

Differentiating and using operator notation we obtain

$$(D + 10)i = 0 \qquad \text{with a solution} \qquad i = c\, e^{-10t} \qquad (2)$$

Setting $t = 0$ in equation (1) gives the initial current $i_0 = 100/5000 = 0·02$ A. Substitute this in (2) and obtain $c = 0·02$. Then the current is

$$i = 0·02\, e^{-10t} \text{ amperes} \qquad (3)$$

and the transient voltages across the circuit elements are

$$v_R = Ri = 5000(0·02\, e^{-10t}) = 100\, e^{-10t} \text{ volts}$$

$$v_C = \frac{1}{C}\int i\, dt = \frac{1}{20 \times 10^{-6}} \int 0·02\, e^{-10t}\, dt = 100(1 - e^{-10t}) \text{ volts}$$

The transients are shown in Fig. 16-24. In the steady state, $v_R = 0$ and $v_C = 100$ V.

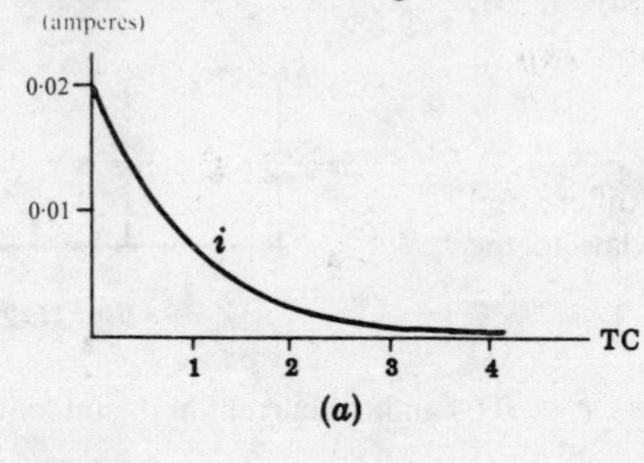

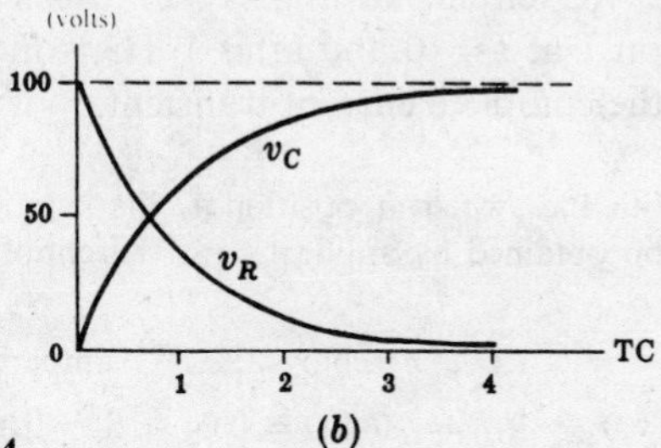

Fig. 16-24

16.6. The 20 μF capacitor in the RC circuit shown in Fig. 16-25 has an initial charge $q_0 = 500$ microcoulombs with the polarity shown in the diagram. At $t = 0$ the switch is closed, thereby applying the constant voltage $V = 50$ volts. Find the current transient.

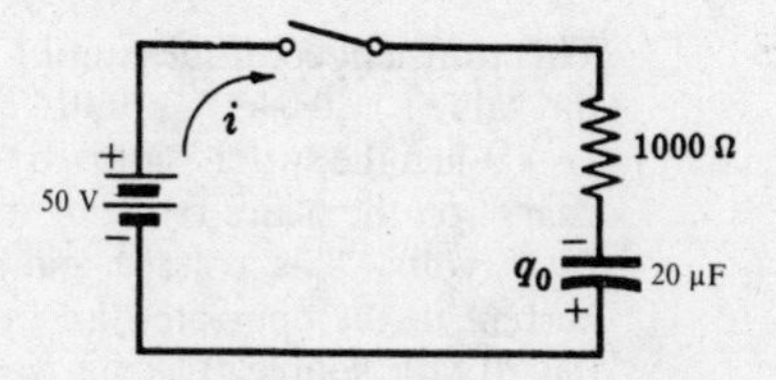

Fig. 16-25

When the switch is closed the equation is

$$1000i + \frac{1}{20 \times 10^{-6}} \int i\, dt = 50 \qquad \text{or} \qquad (D + 50)i = 0 \qquad (1)$$

and the solution is

$$i = c\, e^{-50t} \text{ amperes} \qquad (2)$$

Now the 50 volt source drives a current in the direction shown in the diagram, resulting in + charge on the top plate of the capacitor. The initial charge q_0 on the capacitor has an equivalent voltage $V_0 = q_0/C = (500 \times 10^{-6})/(20 \times 10^{-6}) = 25$ volts which also sends a current in the direction of i as shown. Then at $t = 0$ the initial current is $i_0 = (V + q_0/C)/R = (50 + 25)/1000 = 0·075$ A. Substituting in equation (2), we find $c = 0·075$ and hence $i = 0·075e^{-50t}$ amperes.

16.7. Repeat Problem 16.6 using the charge basis for the transient.

The charge basis equation is

$$1000\frac{dq}{dt} + \frac{q}{20 \times 10^{-6}} = 50 \text{ or } (D + 50)q = 0{\cdot}05 \qquad (1)$$

with solution
$$q = c\,e^{-50t} + 10^{-3} \text{ coulombs} \qquad (2)$$

At $t = 0$ the capacitor has a positive charge of $0{\cdot}5 \times 10^{-3}$ coulombs on the lower plate. The polarity of the charge deposited during the transient is positive on the upper plate. Hence put $q_0 = -0{\cdot}5 \times 10^{-3}$ and $t = 0$ into equation (2) and find $c = -1{\cdot}5 \times 10^{-3}$. Then $q = -1{\cdot}5 \times 10^{-3}\,e^{-50t} + 10^{-3}$ and the current transient is $i = dq/dt = 0{\cdot}075e^{-50t}$ amperes.

The transient in Fig. 16-26(*a*) shows that the capacitor has an initial charge of $0{\cdot}5 \times 10^{-3}$ coulombs, positive on the lower plate, and a final charge of $1{\cdot}0 \times 10^{-3}$ coulombs, positive on the top plate. The current transient $i = dq/dt$ is shown in Fig. 16-26(*b*).

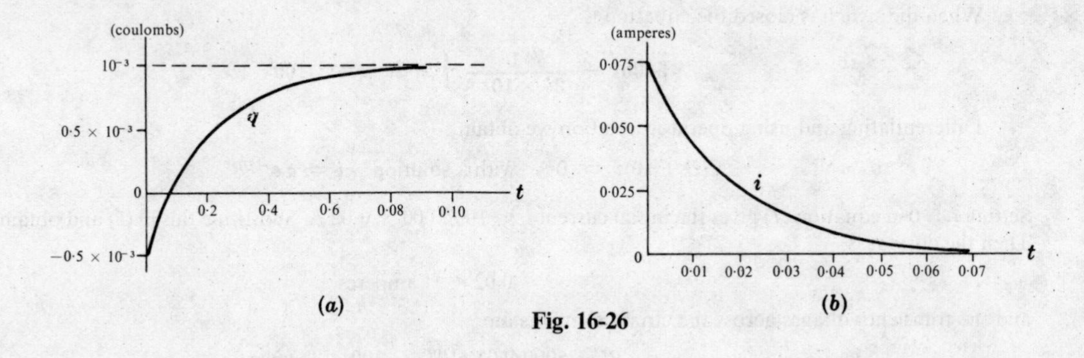

(*a*) Fig. 16-26 (*b*)

16.8. In the *RC* circuit of Fig. 16-27 the switch is closed on position 1 at $t = 0$ and after 1 TC is moved to position 2. Find the complete current transient.

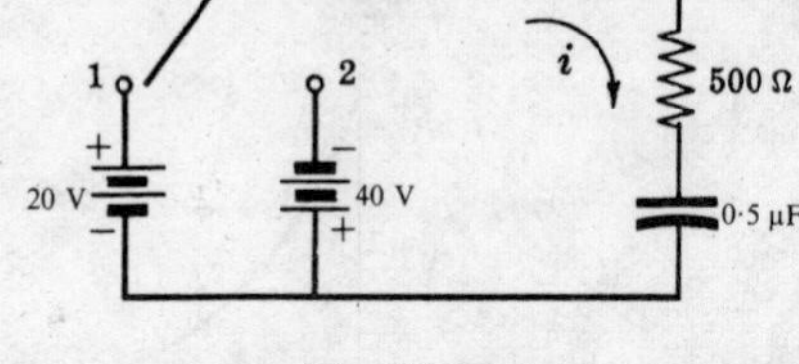

Fig. 16-27

With the switch at position 1, the solution of the differential equation obtained by application of Kirchhoff's voltage law to the circuit is

$$i = c_1\,e^{-t/RC} = c_1\,e^{-4000t} \text{ amperes} \qquad (1)$$

At $t = 0$, $i_0 = V/R = 20/500 = 0{\cdot}04$ A. Substituting into (*1*), $c_1 = 0{\cdot}04$ and the current in the interval $0 < t < 1$ TC is

$$i = 0{\cdot}04\,e^{-4000t} \text{ amperes} \qquad (2)$$

This transient continues until $t = 1$ TC $= RC = 500(0{\cdot}5 \times 10^{-6}) = 250$ microseconds. At this point the current has the value $i = 0{\cdot}04e^{-1} = 0{\cdot}0147$ A.

When the switch is moved to position 2, the capacitor has a charge on the plates resulting in a voltage $v_C = 20(1 - e^{-1}) = 12{\cdot}65$ volts. This voltage and the 40 volt source both drive current in the opposite direction from the current caused by the 20 volt source. Letting $t' = 1$ TC, the current equation for the second transient is

$$i = c_2\,e^{-4000(t-t')} \text{ amperes} \qquad (3)$$

At $t = t'$, $i = -(40 + 12{\cdot}65)/500 = -0{\cdot}1053$ A. Substituting into (*3*), $c_2 = -0{\cdot}1053$ and the current is

$$i = -0{\cdot}1053\,e^{-4000(t-t')} \text{ amperes} \qquad (4)$$

The complete current transient is shown in Fig. 16-28. At 1 TC, the current has a peak value of $-0{\cdot}1053$ A.

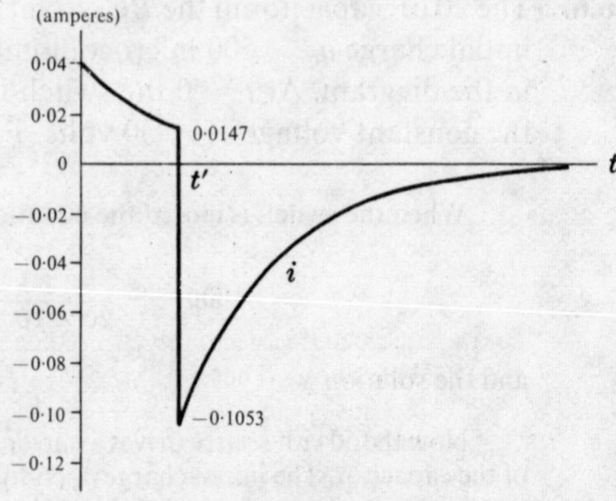

Fig. 16-28

16.9. Determine the charge transient for Problem 16.8 and differentiate to obtain the current.

The equation on the charge basis with the switch in position 1 is

$$500\,\frac{dq}{dt} + \frac{q}{0{\cdot}5 \times 10^{-6}} = 20 \text{ or } (D + 4000)q = 0{\cdot}04 \qquad (1)$$

and the solution is
$$q = c_1\,e^{-4000t} + 10 \times 10^{-6} \text{ coulombs} \qquad (2)$$

At $t = 0$, $q_0 = 0$. Using this initial condition in *(2)*, we obtain $c_1 = -10 \times 10^{-6}$ and thus

$$q = 10 \times 10^{-6}\,(1 - e^{-4000t}) \text{ coulombs} \qquad (3)$$

This equation applies for $0 < t < t'$ where $t' = 1$ TC. At 1 TC the charge on the capacitor is $q = 10 \times 10^{-6}(1 - e^{-1}) = 6{\cdot}32 \times 10^{-6}$ coulombs.

With the switch in position 2 the differential equation is

$$500\,\frac{dq}{dt} + \frac{q}{0{\cdot}5 \times 10^{-6}} = -40 \text{ or } (D + 4000)q = -0{\cdot}08 \qquad (4)$$

and the solution is
$$q = c_2\,e^{-4000(t-t')} - 20 \times 10^{-6} \text{ coulombs} \qquad (5)$$

Now we determine c_2 by substituting the value of q at 1 TC and setting $t = 1$ TC in equation *(5)*. Thus, $6{\cdot}32 \times 10^{-6} = c_2(1) - 20 \times 10^{-6}$ or $c_2 = 26{\cdot}32 \times 10^{-6}$. Then

$$q = 26{\cdot}32 \times 10^{-6}\,e^{-4000(t-t')} - 20 \times 10^{-6} \text{ coulombs} \qquad (6)$$

The complete charge transient is shown in Fig. 16-29. We obtain the corresponding current transient by differentiating equations *(3)* and *(6)*. Thus in the interval $0 < t < t'$ the current is

$$i = \frac{d}{dt}\{10 \times 10^{-6}\,(1 - e^{-4000t})\} = 0{\cdot}04\,e^{-4000t} \text{ amperes}$$

and when $t > t'$,

$$i = \frac{d}{dt}\{26{\cdot}32 \times 10^{-6}\,e^{-4000(t-t')} - 20 \times 10^{-6}\} = -0{\cdot}1053\,e^{-4000(t-t')} \text{ amperes}$$

These same results were obtained in equations *(2)* and *(4)* of Problem 16.8.

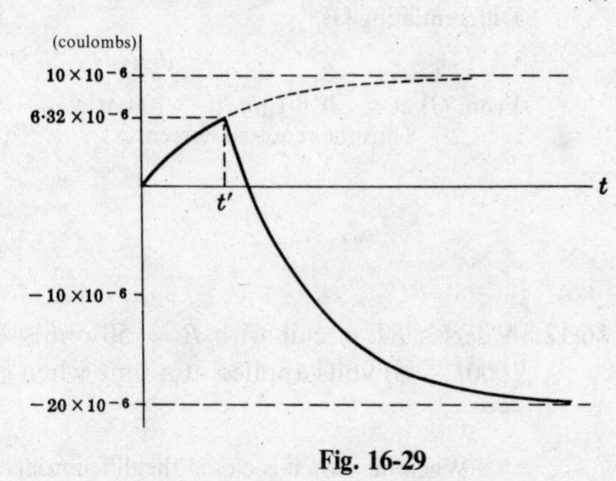

Fig. 16-29

16.10. A series *RLC* circuit with $R = 3000$ ohms, $L = 10$ H and $C = 200\ \mu$F has a constant voltage $V = 50$ volts applied at $t = 0$. Find the current transient and the maximum value of the current if the capacitor has no initial charge.

The equation after the switch is closed is

$$3000i + 10\frac{di}{dt} + \frac{1}{200 \times 10^{-6}}\int i\,dt = 50 \quad \text{or} \quad (D^2 + 300D + 500)i = 0 \qquad (1)$$

The roots of the characteristic equation are $D_1 = -298{\cdot}3$ and $D_2 = -1{\cdot}67$ and

$$i = c_1\,e^{-1{\cdot}67t} + c_2\,e^{-298{\cdot}3t} \text{ amperes} \qquad (2)$$

To evaluate c_1 and c_2 we use two initial conditions. Since the series circuit contains an inductance, the current function must be continuous. Therefore since $i = 0$ at $t = 0-$, i is also zero at $t = 0+$. Then from equation *(1)*, $10\,di/dt = 50$ and $di/dt = 5$. Now write equation *(2)* at $t = 0$: $0 = c_1(1) + c_2(1)$ or $c_1 + c_2 = 0$. Setting $t = 0$ in the first derivative of *(2)* and substituting $di/dt = 5$, we obtain $5 = -1{\cdot}67c_1 - 298{\cdot}3c_2$. Solving the two simultaneous equations relating the constants, we find $c_1 = 0{\cdot}0168$ and $c_2 = -0{\cdot}0168$. Then

$$i = 0{\cdot}0168\,e^{-1{\cdot}67t} - 0{\cdot}0168\,e^{-298{\cdot}3t} \text{ amperes} \qquad (3)$$

To find the maximum current we set di/dt equal to zero and solve for t.

$$di/dt = (0{\cdot}0168)(-1{\cdot}67)e^{-1{\cdot}67t} - (0{\cdot}0168)(-298{\cdot}3)e^{-298{\cdot}3t} = 0 \text{ or } t = 0{\cdot}0175 \text{ s}$$

Substituting this value of t into equation (3), we find 0·0161 A.

16.11. A series RLC circuit with $R = 50$ ohms, $L = 0{\cdot}1$ H and $C = 50$ μF has a constant voltage $V = 100$ volts applied at $t = 0$. Find the current transient assuming zero initial charge on the capacitor.

When the switch is closed the following differential equation is obtained.

$$50i + 0{\cdot}1\frac{di}{dt} + \frac{1}{50 \times 10^{-6}}\int i\,dt = 100 \text{ or } (D^2 + 500D + 2 \times 10^5)i = 0 \qquad (1)$$

The roots of the characteristic equation are $D_1 = -250 + j371$ and $D_2 = -250 - j371$; hence the current is

$$i = e^{-250t}(c_1 \cos 371t + c_2 \sin 371t) \text{ amperes} \qquad (2)$$

The current is zero at $t = 0$. Then from (2), $i_0 = 0 = (1)(c_1 \cos 0 + c_2 \sin 0)$ and $c_1 = 0$. Now equation (2) becomes

$$i = e^{-250t}\, c_2 \sin 371t \text{ amperes} \qquad (3)$$

Differentiating (3),

$$di/dt = c_2\{e^{-250t}(371)\cos 371t + e^{-250t}(-250)\sin 371t\} \qquad (4)$$

From (1) at $t = 0$, $0{\cdot}1(di/dt) = 100$ or $di/dt = 1000$. Substituting into (4) at $t = 0$, $di/dt = 1000 = c_2 371 \cos 0$ and $c_2 = 2{\cdot}7$. Thus the required current is $i = e^{-250t}(2{\cdot}7 \sin 371t)$ amperes.

16.12. A series RL circuit with $R = 50$ ohms and $L = 0{\cdot}2$ H has a sinusoidal voltage source $v = 150 \sin (500t + \varphi)$ volts applied at a time when $\varphi = 0$. Find the complete current.

When the switch is closed the differential equation for the given circuit is

$$50i + 0{\cdot}2\frac{di}{dt} = 150 \sin 500t \text{ or } (D + 250)i = 750 \sin 500t \qquad (1)$$

The complementary function is $i_c = c\,e^{-250t}$.

To find the particular solution we use the method of undetermined coefficients and assume a particular current

$$i_p = A \cos 500t + B \sin 500t \qquad (2)$$

Then

$$i_p' = -500A \sin 500t + 500B \cos 500t \qquad (3)$$

Substituting these expressions for i and i' into equation (1), we obtain

$$(-500A \sin 500t + 500B \cos 500t) + 250(A \cos 500t + B \sin 500t) = 750 \sin 500t$$

Equating the coefficients of sin $500t$ and also of cos $500t$, we obtain

$$-500A + 250B = 750 \text{ and } 500B + 250A = 0 \qquad (4)$$

Solving these simultaneous equations, we find $A = -1{\cdot}2$ and $B = 0{\cdot}6$. Then

$$i_p = -1{\cdot}2 \cos 500t + 0{\cdot}6 \sin 500t = 1{\cdot}34 \sin (500t - 63{\cdot}4°) \text{ amperes} \qquad (5)$$

The complete current is

$$i = c\,e^{-250t} + 1{\cdot}34 \sin (500t - 63{\cdot}4°) \text{ amperes} \qquad (6)$$

At $t = 0$, $i = 0 = c(1) + 1{\cdot}34 \sin(-63{\cdot}4°)$ and $c = 1{\cdot}2$. Then

$$i = 1{\cdot}2\,e^{-250t} + 1{\cdot}34 \sin (500t - 63{\cdot}4°) \text{ amperes} \qquad (7)$$

Fig. 16-30 shows i_c, i_p and their sum i. After the transient is over (approx. at $t = 5$ TC) the current is sinusoidal and lags the applied voltage by $\theta = \tan^{-1} \omega L/R = 63{\cdot}4°$.

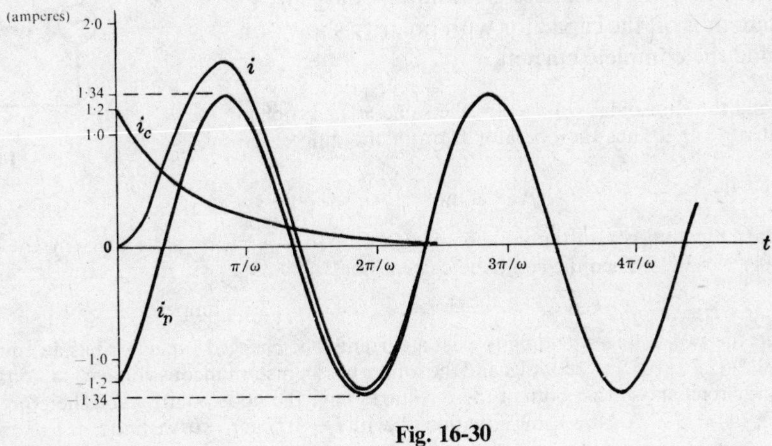

Fig. 16-30

16.13. Referring to the circuit described in Problem 16.12, at what angle φ must the switch be closed in order that the current will go directly into the steady state with no transient?

If $\varphi \neq 0$, we have from equation (*6*) of Problem 16.12,

$$i = c\,e^{-250t} + 1{\cdot}34 \sin(500t + \varphi - 63{\cdot}4°) \text{ amperes}$$

At $t = 0$, $0 = c(1) + 1{\cdot}34 \sin(\varphi - 63{\cdot}4°)$. Now the transient is zero if the constant c is zero; this occurs when $\varphi = (63{\cdot}4° + n\,180°)$, where $n = 0, 1, 2, \ldots$.

16.14. A series *RC* circuit with $R = 100$ ohms and $C = 25$ μF has a sinusoidal voltage source $v = 250 \sin (500t + \varphi)$ volts applied at a time when $\varphi = 0°$. Find the current, assuming there is no initial charge on the capacitor.

When the switch is closed the differential equation for the circuit is

$$100i + \frac{1}{25 \times 10^{-6}} \int i\,dt = 250 \sin 500t \text{ or } (D + 400)i = 1250 \cos 500t \qquad (1)$$

The complementary function is $i_c = c\,e^{-400t}$.

To find the particular current we let the right side of the operator equation be the real part of $1250\,e^{j500t}$ and then assume a particular current

$$i_p = \mathbf{K}\,e^{j500t} \text{ amperes} \qquad (2)$$

Then

$$i_p' = j500\,\mathbf{K}\,e^{j500t} \text{ amperes} \qquad (3)$$

Substituting these values of i and i' into equation (*1*), we obtain

$$j500\,\mathbf{K}\,e^{j500t} + 400(\mathbf{K}\,e^{j500t}) = 1250 e^{j500t} \qquad (4)$$

from which $\mathbf{K} = 1{\cdot}955 \angle -51{\cdot}3°$. This value of **K** is substituted into equation (*2*), but since the driving voltage was the real part of $1250\,e^{j500t}$, the actual current is the real part of (*2*) and $i_p = 1{\cdot}955 \cos(500t - 51{\cdot}3°)$. The complete current is

$$i = c\,e^{-400t} + 1{\cdot}955 \cos(500t - 51{\cdot}3°) \text{ amperes} \qquad (5)$$

At $t = 0$, equation (*1*) is $100i = 250 \sin 0$ or $i = 0$. Now using equation (*5*) with $t = 0$, we find $c = -1{\cdot}22$ and hence

$$i = -1{\cdot}22\,e^{-400t} + 1{\cdot}955 \cos(500t - 51{\cdot}3°) = -1{\cdot}22\,e^{-400t} + 1{\cdot}955 \sin(500t + 38{\cdot}7°) \text{ amperes}$$

16.15. In the *RC* circuit shown in Fig. 16-31, the sinusoidal voltage source $v = 250 \sin(500t + \varphi)$ volts is applied by closing the switch at a time when $\varphi = 45°$. There is an initial charge $q_0 = 5000 \times 10^{-6}$ coulombs on the capacitor with polarity shown on the diagram. Find the complete current.

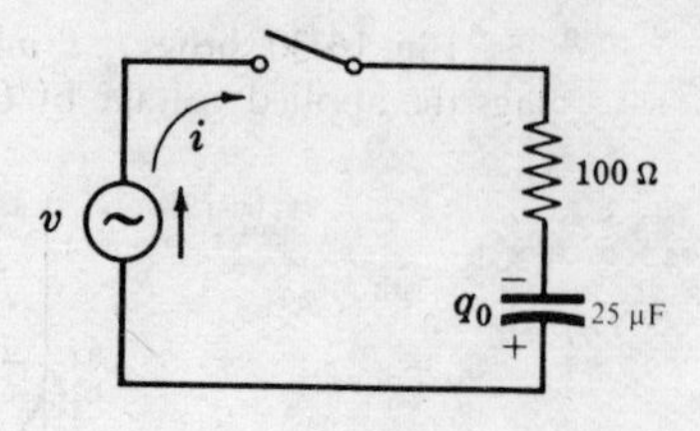

Fig. 16-31

The circuit and the sinusoidal voltage are the same as in Problem 16.14 except that $\varphi = 45°$. Thus the operator form of the differential equation is

$$(D + 400)i = 1250 \cos(500t + 45°) \quad (1)$$

The complementary function is also the same as in Problem 16.14 and the particular current is shifted by 45°, i.e. $i_p = 1{\cdot}955 \sin(500t + 83{\cdot}7°)$. Then the complete current is

$$i = c\,e^{-400t} + 1{\cdot}955 \sin(500t + 83{\cdot}7°) \text{ amperes} \quad (2)$$

At $t = 0$ there are two voltages tending to pass a current. The charged capacitor has an equivalent voltage $V = q_0/C = (5000 \times 10^{-6})/(25 \times 10^{-6}) = 200$ volts and the source has an instantaneous voltage $v = 250 \sin 45° = 176{\cdot}7$ volts. Examination of the circuit shows that both of these voltages have the same polarity and therefore the initial current is $i_0 = (200 + 176{\cdot}7)/100 = 3{\cdot}77$ A. Now using equation (2) with $i = 3{\cdot}77$ at $t = 0$, we find $c = 1{\cdot}83$ and hence the required current is

$$i = 1{\cdot}83\, e^{-400t} + 1{\cdot}955 \sin(500t + 83{\cdot}7°) \text{ amperes}$$

16.16. The series *RLC* circuit shown in Fig. 16-32 has a sinusoidal voltage source $v = 100 \sin(1000t + \varphi)$ volts. If the switch is closed when $\varphi = 90°$, find the current assuming zero initial charge on the capacitor.

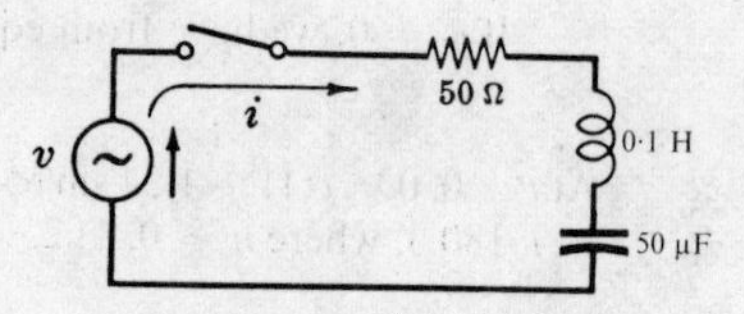

Fig. 16-32

The equation written for the circuit after the switch is closed is

$$50i + 0{\cdot}1\frac{di}{dt} + \frac{1}{50 \times 10^{-6}}\int i\,dt = 100 \sin(1000t + 90°)$$

or

$$(D_2 + 500D + 2 \times 10^5)i = 10^6 \cos(1000t + 90°) \quad (1)$$

The roots of the characteristic equation are $D_1 = -250 + j371$ and $D_2 = -250 - j371$.

The complementary current is $i_c = e^{-250t}(c_1 \cos 371t + c_2 \sin 371t)$ and the particular current, found by the method used in Problem 16.14, is $i_p = 1{\cdot}06 \sin(1000t + 32°)$. Then the complete current is

$$i = e^{-250t}(c_1 \cos 371t + c_2 \sin 371t) + 1{\cdot}06 \sin(1000t + 32°) \text{ amperes} \quad (2)$$

From equation (1) at $t = 0$, $i_0 = 0$ and $di/dt = 1000$. Substituting into (2), we find $c_1 = -0{\cdot}562$. Now differentiating (2), we obtain

$$\frac{di}{dt} = e^{-250t}(-371c_1 \sin 371t + 371c_2 \cos 371t)$$
$$+ (c_1 \cos 371t + c_2 \sin 371t)(-250\, e^{-250t}) + 1{\cdot}06(1000) \cos(1000t + 32°) \quad (3)$$

Substituting $t = 0$, $c_1 = -0{\cdot}562$ and $di/dt = 1000$ into equation (3), we find $c_2 = -0{\cdot}104$. Now equation (2) becomes

$$i = e^{-250t}(-0{\cdot}562 \cos 371t - 0{\cdot}104 \sin 371t) + 1{\cdot}06 \sin(1000t + 32°) \text{ amperes}$$

16.17. A series *RLC* circuit with $R = 100$ ohms, $L = 0{\cdot}1$ H and $C = 50\ \mu$F has a sinusoidal voltage source $v = 100 \sin(1000t + \varphi)$ volts. If the switch is closed when $\varphi = 90°$, find the current assuming no initial charge on the capacitor.

When the switch is closed, the equation for the circuit is

$$100i + 0{\cdot}1\frac{di}{dt} + \frac{1}{50 \times 10^{-6}}\int i\,dt = 100 \sin(1000t + 90°)$$

or

$$(D_2 + 1000D + 2 \times 10^5)i = 10^6 \cos(1000t + 90°) \quad (1)$$

The roots of the characteristic equation are $D_1 = -276{\cdot}5$ and $D_2 = -723{\cdot}5$.

The complementary function is $i c_1 e^{-276{\cdot}5t} + c_2 e^{-723{\cdot}5t}$ and the particular solution, obtained by the method used in Problem 16.14, is $i_p = 0{\cdot}781 \sin(1000t + 51{\cdot}4^\circ)$. Then the complete current is

$$i = c_1 e^{-276{\cdot}5t} + c_2 e^{-723{\cdot}5t} + 0{\cdot}781 \sin(1000t + 51{\cdot}4^\circ) \text{ amperes} \qquad (2)$$

To determine the constants c_1 and c_2, we evaluate i and di/dt at $t = 0$ in equation (1). Substituting the resulting $i_0 = 0$ and $di/dt = 1000$ into equation (2), we obtain

$$i_0 = 0 = c_1(1) + c_2(1) + 0{\cdot}781 \sin 51{\cdot}4^\circ \text{ or } c_1 + c_2 = -0{\cdot}610 \qquad (3)$$

Differentiating (2) and substituting $t = 0$ and $di/dt = 1000$,

$$di/dt = 1000 = -276{\cdot}5c_1 - 723{\cdot}5c_2 + 781 \cos 51{\cdot}4^\circ \text{ or } 276{\cdot}5c_1 + 723{\cdot}5c_2 = -513 \qquad (4)$$

Solving (3) and (4) simultaneously, $c_1 = 0{\cdot}161$ and $c_2 = -0{\cdot}771$. Then

$$i = 0{\cdot}161\, e^{-276{\cdot}5t} - 0{\cdot}771\, e^{-723{\cdot}5t} + 0{\cdot}781 \sin(1000t + 51{\cdot}4^\circ) \text{ amperes}$$

16.18. In the two-mesh network shown in Fig. 16-33 the switch is closed at $t = 0$. Find the transient mesh currents i_1 and i_2 shown in the diagram, and the transient capacitor voltage v_C.

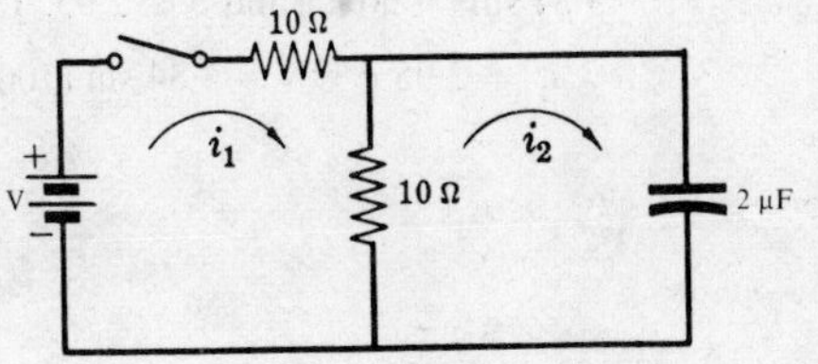

Fig. 16-33

Applying Kirchhoff's voltage law to the two loops, we have

$$20i_1 - 10i_2 = 50 \text{ or } 2Di_1 = Di_2 \qquad (1)$$

$$-10i_1 + 10i_2 + \frac{1}{2 \times 10^{-6}} \int i_2\, dt = 0 \text{ or } -Di_1 + (D + 5 \times 10^4) i_2 = 0 \qquad (2)$$

From equation (1), $Di_1 = \frac{1}{2}Di_2$. Substitute this in (2) and obtain

$$-(\tfrac{1}{2}Di_2) + (D + 5 \times 10^4)i_2 = 0 \text{ or } (D + 10^5)i_2 = 0 \qquad (3)$$

The solution to equation (3) contains a complementary function only, since the equation is homogeneous. Hence

$$i_2 = c\, e^{-10^5 t} \text{ amperes} \qquad (4)$$

Setting $t = 0$ in equation (2), $-10i_1 + 10i_2 = 0$ or $i_1 = i_2$. Then equation (1) at $t = 0$ becomes $20i_1 - 10i_1 = 50$ or $i_1 = i_2 = 5$ amp. Substituting this value of i_2 into (4), we obtain $c = 5$. Thus

$$i_2 = 5\, e^{-10^5 t} \text{ amperes} \qquad (5)$$

Now obtain the transient current i_1 by substituting (5) into equation (1). Thus

$$20i_1 - 10(5\, e^{-10^5 t}) = 50 \quad \text{and} \quad i_1 = 2.5 + 2.5\, e^{-10^5 t} \text{ amperes}$$

The transient voltage across the capacitor, v_C is obtained by the integral of the mesh current i_2:

$$v_C = \frac{1}{C}\int i_2\, dt = \frac{1}{2 \times 10^{-6}} \int 5\, e^{-10^5 t}\, dt = 25(1 - e^{-10^5 t}) \text{ volts}$$

16.19. In the two-mesh network shown in Fig. 16-34 the switch is closed at $t = 0$ and the voltage source is given by $v = 150 \sin 1000t$ volts. Find the mesh currents i_1 and i_2 as given in the diagram.

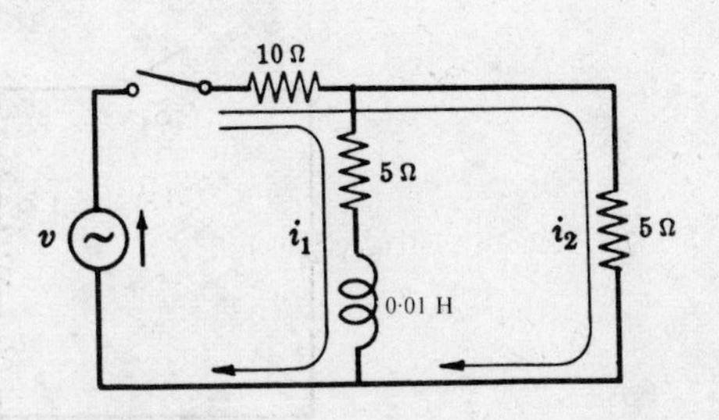

Fig. 16-34

Kirchhoff's voltage law applied to the two indicated loops results in the equations

$$10i_2 + 15i_1 + 0{\cdot}01 \frac{di_1}{dt} = 150 \sin 1000t$$

or

$$(D + 1500)i_1 + 1000i_2 = 15{,}000 \sin 1000t \qquad (1)$$

$$15i_2 + 10i_1 = 150 \sin 1000t \qquad (2)$$

From equation (2) we find that $i_2 = 10 \sin 1000t - \frac{2}{3}i_1$ amperes (3)

Substituting into equation (1), we obtain the differential equation

$$(D + 833)i_1 = 5000 \sin 1000t \qquad (4)$$

The complete solution, found by the method of Problem 16.14, is

$$i_1 = c\, e^{-833t} + 3{\cdot}84 \sin (1000t - 50{\cdot}2°) \text{ amperes} \qquad (5)$$

Now substitute this expression for i_1 into equation (3) and get

$$i_2 = -\tfrac{2}{3}c\, e^{-833t} - 2{\cdot}56 \sin (1000t - 50{\cdot}2) + 10 \sin 1000t$$

$$= -\tfrac{2}{3}c\, e^{-833t} + 8{\cdot}58 \sin (1000t + 13{\cdot}25°) \text{ amperes} \qquad (6)$$

Mesh current i_1 passes through an inductance and must be zero at $t = 0$. Substituting into equation (5), $0 = c(1) + 3{\cdot}84 \sin (-50{\cdot}2°)$ and $c = 2{\cdot}95$. Then the two mesh current equations are

$$i_1 = 2{\cdot}95\, e^{-833t} + 3{\cdot}84 \sin (1000t - 50{\cdot}2°) \text{ amperes and } i_2 = -1{\cdot}97\, e^{-833t} + 8{\cdot}58 \sin (1000t + 13{\cdot}25°) \text{ amperes}$$

Supplementary Problems

16.20. In the series RL circuit shown in Fig. 16-35 switch S_1 is closed at $t = 0$. After 4 ms switch S_2 is opened. Find the current in the intervals $0 < t < t'$ and $t' < t$, where $t' = 4$ ms.
Ans. $i = 2(1 - e^{-500t})$ amperes, $i = 1{\cdot}06\, e^{-1500(t-t')} + 0{\cdot}667$ amperes

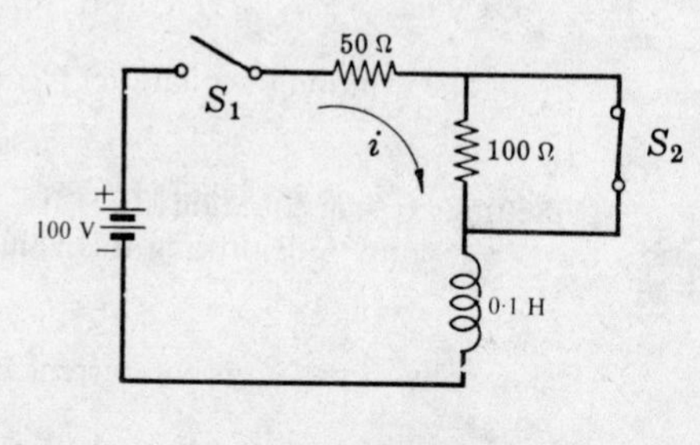

Fig. 16-35

16.21. A constant voltage is applied to a series RL circuit by closing a switch. The voltage across L is 25 volts at $t = 0$ and drops to 5 volts at $t = 25$ ms. If $L = 2$ H, what must be the value of R?
Ans. 128·8 ohms

16.22. In the circuit shown in Fig. 16-36 switch S_1 is closed at $t = 0$ and switch S_2 is opened at $t = 0{\cdot}2$ sec. Find the transient current expressions for the two intervals. *Ans.* $i = 10(1 - e^{-10t})$ amperes, $i = 6{\cdot}97\, e^{-60(t-t')} + 1{\cdot}67$ amperes

16.23. In the circuit shown in Fig. 16-37 the switch is closed on position 1 at $t = 0$, and then moved to position 2 after 1 millisecond. Find the time at which the current is zero and reversing its direction. *Ans.* 1·261 ms.

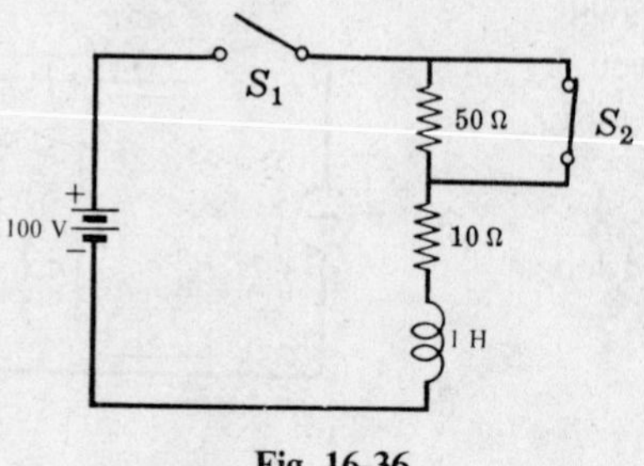

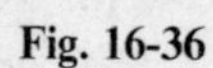
Fig. 16-36

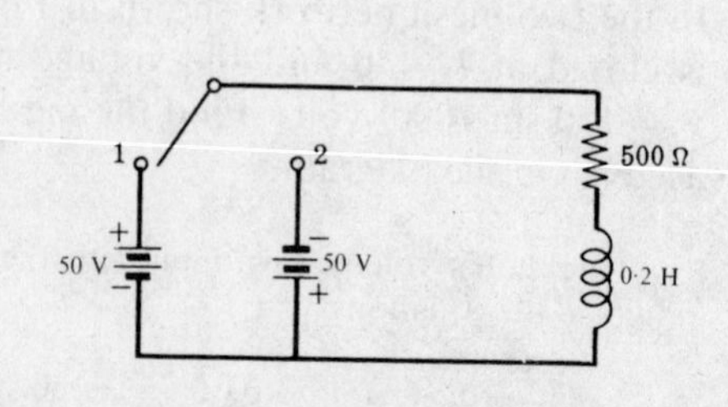

Fig. 16-37

16.24. In the circuit shown in Fig. 16-38 the switch has been closed in position 1 for sufficient time to establish the steady state current. When the switch is then moved to position 2 a transient current exists in the two 50 ohm resistor for a short time. Find the energy dissipated in the resistors during this transient. *Ans.* 8 joules

16.25. The *RC* circuit shown in Fig. 16-39 has an initial charge on the capacitor of $q_0 = 800 \times 10^{-6}$ coulombs with the polarity shown in the diagram. Find both the current and charge transients which result when the switch is closed.
Ans. $i = -10e^{-2.5\times 10^4 t}$ amperes, $q = 400(1 + e^{-2.5\times 10^4 t}) \times 10^{-6}$ coulombs

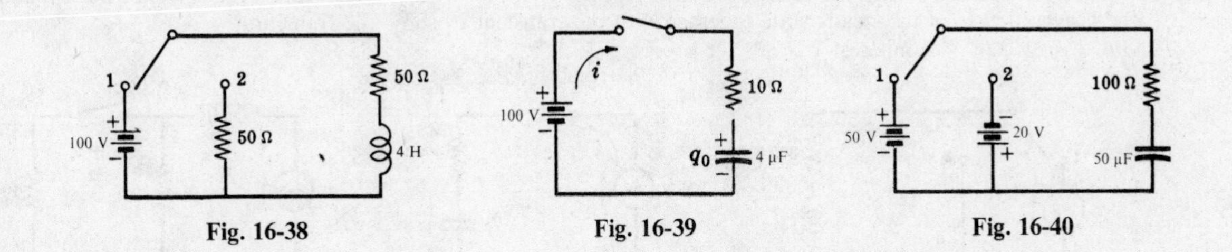

Fig. 16-38 Fig. 16-39 Fig. 16-40

16.26. A capacitor of 2 μF with an initial charge $q_0 = 100 \times 10^{-6}$ coulombs is connected across the terminals of a 100 ohm resistor at $t = 0$. Calculate the time in which the transient voltage across the resistor drops from 40 to 10 volts.
Ans. 277·4 μs

16.27. In the circuit shown in Fig. 16-40 the switch is closed on position 1 at $t = 0$ and then moved to position 2 after 1 TC. Find the transient current expressions for both intervals $0 < t < t'$ and $t' < t$.
Ans. $i = 0{\cdot}5\, e^{-200t}$ amperes, $i = -0{\cdot}516 e^{-200(t-t')}$ amperes

16.28. Referring to Problem 16.27, solve the differential equation on the charge basis. From the transient charge functions obtain the current expressions and compare the results.

16.29. In the circuit shown in Fig. 16-41 the switch is in position 1 for sufficient time to establish the steady state and then moved to position 2. A transient current exists when the switch is moved to position 2 during which energy is dissipated in the two resistors. Find this energy and compare it with that which was stored in the capacitor before the switch was moved. *Ans.* 0·20 joules

16.30. In the circuit shown in Fig. 16-42 capacitor C_1 has an initial charge $q_0 = 300 \times 10^{-6}$ coulombs. If the switch is closed at $t = 0$, find the current transient, charge and final voltage of capacitor C_1.
Ans. $i = 2{\cdot}5\, e^{-2{\cdot}5 \times 10^4 t}$ amperes, $q = 200(1 + 0{\cdot}5\, e^{-2{\cdot}5 \times 10^4 t}) \times 10^{-6}$ coulombs, 33·3 V

16.31. Referring to Problem 16.30, find the transient voltages v_C, v_C and v_R. Show that their sum is zero.
Ans. $v_{C_1} = 33{\cdot}3 + 16{\cdot}7\, e^{-2{\cdot}5 \times 10^4 t}$ volts, $v_{C_2} = 33{\cdot}3(1 - e^{-2{\cdot}5 \times 10^4 t})$ volts, $v_R = -50\, e^{-2{\cdot}5 \times 10^4 t}$ volts

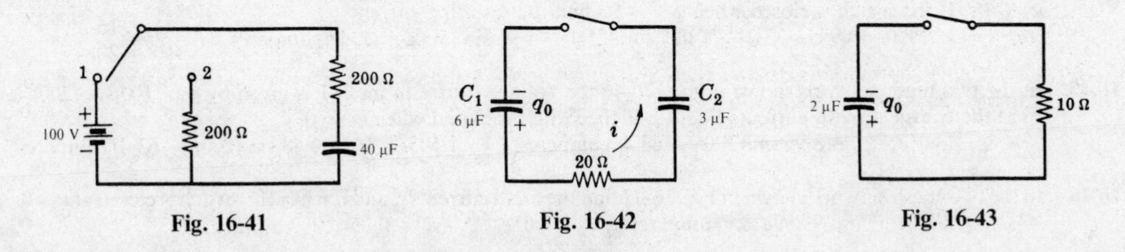

Fig. 16-41 Fig. 16-42 Fig. 16-43

16.32. In the series *RC* circuit shown in Fig. 16-43 the capacitor has an initial charge q_0 and the switch is closed at $t = 0$. Find q_0 if the transient power in the resistor is known to be $p_R = 360e^{-10^5 t}$ watts. *Ans.* 120×10^{-6} coulombs

16.33. A series *RLC* circuit with $R = 200$ ohms, $L = 0{\cdot}1$ H and $C = 100$ μF has a constant voltage $V = 200$ volts applied at $t = 0$. Find the current assuming the capacitor has no initial charge. *Ans.* $i = 1{\cdot}055e^{-52t} - 1{\cdot}055e^{-1948t}$ amperes

16.34. A series *RLC* circuit with $R = 200$ ohms and $L = 0{\cdot}1$ H is to be made critically damped by the selection of the capacitance. Find the required value of *C*. *Ans.* 10μF

16.35. Find the natural frequency of a series *RLC* circuit in which $R = 200$ ohms, $L = 0{\cdot}1$ H and $C = 5$ μF.
Ans. 1000 rad/s

16.36. A series RLC circuit with $R = 5$ ohms, $L = 0{\cdot}1$ H and $C = 500$ μF has a constant voltage $V = 10$ volts applied at $t = 0$. Find the resulting current transient. *Ans.* $i = 0{\cdot}72\, e^{-25t} \sin 139t$ amperes

16.37. An RL series circuit with $R = 300$ ohms and $L = 1{\cdot}0$ H has a sinusoidal applied voltage $v = 100 \cos (100t + \varphi)$ volts. If the switch is closed when $\varphi = 45°$, obtain the resulting current transient.
Ans. $i = -0{\cdot}282e^{-300t} + 0{\cdot}316 \cos (100t + 26{\cdot}6°)$ amperes

16.38. The RL circuit shown in Fig. 16-44 is operating in the sinusoidal steady state with the switch in position 1. The switch is moved to position 2 when the voltage source is $v = 100 \cos (100t + 45°)$ volts. Obtain the current transient and plot the last half-cycle of the steady state together with the transient to show the transition.
Ans. $i = 0{\cdot}282e^{-300t}$ amperes

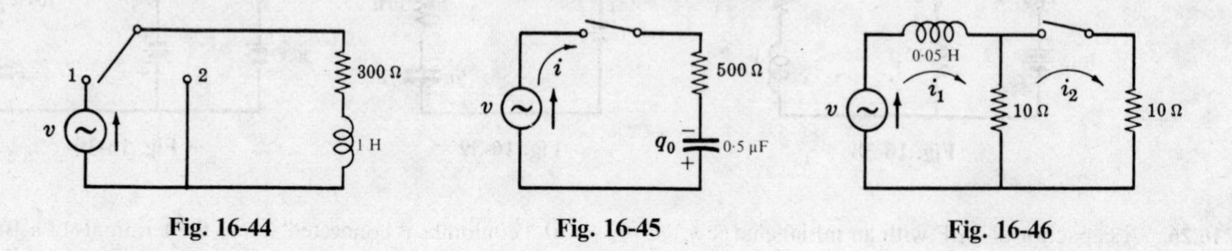

Fig. 16-44 **Fig. 16-45** **Fig. 16-46**

16.39. In the RC circuit shown in Fig. 16-45 the capacitor has an initial charge $q_0 = 25 \times 10^{-6}$ coulombs with polarity shown in the diagram. The sinusoidal voltage $v = 100 \sin (1000t + \varphi)$ volts is applied to the circuit at time corresponding to $\varphi = 30°$. Obtain the current transient. *Ans.* $i = 0{\cdot}1535\, e^{-4 \times 10^3 t} + 0{\cdot}0484 \sin (1000t + 106°)$ amperes

16.40. Referring to Problem 16.39, what initial charge on the capacitor will cause the current to go directly into the steady state without a transient when the switch is closed? *Ans.* $13{\cdot}37 \times 10^{-6}$ coulombs, + on top plate

16.41. Show that a series RLC circuit with a source $v = V_{max} \sin (\omega t + \varphi)$ has a particular solution to its differential equation given by

$$i_p = \frac{V_{max}}{\sqrt{R^2 + (1/\omega C - \omega L)^2}} \sin\left(\omega t + \varphi + \tan^{-1}\frac{(1/\omega C - \omega L)}{R}\right)$$

16.42. A series RLC circuit with $R = 5$ ohms, $L = 0{\cdot}1$ H and $C = 500$ μF has a sinusoidal voltage $v = 100 \sin (250t + \varphi)$ volts applied at a time when $\varphi = 0°$. Find the resulting current.
Ans. $i = e^{-25t}\,(5{\cdot}42 \cos 139t - 1{\cdot}89 \sin 139\,t) + 5{\cdot}65 \sin (250t - 73{\cdot}6°)$ amperes

16.43. A series RLC circuit with $R = 200$ ohms, $L = 0{\cdot}5$ H and $C = 100$ μF has a sinusoidal voltage source $v = 300 \sin (500t + \varphi)$ volts. If the switch is closed when $\varphi = 30°$, find the resulting current.
Ans. $i = 0{\cdot}517\, e^{-341{\cdot}4t} - 0{\cdot}197\, e^{-58{\cdot}6t} + 0{\cdot}983 \sin (500t - 19°)$ amperes

16.44. A series RLC circuit with $R = 50$ ohms, $L = 0{\cdot}1$ H and $C = 50$ μF has a sinusoidal voltage source $v = 100 \sin (500t + \varphi)$ volts. If the switch is closed when $\varphi = 45°$, find the resulting current.
Ans. $i = e^{-250t}\,(-1{\cdot}09 \cos 371t - 1{\cdot}025 \sin 371t) + 1{\cdot}96 \sin (500t + 33{\cdot}7°)$ amperes

16.45. In the two-mesh network shown in Fig. 16-46 the voltage source in mesh 1 is given by $v = 100 \sin (200t + \varphi)$ volts. Find the transient mesh currents i_1 and i_2 if the switch is closed when $\varphi = 0$.
Ans. $i_1 = 3{\cdot}01e^{-100t} + 8{\cdot}96 \sin (200t - 63{\cdot}4°)$ amperes, $i_2 = 1{\cdot}505e^{-100t} + 4{\cdot}48 \sin (200t - 63{\cdot}4°)$ amperes

16.46. In the two-mesh network shown in Fig. 16-47 find the mesh currents i_1 and i_2 when the switch is closed at $t = 0$.
Ans. $i_1 = 0{\cdot}101e^{-100t} + 9{\cdot}899e^{-9950t}$ amperes, $i_2 = -5{\cdot}05e^{-100t} + 5 + 0{\cdot}05e^{-9950t}$ amperes

16.47. In the two-mesh network shown in Fig. 16-48 the switch is closed at $t = 0$. Find the resulting currents i_1 and i_2.
Ans. $i_1 = 1{\cdot}67e^{-6{\cdot}67t} + 5$ amperes, $i_2 = -0{\cdot}555e^{-6{\cdot}67t} + 5$ amperes

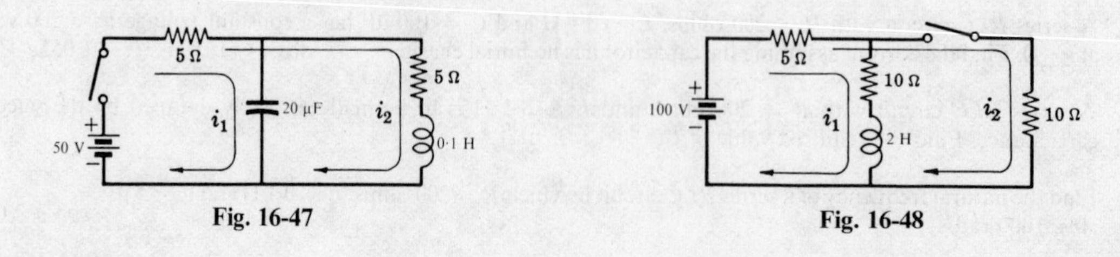

Fig. 16-47 **Fig. 16-48**

CHAPTER 17

Transients by the Laplace Transform Method

INTRODUCTION

In Chapter 16 we analyzed the transient currents in circuits containing energy storage elements. The application of Kirchhoff's laws to such circuits resulted in one or more differential equations in the time domain, depending on the configuration of the circuit. These equations were then solved by classical methods. However, in many cases these methods are not convenient and in this chapter we introduce another method, called the Laplace transform method, which provides more direct solutions to the differential equations. Moreover, some irregular forcing functions cannot be easily solved by classical methods whereas the Laplace method provides a solution to such problems.

This chapter shows only basic applications of the Laplace transform method. The formal mathematical derivations and more complex applications are left to texts devoted primarily to transient analysis.

THE LAPLACE TRANSFORM

If $f(t)$ is a function of t, defined for all $t > 0$, the Laplace transform of $f(t)$, denoted by the symbol $\mathcal{L}[f(t)]$, is defined by

$$\mathcal{L}[f(t)] \quad = \quad \mathbf{F(s)} \quad = \quad \int_0^\infty f(t)\, e^{-st}\, dt \qquad (1)$$

where the parameter $\mathbf{s}$ may be real or complex. In circuit applications we assume $\mathbf{s} = \delta + j\omega$. The operation $\mathcal{L}[f(t)]$ transforms a function $f(t)$ in the *time domain* into a function $\mathbf{F(s)}$ in the *complex frequency domain* or simply the **s** *domain*. The two functions $f(t)$ and $\mathbf{F(s)}$ thus form a transform pair. Extensive tables list such pairs. The transforms shown in Table 17-1, Page 267, are sufficient for the purposes of this chapter.

Sufficient conditions for the existence of the Laplace transform are that the function $f(t)$ must be (*a*) piecewise continuous and (*b*) of exponential order. The function $f(t)$ is of exponential order if $|f(t)| < Ae^{\alpha t}$ for all $t > t_0$, where both A and t_0 are positive constants. When these conditions are met the direct transformation integral is convergent for all $\delta > \alpha$ and $\mathbf{F(s)}$ exists. All functions in circuit analysis satisfy requirements (*a*) and (*b*).

Example 1.

The function shown in Fig. 17-1 is called a *step function* and is defined by $f(t) = A,\ t > 0$. Find the corresponding Laplace transform.

Applying equation (*1*) to the function $f(t) = A$, we write

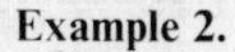

$$\mathcal{L}[A] \quad = \quad \int_0^\infty Ae^{-st}\, dt \quad = \quad \left[-\frac{A}{\mathbf{s}}\, e^{-\mathbf{s}t}\right]_0^\infty \quad = \quad \frac{A}{\mathbf{s}}$$

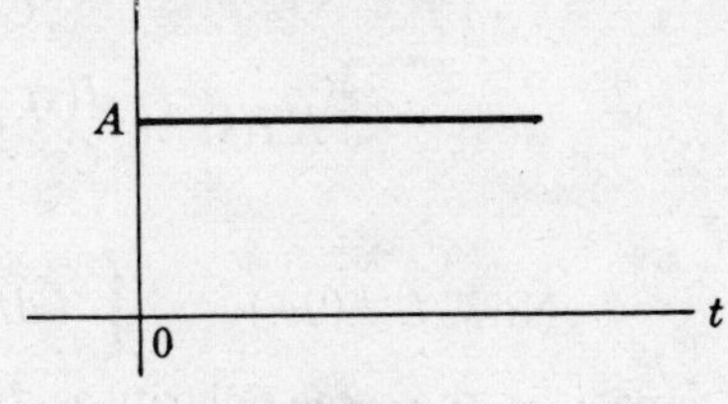

Fig. 17-1

Example 2.

Obtain the Laplace transform of $f(t) = e^{-at}$ where a is a constant.

$$\mathcal{L}[e^{-at}] \quad = \quad \int_0^\infty e^{-at}\, e^{-st}\, dt \quad = \quad \int_0^\infty e^{-(a+\mathbf{s})t}\, dt \quad = \quad \left[-\frac{1}{(a+\mathbf{s})}\, e^{-(a+\mathbf{s})t}\right]_0^\infty \quad = \quad \frac{1}{\mathbf{s}+a}$$

Example 3. Find the Laplace transform of $f(t) = \sin \omega t$.

$$\mathcal{L}[\sin \omega t] = \int_0^\infty \sin \omega t\, e^{-st}\, dt = \left[\frac{-s(\sin \omega t)e^{-st} - e^{-st}\omega \cos \omega t}{s^2 + \omega^2}\right]_0^\infty = \frac{\omega}{s^2 + \omega^2}$$

Example 4. Find the Laplace transform of the derivative df/dt.

$$\mathcal{L}[df/dt] = \int_0^\infty (df/dt)e^{-st}\, dt$$

Integrate by parts, using $\int u\, dv = uv - \int v\, du$ where $u = e^{-st}$, $dv = df$, $v = f$. Now

$$\mathcal{L}[df/dt] = \left[e^{-st} f\right]_0^\infty - \int_0^\infty f(-se^{-st})\, dt = -f(0+) + s\int_0^\infty fe^{-st}\, dt = -f(0+) + s\mathbf{F}(s)$$

where $f(0+)$ is the value of the function as zero is approached from the right, i.e. the value of the function at $t = (0+)$.

Example 5. Find the Laplace transform of the integral $\int f(t)\, dt$.

$$\mathcal{L}\left[\int f(t)\, dt\right] = \int_0^\infty \int f(t)\, dt\; e^{-st}\, dt$$

Integrate by parts letting $u = \int f(t)\, dt$ and $dv = e^{-st}\, dt$. Then

$$\mathcal{L}\left[\int f(t)\, dt\right] = \left[\int f(t)\, dt\left(-\frac{1}{s}e^{-st}\right)\right]_0^\infty - \int_0^\infty \left(-\frac{1}{s}e^{-st}\right) f(t)\, dt$$

$$= \frac{1}{s}\int f(t)\, dt\Big|_{0+} + \frac{1}{s}\mathbf{F}(s)$$

where $\int f(t)\, dt\Big|_{0+}$ is the value of the integral at $0+$, also written $f^{-1}(0+)$. Then the Laplace transform of an integral is

$$\mathcal{L}\left[\int f(t)\, dt\right] = \frac{1}{s}\mathbf{F}(s) + \frac{1}{s}f^{-1}(0+)$$

The transform pairs obtained in these examples appear in Table 17-1, Page 267.

APPLICATIONS TO CIRCUIT ANALYSIS

The series RC circuit of Fig. 17-2 has an initial charge q_0 on the capacitor with the polarity shown in the diagram. When the switch is closed the constant voltage source V is applied to the circuit and the differential equation for the circuit is

$$Ri + \frac{1}{C}\int i\, dt = V \qquad (2)$$

Using $I(s)$ for the s domain current, we take the Laplace transform of each term in equation (*2*).

$$\mathcal{L}[Ri] + \mathcal{L}\left[\frac{1}{C}\int i\, dt\right] = \mathcal{L}[V] \qquad (3)$$

$$R\,I(s) + \frac{I(s)}{Cs} + \frac{f^{-1}(0+)}{Cs} = \frac{V}{s} \qquad (4)$$

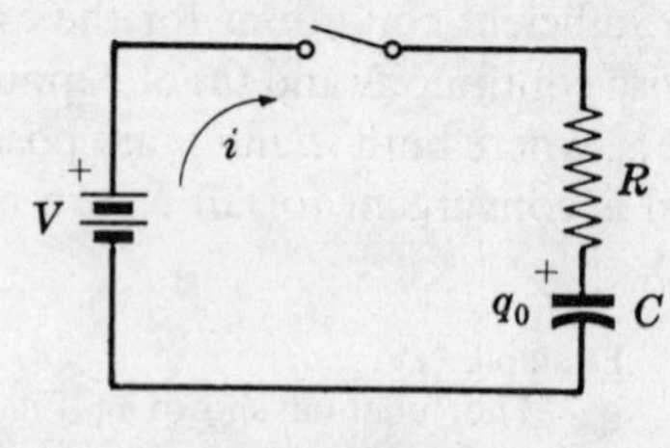

Fig. 17-2

Now $f^{-1}(0+) = \int i\, dt\Big|_{0+} = q(0+)$. The initial charge q_0 is positive on the upper plate of the capacitor, the same polarity as the charge deposited by the source V. Thus the sign is positive. Substituting q_0 in equation (*4*), we obtain

$$R\,I(s) + \frac{I(s)}{Cs} + \frac{q_0}{Cs} = \frac{V}{s} \qquad (5)$$

Rearranging terms and factoring $I(s)$,

Table 17-1

LAPLACE TRANSFORMS

	$f(t)$	$\mathbf{F}(\mathbf{s})$
1.	$A \quad t \geqq 0$	$\dfrac{A}{\mathrm{s}}$
2.	$At \quad t \geqq 0$	$\dfrac{A}{\mathrm{s}^2}$
3.	e^{-at}	$\dfrac{1}{\mathrm{s}+a}$
4.	te^{-at}	$\dfrac{1}{(\mathrm{s}+a)^2}$
5.	$\sin \omega t$	$\dfrac{\omega}{\mathrm{s}^2+\omega^2}$
6.	$\cos \omega t$	$\dfrac{\mathrm{s}}{\mathrm{s}^2+\omega^2}$
7.	$\sin(\omega t+\theta)$	$\dfrac{\mathrm{s}\sin\theta+\omega\cos\theta}{\mathrm{s}^2+\omega^2}$
8.	$\cos(\omega t+\theta)$	$\dfrac{\mathrm{s}\cos\theta-\omega\sin\theta}{\mathrm{s}^2+\omega^2}$
9.	$e^{-at}\sin\omega t$	$\dfrac{\omega}{(\mathrm{s}+a)^2+\omega^2}$
10.	$e^{-at}\cos\omega t$	$\dfrac{(\mathrm{s}+a)}{(\mathrm{s}+a)^2+\omega^2}$
11.	$\sinh \omega t$	$\dfrac{\omega}{\mathrm{s}^2-\omega^2}$
12.	$\cosh \omega t$	$\dfrac{\mathrm{s}}{\mathrm{s}^2-\omega^2}$
13.	df/dt	$\mathrm{s}\,\mathbf{F}(\mathrm{s}) - f(0+)$
14.	$\int f(t)\,dt$	$\dfrac{\mathbf{F}(\mathrm{s})}{\mathrm{s}} + \dfrac{f^{-1}(0+)}{\mathrm{s}}$
15.	$f(t-t_1)$	$e^{-t_1 \mathrm{s}}\,\mathbf{F}(\mathrm{s})$
16.	$f_1(t)+f_2(t)$	$\mathbf{F}_1(\mathrm{s})+\mathbf{F}_2(\mathrm{s})$

$$I(\mathbf{s})\left(R+\frac{1}{C\mathbf{s}}\right) = \frac{V}{\mathbf{s}} - \frac{q_0}{C\mathbf{s}} \tag{6}$$

and

$$I(\mathbf{s}) = \frac{1}{\mathbf{s}}(V - q_0/C)\frac{1}{(R + 1/\mathbf{s}C)} = \frac{V - q_0/C}{R}\,\frac{1}{(\mathbf{s} + 1/RC)} \tag{7}$$

Equation (7) in the $\mathbf{s}$ domain has a corresponding time domain equation i. The operation whereby $\mathbf{F(s)}$ is transformed into $f(t)$ is called the inverse Laplace transform, denoted by $\mathcal{L}^{-1}[\mathbf{F(s)}] = f(t)$. Referring to Table 17-1, we see that $\mathbf{F(s)}$ of transform pair 3 is equivalent to the term $1/(\mathbf{s} + 1/RC)$ of equation (7). Then from the definition of the inverse Laplace transform and from the table, we have

$$\mathcal{L}^{-1}[I(\mathbf{s})] = i = \left(\frac{V - q_0/C}{R}\right)\mathcal{L}^{-1}\left[\frac{1}{\mathbf{s} + 1/RC}\right] = \frac{V - q_0/C}{R}\,e^{-t/RC} \tag{8}$$

Now equation (8) is the current transient in the time domain which results when the switch is closed on the RC circuit containing an initial charge q_0 on the capacitor. The initial conditions were inserted in equation (5) in the $\mathbf{s}$ domain; consequently, upon taking the inverse transform the resulting equation already contains the constants.

Note that by performing algebraic manipulations in (6) and (7), the function $I(\mathbf{s})$ was reduced to a form found in the table, thus enabling us to obtain the inverse Laplace transform.

The time function is shown in Fig. 17-3 with an initial current $(V - q_0/C)/R$. If $q_0/C = V$ there is no transient since the initial charge on the capacitor has an equivalent voltage equal to the applied voltage V. If q_0 is of the opposite polarity the sign on q_0/C changes, resulting in a comparatively large initial current.

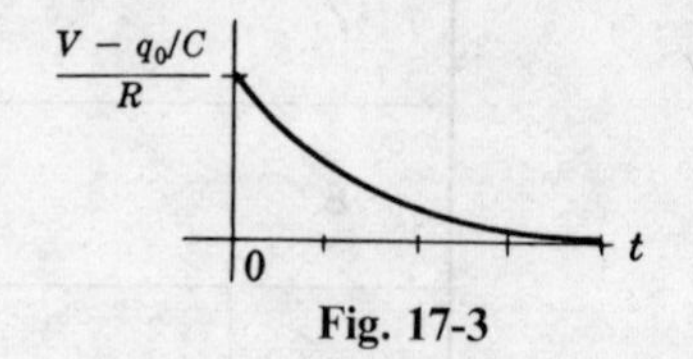

Fig. 17-3

The RL circuit shown in Fig. 17-4 has a constant voltage source V applied when the switch is closed. Application of Kirchhoff's law after closing results in the following equation.

$$Ri + L\frac{di}{dt} = V \tag{9}$$

Now apply the direct Laplace transform to each term and obtain

$$\mathcal{L}[Ri] + \mathcal{L}\left[L\frac{di}{dt}\right] = \mathcal{L}[V] \tag{10}$$

$$R\,I(\mathbf{s}) + \mathbf{s}L\,I(\mathbf{s}) - L\,i(0+) = V/\mathbf{s} \tag{11}$$

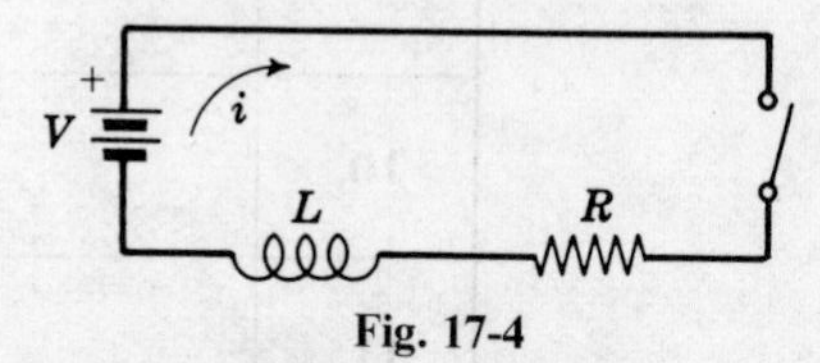

Fig. 17-4

The initial current $i(0+)$ in a series RL circuit which had zero current prior to the closing of the switch is also zero at $t = 0+$. Substituting $i(0+) = 0$ in equation (11), we obtain

$$I(\mathbf{s})(R + \mathbf{s}L) = V/\mathbf{s} \tag{12}$$

and

$$I(\mathbf{s}) = \frac{V}{\mathbf{s}}\,\frac{1}{(R + \mathbf{s}L)} = \frac{V}{L}\left(\frac{1}{\mathbf{s}}\right)\frac{1}{(\mathbf{s} + R/L)} \tag{13}$$

Now the function in equation (13) does not appear in Table 17-1; but if it can be changed to the form $A/\mathbf{s} + B/(\mathbf{s} + R/L)$, transform pairs 1 and 3 can be used on the two parts, and pair 16 indicates that the total time function is the sum of the two time functions, i.e. $\mathcal{L}^{-1}[\mathbf{F}_1(\mathbf{s}) + \mathbf{F}_2(\mathbf{s})] = f_1(t) + f_2(t)$. To obtain the desired sum we set the right side of (13), exclusive of the constant V/L, equal to the sum of the fractions as follows.

$$\frac{1}{\mathbf{s}(\mathbf{s} + R/L)} = \frac{A}{\mathbf{s}} + \frac{B}{(\mathbf{s} + R/L)} = \frac{A(\mathbf{s} + R/L) + B\mathbf{s}}{\mathbf{s}(\mathbf{s} + R/L)} \tag{14}$$

Now from the numerators we obtain the following equation in $\mathbf{s}$:

$$1 = (A + B)\mathbf{s} + AR/L \tag{15}$$

Equating the coefficients of like powers of s, we obtain

$$A + B = 0, \quad A = L/R, \quad B = -L/R \tag{16}$$

Using the indicated partial fractions with A and B as determined above, equation (*13*) becomes

$$I(\mathbf{s}) = \frac{V}{L}\left(\frac{L/R}{\mathbf{s}} + \frac{-L/R}{\mathbf{s}+R/L}\right) = \frac{V}{R}\left(\frac{1}{\mathbf{s}} - \frac{1}{\mathbf{s}+R/L}\right) \tag{17}$$

Applying transforms 1 and 3 of Table 17-1 gives the expression of the inverse transform of the current. Hence

$$\mathcal{L}^{-1}[I(\mathbf{s})] = i = \frac{V}{R}\left\{\mathcal{L}^{-1}\left[\frac{1}{\mathbf{s}}\right] - \mathcal{L}^{-1}\left[\frac{1}{\mathbf{s}+R/L}\right]\right\} \tag{18}$$

and

$$i = \frac{V}{R}(1 - e^{-(R/L)t}) \tag{19}$$

Equation (*19*) is the familiar exponential rise with a steady state current value of V/R.

EXPANSION METHODS

In circuit analysis the expansion of quotients into the sum of several fractions is frequently necessary in obtaining inverse Laplace transforms since the **s** domain current is often the ratio of two polynomials in **s**,

$$I(\mathbf{s}) = P(\mathbf{s})/Q(\mathbf{s}) \tag{20}$$

where $Q(\mathbf{s})$ is of higher degree than $P(\mathbf{s})$. An example of a quotient expansion was shown in equation (*14*).

We now examine the application of the partial fraction expansion method to different cases that occur in the expansion of quotients of polynomials.Another method, called the Heaviside expansion formula, is also introduced here. Its application results in a different approach to the evaluation of the inverse Laplace transform of quotients of polynomials.

1. Partial Fraction Expansion Method.

Equation (*20*) can be written as the sum of fractions each having as denominator one of the factors of $Q(\mathbf{s})$ and as numerator a constant. In expanding the quotient $P(\mathbf{s})/Q(\mathbf{s})$ we must consider the roots of $Q(\mathbf{s})$. They may be real or complex and this gives rise to three cases.

Case 1. The roots of $Q(\mathbf{s})$ are real and unequal.

Consider the following expression for the **s** domain current.

$$I(\mathbf{s}) = \frac{P(\mathbf{s})}{Q(\mathbf{s})} = \frac{\mathbf{s}-1}{\mathbf{s}^2+3\mathbf{s}+2} \tag{21}$$

Factoring $Q(\mathbf{s})$, equation (*21*) can be written as

$$I(\mathbf{s}) = \frac{\mathbf{s}-1}{(\mathbf{s}+2)(\mathbf{s}+1)} = \frac{A}{\mathbf{s}+2} + \frac{B}{\mathbf{s}+1} \tag{22}$$

With $\mathbf{s} = -2$ and $\mathbf{s} = -1$ the expression becomes infinite and *simple poles* are said to exist at these values of **s**. The coefficient of a simple pole $\mathbf{s} = \mathbf{s}_0$ is given by $I(\mathbf{s})(\mathbf{s} - \mathbf{s}_0)\big|_{\mathbf{s}=\mathbf{s}_0}$. Thus to determine the coefficient A, multiply both sides of (*22*) by $(\mathbf{s}+2)$:

$$\frac{\mathbf{s}-1}{(\mathbf{s}+2)(\mathbf{s}+1)}(\mathbf{s}+2) = A + \frac{B}{(\mathbf{s}+1)}(\mathbf{s}+2) \tag{23}$$

Substituting $\mathbf{s} = -2$,

$$A = \left.\frac{\mathbf{s}-1}{\mathbf{s}+1}\right|_{\mathbf{s}=-2} = 3$$

Similarly, $$B = \left.\frac{s-1}{s+2}\right|_{s=-1} = -2$$

Substituting these values in *(22)*, the **s** domain current is

$$I(s) = \frac{3}{s+2} + \frac{-2}{s+1} \qquad (24)$$

The inverse Laplace transform of $I(s)$, obtained from Table 17-1, is $i = 3e^{-2t} - 2e^{-t}$.

Another method. Multiply both sides of *(22)* by $(s+2)(s+1)$:

$$s - 1 = A(s+1) + B(s+2) = (A+B)s + A + 2B$$

Now equating coefficients of like powers of **s**, $A + B = 1$ and $A + 2B = -1$. Then $A = 3$ and $B = -2$, the same values obtained above. This alternate method always leads to simultaneous equations which must then be solved to yield the desired coefficients, whereas the first method results in simple independent equations for each coefficient.

Case 2. The roots of $Q(s)$ are real and equal.

Consider the following expression for the **s** domain current.

$$I(s) = \frac{P(s)}{Q(s)} = \frac{1}{s(s^2+6s+9)} = \frac{1}{s(s+3)^2} \qquad (25)$$

Then $$\frac{1}{s(s+3)^2} = \frac{A}{s} + \frac{B}{s+3} + \frac{C}{(s+3)^2} \qquad (26)$$

Multiplying both sides of *(26)* by **s** and setting $s = 0$,

$$A = \left.\frac{1}{(s+3)^2}\right|_{s=0} = \frac{1}{9}$$

For the case of repeated roots the coefficient of the quadratic term is given by $I(s)\,(s - s_0)^2\,|_{s=s_0}$. Thus

$$C = \left.\frac{1}{s}\right|_{s=-3} = -\frac{1}{3}$$

Then the coefficient of the associated linear term is given by $\left.\frac{d}{ds}\,[I(s)\,(s-s_0)^2]\right|_{s=s_0}$. Thus

$$B = \left.\frac{d}{ds}\left(\frac{1}{s}\right)\right|_{s=-3} = \left.-\frac{1}{s^2}\right|_{s=-3} = -\frac{1}{9}$$

Substituting these values into equation *(26)*, the **s** domain current is

$$I(s) = \frac{\frac{1}{9}}{s} - \frac{\frac{1}{9}}{s+3} - \frac{\frac{1}{3}}{(s+3)^2} \qquad (27)$$

and the inverse Laplace transform is $i = \frac{1}{9} - \frac{1}{9}e^{-3t} - \frac{1}{3}te^{-3t}$.

Another method. Multiplying both sides of *(26)* by $s(s+3)^2$, we obtain

$$1 = A(s+3)^2 + Bs(s+3) + Cs = (A+B)s^2 + (6A+3B+C)s + 9A$$

Equating coefficients of like powers of **s**, $A + B = 0$, $6A + 3B + C = 0$ and $9A = 1$; then $A = \frac{1}{9}$, $B = -\frac{1}{9}$ and $C = -\frac{1}{3}$, the same results obtained above.

Case 3. The roots of $Q(s)$ are complex.

Consider the following expression for the **s** domain current.

$$I(s) = \frac{P(s)}{Q(s)} = \frac{1}{s^2+4s+5} = \frac{1}{(s+2+j)(s+2-j)} \qquad (28)$$

Since $Q(s)$ has complete conjugate roots, the constants in the numerators of the partial fractions are also complex conjugate. Thus

$$\frac{1}{(s+2+j)(s+2-j)} = \frac{A}{s+2+j} + \frac{A^*}{s+2-j} \tag{29}$$

Multiplying both sides of (*29*) by $(s + 2 + j)$ and setting $s = -2 - j$, we obtain

$$A = \left.\frac{1}{s+2-j}\right|_{s=-2-j} = j\tfrac{1}{2} \quad \text{and} \quad A^* = -j\tfrac{1}{2}$$

Substituting these values into equation (*29*), the **s** domain current is

$$I(s) = \frac{j\frac{1}{2}}{s+2+j} + \frac{-j\frac{1}{2}}{s+2-j} \tag{30}$$

The inverse Laplace transform is $i = e^{-2t} \sin t$.

Another method. Multiplying both sides of (*29*) by $(s + 2 + j)(s + 2 - j)$, we obtain

$$1 = A(s+2-j) + A^*(s+2+j)$$

Equating coefficients of like powers of **s**, $A + A^* = 0$ and $A(2 - j) + A^*(2 + j) = 1$; then $A = j\frac{1}{2}$ and $A^* = -j\frac{1}{2}$.

2. Heaviside Expansion Formula.

The Heaviside expansion formula states that the inverse Laplace transform of the quotient $I(s) = P(s)/Q(s)$ is given by

$$\mathcal{L}^{-1}\left[\frac{P(s)}{Q(s)}\right] = \sum_{k=1}^{n} \frac{P(a_k)}{Q'(a_k)} e^{a_k t} \tag{31}$$

where a_k are the n distinct roots of $Q(s)$.

We apply now the Heaviside expansion formula to the expression for the **s** domain current given in Case 1 above.

$$I(s) = \frac{P(s)}{Q(s)} = \frac{s-1}{s^2+3s+2} = \frac{s-1}{(s+2)(s+1)} \tag{32}$$

Now $P(s) = s - 1$, $Q(s) = s^2 + 3s + 2$ and $Q'(s) = 2s + 3$. The roots are $a_1 = -2$ and $a_2 = -1$. Then from (*31*) we have

$$i = \mathcal{L}^{-1}\left[\frac{P(s)}{Q(s)}\right] = \frac{P(-2)}{Q'(-2)} e^{-2t} + \frac{P(-1)}{Q'(-1)} e^{-t} = \frac{-3}{-1} e^{-2t} + \frac{-2}{1} e^{-t} = 3e^{-2t} - 2e^{-t}$$

INITIAL VALUE THEOREM

From Example 4,

$$\mathcal{L}[df/dt] = \int_0^\infty (df/dt)e^{-st}\,dt = s\mathbf{F}(s) - f(0+) \tag{33}$$

Taking the limit in (*33*) as $s \to \infty$, we have

$$\lim_{s\to\infty} \int_0^\infty (df/dt)e^{-st}\,dt = \lim_{s\to\infty} \{s\mathbf{F}(s) - f(0+)\} \tag{34}$$

The integrand contains e^{-st} which approaches zero as $s \to \infty$. Thus

$$\lim_{s\to\infty} \{s\mathbf{F}(s) - f(0+)\} = 0 \tag{35}$$

Since $f(0+)$ is a constant, we can write (*35*) as

$$f(0+) = \lim_{s\to\infty} \{s\mathbf{F}(s)\} \tag{36}$$

Equation (*36*) is the statement of the initial value theorem. Thus we can find the initial value of a time function $f(t)$ by multiplying the corresponding **s** domain function $\mathbf{F}(s)$ by **s** and taking the limit as $s \to \infty$.

Example 6.

In the *RC* circuit of Fig. 17-2 the s domain current $I(s) = \frac{V - q_0/C}{R}\left(\frac{1}{(s + 1/RC)}\right)$ [see equation (*7*)]. Determine the initial current $i(0+)$ using the initial value theorem.

From equation (*36*),

$$i(0+) \quad = \quad \lim_{s \to \infty} \left\{ \frac{V - q_0/C}{R} \left(\frac{s}{(s + 1/RC)} \right) \right\} \quad = \quad \frac{V - q_0/C}{R}$$

This result was shown in Fig. 17-3.

FINAL VALUE THEOREM

From Example 4,

$$\mathcal{L}\,[df/dt] \quad = \quad \int_0^\infty (df/dt)e^{-st}\,dt \quad = \quad s\,\mathbf{F}(s) \; - \; f(0+) \tag{37}$$

Taking the limit in (*37*) as $s \to 0$, we have

$$\lim_{s \to 0} \int_0^\infty (df/dt)e^{-st}\,dt \quad = \quad \lim_{s \to 0} \{s\,\mathbf{F}(s) \; - \; f(0+)\} \tag{38}$$

Since $\lim_{s \to 0} \int_0^\infty (df/dt)e^{-st}\,dt = \int_0^\infty df = f(\infty) - f(0)$ and $\lim_{s \to 0} f(0+) = f(0+)$, equation (*38*)

becomes

$$f(\infty) \; - \; f(0) \quad = \quad -f(0+) \; + \; \lim_{s \to 0} \{s\,\mathbf{F}(s)\} \tag{39}$$

or

$$f(\infty) \quad = \quad \lim_{s \to 0} \{s\,\mathbf{F}(s)\} \tag{40}$$

Equation (*40*) is the statement of the final value theorem. By analogy with the application of the initial value theorem, we can find the final value of a time function $f(t)$ by multiplying the corresponding **s** domain function **F(s)** by **s** and taking the limit as $s \to 0$. However, equation (*40*) may be applied only when all roots of the denominator of **s F(s)** have negative real parts. This restriction excludes the sinusoidal forcing functions since the sine function is indeterminate at infinity.

Example 7.

In the *RL* circuit of Fig. 17-4 the s domain current $I(s) = \frac{V}{R}\left\{\frac{1}{s} - \frac{1}{s + R/L}\right\}$ [see equation (*17*)]. Determine the final value of the current.

From equation (*40*),

$$i(\infty) \quad = \quad \lim_{s \to 0} \frac{V}{R}\left\{\frac{s}{s} - \frac{s}{s + R/L}\right\} \quad = \quad V/R$$

s DOMAIN CIRCUITS

The equation for the series *RLC* circuit shown in Fig. 17-5 below is

$$Ri \; + \; L\frac{di}{dt} \; + \; \frac{1}{C}\int i\,dt \quad = \quad v \tag{41}$$

This integro-differential equation was solved in Chapter 16 by classical methods.

In the sinusoidal steady state, the three circuit elements *R*, *L* and *C* have complex impedances which are given in terms of ω, namely R, $j\omega L$ and $1/j\omega C$ respectively. Then the circuit equation is transformed from the time domain into the frequency domain, and with this transformation the voltages and currents became phasors. Now the equation of the series *RLC* circuit shown in Fig. 17-6 below is

$$R\mathbf{I} \; + \; j\omega L\mathbf{I} \; + \; (1/j\omega C)\mathbf{I} \quad = \quad \mathbf{V} \tag{42}$$

The advantage gained by the transformation is that the transformed equation can be treated algebraically to obtain the phasor current **I**. The various voltage drops are simply products of the phasor current and the impedance of the particular circuit element.

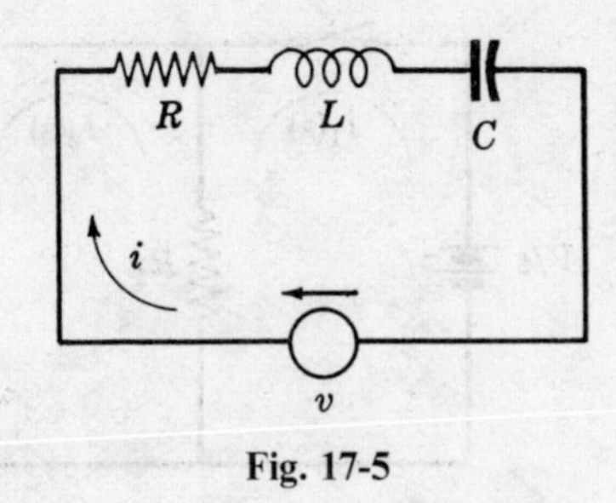

Fig. 17-5

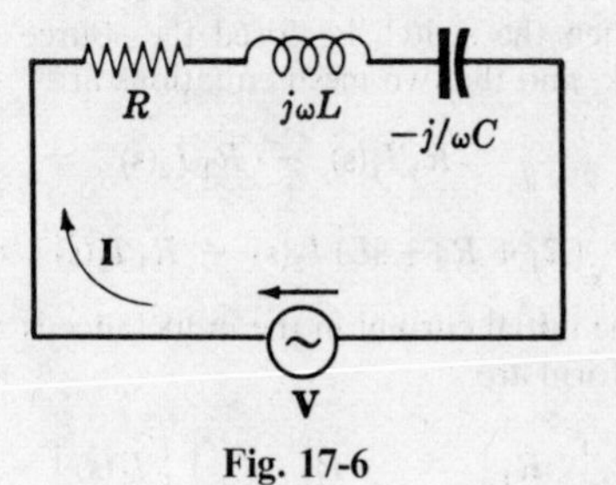

Fig. 17-6

The Laplace transform method results in a transformation of the voltage drop Ri in the time domain to $R\,I(\mathbf{s})$ in the $\mathbf{s}$ domain. Similarly, the voltage across an inductance $L(di/dt)$ becomes $\mathbf{s}L\,I(\mathbf{s}) - L\,i(0+)$ and the voltage across a capacitor $\frac{1}{C}\int i\,dt$ becomes $\frac{1}{\mathbf{s}C}I(\mathbf{s}) + \frac{q_0}{\mathbf{s}C}$. Then the equation for the series circuit shown in Fig. 17-7 is

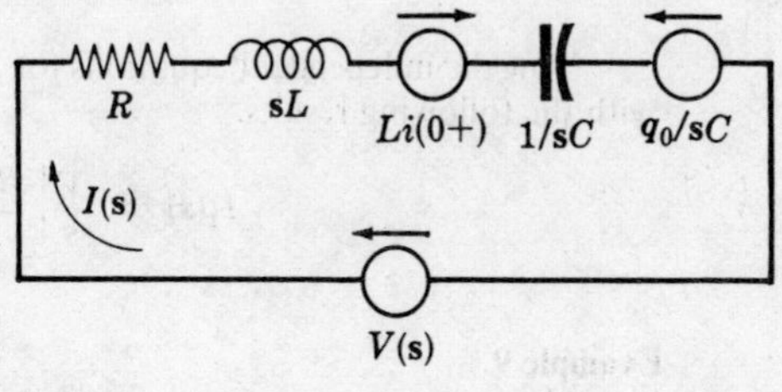

Fig. 17-7

$$R\,I(\mathbf{s}) + \mathbf{s}L\,I(\mathbf{s}) - L\,i(0+) + \frac{1}{\mathbf{s}C}I(\mathbf{s}) + \frac{q_0}{\mathbf{s}C} = V(\mathbf{s}) \tag{43}$$

or

$$I(\mathbf{s})\{R + \mathbf{s}L + 1/\mathbf{s}C\} = V(\mathbf{s}) - q_0/\mathbf{s}C + L\,i(0+) \tag{44}$$

In equation (*44*), $R + \mathbf{s}L + 1/\mathbf{s}C$ is the $\mathbf{s}$ domain impedance $Z(\mathbf{s})$; the ratio of the excitation to response. $Z(\mathbf{s})$ has the same form as the complex impedance of the sinusoidal steady state, $R + j\omega L + 1/j\omega C$. The equations of both the mesh current and the node voltage methods of analysis can be easily applied to $\mathbf{s}$ domain circuits provided the proper signs are used on the initial condition terms $L\,i(0+)$ and $q_0/\mathbf{s}C$.

Consider the circuit of Fig. 17-8(*a*) in which an initial current i_0 exists while the switch is in position 1. At $t = 0$ the switch is moved to position 2, thus introducing in the circuit a constant source V and a capacitor with an initial charge q_0. The assumed positive current i is selected in a clockwise direction as shown in the diagram.

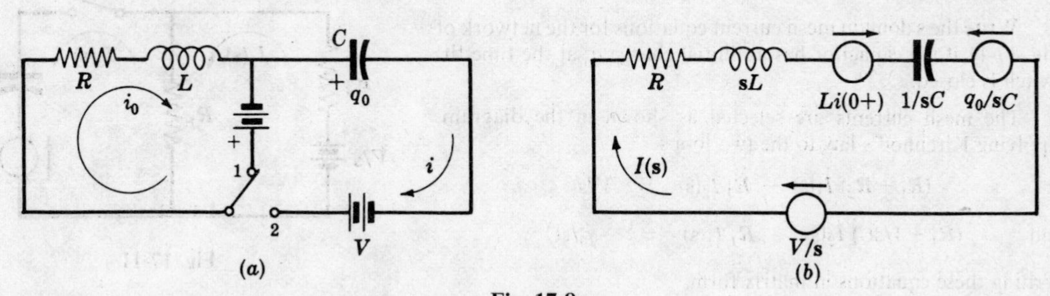

Fig. 17-8

Now the constant source is transformed to $V/\mathbf{s}$ and the resulting current is $I(\mathbf{s})$ as shown in Fig, 17-8(*b*). The initial condition terms are now sources with directions as shown and the corresponding equation would be identical to equation (*44*) above. For an initial current i_0 in the opposite direction or a charge q_0 of opposite sign, the signs of the terms $L\,i(0+)$ and $q_0/\mathbf{s}C$ would change accordingly. The following examples show how the $\mathbf{s}$ domain equations are similar to the phasor equations treated earlier in this text. All of the network theorems developed and applied to the sinusoidal steady state have their counterparts in the $\mathbf{s}$ domain.

Example 8.

In the two-mesh network shown in Fig. 17-9 below the s domain mesh currents are selected as shown in the diagram. If the switch is closed at $t = 0$, obtain the equations for $I_1(\mathbf{s})$ and $I_2(\mathbf{s})$.

When the switch is closed the source V/s is applied to the network, and the two mesh equations are

$$R_1 I_1(s) - R_1 I_2(s) = V/s$$

and

$$(R_1 + R_2 + sL)\, I_2(s) - R_1 I_1(s) = L\, i(0+)$$

Since the initial current in the inductance is zero, the equations in matrix form are

$$\begin{bmatrix} R_1 & -R_1 \\ -R_1 & R_1 + R_2 + sL \end{bmatrix} \begin{bmatrix} I_1(s) \\ I_2(s) \end{bmatrix} = \begin{bmatrix} V/s \\ 0 \end{bmatrix}$$

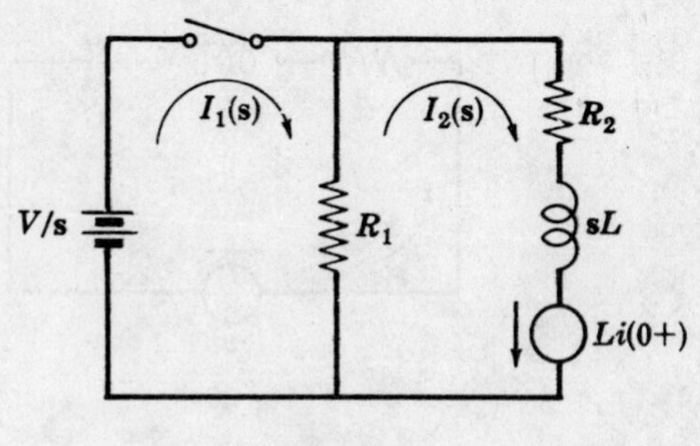

Fig. 17-9

Now the independent equations for I_1 (s) and I_2(s) are obtained either by substitution or by determinant methods, with the following results.

$$I_1(s) = \frac{V}{s}\left[\frac{R_1 + R_2 + sL}{R_1(R_2 + sL)}\right] \quad \text{and} \quad I_2(s) = \frac{V}{s}\,\frac{1}{(R_2 + sL)}$$

Example 9.

Write the s domain node voltage equation for the network shown in Fig. 17-10.

Node *1* and the reference are selected as shown in the diagram and when the switch is closed the nodal equation is

$$\frac{V_1(s) - V/s - L\, i(0+)}{sL} + \frac{V_1(s)}{R_1} + \frac{V_1(s)}{R_2} = 0$$

or

$$(1/sL + 1/R_1 + 1/R_2)\, V_1(s) = \frac{V/s + L\, i(0+)}{sL}$$

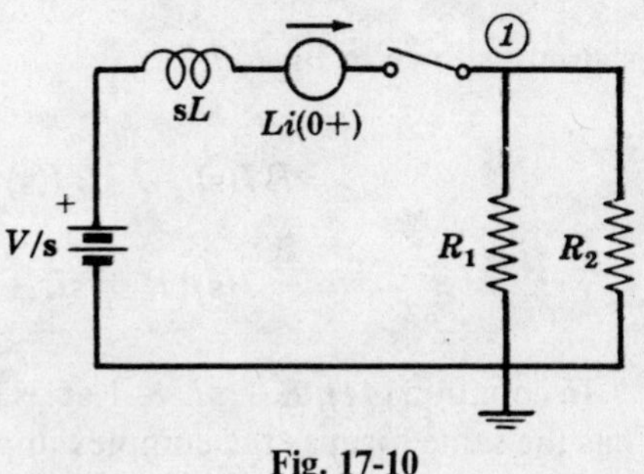
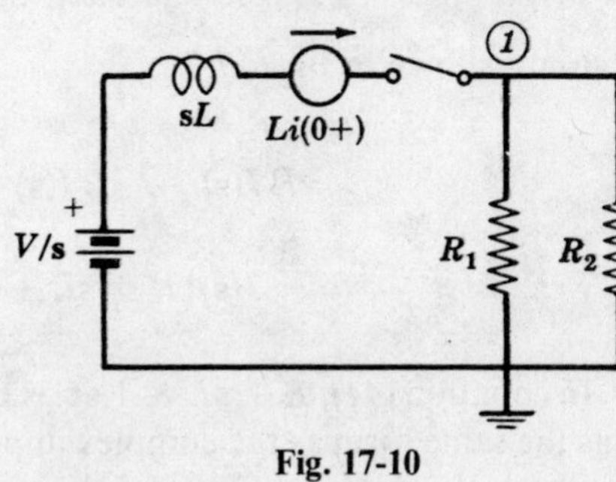

Fig. 17-10

The initial current in the inductance is zero; hence the equation for node voltage V_1(s) is

$$V_1(s) = \frac{V}{s}\left(\frac{R_1 R_2}{R_1 R_2 + sLR_2 + sLR_1}\right)$$

Example 10.

Write the s domain mesh current equations for the network of Fig. 17-11 if the capacitor has an initial charge q_0 at the time the switch is closed.

The mesh currents are selected as shown in the diagram. Applying Kirchhoff's law to the two loops,

$$(R_1 + R_2)\, I_1(s) - R_1 I_2(s) = V/s$$

and

$$(R_1 + 1/sC)\, I_2(s) - R_1 I_1(s) = -q_0/sC$$

Fig. 17-11

Writing these equations in matrix form,

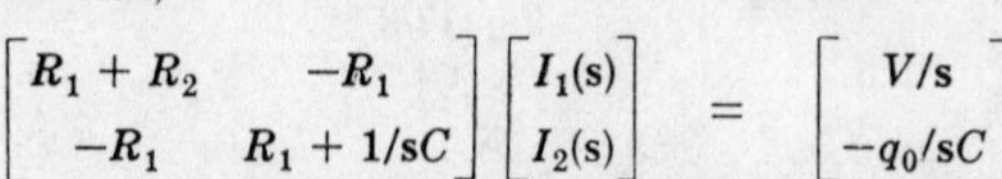

$$\begin{bmatrix} R_1 + R_2 & -R_1 \\ -R_1 & R_1 + 1/sC \end{bmatrix} \begin{bmatrix} I_1(s) \\ I_2(s) \end{bmatrix} = \begin{bmatrix} V/s \\ -q_0/sC \end{bmatrix}$$

Solved Problems

17.1. Find the Laplace transform of $e^{-at}\cos\omega t$, where a is a constant.

Applying the defining equation $\mathcal{L}[f(t)] = \int_0^\infty f(t)e^{-st}\,dt$ to the given function, we obtain

$$\begin{aligned}\mathcal{L}[e^{-at}\cos\omega t] &= \int_0^\infty \cos\omega t\, e^{-(s+a)t}\,dt \\ &= \left[\frac{-(s+a)\cos\omega t\, e^{-(s+a)t} + e^{-(s+a)t}\,\omega\sin\omega t}{(s+a)^2+\omega^2}\right]_0^\infty \\ &= \frac{s+a}{(s+a)^2+\omega^2}\end{aligned}$$

17.2. If $\mathcal{L}[f(t)] = \mathbf{F}(s)$, show that $\mathcal{L}[e^{-at}f(t)] = \mathbf{F}(s+a)$. Apply this result to Prob. 17.1.

By definition, $\mathcal{L}[f(t)] = \int_0^\infty f(t)e^{-st}\,dt = \mathbf{F}(s)$. Then

$$\mathcal{L}[e^{-at}f(t)] = \int_0^\infty e^{-at}[f(t)e^{-st}]\,dt = \int_0^\infty f(t)e^{-(s+a)t}\,dt = \mathbf{F}(s+a) \qquad (1)$$

Since $\mathcal{L}[\cos\omega t] = \dfrac{s}{s^2+\omega^2}$, (see Table 17-1), it follows from (1) that $\mathcal{L}[e^{-at}\cos\omega t] = \dfrac{s+a}{(s+a)^2+\omega^2}$, as determined in Problem 17.1.

17.3. Find the Laplace transform of $f(t) = 1 - e^{-at}$, where a is a constant.

We have

$$\begin{aligned}\mathcal{L}[1-e^{-at}] &= \int_0^\infty (1-e^{-at})e^{-st}\,dt = \int_0^\infty e^{-st}\,dt - \int_0^\infty e^{-(s+a)t}\,dt \\ &= \left[-\frac{1}{s}e^{-st} + \frac{1}{s+a}e^{-(s+a)t}\right]_0^\infty = \frac{1}{s} - \frac{1}{s+a} = \frac{a}{s(s+a)}\end{aligned}$$

17.4. Find $\mathcal{L}^{-1}\left[\dfrac{1}{s(s^2-a^2)}\right]$.

Using the method of partial fractions,

$$\frac{1}{s(s^2-a^2)} = \frac{A}{s} + \frac{B}{s+a} + \frac{C}{s-a}$$

and the coefficients are

$$A = \left.\frac{1}{s^2-a^2}\right|_{s=0} = -\frac{1}{a^2} \qquad B = \left.\frac{1}{s(s-a)}\right|_{s=-a} = \frac{1}{2a^2} \qquad C = \left.\frac{1}{s(s+a)}\right|_{s=a} = \frac{1}{2a^2}$$

Now
$$\mathcal{L}^{-1}\left[\frac{1}{s(s^2-a^2)}\right] = \mathcal{L}^{-1}\left[\frac{-1/a^2}{s}\right] + \mathcal{L}^{-1}\left[\frac{1/2a^2}{s+a}\right] + \mathcal{L}^{-1}\left[\frac{1/2a^2}{s-a}\right]$$

The corresponding time functions are found in Table 17-1:

$$\begin{aligned}\mathcal{L}^{-1}\left[\frac{1}{s(s^2-a^2)}\right] &= -\frac{1}{a^2} + \frac{1}{2a^2}e^{-at} + \frac{1}{2a^2}e^{at} \\ &= -\frac{1}{a^2} + \frac{1}{a^2}\left(\frac{e^{at}+e^{-at}}{2}\right) = \frac{1}{a^2}(\cosh at - 1)\end{aligned}$$

17.5. Find $\mathcal{L}^{-1}\left[\dfrac{s+1}{s(s^2+4s+4)}\right]$.

Using the method of partial fractions, we have

$$\frac{s+1}{s(s+2)^2} = \frac{A}{s} + \frac{B}{s+2} + \frac{C}{(s+2)^2}$$

Then $$A = \left.\frac{s+1}{(s+2)^2}\right|_{s=0} = \frac{1}{4} \quad \text{and} \quad C = \left.\frac{s+1}{s}\right|_{s=-2} = \frac{1}{2}$$

The coefficient of the quadratic term is

$$B = \left.\frac{d}{ds}\left[\frac{s+1}{s}\right]\right|_{s=-2} = \left.-\frac{1}{s^2}\right|_{s=-2} = -\frac{1}{4}$$

Now $$\mathcal{L}^{-1}\left[\frac{s+1}{s(s^2+4s+4)}\right] = \mathcal{L}^{-1}\left[\frac{\frac{1}{4}}{s}\right] + \mathcal{L}^{-1}\left[\frac{-\frac{1}{4}}{s+2}\right] + \mathcal{L}^{-1}\left[\frac{\frac{1}{2}}{(s+2)^2}\right]$$

The corresponding time functions are found in Table 17.1, and

$$\mathcal{L}^{-1}\left[\frac{s+1}{s(s^2+4s+4)}\right] = \frac{1}{4} - \frac{1}{4}e^{-2t} + \frac{1}{2}te^{-2t}$$

17.6. In the series *RC* circuit of Fig. 17-12, the capacitor has an initial charge $q_0 = 2500 \times 10^{-6}$ coulombs. At $t = 0$, the switch is closed and a constant voltage source $V = 100$ volts is applied to the circuit. Use the Laplace transform method to find the current.

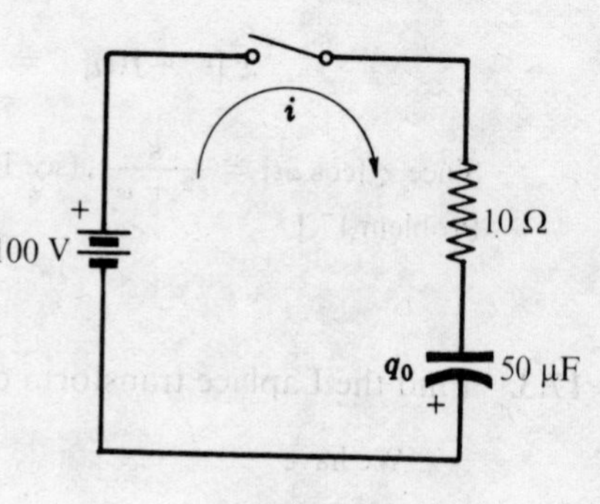

Fig. 17-12

The time domain equation for the given circuit after the switch is closed is

$$Ri + \frac{1}{C}\int i\,dt = V$$

or $$10i + \frac{1}{50 \times 10^{-6}}\int i\,dt = 100 \qquad (1)$$

Taking the Laplace transform of the terms in (*1*), we obtain the **s** domain equation

$$10\,I(s) + \frac{I(s)}{50 \times 10^{-6}\,s} + \frac{q_0}{50 \times 10^{-6}\,s} = \frac{100}{s} \qquad (2)$$

The polarity of q_0 shown in the diagram is opposite to the polarity of the charge which the source will deposit on the capacitor; hence the **s** domain equation is

$$10\,I(s) + \frac{I(s)}{50 \times 10^{-6}\,s} - \frac{2500 \times 10^{-6}}{50 \times 10^{-6}\,s} = \frac{100}{s} \qquad (3)$$

Rearranging, $$I(s)\left\{\frac{10s + 2 \times 10^4}{s}\right\} = \frac{150}{s} \qquad (4)$$

or $$I(s) = \frac{15}{s + 2 \times 10^3} \qquad (5)$$

The time function is now obtained by taking the inverse Laplace transform of (*5*):

$$\mathcal{L}^{-1}[I(s)] = i = \mathcal{L}^{-1}\left[\frac{15}{s + 2 \times 10^3}\right] = 15e^{-2\times10^3 t} \text{ amperes}$$

If the initial charge q_0 is positive on the upper plate of the capacitor, the sign of q_0/sC in equation (*3*) is positive. Then the right side of equation (*4*) becomes $50/s$, thus generating a current transient $i = 5e^{-2\times10^3 t}$ amperes.

17.7. In the *RL* circuit shown in Fig. 17-13 below, the switch is in position 1 long enough to establish steady state conditions and at $t = 0$ is switched to position 2. Find the resulting current.

Assume the direction of the current as shown in the diagram. The initial current is then $i_0 = -50/25 = -2$ A

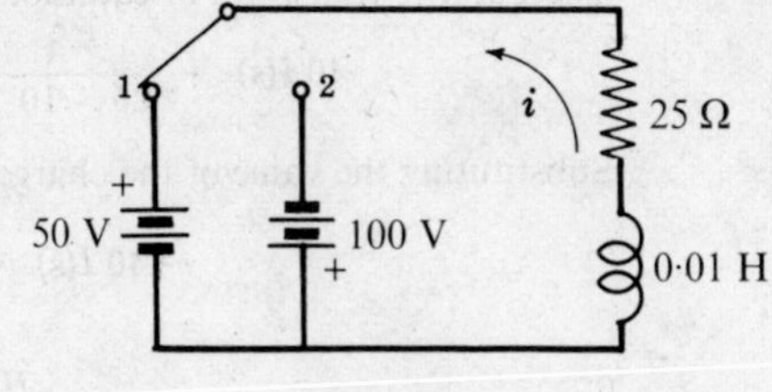

Fig. 17-13

The time domain equation is

$$25i + 0{\cdot}01(di/dt) = 100 \qquad (1)$$

Taking the Laplace transform of (*1*),

$$25\ I(\mathrm{s}) + 0{\cdot}01\mathrm{s}I(\mathrm{s}) - 0{\cdot}01\ i(0+) = 100/\mathrm{s} \qquad (2)$$

Substituting for $i(0+)$,

$$25\ I(\mathrm{s}) + 0{\cdot}01\mathrm{s}\ I(\mathrm{s}) + 0{\cdot}01(2) = 100/\mathrm{s} \qquad (3)$$

and
$$I(\mathrm{s}) = \frac{100}{\mathrm{s}(0{\cdot}01\mathrm{s}+25)} - \frac{0{\cdot}02}{0{\cdot}01\mathrm{s}+25} = \frac{10^4}{\mathrm{s}(\mathrm{s}+2500)} - \frac{2}{\mathrm{s}+2500} \qquad (4)$$

Expanding $\frac{10^4}{\mathrm{s}(\mathrm{s}+2500)}$ of equation (*4*) by the method of partial fractions,

$$\frac{10^4}{\mathrm{s}(\mathrm{s}+2500)} = \frac{A}{\mathrm{s}} + \frac{B}{\mathrm{s}+2500} \qquad (5)$$

Then
$$A = \left.\frac{10^4}{\mathrm{s}+2500}\right|_{\mathrm{s}=0} = 4 \quad \text{and} \quad B = \left.\frac{10^4}{\mathrm{s}}\right|_{\mathrm{s}=-2500} = -4$$

Substituting these values into equation (*4*),

$$I(\mathrm{s}) = \frac{4}{\mathrm{s}} - \frac{4}{\mathrm{s}+2500} - \frac{2}{\mathrm{s}+2500} = \frac{4}{\mathrm{s}} - \frac{6}{\mathrm{s}+2500} \qquad (6)$$

Taking the inverse Laplace transform of equation (*6*), we obtain $i = 4 - 6e^{-2500t}$ amperes.

17.8. In the series RL circuit of Fig. 17-14 an exponential voltage $v = 50e^{-100t}$ volts is applied by closing the switch at $t = 0$. Find the resulting current.

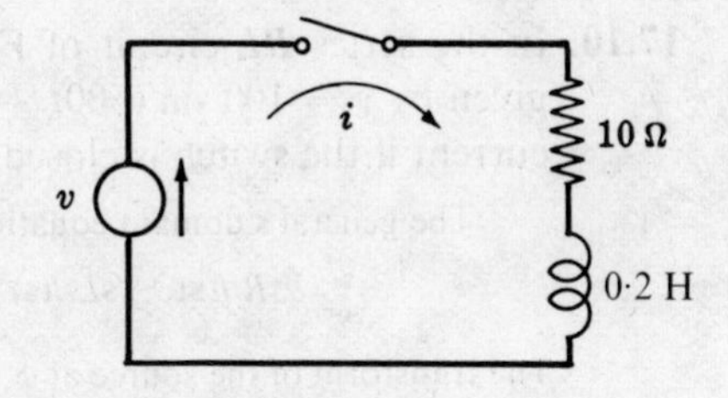

Fig. 17-14

The time domain equation for the given circuit is

$$Ri + L(di/dt) = v \qquad (1)$$

In the s domain (*1*) has the form

$$R\ I(\mathrm{s}) + \mathrm{s}L\ I(\mathrm{s}) - L\ i(0+) = V(\mathrm{s}) \qquad (2)$$

Substituting the circuit constants and the transform of the source $V(\mathrm{s}) = 50/(\mathrm{s}+100)$ in (*2*),

$$10\ I(\mathrm{s}) + \mathrm{s}(.2)\ I(\mathrm{s}) = \frac{50}{\mathrm{s}+100} \quad \text{or} \quad I(\mathrm{s}) = \frac{250}{(\mathrm{s}+100)(\mathrm{s}+50)} \qquad (3)$$

By the Heaviside expansion formula, $\mathcal{L}^{-1}[I(\mathrm{s})] = \mathcal{L}^{-1}\left[\frac{P(\mathrm{s})}{Q(\mathrm{s})}\right] = \sum_{n=1,2} \frac{P(a_n)}{Q'(a_n)} e^{a_n t}$, where $P(\mathrm{s}) = 250$, $Q(\mathrm{s}) = \mathrm{s}^2 + 150\mathrm{s} + 5000$, $Q'(\mathrm{s}) = 2\mathrm{s} + 150$, $a_1 = -100$ and $a_2 = -50$. Then

$$i = \mathcal{L}^{-1}[I(\mathrm{s})] = \frac{250}{-50} e^{-100t} + \frac{250}{50} e^{-50t} = -5e^{-100t} + 5e^{-50t} \text{ amperes}$$

17.9. The series RC circuit of Fig. 17-15 has a sinusoidal voltage source $v = 180 \sin(2000t + \varphi)$ volts and an initial charge on the capacitor $q_0 = 1250 + 10^{-6}$ coulombs with polarity as shown. Determine the current if the switch is closed at a time corresponding to $\varphi = 90°$.

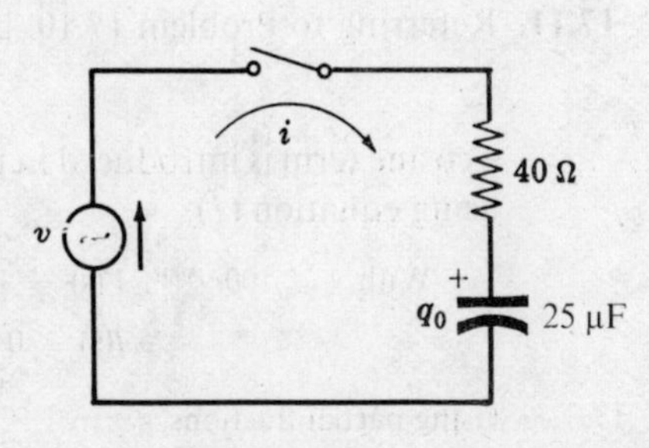

Fig. 17-15

The time domain equation of the circuit is

$$40i + \frac{1}{25 \times 10^{-6}} \int i\, dt = 180 \sin(2000t + 90°)$$

The Laplace transform of equation (*1*) results in the **s** domain equation

$$40\,I(\mathbf{s}) + \frac{1}{25\times10^{-6}\,\mathbf{s}}I(\mathbf{s}) + \frac{q_0}{25\times10^{-6}\,\mathbf{s}} = 180\left\{\frac{\mathbf{s}\sin 90^\circ + 2000\cos 90^\circ}{\mathbf{s}^2 + 4\times10^6}\right\} \tag{2}$$

Substituting the value of the charge q_0 into (*2*),

$$40\,I(\mathbf{s}) + \frac{4\times10^4}{\mathbf{s}}I(\mathbf{s}) + \frac{1250\times10^{-6}}{25\times10^{-6}\,\mathbf{s}} = \frac{180\,\mathbf{s}}{\mathbf{s}^2+4\times10^6}$$

or

$$I(\mathbf{s}) = \frac{4{\cdot}5\,\mathbf{s}^2}{(\mathbf{s}^2+4\times10^6)(\mathbf{s}+10^3)} - \frac{1{\cdot}25}{\mathbf{s}+10^3} \tag{3}$$

Applying the Heaviside expansion formula to the term $\dfrac{4{\cdot}5\,\mathbf{s}^2}{(\mathbf{s}^2+4\times10^6)(\mathbf{s}+10^3)}$ in (*4*), we have $P(\mathbf{s}) = 4{\cdot}5\,\mathbf{s}^2$, $Q(\mathbf{s}) = \mathbf{s}^3 + 10^3\,\mathbf{s}^2 + 4\times10^6\,\mathbf{s} + 4\times10^9$, $Q'(\mathbf{s}) = 3\,\mathbf{s}^2 + 2\times10^3\,\mathbf{s} + 4\times10^6$, $a_1 = -j2\times10^3$, $a_2 = j2\times10^3$ and $a^3 = -10^3$. Then

$$i = \frac{P(-j2\times10^3)}{Q'(-j2\times10^3)}e^{-j2\times10^3t} + \frac{P(j2\times10^3)}{Q'(j2\times10^3)}e^{j2\times10^3t} + \frac{P(-10^3)}{Q'(-10^3)}e^{-10^3t} - 1{\cdot}25e^{-10^3t}\text{ amperes}$$

$$= (1{\cdot}8 - j0{\cdot}9)e^{-j2\times10^3t} + (1{\cdot}8 + j0{\cdot}9)e^{j2\times10^3t} - 0{\cdot}35e^{-10^3t}\text{ amperes} \tag{4}$$

$$= -1{\cdot}8\sin 2000t + 3{\cdot}6\cos 2000t - 0{\cdot}35e^{-10^3t}\text{ amperes}$$

$$= 4{\cdot}02\sin(2000t + 116{\cdot}6^\circ) - 0{\cdot}35e^{-10^3t}\text{ amperes}$$

At $t = 0$ the current is given by the instantaneous voltage, consisting of the source voltage and the charged capacitor voltage, divided by the resistance. Thus

$$i_0 = \left(180\sin 90^\circ - \frac{1250\times10^{-6}}{25\times10^{-6}}\right)\Big/40 = 3{\cdot}25\text{ A}$$

The same result is obtained if we set $t = 0$ in equation (*4*).

17.10. In the series *RL* circuit of Fig. 17-16 the sinusoidal source is given by $v = 100\sin(500t + \varphi)$ volts. Determine the resulting current if the switch is closed when $\varphi = 0$.

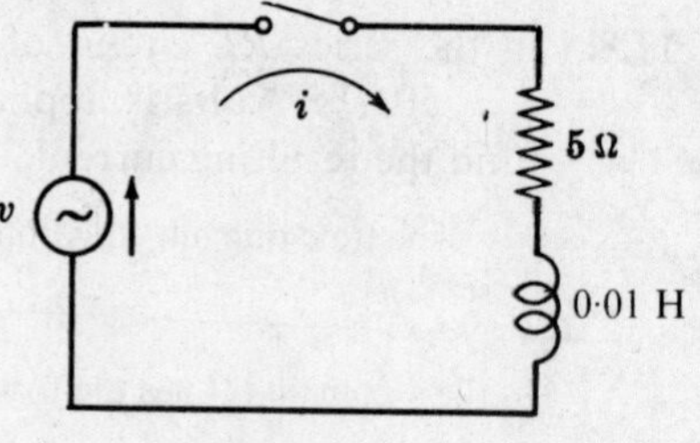

Fig. 17-16

The general **s** domain equation of a series *RL* circuit is

$$R\,I(\mathbf{s}) + \mathbf{s}L\,I(\mathbf{s}) - L\,i(0+) = V(\mathbf{s}) \tag{1}$$

The transform of the source at $\varphi = 0$ is $V(\mathbf{s}) = \dfrac{500(100)}{\mathbf{s}^2 + (500)^2}$.

Since there is no initial current in the inductance, $L\,i(0+) = 0$. Substituting the circuit constants into equation (*1*),

$$5\,I(\mathbf{s}) + 0{\cdot}01\mathbf{s}I(\mathbf{s}) = \frac{5\times10^4}{\mathbf{s}^2+25\times10^4}\text{ and } I(\mathbf{s}) = \frac{5\times10^6}{(\mathbf{s}^2+25\times10^4)(\mathbf{s}+500)} \tag{2}$$

Expanding (*2*) by partial fractions,

$$I(\mathbf{s}) = 5\left(\frac{-1+j}{\mathbf{s}+j500}\right) + 5\left(\frac{-1-j}{\mathbf{s}-j500}\right) + \frac{10}{\mathbf{s}+500} \tag{3}$$

Then the inverse Laplace of (*3*) is

$$i = 10\sin 500t - 10\cos 500t + 10e^{-500t} = 10e^{-500t} + 14{\cdot}14\sin(500t - \pi/4)\text{ amperes}$$

17.11. Referring to Problem 17.10, by writing the voltage function as

$$v = 100e^{j500t}\text{ volts} \tag{1}$$

a cosine term is introduced in the source voltage. Determine the current for the circuit of Problem 17.10, using equation (*1*).

With $v = 100e^{j500t}$, $V(\mathbf{s}) = 100/(\mathbf{s} - j500)$ and the **s** domain equation is

$$5\,I(\mathbf{s}) + 0{\cdot}01\mathbf{s}\,I(\mathbf{s}) = 100/(\mathbf{s}-j500)\text{ and } I(\mathbf{s}) = 10^4/(\mathbf{s}-j500)(\mathbf{s}+500) \tag{2}$$

Using partial fractions,

$$I(\mathbf{s}) = \frac{10-j10}{\mathbf{s}-j500} + \frac{-10+j10}{\mathbf{s}+500} \tag{3}$$

Now taking the inverse Laplace transform of (3), the corresponding time function of the current is

$$\begin{aligned} i &= (10 - j10)e^{j500t} + (-10 + j10)e^{-500t} \text{ amperes} \\ &= 14\ 14 e^{j(500t - \pi/4)} + (-10 + j10)e^{-500t} \text{ amperes} \\ &= 14\ 14\{\cos(500t - \pi/4) + j \sin(500t - \pi/4)\} + (-10 + j10)e^{-500t} \text{ amperes} \end{aligned} \quad (4)$$

Since the source voltage in Problem 17.10 contained only the imaginary part of (1), the resulting current is the imaginary part of equation (4),

$$i = 14{\cdot}14 \sin(500t - \pi/4) + 10e^{-500t} \text{ amperes}$$

17.12. In the series *RLC* circuit shown in Fig. 17-17, there is no initial charge on the capacitor. If the switch is closed at $t = 0$, determine the resulting current.

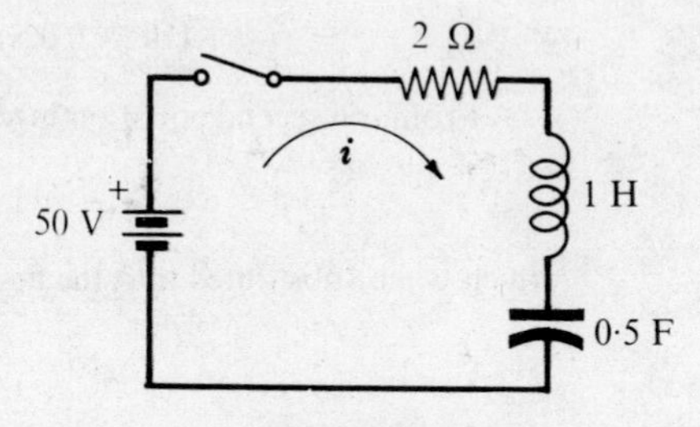

Fig. 17-17

The time domain equation of the given circuit is

$$Ri + L\frac{di}{dt} + \frac{1}{C}\int i\,dt = V \quad (2)$$

The Laplace transform of the terms in (1) results in the s domain equation

$$R\,I(s) + sL\,I(s) - L\,i(0+) + \frac{1}{sC}I(s) + \frac{q_0}{sC} = \frac{V}{s}$$

From the initial boundary conditions, $L\,i(0+) = 0$ and $q_0/sC = 0$. Substituting the circuit constants in (2), we have

$$2\,I(s) + 1s\,I(s) + \frac{1}{0{\cdot}5s}I(s) = \frac{50}{s} \quad (3)$$

or

$$I(s) = \frac{50}{s^2 + 2s + 2} = \frac{50}{(s+1+j)(s+1-j)} \quad (4)$$

Expanding (4) by partial fractions,

$$I(s) = \frac{j25}{(s+1+j)} - \frac{j25}{(s+1-j)} \quad (5)$$

and the inverse Laplace transform of (5) results in the time domain current

$$i = j25\{e^{(-1-j)t} - e^{(-1+j)t}\} = 50e^{-t}\sin t \text{ amperes}$$

17.13. In the two-mesh network of Fig. 17-18, the two mesh currents are selected as shown. Write the s domain equations in matrix form and construct the corresponding circuit.

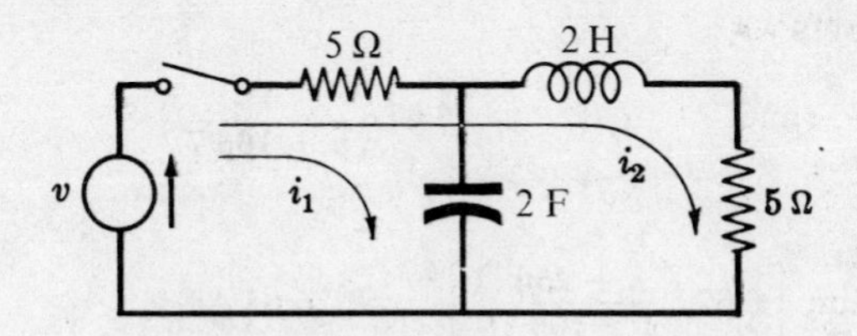

Fig. 17-18

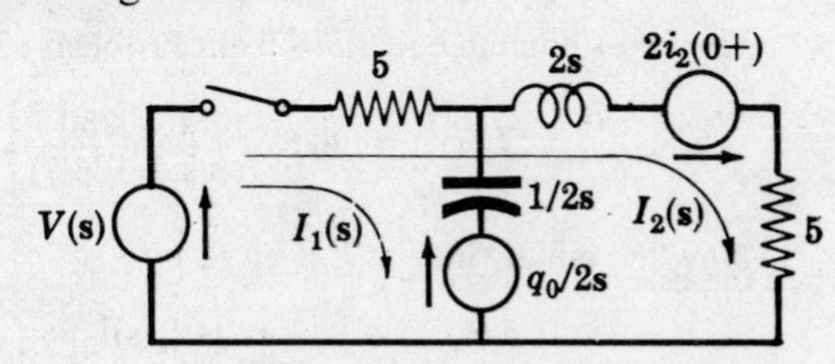

Fig. 17-19

Writing the set of equations in the time domain,

$$5i_1 + \frac{1}{2}\int i_1\,dt + 5i_2 = v \quad \text{and} \quad 10i_2 + 2(di_2/dt) + 5i_1 = v \quad (1)$$

Taking the Laplace transform of (1) to obtain the corresponding s domain equations,

$$5\,I_1(s) + \frac{1}{2s}I_1(s) + \frac{q_0}{2s} + 5\,I_2(s) = V(s) \qquad 10\,I_2(s) + 2s\,I_2(s) - 2\,i_2(0+) + 5\,I_1(s) = V(s) \quad (2)$$

When this set of s domain equations is written in matrix form, the required s domain circuit can be determined by examination of the $Z(s)$, $I(s)$ and $V(s)$ matrices (see Fig. 17-19).

$$\begin{bmatrix} 5 + 1/2s & 5 \\ 5 & 10 + 2s \end{bmatrix} \begin{bmatrix} I_1(s) \\ I_2(s) \end{bmatrix} = \begin{bmatrix} V(s) - q_0/2s \\ V(s) + 2i_2(0+) \end{bmatrix}$$

17.14. In the two mesh network of Fig. 17-20, find the currents which result when the switch is closed.

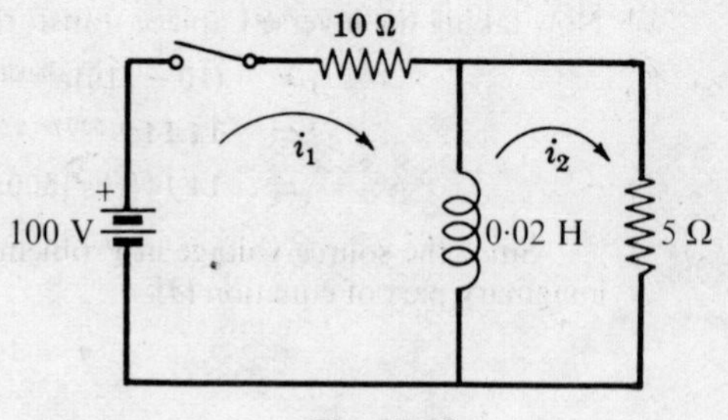

Fig. 17-20

The time domain equations for the network are

$$10i_1 + 0{\cdot}02\frac{di_1}{dt} - 0{\cdot}02\frac{di_2}{dt} = 100$$
$$0{\cdot}02\frac{di_2}{dt} + 5i_2 - 0{\cdot}02\frac{di_1}{dt} = 0 \qquad (1)$$

Taking the Laplace transform of set (*1*),

$$(10 + 0{\cdot}02\mathbf{s})\,I_1(\mathbf{s}) - 0{\cdot}02\,I_2(\mathbf{s}) = 100/\mathbf{s} \quad (5 + 0{\cdot}02\mathbf{s})\,I_2(\mathbf{s}) - 0{\cdot}02\mathbf{s}\,I_1(\mathbf{s}) = 0 \qquad (2)$$

From the second equation in set (*2*) we find

$$I_2(\mathbf{s}) \;=\; I_1(\mathbf{s})\left(\frac{\mathbf{s}}{\mathbf{s}+250}\right) \qquad (3)$$

which when substituted into the first **s** domain equation gives

$$(10 + 0{\cdot}02\mathbf{s})\,I_1(\mathbf{s}) - 0{\cdot}02\mathbf{s}\left\{I_1(\mathbf{s})\left(\frac{\mathbf{s}}{\mathbf{s}+250}\right)\right\} \;=\; \frac{100}{\mathbf{s}} \qquad (4)$$

or

$$I_1(\mathbf{s}) \;=\; 6{\cdot}67\left\{\frac{\mathbf{s}+250}{\mathbf{s}(\mathbf{s}+166{\cdot}7)}\right\} \qquad (5)$$

Now apply the partial fraction method to (*5*) and obtain

$$I_1(\mathbf{s}) = \frac{10}{\mathbf{s}} - \frac{3{\cdot}33}{\mathbf{s}+166{\cdot}7} \text{ and } i_1 = 10 - 3{\cdot}33e^{-166\cdot7t} \text{ amperes} \qquad (6)$$

Finally, substitute equation (*5*) into (*3*) and obtain the **s** domain equation

$$I_2(\mathbf{s}) \;=\; 6{\cdot}67\left\{\frac{\mathbf{s}+250}{\mathbf{s}(\mathbf{s}+166{\cdot}7)}\right\}\frac{\mathbf{s}}{\mathbf{s}+250} \;=\; 6{\cdot}67\left(\frac{1}{\mathbf{s}+166{\cdot}7}\right) \text{ and } i_2 = 6{\cdot}67e^{-166\cdot7t} \text{ amperes} \qquad (7)$$

17.15. Apply the initial and final value theorems to the **s** domain equations $I_1(\mathbf{s})$ and $I_2(\mathbf{s})$ in Problem 17.14.

The two **s** domain equations from Problem 17.14 are

$$I_1(\mathbf{s}) \;=\; 6{\cdot}67\left\{\frac{\mathbf{s}+250}{\mathbf{s}(\mathbf{s}+166{\cdot}7)}\right\} \quad \text{and} \quad I_2(\mathbf{s}) \;=\; 6\;67\left(\frac{1}{\mathbf{s}+166{\cdot}7}\right)$$

Now the initial value of i_1 is given by

$$i_1(0) \;=\; \lim_{\mathbf{s}\to\infty}\,[\mathbf{s}\,I_1(\mathbf{s})] \;=\; \lim_{\mathbf{s}\to\infty}\left[6{\cdot}67\left(\frac{\mathbf{s}+250}{\mathbf{s}+166{\cdot}7}\right)\right] \;=\; 6{\cdot}67\text{ A}$$

and the final value

$$i_1(\infty) \;=\; \lim_{\mathbf{s}\to 0}\,[\mathbf{s}\,I_1(\mathbf{s})] \;=\; \lim_{\mathbf{s}\to 0}\left[6{\cdot}67\left(\frac{\mathbf{s}+250}{\mathbf{s}+166{\cdot}7}\right)\right] \;=\; 6{\cdot}67(250/166{\cdot}7) \;=\; 10\text{ A}$$

The initial value of i_2 is given by

$$i_2(0) \;=\; \lim_{\mathbf{s}\to\infty}\,[\mathbf{s}\,I_2(\mathbf{s})] \;=\; \lim_{\mathbf{s}\to\infty}\left[6{\cdot}67\left(\frac{\mathbf{s}}{\mathbf{s}+166{\cdot}7}\right)\right] \;=\; 6{\cdot}67\text{ A}$$

and the final value

$$i_2(\infty) \;=\; \lim_{\mathbf{s}\to 0}\,[\mathbf{s}\,I_2(\mathbf{s})] \;=\; \lim_{\mathbf{s}\to 0}\left[6{\cdot}67\left(\frac{\mathbf{s}}{\mathbf{s}+166{\cdot}7}\right)\right] \;=\; 0$$

Examination of the circuit in Fig. 17-20 verifies each of the above initial and final values. At the instant of closing, the inductance presents an infinite impedance and the current $i_1 = i_2 = 100/(10 + 5) = 6{\cdot}67$ A. Then in the steady state the inductance appears as a short circuit; hence $i_1 = 10$ A and $i_2 = 0$.

17.16. Referring to the circuit of Fig. 17-20, find the equivalent impedance of the network and construct the circuit using this impedance.

In the s domain the 0·02 H inductance has an impedance $Z(s) = 0{\cdot}02s$ which can be treated the same as $j\omega L$ in the sinusoidal steady state. Therefore the equivalent impedance of the network as seen from the source is

$$Z(s) = 10 + \frac{0{\cdot}02s(5)}{0{\cdot}02s + 5} = \frac{0{\cdot}3s + 50}{0{\cdot}02s + 5} = 15\,\frac{s + 166{\cdot}7}{s + 250} \qquad (1)$$

The circuit containing the equivalent impedance is shown in Fig. 17-21.

The current is

$$I_1(s) = \frac{V(s)}{Z(s)} = \frac{100}{s}\left\{\frac{s + 250}{15(s + 166{\cdot}7)}\right\}$$

$$= 6.67\left\{\frac{s + 250}{s(s + 166{\cdot}7)}\right\}$$

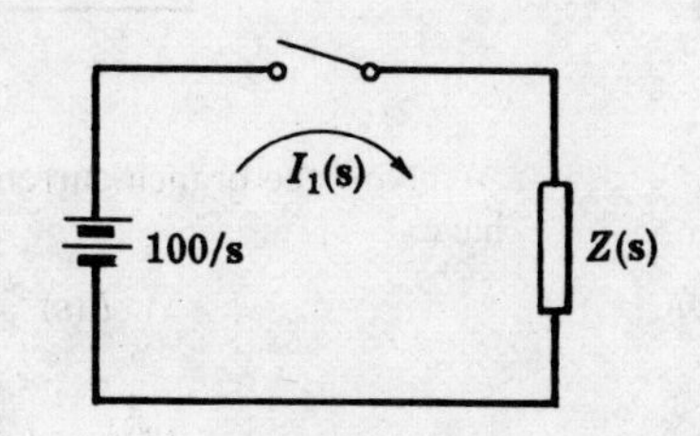

Fig. 17-21

This expression is identical with equation (5) of Prob. 17.14 and thus the time function is $i_1 = 10 - 3{\cdot}33e^{-166{\cdot}7t}$ amperes.

17.17. In the two-mesh network shown in Fig. 17-22 there is no initial charge on the capacitor. Find the mesh currents i_1 and i_2 which result when the switch is closed at $t = 0$.

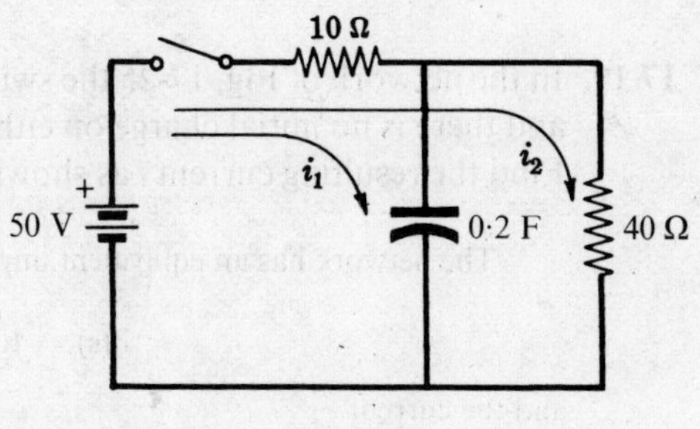

Fig. 17-22

The time domain equations for the circuit are

$$10i_1 + \frac{1}{0{\cdot}2}\int i_1\,dt + 10i_2 = 50$$
$$50i_2 + 10i_1 = 50 \qquad (1)$$

and the corresponding s domain equations are

$$10\,I_1(s) + \frac{1}{0{\cdot}2s}I_1(s) + 10\,I_2(s) = 50/s \quad 50\,I_2(s) + 10\,I_1(s) = 50/s \qquad (2)$$

or in matrix form,

$$\begin{bmatrix} 10 + 1/0{\cdot}2s & 10 \\ 10 & 50 \end{bmatrix}\begin{bmatrix} I_1(s) \\ I_2(s) \end{bmatrix} = \begin{bmatrix} 50/s \\ 50/s \end{bmatrix}$$

from which $I_1(s) = 5/(s + 0{\cdot}625)$ and $i_1 = 5e^{-0{\cdot}625t}$ amperes.

To find i_2, substitute the value of i_1 into the second of the time domain equations (1):

$$50i_2 + 10(5e^{-0{\cdot}625t}) = 50 \text{ and } i_2 = 1 - e^{-0{\cdot}625t} \text{ amperes}$$

17.18. Referring to Problem 17.17, obtain the equivalent impedance of the s domain network and determine the total current and the branch currents using the current division rule.

The s domain equivalent impedance is

$$Z(s) = 10 + \frac{40(1/0{\cdot}2s)}{40 + 1/0{\cdot}2s} = \frac{80s + 50}{8s + 1} = 10\,\frac{s + 5/8}{s + 1/8} \qquad (1)$$

The equivalent circuit is shown in Fig. 17-23 below and the resulting current is

$$I(s) = \frac{V(s)}{Z(s)} = \frac{50}{s}\left\{\frac{s + 1/8}{10(s + 5/8)}\right\} = 5\,\frac{s + 1/8}{s(s + 5/8)} \qquad (2)$$

Expressing equation (2) in terms of partial fractions,

$$I(s) = \frac{1}{s} + \frac{4}{s + 5/8} \quad \text{from which} \quad i = 1 + 4e^{-5t/8} \text{ amperes} \qquad (3)$$

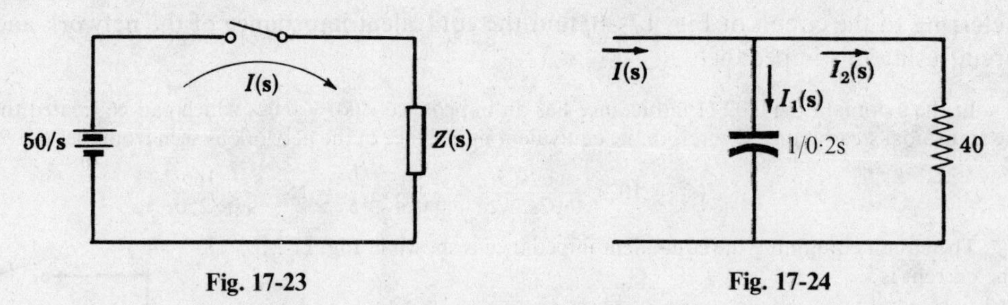

Fig. 17-23

Fig. 17-24

Now the branch currents $I_1(s)$ and $I_2(s)$ can be obtained by the current division rule. Referring to Fig. 17-24, we have

$$I_1(s) = I(s)\left(\frac{40}{40 + 1/0{\cdot}2s}\right) = \frac{5}{s + 5/8} \text{ and } i_1 = 5e^{-0{\cdot}625t} \text{ amperes}$$

$$I_2(s) = I(s)\left(\frac{1/0{\cdot}2s}{40 + 1/0{\cdot}2s}\right) = \frac{1}{s} - \frac{1}{s + 5/8} \text{ and } i_2 = 1 - e^{-0{\cdot}625t} \text{ amperes}$$

17.19. In the network of Fig. 17-25 the switch is closed at $t = 0$ and there is no initial charge on either of the capacitors. Find the resulting current i as shown in the diagram.

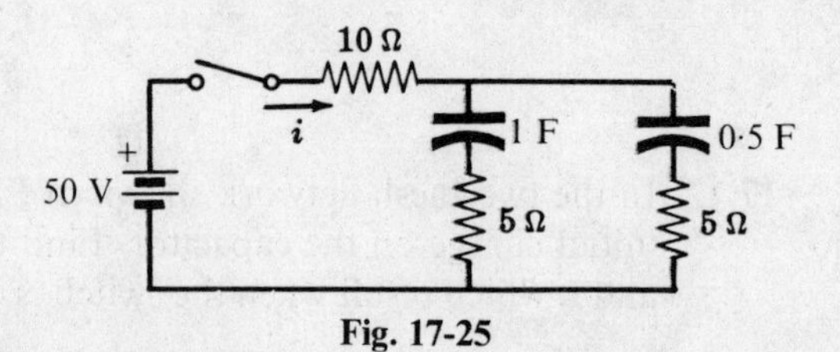

Fig. 17-25

The network has an equivalent impedance in the **s** domain

$$Z(s) = 10 + \frac{(5 + 1/s)(5 + 1/0{\cdot}5s)}{(10 + 1/s + 1/0{\cdot}5s)} = \frac{125s^2 + 45s + 2}{s(10s + 3)} \tag{1}$$

and the current

$$I(s) = \frac{V(s)}{Z(s)} = \frac{50}{s}\frac{s(10s + 3)}{(125s^2 + 45s + 2)} = \frac{4(s + 0{\cdot}3)}{(s + 0{\cdot}308)(s + 0{\cdot}052)} \tag{2}$$

Expressing the **s** domain current in terms of partial fractions,

$$I(s) = \frac{1/8}{s + 0{\cdot}308} + \frac{31/8}{s + 0{\cdot}052} \text{ and } i = \frac{1}{8}e^{-0{\cdot}308t} + \frac{31}{8}e^{-0{\cdot}052t} \text{ amperes}$$

17.20. Apply the initial and final value theorems to the **s** domain current of Porblem 17.19.

Since $I(s) = \dfrac{1/8}{s + 0{\cdot}308} + \dfrac{31/8}{s + 0{\cdot}052}$, the initial current is

$$i(0) = \lim_{s\to\infty} [s\,I(s)] = \lim_{s\to\infty}\left[\frac{1}{8}\left(\frac{s}{s + 0{\cdot}308}\right) + \frac{31}{8}\left(\frac{s}{s + 0{\cdot}052}\right)\right] = 4 \text{ A}$$

and the final current is

$$i(\infty) = \lim_{s\to 0} [s\,I(s)] = \lim_{s\to 0}\left[\frac{1}{8}\left(\frac{s}{s + 0{\cdot}308}\right) + \frac{31}{8}\left(\frac{s}{s + 0{\cdot}052}\right)\right] = 0$$

Examination of the circuit given in Fig. 17-25 shows that initially the total circuit resistance is $R = 10 + 5(5)/10 = 12{\cdot}5$ and thus $i(0) = 50/12{\cdot}5 = 4$ A. Then in the steady state both capacitors are charged to an equivalent voltage of 50 volts and the current is zero.

Supplementary Problems

17.21. Find the Laplace transform of each function.

(a) $f(t) = At$ (c) $f(t) = e^{-at}\sin\omega t$ (e) $f(t) = \cosh\omega t$

(b) $f(t) = te^{-at}$ (d) $f(t) = \sinh\omega t$ (f) $f(t) = e^{-at}\sinh\omega t$

Ans. (a)-(e) See Table 17-1, (f) $\dfrac{\omega}{(\mathbf{s}+a)^2-\omega^2}$

17.22. Find the inverse Laplace transform of each function.

(a) $\mathbf{F(s)} = \dfrac{\mathbf{s}}{(\mathbf{s}+2)(\mathbf{s}+1)}$ (d) $\mathbf{F(s)} = \dfrac{3}{\mathbf{s}(\mathbf{s}^2+6\mathbf{s}+9)}$ (g) $\mathbf{F(s)} = \dfrac{2\mathbf{s}}{(\mathbf{s}^2+4)(\mathbf{s}+5)}$

(b) $\mathbf{F(s)} = \dfrac{1}{\mathbf{s}^2+7\mathbf{s}+12}$ (e) $\mathbf{F(s)} = \dfrac{\mathbf{s}+5}{\mathbf{s}^2+2\mathbf{s}+5}$

(c) $\mathbf{F(s)} = \dfrac{5\mathbf{s}}{\mathbf{s}^2+3\mathbf{s}+2}$ (f) $\mathbf{F(s)} = \dfrac{2\mathbf{s}+4}{\mathbf{s}^2+4\mathbf{s}+13}$

Ans. (a) $2e^{-2t}-e^{-t}$ (d) $\frac{1}{3}-\frac{1}{3}e^{-3t}-te^{-3t}$ (g) $\frac{10}{29}\cos 2t+\frac{4}{29}\sin 2t-\frac{10}{29}e^{-5t}$

(b) $e^{-3t}-e^{-4t}$ (e) $e^{-t}(\cos 2t+2\sin 2t)$

(c) $10e^{-2t}-5e^{-t}$ (f) $2e^{-2t}\cos 3t$

17.23. A series *RL* circuit with $R = 10$ ohms and $L = 0{\cdot}2$ H has a constant voltage $V = 50$ volts applied at $t = 0$. Find the resulting current using the Laplace transform method. *Ans.* $i = 5 - 5e^{-50t}$ amperes

17.24. In the series *RL* circuit of Fig. 17-26 the switch is in position 1 long enough to establish the steady state and is switched to position 2 at $t = 0$. Find the current. *Ans.* $i = 5e^{-50t}$ amperes

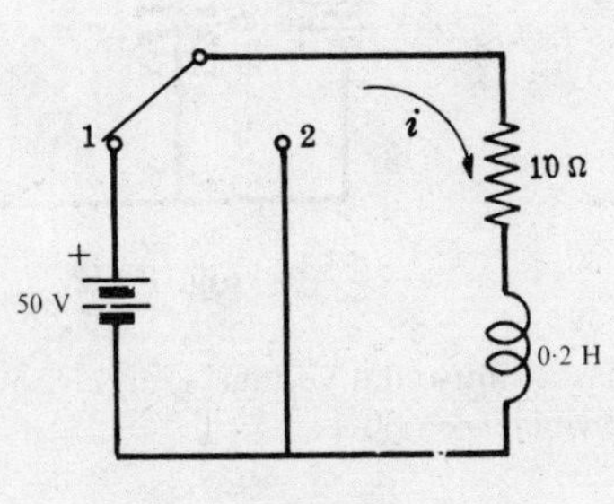

Fig. 17-26

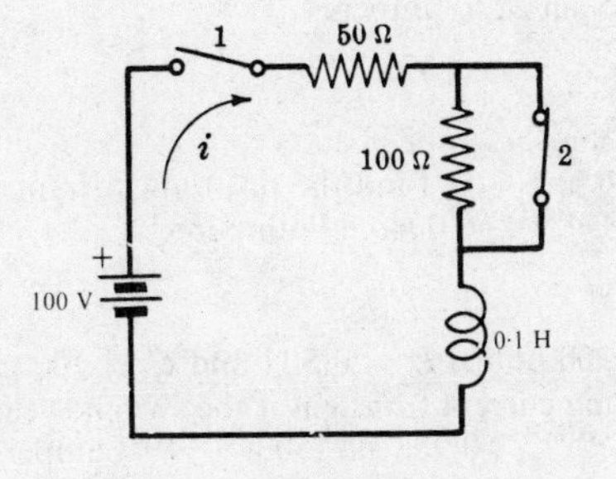

Fig. 17-27

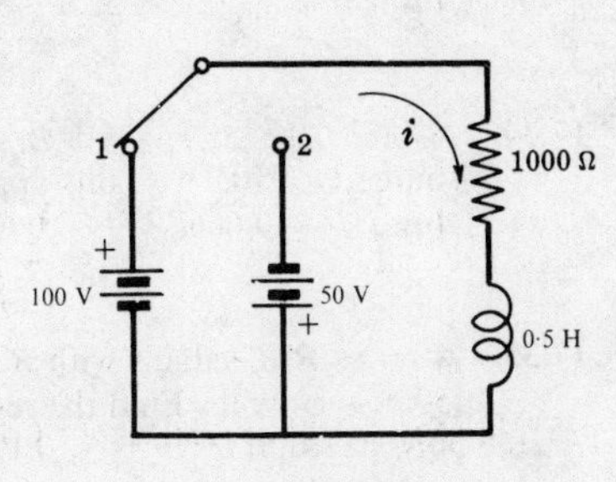

Fig. 17-28

17.25. In the circuit shown in Fig. 17-27 switch 1 is closed at $t = 0$ and then at $t = t' = 4$ milliseconds switch 2 is opened. Find the transient current in the intervals $0 < t < t'$ and $t' < t$.
Ans. $i = 2(1 - e^{-500t})$ amperes, $i = 1{\cdot}06e^{-1500(t-t')} + 0{\cdot}667$ amperes

17.26. In the series *RL* circuit shown in Fig. 17-28 the switch is closed on position 1 at $t = 0$ and then at $t = t' = 50$ microseconds is moved to position 2. Find the transient current in the intervals $0 < t < t'$ and $t > t'$.
Ans. $i = 0{\cdot}1(1 - e^{-2000t})$ amperes, $i = 0{\cdot}06e^{-2000(t-t')} - 0{\cdot}05$ amperes

17.27. A series *RC* circuit with $R = 10$ ohms and $C = 4\ \mu$F has an initial charge $q_0 = 800 \times 10^{-6}$ coulombs on the capacitor at the time the switch is closed, applying a constant voltage $V = 100$ volts. Find the resulting current transient if the charge is (a) of the same polarity as that deposited by the source and (b) of the opposite polarity.
Ans. (a) $i = -10e^{-25\times10^3 t}$ amperes, (b) $i = 30e^{-25\times10^3 t}$ amperes

17.28. A series *RC* circuit with $R = 1000$ ohms and $C = 20\ \mu$F has an initial charge q_0 on the capacitor at the time the switch is closed, applying a constant voltage $V = 50$ volts. If the resulting current is $i = 0{\cdot}075e^{-50t}$ amperes, find the charge q_0 and its polarity. *Ans.* 500×10^{-6} coulombs, opposite polarity to that deposited by source

17.29. In the *RC* circuit shown in Fig. 17-29 the switch is closed on position 1 at $t = 0$ and then at $t = t' = 1$ TC is moved to position 2. Find the transient current in the intervals $0 < t < t'$ and $t > t'$.
Ans. $i = 0{\cdot}5e^{-200t}$ amperes, $i = 0{\cdot}516e^{-200(t-t')}$ amperes

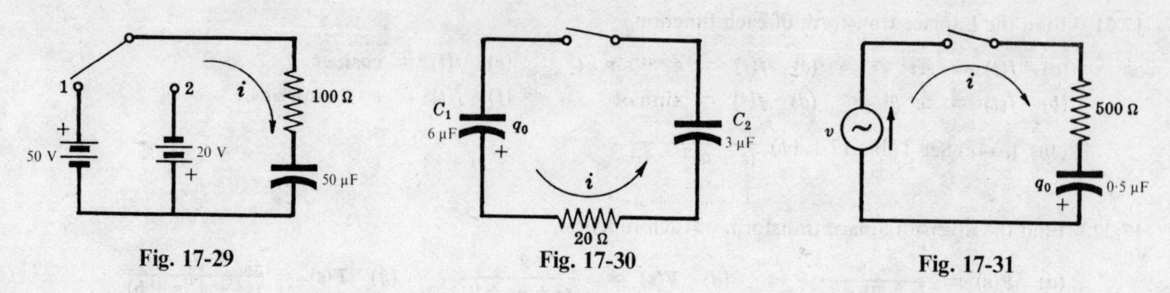

Fig. 17-29 Fig. 17-30 Fig. 17-31

17.30. In the circuit of Fig. 17-30 capacitor C_1 has an initial charge $q_0 = 300 \times 10^{-6}$ coulombs at the time the switch is closed. Find the resulting current transient. *Ans.* $i = 2{\cdot}5e^{-2 \cdot 5 \times 10^4 t}$ amperes

17.31. In the series *RC* circuit shown in Fig. 17-31 the capacitor has an initial charge $q_0 = 25 \times 10^{-6}$ coulombs and the sinusoidal voltage source is $v = 100 \sin(1000t + \varphi)$ volts. Find the resulting current if the switch is closed when $\varphi = 30°$.
Ans. $i = 0{\cdot}1535e^{-4000t} + 0{\cdot}0484 \sin(1000t + 106°)$ amperes

17.32. A series *RLC* circuit with $R = 5$ ohms, $L = 0{\cdot}1$ H and $C = 500\ \mu$F has a constant voltage $V = 10$ volts applied at $t = 0$. Find the resulting current. *Ans.* $i = 0{\cdot}72e^{-25t} \sin 139t$ amperes

17.33. In the series *RLC* circuit of Fig. 17-32 the capacitor has an initial charge $q_0 = 1$ mC and the switch is in position 1 long enough to establish the steady state. Find the transient current which results when the switch is moved from position 1 to 2 at $t = 0$.
Ans. $i = e^{-25t}(2 \cos 222t - 0{\cdot}45 \sin 222t)$ amperes

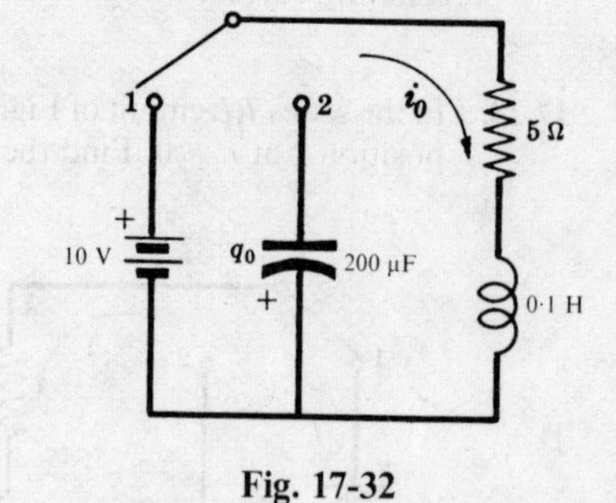

Fig. 17-32

17.34. A series *RLC* circuit with $R = 5$ ohms, $L = 0{\cdot}2$ H and $C = 1$ F has a voltage source $v = 10e^{-100t}$ volts applied at $t = 0$. Find the resulting current.
Ans. $i = -0{\cdot}666e^{-100t} + 0{\cdot}670e^{-24 \cdot 8t} - 0{\cdot}004e^{-0 \cdot 2t}$ amperes

17.35. A series *RLC* circuit with $R = 200$ ohms, $L = 0{\cdot}5$ H and $C = 100\ \mu$F has a sinusoidal voltage source $v = 300 \sin(500t + \varphi)$ volts. Find the resulting current transient if the switch is closed when $\varphi = 30°$.
Ans. $i = 0{\cdot}517e^{-341 \cdot 4t} - 0{\cdot}197e^{-58 \cdot 6t} + 0{\cdot}983 \sin(500t - 19°)$ amperes

17.36. A series *RLC* circuit with $R = 5$ ohms, $L = 0{\cdot}1$ H and $C = 500\ \mu$F has a sinusoidal voltage source $v = 100 \sin(250t + \varphi)$ volts. Find the resulting current if the switch is closed when $\varphi = 0°$.
Ans. $i = e^{-25t}(5{\cdot}42 \cos 139t + 1{\cdot}89 \sin 139t) + 5{\cdot}65 \sin(250t - 73{\cdot}6°)$ amperes

17.37. In the two-mesh network of Fig. 17-33 the currents are selected as shown in the diagram. Write the time domain equations, transform them into the corresponding **s** domain equations, and obtain the transient currents i_1 and i_2.
Ans. $i_1 = 2{\cdot}5(1 + e^{-10^5 t})$ amperes, $i_2 = 5e^{-10^5 t}$ amperes

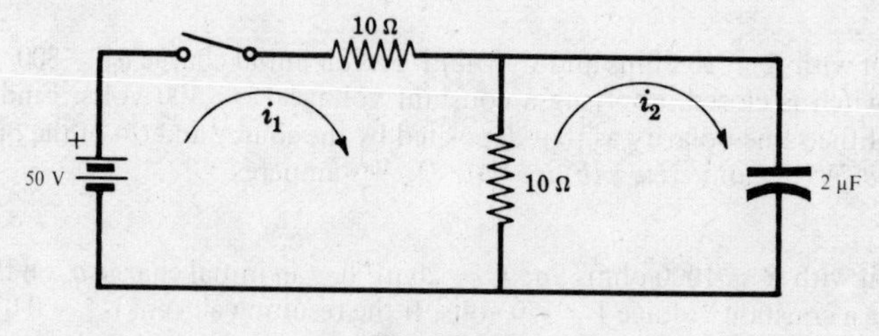

Fig. 17-33

17.38. For the two-mesh network shown in Fig. 17-34 find the currents i_1 and i_2 which result when the switch is closed at $t = 0$.
Ans. $i_1 = 0{\cdot}101e^{-100t} + 9{\cdot}899e^{-9950t}$ amperes, $i_2 = -5{\cdot}05e^{-100t} + 5 + 0{\cdot}05e^{-9950t}$ amperes

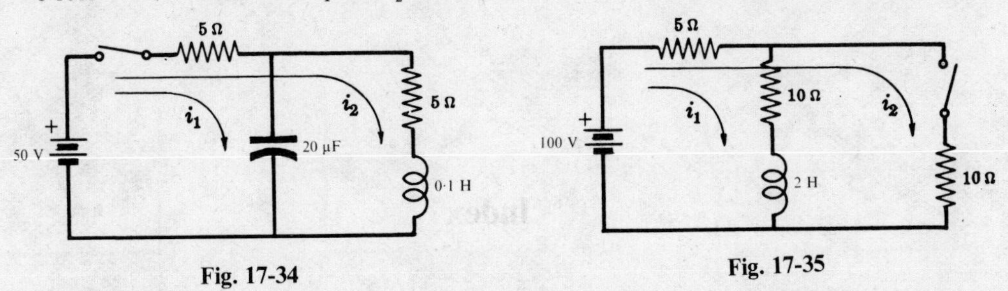

Fig. 17-34

Fig. 17-35

17.39. In the network shown in Fig. 17-35 the 100 volt source passes a continuous current in the first loop and the switch is closed at $t = 0$ placing the 10 ohm resistor in parallel with the branch containing the series combination of $R = 10$ ohm and $L = 2$H. Find the resulting currents. *Ans.* $i_1 = 1{\cdot}67e^{-6{\cdot}67t} + 5$ amperes, $i_2 = -0{\cdot}555e^{-6{\cdot}67t} + 5$ amperes

17.40. The two-mesh network shown in Fig. 17-36 contains a sinusoidal voltage source $v = 100 \sin(200t + \varphi)$ volts. At $t = 0$ the angle $\varphi = 0$ and the switch is closed, placing the second 10 ohm resistor in parallel with the first. Find the resulting mesh currents with directions as shown in the diagram.
Ans. $i_1 = 3{\cdot}01e^{-100t} + 8{\cdot}96 \sin(200t - 63{\cdot}4°)$ amperes, $i_2 = 1{\cdot}505e^{-100t} + 4{\cdot}48 \sin(200t - 63{\cdot}4°)$ amperes

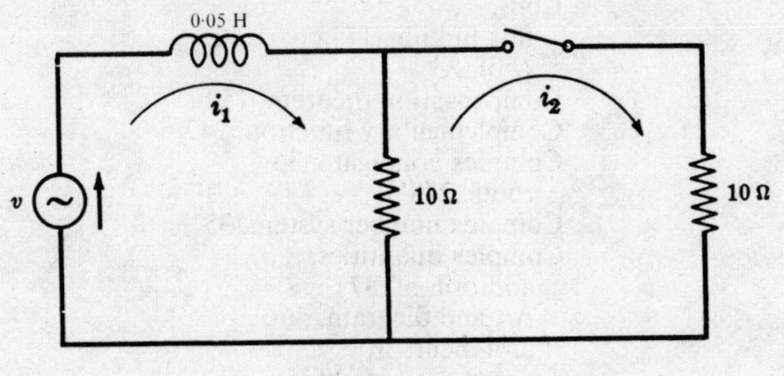

Fig. 17-36

Index

GREYSCALE

BIN TRAVELER FORM

Cut By Dina Lopez Qty 46 Date 10-31

Scanned By__________ Qty______ Date______

Scanned Batch IDs

__________ __________ __________

Notes / Exception